INTRODUCTION TO DESIGN AND ANALYSIS OF EXPERIMENTS

Springer
New York
Berlin
Heidelberg
Barcelona
Budapest
Hong Kong
London
Milan
Paris
Santa Clara
Singapore
Tokyo

GEORGE W. COBB
Mount Holyoke College

INTRODUCTION TO DESIGN AND ANALYSIS OF EXPERIMENTS

Springer

Library of Congress Cataloging-in-Publication Data
Cobb, George W.
Introduction to design and analysis of experiments / George W. Cobb.
p. cm.—(Textbooks in mathematical sciences)
Includes bibliographical references and index.
ISBN 0-387-94607-1 (alk. paper)
1. Experimental design. 2. Analysis of variance. I. Title. II. Series.
QA279.C625 1997
001.4'34—dc21 97-996

Printed on acid-free paper.

Production managed by Lesley Poliner; manufacturing supervised by Jacqui Ashri.
Typeset by Matrix Publishing Services, York, PA, from Microsoft Word files supplied by the author.
Printed and bound by Hamilton Printing Co., Rensselaer, NY.
Printed in the United States of America.

9 8 7 6 5 4 3 2 1

ISBN 0-387-94607-1 Springer-Verlag New York Berlin Heidelberg SPIN 10516493

To my father,
Whitfield Cobb,
teacher and statistician,
and to the memory of my mother,
Polly Cobb,
teacher also, but assuredly not a statistician.

CONTENTS

TO THE INSTRUCTOR

This book provides an applied approach to the design and analysis of experiments for students with no previous background in statistics. It has been used for two kinds of courses: (1) an alternative first course in statistics for undergraduate biology and psychology majors; and (2) an elementary service course for graduate students from a variety of fields that use statistics. It introduces the main elements of statistical thinking in the context of experimental design and ANOVA and thus makes it possible to cover in a single semester material that is traditionally spread over two semesters.

Goals. Because this book is in many ways unconventional in its approach, organization, and exposition, it seems useful to be explicit about goals. A diligent reader of this book will learn

1. *To choose sound and suitable design structures.* This includes evaluating the choice of response variable, identifying and then deciding how to handle extraneous influences, and evaluating the pros and cons of various design strategies such as blocking. (Many books with the word "design" in the title don't really teach design; they teach ways to analyze experiments designed by someone else. This book emphasizes the designs themselves, separate from, and prior to, formal ANOVA.)
2. *To recognize the structure of any balanced design built from crossing and nesting.* This includes not just learning to recognize the factors of a design and how they are related but also learning to recognize when a given data set does *not* have an ANOVA structure. When I first started teaching experimental design, I wasn't prepared for the trouble some students had in recognizing and distinguishing the various designs when they were given only verbal descriptions of the experiments. To make this easier, I have borrowed from modern abstract treatments of ANOVA the idea that a factor is a partition of the observations. Regarding a design as a set of partitions lets students focus on the ways the observations should be grouped to give meaningful comparisons; the set of groupings then tells the kind of design. (I've included several exercises where the task is to go from the words to a standard data format that shows the partitions of the data.)
3. *To explore real data sets using a variety of graphs and numerical methods.* The graphs include the usual parallel dot plots, interaction plots, and scatterplots of residuals versus fitted values, plus within-blocks scatterplots and two kinds of diagnostic scatterplots for choosing a transformation. However, exploratory analysis is more a matter of attitude than a collection of methods, and the spirit of exploration must come from trying to understand real data. Almost all my

examples and exercises are based on real data, some from student honors projects but most from journal articles in psychology and biology. Both the examples and the exercises include many extended explorations of these data sets.

4. *To assess how well the standard assumptions of ANOVA fit a data set and if the fit is poor, to choose a suitable remedy such as transforming to a new scale.* Informal checking of assumptions appears early, in connection with exploring the patterns in a data set. Most chapters contain examples of data sets that fail to fit the assumptions, and Chapter 12 is devoted entirely to assumption checking and remedies for data sets that fail to fit one or more assumptions.
5. *To decompose any balanced data set into "overlays" (components corresponding to the factors of the design) and find the parallel decompositions of the sums of squares and degrees of freedom.* For mathematically sophisticated students, a familiar and effective approach to the ANOVA decomposition is based on the Euclidean geometry of R^n: sums of squares and degrees of freedom are tied conceptually to a linear decomposition obtained by projecting the data onto mutually orthogonal subspaces. Moreover, as Peter Fortini has shown, group representation theory provides a direct route from the partitions of a design to the ANOVA decomposition. Although this approach is beyond the reach of many students, there is an elementary analog in which the perpendicular projections correspond to tables of group averages. That correspondence leads to an elementary recursive algorithm suitable for all balanced designs. The rule has an intuitive rationale based on comparing groups while holding appropriate other conditions constant. An easy parallel gives degrees of freedom.
6. *To construct the interval estimates and F-tests of formal inference.* This includes learning to find appropriate denominator mean squares by finding EMSs, for any combination of fixed and random factors, using both the restricted and unrestricted models for mixed interaction terms.
7. *To interpret numerical patterns and formal inferences in relation to the relevant applied context.* I find that careful interpretation is hard to teach and that, particularly at first, it is not enough just to remind students to interpret their numerical results. Thus I have included many exercises that give concrete suggestions about what to look for, in the hope that hints will make it easier for students to recognize that statistics involves interpretation, and interpretation involves issues of substance.

The sample exam questions that follow this section illustrate the kinds of things a student should be able to do after completing this book.

General approach. In my exposition I have chosen (1) to keep technical demands low, (2) to emphasize examples and context more than mathematical derivations, (3) to write for a course based on active learning, and (4) to do as much as I could with a carefully chosen small set of basic but easily generalizable ideas. These four choices make the book quite different from others on the same subject.

1. *Modest technical demands.* Although I have included the standard algebraic notation in optional sections, I have worked hard to make the book accessible to

students who do not have strong algebraic skills. I use a concrete visual representation of the factor structure of a design instead of the usual heavily subscripted equations. Representing the factors visually as partitions of the data brings with it several simplifications in the exposition. It anticipates the formal analysis, since the partitions correspond one to one to sets of averages and to lines in the ANOVA table. It makes the partial ordering of the factors easy to see, because one factor is "inside" a second "outside" factor if superimposing their factor diagrams gives back the first diagram. The partial ordering is the key both to decomposing a data set and to finding expected mean squares. (The set of estimated effects for a factor equals the set of group averages for that factor minus the sum of estimated effects for all "outside" factors; the EMS for a factor contains a term for every "inside" factor that is random and is restricted in the same way.) Both operations are easier to understand and to justify intuitively using the visual notation.

In the same spirit, I emphasize variability in observed data as more concrete and more accessible than the abstract constructions of formal probability theory.

2. *Examples and context rather than derivations*. Mathematical statistics has a deductive structure, but data analysis is an interpretive activity. In a textbook these two aspects of the subject are unavoidably at odds, as the role of context makes clear. The deductive structure of formal ANOVA is clearest in the abstract and precise language of mathematics, with all applied context stripped away. *In mathematics, context obscures structure*. In data analysis, however, we are ultimately interested in telling the story of individual data sets, and for that the applied context is essential. *In data analysis, context provides meaning*. There are already many books on design and ANOVA that emphasize its deductive structure. In such books examples tend to come after a more abstract exposition, trimmed and tailored to fit the method *du jour*. In this book I rely heavily on examples and context to show how statistical methods are used in meaningful ways. I am less concerned with mathematical structure and more concerned with interpretation.
3. *A book for active learning*. There are two main elements here: style and exercises.

   (a) Style. The adventurous teacher of statistics faces a dilemma. On the one hand, much recent research on how students learn tells us we should use class time for hands-on activities, group problem solving, and other alternatives to lecturing. On the other hand, textbooks are usually written to serve as back-up for a set of lectures, which makes them too hard for many students to learn from until after they have heard the lecture. I have tried to write this book in a way that makes it easier for nonmathematical students to use for primary exposition, so that teachers will have more choices about what to do in class. I talk directly to the student/reader, whom I assume is seeing the material for the first time.

   (b) Exercises. Also in the spirit of active learning, I have worked at least as hard on the exercises as on the text of the chapters. Most sections include several kinds of exercises: quick answer, akin to vocabulary-and-grammar drill in a language; easy mechanical drill for computational skills; and other exercises that don't involve much arithmetic but do require thinking about the material from a different angle. Each section has many exercises suitable for essay questions, class discussion, or group work.

4. *A small set of basic ideas*. My goal is for students to develop a flexible ability to apply a few broad principles. Thus three core chapters (Chapters 5, 6, and 7) are devoted to two principles for assigning treatments to units (random assignment and blocking) and one principle for choosing a treatment structure (factorial crossing). These principles are first applied, in those same three chapters, to construct four basic designs: completely randomized (one-way and two-way); complete block; Latin square; and split plot/repeated measures. Chapter 8 treats all four designs together, with sections on recognizing designs when given a verbal description, on choosing from among the four designs, and on learning rules to help avoid misapplying the ideas to inappropriate structures.

   After this chapter of review and consolidation come three extensions, which enlarge the scope of a student's skill to encompass all balanced designs. In the first extension, Chapter 9, the principle of factorial crossing is applied to generate a family of designs from each of the original four designs. Later, Chapter 10, the second extension, incorporates hierarchical designs so that a student can handle the factor structure and decomposition for any design built of crossing and nesting. By maintaining a distinction between the factor structure, which determines the decomposition, and the distributional assumptions that determine the set of possible inferences, it is natural to put off a systematic discussion of fixed versus random effects until students have become adept at applying basic design principles and recognizing factor structures. Thus it is the third and final extension—Chapter 13—that deals with fixed versus random effects and restricted versus unrestricted models for mixed effects.

Suggestions for teaching from this book.
At Mount Holyoke I use this book in a course for science majors who have not yet studied statistics. We cover most sections from the first fourteen chapters, in order, over the course of a thirteen-week semester, meeting three times per week. We go through Chapters 1–3 quickly, in the first two weeks, and then average one chapter per week after that. However, the topics that constitute design and ANOVA can be presented in a very large number of reasonable orders, and I have written the book to offer the instructor maximum flexibility in choosing whatever sequence seems best.

The purpose of the three introductory chapters is to give the beginner an overview of the whole book, to provide a sense of where things are headed. Many of the explanations in these chapters are deliberately abbreviated or incomplete, because the same ideas will be developed more slowly and thoroughly in subsequent chapters. Nevertheless, I think many students find it helpful to read this extended overview before taking on the main chapters in earnest, and helpful also to return to these chapters as a review later on. However, recognizing (with one reviewer's emphatic help!) that some instructors dislike extended introductions, I have made these chapters optional. With the exception of one two-page appendix on the general rule for decomposing a data set, and the description of the hamster experiment, none of material in Chapters 1–3 is required for later chapters.

This makes Chapter 4, on the response, conditions, and material, another possible starting point. The topics here include scales of measurement, reliability and validity, observational versus experimental studies, placebos and controls, and experimental units. If students have already studied these in another course, it may make sense to skip most or all of Chapter 4, and begin the course with Chapter 5.

Chapters 5–7, on three design principles and four basic designs, form the core of the book, and must be covered, in order, except that the sections on algebraic notation and on computers are optional, and the sections on decomposition and F-tests can be postponed until the end of Chapter 7. After Chapter 7, all the remaining chapters are meant to be usable independently of one another, provided one is willing to refer back, as necessary, for the descriptions of data sets. (The data sets introduced in one chapter keep returning in later chapters if the topic at hand promises to add to one's understanding of the data).

One reviewer of a very early version of this book commented, "I sometimes get the feeling he doesn't trust me to do my own lecturing." Although I disavow any such imperialistic motives toward my colleagues' lectures, I do acknowledge a certain accuracy in the reviewer's perceptions: If we are to encourage active learning in class, then the textbook needs to carry more of the burden of exposition.

Applied statistics lends itself naturally to discussion: for example, on the merits of completely randomized versus complete block designs to see whether carpeting raises levels of airborne bacteria in hospital rooms, on the choice between logarithms and reciprocal roots for analyzing the effect of day length on the concentration of neurotransmitter in the brain of a hibernator, on the way a few mildly deviant observations should qualify conclusions from the F-test in a study of babies learning to walk, or on the relationship between certain three-way interactions and an hypothesis about schizophrenia. The list of possibilities is endless, and students are generally surprised and engaged when they find there is so much more to our subject than just fancy arithmetic.

The main obstacle to using class time to involve students in discussion (and other forms of active learning) is not so much in the nature of our subject as in the way we teachers have been expected to present it. Students get most of the material from a set of lectures, and use their book for repetition, for detail, and as a source of exercises. This pattern often makes it hard to find class time for anything but lecturing. To make it easier for teachers to depart from this pattern if they want to, I have written a book that students can actually read on their own, without first covering the same ground in a lecture. I encourage active reading, at times asking the reader to anticipate where a particular analysis will end up, at times (as another reviewer noticed) inviting the reader to disagree with me on a matter of judgment. I assume that most of my readers are not already used to reading with pencil and paper, working out the details as they read, and so I have included more detail and commentary than is typical in quantitative subjects. (That, and the large number of extended examples, are the main reasons for the book's Dickensian dimensions.) One consequence of these choices is that the student who tries to use the book without actually reading it may find the experience frustrating. As one colleague who has taught from the book several times puts it, "I tell the students it's all there, but they really have to read it."

SAMPLE EXAM QUESTIONS

A. The Effect of Expectations on Biofeedback

Read the following description and then describe three experimental plans: (1) completely randomized; (2) based on Latin squares; and (3) split plot/repeated measures. For each plan, tell how you would use your subjects: tell the units and the conditions; and tell how to assign conditions to units. Draw and label a rectangular diagram showing what the data from your design would look like. Then tell whether you would recommend such an experimental plan. If you would not recommend it, why not? Which of the other plans is better, and why?

Experiments have shown that skin temperature and relaxation are related: the more relaxed you are, the higher your skin temperature. Because of this relationship, doctors sometimes use biofeedback to help tense patients relax. If you were such a patient, the doctor would attach an electronic sensor to the tip of your index finger and connect the sensor to a machine that measures your skin temperature and plays a musical tone that gets higher when the temperature goes up and gets lower when the temperature goes down. Many people who practice regularly become able to control their skin temperature.

Suppose you want to design an experiment to study factors that influence how well people learn to raise their skin temperature. Suppose you have 12 human volunteers, people who know nothing about either biofeedback or the scientific literature on control of skin temperature.

It's reasonable to wonder whether patients who believe they can learn to control their temperature will do better than those who believe they probably can't. One way to study the effect of expectation would be to compare subjects who are told that most people can learn to raise their temperature (high expectation) with subjects who are told that hardly anyone is able to do it (low expectation).

It's also reasonable to wonder about the effect of the biofeedback, the changing tone that the machine makes as your temperature changes. Suppose that in addition to expectation you want to compare three kinds of feedback:

True feedback (TF): The tone goes up when the subject's temperature goes up and goes down when the temperature goes down.

False feedback (FF): The machine plays back tones from someone else's biofeedback session. The tone has no relationship to the subject's own temperature.

No feedback (NF): The machine just records the temperature but doesn't make any sound at all.

B. Hypnosis and Learning (*J. Personality*, 1965)

1. Read the description that follows, and then (a) tell the response; (b) list all structural factors for the design; (c) tell whether each is a nuisance factor or a factor of

interest; (d) tell the number of levels; (e) tell whether each factor of interest is experimental or observational; (f) tell what the units are; and (g) name the design.

The purpose of this study was to compare the effects of two kinds of hypnosis (traditional and "alert" hypnosis) on learning. Fifteen undergraduates served as subjects. Each subject learned two lists of 16 word–number pairs—one list under normal waking conditions, and a second list under one of three randomly chosen conditions: waking; traditional hypnosis; or alert hypnosis. The design was balanced so that for the second set of conditions, there were five subjects under each condition. The two lists were of similar degree of difficulty.

The lists were learned as follows: a subject heard a tape recording of the list and was then asked to repeat the list. If the subject made any errors, the process was repeated, up to a maximum of 15 times, until the subject gave back the list without errors. The following numbers give the total numbers of errors for each list, for each subject.

Subject	1	2	3	4	5	6	7	8	9	10	11	12	13	14	15
Test 1	67	65	12	11	3	106	30	32	17	15	71	65	39	16	4
Test 2	8	85	18	28	3	105	31	29	38	49	36	37	26	29	2
Condition	Waking hypnosis					Traditional hypnosis					Alert hypnosis				

2. (a) Tell how you would run the experiment with the same response, the same set of subjects, and the same set of conditions to compare, but using a randomized complete block design. Tell the units, the blocks, and the number of units per block. (b) Consider how you could use Latin squares to improve on the RCB plan in (a). Draw and label a diagram showing how you would assign conditions to units, and tell the two nuisance factors in your design. There is a third nuisance influence, one that is completely confounded with one of the structural factors in the design actually used in part 1. What is it?

C. The Effect of Darkness on the Extinction of Learned Behavior in Rats

The next several questions are all based on the results of an experiment designed to study the effect of darkness on the learning behavior of rats. The experiment has a split plot/repeated measures design.

Twenty rats were taught to press a bar in order to get pellets of food. After initial training—once the rat had learned that it got a pellet each time it pressed the bar—all 20 rats were tested. The numbers below, in the "Before" column, give 100 × log(# bar presses per minute) during this first test period.

After the first test period, all 20 rats were subjected to "extinction of learned behavior:" they no longer got rewarded with pellets when they pressed the bar. Ten of the rats had been randomly assigned to the Dark condition. They were kept in total darkness during the extinction phase of the experiment. The other 10 rats were kept in conditions that were the same as the first 10, except that they had normal lighting during the extinction phase.

After extinction, all 20 rats were tested a second time. The numbers in the "After" column give 100 × log(# bar presses per minute) during the second test period.

Light			Dark		
RAT	**Before**	**After**	**RAT**	**Before**	**After**
1	271	201	11	275	157
2	231	203	12	150	123
3	176	92	13	201	176
4	249	161	14	284	135
5	240	239	15	234	74
6	290	253	16	225	138
7	175	123	17	244	153
8	236	131	18	227	88
9	256	210	19	271	155
10	239	170	20	264	160

Condition averages

Light	Before	236.3
	After	178.3
Dark	Before	237.5
	After	135.9

1. Write the factor diagram.
2. Compute (a) the grand average, (b) the estimated overall effect for Light, (c) the estimated overall effect for Before, (d) the estimated interaction effect for Light, Before, (e) the block effect for Rat IV, and (f) the residual for Rat I, Before.
3. Using the estimated effects from 2 to compute SSs, give the complete ANOVA table, and mark with an asterisk any factor whose observed effects are too big to be due just to chance error. Use the facts that $SS_{\text{Total}} = 1{,}685{,}914$ and $MS_{\text{Blocks}} = 2601.1$.
4. Make a scatterplot of After = y versus Before = x using *L*s for the rats who got light and *D*s for the rats who got darkness. Use the balloon rule to estimate the correlation between Before and After for the rats who got light.
5. Find confidence intervals for each of the following contrasts. (Don't bother here with adjustments for simultaneous confidence intervals; just use the ordinary t-value.)

   a. Light versus Dark

   b. Light versus Dark for Before

   c. Light versus Dark for After
6. Write a sentence or two explaining what these confidence intervals tell you about the behavior of the rats. Why does it make sense to do two separate comparisons

of Light versus Dark in 5(b) and (c)? How would these comparisons turn out if darkness during the extinction phase has a real effect?

7. Find confidence intervals for each of the following contrasts. (Don't bother here with adjustments for simultaneous confidence intervals; just use the ordinary t-value.)

   a. Before versus After

   b. Before versus After for the rats that got light

   c. Before versus After for the rats that got darkness

8. Draw an interaction graph and use it, together with your answers to 1–7, to discuss what the data tell you about the effect of light and darkness on the number of times the rats pressed the bar.

D. Circadian Rhythms (Part I)

The numbers below come from an observational complete block study (*Steroids*, Vol. 20, pp. 269–273) whose purpose was to determine whether hormone levels in adult males varied randomly or according to a systematic 24-hour cycle (a circadian rhythm). The response is the concentration of plasma testosterone in milligrams per deciliter. Each subject was measured four times during the course of a 24-hour period.

Subj.	9 a.m.	3 p.m.	9 p.m.	3 a.m.	Avg.	SD	ln (avg.)	ln (SD)
1	7	28	18	32	21.25	11.18	3.1	2.4
2	64	31	88	75	64.50	24.39	4.2	3.2
3	83	31	36	36	46.50	24.45	3.8	3.2
4	19	20	29	24	23.00	4.55	3.1	1.5
5	10	28	88	73	49.75	36.77	3.9	3.6
6	56	41	31	56	46.00	12.25	3.8	2.5
7	55	106	15	97	68.25	41.88	4.2	3.7
8	38	100	23	49	52.50	33.41	4.0	3.5
9	23	163	17	44	61.75	68.49	4.1	4.2
10	6	30	10	47	23.25	19.00	3.1	2.9
Avg.	36.1	57.8	35.5	53.3	45.68			

1. *Parallel dot graph*. A partial dot graph appears below (not shown here). Complete the dot graph and discuss the patterns. (I suggest you round the response values to the nearest 10 for your plot.)

2. *Within-blocks scatterplot*. Plot (rounded) response values for 3 p.m. (y) versus (rounded) values for 9 a.m. (x), and discuss what the pattern suggests about the assumptions for ANOVA.

3. *Diagnostic plot*. Use the values of log(avg.) and log(SD) for subjects to construct a diagnostic plot. Then describe the pattern: what is the evidence that a change of scale is called for? Fit a line by eye, and estimate the slope. What transformation does the plot suggest?

4. *Residuals versus fitted values*. A scatterplot of Res vs. Fit is given below (Note: It is not shown here). (a) Describe the pattern in words, and tell what the pattern suggests about the fit of the model and the standard assumptions for ANOVA. (b) Briefly discuss what you have found from the four plots in 1–4: are the suggestions about fit and assumptions consistent, or at odds with each other?
5. ANOVA. For the untransformed data, $SS_{Total} = 126.935$ and $MS_{Res} = 1047.5185$. Use this information, together with the column averages, to give a complete ANOVA table. What do you conclude from the F-ratios for subjects and times?
6. *Confidence intervals*. Find ordinary 95% confidence intervals for the four condition averages. Sketch the four intervals on a graph, and comment briefly on the pattern: does this evidence support the hypothesis of a 24-hour cycle for testosterone levels?

E. Circadian Rhythms (Part II)

The numbers below come from the study described in Section D. The first row gives condition averages from the data set in Section D. Rows 2 and 3 give condition averages for the same 10 subjects, at the same four times, but for concentrations of two other hormones. (Test = testosterone, And = androstenedione, Deh = dehydrocepiandrosterone.)

	9 a.m.	3 p.m.	9 p.m.	3 a.m.	Avg
Test	33.0	56.5	35.5	55.0	45.0
And	12.0	32.0	12.0	24.0	20.0
Deh	15.0	31.5	18.5	35.0	25.0
Avg	20.0	40.0	22.0	38.0	30.0

Although the cell entries are in fact averages, the table has the same structure as a two-way factorial with one observation per cell. The questions that follow concern that two-way structure.

7. Draw an interaction graph, and comment on the pattern and its meaning.
8. Decomposition. Write out a decomposition of the data into grand average, hormone effects, time effects, and "residuals" = interaction effects.
9. "Residuals" versus "fitted values." Use the partial fit for interaction as fitted values to scatterplot Res vs. Fit. Discuss what the pattern in the plot suggests: to what extent does the evidence indicate that transforming might improve the fit of the no-interaction model?
10. Residuals versus comparison values. (a) Construct a table of comparison values and scatterplot Res vs. Comp. (b) Fit a line by eye, estimate its slope, and tell what transformation you would use, if any, for these data. (c) What does the shape of the plot tell you about how effective the transformation is likely to be?

TO THE STUDENT

Once, years ago I saw a television interview with Linus Pauling, who is famous for winning two Nobel prizes, one for his research on the structure of large molecules and the other for his efforts on behalf of nuclear disarmament and world peace. I've forgotten most of the interview, but I still remember one short question and answer. The interviewer wanted to know, "How did you manage to win *two* Nobel prizes?" Pauling thought a minute and then answered, "You have to have lots of ideas, and throw away the bad ones." Part of the reason I still remember what he said is that, even though he didn't plan it this way, his answer tells a lot about why statistics is worth learning.

I'll try to explain. Many students tend to think of statistics as a branch of mathematics; many mathematicians do, too. However, at least some of the people who think of statistics as a branch of mathematics are really thinking about arithmetic, which is not at all the same as mathematics. Statistics sometimes gets the reputation of being terribly boring, a subject where you do nothing but memorize formulas and practice long computations. Statisticians, however, think of statistics as a subject that involves much more than just computation, and much more than mathematical theory, too. For example, to plan a good scientific experiment, or to decide whether an experiment is well planned, you need to know quite a bit of statistics but not very much math. Even when you analyze the results of an experiment, computing is only part of the work. Deciding *what* to compute, and deciding what the numbers mean, are crucial. After all, there's not much point in doing a lot of arithmetic (or trying to persuade a computer to do the arithmetic for you) unless your effort ends up telling you something interesting about the experiment that gave you the numbers. At its best, statistics is like detective work with numbers: the work you do uncovers clues and helps you run down hunches about what the data have to tell you. Sometimes the process can get quite exciting, because statistics can lead you to discover things that you can't see just from looking casually at the data. At the same time, you have to be careful, because sometimes what looks like a message from the data is just a coincidence caused by chance fluctuations. To summarize, part of what a statistician does is to use various methods to get hunches about the data; another part is to use other methods to decide which hunches can be turned into sound conclusions. In the spirit of Linus Pauling's approach, good statistical work can help you have lots of ideas about the data and can help you throw away the bad ones.

I don't want to mislead you. Even though statistics deals with ideas, you still have to be willing to learn how to compute things. It's important to remember, however, that statistics doesn't start with computing, and it doesn't stop there either. Somebody has to choose *what* to compute, and if he or she makes a wrong choice, the computations may be worse than worthless. Somebody has to figure out what the computations mean; otherwise they mean absolutely nothing.

Maybe all this strikes you as obvious. I hope it does. But I know from experience that, in the middle of a statistics course, it is all too easy to get so involved in computing that you lose touch with the ideas. That's why I've written this book in a way that doesn't depend on formulas, even though symbols and formulas can provide a powerful way of expressing ideas. Although I believe that many important ideas can be expressed more compactly, and sometimes more precisely, using carefully chosen symbols, I do not accept the defeatist attitude that you can't learn to design and analyze experiments unless you use a lot of algebra. I see an important distinction between logical thinking and the language you use for it. In this book I have chosen not to rely on algebraic language, but I nevertheless believe that if you are to learn to use statistical methods appropriately, you must be willing to work hard at understanding their logic.

Although I believe that statistics has a logical structure that exists almost independently of the people who use it, I also believe it is important to recognize that, in the end, it is individual people who try to apply statistical ideas. Good statistical work calls for judgment, and judgments are often influenced by personal experience and feeling. From time to time in this book, I say things that represent my own view, things that others might see differently. I don't do this often; where I do, I try to make it clear that what I say is a personal opinion. Many of the statisticians I have learned from, first at the Medical College of Virginia, and later at Harvard, taught me that hunches, guesses, and opinions are often important but also that it is equally important to be clear in identifying opinions as such. I have tried to follow their example.

Finally, I hope that as you work through this book, you'll keep reminding yourself there's no point in learning the methods unless you also learn when to use them and what they tell you. Keep this in mind especially when you work the problems. You can't learn to think like a statistician just by reading a book, anymore than you can learn to swim just by lying in the sun next to a pool. You've got to get in the water and move your arms and legs.

ACKNOWLEDGMENTS

When I was an undergraduate at Dartmouth, I heard John Kemeny tell a class, "There are two kinds of mathematicians: those who use examples (to help them learn mathematics), and those who use examples but don't admit it." Although Kemeny was referring to printed examples, not living ones, and although I no longer think of myself as a mathematician, it is nevertheless a pleasure to admit in print that this book owes much to the many teachers who have served as my examples over the years. Two teachers from high school, Kenneth Walker at Guilford (NC) High School and Grant Fraser at George School (PA) inspired me to study mathematics. As an undergraduate at Dartmouth, I was influenced not only by mathematicians John Kemeny, Laurie Snell, Reese Prosser, and David Kelly, but also by poet, polyglot, and professor of Russian, Walter Arndt. In part due to his influence, I nearly went to graduate school to study Russian literature. (Textual exegesis is, after all, just another form of data analysis!) At the Medical College of Virginia, I had my first serious encounter with applied statistics, thanks to James Kilpatrick, Ray Myers, Hans Carter, Roger Flora and Van Bowen. But for their example and influence, I might still be analyzing Pushkin and Dostoevsky instead of hamster enzymes and shrews' brain waves. (Should I really be thanking those teachers?) Earning my Ph.D. in statistics at Harvard, I had three superb models to teach and inspire me: William Cochran, Frederick Mosteller, and Arthur Dempster, my dissertation advisor. The list of inspiring examples didn't stop there, however: perhaps the best thing about learning is that it doesn't have to end just because someone hands you a piece of paper. In recent years I've continued to be influenced by the wonderful example of others, especially David Moore and Richard Scheaffer. All these teachers have indirectly helped me make this book what it is, and I thank them.

Many others have made much more direct contributions. John Kimmel and Alexander Kugushev served as editors for earlier versions of this book. Although neither John nor Alex has been in on the final stages of turning the manuscript into a book, I am grateful for their earlier contributions, and for our continuing association. In addition, I wish to thank a large number of anonymous reviewers of early versions of the manuscript. Some, who didn't like what I was trying to do, at least deserve thanks for the pain they must have endured reading something that struck them as so hopelessly wrong-headed, while others made such immensely useful suggestions that I'm sorry not to be able to thank them by name. I also thank Edward Stanek, of the University of Massachusetts, for his thorough and thoughtful comments on my treatment of mixed models in Chapter 13. Even though Ed may not agree with all that I've written there, his comments have given me new insights.

A handful of colleagues taught from various versions of the book during its long gestation. Over the years Loveday Conquest (University of Washington), Robert Wardrop (University of Wisconsin at Madison), and Jeffrey Witmer (Oberlin College) have all been extremely helpful. At Mount Holyoke College, I've had many helpful conversations with Alan Siegel, Barbara Peskin, Margaret Robinson and especially Janice Gifford. In addition, Francis Anscombe of Yale University earned an enduring debt of gratitude for the long lists of helpful suggestions he sent me over the course of the semester he used the manuscript. Many of his suggestions appear in the book almost exactly as he sent them, and the book is much the richer for it.

Before I had ever put pencil to paper, a version of this book had been evolving slowly, for more than a decade, inside my head. Any such evolutionary process requires an ecosystem as context, and I am grateful to the many inquisitive, thoughtful, and articulate students whose comments and reactions provided the forces of natural selection that drove some early ideas into oblivion while encouraging others to survive. In this context, I particularly thank Anne Weaver, M.D., for her careful and copious comments.

Jeremiah Lyons, Editorial Director for the Physical Sciences at Springer-Verlag New York is an author's dream. From the moment he decided Springer should publish my book, I could not have asked for a more supportive publisher. A key element of that support has come from the production team. Production of this book posed a real challenge, demanding more than the usual amount of skill, imagination, flexibility, effort and patience, because of my reliance on a large number of diagrams and other visual substitutes for the equations traditionally used to present analysis of variance. I am especially grateful to my copyeditor, Heather Jones, to Pam Whiteley, production coordinator at Matrix Publishing Services, to Karen Phillips, assistant manager for design, and to Lesley Poliner, senior production editor, both at Springer-Verlag.

Finally, although the greatest debts are often the hardest to put into words, I want to thank my wife Cheryl and ten-year-old daughter LeeTae. Lee has sometimes come to one of my statistics classes to help with an activity, but neither she nor Cheryl is a statistician, and neither she nor Cheryl feels even the tiniest bit deprived as a consequence. That makes me all the more grateful to them, first for tolerating my devotion to this project, and above all for being there all these years, enriching the non-statistical part of my life.

George W. Cobb
Mount Holyoke College
December, 1997

CHAPTER 1

INTRODUCTION TO EXPERIMENTAL DESIGN

1. The Challenge of Planning a Good Experiment

Several years ago, at Mount Holyoke College, a biology major named Kelly Acampora designed a clever experiment as part of her senior honors project. She was interested in what happens when animals hibernate, and in particular, she wanted to study how changes in an animal's environment cause the animal to start hibernating. The decisions she had to make in planning her experiment are fairly typical of the decisions involved in many kinds of experimental work in biology and psychology: decisions about what questions to ask, and decisions about how to gather information as efficiently as possible. By making sound decisions, Kelly was able to make her experiment very efficient: using only 8 hamsters, and only 2 measurements per hamster, she was able to collect enough information to study five different questions related to her experiment.

In planning her work, Kelly relied on a number of key ideas from the branch of statistics called experimental design. In the next three chapters, I'll use the story of Kelly's experiment as a way to introduce you to some of those key ideas. This first chapter introduces the main ideas you need for planning an experiment. Then the next two chapters deal with how you analyze the results. As you read these chapters, don't feel you have to understand everything the first time through; instead, read to get a general sense of the main ideas. Keep in mind that anything really important will reappear in later chapters. If you start to feel swamped, you may find that the summaries I've put at the ends of the chapters can help you tell the forest from the trees.

Here's a summary of this section, the first of three in this chapter:

- The content of an experiment: three decisions
- Three sources of variability: one we want and two we don't
- Three kinds of variability: one we want, one we can live with, and one that threatens disaster
- Chance error and bias compared
- Design in a nutshell: isolate the effects of interest—control what you can, and randomize the rest

The Content of an Experiment: Three Decisions

In planning almost any experiment, you have to decide
1. **what** measurement **to make (the** response**),**
2. **what** conditions **to study (the** treatments**), and**
3. **what** experimental material **to use (the** units**).**

At the start of almost any scientific study, you need to decide what questions to ask and what kind of information to gather. What questions you ask should depend more than anything else on your sense of what is important. However, it's not enough just to choose a general question, like what makes animals hibernate. Your question must be specific enough to allow you to decide the kind of information you'll need to answer it. In other words, there's a tradeoff. Questions that are really important tend to be quite general and hard to answer. It takes a skillful scientist to choose a question that is both general enough to be important and specific enough that an experiment can stand a chance of answering it. (Of course, there are lots of important questions that can't be answered by experiments, but this is a book about the ones that can.) All this may seem obvious, but there's a reason for my mentioning it: statistical ideas can't help you very much in deciding what questions to study, and since this is a book about statistics, I won't have much to say about choosing the subject for an experiment. As you read the rest of the book, it is important to keep in mind that the ideas of experimental design deal with turning a good question into a good experiment; they won't help you find the good question in the first place.

Once Kelly had decided on her general question—How do changes in the environment influence an animal to start hibernating?—she had to recast the question in a form that an experiment might be able to answer. What changes in the environment would she study? If you think about what happens to the environment as winter arrives, you'll no doubt think of two kinds of changes: it gets colder, and the days grow shorter. Kelly decided to focus on the second kind of change—changes in photoperiod (day length).

Another decision Kelly faced in making her question more specific was what kind of measurement to make in order to study the effect of day length. One possibility would be to measure something like heart rate or breathing rate, since the hearts and lungs of animals slow down when they hibernate. To a statistician who knows no biology, that possibility might seem reasonable: it could be turned into an experiment that would be statistically sound. Such an experiment, however, would be of no interest to biologists. They already know that when days grow short, the heart of a hibernator slows down. So Kelly decided to look instead for a measurement related to the level of nerve activity in an animal. The measurement she chose was the concentration of a particular enzyme (Na^+K^+ATP-ase) that is known to be part of a biochemical reaction, called the *sodium pump*, that transmits nerve impulses. She knew that when an animal shows high concentrations of the sodium pump enzyme, it has a high capacity for transmitting nerve impulses. If the concentrations of enzyme were to change as days grew shorter, that change would tell her about a change in the animal's capacity for nervous activity.

Kelly also had to decide what animal to study. For a variety of reasons, she chose to work with golden hamsters. Obviously, she had to choose an animal that is able to hibernate, and the golden hamster is. Almost as obviously, some hibernators are more suited to a small lab experiment than others. Bears, for example, would present a number of problems. Golden hamsters are easy to get from a biological supply house; they don't cost very much; and they are small enough not to take over a whole lab, at least as long as you keep the males and females apart.

At this point, Kelly had made three decisions: to study the effect of day length (the treatments); to measure the concentration of sodium pump enzyme (the response); and to use golden hamsters (the material). As a result of these three decisions, she could replace her general question—How do changes in the environment influence an animal to start hibernating?—with a specific version of it—What is the effect of changing day length on the concentration of the sodium pump enzyme in the golden hamster?

Kelly's three decisions about content are typical of most experiments. They can make the difference between an interesting experiment and one nobody really cares about. For the most part, however, these are not decisions that statistics can help you with. To make such decisions wisely, you have to know your biology, or psychology, or whatever.

Don't get me wrong, though. Statistical thinking is important, too. You can start with a good question; then make good decisions about the conditions, the measurement, and the experimental material; and still come up with a worthless experiment. It happens all the time, but knowing the key ideas of experimental design can help you make your experiment a good one.

Three Sources of Variability: One We Want and Two We Don't

In planning an experiment, try to identify all major sources of variability. Most data sets contain variability from three main sources, which correspond to the three decisions about content:

1. **variability due to the conditions of interest (wanted),**
2. **variability in the measurement process (unwanted), and**
3. **variability in the experimental material (unwanted).**

A good design will let you estimate the amount of variability due to each source.

Variability is everywhere in science, and is the main reason scientists need statistical thinking, both for planning studies and for analyzing the results. A good design and analysis make it possible to learn about the variability due to the conditions of interest even though variability from other sources affects your data. In planning any experiment, you should think about variability as you look ahead to the data you expect to get and to the ways you hope to use that data. To help you do this with Kelly's experiment, I'll show you eight of her actual measurements:

TABLE 1.1 Enzyme concentrations (mg/ml) in the hearts of eight hamsters.

Hamsters raised with long days:	1.49	1.53	1.56	1.79
Hamsters raised with short days:	1.39	1.49	1.25	1.38

On the surface, it looks as though day length does affect the enzyme concentration. The average for the long-day hamsters is 1.59 mg/ml, which is higher than the short-day average of 1.32 mg/ml. Notice, however, that the eight numbers show a lot of variability. The long-day measurements range from 1.49 mg/ml to 1.79 mg/ml, for example, and the difference of 0.30 mg/ml is bigger than the difference between the long- and short-day averages. In order to analyze the data, we need to estimate how much of the difference between the averages is due to day length and how much is due to other sources.

For Kelly's hamster experiment, we can list three sources that may contribute to the variability in the numbers.

1. *Variability in the conditions of interest*: the effects of day length. This is the source of variability we want, and estimating the size of the day-length effects is the main goal of the experiment. However, Kelly's choice of conditions was in fact more ambitious than I've told you. If you think about the nervous system, you might guess that some organs of the body are more involved with nerve activity than others. Not surprisingly, the concentration of enzyme you find depends on the organ you measure. Kelly decided to study two organs, the heart and the

brain. That decision led to two new questions she wanted her experiment to answer. First, and more obvious, is there a difference in enzyme concentration between the hearts and brains of hamsters? This question was no doubt once of interest to biologists, but not so much anymore, because they know the answer. Brains have higher concentrations than hearts. There's a second, more interesting question, though: does day length affect the enzyme concentrations of hearts and brains in the same way? After all, hearts and brains have different jobs to do, so you might expect them to behave differently as they get ready for hibernating. This kind of question, about **interaction**, is extremely important in experimental work, so it's worth looking more carefully at what it asks. The key idea is that there is variation from *two* sets of conditions involved: long days versus short, and hearts versus brains. The question asks whether the difference in response due to one set of conditions (day length) is the same for each of the other conditions (heart and brain). It takes special planning if an experiment is to be able to answer questions about interaction, as we'll see later on.

2. *Variability in the response*: measurement error. This is not a source of variability we want, but it's a fact of life in science that if you make two measurements under conditions that are as nearly the same as you can make them, the numbers you get are almost never exactly equal. Our experiment should let us estimate the typical size of differences due to such unavoidable variability in the measurement process. To find the concentration of enzyme, Kelly had to sacrifice the hamster, dissect the tissue she was interested in, then use a difficult sequence of steps to suspend the enzyme in liquid, and, in the final step, measure the concentration of enzyme using a spectrophotometer to measure the amount of light absorbed by the suspended particles. At each stage of the whole process, things might have gone a little differently from the way they did. The dissection is too hard to produce pure heart tissue. There's no way to get chemicals that are 100% pure to use in the tissue preparation; even if you could, it's impossible to measure amounts that are exactly right. The spectrophotometer has to be calibrated against a standard before you use it to make the measurements you're interested in; each time you recalibrate it, you introduce variability. Ultimately, each measurement depends on the chancelike behavior of individual suspended particles.

3. *Variability in the experimental material.* Like variability in the response, variability in the material is unwanted but unavoidable. No two hamsters are biologically the same; some just naturally have higher enzyme concentrations than others. Moreover, we can expect the hamsters' environment and behavior to have some effect on enzyme concentration. The ability to make the enzyme depends to some extent on diet. The amount of nerve activity will depend to some extent on the level of stimulation in the environment.

The following table summarizes some of the sources of variability in the hamster experiment:

TABLE 1.2 Some sources of variability in the hamster experiment

1. Conditions of interest
   a. Day length—long versus short
   b. Organ—heart versus brains
2. Measurement process
   a. Preparation of the enzyme suspension
   b. Calibration of the spectrophotometer
3. Experimental material
   a. Biological differences among hamsters
   b. Environmental and behavioral differences

So far, we've looked at variability in terms of where it comes from. Variability from two of the three sources is unwanted—it just gets in the way when we try to study the variability that interests us. In order to develop strategies for handling the unwanted variability, we'll look next at different ways that variability can behave.

Three Kinds of Variability: One We Want, One We Can Live With, and One That Threatens Disaster

Any experiment is likely to involve three kinds of variability:

1. **Planned, systematic variability—the kind we want**
2. **Chance-like variability—the kind we can live with, and**
3. **Unplanned, systematic variability—the kind that THREATENS DISASTER.**

The practical value of these three categories is that they can (1) help you spot "bad" (unplanned, systematic) variability in time to look for ways to convert it into "tolerable" (chance-like) or even "good" (planned, systematic) variability, and (2) sometimes help you reduce the amount of chance-like variability.

Planned, systematic variation: the kind we want. This kind always includes the differences due to the conditions of interest. In the hamster experiment, for example, differences in enzyme concentration due to day length and organ are both planned and systematic. When systematic variability of this sort is present, we expect groups of response values to form clusters, like the two clusters we found in the eight measurements in Table 1.1: one cluster for long days, one for short. A key idea is that when systematic variability is planned, we can tell *before we make the measurements* which ones we expect to cluster together. For example, Kelly knew before she actually measured the enzyme concentrations which measurements would belong to the long-day clusters. She could tell, because she could *assign* a day length, long or short, to each hamster. Similarly, she could *select* a particular organ, heart or brain, so she knew in advance which measurements would belong to the heart cluster, and which to the brain cluster.

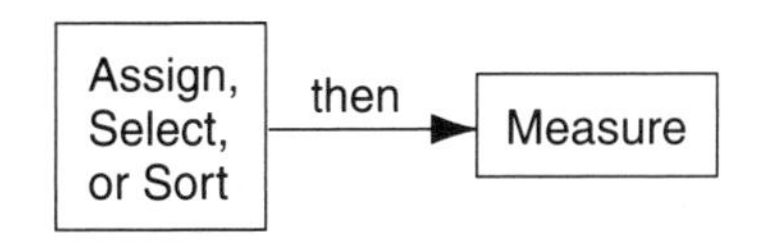

PLANNED MEANS PRE-SORTED

Unfortunately, systematic variation can be unplanned as well as planned, and the conditions of interest are almost never the only source of systematic variation. For Kelly, there would be some systematic variation in the material (differences among hamsters) and in the measurement process (linked to calibration). As you'll see later, one of the most useful design strategies—blocking—can sometimes turn such differences into planned, systematic variation.

Chance-like variation: the kind we can live with. This kind always includes an unplanned component from the measurement process, called **measurement error**. (As you'll see later, many good experiments include a *planned* chance-like component as well.)

Chance-like variation behaves like drawing at random from a box of numbered tickets. You can't tell ahead of time what number you'll get on any one draw, and if you draw again you're likely to get a different number, but underneath the surface chaos, there's a regular pattern.

When you toss a fair coin several times in a row, and think about predicting what will happen, two things stand out. One, you can't tell ahead of time whether a particular toss will land heads or tails, and two, you can tell that, over the long run, about half the tosses will land heads. Drawing from a box of tickets with numbers on them is similar: There will be chance-like variation from one draw to the next, but in the long run, the average will "settle down" just like the percent heads settles down to 50% in a long string of tosses.

There's an important connection between random draws and scientific measurement, one that scientists first began to understand in the 1700s. While trying to figure out how our solar system works (among other things), they found that repeated measurements of the same thing, measurements they had expected to be equal, kept turning out to be different. You've probably run into this yourself: you repeat a certain measurement in the lab, and what you get the second time isn't exactly equal to what you got the first time. However, even though individual measurements may differ, they tend to fluctuate around some central value.

Engineers sometimes say that a measurement is part signal, part noise: part message, and part chance fluctuation. For coin tossing, the hidden signal is that about half the tosses land heads, half land tails. The noise comes from never getting a single toss to land half heads and half tails. For random draws, the hidden signal is that the average of the draws tends to cluster near the average of the tickets in the box. The "noise" corresponds to the chance differences between one draw and the next.

A useful name for the random noise is **chance error**: chance, because it is unpredictable, like a coin toss; and error, because it keeps the measurement process from telling you the true value. In getting used to the term "chance error," it's important to realize that "error" doesn't mean "mistake." It's not a mistake if a single coin toss lands heads instead of half heads, half tails. Instead, it's something built into the process, and not at all a surprise. The same is true of scientific measurements. Every measurement, no matter how careful, has a certain amount of chance error built into the process.

Fortunately, chance-like variation has two properties that make it possible for us to learn from an experiment even when this unwanted kind of variation is present. First, just by the luck of the draw, some chance errors will be positive, others will be negative, and on balance they will tend to cancel each other out, at least partially, when you compute an average. The more measurements that go into your average, the smaller the chance part of the average will tend to be, and the more accurate the average itself will be. Second, if your experiment is well-planned, you will be able to use your measurements to estimate the size of the chance-like variation, and that will make it possible to figure out how accurate your average is likely to be.

What does all this tell us about designing an experiment?

In order to estimate the typical size of chance errors, you need to make more than one measurement with conditions held constant.

Any reading of the concentration of sodium-pump enzyme will have chance-like measurement error built into it, like it or not. This means that you need to measure things more than once, in order to get a good idea of how big chance error is. To see how important this is, imagine that Kelly only made two measurements: one enzyme concentration for long days, and one for short days. Here are two of her actual measurements:

Long days: 1.53 mg/ml
Short days: 1.39 mg/ml

We'd like to be able to say that the difference in enzyme concentration is due to the difference in day length, but we can't: since we don't know how big the chance errors are, we don't know how much of the difference is due to day length and how much is due just to chance. The only way around the problem is to make several measurements of the same thing and see how different they turn out. If all the measurements for long days turn out close to 1.53, and all those for short days turn out close to 1.39, then we're in business: we can safely conclude that day length makes a "real" difference in enzyme concentration.

Unplanned, systematic variation: the kind that threatens disaster. This kind of variation is one of the main causes of wrong conclusions and ruined studies. It can lurk almost anywhere. If you're not careful, your experiment can end up with unplanned, systematic variation from your material, from your measurement process, or even built into the conditions you want to study. The main threat is that such variation can **bias** your results by systematically raising or lowering a set of measurements for reasons you never recognize. There are two main strategies—blocking and randomization—for handling this kind of variation. Blocking turns possible bias into planned, systematic variation. Randomization turns it into planned, chancelike variation.

Selection bias: An example. One of my statistics teachers was the late W. G. Cochran, a major contributor to the theory of experimental design. In one of his courses, he told a story that illustrates how systematic, unplanned variation can really ruin an experiment. The experiment was part of a study of rat nutrition, with two diets, ordinary lab chow, and lab chow with a special supplement, which was thought to be better. The two people running the experiment grabbed rats out of a cage, one at a time, and assigned them to get either lab chow or the special diet. Later, when the results were in, there was a surprise: the special diet hadn't done as well as expected. A closer look at the data showed what seemed to have happened. For the rats chosen by one of the two experimenters, the results were okay: the rats getting the special diet did better. Only the rats chosen by the other experimenter, my teacher, seemed out of line. Cochran and his co-worker finally decided that, without realizing it, Cochran had acted almost as if he had felt sorry for the scrawnier rats: he had assigned most of them to the better diet. So his half of the rats on the better diet had started out less healthy, and that had made the diet look bad. Notice what might have happened if there had only been one person, my teacher, assigning the rats to diets. Without the other set of rats, it would have been impossible to figure out what had happened, and the study would have appeared to show that the special diet wasn't as good as they thought. That's bias.

There are various ways bias might be involved with Kelly's hamster experiment. Think about the average enzyme concentrations for the hamsters raised with long days (1.59 mg/ml) and those raised with short days (1.32 mg/ml). Although it looks like day length may have a "real" effect, part of the difference in the averages could be due to unplanned but systematic variation in her material or in her measurement process. Here's one story I invented to show how bias might creep in. Suppose that the four hamsters raised with long days all came from the same litter, and the four hamsters raised with short days came from another litter. Suppose, too, that the first litter just naturally had higher enzyme concentrations to begin with. Then, regardless of day length, those hamsters might tend to show higher concentrations at the end of the experiment. If bias of that sort were present, the results would *seem* to show a real effect of day length even though part of the effect would be due to unplanned but systematic variation in the way the hamsters were chosen for the two groups.

With a little imagination, you can think of all sorts of ways bias can creep into an experiment. Suppose Kelly Acampòra had measured all the hamsters that got long days at one time and then measured all the hamsters that got short days at another time. It's not hard to think of things other than day length that could make the long- and short-day averages different: perhaps the lab used up one batch of chemical reagents during the first set of measurements and had to switch to a new batch for the second set; or perhaps, for whatever reason, the measuring instrument got set a bit too high for the first set of measurements.

Chance Error and Bias Compared

Bias and chance error are similar in that both prevent your observations from being equal to their "true" values, but they are different in that whereas chance error is random, bias is not. Like the toss of a fair coin, chance errors sometimes go one way, sometimes the other, and in the long run they tend to cancel each other when you compute an average. Bias, on the other hand, tends to push every measurement in the same direction, there's no helpful canceling, and you end up with a distorted message from the data.

Target metaphor. One way to think about chance error and bias is to imagine target shooting with a rifle. If there's nothing wrong with the rifle, a series of 10 shots might look like the picture on the left in Fig. 1.1. If on the other hand, something is wrong with the scope on the rifle, so that, unknown to the marksman, the aim is always high

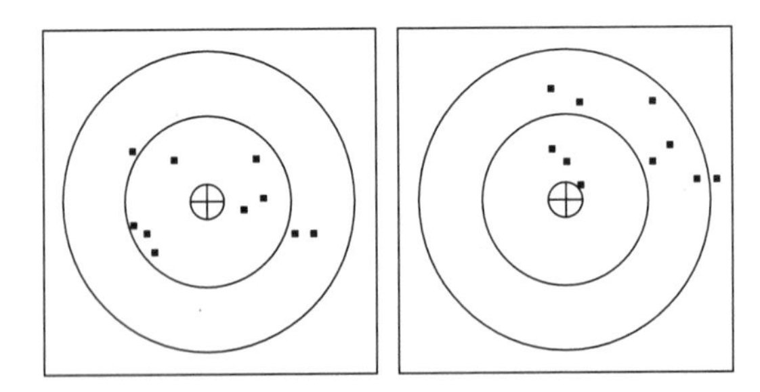

FIGURE 1.1 Chance errors in shooting at a target, without bias (left) and with bias. The bull's-eye is the true value, and each shot is a measurement. In the left-hand picture (no bias), the difference between where the shot hits and the bull's-eye is due entirely to chance error. There are about as many hits to the left of the bull's-eye as to the right and about as many above as below. Overall, the errors tend to balance. In the right-hand picture, chance error is once again present, in exactly the same pattern as on the left. Here, however, bias tends to make all the measurements off in the same direction, in this case high and to the right, as though the scope of the rifle were badly set, just as an instrument might be badly calibrated.

In scientific work, the bull's-eye is always invisible. You only get to see where the shots land. So mentally erase the target and bull's-eye from both pictures, and notice two things. First, provided you have several observations, you can judge the size of the chancelike variation from their scatter. Second, *there's no way to tell from the pattern of the observations which picture is the one with the bias.* That's why bias is such a threat.

and to the right, then the shots still show a scattering (chance error), but they also all tend to be off in the same direction, as in the picture on the right.

Sampling from a population

	Collection of possible values	Numerical summary
What you want to know about	Population	Parameter
What you actually get to see	Sample	Statistic

Doing statistics is a bit like eating a sausage: you know there's got to be a connection between the observable data on your plate and the unseen process that was responsible for creating it, but the connections are not always obvious. Thinking about drawing tickets from a box, and choosing a *sample* from a *population*, can help clarify those connections.

A box model for measurement error. For the moment, we'll pretend there are no hamster differences and that all hamsters respond to day length in exactly the same way. If we ignore brains and just work with hamster hearts, then the only variation is due either to day length or to measurement error. Imagine that you have two boxes of numbered tickets, one box labeled "Long Days" and one "Short Days." The numbers on the tickets in the box for Long Days represent the hypothetical set of all possible values you could get when you measure the enzyme concentration for a hamster raised with long days. The box for Short Days is similar. Each box represents a **population**, the set of all possible outcomes of interest, for one day length or the other. The average value of all the tickets in a box tells the enzyme concentration for a hamster raised with so many hours of light per day. Each average is a numerical property of the whole population and is called a **parameter**.

We want to know whether there is a difference between the two population averages, for long and short days. We can't measure the population averages directly, because we can't get a complete set of all possible outcomes for each day length. Instead, we have to choose a **sample** from each population and use our samples to **estimate** the unknown population parameters. Each of our observed values is one of the possible values, chosen from one of the populations, and each set of observed values is a sample. Briefly: the population is what you want to know about; the sample is what you get to see. We can compute the average of the numbers in each sample and use those averages to estimate the unknown population averages. Each sample average is an example of a **statistic**, a numerical property of a sample.

"True value" = parameter. Statisticians sometimes refer to population parameters as "true values." I find this informal language useful. The observations in a sample aren't "false," but their purpose is almost always to tell us about a more important truth that remains hidden in the unseen population. In the same spirit, many statisticians talk about whether an observed difference is "real." Here again, the informal language can be a useful shorthand reminder that we are ultimately more interested in the unseen population than in the sample. Obviously, if our sample averages for long and short days show a difference, that difference is real, but as statisticians use the word, a "real" difference is one that is judged too big to be due to chance error alone. In other words, a "real" difference is a difference in true (population) values.

Chancelike variation. How can the ideas of population and sample help explain the way measurement errors behave? I'll first describe the ideal situation, called **random sampling**, and then say just a little about the connection to measurement errors.

Think of your population as a box of numbered tickets, and suppose you get your sample by mixing the tickets thoroughly, then drawing out one ticket at random for each observation. The chance is in the process of choosing, and each ticket has the same chance of getting chosen. For each draw, the chance process is responsible for the particular value you get, and the chance error is the difference between the observed value on the ticket and the "true value," the population average.

Chance error = [Number on ticket] − [Average for the box]
= Observed value − "True" (population) value
Observed value = "True" value + Chance error

In practice, the actual situation is almost always less than ideal. When Kelly made her measurements, for example, she didn't actually draw at random from a box. She hoped that her process of measuring behaved a lot like the process of drawing tickets at random, because her statistical analysis depended on being able to assume that the actual process behaved a lot like the ideal one. Fortunately, a lot of evidence suggests that such an assumption is often reasonable, provided your study is well designed.

Bias and sampling. You can make the idea of bias more precise by relating it to sampling from a population. If you choose your sample strictly at random, so that each value has the same chance of getting into your sample, then the process of sampling doesn't tend to favor one value over another: the process is unbiased. On the other hand, imagine that values larger than the population average were written on big tickets and that numbers smaller than the average were written on tiny tickets. You'd then be more likely to get the bigger numbers in your sample, because you'd be more likely to choose the bigger tickets. In the same way, a biologist using a net to collect a sample of fish might be more likely to get the bigger, older fish because the smaller ones would have a better chance of slipping through the net. A psychotherapist may get a biased view of human nature because any group of therapy patients is biased if you regard it as a sample from the population as a whole. Similarly, if Kelly weren't careful about how she assigned day lengths to her hamsters, she could end up introducing bias. As you'll see in Section 2 of this chapter, drawing strictly at random to protect against bias will be one of Kelly's design strategies.

Design in a Nutshell: Isolate the Effects of Interest—Control What You Can and Randomize the Rest

In all, we can list five sets of questions that Kelly wanted her experiment to answer.

1. *Long versus short days.* Does day length affect enzyme concentrations? If so, how big is the effect?
2. *Hearts versus brains.* Do the hearts and brains of hamsters have different concentrations of the enzyme? If so, how big is the difference?
3. *Interaction.* Do hearts and brains react differently to day length? More precisely, is the difference in enzyme concentrations between long and short days different for hearts and brains? If so, what is the pattern of the difference?
4. *Hamsters.* Is the variability from one hamster to the next big enough to detect? If so, how much variability is there?
5. *Measurement error.* How big is the chance error built into the process of measuring enzyme concentrations?

Kelly was most interested in the first three questions. The last two were forced on her—they deal with unavoidable nuisance variation. Kelly's challenge in planning her experiment was to *isolate the effects of interest* (the effects of day length on hearts and brains) from each other and from the nuisance variation. Some things were under her control: for each of the hamsters she could choose how much light it would get in each 24-hour period, and she could choose which of its organs to measure. Her general strategy was to *control what she could, and randomize the rest.* When possible, she would try to turn unplanned, systematic variation into planned, systematic variation. When that was not possible, she would use a chance device to turn unplanned, systematic variation into planned, chancelike variation.

The next section will explain the details, but here's a quick sketch of her actual plan: she decided to raise four hamsters with each day length, keeping all other conditions as constant as possible. Half of the eight hamsters, chosen by a chance device, were raised with long days (16 hours of light, 8 hours of darkness), and the four that were left were raised with short days (8 hours light, 16 hours darkness). Each hamster gave two measurements: one for the heart and one for the brain.

2. Three Basic Principles and Four Experimental Designs

Overview. In planning any experiment, once you have chosen your measurement, conditions, and material, you face a new set of choices—about how to assign conditions to material. The three design principles of this section—random assignment, blocking, and factorial crossing—deal with these choices.

In thinking about how to design experiments, I find it useful to recognize two different points of view, which I call "what you do" and "what you get." "What you do" refers to the choices you have to make in order to carry out your experiment: how you choose your conditions and how you assign them to your material. "What you get" refers to the structure of the data set you get from your design. More specifically, by "structure" I mean the sets of averages and comparisons (basically, ways to group the observations) that make sense for a particular data set. Your choices for "what to do" will determine the structure that you get, and vice versa: the comparisons you want to make should determine your choices of what to do.

I start this section with "what you do": two principles for assigning conditions to material (random assignment, blocking) plus one principle for choosing the conditions (factorial crossing). For each of the three principles, I give the simplest possible design that uses it. Then I show how you can use all three principles at once to get a fourth design, the one Kelly chose for her experiment. Section 3 deals with "what you get." For each of the four designs from Section 2, I describe the structure of the design in terms of the comparisons that the design lets you make.

Random Assignment and the One-Way Randomized Basic Factorial (RBF[1]) Design

FIRST DESIGN PRINCIPLE: RANDOM ASSIGNMENT

In planning an experiment, any assigning that would otherwise be haphazard should be done using a chance device.

Suppose that Kelly cared only about the hearts of her hamsters and wanted to know whether day length affects the enzyme concentration in hamsters' hearts. If she didn't have to worry about measurement error, and differences between hamsters, and bias, she could simply raise one hamster with long days, raise one other with short days, and compare the two enzyme concentrations. Of course, life isn't that simple. In order to estimate the size of the differences due to measurement error and to variability in her material, Kelly needed to assign several hamsters to each day length. To protect against bias, and to make the hamster differences behave in a chancelike way, she should randomize the assignment: *use a chance device to decide which hamsters get long days and which get short.*

"Random" has a precise meaning. Here's one way it might work. Kelly had eight hamsters in all, four for long days, four for short. So she could put eight slips of paper into a box, four that said "Long" and four that said "Short." One at a time, she could take a hamster from the set of eight, then mix up the slips of paper in the box and draw one out. The hamster would then be assigned to the condition written on the slip of paper. It's important here to pay attention to the difference between haphazard and random. When my teacher who did the rat experiment grabbed rats out of a cage, he thought he was choosing rats at random, but he found out later that there was a pattern to his choices. His choices were haphazard, not planned, but they weren't really random. To get a random selection, you have to use a chance device, like drawing slips of paper out of a box, or working with a table of random numbers, created especially for scientists to use as a chance device. (In actual practice, drawing from a box doesn't work so well because it's hard to get the tickets thoroughly mixed. Use the box for understanding, but use a table of random numbers for a real experiment.)

Why randomize? *Randomizing converts unplanned, systematic variability into planned, chancelike variability.* This brings two main advantages. First, randomizing protects against possible bias. Suppose some of the hamsters naturally have higher enzyme concentrations than others. If you randomize the assignment, then just by chance it is likely that some will be assigned to long days and some to short. The same would be true of the hamsters with lower enzyme concentrations: by chance some would be assigned to long days and some to short. Within each group, long and short, the higher and lower concentrations will tend to balance each other. The effects would go in opposite directions like chance error and would tend to cancel out when Kelly computed her averages.

Second, randomizing makes statistical analysis possible. Randomizing means that we'll have two different kinds of chance error, from two different sources, both built into Kelly's experiment. The first is measurement error, which usually behaves as though it were chancelike. The second is the "error" due to variability in the material, which is made to behave in a chancelike way by randomizing the assignment of day lengths to hamsters. When we get to the analysis of Kelly's data, we will need to rely on the predictable regularities present in chancelike behavior. Statistical analysis *assumes* that those regularities are present. If Kelly had not randomized the assignment of conditions to material, the assumption might not be appropriate.

A randomized basic factorial design. Random assignment leads directly to the simplest experimental plan: using a chance device, randomly assign a day length (long or short) to each hamster. For **balance**, make sure that half the hamsters get long days and half get short.

Here's the way the list of measurements for this simple plan might look, using actual enzyme concentrations:

Long Days		Short Days	
Hamster	**Conc. (mg/ml)**	**Hamster**	**Conc. (mg/ml)**
#4	1.490	#3	1.385
#1	1.525	#8	1.485
#7	1.555	#5	1.255
#2	1.790	#6	1.285
Average	1.590		1.353

EXAMPLE 1.1 RANDOMIZED BASIC FACTORIAL EXPERIMENT TO COMPARE LONG AND SHORT DAYS

Treatments: long and short days
Experimental units: hamsters (8 in all)
RBF design: each of the eight hamsters is randomly assigned a day length, long or short. For balance, each day length is assigned to half the hamsters. ■

Treatments and units. Although this plan is very simple, it illustrates several important ideas. First, the conditions we want to compare, the long and short days, are assigned to material: we can control the process of assignment. Statisticians often use the word **treatments** to refer to conditions that can be assigned. (For some situations, the conditions you want to compare—male and female, for example—come already built into the material and can't be assigned.)

Second, it's an important feature of the plan that the treatments get assigned to hamsters individually rather than to groups of hamsters. Each hamster is an **experimental unit**, the chunk of material that is assigned a treatment. (If you've never run into this meaning of "units" before, notice that the word means something very different from the more familiar "units of measurement," like grams or liters.) *The amount of information you get from an experiment depends on the number of experimental units: the more units, the better.* (I discuss this in more detail in Section 3 of Chapter 4.)

If your experimental material is reasonably uniform, then the strategy of completely random assignment is likely to be a good choice. Often, though, if there is a lot of variability in your experimental units, you can improve on your design by using the second basic design principle, blocking.

Blocking and the One-Way Complete Block (CB[1]) Design

SECOND DESIGN PRINCIPLE: BLOCKING

First sort (or subdivide) your experimental material into groups (blocks) of similar units; then assign conditions to units separately within each block.

Blocking converts unplanned, systematic variation into planned, systematic variation. To make the idea concrete, consider an example of how Kelly might have used blocking as an alternative to completely random assignment. First remind yourself of the RBF[1] design of Example 1.1: we randomly assigned a day length to each hamster, regarding all eight hamsters as interchangeable, without making any distinctions among them.

Now suppose that we know that our hamsters came from four different litters, exactly two hamsters per litter. (More generally, suppose we know of a meaningful way to sort them into four pairs of similar hamsters before we assign day length.) Then we can take each pair of hamsters as a **block** and assign day length to hamsters separately within each pair: one randomly chosen hamster from each pair will get long days, the other short. This plan gives an example of a complete block design, which takes the hamsters within each pair as interchangeable: first sort the hamsters into four pairs; then within each pair, randomly assign long days to one hamster and short days to the other.

EXAMPLE 1.2 RANDOMIZED COMPLETE BLOCK EXPERIMENT TO COMPARE LONG AND SHORT DAYS

Treatments: long and short days
Experimental units: hamsters (8 in all)
Blocks: pairs of similar hamsters (4 pairs)
CB design: randomly assign long days to one hamster in each pair, short days to the other. ■

How to choose blocks. Blocking is a good strategy provided you can sort your material into groups of similar units. Here "similar" has a specialized meaning: units are similar if they are likely to give similar values for your measurement. To illustrate this specialized meaning, compare the following three ways to choose blocks for the hamster experiment.

(1) If our hamsters should actually happen to come from four different litters, two hamsters per litter, we could use pairs of littermates as blocks.
(2) We could weigh the hamsters and then put the two heaviest together in a pair, the next two together, and so on.
(3) Instead of weighing the hamsters, we could measure how fast they use oxygen, and put the two with the fastest rates together, and so on.

For (1), with four pairs of littermates, blocking would be a good strategy, because hamsters with similar genes would tend to have similar enzyme concentrations. However, you're not likely to get hamsters that come to you in genetically similar pairs like that. For (2), sorting by weight, blocking is probably not so good: there's no reason to expect hamsters that weigh about the same to have similar enzyme concentrations. For (3), sorting by oxygen rate, blocking might be worthwhile, depending on whether enzyme concentrations are related to rate of metabolism.

Nonrandomized block designs. Sometimes the two design principles of random assignment and blocking are in conflict. To see this, imagine that Kelly was interested only in comparing hearts and brains, using four hamsters, all raised with long days.

Think first about using something like a completely randomized design: use a chance device to assign the conditions to equal numbers of hamsters. Here, the two treatments are "Heart" (measure the enzyme concentration in the heart) and "Brain." For a balanced design, you would assign "Heart" to two hamsters chosen at random and assign "Brain" to the other two.

I hope you see that this design is really very inefficient. After all, each hamster could provide two measurements: one heart and one brain. Why not use them both? That way we'll get a total of eight measurements from the four hamsters. This logic gives us a complete block design, but one with an important shortcoming: there is no randomization.

Long-days Hamster	Heart	Brain	Hamster Average
#4	1.490	6.625	4.058
#1	1.525	10.375	5.950
#7	1.555	9.900	5.728
#2	1.790	8.800	5.295
Average	1.590	8.925	5.258

Example 1.3 Nonrandomized Complete Block Design for Comparing Hearts and Brains

Conditions: the organ chosen—heart or brain
Units: chunks of hamsters (8 in all, 2 chunks from each hamster)
Blocks: hamsters (4 in all; all raised with long days)
CB design: each block (hamster) provides one heart measurement and one brain measurement, *but* the conditions (heart and brain) are *not* assigned to the units (chunks of hamsters), and there is no randomization. ■

This block design has two advantages over the design that randomly assigns two hamsters to "Heart" and two to "Brain." First, we have twice as many measurements for the same number of hamsters. Second, the comparisons we can make using this design are more efficient. Each row corresponds to a single hamster; thus each row allows us to compare a heart concentration with a brain concentration, without the interference of hamster differences.

Observational versus experimental conditions. In an important way the last example (for comparing hearts and brains) differs from the first two (for comparing long and short days). In Examples 1.1 and 1.2, we were able to assign conditions: day length was a treatment. In the last example, we could not assign the conditions of interest: the hamsters came with their hearts and brains already built in. Because the difference between the two kinds of conditions is often important, statisticians use different words for the two kinds. Those you can assign, like day length in the first two examples, are called **experimental** conditions; those that come built in, like heart and brain in the last two examples, are called **observational**. If you have a choice, it is more scientifically sound to make the conditions experimental rather than observational, so that you can randomize.

Looking ahead. Why is this important? In large part because of the way statistical analysis of data relies on the assumption that errors are chancelike. It is usually safe to assume that measurement errors are chancelike, and, provided you randomize properly, it is also safe to assume that variability in your material will behave in a chancelike way. If you can't or don't randomize, however, you still need the assumptions in order to do parts of the analysis, but the assumptions may not fit your data. Chapter 2 will spell out these assumptions and introduce some of the ways you can use your data to check them. Chapter 12 will go into this in more detail.

The RBF and CB examples illustrate two of the three basic principles Kelly used in planning her experiment: random assignment and blocking. The two designs give two alternative strategies for assigning treatments to units. So far, however, I haven't said much about the nature of the treatments that get assigned. The third basic design principle, factorial crossing, deals with situations where the treatments themselves have a particular kind of structure.

Factorial Crossing and the Two-Way Basic Factorial (BF[2]) Design

Third design principle: Factorial crossing

If you want to compare the effects of two or more sets of conditions in the same experiment, cross them: take the set of all possible treatment combinations as your conditions.

Of all the questions Kelly's actual experiment was designed to answer, the one I find most interesting is the one about interaction: is the difference in enzyme concentrations between long and short days the same for hearts as for brains? I wasn't surprised to find that brains have a higher concentration of sodium pump enzyme than hearts: it makes sense that the brain should have a higher capacity for nerve activity than the heart. Nor was I surprised that the data seem to show that when days are shorter, the enzyme concentrations in hamster hearts are lower. We know that days grow short as winter arrives and that hearts slow down when an animal hibernates, so it makes sense that when days are short, an animal's heart would not need as big a capacity for nerve activity. What about brains? They are obviously not the same as hearts, so there's no reason to expect them to react to a change in day length in the same way that hearts do. Perhaps, since brains have such huge concentrations of enzyme when days are long, they will react to short days by showing a huge drop in enzyme concentrations, a much bigger drop than hearts show. Or perhaps brain work doesn't slow down much at all when an animal hibernates. In that case the enzyme concentrations for short days might not be much different from the concentrations for long days. Perhaps, during hibernation, the capacity for nerve activity, as measured by enzyme concentration, is not directly related to the immediate need for nerve activity, so that the brain stores up enzyme that it doesn't need right away.

Imagine planning an experiment to study these issues. Until now, each of the designs we've looked at has been for comparing just one set of conditions, either long days versus short, or hearts versus brains. Here we want a design that lets us compare both sets of conditions at once. The basic idea is quite simple. In order to study both day length and organ, we include all possible combinations of day length and organ as conditions in our experiment: (1) long days/heart; (2) long days/brain; (3) short days/heart; and (4) short days/brain.

Once we've decided to use this set of combinations as our conditions, planning the rest of the experiment involves no new ideas. As before, we need to choose experimental units and then choose a strategy for assigning conditions to units. As always, the simplest strategy is complete randomization: we take the eight hamsters as a set of experimental units, and randomly assign one of the four conditions to each hamster. For balance, each condition is assigned to two hamsters. This design is called a randomized **two-way** basic factorial (RBF[2]) design.

Condition Day Length/Organ	Hamster	Enzyme Conc. (mg/ml)	Condition Average
Long days/heart	#7	1.555	1.673
	#2	1.790	
Long days/brain	#4	6.625	8.500
	#1	10.375	
Short days/heart	#3	1.385	1.320
	#5	1.255	
Short days/brain	#8	11.625	12.425
	#6	13.225	

EXAMPLE 1.4 RANDOMIZED TWO-WAY BASIC FACTORIAL EXPERIMENT FOR THE HAMSTERS

The four conditions are the four possible combinations of (1) day length (long or short) and (2) organ (heart or brain). As in any balanced RBF design, each condition is randomly assigned to the same number of units. ■

The two-way treatment structure is easier to see if we rearrange the treatment combinations and the data into two-way tables:

Treatment Combinations

	Heart	Brain
Long	Long days/heart	Long days/brain
Short	Short days/heart	Short days/brain

Data

	Heart	Brain
Long	1.555	6.625
	1.790	10.375
Short	1.385	11.625
	1.255	13.225

The table on the left shows the four possible treatment combinations. The first treatment, day length, corresponds to rows. The second treatment, organ, corresponds to columns. Two treatments are **crossed** if all possible treatment combinations occur in the design. In terms of the table, the design should include a condition for each little box, or **cell**, where a row crosses a column. The table on the right shows the eight observed values, which you can use to check for interaction. Look first at the heart column, and make a rough estimate of the difference in enzyme concentrations between long and short days. Then do the same for the brains. Is the difference between long and short days about the same for hearts as for brains?

Factors and levels. In thinking about factorial crossing, remember that statisticians reserve the word "treatments" for experimental conditions, that is, for conditions you can actually assign to your units. The third design principle also applies to observational conditions, those like "male" and "female" that you can't assign because they come already built into your material. It's useful to have a substitute word for "treatment" that we can use for observational as well as experimental conditions; statisticians use the word **factor**. For the design of Example 1.4, Day Length and Organ are factors. Notice that the factor is Day Length, not long days or short days; these are called **levels** of the factor. (See the Review Exercise 7 for why statisticians chose the word "level.") In this language, heart and brain are levels of the factor Organ. Two factors in a design are **crossed** if every possible combination of the form "level of 1st factor, level of 2nd factor" occurs as a condition in the design.

Remind yourself by way of review that whereas complete randomization and blocking are strategies for assigning conditions and say nothing about how you choose the conditions, factorial crossing is a strategy for choosing conditions and says nothing about how you assign them. You can use factorial crossing together with complete randomization, or with blocking, or with various combinations of the two.

Kelly's Experiment: A Split Plot/Repeated Measures Design, or SP/RM

At this point you've seen three design principles and a simple design based on each principle. With all three principles in hand, we can now go back to Kelly's actual experiment to see how it uses them all. Her plan is called a **split plot/repeated measures** design, or SP/RM. (Chapter 9 discusses such designs in detail.) First, we randomly assign long days to four of the eight hamsters and short days to the other four. Then we use each hamster twice, once to get a heart concentration and once to get a brain concentration.

	Hamster	Heart	Brain
Long days	#4	1.490	6.625
	#1	1.525	10.375
	#7	1.555	9.900
	#2	1.790	8.800
Short days	#3	1.385	12.500
	#8	1.485	11.625
	#5	1.255	18.275
	#6	1.285	13.225

Example 1.5 Kelly's Split Plot/Repeated Measures Design

- There are two sets of conditions to compare: long days versus short, and hearts versus brains. All possible combinations of long or short, heart or brain, occur as conditions, as in a two-way factorial design.
- There are units of two different sizes: hamsters and chunks of hamsters. Each larger unit (hamster) is a block of smaller units.
- One set of conditions (day length) gets assigned to the larger units (hamsters) completely at random, as in a RBF design.
- The other set of conditions (organ) gets associated with the smaller units (chunks of hamsters) separately within each block, as in a CB design. ■

A key feature of the SP/RM design is that it has units of two different sizes, one size for each factor of interest. The larger unit is a hamster; day lengths are assigned to hamsters completely at random, as in an RBF design. The smaller unit is a chunk of a hamster; these units come grouped in blocks of similar units, two chunks per hamster. Levels of the observational factor (organ) are associated with these smaller units, separately within each block, as in a CB design. (If "Heart" and "Brain" were experimental treatments that we could assign instead of observational conditions, we would use a chance device to randomize the assignment separately within each block of units.)

Notice how Kelly's design uses all three design principles:

(1) **Factorial crossing**: all possible combinations of day length and organ occur as conditions in the design.

(2) **Randomization**: long and short days are assigned to hamsters completely at random.

(3) **Blocking**: the chunks of hamsters associated with hearts and brains come (pre)sorted into blocks, two chunks per hamster.

So far, this chapter has emphasized "what you do": how you choose your conditions and how you assign them to units. The next section shifts the emphasis to "what you get," that is, to the structure of the data sets that come from the designs. Each design has its own set of meaningful averages, and it's worth learning to think about the different designs in terms of which sets of averages have meaning.

3. The Factor Structure of the Four Experimental Designs

Every experimental design has its own characteristic set of meaningful averages, or, what amounts to the same thing, its own set of meaningful ways to sort the observed values into groups.

Now that you've seen the ideas Kelly used to choose the design for her experiment, it's time to look more carefully at the relationship between her design and the list of five questions she wanted to study. What are the effects of (1) day length and (2) organ? (3) Do day length and organ interact? (4) How big are the hamster differences? (5) What is the typical size of the measurement errors?

Each question involves one or more comparisons. To make each set of comparisons, we'll first sort the observed values into groups, then compute an average for each group, and compare the group averages. It is important to understand that whenever you compute an average, you work with a group of observed values—whichever ones you add up to get the average.

It may help to think of each measurement written on a 3 × 5 file card together with the conditions for the measurement. For example, the card for the measurement in the upper left corner in Example 1.5 might look like this:

	Conc.:	1.490	mg/ml
Conditions:			
Hamster:	#4		
Day length:	long		
Organ:	heart		

FIGURE 1.2

As we come to each question and the comparison it involves, you can think of sorting the 16 cards into piles, computing the average concentration for each pile, and comparing the averages. Consider day length, for example. If you look over the 16 measurements in Example 1.5, you'll see that there are 8 measurements (top half of the table) for long days and 8 (bottom half) for short days. To compare long days with short, we would sort the 16 cards into two piles of 8, one for long days and one for short, then compare the average of the 8 long-day measurements with the average of the 8 short-day measurements. The 8 and 8 balance, but there is more involved than just that. Exactly half of each pile of 8 cards is for hearts, and the other half is for brains. So for long days, the 4 heart measurements (left side of the table) balance the 4 brain measurements (right side), and the same is true for short days. There's yet another kind of balance as well. Ideally, we'd like to have each hamster contribute two measurements, one for long days, and one for short, but we can't have that, because you can't measure the enzyme concentration without sacrificing the hamster. Remember, though, that the hamsters were assigned a day length using a chance device. That means that each hamster has the same chance as each other hamster of ending up with long days. The beauty of using the chance device is that we get a kind of balance after all: some hamsters with higher enzyme concentrations are assigned to long days and some to short, and so the hamster differences tend to balance each other.

For each of the four designs of Section 2, some sets of averages will be worth looking at because they will tell you something meaningful about the data, but other sets of averages, although you could compute them, wouldn't tell you anything meaningful.

Each way of sorting the observations into meaningful groups is called a factor; each of the groups is called a level of the factor.

Notice that here the meaning of the word "factor" is not the same as it was in the last section although the two meanings are very closely related. It's too bad that statisticians use the same word to mean two somewhat different things, but that's how it is. Until you get used to both meanings, you may find that it helps to remind yourself that the earlier meaning is for "what you do," and the new meaning is for "what you get."

From the point of view of "what you get," you can tell various designs apart using their sets of factors, or **factor structure**.

Factor Structure of the One-Way Basic Factorial (BF[1]) Design

In thinking about various experimental plans, it is useful to have a simple shorthand notation for summarizing and comparing their structure. This shorthand is simply a way of using pictures to record the ways the observations are sorted into groups.

Turn back to Example 1.1, and look over the data for that one-way basic factorial design: which averages are worth computing? The eight observed values are arranged in two columns, four long-day observations in the first column and four short-day observations in the second column. The purpose of the experiment was to compare long and short days, so it makes sense to compute a long-day average (add the four numbers in the first column, then divide by 4) and a short-day average. Each of these two averages is based on a group of observed values; together, the two averages correspond to sorting all eight observations into two groups, column 1 and column 2:

	Long	Short
	1.490	1.385
	1.525	1.485
	1.555	1.255
	1.790	1.285
Avg.	1.590	1.353

Shorthand

FIGURE 1.3 Factor for day length, with the observations sorted into two groups.

Notice two things about this factor: every observation belongs to a group, and no observation belongs to more than one group. (These two properties will be true of every factor considered in this book.) Notice also the shorthand I have used at the right side of the picture to represent the factor. The picture tells you that this factor, Day Length, sorts the observations into two groups that correspond to the left and right columns of the table of data. The two groups, for long and short, are the two levels of the factor.

Factor Structure of the One-Way Complete Block (CB[1]) Design

Turn back to the complete block design of Example 1.3, which used four hamsters as blocks in order to compare hearts and brains. Look over the data, and think about which sets of averages have meaning. How many factors are there for this design?

Here, just as for the basic factorial example, it makes sense to compute an average for each column, this time in order to compare hearts (column 1) and brains (column 2). Thus one factor, for organ, sorts the observations into two columns. In addition, for this design, it makes sense to compute row averages. Each row corresponds to a different hamster; comparing row averages tells us about the differences between hamsters (blocks). Thus another factor, for hamsters, sorts observed values into four rows. Notice that for the basic factorial example, row averages do not make sense. For the block design, the two numbers in a row have something in common: they belong to the same hamster. For the factorial design, the two numbers in the first row belong to different hamsters; in fact, it's purely a matter of chance that the two numbers happened to end up in the same row.

For the block design, we've found two sets of meaningful averages: organs (= columns = conditions) and hamsters (= rows = blocks). (See Fig. 1.4.) Every one-way complete block design will have these same two ways of sorting the observations—conditions and blocks.

Heart	Brain
1.490	6.625
1.525	10.375
1.555	9.900
1.790	8.800

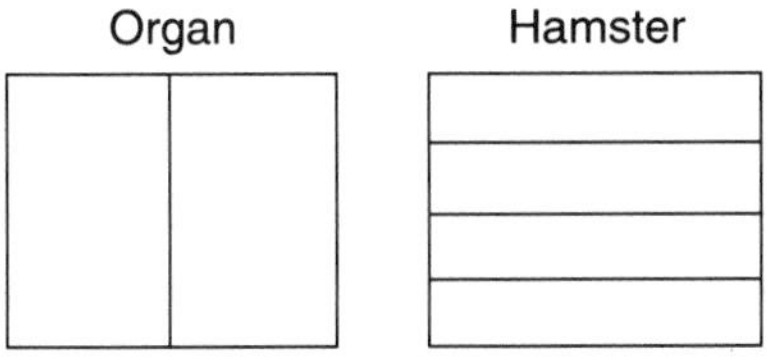

Organ

Hamster

FIGURE 1.4 Data from the complete block design, showing two meaningful sets of averages. The two factors, organs and hamsters, correspond to columns and rows. Every CB[1] design has these same two factors, for conditions (or treatments) and for blocks.

Nuisance factors and factors of interest. Notice that the goal of this experiment—to compare hearts and brains—involves one of the two factors (organ) but not the other (hamsters). This is typical of an experiment in blocks. One of the two factors is directly related to the goal of the experiment and is sometimes called a **factor of interest**. The other factor, hamsters, became a factor for us only because we chose to make it one in order to deal efficiently with the natural variation in the experimental material, in our case natural differences between hamsters. Such a factor is sometimes called an extraneous factor, but I prefer to use the less formal name **nuisance factor**.

Finally, notice that the factor of interest has two levels (heart and brain) and that each hamster, each block, provides one measurement of each kind. In a **complete** block design, each block provides a measurement for each and every level of the factor of interest. In other words, the nuisance factor and the factor of interest are **crossed**; all possible combinations of organ and hamster number occur in the design.

Factor Structure of the Two-Way Basic Factorial (BF[2]) Design

Turn back to the data given in Example 1.4, and see if you can find three sets of meaningful averages. It may help you to ask which averages you would compare to judge the overall difference between long and short days, the overall difference between hearts and brains, and whether the difference between long and short days is the same for hearts as for brains.

1. *Day length.* To judge the overall effect of day length, we first compute the average of all four long-day observations (first two rows of data), lumping the hearts and brains together. This average is 5.086 mg/ml. In the same way, we compute an average of 6.873 for the four short-day observations (last two rows). This factor sorts the observations into groups of rows: long days (rows 1 and 2) and short (rows 3 and 4).
2. *Organs.* Here we first compute the average of the four heart measurements (column 1), lumping long and short days together, then compare them with the average of the four brain measurements (column 2). This second structural factor sorts the observations into columns. The averages work out to 1.496 mg/ml (hearts) and 10.463 mg/ml (brains).
3. *Interaction of day length and organs.* Here, as always, interaction refers to a difference of differences. We first find the difference between long and short days for the hamster hearts and then find the difference for brains. To do this we work with the four condition averages, or cell averages, as Fig. 1.5 shows. (Each little box in the two-way table is called a **cell**.)

	Heart	Brain
Long	1.673	8.500
Short	1.320	12.425

FIGURE 1.5 Cell averages for the two-way BF design. For hearts, the long-day average is 0.353 mg/ml *higher* than the short-day average. For brains, the long-day average is 3.925 mg/ml *lower* than the short-day average. The difference due to day length is not the same for hearts as for brains: interaction is present.

	Heart	Brain	Day Length	Organ		Interaction	
Long	1.555	6.625					
	1.790	10.375					
Short	1.385	11.625					
	1.255	13.225					

FIGURE 1.6 Three meaningful sets of averages for the two-way factorial experiment. In any two-way factorial design, the factor for interaction comes from crossing the two factors of interest (here, Day Length and Organ), placing one on top of the other to get the interaction groups, or cells.

Two additional factors present for every design. Although the factors we have identified are the most useful for telling one design from another, each of the designs we have looked at also has two more factors, one for the **grand average** and one for what we shall call **residual error**.

1. *Grand average.* It makes sense to compute an overall average of all the observations, as a kind of benchmark value that tells us the average enzyme concentration for the experiment as a whole. The factor for the grand average puts all the observations together in one group; or, in other words, this factor has only one level.
2. *Residual error.* This factor corresponds to the differences between individual observed values; the differences are due in part to measurement error, which is chancelike; in part to variability in the material, which is chancelike if we have randomized; and in part, perhaps, to other sources. In terms of sorting, the factor for residual error always puts each observation in a group by itself; the factor has as many levels as there are observations. To get a sense of how big the residual errors are for this (or any) experiment, we compare individual observed values with each other, one against one. For these comparisons, each "average" is just a single observed value.

These two factors, for the grand mean and for residual error, will always be present, automatically, as part of the structure of every design in this book. So, in comparing designs, we can ignore these two factors, because they give no information about what makes one design different from another. It will be handy to have a name to distinguish these two factors, which I'll call **universal**, from the other, **structural** ones. In this language, the one-way basic factorial design has only one structural factor (treatments), the complete block design has two structural factors (treatments and blocks), and the two-way basic factorial design has three (two factors of interest and their interaction).

Factor Structure of the Split Plot/Repeated Measures (SP/RM) Design

What you do: comparisons with simpler designs. Remember that the SP/RM design combines features of the other three designs of this chapter. Like the two-way basic factorial, this repeated measures design has two factors of interest. All possible combinations of the two factors are present (the two factors are crossed), and there is more than one measurement for each combination (4 per combination, in this case) so it is possible to measure interaction. For one of the two factors of interest—day length—the design is like a randomized one-way basic factorial design: we randomly assigned levels of the factor, long and short days, to the hamsters. For the other factor of interest—organ—the design is like a complete block design, with the nuisance factor, hamsters, providing the blocks.

What you get: the factor structure. To get the list of factors for Kelly's experiment, we begin with the list of comparisons from page 13:

1. Day length (long versus short)
2. Organs (heart versus brain)
3. Interaction of day length with organs
4. Hamsters
5. Residual error

1. *Day length (between blocks)*. The first structural factor corresponds to sorting the 16 observations into two groups of eight. The first four rows of data contain the eight numbers from the long-day hamsters, and the last four rows contain the numbers for short days, so the two groups for this factor correspond to rows 1–4 and rows 5–8.

   Statisticians call day length a between-hamsters factor, because when you sort the data into piles for long and short days, there is one set of hamsters in the long pile (numbers 4, 1, 7, 2) and a different set of hamsters in the short pile. To compare long with short, you compare two different groups of hamsters.

2. *Organs (within blocks)*. The second structural factor sorts the observations into two columns, one for hearts and one for brains. Organ is a within-hamsters factor, because you can compare levels of that factor, hearts and brains, using the same hamster. If you sorted the data cards into a pile for hearts and a pile for brains, each hamster would contribute two cards, one to each pile.

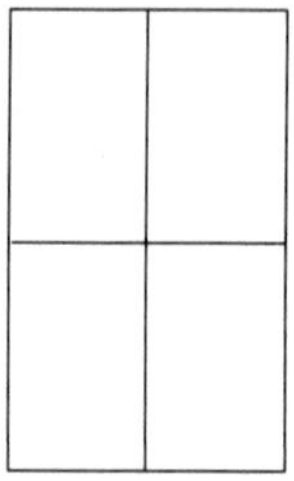

3. *Interaction of day length and organs*. This factor sorts the numbers into four groups. Remember that here the comparison we want is a two-layer comparison: we start with the separate effects of day length for hearts and for brains, then compare these effects. To make the comparison, we need to sort the eight numbers in the heart column into two groups, for long and short days, and then do the same for the eight numbers in the brain column.

4. *Hamsters*. This fourth factor gives us eight groups of two numbers each, one group for each hamster. Because hamsters correspond to rows in the table, the hamster factor sorts observations into rows.
5. *Residual errors*. The fifth and last set of comparisons, for judging the sizes of the residual errors, corresponds to one of the two universal factors, and puts each observation in a group by itself.

In all then, we have four structural factors, plus the two universal ones, as summarized in Fig. 1.7.

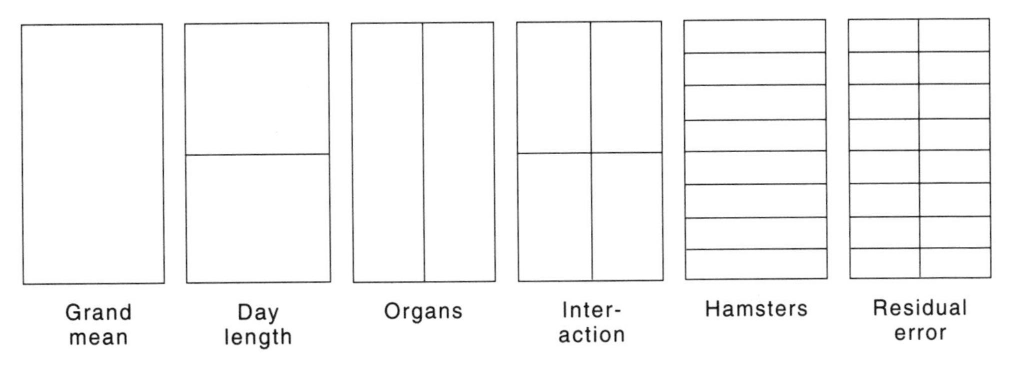

FIGURE 1.7 Factor structure of Kelly's split plot/repeated measures design. Every such SP/RM design has four structural factors:

1. Between-blocks factor of interest (day length)
2. Within-blocks factor of interest (organs)
3. Interaction of the two factors of interest
4. Blocks (hamsters), a nuisance factor

Looking ahead. The hamster experiment is very efficient, but the efficiency of the design carries a cost of added complexity. The way we've planned things, each measurement contains several pieces of information, all neatly packaged in a single number, the enzyme concentration. For example, measurement #1 in the list contains pieces of information about long days, about the heart, about interaction, about hamster #4, and about the chance error built into the measurement process. How can we possibly split those pieces of information apart when all we have is the measurement? For now, I hope you'll be satisfied just to know that it is possible and, in fact, not terribly hard if you realize that the alternative is to buy many more hamsters, take care of them in the lab, do additional dissections, and so on. The next two chapters introduce a way to split the pieces of information apart and a way to draw sound conclusions from the results. Basically, what's involved is like sorting the cards into piles, except that it's done using arithmetic.

APPENDIX: OTHER WAYS TO GUARD AGAINST BIAS

Bias can result from the natural variability in your material, especially if you don't randomize, but it can also arise from other sources as well, particularly in two kinds of situations: those in which you do something to living organisms, and those in which your measurement depends on human judgment. Three useful ideas for avoiding bias in these situations have shorthand names: **placebos**; "**blind**" **subjects**; and "**blind**" **evaluation**. I'll take them in order.

Guarding Against Unplanned, Systematic Variation in the Conditions

Placebos. Doctors and medical researchers have known for a long time that you can often make sick people feel better, or keep healthy people from getting sick, just by giving them what they think is medicine, even if there are no active ingredients in what you give them. This fact can easily make a drug look good when the only effect comes from taking something that a person in a white coat gives you, regardless of what that something really is. The bias that comes from the reassurance of taking an otherwise worthless drug substitute is called the **placebo effect**, and the drug substitute itself is called a **placebo**. If you want to test a new drug, you should plan your study so that if the people who get the drug get better, or don't get sick, you can be sure it was the drug and not just the placebo effect that was responsible. One standard plan based on this idea uses a chance device to divide subjects into two groups. Half get the real drug, and half get the placebo.

"Blind" subjects. Of course, using a placebo won't avoid bias if the people who get the placebo know what they're getting. To avoid bias from the placebo effect, you have to make the placebo so much like the real thing that your subjects can't tell the drug from the placebo, and you have to make sure the subjects don't find out which one they get. If the people in your study are kept in the dark like this, they are called **blind subjects**. Using blind subjects is absolutely necessary if you are to avoid the bias of the placebo effect. However, here, as so often, there's more to the story than just statistics. Secretly giving a placebo to people who need treatment is unethical. (Standard practice is to tell all your subjects that some of them, chosen by a chance device, will get the placebo and others will get the drug.)

Control groups. What if you don't plan to do medical research? Although the word "placebo" is most often used in connection with medical studies of human subjects, the basic idea—using a substitute treatment to avoid bias—is much more general and can be useful in any study of living organisms, plants as well as animals. For example, a colleague at Mount Holyoke, a plant ecologist, used the idea in designing a study to see whether buttercups that grew naturally in damp woods would grow equally well in a sunny field. Her plan was to transplant some of the buttercups from the woods to a field and later compare them with other buttercups she'd left in the woods. She knew, however, that plants don't like to be dug up and that to compare transplants with other buttercups that had never been dug up would bias her results by putting the transplants at a disadvantage. So she dug up twice as many plants as she planned to put in the field, and she used a chance device to sort them into two groups. One group went to the field and the other, the **control group**, went back to the woods. By making the control group as much like the treatment group as possible, except for ending up in the field, she avoided bias that would otherwise have made her results worthless.

Guarding Against Unplanned, Systematic Variation in the Response

Blind evaluators. Another sort of bias can undermine experiments for which the measurement depends on human judgments, as many studies in medicine and psychology do. For example, in the 1950s a large study involving thousands of school children was done to test a new vaccine that was supposed to prevent polio. Some of the children got the vaccine in a shot, and some got a placebo shot. The subjects were divided into treatment and control groups using a chance device, and none of them knew which group they were in, so that part of the experiment was sound: the design would prevent the bias of the placebo effect. However, there was another source of bias to worry about. Someone had to decide who among the children who got sick had polio and who had some other disease. Medical diagnosis is not easy, and a doctor is almost certain to be influenced by what he or she expects to find. In many situations, it may be quite proper for expectations to influence a diagnosis, and impossible to avoid, as well, but the doctors in the polio study didn't want to be influenced by knowing which children got the real vaccine. To avoid this kind of bias, sometimes called evaluator bias, the study was set up so that the doctors who made the diagnoses didn't know who had gotten the vaccine and who had gotten the placebo: they were **blind evaluators**. A study like this one, with both the subjects and the evaluators kept in the dark about who got the placebo, is called **double blind**.

SUMMARY

1. *Planning an experiment: General concepts and strategies*

   a. **Content**: In planning almost any experiment, you first have to decide (1) what **measurement** to make (the **response**), (2) what **conditions** to study (the **treatments**), and (3) what **experimental material** to use (the **units**).

   b. **Sources of variability**. Try to identify all major sources. Most data sets contain variability from three sources: the conditions; the material; and the measurement process. A good design will let you estimate the amount of variability due to each source.

   c. **Three kinds of variability**: (1) planned, systematic variability we want; (2) chancelike variability we can live with; and (3) unplanned, systematic variability that threatens disaster.

   d. **Measurement error**. Every scientific measurement contains error. In order to estimate the typical size of such measurement errors, you need to make more than one measurement with conditions held constant. The measurement errors typically behave in a chancelike way, and statistical analysis relies on certain predictable regularities of chancelike behavior. (In particular, chancelike errors tend to go in opposite directions and thus cancel each other when you take averages.)

   e. **Bias**. Unlike measurement error, which is chance-like, bias tends to go all in the same direction, to make all your measurements larger than the true value, or all of them smaller.

   f. **Sampling from a population**. The **population** is the entire set of values you want to know about; the **sample** is the set you observe. A **parameter** is a numerical property of the population; a **statistic** is a numerical summary computed from the sample. Ideally, you choose your sample from the population at random, so that each value has the same chance of getting chosen. (The process of random sampling is unbiased.)

   g. **Isolate the effects of interest: Control what you can, and randomize the rest**. As a rule, you should use a chance device to randomize the assignment of conditions to your experimental material. There are two main reasons: (1) randomizing reduces the risk that the variability in your material will bias your results; (2) randomizing makes the variability behave in a chancelike way, which allows you to rely on the regularities of chance behavior when you analyze your data.

2. *Factor structure*

   a. A **factor** is a way of sorting the observed values into groups so that every observation belongs to a group, and no observation belongs to more than one group. The groups are called **levels** of the factor. The complete list of factors for a design is called its **factor structure**.

   b. Every design has two **universal** factors, one for the grand mean (all observations in the same one group) and one for residual error (each observation is by itself in its own group). The other, **structural**, factors are what make the structure of one design different from another.

3. *Treatments, units, and four experimental designs*

   a. In a true **experiment**, the conditions of interest, or **treatments**, get assigned to chunks of experimental material, or **units**, using a chance device. In an **observational study**, the material comes with the conditions already built in. (For designs with more than one factor of interest, some factors may be experimental, others observational.)

b. For a **randomized basic factorial** (RBF) experiment, you randomly assign treatments to units, regarding the units as interchangeable. In a balanced RBF, each treatment is assigned to the same number of units. The simplest (one-way) basic factorial, or BF[1], has only one structural factor, treatments.

c. For a **complete block** (CB) experiment, you first sort your experimental material into blocks of similar units and then randomly assign treatments to units separately within each block, regarding the units in each block as interchangeable. Within each block, each of the treatments is assigned to one unit. The CB[1] design has two structural factors: one factor of interest and one nuisance factor (blocks).

d. Two factors in a design are **crossed** if all possible combinations of levels of the two factors occur in the design. The simplest design with crossed factors is the **two-way RBF**: you randomly assign treatment combinations to units just as for the one-way RBF. The two-way RBF has three structural factors: two factors of interest and their interaction. (In order to measure interaction, you need to have more than one measurement for each combination of factor levels.)

e. The **split plot/repeated measures** (SP/RM) design has units of two different sizes. The larger units are blocks of smaller units. Levels of one factor of interest, the **between-blocks** factor, get assigned to the larger units as in a RBF. Levels of the other factor of interest, the **within-blocks** factor, are assigned to the smaller units separately within each block, as in a CB. The simplest SP/RM has four structural factors: two (crossed) factors of interest, their interaction, and one nuisance factor (blocks).

NOTES AND SOURCES

1. ***Data sets.*** Throughout the book all data are real unless noted otherwise. Numbers marked * in a data set were changed slightly to simplify arithmetic. Sources are given in the index to data sets at the end of the book.

2. ***Animal experiments.*** I'm not automatically against research that sacrifices animals. Nevertheless, I urge you to think not only about whether the results of an experiment are important enough to justify taking the hamsters' lives but also about whether there might be other ways to get the results.

3. ***Interaction.*** The word "interaction" has both an informal, everyday meaning and a more precise, specialized statistical meaning. The informal meaning can help you recognize situations where the statistical meaning might apply, but it is often too vague for scientific work. In statistics, interaction refers to a *difference of differences*, which you get by subtracting averages of observed values. For the hamster data, the question "Is there interaction?" is shorthand for "Is the difference between average enzyme concentrations for long and short days the same for hearts as for brains?" (I'll illustrate this meaning using numbers in the next chapter.)

4. ***The box model*** is, in a different guise, almost as old as probability theory, but my particular description of it owes much to David Freedman, Robert Pisani, and Roger Purves (1978), *Statistics*. New York: W. W. Norton & Company. There's now a third edition (1998).

5. ***"What you do" versus "what you get."*** Both points of view are important, but in general statistics books don't make a clear distinction between them, which is unfortunate, I believe, because some statistical terms like "factor" have somewhat different meanings, depending on the point of view. (1) Notice that in terms of "what you get," the two structural factors Day Length and Organs correspond exactly to the two treatment factors that we crossed in Section 3 to get the BF[2] design. Even though the two meanings of "factor"—for "what

you do" (Section 2) and for "what you get" (Section 3)—are not the same, the meanings *are* closely related. (2) Recall the two examples, one experimental and the other observational, of CB[1] designs from pages 17–18. They are clearly different in terms of "what you do"; but looked at in terms of "what you get," the two designs have the same factor structure. (3) Finally, note that one of the everyday meanings of "factor" is "something that contributes to a result," which is pretty close to the more specialized statistical meaning.

6. ***Whole plot error and subplot error.*** Remember that the SP/RM design has units of two sizes, and notice that the last two factors in its factor list on page 28 correspond to the two sizes of units: hamsters (larger units) and residual error (smaller units). The variability from the larger units is often called **whole plot error**, and the variability from the chunks of hamsters is **subplot error**. For this particular example, the assignment of day lengths to hamsters was randomized, and we can safely assume that the whole plot error will be chancelike. For the subplots, however, conditions were not assigned, and so the subplot (= residual) errors may well not be chancelike.

7. ***Is Kelly's design really an SP/RM?*** There are sound reasons why some statisticians would disagree with my calling Kelly's design an SP/RM. I take these reasons up in Chapter 2, Section 4, and then briefly consider a different model (a bivariate RBF[1]) for the data set. I mention the possibility of disagreement here mainly to emphasize that statistical work is rarely automatic. This book does not attempt to explain multivariate methods, although in Chapters 4, 8, and 12 I do discuss ways to recognize data sets for which you need these methods. As for Kelly's data set, it passes tests of the assumptions that must be satisfied in order to use methods appropriate for the SP/RM design. In my judgment, regarding the design as an SP/RM provides one, but not the only, reasonable way to think about her data.

8. ***Notation for the designs.*** It is convenient to have quick ways to refer to the various designs. We use an R to indicate that all treatments are randomly assigned. Thus BF is for any basic factorial, and RBF is for a randomized BF. (Many books call the RBF a completely randomized design, abbreviated CR.) Similarly, CB refers to any complete block design, and RCB refers to a randomized CB. A number in square brackets after the design tells the number of crossed factors of interest. So far, we have seen a BF[1] and a BF[2], but we'll eventually see many more variations. For the SP/RM design, we need two numbers: the first tells the number of between-blocks factors; the second tells the number of within-blocks factors. In this notation, Kelly's design is an SP/RM[1,1].

REVIEW EXERCISES

1. List the three decisions Kelly made as the first step in planning her experiment.

2. This chapter listed three sources of variability present in almost every data set. Which source is for the variability that

a you can nearly always expect to be chancelike?

b. will be chancelike if you assign conditions using a chance device?

c. will nearly always not be chancelike?

3. Fill in the blanks, choosing between "bias" and "chance error."

a. A major reason for wanting several measurements under the same conditions is to be able to estimate the size of ____________.

b. When ____________ is present, all your measurements tend to be off in the same direction, either too high or too low.

c. ____________ tends to go in different directions for different measurements and to cancel when you compute averages.

4. True or false:

a. An important reason why large samples are better than small ones is that the bigger the sample, the smaller the bias.
b. If you have a sample of several measurements, all made under the same conditions, you can usually tell, just from looking at the numbers, whether the sampling process is biased and whether the measurement process was biased.
c. If you have a sample of several measurements, all made under the same conditions, you can usually get some idea, just from looking at the numbers, about how big the measurement errors are.
d. An important reason why large samples are better than small ones is that chance-like errors tend to cancel each other when you compute averages.
e. A statistic is a number you compute from a sample.
f. The population average is an example of a statistic.
g. Parameters are used to estimate statistics.
h. For a random sample, each individual in the population has the same chance of being in the sample.

5. For this exercise, the population is the box of numbered tickets shown below:

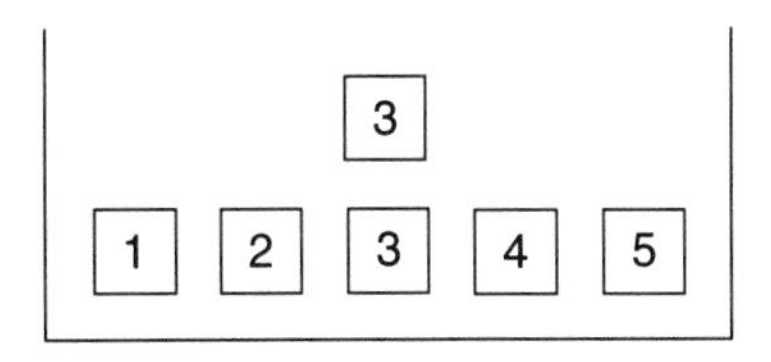

a. The population average is equal to ________.
b. If you draw a ticket at random and define chance error to be (# on ticket) − (population average), then there are five different possible values for chance error. List them.
c. Which possible chance error is most likely?
d. Suppose you get a sample by drawing three tickets at random, and that just by the luck of the draw, your sample turns out to be:

i. The average for the sample is ________.
ii. Which of A or B is correct?
A. The fact that the average for this sample is not equal to the population average proves that the sampling method is biased.
B. The difference between the sample average and population average is due to chance errors that come from the method of sampling.

6. Each of the following sampling methods is biased. For each, tell which individuals in the population are more likely to be chosen, tell which are less likely, and explain why.

a. Population: all fish in Moosehead lake, a large lake in northern Maine
Sampling method: drag a net with one-inch mesh (hole size) behind a motor boat, up and down the length of the lake.
b. Population: all people over age 18 in New York City
Sampling method: run an advertisement in the *New York Times*, asking for volunteers; then choose at random from the list of people who volunteer.
c. Population: all trees in a one-acre lot
Sampling method: assume you have a map of the lot, showing the location of each tree. To get the first tree in your sample, you first choose a point at random on the map,

then take the tree closest to that point for your sample. To get a second tree, you repeat the process: choose a new point at random, and take the tree closest to that point. Continue choosing random points and closest trees until you have a sample of 20 trees. (This method was once in common use, until someone discovered the bias. If you don't see the bias at first, it might help to draw a little map with about 10 trees. Put in several younger trees, which grow close together, and two or three older trees, whose large leaf canopies discourage other trees from growing nearby. Which trees are more likely to be chosen by the sampling method?)

7. List three fundamental principles of experimental design. For each, give the name of two designs that use the principle.

8. Give two reasons for randomizing the assignment of conditions to material.

9. Define a factor, and tell the difference between a universal factor and a structural factor.

10. Suppose your data set consists of one set of measurements, all made under the same conditions. True or false:

a. The design for this data set does not have both of the usual two universal factors.

b. The design for this data set has no structural factors.

11. Short answer:

a. Which two of the designs of this chapter have a nuisance factor as one of the structural factors?

b. True or false: In a complete block design, the factor blocks is crossed with the factor of interest.

c. Why can't you use the complete block design described in this chapter to measure interaction between blocks and the factor of interest?

12. Several years ago a graduate student I knew did a study of the shape of sponge cells. He wanted to see if the shape depends on whether the cells came from a white or a green sponge and on whether the cells came from the tip or the base of the sponge.

a. List a complete set of conditions for such a study. Which of the three design principles applies here?

b. Is sponge color an experimental or observational factor? Is site (tip or base) observational or experimental?

13. A lot of the language of experimental design comes from agriculture. Here's a simple but typical example: you want to study the effect of two fertilizers, sulfate of ammonia (for nitrogen) and superphosphate (for phosphorus), on the yield of Brussels sprouts. You decide to use each fertilizer at four **levels**: 0, 30, 60, and 90 pounds per acre.

a. List a complete set of treatment combinations in a two-way table, as on page 20.

b. Suppose you have a bunch of similar square plots, 64 in all, with the same size and similar soil conditions. Which design principle would you use for assigning treatments to units?

14. (The purpose of this exercise is to illustrate two different ways to use the principle of blocking.) Suppose you want to design a simple experiment in cognitive psychology to study the effects of a distracting noise on short-term memory. You decide to compare two conditions: noise present (treatment) and noise absent (control). You have 20 student volunteers to use as subjects. For each of the following two blocking strategies, tell what the blocks are and what the units are.

a. (**Sorting**: a matched subjects design) Give your subjects a pretest to measure how well they remember words, and use the pretest scores to sort the 20 subjects into 10 pairs. One randomly chosen member of each pair then gets the treatment, and the other gets the control.

b. (**Reusing**: a within-subjects design) There's no pretest, and each subject performs twice, once under each condition. You toss a coin to decide the order of the conditions, and do this separately for each subject.

15. *Needle threading.* No one is going to win a Nobel prize for a study of the factors that affect the speed of needle threading. Nevertheless, please bear with me: this is an exercise that gives you a chance to apply many important principles of design, and I don't have to assume that you have the sort of specialized knowledge that you might need to plan other experiments.

It is reasonable to think that how fast you or I can thread a needle depends both on the color of the thread and on the background behind it. Black thread against a white background would certainly be easier than black thread against a black background. Suppose your goal is to measure the relationship between thread color (white, black, red) and background color (white, black) on the speed of needle threading.

a. Measurement. There are many different ways to measure how fast a person can thread a needle; some of these ways are much better than others. (i) Think of two different ways, and describe each one as if you were writing out instructions for an assistant; assume your assistant is a bit like a computer: very conscientious, but not terribly bright, so that you have to spell everything out in detail. (ii) Which of your two ways is better? Why?

b. Two-way completely randomized factorial design. Tell how to run the experiment as a RBF[2]. List what you regard as any important advantages or disadvantages of the design.

c. Complete block design. List any nuisance influences that you consider likely to be important. Pick the one you regard as most important, and describe two different ways to use that nuisance influence to define blocks as a nuisance factor.

d. Comparison. Which of designs (b) or (c) do you judge better? Give a reason for your answer.

e. Split plot/repeated measures design with gender as another factor of interest. Suppose you want to compare males and females. Tell in detail how you could run the experiment with units of two sizes: describe your units, the between-blocks and within-blocks treatments, and how you would assign them. (Pause for a minute to think about how a careless plan for choosing subjects in design (b) could have introduced a fatal selection bias.)

Recognizing factors. For each of the following data sets, first decide which sets of averages are meaningful. Then draw the factor diagram.

16. *Sleeping shrews.* The numbers below give heart rates, in beats per minute, for four tree shrews, during each of three kinds of sleep: light slow-wave sleep (LSWS), deep slow-wave sleep (DSWS), and rapid eye movement sleep (REMS). (Observations marked * were changed slightly to simplify arithmetic in a later exercise.)

Block(shrew)	**LSWS**	**Condition DSWS**	**REMS**
I	14.0	11.7	15.7
II	25.8	21.1	21.5
III	20.8	19.7	18.3
IV	19.0	18.3*	17.3*

17. *Pigs and vitamins.* For this data set, each of 12 piglets was randomly assigned to get one of four diets. The numbers in the table give weight gains, in pounds per day:

		Antibiotics					
		0 mg			**40 mg**		
B12	0 mg	1.30	1.19	1.08	1.05	1.00	1.05
	5 mg	1.26	1.21	1.19	1.52	1.56	1.55

CHAPTER 2

INFORMAL ANALYSIS AND CHECKING ASSUMPTIONS

OVERVIEW

This chapter has a promise to keep: I've told you that Kelly's split plot/repeated measures design lets you study five different questions, but so far we've hardly looked at Kelly's actual measurements. Instead, we've been looking mainly at the structure of her plan. In this chapter we'll begin to see what the numbers can tell us about the hamsters. We won't do much computing, barely more than an average or two, but I'll show you three useful graphs. It's really quite remarkable how much you can learn just from plotting your data. The three plots in this chapter will help us prepare for the more formal and fancy methods of analysis coming in Chapter 3. Ultimately, in order to separate signal from noise, we'll need to split our data into pieces and use some of the pieces to estimate chance error sizes.

This chapter comes in four sections. The first is a general introduction to the idea of splitting the variation in a data set into pieces, one piece for each factor in your design. Although you can do the arithmetic to split apart any data set in this way, the results will have meaning only if your data set fits a particular set of six assumptions. Section 2 introduces those assumptions. Sections 3 and 4 then describe three different kinds of graphs and begin the important process of checking to see how well the assumptions fit the hamster data. It turns out that for Kelly's data, as for many data sets, you have to transform the numbers to a new scale in order to make one of the assumptions reasonable.

1. What Analysis of Variance Does
2. The Six Fisher Assumptions
3. Informal Analysis, Part 1: Parallel Dot Graphs and Choosing a Scale
4. Informal Analysis, Part 2: Interaction Graph and Scatterplot for the Log Concentrations

1. What Analysis of Variance Does

All the important questions from Chapter 1 involve the idea of residual error, at least indirectly. In order to draw conclusions from the data, we have to find some way to answer the question, How big are these errors? For example, consider the first of Kelly's five questions, asking about the effect of day length. Of the 16 observed values, 8 came from long-day hamsters, and 8 came from short-day hamsters. To summarize the effect of day length on enzyme concentration, we can compute the average for each group of 8 numbers and then ask how far each average is from the grand average of all 16 observations lumped together.

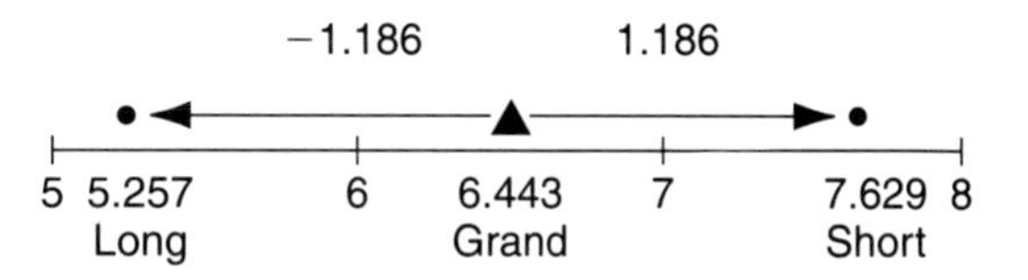

FIGURE 2.1 Estimating the effect of day length. The distances from the two day-length averages to the grand average estimate the effects of long and short days.

If we didn't have residual error to worry about, we might be able to stop with this summary. However, we assume that each of the observed numbers is part true value, part error, and so each of the two averages also contains some error. It's not enough to say, "The long day average is 1.186 mg/ml smaller than the grand average." Part of that 1.186 mg/ml is "true" distance—the "real" effect of long days—but part is error. At this point, we can't yet say how much of the distance might be due to error. However, if it is reasonable to assume that the errors are chancelike, we can then use the predictable regularities of chance behavior to estimate a typical size for the chance errors (a number called the **standard deviation**) and use that size as a yardstick for judging how big the "true" effect of day length is.

Unfortunately, the problem of measuring the errors is made difficult by an obvious but annoying fact: we can't see the errors themselves; we can only see the measurements. Difficulties of this sort are a frequent and familiar challenge in science. For example, you can't see how much enzyme is in a hamster's brain just by looking at the hamster: someone has to invent a method for measuring the concentration. The method we'll use to measure error size is called **analysis of variance**, or **ANOVA** for short. It was invented during the early part of this century by Ronald Fisher, a British geneticist and statistician who also invented much of the theory of experimental design. His method is based on a key idea, associated with the word "analysis," whose story can be traced a long way back in the history of science.

Analysis and splitting: a bit of history. Over 300 years ago, in a darkened laboratory chamber, a man carried out a series of experiments that were to disprove a theory scientists had thought true for over 1000 years. The year was 1665 and the man was Isaac Newton, co-inventor of the calculus and founder of modern physics. Ever since the time of Aristotle, around 350 B.C., scientists had believed that colored light, of the sort you see in rainbows, came about when pure white light was modified in some way. Newton proved the scientists wrong by projecting a narrow beam of white light onto the wall of his darkened chamber in such a way that the white light split into lights of different colors. Thus he showed that white light was, in fact, a mixture made by mixing all the colors of the rainbow.

You can look at Newton's discovery as an example of the very general idea that some "thing" once thought to be pure and unbreakable can actually be split apart and understood in terms of its pieces. For example, from the time of the ancient Greeks, water was considered to be one of nature's four elements, or unbreakable building blocks. Then, around 1775, work by Cavendish, Priestly, and Lavoisier showed that water was made up of hydrogen and oxygen, which came to be called *elements* because now they were considered unbreakable. Eventually other scientists showed that hydrogen and oxygen could themselves be split apart into protons, neutrons, and electrons. A similar but more recent example: the method of chromatography is used to analyze proteins by splitting them apart into amino acids and separating the amino acids in order to understand the structure of the proteins.

Social scientists also use the idea of splitting something into simpler parts, but because the things they study are often more abstract, the splitting is harder and more controversial. For example, some psychologists who study intelligence try to split it into pieces having to do with verbal ability, quantitative ability, and so on. Because you can't hold intelligence in your hand the way you can a bit of protein, there's less agreement about what the splitting really means, but the idea of splitting is there, nevertheless.

The assembly line metaphor. Just as Newton analyzed light by splitting it into components, Fisher's method analyzes the variation in a data set by splitting it into components. To understand what this splitting process does, I find it helpful to rely on a metaphor for the mechanism that creates the observations. The metaphor is not something you should take literally; it's something I made up. Nevertheless, if you use the metaphor as a way of thinking about the way observations *behave*, I think you'll find it helpful.

Imagine an assembly line of the sort used to build cars, except this one builds observed values. Each observation starts out with just a benchmark number, like the bare frame of a car, and then moves down the assembly line, getting new pieces added on at each of several workstations along the way, until at the last station a residual error gets added on, and the sum of all the pieces rolls off the line in the form of an observed value. The size of the final product—the observed value—depends on the sizes of the pieces that are added on at each work station. Between the Benchmark (first station) and the Residual Error (last station) will be one station for each set of conditions that influence the measurement process. For the hamster data, for example, there would be workstations for Day Length, Organ, Interaction, and Hamsters.

Now imagine yourself present at the creation of the first observed value, the 1.49 mg/ml. At the Benchmark station, the first worker takes the benchmark amount, so many milligrams of enzyme, and puts it on the conveyer belt: this starting amount will be the same for all the measurements in the experiment. When the enzyme gets to the next station, Day Length, the worker there checks to see whether the instructions say "Long" or "Short," and either adds or subtracts enzyme according to the instructions. Just like a worker who must always choose between a Dodge or Plymouth body to bolt onto the car frame, the day-length worker has only two choices about the amount of enzyme to add or subtract: one choice for long and one for short. The pile of enzyme now moves to the next workstation, Organ, and then on down the line. At each station a worker checks the instructions and either adds or subtracts enzyme. Finally, the enzyme reaches the last station, Residual Error, where a worker adds or subtracts a final amount. The completed pile of enzyme then rolls off the line, where a scientist measures how much is there and records the observed value.

Of course, the actual process in the lab is quite different, but just the same, scientific measurements very often behave pretty much the way you'd expect if they'd been built on the assembly line as described here, and so the metaphor gives a concrete way to think about the factor structure of an experiment. From the point of view of the metaphor, each factor corresponds to a workstation on the assembly line. The list of factors tells how many workstations there are, how many choices there are at each station, and how to decide which choice to make for each observation as it comes down the line.

Decomposing observed values. The idea of splitting behind Fisher's analysis of variance is essentially the assembly line model: you can think of the numbers you get from a scientific experiment as made up of several pieces, all added together. To understand the results of an experiment, you should split the numbers apart and study the pieces. To take a very simple case, you can think of an observed value as made up of two pieces: a true value plus an error:

$$\text{Observed number} = \text{True value} + \text{Residual error.}$$

Of course, the numbers you get from the lab don't come to you already split apart, so when you look at a measurement you can't tell how much is true value and how much is error. Nevertheless, if you make several measurements under the same conditions, so that the true value is the same in each case (or at least approximately the same), then you can use the average of the measurements as an estimate of the true value and you can look at how far each measurement is from the average in order to get a sense of how big the errors are.

Most experiments give you measurements for more than just one fixed condition because the goal of the experiment involves one or more comparisons. The numbers from such comparative experiments have more than two pieces, more than just true value plus error: the true value itself can be split apart into pieces. For example, take the hamster data, and consider the concentration of 1.49 mg/ml found for the heart of hamster #4, who was raised with long days. The following equation summarizes the idea behind Fisher's analysis of variance, applied to the 1.49, writing it as a sum of pieces, with one piece for each of the factors we listed in the last section of Chapter 1. (I've already explained how I got the piece for day length. In the next chapter I'll show you more about how I got the numbers.)

1.49 =	piece common to all 16 observations	(+6.44)
+	piece common to all long-day observations	(−1.18)
+	piece common to all heart measurements	(−4.97)
+	interaction piece common to all long-day heart observations	(+1.30)
+	piece common to the two observations for hamster 4	(−1.20)
+	estimate of residual error.	(+1.10)

If we can find a way to estimate each of the first five pieces, then whatever is left over will be an estimate of the residual error. We can carry out the same kind of splitting on all the observed values to get estimates for each of the residual errors. Then we can look at all these estimates together and decide how big a typical error is. Finally, we can compare the size of the piece due to day length, for example, with the size of residual error, and use the assumption that the errors are chancelike to decide whether the effect of day length is real and, if so, how big the true effect might be. That's exactly what analysis of variance does, *provided certain key assumptions hold.*

2. The Six Fisher Assumptions

In order to make it easier to decide when Fisher's method is appropriate to use, statisticians have restated his ideas as a set of specific assumptions about the data—two assumptions about the unknown true values and four assumptions about the residual errors.

Two Assumptions About Unknown True Values

It is the first pair of assumptions, about the true values, that justify computing averages to use as estimates. To see this, consider once again the simple situation where you have made several measurements, all under the same conditions, and you assume that each observed value equals the unknown true value plus an error. For example, here are the four enzyme concentrations (in mg/ml) for the hearts of the hamsters that got long days:

1.490	1.525	1.555	1.790
True + Error_1	True + Error_2	True + Error_3	True + Error_4

We can think of each number as the sum of the unknown true concentration plus an error. There are four different errors, one for each observation, but the true value is the same for all four observations.

To get the average of these four measurements, we first compute their sum, which we can rearrange so that all the "true" parts are together at the start and all the error terms are at the end. Then we divide by the number of observations:

$$\begin{aligned}\text{Sum} &= (\text{True} + \text{True} + \text{True} + \text{True}) + (\text{Error}_1 + \text{Error}_2 + \text{Error}_3 + \text{Error}_4).\\ \text{Avg} &= \text{True} + (\mathit{Error}_1 + \mathit{Error}_2 + \mathit{Error}_3 + \mathit{Error}_4)/4\\ &= (1.490 + 1.525 + 1.555 + 1.790)/4 = 1.590 \text{ mg/ml.}\end{aligned}$$

This algebra shows that *the average of the measurements equals the true value plus the average of the errors*, which is what allows us to regard the 1.590 as an estimate of the underlying true enzyme concentration for the heart of a hamster raised with long days. In order for this to work, two assumptions must be true.

C: The unknown true values are CONSTANT. In other words, all the observed values made under the same conditions contain the same true value.

The assumption of constant effects (C) says the true value is the same for all four measurements. If the true values were not all the same, then it wouldn't make sense to compute one estimate of the true value.

A: The pieces that go together to make the observed value are combined by ADDING them, not by some other process, such as multiplying.

The additivity assumption (A) says that each observed value equals true value *plus* error. Notice how our algebra relied on this assumption. Because the pieces of the individual observations are combined by adding, we can reorder the terms that make up the sum of the observations to form two pieces, with true values grouped together at the start and errors grouped together at the end. Moreover, the additivity assumption guarantees that the first, "true" part and the second, "error" part are combined by adding.

The next set of assumptions guarantee that, among other things, the average of the errors will tend to be near zero, so the average of the observations will tend to be near the true value.

Four Assumptions About Residual Errors

Fisher's four assumptions about the residual errors take some time and work to understand; Chapter 12 deals with these assumptions in detail. For now, you can make a good start by using a metaphor I borrowed from a wonderful book by three Berkeley statisticians.

The box model for chance errors. Imagine that your first measurement is being built. All the appropriate true values have been added together, and your measurement-to-be sits at the end of the assembly line, waiting to have its residual error added on. Now for the Berkeley metaphor: next to the workstation for residual error sits a box filled with tickets; each ticket has a number printed on it. To get the residual error for your measurement, the worker reaches into the box, mixes up the tickets, and draws one out at random. The number on that ticket tells the residual error for your measurement; the worker adds this chance error to the waiting sum of true values, and the total comes off the line as your measurement. Now the second sum of true values comes along. The worker puts back the last ticket, to make sure the contents of the box will be exactly the same as before, then mixes the tickets again, and draws one out to get the next error. The same thing happens over and over, once for each measurement.

True, the box is imaginary, just like the assembly line, and we almost never know much about how residual errors actually get created, but years of experience have shown that for many kinds of scientific work, residual errors, in fact, behave as if they were created in this way. If your goal is to learn how residual errors often behave, the box metaphor is a helpful thing to keep in mind.

Four assumptions based on the box model

Z: The average of the tickets in the box equals ZERO.

This means that the average of a set of observed values will tend to fall near the underlying true value, and also, the more observations you make, the more sure you can be that (Avg − True) will be small. The likely size of (Avg − True) will depend on the typical size of the chance errors in the box. In Chapter 3 I'll explain how to measure that typical size by computing the **standard deviation**. The next assumption, when true, guarantees that the computation is meaningful.

S: All the errors are like draws from the *SAME* error box. Thus there is one typical size, or STANDARD DEVIATION (SD), for all the errors.

Different experiments will have different error boxes, but analysis of variance assumes that for each experiment there is only one box. This assumption of same standard deviation (S) is important because if it is correct, then it makes sense to estimate one typical size for residual error to use as a yardstick for judging comparisons based on the data.

I: The errors are INDEPENDENT, that is, they are not related by any patterns.

In terms of the metaphor, the independence assumption (I) says that the errors behave as if you always put the last ticket back, and then thoroughly mix the tickets, before you draw the next one. The assumption is important because it guarantees that individual errors will tend to go in opposite directions—some positive, others negative—so they will tend to cancel when you compute averages.

N: The distribution of error sizes follows a NORMAL curve.

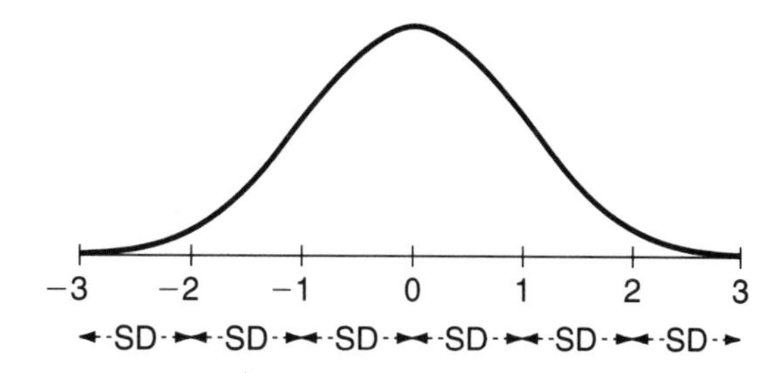
−3 −2 −1 0 1 2 3
SD SD SD SD SD SD
For a set of chance errors,
roughly 68% will be within 1 standard deviation of 0,
roughly 95% will be within 2 standard deviations of 0,
almost 100% will be within 3 standard deviations of 0.

FIGURE 2.2 N: The distribution of error sizes follows a **NORMAL** curve.

If you aren't familiar with the normal distribution or the bell-shaped curve, it will be easier to understand this assumption later, after I've built up an example to illustrate it. For now, just focus on how the 1-2-3-SD rule relates error sizes to the standard deviation: two-thirds are within one SD of zero; 95% are within two SDs of zero; and essentially all chance errors are within three SDs of zero (Fig. 2.2). Thus the normality assumption (N) says that it's unusual to get really extreme observations, very high or very low. Such extreme values are called **outliers**, and the rule helps you identify them and tells you to give them special attention.

Some of the connections between Fisher's assumptions and his method of analyzing experiments are not easy to see right off the bat, so I will come back to those connections as they arise in the context of particular examples. The first of these examples comes up in the next section: one of the assumptions doesn't fit the hamster data. Part of the reason I chose the hamster experiment as an example is to emphasize that statistical work is rarely automatic. I wanted you to see, early in the book, that to do a good analysis you have to make thoughtful choices. There may be more than one model to choose from, and checking assumptions is an important part of finding a good model.

3. Informal Analysis, Part 1: Parallel Dot Graphs and Choosing a Scale

In general, when you analyze a data set you don't have the advantage of being told ahead of time whether the assumptions fit your data. To make your experience with this example more realistic, put aside for the moment the hint that I've given about what to look for, and run your eyes over the data, looking for patterns and thinking about what they might mean.

	Hamster	Heart	Brain
Long days	#4	1.490	6.625
	#1	1.525	10.375
	#7	1.555	9.900
	#2	1.790	8.800
Short days	#3	1.385	12.500
	#8	1.485	11.625
	#5	1.255	18.275
	#6	1.285	13.225

FIGURE 2.3 Hamster data: Enzyme concentrations in mg/ml. Perhaps the single most striking thing about the numbers is that the heart concentrations are all low, between 1 and 2 mg/ml, whereas the brain concentrations are roughly 10 times as big. Now isolate the column for hearts, and compare the enzyme concentrations for long and short days; do the same for brains. For hearts, the effect of day length doesn't seem very big, although all four of the concentrations for long days are higher than the concentrations for short days: there's a suggestion that as days grow short, enzyme concentrations in the hearts of hamsters tend to drop. For brains, the reverse is true. The concentrations for long days are quite a bit lower than the concentrations for short days, suggesting that as days grow short, the enzyme concentrations in the brain tend to go up.

Averages and Parallel Dot Graphs

One way to summarize these patterns is to compute averages.

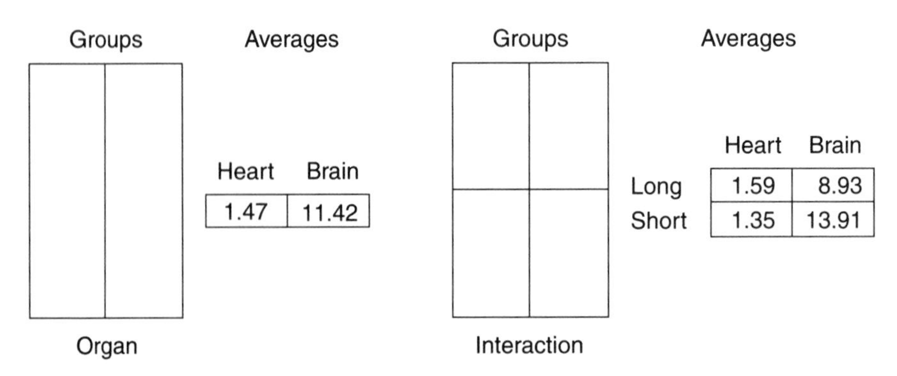

FIGURE 2.4 Two sets of average enzyme concentrations (in mg/ml) for the hamseter data. Each set of averages is based on sorting the observed values into groups, with one average per group.

The averages have two advantages here. Each average summarizes a group of observations with a single number, getting rid of extra detail to make patterns easier to see, and the averages also make it easier to state comparisons quantitatively. For example, for hearts the average for short days is 1.35 mg/ml, which is 0.24 mg/ml lower than for long days; for brains the average for short days is 13.91 mg/ml, which is 4.98 mg/ml higher than for long days.

There are sometimes disadvantages to using averages. The detail that gets washed out when you compute an average can be important. Also, when you have more than just a few groups to compare, it can be hard to recognize and absorb all the information in the averages. You can often get around those disadvantages with a good graph or two. One is the **parallel dot graph**.

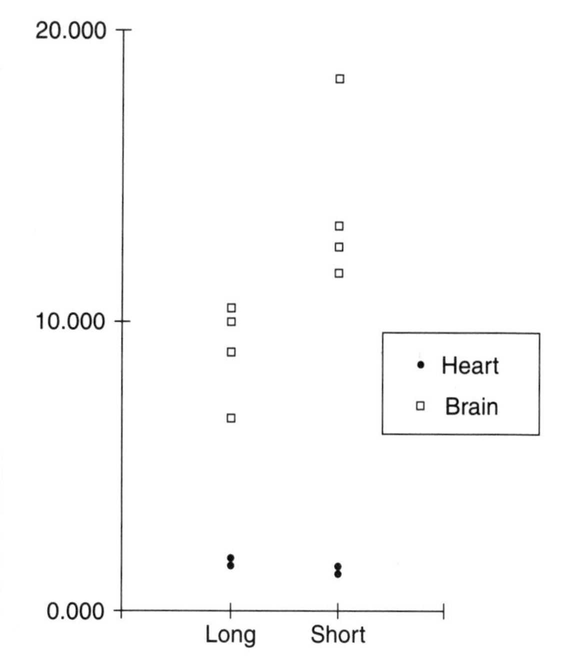

FIGURE 2.5 Parallel dot graph for the hamster data. Each observed value is plotted as a dot, with its height equal to the observed value. The dots are arranged in parallel columns, with one column for each of the groups you want to compare. (Chapter 5 gives more detailed instructions and many examples.)

Figure 2.5 shows the same patterns we found using averages: brains have a lot more of the sodium pump enzyme than hearts do, and going from long days to short seems to lower the heart concentrations a little while raising the brain concentrations a lot. Figure 2.5 also shows some important patterns that you can't see from the averages alone. Look over the graph again, keeping in mind our goal of finding the typical size of a residual error. We'd like to be able to compute a single number and say, "Here. This is our estimate for the size of a typical error." A careful look at the graph warns of trouble ahead.

Measuring spread. Look again at each group of four dots, paying particular attention this time to the range of values in each group. For example, the four measurements for Long Days/Heart range from a low of 1.49 mg/ml to a high of 1.79 mg/ml. The **range** is the distance from low to high, in this case, 0.30 (0.30 = 1.79 − 1.49).

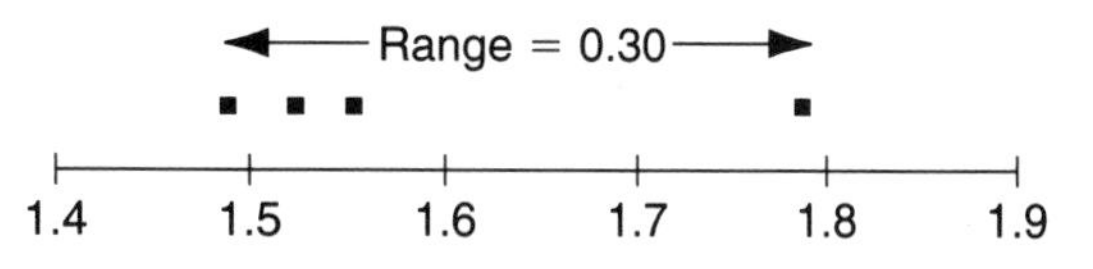

FIGURE 2.6 Range of enzyme concentrations in the hearts of the long-day hamsters (in mg/ml).

Because all four measurements were for the hearts of hamsters raised with long days, the range of values gives an indication of the variability in the data when conditions are held constant. Here are the ranges for all four sets of conditions:

TABLE 2.1 Ranges for enzyme concentrations for four groups of observations (in mg/ml)

Day length	Organ	Low	High	Range = high − low
Long	Heart	1.49	1.79	0.30
	Brain	6.63	10.38	3.75
Short	Heart	1.25	1.48	0.23
	Brain	11.63	18.28	6.65

The largest range is 6.65 (Short Days/Brain), while the smallest range is only 0.23 (Short Days/Heart). This kind of situation, with the largest range almost 30 times as big as the smallest range, is like a monkey wrench in the machinery of analysis of variance. Why? Because one goal is to compute a single number that estimates the typical size of residual error, and we seem to have not one typical size but two. For heart measurements, the ranges are small (.30 for long days, .23 for short), while for brain measurements the ranges are large (3.75 and 6.65). In fact, our data violate one of the Fisher assumptions that are needed to make analysis of variance work:

S: All the errors are like draws from the same error box.
Thus there is one standard deviation for all the errors.

There's an important lesson here, which is part of the reason I chose this data set for my example:

Statistical methods are not idiot proof.

If you use a method designed to give you a single number as the typical size of residual error, that method will give you a single number, even if there are really two typical sizes. If you want lobster for dinner, don't order pizza.

Choosing a New Scale

This kind of data set, with more than one size for residual error, is really quite common, as you can easily convince yourself. Notice that the small errors go with small measurements, and large with large:

TABLE 2.2 Pattern relating average and range for four groups of observations.

Day length	Organ	Average	Range
Long	Heart	1.59	0.30
	Brain	8.93	3.75
Short	Heart	1.35	0.23
	Brain	13.91	6.65

The match is perfect. The smallest average (1.35) and smallest range (0.23) go together, and so on, up the line to the largest average (13.91) with the largest range (6.65). The pattern suggests, for this data set at least, that the size of the error depends on the size of the measurement.

Average and spread are often related. To see how common it is for the size of the error to depend on the size of the measurement, just do a small thought experiment. First, estimate the height of the letter E, and think about how far off your estimate is likely to be. Then do the same for the diagonal of this page: estimate, then think about the likely size of your error. Finally, estimate the distance to the nearest post office.

The pattern holds not just for lengths. Try weights: think about estimating the number of grams in a penny and the number of grams in a Cadillac. Which estimate will have the bigger error? Try times: how many seconds would it take you to flip a coin? To build a house? Which estimate has the larger error? Because the pattern (large errors with large observed values, small with small) often holds for lengths, weights, and times, it's not really surprising that it's quite commonly true of all sorts of scientific measurements.

Have I been leading you down the primrose path then? I promised you a method for estimating the typical size of residual error, only to give you a data set with two sizes, not one. Have faith. A solution is at hand.

Transforming to logs. If you've taken a chemistry course, you're already familiar with the kind of thinking we need to get around the problem of two very different sizes for chance errors. Back when I was in high school, learning about acids and bases, I had to memorize the definition of pH: "pH is the negative logarithm (to the base 10) of the hydrogen ion concentration in moles per liter." Only years later did I come to understand why chemists wanted to use logarithms: they make it much easier to think about numbers of very different sizes. For example, think about comparing the concentration of hydrogen ions (measured in moles per liter) for water, milk, apples, and stomach acid. For pure water, the concentration is .0000001; for milk, it's .000001; for apples, .001; and for stomach acid, about .1. Measured in moles per liter, these numbers are hard to compare, but pH makes it easy. It's not a matter of pH being right and moles per liter being wrong. It's just that for some purposes pH is easier to work with.

TABLE 2.3 pH makes it easier to compare numbers of very different sizes.

Substance	Moles per liter	pH
Water	$.0000001 = 1 \div 10 \times 10 \times 10 \times 10 \times 10 \times 10 \times 10$	7
Milk	$.000001 = 1 \div 10 \times 10 \times 10 \times 10 \times 10 \times 10$	6
Apples	$.001 = 1 \div 10 \times 10 \times 10$	3
Stomach acid	$.1 = 1 \div 10$	1

The pH column on the far right tells the number of 10s in the denominator.

Now see what happens when we try the pH idea on the hamster data. Instead of looking at the concentration of the enzyme, let's compute the logarithm of each concentration. For hamster #4, the concentration in the heart is 1.49. The logarithm button (base 10) on my calculator gives log(1.49) = .17, rounded off to two decimal places. To make the numbers easier to work with, I've decided to multiply each number by 100, so the .17 becomes 17. Doing the same for the rest of the data (taking logs, rounding off, then multiplying by 100) gives me the same information as before, but expressed in a new scale.

	Hamster	Heart	Brain
Long days	#4	17	81
	#1	19	101
	#7	19	99
	#2	25	95
Short days	#3	14	110
	#8	17	105
	#5	10	126
	#6	11	111

FIGURE 2.7 Transformed hamster data: $100 \times$ log enzyme concentrations (mg/ml).

Notice how logarithms have gotten around the problem of grossly unequal ranges. For log concentrations, the ranges of values for the four conditions are much more nearly equal. For Long Days, Heart, the largest value is 25 and the smallest is 17; so the range is 8. The ranges for the other conditions are given below:

TABLE 2.4 Ranges for the transformed enzyme concentrations.

Day length	Organ	Low	High	Range
Long	Heart	17	25	8
	Brain	81	101	20
Short	Heart	10	17	7
	Brain	105	126	21

Before logs, the largest range was almost 30 times the smallest range. After logs, the largest range is only 3 times the smallest, a big improvement.

Transforming the data to a new scale has made the ranges for the four groups close enough to each other that it is no longer unreasonable to assume that the errors behave as if drawn from the same box.

Is it cheating to change scales? I can imagine you complaining: "The original data showed ranges of two different sizes. You didn't like that, and so you fiddled with the data until you got ranges that you did like. If that's how you play the game, you can prove anything with statistics, so I'm better off not trusting statisticians, at least not statisticians like you."

If that's more or less how you feel, let me try to explain why taking logarithms is not cheating. It's worth spending a couple of paragraphs on this because the situation illustrates something quite common in science: a great many numerical descriptions are not absolute, but depend on the scale you use. In transforming to logarithms, I haven't really changed the results of the experiment at all; I've just found a different way to describe those results.

The basic idea is easier to think about in a more familiar setting. If I want to compare my car's fuel efficiency with a friend's, I use miles per gallon as my scale and report the number 40. But if I'm lending my car to the friend for a trip to Boston, it's more useful to report that a round trip takes 4.5 gallons. In effect, I've turned my units of measurement upside down and used gallons per mile as my scale. Whichever scale I use, it's the same car and the same fuel efficiency. The same is true for enzyme concentrations. If I report milligrams per milliliter, each number tells how many milligrams of enzyme there were for each milliliter of tissue preparation. But it is just as correct to report milliliters per milligram. Then each number would tell how many milliliters of tissue preparation it took to get one milligram of enzyme.

There are two important points here. First, while milligrams per milliliter may seem like a natural way to measure enzyme concentrations, milliliters per milligram gives another scale that is every bit as natural, so there is not one God-given scale for enzyme concentrations. Second, the size of variability depends on your scale. To see this, consider three cars with very different fuel efficiencies: 20 mpg, 40 mpg, and 60 mpg. Notice that in the scale of miles per gallon, the numbers are evenly spaced. If we change scales to the number of gallons for the 180-mile round trip to Boston, the cars take 9 gallons, 4.5 gallons, and 3 gallons: in this scale the numbers are not evenly spaced. In the same way, for Kelly's data the size of the range (and the sizes of the errors) is not a property of the hamsters and their enzymes, and nothing else; the range also depends on the scale of measurement. There is not just one natural set of ranges that we have to use in order to be honest about our analysis. Because there are several reasonable scales to choose from, and because we're getting ready to use a method that starts by assuming the errors are of roughly the same size, we should feel virtuous, not devious, if we can choose a scale of measurement that makes the assumption approximately true. From here on, we'll work with the data in our new scale.

4. Informal Analysis, Part 2: Interaction Graph and Scatterplots for the Log Concentrations

Graphs and averages in the new scale. We can start with a new copy of the parallel dot graph, this time using the transformed data:

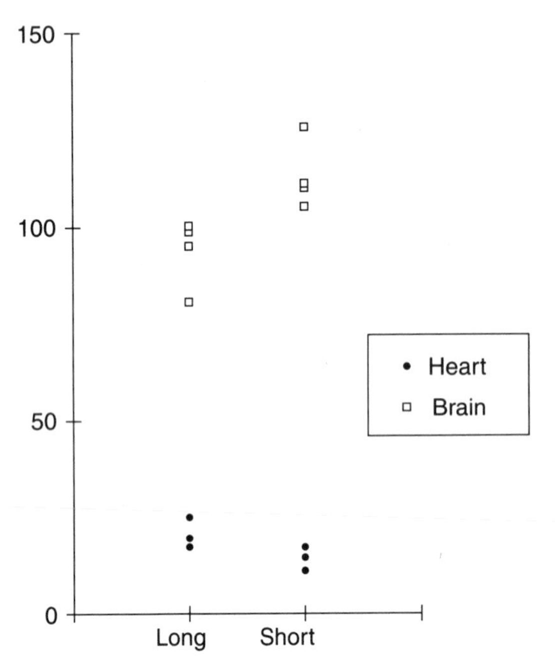

FIGURE 2.8 Parallel dot graph for the transformed hamster data. Note that, in comparison with the earlier graph, the ranges for the four groups are much more nearly equal.

This graph for the log concentrations shows a number of patterns:

1. Heart concentrations are a lot lower than brain concentrations, just as they were before we took logs: transforming hasn't changed that pattern.
2. The interaction pattern remains the same as before, too. For hearts the concentrations for short days are lower than concentrations for long days, while for brains the opposite is true: concentrations for short days are higher than concentrations for long days.
3. After transforming to logs, the enzyme concentrations for the four conditions show roughly the same spread, although the concentrations for brains are still more spread out than the concentrations for hearts.
4. One of the values for Long Days, Brain is a bit lower than the other three values; and one of the values for Short Days, Brain is a bit higher than the other three values. However, none of the values seems wildly out of line.

We compute averages in Fig. 2.9 to summarize the first two patterns quantitatively:

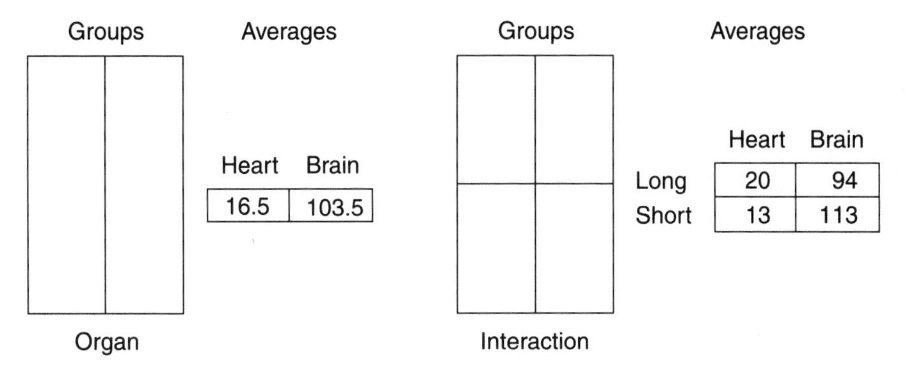

FIGURE 2.9 Two sets of average enzyme concentrations for the transformed hamster data. There are several things to notice here: (a) Each set of averages corresponds to a factor of the SP/RM design. (b) Transforming to logs hasn't changed the factor structure of the experiment: the ways of sorting observations into groups are the same as before. (c) For completeness, we ought to compute one set of averages for each of the other factors in the design, apart from residual error, although I won't do that here.

Interaction Graphs

The most interesting pattern to me is number 2 in the previous list, that the effect of day length seems to be different for hearts and brains. There's a nice way to show this pattern in a picture called an **interaction graph**:

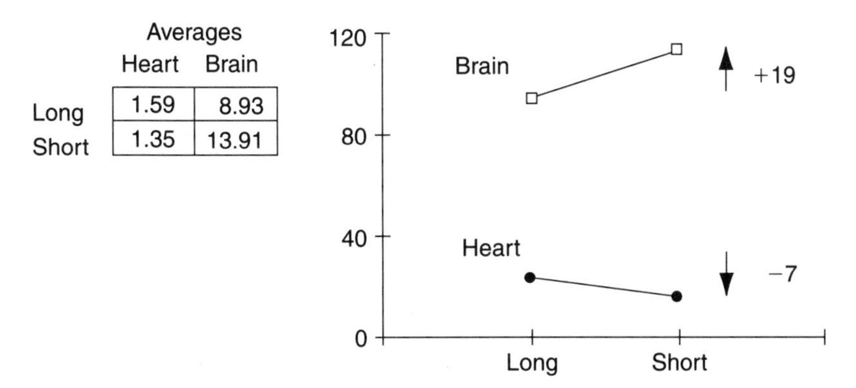

FIGURE 2.10 Interaction graph for the interaction of day length with organ. Each average is plotted as a dot. The two columns of dots correspond to Long and Short, the two levels of the day-length factor. The second factor involved in the interaction is shown by connecting the dots that go with each level of that factor. Here there are two lines: one for hearts and one for brains. (Chapter 6 gives detailed instructions for drawing and interpreting interaction graphs, as well as many examples.) For hearts, the line slopes down: the difference due to Day Length, Short minus Long, is −7. For brains, the line slopes up: the difference is +19.

Figure 2.10 shows quite clearly—more clearly than the averages alone—that brain concentrations are a lot higher than heart concentrations, and that in going from long days to short, the brain concentrations go up, while the heart concentrations go down. If there were no interaction, the lines would be parallel: heart and brain concentrations would go up (or down) by the same amount.

What the graph can't tell us is how much of the pattern might be due to the errors; that's why we need a more formal analysis, based on the assumption that the errors show the regularities of chancelike behavior. However, the graph does summarize the questions we want the formal analysis to answer:

1. How much of the apparent huge difference between the brain concentrations and the heart concentrations is "real," and how much is just error?
2. Does day length have a "real" effect? If so, what can we say about the likely size of the "true" effect?
3. Is the effect of day length "really" different for hearts and brains? If so, how big is the "true" difference?

The formal analysis begins in the next chapter, where I will show you how to split the observed values apart and how to find a typical size for the residual errors. First, though, we need to examine another assumption about these errors.

Scatterplots

So far I've shown you three tools for informal analysis of data: averages; parallel dot graphs; and interaction graphs. There's one more tool I want to show you: the **scatterplot**. The scatterplot has many uses, just as a parallel dot graph does, but it is designed for a particular situation—when your numbers come in pairs. The scatterplot is often used to answer a particular question: Do the number pairs show a pattern of high with high, medium with medium, low with low? Or the opposite, high with low, medium with medium, low with high? Or neither? In this section we'll use a scatterplot to check the assumption of independence for the residual errors in Kelly's data set.

Drawing a scatterplot. The hamster data give us eight pairs of numbers—one heart concentration and one brain concentration for each hamster—so our scatterplot will show eight points, one for each hamster. Drawing the scatterplot is a lot like drawing a parallel dot graph, but with one major difference. Just as with the dot graph, the height of a dot (y-value) tells the observed value; in this case, the enzyme concentration in the brain. Scatterplots and dotplots differ in the way they use the x-axis: for a dot graph, the x-value tells you which group an observation comes from; for a scatterplot, the x-value tells you the other number in the number pair; in this case, the heart concentration of the hamster. To get the entire scatterplot, you go through pretty much the same steps as you would for a parallel dot graph, plotting points one at a time (Fig. 2.11). For the dot graph, you have only a few columns of dots, one for each group, and the groups may not have a logical order. For a scatterplot, there are lots of columns, one for each possible value of the heart concentration (or whatever), and these numerical values correspond to points on a number line.

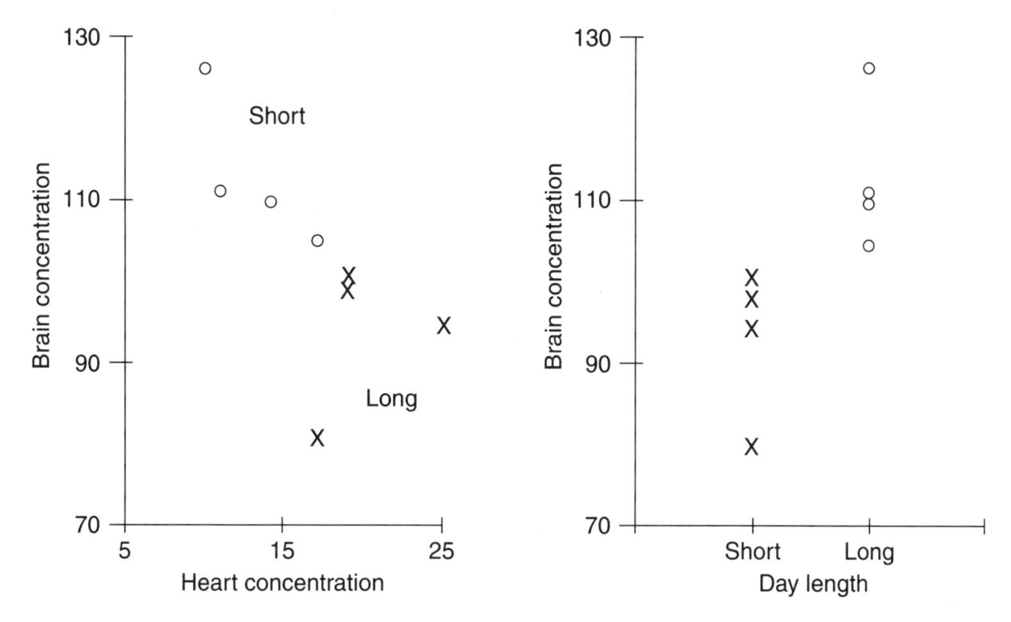

FIGURE 2.11 Scatterplot and dot graph for the hamster data. For both graphs, the y-value (vertical scale) equals the enzyme concentration in the brain. For the dot graph on the right, the x-value tells the group, long days or short. For the scatterplot on the left, the x-value tells the enzyme concentration in the heart. The scatterplot shows the pattern of how heart and brain concentrations are related. Points that are high on the vertical scale represent hamsters with high brain concentrations; low points represent low brain concentrations. Points to the left are low on the horizontal scale (low heart concentrations); points to the right represent high heart concentrations. With this information as background, what does the picture tell you about the hamsters?

Reading the scatterplot. For me, three things stand out as I look at the scatterplot. First, the Xs are in the lower right, and the Os are in the upper left. Short-day hamsters have lower heart concentrations and higher brain concentrations, whereas the long-day hamsters have higher heart concentrations and lower brain concentrations. Second, the X for the hamster with the lowest brain concentration seems peculiar. If you cover up that one point, the other seven points all lie quite close to a line that slopes down, from the northwest to the southeast. In the context of this pattern, the one odd hamster is an **outlier**, an observation far enough from the rest to seem suspicious. Third, if you ignore the outlier, the other points are so close to a line that if you know the enzyme concentration for one organ, you could come pretty close to predicting the concentration for the other organ, just by pretending the point for the hamster was on the line. The downward slope to the line tells you that, hamster for hamster, the higher the heart concentration, the lower the brain concentration. (If you wanted to be more precise, you could sketch in a line and use it to discover a rough quantitative relationship: an increase of 10 in the heart concentration corresponds to a drop of 20 in the brain concentration.)

A challenge to the SP/RM model. If we ignore the one questionable long-day hamster, the pattern in this scatterplot shows what is called a **negative** relationship between the heart and brain concentrations: higher heart values tend to go with lower brain values, and vice versa. The numbers behave roughly as though inside each hamster, the heart and brain were competing for a fixed amount of enzyme: the more the brain gets, the less there is for the heart, and vice versa. This behavior is at odds with one of the standard assumptions of the SP/RM model: the pattern tells us that the two measurements within a block are *not independent.*

Should we abandon the analysis and start over? Fortunately, we can still do a reasonable analysis using the factor structure of the SP/RM model from before, with but one change to the Fisher assumptions. Instead of assuming the residual errors are independent, we'll assume they come in pairs, one pair per hamster, and that these pairs of errors have a negative relationship: if one error is positive, the other tends to be negative, and vice versa. This modified model does fit the pattern in the scatterplot. Moreover, we can carry out our analysis of variance in the same way we would if we'd assumed the errors to be independent.

A model, just like a metaphor, oversimplifies and so is never completely "true." But a good model, like a good metaphor, helps you focus on crucial aspects of some more complex situation. Our modified SP/RM model will lead to a useful analysis, which I'll sketch for you in the next chapter.

SUMMARY

1. *The six Fisher assumptions (ACZSIN)*

   A: The pieces that go together to make the observed value are combined by **adding** them, not by some other process, such as multiplying.

   C: The unknown true values are **constant**. In other words, all the observed values made under the same conditions contain the same true value.

   Z: The average of the tickets in the error box is **zero**.

   S: All the errors are like draws from the **same** error box. Thus there is one **standard deviation** for all the errors.

   I: The errors are **independent**, that is, not related by any patterns.

   N: The distribution of error sizes follows a **normal** curve, at least approximately:

   For any set of errors,

   roughly 68% will be within 1 standard deviation of 0,
   roughly 95% will be within 2 standard deviations of 0,
   nearly 100% will be within 3 standard deviations of 0.

2. *Four tools for informal analysis*

   a. **Averages**: for any factor in the design, you can compute a set of averages, one for each level (group) of the factor.

   b. **Parallel dot graphs**: each observed value gets plotted as a dot, with its height equal to the observed value. The dots are arranged in parallel columns, with one column for each of the groups you want to compare. Dot graphs show spread, gaps, and outliers.

   c. **Interaction graphs**: compute averages corresponding to the levels of the interaction factor, and plot the averages in columns, one column for each level of the first factor of interest. Then join the dots by lines, one line for each level of the second factor of interest. If the lines are nearly parallel, there is no interaction.

d. **Scatterplots**: plot pairs of numbers as points. The first number (x-value) in the pair locates the point along the horizontal axis; the second number (y-value) gives the height of the point. The plot shows how the numbers in the pairs are related. In a negative relationship, higher x-values go with lower y-values, and vice versa. In a positive relationship, high goes with high, low with low.

3. *Choosing a scale.* For the hamster data, a parallel dot graph showed that the residual errors seemed to be of two very different sizes. To make the data more in line with Fisher's assumption S (same error box, same standard deviation), we transformed the data to a new scale by taking logarithms.

4. *Graphs for the transformed data.* These showed several patterns:

- Heart concentrations are a lot lower than brain concentrations.
- Interaction pattern: for hearts enzyme concentrations for short days are lower than concentrations for long days; while for brains the opposite is true: concentrations for short days are higher than concentrations for long days.
- After transforming to logs, the enzyme concentrations for the four conditions show roughly the same spread, although the concentrations for brains are a bit more spread out than the concentrations for hearts. One of the values for Long Days, Brain is a bit lower than the other three values; and one of the values for Short Days, Brain is a bit higher than the other three values. However, none of the values seems wildly out of line.
- Scatterplot: with the exception of one outlier (hamster #4), the points all lie quite close to a downward-sloping line—heart and brain values have a negative relationship, whereas data from an SP/RM model tend to show a positive relationship. We can either change to a bivariate RBF[1] model or keep the SP/RM factor structure and change the assumption of independent errors.

REVIEW EXERCISES

Note: These exercises cover general ideas. There are additional exercises on these same topics at the ends of sections in Chapters 5 to 9 and 12.

The data set in Fig. 2.12 also came from Kelly Acampora's experiment. The design is the same as before; in fact, the hamsters were the same ones. Each number is the specific activity of the sodium pump enzyme. The specific activity of an enzyme measures its effectiveness as a catalyst, in units per milliliter. Essentially, the concentration tells how much is there, the specific activity tells how potent it is.

1. List the structural factors for the data set in Fig. 2.12.

2. Draw a parallel dot graph, using the graph from Section 3 as a guide.

3. Compute the average and range using the same four groups as in Section 3 and comment on what you find. Are the four ranges roughly equal? Does the range seem to be related to the average by a pattern?

4. Draw an interaction graph, using the one from Section 4 as a guide. Does the effect of day length, as measured by specific activity, seem pretty much the same for hearts and for brains?

5. I find it helps me remember the Fisher assumptions if I remember the nonsense word ACZSIN; each letter stands for an assumption: Additivity; Constant effects; Zero average for

	Hamster	Heart	Brain
Long days	#4	246	394
	#1	169	297
	#7	186	216
	#2	183	373
Short days	#3	468	216
	#8	390	198
	#5	314	194
	#6	508	192

FIGURE 2.12 Hamster data: Specific activity of the enzyme.

the box; Same standard deviation; Independence; and Normality. List the letters of the four assumptions about the residual errors.

6. Explain how the assumption of constant effects (C) is part of the justification for computing averages.

7. Here's a simple example for which the additivity assumption (A) is false. The purpose of the exercise is to illustrate a relationship between assumption A and the use of averages as summaries. Suppose each observed value equals a Benchmark value of 100 *times* a chance error and that chance errors get chosen at random from the box of four (equally likely) numbered tickets shown below:

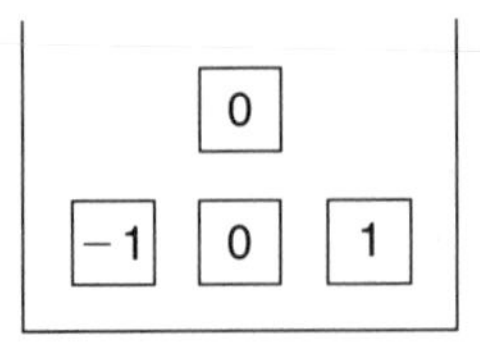

a. Use the equation(Observed value) = (Benchmark value) × (Chance error) to compute observed values corresponding to each of the four chance errors.

b. Compute the average of the four observed values from part a, and compare your result with the true Benchmark value of 100.

c. Now repeat parts a and b, this time using the model (Observed value) = (Benchmark value) + (Chance error) for which the additivity assumption is true.

8. Draw a scatterplot for the specific activity data, using the example in Section 4 as a guide.

a. Are there any obvious outliers?

b. Which of the following best describes the plot?

i. All eight points lie near a single line.

ii. All eight points form a single balloon-shaped cloud.

iii. The points form two distinct clusters, one for long days, one for short.

iv. None of the above.

9. Write a half-page summarizing what Exercises 2 to 4 and 8 tell you about the specific activity data.

NOTES AND SOURCES

1. ***Decomposition.*** The beginning of the chapter drew a parallel between Newton's decomposition of white light into components of various colors and Fisher's decomposition of variation into components. This parallel is, in fact, quite deep: both decompositions are, abstractly, instances of harmonic analysis, a mathematical method for splitting a collection of numbers into components whose structure comes from a group of symmetries. Chapter 10 discusses this in more detail.

2. ***The error box.*** This metaphor is from David Freedman, Robert Pisani, and Roger Purves (1978), *Statistics*. New York: W. W. Norton & Company.

3. ***Transforming.*** Chapter 12, which deals with assumption checking and remedies, tells ways you can find a transformation when you need one.

4. ***Outliers.*** Different statisticians use the word "outlier" to mean slightly different things. Many use it as I use it in this book, to mean any observation that lies far enough away from other observations to seem suspicious and worthy of special attention. Other statisticians reserve the word to mean only those suspicious-looking observations that have passed some sort of screening test to certify their deviance. Sometimes an outlier comes from a recording error—someone misread an instrument or wrote a number down wrong. Sometimes an outlier comes from unusual circumstances—an instrument malfunctioned, a plant got overwatered, or a subject misunderstood the instructions. Sometimes, though, it's just part of the way things go that an occasional observation will be extreme. If you know an observation is bad, you should throw it out, of course. But much of the time, as with the outlier in the hamster experiment, there's no way to know. In such instances, there are various strategies for dealing with outliers, some of them quite elaborate. One general approach is simply to redo your analysis without the outlier, regarding that observation as missing, to see how much the results of your analysis change. If they change a lot, you know that what you conclude from the analysis depends on what you decide about the outlier. On the other hand, if the results don't change much when you leave out the outlier, you know that it doesn't really matter whether you regard that observation as "bad." Chapter 12 deals with this issue in more detail.

5. ***Negative or positive relationship?*** The scatterplot in Section 4 showed a negative relationship between heart and brain concentrations, and I said that the usual SP/RM model predicts a positive relationship. Here's some more detail; in particular, why the assumptions of the SP/RM model lead to data showing a positive relationship (Fig. 2.13). (I'll return to this in more detail in Chapters 7 and 12.)

Think of the four long-day hamsters, each with a heart and a brain concentration. Suppose the average concentrations are the same as for the actual data: 20 for hearts, 94 for brains. These averages represent the contributions from the first four factors of the SP/RM model: Benchmark; Day Length; Organ; and Interaction. The last two factors, for Hamsters and Residual Error, are responsible for what makes one hamster's heart and brain concentrations different from those of another hamster. The assumption of independence tells us the residual errors have no relationship, and so these errors do not contribute to a positive or negative relationship between heart and brain values. However (according to the assumption of constant effects), each hamster effect is the same for both the heart and the brain values of that hamster. A positive hamster effect makes both values higher by the same amount; a negative effect makes both values lower. Thus higher heart values tend to go with higher brain values, lower with lower. This is the opposite of the negative relationship we find in the scatterplot for the actual data.

6. ***An alternative model.*** Example 1.1 on page 16 illustrates the simplest design based on random assignment, the one-way completely randomized design, or RBF[1] for short. Compare Kelly's actual experiment with that example and notice that the way day lengths (treatments)

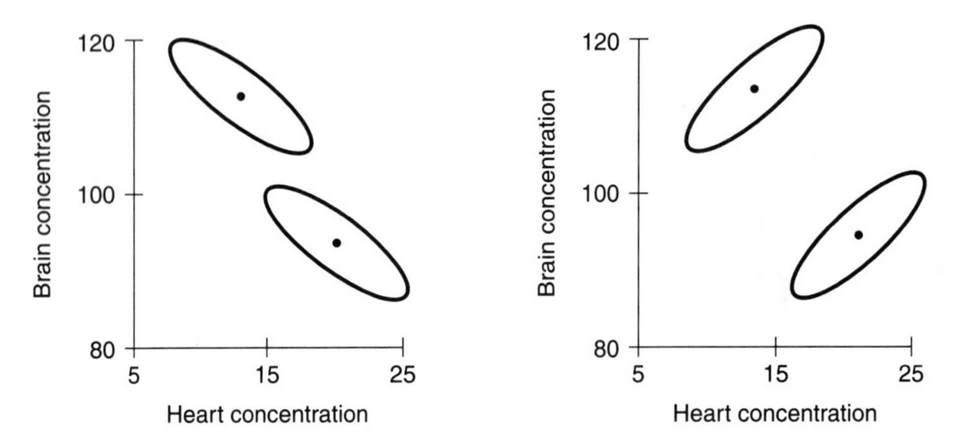

FIGURE 2.13 Comparison of actual data with data from the SP/RM model. The actual data (left) show a negative relationship within each group of hamsters (i.e., for each day length): higher heart values with lower brain values within each group, and vice versa. Data from a properly randomized SP/RM model (right) would ordinarily show a positive relationship within each group: high with high, low with low.

are assigned to hamsters (units) is the same for both. The only difference is that for the RBF[1] there is just one measurement (heart concentration) per hamster, whereas for the actual experiment, there were two: heart and brain concentrations. You can think of Kelly's experiment as a **bivariate** RBF[1] design, that is, an RBF[1] with two kinds of measurements per unit. Many statisticians would consider the two-variable RBF[1] a better model for the hamster data than the SP/RM, partly because of the scatterplot. However, ANOVA for bivariate data goes beyond what this book covers. (See any book on multivariate analysis.)

CHAPTER 3

FORMAL ANOVA: DECOMPOSING THE DATA & MEASURING VARIABILITY, TESTING HYPOTHESES, & ESTIMATING TRUE DIFFERENCES

OVERVIEW

There are several goals for this chapter. The first is to split Kelly's observed values into pieces, one piece for each factor, including the factor for residual error. As long as the Fisher assumptions fit the data, we can think of the numbers we get from splitting the observations as estimates of the "true" amounts added or subtracted at each of the work stations on the assembly line. A second goal is to compute a summary number, called a mean square, for each factor in the design, to measure the variability due to that factor. In particular, in the third section, we'll use the mean squares for the 16 residuals to compute the standard deviation, which gives a typical size of the residual errors. In the last two sections of the chapter, we'll assume that the residual errors are in fact chancelike, and we'll compare our estimates of the true values with the chance error sizes, first to see which of the observed effects (for day length, organs, and so on) are too big to be due just to chance error and then to make inferences about the sizes of the underlying true values.

1. Decomposing the Data
2. Computing Mean Squares to Measure Average Variability
3. Standard Deviation = Root Mean Square for Residuals
4. Formal Hypothesis Testing: Are the Effects Detectable?
5. Confidence Intervals: The Likely Size of True Differences

1. Decomposing the Data

Goal of the decomposition. Our first goal is to split each observed value into a sum of pieces, one for each factor. Here's what we'll end up with for the observation in the upper left corner of the data table in Figure 2.7—the 17 for the heart of hamster #4, who was raised with long days:

17	=	(+60)	+ (−3)	+ (−43.5) +	(+6.5)	+	(−3)	+	(+5)
Observation		Benchmark	Long days	Heart	Interaction		Hamster #4		Residual error
OBS	=	FIT						+	RES

FIGURE 3.1 Decomposition of the first observation. The observed value of 17 is written as a sum of six pieces, one piece for each factor. Once you understand where these numbers come from, you can use them to see, for example, that on average, long days lower the enzyme concentration by a little bit (−3), that concentrations in the heart are a lot lower than the benchmark (−43.5), that the estimated interaction effect is about twice as big as the effect for long days, and so on.

The sum of the first five pieces—everything but the estimate for residual error—is called the *fitted value* (fit for short), which in this case equals 12. The difference between the observed value and the fitted value is the residual, our estimate for the residual error: Res = Obs − Fit = 17 − 12 = 5.

The estimates in the decomposition are all based on averages computed from the observations.

We begin with a complete set of averages, one set for each factor (except for residual error):

Observed values Heart	Observed values Brain	Hamster Averages
17	81	49
19	101	60
19	99	59
25	95	60
14	110	62
17	105	61
10	126	68
11	111	61

	Condition Averages Heart	Condition Averages Brain	Day Length Averages
Long	20	94	57
Short	13	113	63
Organ Averages	16.5	104	60 (Grand Average)

FIGURE 3.2 A complete set of averages for the transformed hamster data. There is one set of averages for each factor, except for chance error. The eight hamster averages are the eight row averages shown next to the data table on the left. Averages for the other factors appear in the table on the right. The grand average is 60. The day length averages, 57 (long) and 63 (short), are both quite close to the grand average. The Organ averages, on the other hand, 16.5 (hearts) and 103.5 (brains), are quite far from the grand average. The four condition averages, 20 for long days/heart, and so forth, correspond to the groups for the Interaction factor. The pieces in the decomposition come from adding and subtracting these averages.

From averages to estimated effects. To illustrate the general idea of the decomposition, I'll show you a few estimates that I hope you'll find reasonable just from common sense. There's also a systematic way to get the estimates, based on a rule that I'll explain in an appendix at the end of this chapter.

Benchmark. We want an estimate of the piece common to all 16 observations, an estimate of the value "everybody starts out at" before the pieces due to day length, organ, and so on get added on. Our estimate here is just the grand average of all 16 observations, which works out to 60 for this data set.

Day length. Think of the estimated effect for long days as the answer to the following question: how far is the long-day average from the grand average? Answer: there are eight observed values for the four hamsters who got long days. The average of these eight long-day numbers is 57, which is 3 lower than the estimated benchmark of 60.

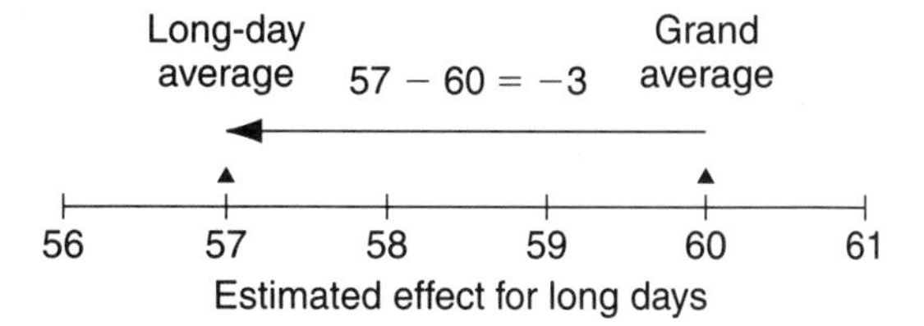

FIGURE 3.3 Estimated effect for long days = long-day average − grand average. For short days, the idea is the same: the short-day average is 63, and the estimated effect equals the short-day average minus the grand average: 63 − 60 = +3. Notice that the estimated effects for long and short days add to zero.

Organs. The estimated effects for hearts and brains work the same way as for day length: estimated effect equals average minus grand average. Question: how far are the heart and brain averages from the grand average? Answer: the average of the eight heart measurements is 16.5, and the average of the eight brain measurements is 103.5. The heart average is 43.5 lower than the grand average of 60, and the brain average is 43.5 higher than the grand average. In other words, the typical heart measurement equals grand average plus (−43.5); the typical brain measurement equals grand average plus (+43.5). Notice that here again, just as for day length, the estimated effects add to zero.

Hamsters. What's the estimated effect for hamster #4? The average of the two measurements for that hamster equals (17 + 81)/2 = 49. If you go by pattern alone, you'd probably guess that the estimated effect is "average minus grand average," or 49 − 60 = −11. However, if you look back at the decomposition, you'll find −8, not −11.

There's a moral here, one that applies to any mathematical subject: it's risky to go by pattern alone and ignore the meaning. So think about the meaning in this case. We want the hamster effects to tell us about the differences between individual hamsters, other things being equal. Common sense suggests that because hamster #4 got long days, we ought to compare him with the other long-day hamsters. In other words, the estimated effect for #4 equals his average (49) minus the average of all the long-day hamsters (57): 49 − 57 = −8. This number tells us that #4 was, on average, 8 lower than the average of all the hamsters that got the same treatment.

Interaction. I'll skip over the interaction effects for now, since they're more complicated than the others.

Residual error. To get the residual (our estimate for the error), we add up all the other estimated effects to get the fitted value, which is an estimate of what the observed value would be without its residual error. To estimate the error, we subtract:

$$\begin{aligned} \text{Obs} &= \text{Fit} + \text{Res}, \qquad \text{so} \\ \text{Res} &= \text{Obs} - \text{Fit} \\ &= 17 - 12 = 5. \end{aligned}$$

I hope this quick overview has suggested that we can use a decomposition to compare the effects of the various influences on the enzyme concentrations and, moreover, that many of the estimates are just a matter of common sense. At the same time, it's not always obvious how to combine the various averages to get estimated effects, and you will eventually need to learn a general rule (explained in the appendix) that works all the time. For now, I want to skip over the computing rule in order to focus on what the results of the decomposition tell us about the hamsters.

The complete decomposition takes each of the 16 observations we started with and writes it as a sum of six pieces, one piece for each factor. How can we display all this information in a way that shows the important patterns? With so much to show, any display will take some practice before you can read it easily, but the display in Fig. 3.4 shows the form I think is most useful, once you get used to it. It's basically the factor diagram, with numbers filled in. There are two useful ways to read the display: one observation at a time or one factor at a time. I urge you to practice until you feel at home with both. Here's a quick exercise (no arithmetic required) that I suggest you do before starting the next section: for each box of the decomposition, write down a single number that you consider reasonable as a typical size for the numbers in that box.

2. Computing Mean Squares to Measure Average Variability

Preview: comparing mean squares. Each table in a decomposition, except for the grand average, shows variability. The variability is due to some mix of planned variability linked to the factor in question plus other variability that ANOVA assumes is chancelike. Our goal is to decide how much of the variability in each table is linked to its factor and how much is due to chance. To do this, we'll develop a measure of the average size of the variability in a table. Then we'll compute the average size for each table in the decomposition and compare the average sizes. In particular, we'll be able to compare the variability for each factor with the variability due to chance error in order to decide which factors have effects that are big enough to be declared

Observed values			Benchmark			Day length			Organs	
17	81		60	60		−3	−3		−43.5	43.5
19	101		60	60		−3	−3		−43.5	43.5
19	99		60	60		−3	−3		−43.5	43.5
25	95	=	60	60	+	−3	−3	+	−43.5	43.5
14	110		60	60		3	3		−43.5	43.5
17	105		60	60		3	3		−43.5	43.5
10	126		60	60		3	3		−43.5	43.5
11	111		60	60		3	3		−43.5	43.5

	Interaction			Hamsters			Residual error	
	6.5	−6.5		−8	−8		5	−5
	6.5	−6.5		3	3		−4	4
	6.5	−6.5		2	2		−3	3
+	6.5	−6.5	+	3	3	+	2	−2
	−6.5	6.5		−1	−1		2	−2
	−6.5	6.5		−2	−2		6	−6
	−6.5	6.5		5	5		−8	8
	−6.5	6.5		−2	−2		0	0

FIGURE 3.4 Complete decomposition of the hamster data. *One observation at a time*: The numbers add up according to where they sit in the boxes. For example, the numbers in the upper left corners together give the decomposition of the observed value of 17 for the heart of hamster #4. To see this, start with the box of observed values and go from one box to the next, picking out the numbers from the upper left corners. Check that these numbers give you the decomposition we started with in Fig. 3.1:

$$17 = (60) + (-3) + (-43.5) + (6.5) + (-8) + (5).$$

One factor at a time: Each box corresponds to a factor. For example, the box for day length shows that the eight observations for long days (top half) all get a -3; the eight for short days all get a $+3$. By comparing one box with another, you can get a general sense of which factors have the biggest effects and how the sizes compare to the estimated errors. Here, for example, a typical residual seems to be roughly ± 5. The day-length effects are about half as big as this; the organ effects are more than eight times that big; and the interaction effects are between one and two times as big.
For various reasons, in the next section it will turn out to be important to be able to see the patterns of repetitions and the patterns of adding to zero within each of the boxes. For practice, check these kinds of patterns for Day Length and for Hamsters. Day Length: there are two effects; each occurs eight times and the effects add to zero over the whole box. Hamsters: There are eight effects; each occurs twice and the effects add to zero in two sets—the top four rows add to zero, as do the bottom four rows.

"real." Consider, for example, the interaction effects for the hamster data, ± 6.5. Are these observed effects too big to be due just to chance error? It will turn out that the average variability in the interaction table of the decomposition is almost 13 times as big as the average variability in the residual table. We'll end up concluding that the interaction effects are "real."

The comparison begins by computing a summary number, called a **mean square**, for each table. The mean square serves as our measure of the average variability in the table. Here's how you compute one using three steps labeled as SS, df, and MS:

1. **SS = sum of squares**. Square all the numbers in the table, and add them up. This total is a measure of the overall variability in the table.
2. **df = degrees of freedom**. This number counts the number of units of information about residual error contained in the table.
3. **MS = mean square**. Divide the SS by the df: MS = SS/df. This number measures the average variability per unit of information.

In equation form,

$$\begin{matrix}\text{Variability}\\ \text{per unit}\\ \text{of information}\end{matrix} = \frac{\text{Overall variability}}{\#\text{ units of information}} = \frac{\text{Sum of squares}}{\text{Degrees of freedom}}.$$

With this preview as background, we'll now go through the three steps, computing sums of squares, degrees of freedom, and then mean squares for all six factors.

Sum of Squares (SS) = Measure of Overall Variability

We want to compute a number for each factor that tells us how much variation is in the box for that factor in the decomposition. The number Fisher chose is the sum of squares: square each number in the box and then add up all the squares. Let's do some arithmetic first and then see what the sums of squares tell us.

Day length. Turn back to the decomposition at the end of the last section, and look over the box for Day Length. (Notice that the numbers in the box add to zero.) To get the sum of squares, just square each of the 16 numbers in the box and then add them up:

$$\begin{aligned}SS_{\text{Days}} &= (-3)^2 + (-3)^2 + \cdots + (-3)^2 + (+3)^2 + (+3)^2 + \cdots + (+3)^2\\ &= 9 + 9 + \cdots + 9 + 9 + 9 + \cdots + 9\\ &= 9 \times 16\\ &= 144.\end{aligned}$$

As I hope you see, there will often be shortcuts for sums of squares.

Organs. This box has eight −43.5s and eight +43.5s. (These 16 numbers add to zero.) The square of ±43.5 is 1892.25, and there are 16 such squares in all:

$$SS_{\text{Organs}} = 16 \times (\pm 43.5)^2 = 16 \times 1892.25 = 30{,}276.$$

You should always feel free to use any shortcuts you can find; of course, you could choose to compute every sum of squares "from scratch": square every number and then add. The basic idea is so simple that I won't show any more of the computations; before I show the other sums of squares, however, look over the remaining boxes of the decomposition, and try to roughly guess how the other SSs will turn out. Can you guess the rank order, from largest to smallest? (I'm quite serious about this suggestion. When you make rough guesses of this sort, your mind tends to focus less on the mechanics and more on the meaning.) If you did the little exercise I suggested at the end of Section 1, you should find that for each box, 16 × (typical size)^2 is a rough estimate for the SS.

The meaning of the SS as a measure of overall variability. What do the sums of squares tell us? Fisher's idea is that the SS for a factor is a measure of the overall variability in the numbers for that box of the decomposition. For example, consider the sum of squares for the residuals, which comes to 316. Here's one way Fisher might have thought about it:

> "We want to measure how spread out the residuals are. Notice that they add to zero. Since we know they add to zero, we also know that their average is zero, so "spread out" is the same as "far away from zero." If a number is close to zero, like +1 or −1, its square will be small; if a number is farther away from zero, like −3 or +5, its square will be bigger. If the residuals aren't spread out much, they will all be close to zero. So their squares will all be small, and when we add the squares, we'll get a small number:
>
> $$\text{Residuals close together} \longleftrightarrow \text{Small sum of squares.}$$
>
> On the other hand, if the residuals are really spread out, at least some of them will be far away from zero, so their squares will be big, and when we add the squares, we'll get a big number:
>
> $$\text{Residuals spread out} \longleftrightarrow \text{Big sum of squares.}$$

The same thinking applies to the other sums of squares as well: the sum of squares measures the overall variability for a set of numbers, provided those numbers add to zero. (You can check that the numbers in each of the boxes for Day Length, Organ, Interaction, Hamsters, and Residuals do in fact add to zero.)

Although the observed values don't add to zero, and the grand average is not zero, there's a reason for computing a sum of squares for these two boxes as well: the sums of squares for the six factors of the decomposition add up to give the sum of squares for the observed values.

Source	Sum of Squares
Grand avg	57,600
Day/length	144
Hamsters	240
Organs	30,276
Interaction	676
Residuals	316
Observed values	89,252

FIGURE 3.5 Sums of squares for the hamster data.

The sums of squares for the six boxes of the decomposition add up to give the sum of squares for the table of observed values. Each sum of squares (other than those for the grand average and the observed values) measures the overall variability due to the factor in question. The rows for Day Length and Hamsters go together because day length was assigned to whole hamsters, unlike organs, for which the "unit" is a part of a hamster.

The numbers in Fig. 3.5 show that the variability due to organs is huge in comparison to the other factors. The variability due to interaction comes next; the SS for interaction is more than twice as big as the SS for residuals. Remember, though, that SSs measure overall variability, not average variability.

From overall to average variability. Our next goal is to go from these sums of squares, which measure overall variability, to mean squares, which measure a kind of average variability. Consider the residuals, for example, with sum of squares equal to 316. The basic idea goes like this: the 316 measures overall variability from 16 numbers at once. Since 16 numbers contributed to the 316, we should divide by 16 to get average variability. This is the right general idea, but there's an extra wrinkle to worry about. Even though 16 residuals are in the box, it turns out there are only 6 units of information about chance error. Thus according to Fisher's theory, we should divide by 6, not 16:

$$\begin{array}{c}\text{Average variability}\\ \text{per unit of info.}\end{array} = \frac{\text{overall variability}}{\#\text{ units of information}} = \frac{316}{6} = 52.67.$$

In the next two steps, I'll explain what I mean by units of information and why there are only 6 of them for the residual box.

Degrees of Freedom (df) = Units of Information About Residual Error

There are two useful ways to think about degrees of freedom: in terms of free numbers and in terms of information about residual error. The first way (free numbers) will show how to count degrees of freedom for each factor and will lead to a general rule for all balanced designs. The second way (units of information) is closer to the logic of mean squares and testing whether the effects of day length or organs are too big to be due just to chance. I'll start with free numbers and come back to units of information at the third step, the mean squares.

The df for a table equals the number of free numbers, the number of slots in the table you can fill in before the patterns of repetitions and adding to zero tell you what the remaining numbers have to be.

Here's how the degrees of freedom will turn out: one each for the Grand Average, Day Length, Organs, and Interaction; six each for Hamsters and Residuals. (Notice that the degrees of freedom for the six factors add up to 16, the number of observed values: 1 + 1 + 1 + 1 + 6 + 6 = 16. This will always happen.)

In the next example I'll show you how I got the degrees of freedom for four of the factors. As you read what I've written about each factor, turn back to the decomposition in Fig. 3.4 so you can see the patterns of repetition and adding to zero. I know it's a nuisance to have to flip pages back and forth, but it's important that you see the patterns yourself.

Example 3.1 Finding the Hamster DF by Counting Free Numbers

Grand Average: df = 1

This box has only one number (60) repeated 16 times. Since you know the box is for the grand average, as soon as you know one number in the box, you can figure out all the others because they have to be the same. There's only one free number, thus only one degree of freedom here.

Day Length: df = 1

There are two numbers in this box: a −3 repeated eight times in the top half and a +3 repeated eight times in the bottom half. It's no coincidence that the numbers add to zero; rather, it's a built-in property of the decomposition. (I'll say more about why a little later.) Because the numbers must add to zero, as soon as we know the first −3 in the top half of the box, we can fill in the rest: the other seven numbers in the top half must all be −3s, and each of the eight numbers in the bottom half has to be a +3; otherwise, the box wouldn't add to zero. There's only one free number here, and only one degree of of freedom for Day Length.

Organs: df = 1

There are two numbers in this box: a −43.5 repeated eight times in the Heart column and a +43.5 repeated eight times in the Brain column. The numbers add to zero, just as they did for Day Length, so we only need to know one number to fill in the whole box: one free number, and so one degree of freedom for Organs.

Hamsters: df = 6

The eight numbers in this box (one per row) add to zero in two sets of four: the top four rows add to zero, and so do the bottom four. If we're told the numbers in the first three rows (−8, +3, +2), we know what the number in the fourth row must be (+3), so the first four rows give us three free numbers. The bottom four rows give us three more, for a total of six free numbers, six degrees of freedom. ■

Patterns of adding to zero. The patterns of repetitions are usually easy to see directly, but you can always get them from the factor diagram. For example, the factor for Hamsters has eight groups, the eight rows. Within each of these groups, the numbers will be repetitions of each other, so there can be at most one free number per group, at most eight in this case. In fact, of course, there are only six free numbers because of the patterns of adding to zero. It is these patterns of adding to zero that can get tricky sometimes. The basic idea, however, is simple:

> **Whenever you subtract the average from a set of numbers, the leftovers from the average, called deviations, will add to zero.**

EXAMPLE 3.2 PATTERNS OF ADDING TO ZERO IN THE HAMSTER DATA

Day Length. The two averages for this factor are 57 (Long days) and 63 (Short). To get the effects, we subtract the grand average (60), which equals the average of 57 and 63. The two effects, −3 and +3, add to zero.

Organs. The situation here is the same. The there are two organ averages, 16.5 (Hearts) and 103.5 (Brains). The average of these two numbers equals 120/2 = 60 (the grand average), which we subtract to get the effects, −43.5 (Hearts) and +43.5 (Brains). The two deviations from their average add to zero.

Hamsters. Here things are a bit more complicated. The hamster averages for the four long-day hamsters are 49, 60, 59, and 60. The average of these four numbers equals 57, the long-day average, and the corresponding hamster effects are the deviations from the long-day average: −8 (= 49 − 57); +3 (= 60 − 57); +2 (= 59 − 57); and +3 (= 60 − 57). These four long-day hamster effects add to zero. In the same way, you can check that the four short-day hamster effects add to zero because we got them by subtracting the short-day average.

For some factors, like Interaction, the patterns of adding to zero are more complicated because you have to subtract more than one average to get each effect. Fortunately, there is an easy way to keep track of which averages get subtracted, but because the rule is secondary for now, I've put it in the appendix. ■

Mean Square (MS) = Average Variability per Unit of Information

To get the mean square (average variability) for a factor, just divide the sum of squares (overall variability) by the degrees of freedom (units of information): MS = SS/df.

Source	SS	df	MS
Grand average	57,600	1	57,600
Long/short	144	1	144
Hamsters	240	6	40
Heart/brain	30,276	1	30,276
Interaction	676	1	676
Residual	316	6	52.67
Total	89,252	16	

FIGURE 3.6 Mean squares for the hamster data. Each mean square summarizes the average variability for the numbers in a box of the decomposition and serves as a measure of how much variability in the observed values is linked to each factor. To take the most extreme example, compare the mean square for organs with the mean square for residuals, and consider whether the observed difference between hearts and brains is too big to be due just to chance error. The mean square for organs is huge, more than 500 times the mean square for residuals. In the next chapter, we'll end up concluding that the organ effects are real.

The logic here depends on the meaning of degrees of freedom as units of information about chance error. I hope you agree that the sum of squares gives one reasonable way to measure overall variability and that it makes sense to want to compute average variability. However, I haven't yet told you why I divide by degrees of freedom to get our measure of average variability.

Here's an informal explanation. According to our model, each observed value contains a chance error, so our data set provides one unit of information about chance error for each observation—16 in all. When we split these 16 observed values into pieces, we don't create new information about chance errors, nor do we destroy information. Instead, we parcel out the 16 units we started with, so many to this factor, so many to that one. Taken together, the 6 boxes of the decomposition still contain 16 units.

Thinking about free numbers tells us how many units go to each factor: each free number provides one unit of information about chance error. For example, consider the box for the grand average, which has the number 60 repeated 16 times. As soon as you know the one number 60, you know all the information that box has to tell you. Similarly, for Day Length, or for Organs, as soon as you know any one number in the box for the factor, you can fill in the whole box, so there's only one unit of information.

I'll come back to this later in the chapter, and again in other chapters as well. First, however, I'll tell you how to compute a standard deviation and show you how to use it to check two of the Fisher assumptions.

3. Standard Deviation = Root Mean Square for Residuals

We've already computed a mean square for residuals (52.7), which measures average variability due to chance error. Remind yourself that this number is a kind of average of the squares of the residuals: we squared each residual, added them up, then divided by how many units of information we had. To get the standard deviation, we simply take a square root to undo the squaring from before:

$$\mathrm{SD} = \sqrt{\mathrm{MS_{Res}}}.$$

For the hamster data, $\mathrm{SD} = \sqrt{52.7} \approx 7.3$. Roughly, this number tells us that for the hamster data, each observed value equals its true value, give or take about 7 or so.

Observed value = True value ± about one standard deviation.

If the normality assumption (N) fits our data, that is, if our residual errors follow the normal curve, then we can be more specific: about 2/3 of our errors will be no bigger than ±7.3 (one SD); roughly 95% will be at most ±14.6 (two SDs); none, or almost none, will be bigger than ±21.9 (three SDs).

EXAMPLE 3.3 CHECKING THE NORMALITY ASSUMPTION FOR THE HAMSTER DATA

If you turn back (yet again!) to the decomposition in Fig. 3.4 you can check whether and how well the residuals fit this ideal pattern. The dot graph in Fig. 3.7 shows each of the 16 residuals from the decomposition plotted as a point.

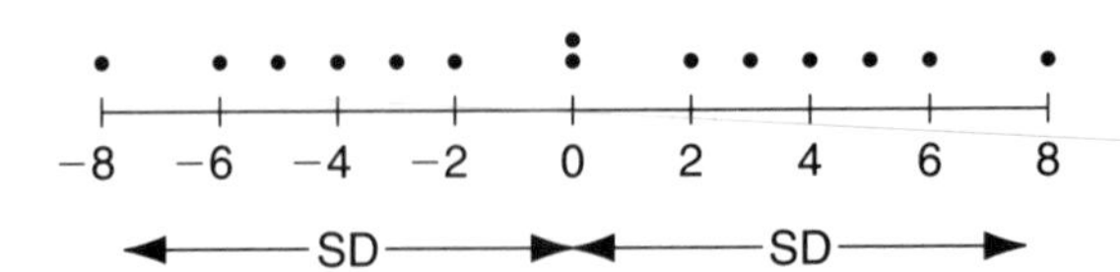

FIGURE 3.7 Using residuals and standard deviation to check the normality assumption. If the assumption fits, roughly 2/3 of the chance errors will be within one SD of zero, 95% within two SDs. However, when the number of residuals is quite a bit bigger than the df, the residuals themselves tend to underestimate the individual chance errors. Since we have 16 residuals, with only 6 df, we should expect to find more than 2/3 of the residuals within one SD of zero and more than 95% within two SDs. In fact, 14 of 16, or 87.5%, are within one SD, and 100% are within two SDs. The normality assumption fits quite well. ■

We can also use SDs to check the assumption (S) of same standard deviations. Remember that I first showed you a way to check this assumption back in Chapter 2, page 48. I divided the observed values into four groups, one group for each combination of Day Length and Organ; then I computed the range (largest − smallest) of the observations in each group and compared ranges. We found that the largest range was almost 30 times as big as the smallest: there were errors of very different sizes.

Back then, I hadn't yet told you how to compute standard deviations, and so I used the range to measure how spread out the observations were. The range has the advantage of being simple and quick to compute, but the SD is almost always a more reliable way to measure error size. Now that you know how to find SDs, we can use them to compare error sizes for the four groups. Doing this will also give you a review of true values, residual errors, decompositions, and standard deviations.

EXAMPLE 3.4 CHECKING THE ASSUMPTION OF SAME SDs

Let's pick one of the four conditions, Long Days/Heart. The four log concentrations for this condition are 17, 19, 19, and 25, with an average of 20. Although all four numbers seek to measure the same thing (the log concentration of sodium pump enzyme in the heart of a hamster raised with 16 hours of light per day), the numbers are not equal. The differences are due to many influences, such as genetic differences between hamsters, perhaps differences in their environments (where the cage was put) and behavior (how much the hamster ate, how much it exercised), and uncontrolled variability in the method of extracting and measuring the enzyme. We lump all these sources of variability together with the label "residual error" and think of each observation as composed of two pieces, true value and error:

$$\text{Observed value} = \text{True value} + \text{Residual error}.$$

The true value is the (unknown) enzyme concentration that is typical of all hearts of hamsters raised with 16 hours of light. Different hamsters have different enzyme concentrations, but for now we will regard those differences as part of residual error, not part of what we call the true value. The true value is a single number, unknown, but the same for all the hamsters.

In terms of factors, we're assuming the simplest possible model, with no structural factors, just the two universal factors: the benchmark (true value) plus residual error. The grand average of the four observations equals 20. Each residual equals Obs − Avg.

	Obs		Avg		Res
	17		20		−3
	19	=	20	+	−1
	19		20		−1
	25		20		5
SS	1636	=	1600	+	36
df	4	=	1	+	3

Residual box
SS = 36
df = 3
Residuals add to zero

FIGURE 3.8 Decomposition of the data for the hearts of the long-day hamsters. For this way of looking at the data, there are only two factors: the benchmark (true value) and residual error. The average of the four observations (20) is an estimate of the true value, and each Res = Obs − Avg is an estimate of residual error.

Notice how the decomposition splits three things in parallel: the pieces of the observed values add up (linear decomposition); the sums of squares add up (quadratic decomposition); and the degrees of freedom add up.

To get the SD for this group of observations, we first compute $MS_{Res} = SS_{Res}/df_{Res} = 36/3 = 12$ and then take the square root: $SD = \sqrt{MS_{Res}} = \sqrt{12} \approx 3.5$. Notice that 3.5 seems about right for the typical size of residual error. One of the residuals (the −3) is quite close to 3.5, apart from the minus sign; two of the residuals (the −1s) are a bit smaller than 3.5; and one (the +5) is a bit bigger. So 3.5 seems like a reasonable summary number for residual error. If you told someone the enzyme concentrations were roughly equal to 20, give or take about 3.5 or so, they'd have a pretty good idea of what the data actually look like.

The SDs for the other groups work out to 3.2 (Short Days/Heart), 9.0 (Long Days/Brain), and 9.1 (Short Days/Brain). Even in the log scale, the errors for the brain concentrations are bigger than those for the hearts, but the two sizes aren't so very far apart. The largest SD is 9.1; the smallest is 3.2; and the ratio, SD_{Max}/SD_{Min}, is less than 3.

Contrast all this with what we get for the untransformed concentrations, in mg/ml:

Group		Avg	SD
Long Days	Heart	1.590	0.136
	Brain	8.925	1.669
Short Days	Heart	1.353	0.104 ← SD_{Min}
	Brain	13.906	2.985 ← SD_{Max}

FIGURE 3.9. Two patterns say to transform the raw data (mg/ml).
1. The largest SD is almost 30 times the smallest SD. As a rough guide, you should try to transform if SD_{Max}/SD_{Min} is bigger than 3.
2. The bigger the average, the bigger the SD. In other words, the typical size of residual error depends on the size of the measurement. In such situations, you can usually find a new scale that makes your SDs roughly equal. (See Chapter 12 for more on both these patterns.)

These computations back up the conclusion we reached before: the raw concentrations, in mg/ml, do not satisfy the assumption of same SDs, and so the raw data ought to be transformed before we do formal tests to see whether the observed differences are "real." The transformed data, in the log scale, satisfy the assumption reasonably well, and so we can go ahead with our analysis of variance. ■

4. Formal Hypothesis Testing: Are the Effects Detectable?

From description to inference: conclusions depend on the Fisher assumptions. Remind yourself (from Chapter 1, pages 11–12) of the difference between a sample and a population, between estimates and true values. The sample is what we get to see and what we use to compute estimates, as in the last chapter. We never get to see the population, and we never know the true values—but, provided the Fisher assumptions fit our data, we can use our estimates to make inferences about the true values. This section and the next illustrate some of those inferences. Until now, we've not tried to draw conclusions about the true values, and so we've not really needed to rely on the Fisher assumptions. In what follows, the methods I'll show you do depend on those assumptions. In particular, from here on, we'll assume the residual errors are chancelike. To emphasize this assumption, I shall sometimes call them "chance errors."

Consider once again the question of interaction: whether the true effect of day length on enzyme concentration is different for the hearts and brains of golden hamsters. The method of hypothesis testing seeks to answer questions like this one, in effect by holding a jury trial. The data set stands "accused" of having real differences. The evidence is provided by the observed values, which we will combine to form a special one-number summary called an **F-ratio**. The jury takes the form of a table of **critical values**, created by Fisher, that tell us how to judge the *F*-ratio. If the *F*-ratio is bigger than Fisher's critical value, we declare the data "guilty": our verdict is that the differences are "real," not due to chance error alone. On the other hand, if the *F*-ratio is less than the table value, our verdict is "innocent": the differences might reasonably be due to chance and nothing more.

For a data set like ours, we have what the newspapers might call a "multicount indictment." The data are suspected of several kinds of possible true differences: differences due to day length, to organ, to interaction, and to hamsters. Analysis of variance provides a method for reaching a verdict on each count. We compute an *F*-ratio, our one-number summary of the evidence, for each set of differences. Each *F*-ratio summarizes a comparison: we compare the average variability due to interaction, or to organ, or whatever, with the average variability due to chance error:

$$F\text{-ratio} = \begin{array}{c}\text{Summary}\\ \text{of the}\\ \text{evidence}\end{array} = \frac{\text{Average variability due to interaction}}{\text{Average variability due to chance error}}.$$

If this ratio is a lot bigger than 1, it means the average variability due to the difference in question is a lot bigger than the average variability due to chance error. In the case of interaction for the hamster data, the *F*-ratio turns out to be 12.84. This number tells us that our estimate of the variability due to interaction is more than 12 times as big as our estimate of the variability due to chance error. To judge the strength of this evidence, we take our 12.84 to the jury: Fisher's table value equals 5.99, which tells us our evidence (12.84) is stronger than what we need (5.99) to return a verdict of guilty. We conclude that there is a real interaction effect.

The rest of this section is divided into two parts. The first gives *F*-ratios for the hamster data and tests the corresponding hypotheses. The second presents some of the reasons for using *F*-ratios to do the tests.

The Analysis of Variance Table: Summarizing the Evidence

Fisher invented a handy table for keeping track of our work, called an **analysis of variance table**, or ANOVA table. The table has one row for each factor in the design and one column for each step in the analysis. You've already seen most of the ANOVA table for the hamster data. Here's the complete table:

TABLE 3.1 ANOVA Table for Log Concentration of Sodium Pump Enzyme

Source	SS	df	MS	F-ratio	Critical value
Grand avg	57,600	1	57,600		
Long/short	144	1	144	3.60	5.99
Hamsters	240	6	40	0.76	4.28
Heart/brain	30,276	1	30,276	574.86	5.99
Interaction	676	1	676	12.84	5.99
Residual	316	6	52.67		
Total	89,252	16			

The table has one row for each factor in the design and one column for each step in the analysis. Filling in the table means (1) computing, for each box of the decomposition, the sum of squares, degrees of freedom, and mean square, (2) dividing mean squares to get F-ratios, and (3) comparing them with Fisher's critical values.

The F-ratios for Long/short and for Hamsters are smaller than the corresponding table values: we conclude that the observed differences due to Day Length and Hamsters may be due to chance error and nothing more. The F-ratios for Heart/brain and Interaction are bigger than their critical values: these effects are "real."

Computing the F-ratios. The basic idea is that for each factor you want to test, you take the mean square for that factor and divide by the mean square for (usually) residuals. Since the Interaction factor interests me the most, I'll use it as an example:

$$\begin{matrix}\text{F-ratio}\\ \text{for}\\ \text{Interaction}\end{matrix} = \frac{\begin{matrix}\text{Mean square}\\ \text{for Interaction}\end{matrix}}{\begin{matrix}\text{Mean square}\\ \text{for Residuals}\end{matrix}} = \frac{676}{52.67} = 12.84.$$

F-Ratio for testing interaction. The F-ratio tells us that the average variability in the Interaction box is about 13 times as big as the average variability in the Residual box. Since the 12.84 is bigger than the table value, we conclude that the interaction effect is real. (In more formal language, we *reject the null hypothesis*, which says there are no true differences due to interaction.)

You get the F-ratios for Hamsters and for Organ in the same way as for Interaction: take the mean square for the factor you want to test, and divide by the mean square for Residuals. The Hamster F-ratio (0.76) is less than the table value (4.28): the observed hamster differences may be just chance error. The Organ F-ratio (574.86) is gigantic, much bigger than the table value (5.99): these differences are "real."

The F-ratio for Day Length is 3.6, but I didn't get it in the same way as I got the F-ratios for the other factors. I didn't divide by the mean square for Residuals; instead, I divided by the mean square for Hamsters. There's a reason for this exception to the usual rule, but I'll save that for later. For now just remember that for the SP/RM design, there is one F-ratio that you don't get in the usual way.

The F-ratio for Day Length is less than the table value, so we conclude that the overall effect of day length (when hearts and brains are averaged together) may be just chance. However, the conclusion here doesn't mean much, because we concluded there's a "real" interaction: for hearts alone, or for brains alone, day length does make a "real" difference.

So far, you've been reading mainly about the mechanics of the ANOVA table: what steps you go through to get the numbers, and what conclusions you get. Next I'll explain some of the logic behind the steps. The better you understand the reasons for the steps, the easier it will be to remember them and to know what they tell you (and what they can't tell you).

The Logic of the F-Test

Note: If you've never seen statistical tests of hypotheses before, you can read Appendix 2 at the end of this chapter for more background on the general logic of hypothesis testing. What follows tends to focus more narrowly on how the F-test works.

If there are no true differences, both numerator and denominator mean squares will be estimates of the same number, and the *F*-ratio will tend to be near 1. If there are real differences, the numerator will tend to be larger than the denominator, and the *F*-ratio will tend to be larger than 1.

Every mean square is part chance error. Since you always compute an F-ratio by dividing one mean square into another, it makes sense that to understand F-ratios you need to look more closely at mean squares. I've told you that each mean square is a summary measure of average variability per unit of information. While that is true, there's more to the story. Just as each observed value is a sum of two or more pieces, so each mean square—though a single number—is, underneath, a sum of pieces. For example, consider the mean square we got from the box for hearts versus brains by squaring, adding, and then dividing by the degrees of freedom. The numbers in the heart/brain box are estimates, not true differences: each number is part true difference but also part chance error, because the splitting we carried out could not be perfect. As a consequence, the mean square is partly a measure of variability due to organ differences, but there's a chance error piece as well. The degrees of freedom for a box tell us how many units of information there are, but these are quite specifically units of information about chance error. When we divide the sum of squares by the degrees of freedom, we divide by exactly the right number to make the chance error piece an estimate of the variability per unit of information about chance error. In other words, this piece of the mean square for Organ is an estimate of the same number that is estimated by the mean square for Residuals. The rest of the Organ mean square is due to true heart/brain differences. When we divide the mean square for Organ by the mean square for Residuals to get an F-ratio, the number we get is itself a sum of two parts:

$$F\text{-ratio} = \frac{\text{Mean square for Organ}}{\text{Mean square for Residuals}} = \frac{\text{Estimate 1 for chance error} + \text{Piece from true organ differences}}{\text{Estimate 2 for chance error}}$$

$$= \underbrace{\frac{\text{Estimate 1 for chance error}}{\text{Estimate 2 for chance error}}}_{\text{First part}} + \underbrace{\frac{\text{Piece from true organ differences}}{\text{Estimate \#2 for chance error.}}}_{\text{Second part}}$$

Look again at the first part. Both the numerator and denominator estimate the same number, and so the first part is just an estimate of 1. The second part estimates a number that will equal zero if there are no true organ differences but will be bigger than zero if there are true differences. So the two pieces added together estimate a number that equals $1 + 0 = 1$ if there are no true organ differences and equals some number bigger than 1 if there are true differences. We can summarize what we've found like this:

Relation Between the True Condition and the F-Ratio

True condition	*F-ratio*			*Tends to be*
1. No real differences	Estimate of 1	+	Estimate of 0	Near 1
2. Real differences	Estimate of 1	+	Estimate of some number bigger than 0	Bigger than 1

There are two possible true conditions: real differences, and no real differences. For each possible condition, there is a corresponding result we expect for the F-ratio: if there *are* real differences, we expect the F-ratio to be quite a bit bigger than 1; if there are *no* real differences, we expect the F-ratio to be pretty close to 1.

When we use an F-ratio to draw conclusions about the data, we work backwards from the value of the F-ratio to a conclusion about the conditions that might have produced that value. If the F-ratio we compute turns out to be near 1, what we see is just what we should expect to see if there were no real differences. We can't be sure there are no differences, but we do know that the data have given us no evidence of differences. If the F-ratio we compute is a lot bigger than 1, what we see is what we expect to see if the differences are real. Here again, we can't be sure the differences are real, but if the F-ratio is big enough, we know it is not at all like what we expect if there are no differences, and we conclude that the differences are real.

Limitations of the* F*-test. Although it's easy to summarize what a big *F*-ratio tells you (that the observed differences are too big to be due just to chance error), if you're not careful, you can fall into the trap of thinking the *F*-ratio tells you more.

A large F-ratio does *not*
- **check the design of your experiment,**
- **check whether your model is good, or**
- **check whether the six key assumptions fit your data.**

All these things are crucial. If your design is a bad one, or your model is wrong, or one of the assumptions doesn't fit your data, you can get a large *F*-ratio and be misled about what it means. *Statistical analysis cannot be made idiot-proof.* The only way to become good at statistical thinking is to develop your judgment.

For the hamster data, there are various reasons to be cautious about conclusions from the *F*-ratios. Probably the two most important are these:

1. The subplot factor (Organ) could not be assigned, and so there could be no randomization. Our scatterplot of heart and brain readings showed a negative relationship.
2. According to the scatterplot, the data contain an outlier.

Suppose you do have a sound design, an appropriate model, and data that fit the Fisher assumptions, so that your *F*-ratios are not misleading. Even so, they are limited in what they tell you:

An F-ratio tells you whether the true differences are big enough to be *detected* by your experiment, but not what size the differences are—*not* whether they are big enough to be scientifically interesting.

Too many people who use ANOVA look only at the results of the *F*-tests. If we were to adopt that narrow approach, we could put the results of all our work into a thimble. The *F*-tests only give us a set of four yes/no conclusions:

no: overall day length effects are not detectable;

yes: overall organ effects are detectable;

yes: interaction effects are detectable;

no: hamster differences are not detectable.

The *F*-tests are always limited in what they can tell you. Worse yet, if you use them in a lazy and mechanical way, without thinking about the patterns in your data, they can sometimes be misleading. For example, in the case of the hamster data, the first *F*-test—by itself—tells us that overall, the effects of day length are not detectable. In fact, as we know from the test for interaction, day length has detectable but opposite effects on the enzyme concentrations of hearts and brains. The first *F*-test misses this effect because it averages hearts and brains together. Similarly, the fourth *F*-test tells us that differences due to hamsters are not detectable. Here, too, the conclusion is misleading if taken by itself. We estimated hamster effects by lumping together the two readings—one heart, one brain—for each hamster. We know from a scatterplot that if one of the two readings tends to be higher than average, the other tends to be lower than average. Lumping the two together leads us to underestimate the variability between hamsters.

Clearly, then, the data contain much more information than a summary based on *F*-tests can tell us. In particular, the data permit us to estimate the likely size of true differences. That's the topic of the last section of this chapter.

5. Confidence Intervals: The Likely Size of True Differences

Intervals for the Effect of Day Length on Hearts and Brains

The *F*-tests of the last section led us to conclude that there is a real interaction: the average difference between long- and short-day enzyme concentrations is not the same for hearts and brains.

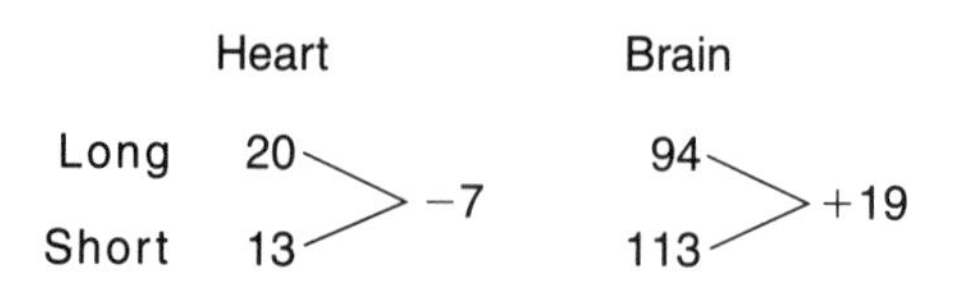

FIGURE 3.10 Average enzyme concentrations showing the effect of day length for hearts and brains. If we take the observed differences at face value, then going from long to short days lowers the heart concentrations by 7 and raises the brain concentrations by 19.

We know that the differences in the averages (the −7 and +19) are estimates: we assume each number equals true value plus chance error. It is natural at this point to ask how good the estimates are. How much of the −7 and +19 is chance error? Could the true effect for brains be as low as 5? As high as 40? You might think we could split our estimates apart, just as we did with the raw data, but unfortunately we can't. We've already done all the splitting possible for this data set. One way to see this is to think about the chance error part of the +19 for brains. Part of the +19 is chance error, but there's no way to know which direction it goes, whether the error part is negative, making the observation smaller than the true value, or the reverse.

Although we can't tell the direction of the chance error part, it is nevertheless possible to estimate the likely size of it. This number, the **standard error**, tells for estimates what the standard deviation tells for a single observation: the typical size of the chance part. For the effect of day length on hamster brains, this number will turn out to be roughly 6.1, which tells us that the +19 equals the true difference, give or take about 6. Once we have the standard error, we can use a table to convert it to what I call the **95% distance** for the estimate, which we can combine with the estimate itself to get a **95% confidence interval**, a range of values that we think contains the true value somewhere within it. The 95% distance will turn out to be 14.9, which gives a 95% confidence interval with 19 at the center and stretching 14.9 on either side, from 4.1 (= 19 − 14.9) to 33.9 (= 19 + 14.9). Can we be sure the true value is somewhere in that interval? No: because of the uncertainty that comes from chance error, no statement about the unknown true value can be 100% sure. The best we can say is this: the interval was created by a method that works 95% of the time. In 95% of the cases where we use this method to construct an interval, it will contain the true value.

The precise meaning of the confidence interval is a bit tricky. It would be simpler if we could say, "There's a 95% chance that the true value lies between 4.1 and 33.9," but, strictly speaking, that's not true. The 95% is a property of the method, not of the particular interval. In a way, the 95% is like a lifetime batting average—it refers to a long sequence of hits and misses. Some confidence intervals score a hit—they capture the true value. Others miss. Since we never know the true value, we can't tell whether or not it belongs to the interval we just computed. But 95% is not a bad batting average to rely on. (Appendix 2 to this chapter gives two detailed analogies and two simple examples to help explain the main ideas of how confidence intervals work.)

We can compute a second interval for the effect of day length on hamster hearts. Because the heart readings show less variability than the brain readings, the standard error turns out to be quite a bit less, about 2.5, which tells us the likely size of the chance piece of the estimate: The true difference (Long − Short) equals −7, give or take about 2.5 (Fig. 3.11). The 95% distance equals 6.2, which gives a 95% confidence interval from −13.2 (= −7 − 6.2) to −0.8 (= −7 + 6.2).

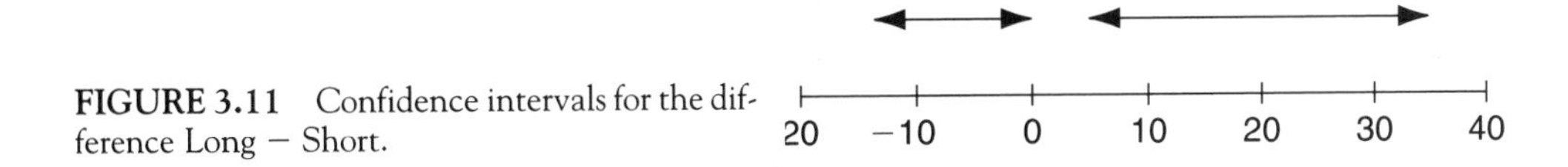

FIGURE 3.11 Confidence intervals for the difference Long − Short.

Transforming back to the original scale. Remind yourself that our estimates of −7 and +19 are measured in units of 100 × log(conc.). The scale raises a question: what does it mean to say that for brains the short-day average is 19 units higher than the long-day average? To get back to the old units of milligram per 100 milliliters, we have to reverse our steps: divide by 100 and then undo the logs:

$$\begin{aligned} 100\log(\text{short conc.}) - 100\log(\text{long conc.}) &= 19 \\ \log(\text{short conc.}) - \log(\text{long conc.}) &= 0.19 \\ \log(\text{short conc./long conc.}) &= 0.19 \\ \text{short conc./long conc.} &= 10^{0.19} = 1.55 \end{aligned}$$

So short conc. = 1.55 long conc.

Using the same steps, you can translate lower and upper points of the confidence interval, and then do the same for hearts:

Brains: Estimate: short conc. = 155% long conc., with a 95% confidence interval going from 110% to 218%;

Hearts: Estimate: short conc. = 85% long conc., with a 95% confidence interval going from 74% to 98%.

Using the 95% Distance to Think About Design Issues

Let's now return to the 95% confidence interval (roughly 19 ± 15) for the effect of day length on the enzyme concentration of hamster brains. The interval, from 4 to 34, is quite wide: the experiment hasn't given us very precise information. (Of course, things could have been much worse, say a 95% distance of 30 and an interval from −11 to 49.) What kind of experiment would it take to give us a 95% distance of 5, with an interval of 19 ± 5, or 14 to 24? My goal in what follows is to say enough about how I computed the 95% distance for you to begin to get a sense of what it is about an experiment that makes the distance large or small.

The standard error, or SE. Our estimate of 19, like any number you compute from the data, is a sum of two pieces, true value plus chance part. Just as the SD tells the typical size of the chance part for an observed value, the standard error, or SE, tells the typical size of the chance part for the estimate. To get the SE you multiply the SD by the leverage factor. The 95% distance equals the SE times a *t*-value from a table.

> **The 95% distance comes from multiplying three numbers together:**
> **(1) $\sqrt{MS_{Res}}$ (or a similar measure of error size)**
> **(2) a leverage factor based on the number of observations**
> **(3) a *t*-value from a table (or computer)**
> **The standard error equals (1) × (2).**

For example, here's the arithmetic for the hamster brains. I'll discuss each of the three terms one at a time.

Estimate ± ($\sqrt{MS_{Res}}$) × (Leverage factor) × (t-value)

19 ± 8.61 × 0.71 × 2.45

Standard Error = 6.1

95% distance = 14.9

FIGURE 3.12 Form of a 95% confidence interval. The 95% distance equals error size ($\sqrt{MS_{Res}}$) times a leverage factor based on the number of observations times a *t*-value from a table. The product of the first two numbers is the standard error (SE), and the interval takes the form Estimate ± SE × *t*-value. Just as 2 × SD works as a 95% distance for the normal curve, *t* × SE gives the 95% distance for the confidence interval.

SD = $\sqrt{MS}$ = typical size of chance error. This first number (8.61 in the example) will always be the square root of some mean square, although which mean square to use will depend on your design and on your estimate. For the BF design, for example, you always use the mean square for residuals, and so the first part of your SE is just $\sqrt{MS_{Res}}$, the usual estimate of the SD. Here's how I computed MS_{Res} for the confidence interval:

	Observations				Ave	Residuals				df	SS
Long	81	101	99	95	94	−13	7	5	1	3	199
Short	110	105	126	111	113	−3	−8	13	2	3	246

$SS_{Res} = 199 + 246 = 445$.
$df_{Res} = 3 + 3 = 6$.
$MS_{Res} = 445/6 = 74.167$.

For some designs and estimates, the choice of $\sqrt{MS}$ may not be so simple, but whatever the mean square, its role is always the same: $\sqrt{MS}$ measures the typical amount of uncertainty in the observed values that go into your estimate. If $\sqrt{MS}$ is large, it will make the SE large, too, and the confidence interval will be wide; if $\sqrt{MS}$ is small, it will make the SE small and the interval narrow. Roughly, the logic goes like this. A large MS means each observed value tends to include a large chance error: the data are "noisy," and the estimate will not be very precise. On the other hand, a small $\sqrt{MS}$ means good data, precise estimates, and a narrow interval.

In practice, there are two strategies you can consider to try to make MS smaller: better design and better lab techniques.

Design: Sometimes you can use blocking to link variability in your measurement process or you material to a factor in the design. For example, Kelly used hamsters as blocks, which reduced the error size for comparing hearts with brains.

Lab technique: You may be able to change to a more precise instrument or procedure. Even when that's not possible, you may be able to reduce variability just by practicing the procedure several times before you begin your experiment.

Leverage factor for the estimate. Our estimate is based on the combined information from several observations, which means that the chance error part of our estimate is smaller than for one observed value. (Most people intuitively feel that the average of several values is more reliable than a single value, and they are right. The leverage factor tells how much more reliable.) In our case, the leverage factor equals .71: the standard error is 71% as big as the $\sqrt{\text{MS}}$. I got the .71 from

$$\text{Leverage factor} = \sqrt{(1/4) + (1/4)} = 0.71.$$

There's a logic behind this number, but for now I'll just give you the bare rules in order to focus on what matters most for planning an experiment.

1. Each 1/4 comes from one of the averages we computed to get the estimate of +19. There was one average for Long Days/Brain, based on four observations, and another average for Short Days/Brain, also based on four observations. The 4 in each denominator is the number of observations in the average. If we'd had 8 observations per average, each denominator would be 8, which would make the leverage factor quite a bit smaller, equal to .5. In general, the more values that go into the averages, the smaller the leverage factor, and hence the smaller the standard error for the estimate. This is of great practical importance because it tells you the numerical relationship between how many observations you make and how precise your estimates will be.
2. The square root: statisticians have discovered that every leverage factor always has a square root. It comes from a very basic pattern in the way chance errors tend to cancel each other when you compute averages. The square root is important to know about in planning experiments because it tells you that to cut the SE in half, you have to use four times as many experimental units. Going from 8 hamsters to 32 hamsters might be a realistic plan. Going from 32 to 4 times 32 is probably not realistic: you'd need almost 100 extra hamsters!

t-value = standard 95% distance from a table. If your chance errors follow the normal curve and you know the exact value of your SE, then the standard 95% distance equals 2, and your confidence interval has the form Estimate ± SE × 2. In real life, however, you have to estimate the SE because you don't know the exact value. This estimation brings in more uncertainty, which means your interval ought to be wider to adjust for the extra uncertainty. Instead of SE × 2, you use SE × *t*-value, to make this adjustment. For us, the *t*-value is about 2.45.

The *t*-values come from the work of two statisticians, William S. Gossett, who did the empirical work, and Ronald Fisher, who provided a theoretical framework. What they discovered was that (provided the six key assumptions hold) every 95% distance equals the standard error of the estimate times a standard 95% distance, and they published a table of these standard distances. The table of these standard distances is called a ***t*-table**, and the distances themselves are called *t*-values. The *t*-value depends on two things: on the degrees of freedom for the mean square you use to get your SE and on the confidence level, usually 95%, as in our example.

The larger the df, the smaller the t-value and the narrower your interval will be. For example, here are a few 95% t-values:

df	1	2	3	4	5	6	10	20	50
t	12.7	4.3	3.18	2.78	2.57	2.45	2.23	2.09	2.01

When you are planning an experiment, the most important lesson from this table of t-values is that when df_{Res} is tiny, adding just a few more observations can reduce the t-value a lot, and so increase your precision a lot. If df_{Res} is around 6 or more, adding extra observations won't change the t-value by much. Additional observations will mainly improve precision by reducing the leverage factor.

Here is a summary of the main points:

1. If df_{Res} is less than 5 or 6, it's almost always worth adding experimental units in order to reduce the t-value.
2. Once df_{Res} is up to 6 or more, the square root effect in the leverage factor takes over: in order to cut the 95% distance in half, you'll need four times as many units.
3. You can sometimes increase precision by reducing the SD, either through design (blocking) or through improved lab technique.

To see these points in action, think about Kelly's decision to use four hamsters for each day length. If she'd used only three, she would have has only four degrees of freedom for Residual and a t-value of 2.78. Her leverage factor for comparing long- and short-day averages would have been 0.82, and the resulting 95% distance would have been 19.6. Had she used five hamsters per group, she'd have gotten a t-value of 2.31, a leverage factor of 0.63, and a 95% distance of 12.5. The two extra hamsters it takes to go from three per group to four per group reduces the 95% distance from almost 20 to just under 15. However, adding another two hamsters reduces the distance only about half as much, from 15 to 12.5.

Our two confidence intervals, together with all the various graphs from Chapter 2, provide a good record of our analysis. In sum, enzyme concentrations are very different for hearts and brains, and the effect of day length on those concentrations goes in opposite directions for the two organs. The center of each confidence interval tells us our best estimate for the true effect of day length on the heart or brain concentration, and the width of the interval tells us how precise the estimate is.

Is our analysis finished? No. In preparing these introductory chapters, I did quite a bit of work with the data that I haven't told you about, things I think should be part of any careful analysis of a data set. Most particularly, we should not accept the results of these chapters until we have explored the possible effect of the outlier that shows up so clearly in the scatterplot of Chapter 3, Section 4. I'll leave that for Chapter 12, however.

Another question I've not addressed is "What do the results suggest about the next experiment?" This question would take us back to biology, away from statistics, and so away from what this book is about, but it's an important question nevertheless.

I raise these points because I believe, and hope you will come to agree, that in a sense you can never finish an analysis. A good experiment answers important questions and in answering them raises others. These others lead naturally to ideas for a new analysis, or a new experiment, and the cycle of design and analysis continues. The process goes back and forth, between design and analysis, between statistics and its subject of application. Throughout the whole process, there are tools to get you from one stepping stone to the next, but it takes judgment, understanding, and a good sense of direction to find the path ahead.

Appendix 1. The General Decomposition Rule: Estimated Effect Equals Average Minus Partial Fit

The title of this section gives a shorthand version of the general rule for finding the pieces in the decomposition of any balanced data set: effect equals average minus partial fit. Before you try to understand the "partial fit" part, first focus on the rest, which says that to estimate the effect for a factor, you start with the average for that factor and subtract some other number. This is just what I did for the hamster data in Section 1: the long-day effect (−3) was the long-day average (57) minus the grand average (60); the heart effect (−43.5) was the heart average (16.5) minus the grand average (60); and the effect for hamster #4 (−8) was the average for #4 (49) minus the long-day average (57). You can even think of the grand average and residual as part of the same pattern. The "effect" for the benchmark (60) equals the grand average (60), which you can think of as the average for the benchmark factor (60) minus zero. The factor for chance error puts each observed value in a group by itself, so the "average" for the group containing the observed value of 17 is just the 17 itself. This makes the residual (5) equal to the "average" for the chance error factor (17) minus the fitted value (12).

If each "effect equals average minus partial fit," how then do you figure out the right partial fit to subtract from each average? We want the estimated effect for long days, for example, to tell us the change in average response we get as a result of assigning the treatment "long days" to a group of observations. More generally, we want the effect of a factor on a group of observations to tell us the change in average response that comes from assigning that factor to the observations. The "change" in response refers to a subtraction—"after the factor" minus "before"—that is, the actual average for the group minus what the average would have been without the factor. The "partial fit" is what the average "would have been." To find it we need to find all the conditions that are the same for all the observations in the group and add up their effects. So finding partial fit boils down to making a list of all the factors that are the same for all the observations in the group. In terms of the assembly line metaphor, we need to combine the pieces that are added to *all* the measurements that go into the average. The idea turns out to be much easier to understand using the factor diagram, which will let us recognize "inside" and "outside" factors. The partial fit for a factor equals the sum of effects for the outside factors. In what follows, I'll give the rule, illustrate it, and then say a little more the logic behind it.

Inside and Outside Factors

Definition: One factor is *inside* another if each group of the first (inside) factor fits completely inside some group of the second (outside) factor.

I'll take a few simple cases first. It may help you to refer to the factor diagram as you read this next set of examples.

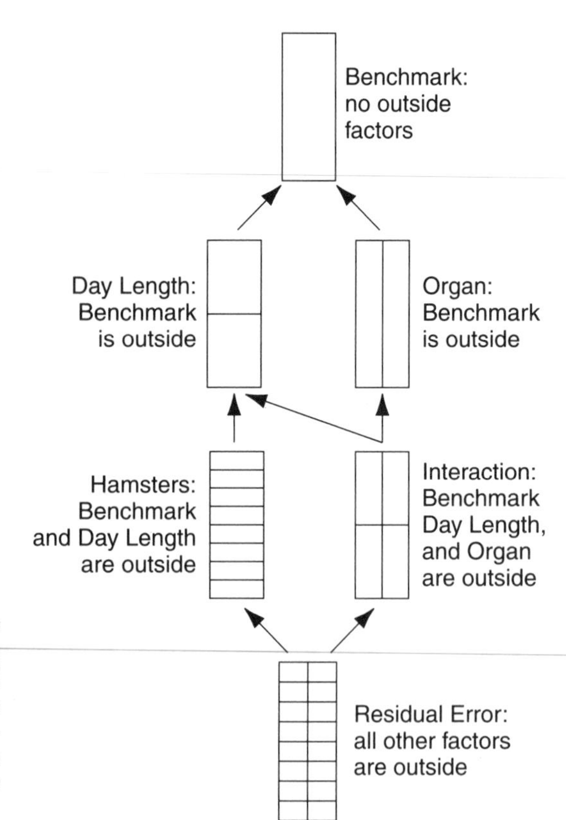

FIGURE 3.13 Inside and outside factors for the hamster experiment. An arrow joining two factors means that the lower factor (tail of the arrow) is inside the upper (head of the arrow). Moreover, if you can get from any factor to any other by following the arrows, the first factor is inside the last (e.g., Hamsters is inside Benchmark).

EXAMPLE 3.5 INSIDE AND OUTSIDE FACTORS FOR THE HAMSTER DATA

- Day length is inside the benchmark factor. The Day Length factor has two groups, long and short. Each group fits completely inside the one group for the benchmark. In fact, *every* factor is inside the benchmark factor, or, turned around, the benchmark is outside every other factor. (Look at the factor diagram so you can picture this.)

- The two factors for Hamsters (rows) and Organs (columns) do not have an inside/outside relationship. Although rows and columns overlap, so that a *part* of each row fits inside a column, and vice versa, you can't fit a whole row completely inside a column, nor can you fit a whole column inside a row. (Notice, by way of review, that Hamsters and Organs are crossed.)
- Residual error is inside Hamsters; Hamsters is outside Residual Error. Each group for the Residual Error factor corresponds to a single observation, and so each group fits completely inside a row, which is a group for the hamster factor. In fact, Residual Error is inside every other factor. ■

Figure 3.13 gives the complete list of inside and outside factors for the hamster data.

Partial Fit Equals the Sum of Estimated Effects for All Outside Factors

Once you can recognize inside/outside relationships among factors, estimating the effects is just a matter of adding and subtracting averages.

Estimated effect for a factor = Average for the factor − Sum of estimated effects for all outside factors.
Effect = Average − Partial fit.

To see why the rule is reasonable, think once again about how we got the effect for hamster #4 in the introductory section: we started with the average of all the measurements for #4, then subtracted the average of all the long-day measurements. Now consider the rule for partial fit and ask yourself, "In terms of the assembly line metaphor, which pieces get added on to all of the measurements (in this case, both measurements) that go into the average for #4?" These are precisely the pieces we want to subtract from the #4 average to get the estimated effect; these pieces correspond to the outside factors.

Example 3.6 The Partial Fit for the Effect of Hamster #4

Go through the list of factors, one at a time, asking "Do both of the # 4 observations get the same piece added on for this factor?" The answer will be yes for the factors that are outside the Hamster factor, and no for the other factors.

Benchmark? Yes, both measurements for #4 get the benchmark amount.

Day Length? Yes, both measurements get the effect for long days.

Organs? No. One measurement gets the heart effect, the other gets the brain effect.

Interaction? Also no, for a similar reason as above.

Hamsters? Yes, both measurements get the effect for hamster #4.

Residual Error? No, the two measurements for #4 don't get the same residual error; each measurement gets its own. ■

In sum, then, all the measurements that go into the average for hamster #4 contain three pieces in common—the hamster #4 effect, which we want to estimate, plus two others: the benchmark and the long-day effect. To estimate the effect for hamster #4, we take the average for #4 and subtract estimates for the benchmark plus the long-day effect.

Notice that the two factors for benchmark and day length are precisely the factors that are outside the Hamster factor. To get the partial fit, we add the estimates for the effects of these outside factors. The sum of these two estimated effects turns out to be the same as the long-day average, so the estimated effect for hamster #4 equals the average for #4 minus the long-day average.

Here's how to apply the general rule to all the factors, one at a time. We start with the outermost factor (Benchmark) and then go on to factors with smaller groups, ending with the innermost factor (Residual Error).

Example 3.7 Estimated Effects for the Hamster Data

Benchmark. There are no outside factors, so the partial fit equals zero. The estimated effect equals the average for the factor, that is, the grand average of 60.

Day Length. Only the benchmark is an outside factor, so the partial fit equals the grand average. The long-day effect (−3) equals the long-day average (57) minus the grand average (60). Similarly for short days: effect (+3) equals average (63) minus partial fit (60).

Organs. Here also, the only outside factor is the benchmark, and the partial fit equals the grand average. The heart effect (−43.5) equals the heart average (16.5) minus the grand average (60). For brains, the average is 103.5, and the estimated effect is 43.5.

Interaction. Here there are three outside factors: benchmark, day length, and organs. The partial fit is the sum of the estimated effects for these three factors.

	Benchmark H	Benchmark B		Day Length H	Day Length B		Organs H	Organs B		Partial fit H	Partial fit B
L	60	60	+	−3	−3	+	−43.5	43.5	=	13.5	100.5
S	60	60		3	3		−43.5	43.5		19.5	106.5

FIGURE 3.14 Partial fit for estimating the interaction effects. The numbers add according to where they sit in the tables. Thus, for example, the numbers in the upper left corner go together for Long Days/Heart:

$$60 + (-3) + (-43.5) = 13.5 \quad \text{(partial fit)}.$$

The partial fit tells us what we ought to expect for the condition average if there were no interaction. If the effect of day length were the same for hearts as for brains, there would be only three pieces common to all four observations that go into the average for Long Days/Brain: benchmark, long-day effect, and heart effect; these are precisely the pieces whose estimates we add to get the partial fit. To estimate the interaction effects, we start with the condition averages and subtract the partial fit (Fig. 3.15).

FIGURE 3.15 Estimated interaction effects.

	Average H	Average B		Partial fit H	Partial fit B		Effect H	Effect B
L	20	94	−	13.5	100.5	=	6.5	−6.5
S	13	113		19.5	106.5		−6.5	6.5

Hamsters. Although there are two outside factors (Benchmark and Day Length), the partial fit simplifies to just the day-length average. From before, the estimated day-length effect equals the day-length average minus the grand average, so

$$\text{Fit} = [\text{Gr. Avg.}] + [(\text{Day Avg.}) - (\text{Gr. Avg.})] = \text{Day-length average.}$$

The estimated effect equals the hamster average minus the day-length average, just as we had before.

Residual error. *All* the other factors are outside, so here the fit equals the sum of all the estimated effects. Since the factor for residual error puts each observation in a group by itself, each "average" here is just an observed value, and the rule gives us "estimated effect = obs − fit." The arithmetic is messy and best left to a computer, but the idea is not nearly as complicated as the number crunching.

Our decomposition is essentially complete. True, we've estimated only one hamster effect, and there are seven others, but the steps are the same as for #4. We also need estimates for the 15 other residual errors, but here, too, finding the estimates involves no new ideas, so I won't give the details. ■

Using the General Rule to Count Degrees of Freedom

The rule for estimating the effects for a factor says to start with the group averages for the factor and then subtract the sum of the effects for all outside factors. The averages for the outside factors are precisely the ones that are subtracted. This fact leads to a shortcut rule for finding degrees of freedom:

df for a factor = # levels for the factor − Sum of df for all outside factors.

EXAMPLE 3.8 USING THE SHORTCUT TO FIND DF FOR THE HAMSTER DATA

Grand average. Number of levels = 1. No outside factors. $df_{\text{Grand}} = 1 - 0 = 1$.

Day Length. Number of levels = 2. Benchmark (df = 1) is outside. $df_{\text{Day length}} = 2 - 1 = 1$.

Organs. Number of levels = 2. Benchmark (df = 1) is outside. $df_{Organs} = 2 - 1 = 1$.

Interaction. Number of levels = 4. Outside factors are Benchmark (df = 1), Day Length (df = 1), and Organs (df = 1). $df_{Interaction} = 4 - (1 + 1 + 1) = 1$.

Hamsters. Number of levels = 8. Outside factors are Benchmark (df = 1) and Day Length (df = 1). $df_{Hamsters} = 8 - (1 + 1) = 6$.

Residuals. Number of levels = 16. All five other factors are outside; their dfs add to 10 (10 = 1 + 1 + 1 + 1 + 6). $df_{Residuals} = 16 - 10 = 6$. ■

Appendix 2. Introduction to the Logic of Hypothesis Tests and Confidence Intervals

Hypothesis Testing.

Working backward from the evidence. In 1963, after President Kennedy was assassinated, Lyndon Johnson appointed the Warren Commission to decide whether the assassination had been part of a conspiracy. There were two competing explanations: that Lee Harvey Oswald was the lone assassin, and that Oswald had been part of a conspiracy. A particularly important piece of evidence was a tape recording of the shots that were fired. Some people claimed that this evidence proved Oswald had not acted alone because a cheap rifle like the one he used could not have fired all the shots in so short a time. Part of the Warren Commission's job was to work backward from the evidence to reach a conclusion about what had actually happened.

In most court cases, the jury, like the Warren Commission, cannot know for sure what actually happened. Its job is to examine the available evidence and try to reconstruct the invisible past. Typically there are two competing explanations, one advanced by the prosecution and one by the defense. Nevertheless, there is only one body of evidence, and there is also only one verdict.

The same holds true for testing hypotheses. There are two competing explanations for what actually happened on the assembly line as the observed values were put together:

i. **The null hypothesis**: The data set is "innocent"; at the work station in question, the worker did nothing; there are no "real" differences; the observed differences are due to chance error.
ii. **The alternative hypothesis**: The data are "guilty"; there are "real" differences, and they account for at least part of the observed differences.

Although there are two possible explanations, there is only one body of evidence, summarized by the *F*-ratio. Our goal is to work backward from this visible evidence to a conclusion about which hypothesis correctly describes the invisible process that gave rise to the data.

Two kinds of errors are possible. In trying to reach our verdict, we never get to see the "true" differences. Because we are forced to work with estimates, it is possible that our conclusion will be wrong, just as a jury's verdict can be wrong. For each of the possible true conditions, there are two possible verdicts, only one of which is right (Fig. 3.16).

True conditions	Verdict: Guilty: Differences real	Verdict: Not guilty: Case not proven
Real differences	Correct	Type II error (miss)
No real differences	Type I error (false alarm)	Correct

FIGURE 3.16 The two types of errors in hypothesis testing. Each row corresponds to one of the two possible true conditions. For any given test, only one of the two rows applies although we cannot be certain which one. For each condition, there are two possible verdicts, one correct, and the other an error. A **miss** occurs if there are "real" differences and the hypothesis test fails to detect them; a **false alarm** occurs if there are no "real" differences, but the test says there are.

Built-in conservatism: innocent until proven guilty. Because wrong verdicts are always possible, jury trials are set up to be conservative: a person is presumed innocent unless the evidence of guilt is very strong. The idea is that it is much worse to convict an innocent person than to let a guilty person go free. Fisher realized that the same conservative approach was best for science, too: it is better to require very strong evidence to conclude there are "real" differences, even if this means that some "guilty" data sets are declared "innocent" because the evidence is not strong enough.

This conservative approach to science is in fact much older than Fisher and goes at least as far back as about 1325, when William of Occam wrote that scientists should avoid complicated explanations unless there was strong evidence that a simple explanation would not work equally well. Occam (1285–1349 A.D.) was an influential philosopher and theologian who helped shift the current of thought in his time away from the older ideas of Aristotle (384–322 B.C.) and Saint Thomas Aquinas (1225–1274 A.D.), both of whom tended to trust logic more than direct observation. (For example, Aristotle, one of the most brilliant thinkers of all time, once said that men had more teeth than women.) Sixteen centuries after Aristotle's failure to count teeth, William of Occam was one of the people who urged scientists, at that time still called natural philosophers, to pay more attention to what they could observe first-hand. In urging a conservative approach to explaining, what he actually wrote, at least according to one version, is that "what can be done with fewer (assumptions) is done in vain with more." His principle is known as **Occam's razor** because it is used to justify carving away unnecessary complications in favor of simpler explanations.

In science, "no differences" is a simpler explanation than "real differences" because if there are real differences a complete explanation would ordinarily have to include some account of the reasons for the differences, whereas if there are no differences you don't need to find reasons to explain them. Fisher chose his table values to be conservative so that you won't conclude there are real differences unless there is strong evidence. This means scientists won't often waste effort looking for reasons to explain differences that don't really exist.

Critical values. Fisher's table value tells us "how big is big enough" to conclude that the differences are real. He chose these values following the conservative principle of jury trials (and Occam's razor) that it is better to let a few real differences get away undetected than to make the other error of declaring differences real when in fact there are none. Thus Fisher's table values correspond to strong evidence in order to protect scientists against the error of declaring an "innocent" data set (no real differences) "guilty." To decide what strong evidence would look like, Fisher carried out a careful study of "innocent" data sets. In effect, he said to himself, "Suppose I look at lots and lots of data sets for which there are no real differences. For each one, I can compute the F-ratio and gradually build up a large catalog of F-ratios, all from data sets with no differences. A few of these data sets will look guilty (large F-ratios) just because of chance error. However, I can choose my table value so large that only 5% of the innocent data sets will have F-ratios that big".

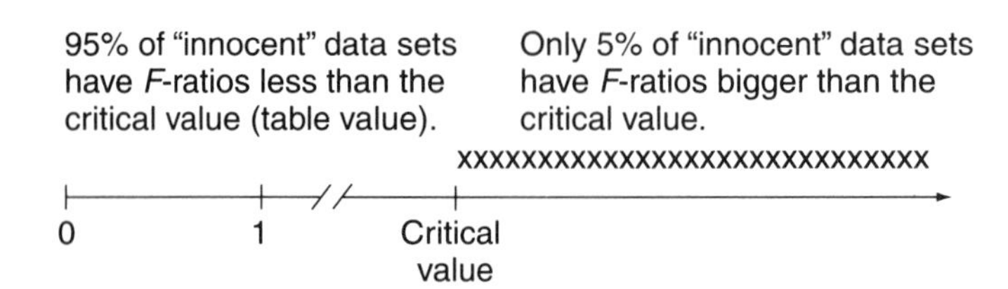

FIGURE 3.17 Possible values for the F-ratio. If the F-ratio is greater than the critical value, the observed differences are declared "real".

Fisher didn't actually have to create all those data sets—he discovered a way to do the same thing using mathematical theory. Using that theory, he chose critical values to have the property that only 5% of the data sets with no real differences would produce F-ratios bigger than the critical value.

Critical values: Fisher's critical values are chosen so that only 5% of the data sets with no real differences will have F-ratios bigger than the critical value.

With this as background, we can now go back and look more carefully at what happens when we use Fisher's table values for a real data set. In practice, with a real data set there are two possible true conditions: real differences or no real differences. For each condition, there is a typical way the *F*-ratio behaves.

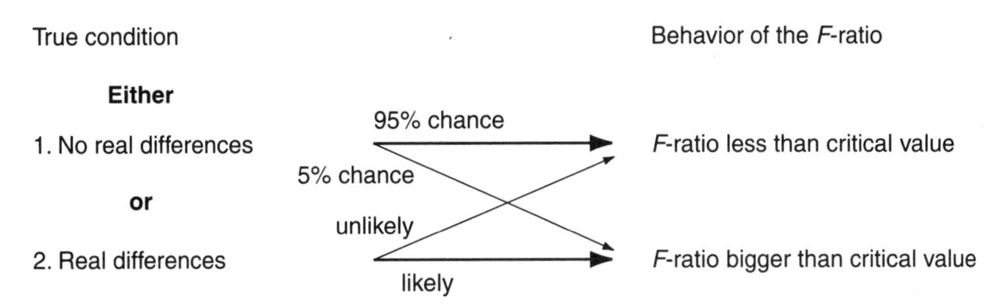

FIGURE 3.18 Relation between the true condition and behavior of the *F*-ratio. This diagram goes "forward," from cause to effect, *from* the true condition *to* the behavior of the *F*-ratio that is typical for a data set under that condition. Our goal in analysis of variance is to go backward, from effect back to cause, *from* the value of the *F*-ratio we compute *to* a conclusion about the true condition.

For each true condition, the heavier arrow points to the more likely outcome and the lighter arrow points to the less likely outcome. When we draw conclusions, we act in effect as if the lighter arrows don't exist, and we follow the heavy arrows backward from observed outcome to the true condition.

***Either/or meaning of a large* F.** Think again about heart/brain differences for the hamster data, which gave us a large *F*-ratio, bigger than the critical value; and think about working backward from the *F*-ratio to a conclusion about the true conditions that might have produced that *F*-ratio. There are two possibilities:

Either/Or Meaning of a Large *F*-Ratio

Either **the heart/brain differences are real, and our F-ratio is about the size we should expect,**

or: **there are no real differences, and we got an unusually large F-ratio just by chance.**

Given this choice of two possible explanations for the large *F*-ratio, Fisher said we should choose to conclude that the differences are real. In more formal language, we reject the null hypothesis of no real differences.

A useful way to fix this logic in your mind is to imagine an argument about what a large *F*-ratio means. On one side of the argument, there is Ronald Fisher, who claims the large *F*-ratio for heart versus brain means that the heart/brain differences are real. On the other side of the argument is a person I'll call the Chance Skeptic, who claims that there are no real differences, and that we got the large F-ratio just by chance. (I got the idea for the dialogue, and the idea of the Chance Skeptic, from the same book with the box model for chance error.)

Chance Skeptic: You say the large F-ratio is evidence of a real difference in enzyme concentrations between hearts and brains, right?

Ronald Fisher: Right.
Chance Skeptic: But it's possible to get an F-ratio that large just by chance, even if there are no real differences. You have to admit that.
Ronald Fisher: It's possible, I agree. But it's not very likely. I've done a theoretical analysis of all the possible data sets with no real differences, and only 5% of them, only 1 out of 20, give F-ratios bigger than my critical value.
Chance Skeptic: But some of those data sets, with no real differences, really did give big F-ratios.
Ronald Fisher: True, but only the unusual ones. If you insist that our hamster data have no real heart/brain differences, you have to believe that our data set is one of those very unlikely ones. Do you really believe that?
Chance Skeptic: I see your point. Unless I'm willing to claim that our data set is something of a freak, I'm forced to conclude that the differences are real.
Ronald Fisher: Precisely.

Confidence intervals.

Built-in shortcomings of the Yes/No approach. The conservatism that Fisher built into his method of testing means that "innocent" doesn't necessarily mean what the word might suggest. The evidence in a statistical trial, just as in a jury trial, may not be conclusive. The jury verdict by itself provides an extremely limited summary of the evidence. Whenever you try to summarize numerical evidence by choosing one or the other of two categories—big or not big, rich or not rich, young or not young—you give up a lot of information. This kind of loss is built into the method of hypothesis testing, which Fisher designed to lead to a conclusion that amounts to choosing from "detectable" or "not detectable". For this reason, you should not regard hypothesis testing by itself as the only goal of a statistical analysis, or even necessarily the main goal. Estimating the likely size of the true effects is also important, because such estimates provide a lot more information than a one-word verdict.

Dart-gun metaphor: the chance is in the process, not the outcome. Imagine a child's dart gun, the kind whose darts have suction cups on the ends, and think about shooting the bullseye of a target. The bullseye is like the true value, and the circle covered by the suction cup where it sticks to the target is like a confidence interval. Suppose you stand fairly close to the target, at just the right distance to give you a 95% chance of covering the bullseye with the suction cup. Here, just as for a confidence interval, the 95% chance is for the *process*, not the target. The bullseye doesn't move, and once your dart has landed, the circle doesn't move: either the suction cup covers the bullseye, or it doesn't. The 95% doesn't refer to a particular shot (or a particular interval), but to the process of shooting (or constructing intervals).

Our confidence interval for the effect of day length on hamster brains is like one shot with the dart gun. We know 95% of the shots cover the true value, but in a case like this one, when we don't know the true value, we can't know whether it's inside the interval from 4 to 34 or not. If we're careful about the meaning of the interval, instead of saying, "There's a 95% chance the true value is between 4 and 34," we should say "The interval from 4 to 34 was constructed using a method that covers the true value 95% of the time." The language is somewhat awkward, and it's easy to understand why people sometimes get a bit careless, but it is important to know the real meaning of the interval. A slight variation on the metaphor may help solidify your understanding and will also get us ready to look at how you construct a confidence interval.

Target-shooting metaphor: finding the 95% distance. Often people who practice target shooting use printed paper targets stapled to a wooden board so they can start each round with a fresh target and so they can take down the target afterwards as a record of how well they've done. To make this metaphor work, imagine a fresh target posted on a board that's never been used before. The true value is the point on the board right under the bull's-eye on the target. The marksman fires several shots, takes down the paper target, and walks away, leaving a pattern of bullet holes in the wooden board. Now you become the statistician. The bullet holes are your observations, and your goal is to locate the true value, the point on the board that was directly under the bull's-eye before the target was removed. It's there somewhere, it doesn't move, but you can't see it. All you've got to go on are the bullet holes.

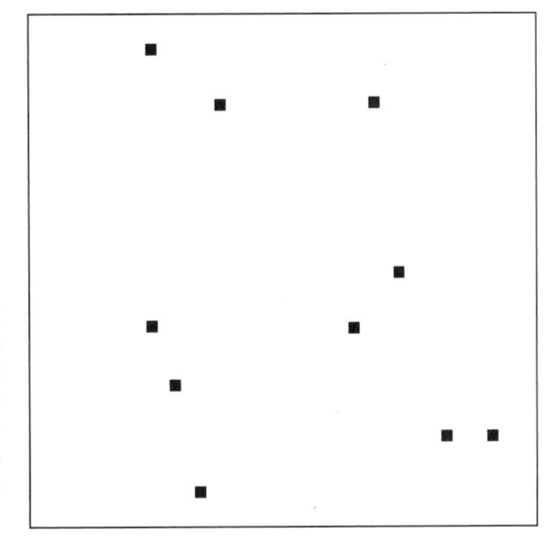

FIGURE 3.19 Target-shooting metaphor for finding a confidence interval. The square is the board the target was stapled to, and the dots are the bullet holes that were left after the paper target was taken down. The statistician's job is to use the observed pattern of bullet holes to guess the likely location of the point that was directly under the bull's-eye of the target.

I'm hoping your intuition gives you some sense of how you'd try to solve the problem. First, you might try to guess where the point is by combining the information from the bullet holes to find a point somewhere in the middle of them, perhaps their center of gravity. In effect, this is what we do when we combine observed values to get an average as our estimate of the true value. Next, you might try to measure how far off your estimate is likely to be. Here's one way to think about your goal: you want to find the right size circle to draw around your point estimate in order to feel pretty sure the unknown true value would be somewhere within the circle. This amounts to finding a distance; your circle will be the set of points within that distance of your estimate.

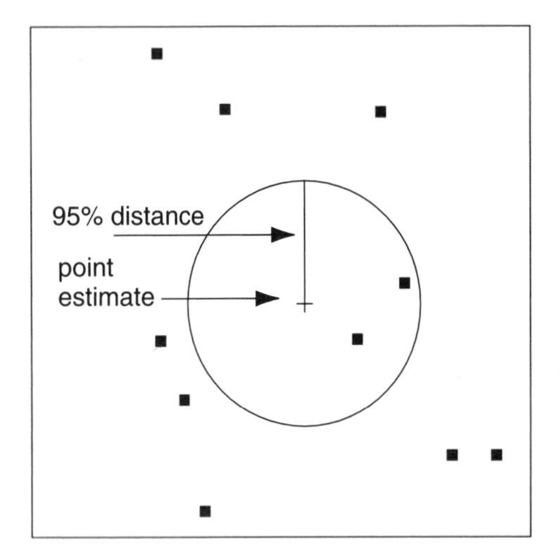

FIGURE 3.20 95% confidence disk for the bull's-eye. We want to find a 95% distance with the property that a circle with center at our point estimate and radius equal to that distance will contain the point that was under the bull's-eye for 95 targets out of every 100.

How do you find the distance? Suppose you have available to you a collection of the last 100 paper targets used by the marksman. Each target shows not only a round of bullet holes but also the bull's-eye, so you can use the targets to check out various distances, trying first one, then another, until you find the one that works, the 95% distance: for 95 out of the 100 targets, the circle you draw using that distance actually includes the bull's-eye; for the other 5 out of 100 it doesn't. Then you can go back to the wooden board, draw a circle around your point estimate using that 95% distance, and claim, "I got this circle using a method that covers the bull's-eye 95 times out of 100."

This is pretty much what statisticians do to find the 95% distance for a confidence interval. The main difference is that they use the Fisher assumptions in place of the marksman, and they use theory based on those assumptions instead of the collection of paper targets. In all other respects, the logic is pretty much the same.

For now, I'll explain only the rough outline of the theory—enough, I hope, to give you some of the main ideas without swamping you with details. I'll start with a simple made-up example, one where I'll tell you exactly how the chance errors behave. If you can see how the logic works in this case, it should help you think about a more realistic example where you can't see the chance errors.

EXAMPLE 3.9 CONFIDENCE INTERVAL FOR A VERY SIMPLE ESTIMATION PROBLEM

For this example you are a statistician, and you agree to play the following game with me: I'm thinking of a number, the secret "true" value. Your job is to find a confidence interval for that true value. I won't tell you the number itself, but I will tell you the number I get by adding together my secret true value plus a chance error. To get my chance error, I'll draw at random from a box with four tickets:

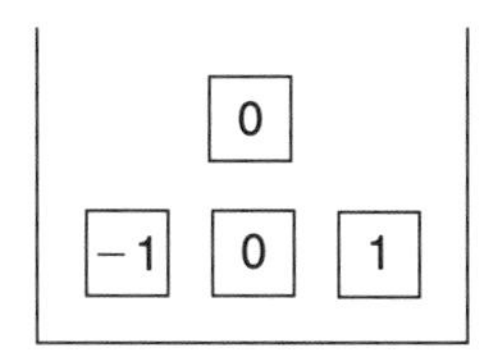

Suppose I tell you that I drew one chance error, that I added it to my true value to get an observed value, and that this observed value equals 19. Find a confidence interval for the true value.

SOLUTION. You might reason like this. "There was a 50% chance (2 out of 4) that the chance error would be 0, which would make the observed value equal to the true value. So if I guess 'True value equals 19', I'd be using a method that's right 50% of the time."

"But there was a 25% chance that the ticket said −1, which would mean the true value was 20 (19 = 20 + [−1]). There's another 25% chance the ticket said +1, which would mean the true value was 18 (19 = 18 + [+1]). So if I guess 'True value is between 18 and 20', I'd be using a method that's right 100% of the time."

In the language of confidence intervals, "18 to 20" is a 100% confidence interval for the true value, and "19 to 19" is a 50% "interval." (For this example, a 95% confidence interval doesn't exist.) ■

Now I'll change the example a bit, to bring us closer to the hamster data:

EXAMPLE 3.10 CONFIDENCE INTERVAL WHEN CHANCE ERRORS FOLLOW THE NORMAL CURVE

This example has two versions, A and B.

Version A. The basic set-up is the same as before. I have a secret true value, and your job is to find a confidence interval. I'll draw one chance error at random, add it to the true value, and tell you the result. This time, however, I won't show you the box of chance errors. Instead, I'll tell you that there are 1000 tickets in the box and 950 of them are spread out between −10 and +10; the other 50 tickets are either less than −10 or bigger than +10. If my observed value (= true value + chance error) equals 19, find a 95% confidence interval for the true value.

SOLUTION. Even though you can't see the tickets, you know from what I've told you that 95% of them are between −10 and +10. Consider a few possible ways the draw from the box might turn out:

i If the chance error I draw is 10, then Obs = True + 10 and since Obs = 19, we have 19 = True + 10, and the true value must be 9.

ii. If the chance error I draw is −10, then Obs = True − 10 and since Obs = 19, we have 19 = True − 10, and the true value must be 29.

iii. If the chance error I draw is anywhere in between −10 and 10, then the true value must be somewhere in between 9 and 29.

Now think about the interval Obs ± 10 as a possible confidence interval. What is the chance this interval scores a "hit," that is, contains the true value? As long as the chance error I draw is one of the 950 tickets between −10 and 10, the distance from Obs to True will be at most 10, and Obs ± 10 will cover the true value. If, on the other hand, I draw one of the other 50 tickets, the distance from Obs to True will be more than 10, and Obs ± 10 will miss the true value. Thus each time you play this game with me, there's a 95% chance that your interval Obs ± 10 will cover the true value.

My actual observed value was 19. Does the interval from 9 to 29 contain the secret number I was thinking of? You'll never know, because I'm not going to tell you either the secret number or the chance error I got. That's how it is with confidence intervals in real life, too.

Version B. Now consider one last example. This set-up is still basically the same: I tell you Obs (the sum of True + Chance error), and your job is to find a 95% confidence interval for the true value. This time, though, I won't even tell you how many tickets there are in the chance error box. All I'll tell you is that the numbers follow the normal curve and have an average of 0 and a standard deviation of 5.

SOLUTION. I hope you recognize that this version of the problem leads to the same 95% confidence interval as before. If the chance errors follow the normal curve, then 95% of them are spread out within two SDs on either side of zero. Since the SD is 5, SD × 2 is 10, so 95% of the tickets are between −10 and 10. An interval of the form Obs ± 10 (or in general, Obs ± SD × 2) has a 95% chance of scoring a hit. ■

SUMMARY

1. **Decomposition**. Decomposing the data means splitting the observed values into pieces, one piece for each factor, in order to estimate the amounts added at each work station of the assembly line metaphor. The general rule (Appendix 1) is that each effect equals average minus partial fit (partial fit = Sum in the box below):

> **Estimated effect for a factor = Average for the factor − Sum of estimated effects for all outside factors.**

[One factor is *inside* another if each group of the first (inside) factor fits completely inside some group of the second (outside) factor.]

2. **Mean squares** summarize the variability within the boxes of the decomposition. Each MS = SS/df and measures average variability per unit of information about residual error.

   SS = sum of squares. Square all the numbers in the table, and add them up. This total is a measure of overall variability in the table.
   df = degrees of freedom. This number counts the number of units of information about residual error contained in the table.

The df for a table equals the number of free numbers, the number of slots in the table you can fill in before the patterns of repetitions and adding to zero tell you what the remaining numbers have to be. (Whenever you subtract the average from a set of numbers, the leftovers from the average, called **deviations**, will add to zero.) Although you can count free numbers directly from boxes of the decomposition, there's a general rule as well (see Appendix 1):

df for a factor = # levels for the factor − Sum of df for all outside factors.

3. The **standard deviation (SD)** gives the typical size (standard) of residual error (deviation). Each observed value equals true value, give or take about one SD. (For the hamster data, the SD is 7.3: each observation equals true value plus or minus about 7.) For data that fit the normality assumption, roughly 2/3 of the errors will be within one SD of 0, roughly 95% within 2 SDs, and all or almost all within 3 SDs.

We compute the SD from the box of residuals in the decomposition:

$$\text{SD} = \text{root mean square for residuals} = \sqrt{\text{MS}_{\text{Res}}}.$$

4. **Testing hypotheses**
   a. Each hypothesis test is like a jury trial, in which you work backward from the evidence to decide which of two explanations correctly describes the invisible process that gave rise to the data. The null hypothesis (innocent) is that there are no true differences; the alternative hypothesis (guilty) is that there are real differences, not just chance error. The test is conservative, set up so that only 5% of "innocent" data sets will be declared to have real differences.
   b. There is one test for each structural factor in the design. Each is based on comparing an *F*-ratio you compute from the data with a table value: if the value you compute is bigger than the one from the table, you declare the differences real. To get the *F*-ratios, you compute sums of squares, degrees of freedom, and mean squares for each box in the decomposition of your data. Each *F*-ratio equals the mean square for the factor you want to test divided by the mean square for (usually) residuals.
   c. A large *F*-ratio tells you the differences from the factor in question are "real," not just chance error, that is, the "true" differences are big enough to be detected by your experiment. An *F*-ratio does not check the design of your experiment, does not check whether you found the right factor structure, and does not check whether the six assumptions fit your data.
5. **Confidence intervals**
   a. The standard error is like an SD for estimates: it tells the typical size of the chance error part of the estimate. Each estimate equals the true value, give or take about one standard error. The standard error equals the square root of some MS (which one depends on the design) times a leverage factor for the estimate. The more observations there are that go into the averages in your estimate, the smaller the leverage factor (and thus the SE) will be.
   b. Each 95% confidence interval is a range of values, on either side of your estimate, constructed by a method that captures the true value in 95 cases out of 100. The interval equals estimate ± 95% distance, and the 95% distance equals the standard error times a standard 95% distance from a *t*-table.

REVIEW EXERCISES

1. *SDs and transforming*. (You might want to review Example 3.4 (p. 73) as you do this exercise. Remember that one of the main reasons for transforming your data to a new scale is to make the assumption of same SDs more appropriate, and recall that the SD is generally more reliable than the range as a way to measure error size.)

The specific activity data given in the Review Exercises of Chapter 2 had observations made under four conditions:

Long days	Heart:	246, 169, 186, 183
	Brain:	394, 297, 216, 373
Short days	Heart:	468, 390, 314, 508
	Brain:	216, 198, 194, 192

a. Look over the numbers and estimate (do not compute):

i. Which group has the largest average? Next largest? Smallest?

ii. Which two groups have the largest SDs? Which groups has the smallest SD?

iii. Do larger SDs tend to go with larger averages, that is, does the error size seem to depend on the size of the measurement?

b. Compute the average and SD separately for each group, following Example 3.4 as a model, and put your results in a little table like the one at the end of Example 3.4. Then find SD_{Max}/SD_{Min}. Should this data set be transformed? Why, or why not?

2. *Informal analysis of the specific activity data.* The table in Fig. 3.21 gives values of 10,000/(specific activity). (Numbers marked * were changed by ±1 or ±2 to simplify the arithmetic.) Notice that the change of scale has reversed directions: large values of the specific activity correspond to small values of 10,000/(specific activity), and vice versa.

Observed values Heart	Observed values Brain	Hamster Average
41	25	33
58*	36*	47
54	46	50
55	29*	42
20*	46	33
24*	50	37
32	52	42
20	52	36

Condition Averages	Heart	Brain	Day Length Averages
Long	52	34	43
Short	24	50	37
Organ Averages	38	104	40 (Grand Average)

FIGURE 3.21 A complete set of averages for the transformed hamster data.

a. Construct a parallel dot graph, and compare it with the one from Review Exercise 2 in Chapter 2. Are the spreads more nearly equal now?

b. Use the four condition averages to construct an interaction graph, and show the differences due to day length for each organ. Is interaction present?

c. Scatterplot brain values versus heart values using separate symbols for long and short days, as in Chapter 2, page 55. Does your plot show positive or negative relationships?

3. *Decomposition of the data*

a. Estimated effects for the transformed data.

i. Effect of day length: How far is it from the long-day average to the grand average? From the short-day average to the grand average?

ii. Effect of organ: How far is it from the heart average to the grand average? From the brain average to the grand average?

iii. Hamster effects: How far is it from the average for the first long-day hamster to the average for all four long-day hamsters? From the average for the second long-day hamster to the average for all four long-day hamsters? From the average for the first short-day hamster to the average for all four short-day hamsters?

iv. Interaction effect: consider the hearts of the long-day hamsters, and suppose there were no interaction. Then the condition average for Long Days, Heart would be the sum [Grand average] + [Distance from long-day average to grand average] + [Distance from heart average to grand average]. Find this sum, which is the partial fit for Long Days, Heart. How far is the actual condition average from this partial fit (which you got by assuming there was no interaction)?

b. *Decomposition of a single observation.* Write the 41 in the upper left corner of the data table as a sum of six pieces, following the format at the beginning of the chapter. (The residual will be equal to whatever number makes the right-hand side add up to 41.)

c. Write out the complete decomposition. The residuals are (by rows)

[−1,1], [2,−2], [−5,5], [4,−4], [0,0], [0,0], [3,−3], [−3,3].

4. *Sums of squares and mean squares for the transformed data*

a. Compute SSs and MSs based on your decomposition, and put them in a table like the one in Section 2, page, 68.

b. Which factors have the largest variability per unit of information?

c. Compute the SD, and use it to check the based on the normal curve, as in Example 3.3.

The bivariate BF[1] model. Think of the enzyme concentration data in the original scale (mg/ml) as coming from a randomized basic factorial design with one factor of interest—Day Length—and two measurements, heart concentration and brain concentration. For this exercise, ignore the brains and work just with the eight heart concentrations given below. (I changed the measurement marked * to make the arithmetic cleaner.)

Long days: 1.490, 1.525, 1.555, 1.790
Average = 1.590

Short days: 1.375*, 1.485, 1.255, 1.285
Average = 1.350

5. Draw and label a factor diagram using arrows to show inside and outside factors, as in Fig. 3.13.

6. Use the general rule from Appendix 1 to estimate the effects for day length. Then find the complete decomposition of the data, and write it in the form shown in Fig. 3.4. (Your decomposition should have one box for the observed values plus three others.)

7. The estimated effect for long days tells how far it is from ________ to ________. The residual for the first observation tells how far it is from ________ to ________. The variability in the residuals is due in part to ________ and in part to ________.

8. Find the degrees of freedom for each box of your decomposition using the general rule in Appendix 1. Then compute sums of squares and mean squares, and put your results in a table.

9. The mean square for day length is ________ times as big as the mean square for residuals.

10. Compute the SD and use it to check the normality assumption as in Example 3.3.

11. Why isn't it necessary to transform this data set?

Sleeping Shrews: A CB[1] Design The heart rates of sleeping shrews given in Review Exercise 16 of Chapter 1 (p. 37) has the structure of a complete block design, with shrews as blocks and kinds of sleep as the factor of interest. Before you begin the written part of the exercise, look over the data set. For which kind of sleep do the shrews' hearts beat fastest? Slowest? Which shrew had the fastest heart? The slowest? Is the variability greater from one shrew to the next, or from one kind of sleep to the next?

12. Draw and label a factor diagram, using arrows to show inside and outside factors.

13. Use the general rule from Appendix 1, together with the averages below, to estimate the effects for the three kinds of sleep and for the four shrews.

Sleep	LSWS	DSWS	REM	
averages:	19.9	17.7	18.2	
Shrew	I	II	III	IV
averages:	13.8	22.8	19.6	18.2

14. The estimated effect for REM sleep tells how far it is (in beats per minute) from __________ to __________. The estimated effect for Shrew I tells how far it is from __________ to __________.

15. Use the rule from Appendix 1 to find the degrees of freedom for each factor in your factor diagram. Then compute SSs and MSs, and summarize your results in a table. (The residual sum of squares is 15.6.)

16. The MS for kinds of sleep is __________ times bigger than the MS for residual error; the MS for shrews is __________ times bigger than the MS for residual error.

ANOVA: F-tests and confidence intervals

17. Here is an incomplete ANOVA table for the specific activity data given in the Review Exercises for Chapter 2.

a. Compute the missing mean squares.

b. Then compute *F*-ratios for Organ, for Interaction, and for Hamsters using the mean square for residuals in the denominator; compute the *F*-ratio for Day Length using the mean square for hamsters in the denominator.

c. What do you conclude from the *F*-ratios?

Source	SS	df	MS	F-ratio	Table value
Grand avg	1,290,496	1			
Long/short	10,816	1			5.99
Hamsters	28,268	6			4.28
Heart/brain	9,216	1			5.99
Interaction	118,336	1			5.99
Residual		6			
Total	1,474,536	16			

18. Use your results from Exercise 4 to construct a complete ANOVA table for the transformed specific activity data. (The table values for the *F*-ratios are the same as the ones above.) Are your conclusions based on these *F*-ratios substantially different from your conclusions in part c?

19. Which of the following are reasonable interpretations of the results of your analysis of the specific activity data? Remember that transforming reversed directions: low values in the new scale correspond to high specific activity, and vice versa.

a. Day length has no detectable effect on the specific activity of the sodium pump enzyme. The observed differences could be due just to natural variability in the material and measurement process.

b. The specific activity level of the enzyme is pretty much the same whether you look at the hearts or brains. The observed differences could be due just to chance.

c. As days grow shorter, the specific activity of the sodium pump enzyme goes up in the hamster hearts but goes down in the hamster brains. These observed differences are too big to be due just to chance variability in the material and the measurement process.

20. Suppose you want to construct a confidence interval based on a single observed value and you know that the observed value equals an unknown true value plus a chance error drawn at random from the following box:

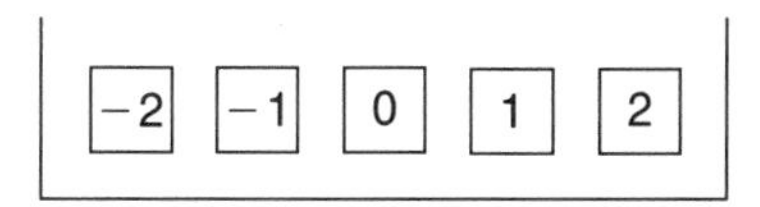

Suppose the observed value turns out to be 21.

a. For this artificially simple example, you can be 100% sure, after observing the 21, that the true value is one of five numbers. List them.

b. What is the confidence level (%) for the "interval" 21 to 21?

c. What is the confidence level (%) for the interval 20 to 22?

21. *The BF[1] model: F-tests and confidence intervals*

a. Use your results from Exercises 5–11 to find the F-ratio for day length. The table value for this F-ratio is 5.99. Does day length have a detectable effect on the enzyme concentration in the hamster hearts?

b. A 95% confidence interval for the "true" effect of day length has the form

[Long-day avg. − short-day avg.] ± (SE)(Standard 95% distance).

Here the S = SD × (leverage factor), the leverage factor is $\sqrt{.5}$, and the standard 95% distance is 2.447. Find the interval.

c. In part b, the observed difference between the long- and short-day averages is like a random draw from a box of numbered tickets. (Refer to Example 3.9.) The average of the tickets equals ____________, and ____________% of the tickets are within ____________ of the true value.

22. *The CB[1] model: F-tests and confidence intervals*

a. Use your results from Exercises 12–16 to find the F-ratios for shrews and for kinds of sleep. The table values are 4.76 (shrews) and 5.14 (sleep). What do you conclude?

b. For this data set, you can find 95% confidence intervals for the unknown "true" heart rates for each of the three kinds of sleep. Each interval has the form [Avg.] ± (SE)(Standard 95% distance). Here the SE = SD × (leverage factor), the leverage factor is 0.5, and the standard 95% distance is 2.447. Find the intervals for the three kinds of sleep, and show all three on a graph like the one in Fig. 3.12.

c. The width of the intervals tells how precisely your averages estimate the unknown true heart rates. Judging from these widths, should the investigators have used more than four shrews?

Quick review

23. Complete the sentence below in two different ways, each time choosing from the words in parentheses to fill in the blanks.

If the null hypothesis is in fact ____________ (true/false) and the *F*-test ____________ (rejects/fails to reject) it, the resulting error is called a ____________ (miss/false alarm).

24. True or false:

a. If the null hypothesis is true, two kinds of errors are possible, miss and false alarm.

b. Hypothesis testing is conservative in that you declare the null hypothesis false unless there is strong evidence to the contrary.

c. If the null hypothesis is true, the numerator and denominator mean squares for the corresponding *F*-ratio will be estimates of the same number, and the *F*-ratio will tend to be near 1.

d. There are two possible meanings for a large *F*-ratio: either the corresponding null hypothesis is true, or an unlikely event has occurred.

e. The critical values for *F*-ratios have the property that if the null hypothesis is false, the chance of a miss is 5%.

f. The critical value for an *F*-ratio is chosen so that 95% of all data sets for which the null hypothesis is true will have *F*-ratios bigger than the critical value.

25. Why is hypothesis testing by itself, without confidence intervals, rarely enough of an analysis for a data set?

26. Fill in the blanks:

a. If you increase the number of observations that go into an average, the standard error for the average goes ____________ (up/down).

b. If you compare two averages, one based on 10 observations with a small standard deviation, the other based on 10 observations with a large standard deviation, the ____________ (first/second) will have the smaller standard error.

c. The larger the standard error for an average, the ____________ (wider/ narrower) the 95% confidence interval.

NOTES AND SOURCES

1. ***Inside and outside factors***. If you already know about nested factors, you'll recognize that "A is nested within B" and "A is inside B; B is outside A" mean almost the same thing. However, for example, every factor is inside the Benchmark factor (grand mean) although one doesn't ordinarily say that factors are nested within the grand mean. Similarly, the factor for chance error is inside every other factor.

2. ***Degrees of freedom.***

a. ***Free numbers.*** Strictly speaking, no single number is either "free" or fixed. The number of "free" numbers is a property of the table as a whole.

b. ***Units of information about chance error.*** Notice that you can't use the estimated organ effects (± 43.5) to estimate the size of chance error. What does it mean, then, to say the box for organs has one unit of information about chance error? It's true, the information about chance error that's contained in the organ effects isn't directly available to us. The estimated effects, ± 43.5, are a mixture of true value plus chance error, and we can't split the two apart. However, and this is crucial: *suppose* the true organ effects are *zero*, that is, suppose there is

$$\text{F-ratio for day length} = \frac{\text{Mean square for day length}}{\text{Mean square for hamsters}} = \frac{144}{40} = 3.6.$$

FIGURE 3.22 F-ratio for day length uses the hamster mean square in the denominator. Day length is a treatment that is assigned to hamsters, the larger of the two kinds of experimental units. Use the mean square for these larger units (hamsters) in the denominator of the F-ratio for day length.

no real difference between hearts and brains. Then the estimated effects contain nothing but chance error. The SS for organs is just the variability (in that box) due to chance error, the df tell us how many units of information about chance error there are, and the mean square for organs tells us the average variability *due to chance error*. To summarize: *if* the "true" effects for organs are zero, then MS_{Organs} measures average variability due to chance error, so MS_{Organs} ought to be about the same size as $MS_{Residuals}$, which also measures average variability due to chance error. (In fact, MS_{Organs} is not at all like what we'd expect if the true organ effects were zero. We conclude there must be "real," nonzero effects due to organs.)

3. ***Computing the standard deviation***

Question: Why do we work with the squares of the residuals? Why not just use the ordinary average of the residuals?

Answer: The residuals always add to zero, so their average will be zero.

Question: Then why not first get rid of the minus signs, then take the regular average?

Answer: Provided you adjust for the degrees of freedom, that's not a bad idea at all, and it works quite well for some situations. However, for reasons I won't go into here, for many situations, using $MS_{Residuals}$ works better or at least as well.

4. ***Unknown true values.*** Since the true value is unknown, we have no way to be sure that the true value is the same for all four hamsters, but "unknown true value" means the part that *is* the same for all the hamsters. In a sense, the true value doesn't really exist—you can't see it under a microscope or hold it in your hand. It's an imaginary quantity, something we made up as part of our attempt to describe the structure of the data. Inventing useful concepts in order to understand the world is in fact quite common in science. "Intelligence" is an invented concept, not a real thing; for that matter, so is "electron."

5. ***The F-ratio for day length.*** Remember that day length is a *between-hamster* factor: different hamsters were used to compute the averages for long days (hamsters 1–4) and short days (hamsters 5–8). In terms of the design of the experiment (split plot/repeated measures), day length was a treatment applied to whole hamsters, which were the larger of the two kinds of experimental units. Thus as I pointed out in Section 2 of Chapter 1, day length is assigned to hamsters as in a randomized basic factorial design with hamsters as units. If the design had in fact been an RBF, the factor for chance error would correspond to hamsters; "mean square for residuals" and "mean square for hamsters" would be different names for the same number; and we would test day length using the mean square for hamsters in the denominator. This logic suggests we should do the same here for our SP/RM design: to test the effect of day length, use the mean square for hamsters in the denominator. In fact, for any SP/RM design, to test the treatment that is assigned to the larger units, the denominator in your F-ratio should be the mean square for those larger units. (see Fig. 3.22.)

CHAPTER 4

DECISIONS ABOUT THE CONTENT OF AN EXPERIMENT

Review and Preview

This chapter is about the three key decisions that together determine the content of an experiment: what measurement to make (the response, Section 1); what conditions to compare (Section 2); and what material to use (Section 3). Think of these three decisions as a bridge from a particular theory or hypothesis you want to study to a plan for gathering and analyzing data, either to test your hypothesis or to focus it by making it more specific.

Example 4.1 Jung's Word Association Test

Identify the response, conditions, and material in the following experiment. Early in this century, the psychologist Karl Jung conducted an important series of experiments based on word associations. His results led him to infer the existence of what he called psychological complexes. Part of his hypothesis was that for each of us certain words belong to a web of ideas with emotional associations—a *complex*—in a way that other words do not. In each of Jung's experiments, he read a list of stimulus words to his subjects, who were to respond to each stimulus with the first word or phrase that came to mind. For each stimulus, Jung measured the time it took his subject to respond.

Solution

Response: Reaction time, in fifths of a second

Conditions: Stimulus words with various meanings

Material: Human subjects

Jung couldn't observe the complexes directly. He hypothesized that a stimulus word that was part of a complex would have a longer reaction time than one that was not, and he chose as his response a quantity that was directly measurable but only indirectly (by way of hypothesis) an indicator of a complex. ■

Exercise Set A

For each of the following situations, give the response, conditions, and material.

1. For a field study of plant competition, an ecologist finds 12 habitats where there are no primrose plants growing, puts in 10 plants each of 2 species of primrose, and returns 5 years later to count the number of plants of each species.

2. For a lab experiment to study the effects of water temperature on the hatch rate of trout eggs, a fishery places batches of eggs into water of several different temperatures and records the percent that hatch for each batch.

3. To see whether concrete words are easier to remember than abstract words, a psychologist reads to subjects a list of 20 words, 10 of each kind, in random order, and for each subject counts the number of words of each kind correctly recalled.

4. To see whether people's positions in a waiting line influence their reaction to someone who breaks in line, a social psychologist arranges for a confederate to break into ticket lines at a commuter rail station. The psychologist compares the reactions of those just behind and just ahead of the line crasher, recording whether or not each person speaks to the offender.

5. *Crabgrass*. If you were a crabgrass plant and someone offered you the choice, would you rather be surrounded by plants of your same species or by plants of a different species? This question is one narrow instance of an important cluster of questions about competition that ecologists study. Here's a simple experimental plan, sometimes called a **de Witt replacement series**, that could be modified and extended to fit a variety of situations. Suppose you have a bunch of seeds for two species of crabgrass, *Digitaria sanguinalis* (D.s.) and *Digitaria ischaemum* (D.i.), and you have prepared 10 styrofoam cups for growing crabgrass. Two of the cups get 20 seeds each of the D.s., randomly chosen from your batch of seeds; two more cups each get 15 seeds of D.s. and 5 of D.i.; each of the next two gets 10 seeds of each species; the next get 5 of D.s. and 15 of D.i.; and the last two each get 20 of D.i. At the end of the experiment, you harvest the plants and compare the dry weights of the plants from each of the 10 cups.

In Exercises 6 to 9, I tell the purpose of an investigation. For each, give a reasonable set of choices for response, conditions, and material.

6. To study the effect of anxiety on memory.

7. To compare long-term and short-term memory for verbal and visual information.

8. To compare the short-term effectiveness of two methods for controlling gypsy moths: a chemical spray (BTU) and a tiny parasitic wasp.

9. To study the effects on the feeding behavior of birds of displaying a model of a small hawk near a backyard feeder.

1. The Response

Statisticians use the word **response** (or **dependent variable**) instead of "measurement" because an experiment may involve several different kinds of measurements, with different purposes, but the response is special.

The *response* is the measurement you use to judge the effect of the conditions.

To see what I mean, think about a learning experiment in which you use a pretest to sort your subjects into groups of roughly equal ability before you conduct the experiment. The pretest is a measurement but not the response. The response might be the score on an achievement test: it is the measurement you use to judge the various learning conditions. Similarly, in a nutrition experiment using rats, you might divide the rats into groups according to their initial body weight, then measure the effectiveness of various diets using weight gain as a percent of initial weight. Initial weight, final weight, and weight gain are all measurements, but for this experiment the response is percent weight gain, the measurement you use to decide which diets are better.

The choice of response depends most of all on what it is you want to study, which means the choice can't be cut free of its context. As a consequence, there are few rules that are both general and practical. There are, however, two general restrictions that are imposed by the nature of the methods in this book and two criteria that are often useful in choosing a good response.

Two Restrictions

One response or many?

Almost all the methods in this book use only one response at a time.

On the surface, this may seem like a severe limitation. Plant biologists often judge the effect of various conditions using a collection of measurements, such as the weight (1) of a plant's root system, (2) of the stem and leaves, (3) of the flowers and seeds, and so forth. Psychologists might record (1) a subject's reaction time, (2) a subject's skin conductivity (GSR), and (3) whether or not the subject's answer was correct. In either of these examples, to single out one of the measurements as the response might sacrifice important information. In such situations, you have essentially two options. You can do several analyses using the methods of this book—one analysis for each choice of response—or you can learn more advanced **multivariate** methods, which allow you to analyze several responses at once. The multivariate option offers the advantage of formal methods that deal simultaneously with two or more response variables. For some experiments multivariate methods are best, but the methods are harder to learn, to use, and to interpret. Even in those cases where statisticians would prefer to use multivariate methods, you can nevertheless get a lot of useful information about each response by using the methods of this book to do several analyses in parallel, one analysis for each choice of response. What you give up is information about how your several response variables are related to each other and how the conditions of the study affect those relationships. For many experiments, that information may be less important than the information you get from the one-at-a-time approach.

Steven's four types. The second general restriction imposed on the response by the methods of this book is less clear cut than the first, but the basic idea is that analysis of variance works better for some kinds of response measurements than for others. Measurements like reaction times, weight gains, and enzyme concentrations are generally OK, but you can sometimes run into trouble if your response has only a small number of possible values (like yes/no, right/left, or high/middle/low), and you can't use the methods at all if the response has values like red/blue/green, which have no natural order.

About 70 years ago, the psychologist Stanley Smith Stevens published a classification system for measurements, sorting them into four categories: nominal, ordinal, interval, and ratio (Fig. 4.1). His system is worth learning for two practical reasons. First, it can help make clearer which kinds of measurements are suitable as the response for an analysis of variance; second, it can help you recognize additional possibilities for your choice of response. Here are his four categories:

1. **Nominal**: The measurement is the name of a category, that is, the measurement tells which of several groups an item belongs to; the groups have no meaningful order. Examples: eye color, ABO blood group.
2. **Ordinal**: The measurement sorts items into categories that are ordered, but there is no meaningful way to measure the distance between categories. Example: color of tiger beetles (brown, brownish red, bright red).
3. **Interval**: The measurement is a number, and the distance (interval) between numbers has meaning, but there is no natural zero (no lowest possible value), so that "twice as big" doesn't make sense. Example: year of birth.
4. **Ratio**: The measurement is a number on a scale that has both a meaningful distance and a natural lowest value, so that "twice as big" (a ratio) makes sense.

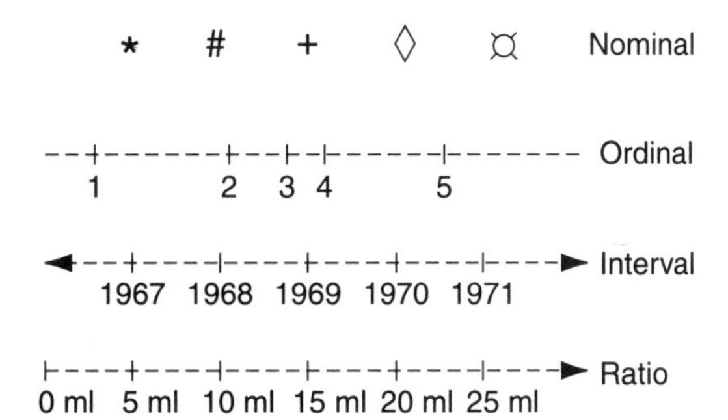

FIGURE 4.1 Examples of Stevens' four types of scales.

Examples: weight, concentration of enzyme, number of test items answered correctly.

Figure 4.2 lists a series of questions you can use to decide the type of measurement.

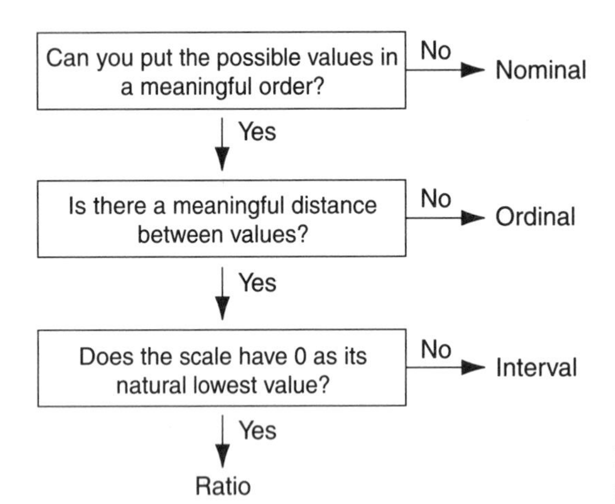

FIGURE 4.2 What kind of measurement scale do you have?

EXAMPLE 4.2 STEVENS' FOUR TYPES

Classify the following measurements using Stevens' system: (a) the weight of a thyroid gland; (b) the body temperature of mouse; (c) a very rough psychiatric classification with four categories, normal/neurotic/borderline/psychotic; and (d) the genotype a plant, animal, or human.

SOLUTION.

a. *Weight:* ratio. Distances make sense ("This gland weighs 50 mg more than that one"), and there is a natural lowest value (0), so ratios make sense ("This gland is twice as heavy as that one").

b. *Temperature*: interval or ratio, depending on which scale you use. Much of the time we think of temperatures in degrees Fahrenheit or Centigrade, as for air temperatures or body temperatures. Distances make sense ("Today is 10 degrees cooler than yesterday"), but ratios don't. You don't hear people say, "Today was twice as hot as yesterday," except as a figure of speech. However, temperature does have a natural zero on the absolute (Kelvin) scale, and if you think of temperature in that way, essentially in terms of kinetic energy, ratios do make sense.

c. *Psychiatric classification*: ordinal. You can order the categories from least disturbed to most disturbed, but there's no meaningful distance between categories: you can't subtract borderline from psychotic. It is quite common with ordinal data to use numbers for the categories (1 = normal, . . . , 4 = psychotic), but there's a big difference between arithmetic that is possible in a mechanical sense and arithmetic that is meaningful. For the numbers themselves, $2 - 1$ is the same as $4 - 3$, but "neurotic minus normal," whatever that means, is definitely not the same as "psychotic minus borderline."

d. *Genotype*: nominal. For example, the ABO system of blood types classifies genotypes into AA/AO/BB/BO/AB/OO. There's no meaningful way to say that one genotype is bigger or smaller than another. ■

Analysis of variance is:

- **impossible for nominal data,**
- **sometimes appropriate for ordinal data, but at the risk of losing information,**
- **generally appropriate for interval and ratio data.**

Analysis of variance works best for interval and ratio data; it is sometimes used for ordinal data but at the risk of losing information; it can't work for nominal data. Remind yourself that analysis of variance works by computing averages, and then ask, for each of Stevens' four types, what an average would tell you. For nominal data, you can't even compute an average because you can't add things such as blood groups or eye colors. ANOVA is out; for nominal data, you need special methods, which are described in many introductory statistics books under the heading of contingency tables or chi-square tests. For ordinal data, if you assign numbers to the categories (1 = normal, . . . , 4 = psychotic), you can at least do the arithmetic, which means that ANOVA is possible in a purely mechanical sense. Still, there may be a real question about what it means. I wouldn't go so far as to call it worthless, but the method does treat $2 - 1 = 1$ the same as $4 - 3 = 1$, and to interpret the results of an analysis you'd have to judge how much the method may have distorted the meaning of your data. For both interval and ratio data, distances make sense, and so do averages; generally analysis of variance works well.

Additional options for choice of response:

- **interval and ratio data: *change* in measurement;**
- **ratio data: *percent change* in measurement.**

One practical value of Stevens' system that gets too little attention is that it can make you more aware of possible choices for a response that you might otherwise overlook. If distance makes sense (interval and ratio data), then you have the option of making before-and-after measurements in order to use the change in the measurement (a distance) as your response. Biologists often use change in weight or change in concentration as a response; psychologists often use change in test score. If ratios make sense, as they do for weights, concentrations, and number right on a test, you have yet another option for the response:

$$\text{percent change} = 100\% \times (\text{new value} - \text{old value})/(\text{old value}).$$

Using percent change is often an effective way to adjust for variability in experimental material, putting everybody on a more equal footing. (There's sometimes a trap to avoid in computing percent change; see Exercise B.7.)

Reliability and Validity: How Good is the Response?

Reliability and validity both refer to the soundness of the connections between the goal of an experiment and your choice of response, that is, between the kinds of conclusions you hope to draw and the kind of evidence you plan to gather.

Reliability is concerned with repeatability.
Validity is concerned with relevance.

In a good experiment, the three decisions about content (response, conditions, material) construct a sound bridge between theory/hypothesis and a means of either testing the hypothesis or gathering data to make it more specific. The decisions reduce a variety of general scientific questions to a common form: what is the effect of the conditions on the material, as measured by the response?

Although rules for making these decisions are in short supply, two clusters of concepts are particularly useful for judging how well the decisions have been made. The technical names for the two clusters are *reliability* and *validity*. Some authors define several different kinds of each, but I prefer to emphasize the general idea and to regard the different kinds as variations on a theme.

Reliable = repeatable.

EXAMPLE 4.3 DISPUTED AUTHORSHIP: THE FEDERALIST PAPERS

You might think that something as simple as counting words would not involve problems of reliability. Two well-known statisticians, Frederick Mosteller and David Wallace, thought so too, until they discovered otherwise, at their own expense. In the late 1780s, Alexander Hamilton, John Jay, and James Madison wrote a series of articles for *The Federalist* in an attempt to persuade the citizens of New York State to ratify the Constitution. Although the articles were published anonymously, historians agree about who wrote each of 70 articles and that each of another 12 was written either by Hamilton or by Madison.

Almost 200 years after the articles appeared, Mosteller and Wallace used a systematic study of prose style to conclude that Madison probably wrote all 12 of the disputed papers. The statisticians based their conclusions on word counts. For example, they chose a common word like "of" and counted how often each of Hamilton and Madison used the word in those articles whose authorship was considered known. Then they compared those results with word counts for the disputed articles. After their study was well under way, Mosteller and Wallace decided to check some of their counts. They discovered, to their chagrin, that some of their recounts gave quite different answers: their method of counting was not nearly as reliable as they had assumed.

They solved the problem by changing their method. Initially they had simply read through the articles, counting words in the same way you or I might. For their new method, they cut the words out of the articles with scissors, put all the "of"s from each article in a separate pile, and then counted. The new method proved highly reliable, except, as they pointed out, when somebody sneezed. ■

As this example illustrates, reliability is often easy to check: make repeated measurements under fixed conditions, and compare the values you get.

Valid = relevant. Unlike reliability, which is often easy to check, validity is often hard to establish and has been the focus of debates about many research studies. *Validity* refers to the relevance of your data (response) to the purpose of your study.

EXAMPLE 4.4 MORTON'S SKULLS

In the mid-1800s, a Philadelphia physician named Samuel Morton, who had amassed a large personal collection of more than 1000 human skulls, published data that he claimed as evidence that blacks and women were less intelligent than white males.

Material: Human skulls

Conditions: Ethnic origin and gender of the skull's original owner

(Note that this study is observational. The material [skulls] comes with the conditions [sex and race] already built in.)

Response: Volume of the skull, in cubic centimeters

Morton concluded that white males tended to have bigger skulls than everyone else and that black females tended to have smaller skulls than the rest: clearly, white males were the smartest.

These days it's easy to see one major problem of validity in Morton's choice of response: whatever intelligence may be, it doesn't depend on the size of your head. Morton had jumped to a wrong conclusion by trying to leap an unbridgeable chasm between the measurable response and the intended focus of his study, between skull size and intelligence. It may be hard to believe that Morton, who in many respects was a careful and conscientious investigator (but also a white male), could have made such a blunder, and it may also be hard to believe that no one called him on it. Yet it is sobering to read (in Stephen Jay Gould's *Mismeasure of Man*) how highly Morton's work was regarded at the time and how many other respected scientists published work of the same sort. Gould argues that the right lesson to take from all this is not how dumb people were in those days but rather that if it was so easy then for careful scientists to be blinded by their expectations, we must assume the same sort of thing happens today as well.

Notice that validity does not refer to the data alone but to the data together with the way they are used. Morton's conclusions from his data would have been valid if he'd stopped at the results that the whites *in his sample* had slightly bigger skulls than the blacks in his sample and that the men had larger skulls than the women. He might even have gone on to recognize that skull volume is closely related to overall body size and that apparent differences due to race and sex go away if you adjust for stature. (There were other serious flaws in Morton's work. See Exercise B.14.) ■

See also Example S4.1: "Freud's Stages of Psychosexual Development."

I think it is clear from these examples why no one has found a comprehensive set of rules for making sound decisions about the response: the decisions depend on the particular context. To see the flaw in Morton's argument that white males are superior, you have to know that skull size is not related to intelligence. To accept the validity of many experiments in cognitive psychology, you have to regard error rates and reaction times as reasonable indicators of how much attention a task requires. To evaluate an experiment in plant ecology, you have to know the difference between a measure of a plant's size, like total biomass, and a measure of reproductive fitness, like total seed set or germination rate.

Here, as in so much of life, it can be hard not to assume that something is unimportant just because it is inaccessible.

See also "The Response in Psychology Research: Special Considerations" (following Example S4.4 in the supplement).

Exercise Set B

Tell which of Stevens' four scale types fits each of the following choices for a response:

1. For seeds of wild grasses: (a) Weight; (b) Species; (c) Percent (out of 10) that germinate; (d) Size category: small, medium, or large.

2. For chickadees: (a) Rank in the social hierarchy; (b) Sex; (c) Number of visits to a feeder in a period of two hours.

3. For a set of five weights in an experiment on perception and labeling: (a) Actual weight in grams; (b) Rank, from 1 = lightest to 5 = heaviest, as judged by a human subject; (c) Label, chosen by a human subject, from "light," "medium," and "heavy."

4. Suppose you want to compare the shapes of eggs for a variety of bird species. What would be a reasonable choice for your (one) response variable? Which of Stevens' four types is it?

5. How can you tell the difference between an interval scale and a ratio scale?

6. Fill in the blanks with one of Stevens' four types. (a) If your response is ____________, then analysis of variance is impossible. (b) If you plan to use the change in a certain measurement as your response, then the measurement itself must be either ______________ or ______________.

7. *Pitfalls of percent change*. Whenever you have before-and-after measurements on a ratio scale, you have the option of using percent change as your response:

$$\text{percent change} = 100\% \times (\text{new value} - \text{old value})/(\text{old value}).$$

Often this choice for a response provides a good way to adjust for initial differences in subjects or material, although there are alternatives: you can use a block design (see Chapter 7), using initial measurements to sort subjects into uniform groups, and/or you can use a general method of adjustment called analysis of covariance (see Chapter 14, Section 3). If you decide to use percent change as your response, there are two traps to avoid:

a. Initial values cross zero. A ratio scale, unlike an interval scale, has a natural zero, which is the lowest possible value. **Do not use percent change as your response if some of your initial values are negative.** The following artificial data set shows before-and-after measurements for four subjects. First notice the pattern: all four subjects increased their scores. Then compute the percent change for each subject, and comment on the pattern in the numbers you get. In what way does converting to percent change distort the original pattern?

Subject	Before	After
I	−10	0
II	−5	5
III	5	10
IV	10	20

b. Some initial values near zero. Choosing percent change as a response may be a good idea if the data suggest that the percent changes are all in the same ballpark, that is, the larger percent change values are not several times larger than the smaller ones. **Don't use percent change if some of your initial values are near zero.** The following artificial data set shows what can happen if you ignore the rule. First note that the changes themselves are all in the same ballpark and might be a suitable choice for your response. Then compute the percent changes and comment on the pattern. In what way does converting to percent change distort the original pattern?

Subject	Before	After
I	0.1	5.1
II	1.0	12.1
III	10.0	23.0
IV	100.0	90.0

Validity: Extreme examples. Each of the following situations describes a purpose and two responses. For each purpose, decide whether response 1 or response 2 is more valid.

8. Purpose: To judge the value of dairy cows
   Response 1: Weight of the cow
   Response 2: Average milk yield, in pounds per week
9. Purpose: To tell the age of trees
   Response 1: Number of rings
   Response 2: Diameter of the tree, measured 4 feet off the ground
10. Purpose: To compare the reproductive fitness of columbine plants
    Response 1: Number of seeds produced per plant
    Response 2: Dry weight of the plant
11. Purpose: To measure how much you learn from this chapter
    Response 1: IQ test
    Response 2: Percent of the exercises you get right
12. Purpose: To measure how fast subjects can read text printed upside-down

Response 1: Time to read and answer two simple questions about an upside-down version of The Lord's Prayer.

Response 2: Time to read and answer two simple questions about an upside-down version of a paragraph of text made up by the investigator.

13. *Deaths and health*

a. In 1996 there were 153,746 recorded deaths in Florida and only 2,562 in Alaska. Is the number of recorded deaths a valid measure of how safe and healthy a state's environment is? Give a reason for your answer.

b. The death rate is the number of recorded deaths per 1,000 people. In 1996 the death rate for Florida was 10.7 (about 1 out of every 100 people died during 1996); the death rate for Alaska was 4.2 (fewer than 1 out of every 200 people). Is the death rate a valid measure of how safe and healthy a state's environment is?

14. *Morton's skullduggery*. In Example 4.4 I pointed out a major problem of validity in the way Morton tried to use skull volume as a measure of intelligence. I also tried to emphasize that Morton was not so much deliberately trying to be misleading as just misled by his own expectations. He himself had discovered and corrected a different problem of validity in his early work: his first sets of measurements were not even valid as a measure of skull volume.

His first method was to fill each skull with mustard seed, then measure how much seed it had taken to fill the skull. He did this by turning the skull upside-down so he could pour the seeds through the hole at the base of the skull; when the skull was full, he transferred the seeds to a graduated cylinder to measure the volume. Before long, he became suspicious of his method based on mustard seed, because the seeds were light and easily compressed. He switched from mustard seed to lead shot, and sure enough, when he remeasured using the new method, he got different numbers. In particular, the new method gave average volumes for white male skulls that were not quite as big as before.

a. Which method—mustard seed or lead shot—would you expect to be less reliable (more variable)?

b. Why was the mustard seed method less valid as a way to measure skull volume? (Why do you suppose the lead shot method gave somewhat lower values for the white male skulls?)

2. Conditions

Isolate the Effects of Interest

An extreme example is often a good tool for understanding and remembering a general idea. Here's one in which surgeons reached the wrong conclusion about an operation because they failed to isolate the effects of interest.

Example 4.5 Surgery as Placebo: Mammary Ligation

Angina pectoris is chest pain associated with disease of the coronary arteries. Around the mid-1950s surgeons in the United States began treating angina patients by performing an operation called internal mammary ligation: give the patient a local anesthetic; make a shallow chest incision; tie off the mammary artery; then sew things back together. Doctors who performed the operation published enthusiastic reports telling how many of their patients felt better following the operation. They had less pain, could exercise more, and took fewer nitroglycerin pills. Even *Reader's Digest* jumped on the bandwagon, publishing an article "New Surgery for Ailing Hearts." All this enthusiasm, however, was based on studies that were not true experiments.

Then, in the late 1950s, two independent investigators published reports of genuine experiments, and opinion about the operation suddenly turned around. For the two new studies, as for the earlier ones, the subjects were angina patients, and the response variables included the patients' own reports of how much pain they felt and how many pills they took. *But*, whereas for the earlier studies there had been only one condition, for the two genuine experiments there were two conditions. Half the patients, chosen by a chance device, got the operation. The other half got a *placebo* operation: give the patient a local anesthetic; make a shallow chest incision, but *don't* tie off the mammary artery; then sew things back together. In both studies the patients were told only that they would be getting an unproved procedure and were part of an effort to evaluate it; they were not told which operation, real or placebo, they got.

Both studies found that the placebo operation did at least as well as the real one, and in some respects better. Internal mammary ligation quickly joined the still-growing crowd of discarded surgical procedures. (For a discussion of ethical issues related to giving patients a fake operation, see Bok.) ■

The early and misplaced enthusiasm that some surgeons felt for mammary ligation can be traced to their failure to isolate the effects of interest. In hindsight we can recognize that patients got better not because their arteries were tied off, but (presumably) because of their faith in the doctors and the surgery. This became clear only after the two experiments, which included a *control group* (the patients who got the *placebo*) designed to isolate the effect of interest (tying off the arteries) by being as much like the real operation as possible except for the "active ingredient." Once the control group made it possible to isolate the effect of tying the arteries, that effect was quickly seen to be tiny or nonexistent.

As you read through the following example, notice that the overall pattern in the numbers would be easier to see if you had the data arranged in a suitable table. Exercise C.5 asks you to construct such a table.

Example 4.6 The Rise and Fall of the Portacaval Shunt

Another operation, the portacaval shunt, was invented to help patients with bad livers by shunting some of the usual blood flow to the liver through an artificial bypass, thus decreasing the workload on the sick liver. The operation was done so often that more than 50 papers were published reporting surgeons' evaluation of it.

Then in 1966 the journal *Gastroenterology* published a study that summarized the earlier reports. Overall, they found that almost 70% of these earlier studies were markedly enthusiastic about the operation, another 20% were moderately enthusiastic, and only about 10% were unenthusiastic: on the surface, the operation looked pretty good. However, they also found that more than 60% of the studies had no control group, about 30% had a control group with serious flaws, and only four studies (fewer than 10% of the 51) had sound controls: underneath the surface, the studies looked pretty bad. Things were actually even worse than that. For the 32 studies with no control group, enthusiasm was high: 24 of the 32 were markedly enthusiastic; only 1 was unenthusiastic. For the 15 studies with poor controls, the results were almost as positive: 10 of the 15 were markedly enthusiastic; only 2 were unenthusiastic. However, for the studies with good controls, the story was quite different: none showed marked enthusiasm; only one showed even moderate enthusiasm, and three (or 75% of the well-controlled studies) showed no enthusiasm. After the article came out, the operation itself was quickly shunted to the same dustbin as mammary ligation. ■

Although control groups and placebos are important concepts, I regard them as particular tools for achieving the more general goal, which is to isolate the effects you want to study.

EXAMPLE 4.7 MEMORY AND MEANING

In remembering what we hear, how much do we rely on meaning? To study this question, you have to find a way to isolate the effect of meaning from other things we may rely on to remember things. A Ph.D. candidate at Harvard, H. E. Wanner (1968), found a way to do this, building the conditions for an experiment from a set of sentences made up to be identical except for small changes in word order. Each subject heard one of these sentences as part of a tape-recorded set of instructions and afterward was shown a pair of sentences—one that had been part of the instructions plus a decoy chosen from the other sentences. The subject was asked to tell which sentence he or she had actually heard.

Here are three of Wanner's sentences:

1. When you score your results, do nothing to correct your answers but mark carefully those answers which are wrong.
2. When you score your results, do nothing to correct your answers but carefully mark those answers which are wrong.
3. When you score your results, do nothing to your correct answers but mark carefully those answers which are wrong.

Some subjects heard either 1 ("mark carefully") or 2 ("carefully mark") and then later were shown both and asked which of the two they actually heard. For this pair of sentences, the difference is purely stylistic: both mean the same. Other subjects heard either 1 ("do nothing to correct your answers") or 3 ("do nothing to your correct answers"); the two sentences in this pair have somewhat different meanings.

Although Wanner used more sentence pairs than I have described here, the basic plan was to have two conditions corresponding to two kinds of pairs, style difference (1 vs. 2) and meaning difference (1 vs. 3). This plan successfully isolated the effect of interest by making the two conditions as alike as possible in all other respects.

Wanner found that when the sentences differed in meaning, subjects could identify the one they had heard almost 100% of the time, but when the only difference was stylistic, the subjects might just as well have tossed a coin: they were right only about 50% of the time. ■

See also Example S4.2: "Sperling's Partial Report Procedure."

Experiments versus Observational Studies

In an *experiment*, you start with one set of subjects or material to which you assign *treatments* (the conditions you want to compare).

In an *observational study*, you start with several *populations* of subjects or objects (with the conditions already built in); you take samples from populations.

An experiment	**compares**	**treatments.**
An observational study	**compares**	**populations.**

Example 4.6 on the portacaval shunt described some of the studies as having good controls and others as having poor controls. A major part of the difference came from the way subjects got into the control group. For the sound studies, the subjects were assigned to the treatment or control group (and the assignment was done using a chance device); for the other studies, the subjects got into the control group by default: for one reason or another, they weren't eligible for the treatment.

It is much easier to isolate the effects of interest if you can assign conditions. In an observational study the conditions you want to study will almost never be the only thing that makes one population different from another. This makes it hard to identify the effects responsible for observed differences.

The main difference between the analysis of a genuine experiment and an observational study comes not from the computations you do—these tend to be the same for both—but from the way you interpret the computations. Observational studies tend to be trickier to interpret than experiments because they are often not as good at isolating the effects of interest. For example, the poorly controlled studies of the portacaval shunt were observational. Patients were not assigned to the control group by choosing from the set of patients who wanted the operation. Instead, the controls were chosen from among those patients who, for one reason or another, were not going to have the operation. Many of them were simply too sick to risk major surgery. Their presence in the control group made the treatment group look better than it otherwise would have.

Because the difference between a true experiment and an observational study is so extremely important, it would be nice if you could count on people always reserving the word *experiment* for studies that assign conditions; unfortunately, people tend to be careless. Often they call a study an experiment when in fact it's observational. To be safe, you shouldn't rely on the words; you have to read the description carefully. If the conditions are not assigned, it's not an experiment, regardless of what it may be called.

EXAMPLE 4.8 THYMUS SURGERY

Most people take it for granted that as we grow up, our arms, legs, and other body parts get larger. As a rule, that's true, but the thymus gland is an exception: it gets smaller as we get older. Ignorance of this fact led turn-of-the century surgeons to adopt a worthless surgical procedure on the basis of observational data. Imagine yourself in their situation: you know that many infants are dying of what seem to be respiratory obstructions and in your search for a cure you begin to do autopsies on infants who die with respiratory symptoms. You're an experienced surgeon, you've done many autopsies in the past on adults who died of various causes, so you decide to rely on these autopsy results as a kind of control. What stands out most when you autopsy the infants is that they all have thymus glands that are much bigger than normal in comparison to body size. Aha! That must be it: the respiratory problems are caused by an enlarged thymus.

It became quite common in the early 1900s for surgeons to treat respiratory problems in young children by cutting out the thymus. In particular, in 1912 Dr. Charles H. Mayo, one of two brothers for whom the Mayo Clinic was named, published an article recommending removal of the thymus. He made this recommendation even though one-third of the children he operated on had died.

Looking back, it's easy to see the problem: the observational studies failed to isolate the effect of interest. A comparison of infants dying of respiratory problems with *infants* dying of other causes would have made it easier to isolate the effect and might have protected doctors from jumping to conclusions. ■

Two related ideas that statisticians use in discussing situations like this are **confounding** and **selection bias**. Both refer to flaws in the design of a study.

> **Two influences on the response are *confounded* if the design makes it impossible to isolate the effects of one from the effects of the other.**

In the surgery example, the effect of age (infant vs. adult) and the effect of illness (respiratory obstruction vs. other) were confounded: the kind of comparisons the doctors chose to do made it impossible for them to tell which was the cause of the differences in the size of the thymus glands. A design that compared infants with infants would have isolated the effect of respiratory obstruction by keeping the other influence (age) constant.

Confounding can occur in many ways, but one common source of confounding, called selection bias, often ruins poorly planned observational studies. In a true experiment, the groups you compare start out more or less uniform—essentially alike except for the differences that you assign. In a good experiment, the conditions you assign are the only important source of differences. In an observational study, however, the groups you compare are different from the very beginning. Although you may choose the groups on the basis of one set of differences, often you can't avoid getting other differences as well. These unwanted effects are confounded with the effects of interest, and the bias that results is called selection bias.

***Selection bias* occurs in observational studies when the process of selecting groups to be compared confounds the effects of interest with other effects.**

In any observational study, it is important to try to choose groups that are as alike as possible except for the effects of interest, but practical considerations may restrict your options, allowing selection bias to creep in. The following example is one where selection bias results from sloppy planning; it would have been possible to avoid the bias by using a more elaborate design. (You might want to read the example as though it were one of those "What's wrong with this picture?" puzzles.)

EXAMPLE 4.9 BOOST YOUR SATS?

Because SAT scores have come to have such an influence on which college a person goes to, there are now lots of special courses that claim to prepare you for the test. Suppose you want to measure the effect of a particular course, using SAT scores as your response. You might first find a reasonably large high school where students are offered a chance to take that course, then randomly select 10 or 20 students who completed the course, together with an equal number randomly selected from the students who took the SATs but had chosen not to take the course.

Suppose you find that the average SAT score for the students who took the course was 30 points higher than for the students in your control group. Do you conclude that the course was effective? Or did you spot the selection bias?

There are several flaws in the plan I just described, but perhaps the biggest one is this: there's likely to be a big difference between the students who chose to take the course and those who chose not to. The course demands a commitment of time and probably costs money too. That means the students who choose it are likely to be more motivated than those who don't. They may be more concerned about getting into the kind of college that requires high SAT scores, they may feel more pressure from their parents, and they may come from homes that can more easily afford the extra cost of the course. It seems very likely that all these differences would have an effect on the SAT scores, and the design of the study gives you no way to separate these effects from the effect of the course itself. (I will come back to this example more than once in the next few chapters, using it to illustrate both good and bad features of various designs. You might want to try to get ahead of me a bit: can you think of a good plan for measuring the effect of the course?) ■

The last example was one where selection bias could have been avoided, but there are many studies where the bias is harder to spot and harder to get around. For example, for over a hundred years anthropologists, psychologists, and biologists have been measuring sex and race differences (and usually concluding that white males are superior). Time and again, though not always, when someone finds a measurable difference that difference gets misinterpreted. For example, remember the observational study (Example 4.4) by Samuel Morton, who took a small but real difference in the sizes of human skulls (a difference due largely to differences in body size) as proof that whites were smarter than blacks. That mistake is now ancient history, but the same sort of thing still happens. In many recent examples where scientists have found a measurable difference between men and women, or whites and blacks, much of the debate about what the differences mean has focused on issues of confounding and selection bias, although few people actually use those words. (See also Example S4.3: "Smoking and Lung Cancer" and Example S4.4: "Gender and Spatial Ability.")

Exercise Set C

1. For each of the following examples, first list the conditions the study was designed to compare; then tell whether the study is observational or experimental. For the purpose of this exercise, give people the benefit of the doubt: if it was possible to assign conditions, assume they actually did.
   a. Example 4.5: Mammary Ligation
   b. Example 4.7: Memory and Meaning
2. A 1970s study of oral contraceptives (birth control pills) gathered data on thousands of women. Among other things, the researchers recorded each woman's blood pressure and how many children she had. The study showed that if you compared blood pressures for women with two children and women with four children, the average for those with four children was about 30 points higher than for those with only two. The careless conclusion is that having children causes your blood pressure to go up. Actually, though, selection bias is at work. Mothers in the first group (two children) tend to be ______________ than mothers in the second group, and as you get ______________, your blood pressure tends to go up. So for this data set, number of children is confounded with ______________.
3. To study the effects of climate on health, you could compare the death rates of two states that have very different climates, like Florida and Alaska. (The death rate for a given year tells how many people out of every 1,000 died during that year.) For 1996 Florida's death rate was more than twice as high as Alaska's.
   a. The reason for using death rate instead of total number of deaths is to avoid con-

founding. If you looked at total number of deaths, the effect of climate would be confounded with ______________. States that were ______________ would tend to have higher values.

b. Even death rate is not a good choice for a response, mainly because the effect of climate would still be confounded with __________. (Think about the kinds of people who move to the two states.)

4. The main reason for using a placebo in the study of the polio vaccine (page 31) was to avoid confounding the effect of __________ with ______________.

5. *Mammary ligation*

a. Reread Example 4.6 on mammary ligation, and list the three possible choices for a response variable mentioned in the description.

b. Of the three, which is the easiest to measure?

c. How would you measure the other two?

d. For the two genuine experiments, how many conditions were compared?

e. For the earlier nonexperimental studies, how many conditions were compared?

f. In studies that use a placebo, the main purpose of the placebo is (choose one)

i. to make the experimental material more uniform

ii. to make the material more representative

iii. to better isolate the effects of interest

6. The response in the polio study was yes/no: either a child got polio or not. The doctors who did the diagnoses didn't know which children had gotten the vaccine and which had gotten the placebo. The reason for keeping them "blind" was to avoid confounding ______________ with ______________.

3. MATERIAL

Units

In an experiment, the *units* are the batches of material that receive the treatments.
In an observational study, the *units* are the items or individuals that make up the population.
The amount of information you get from a study depends much more on how many units you use than on how many response values you have.

Here's an example of a design blunder that comes from not recognizing the units. Suppose you want to compare two textbooks for third-grade arithmetic. Two third-grade teachers have "volunteered" their students for your study, so you (randomly) assign one teacher's class of 30 students to use the first book and the other teacher's 30 students to use the second. You test all 60 students at the beginning of the year and again at the end in order to use the change in test score as your response.

You might think that the units in this design are the third graders, with 60 units in all, but the treatments (textbooks) get assigned to entire classes, not to individual students. Even though there are 60 response values, there are only two units, one for each treatment. If one class does better than the other, there is no way to know whether the difference is due to the textbooks or to other differences between the two classes, such as differences in the effectiveness of the two teachers. A proper design would assign each book to several classes.

A sad experience I had the first time I taught experimental design illustrates the same kind of blunder, this time in an observational study.

EXAMPLE 4.10 THE SHAPE OF SPONGE CELLS

A graduate student was studying the shapes of sponge cells to see whether the shapes of the cells depended on the color of the sponge (white or green) and the location of the cells (tip or base). He painstakingly measured hundreds of cells under a microscope, but, unfortunately, he had based his experimental plan on bad advice: all his hundreds of cells came from just two sponges, one white and one green.

The goal of his study was to compare two hypothetical populations: all the white sponges and all the green sponges of a particular type. Thus the sampling unit was a sponge, not a cell, and he had only two units even though he had almost a thousand response values. ■

Material Should Be Representative and Uniform

Representativeness. Choosing subjects for a psychology experiment usually involves a tradeoff between who's available and how well they represent the group you really want to study. An ideal method for choosing subjects is **random sampling.** If you want to draw conclusions about the adult population of the United States, for example, then the ideal way to get subjects would be to choose them at random (using a chance device) from a complete list of all the adults in the United States. Of course this isn't practical, and you have to make some sort of compromise based on who you can get as subjects. Perhaps you can use paid volunteers from your local area, or if money is tight you might use lower-paid student volunteers from where you work. Volunteers may not be representative of the entire population, and college students may be still less representative, but you can't study the nonvolunteers. All this means that you need to be careful not to assume that the results of your experiment apply to people you haven't studied.

Biologists face a similar tradeoff: lab experiments may be easier than field studies, but in some ways plants raised in a greenhouse or rats bought from a supply house will not be representative of their counterparts in the field.

EXAMPLE 4.11 ABUNDANCE OF PHYTOPLANKTON

For many field studies in ecology, the problem of representativeness is particularly acute. Hulburt gives a striking example: suppose you want to study the abundance of various species of marine plankton, using as your response the density, measured by number of cells per milliliter of seawater. Hulburt's data show quite convincingly that the density of the plankton is highly variable from one place to another, from one time to another, and from one depth to another. Unless you are careful to choose a sample that represents many locations, times, and depths, you are almost certain to be misled by the data you collect. Notice that here, although it is possible to describe the population of interest in a general way—all locations, times, and depths—it is difficult to construct a reasonable list of possibilities from which to choose your sample. (See Exercises D.4–6.) ■

A similar problem is often unavoidable in certain kinds of research on psychotherapy. (See Weiss and Sampson for some interesting work of this sort.) Here the "material" for the study might be a transcript of recorded therapy sessions for a single patient. Because the goal of the research depends on a close and detailed examination of dozens, or even hundreds, of hours of therapy sessions, it would take too long to study several patients. The value of the research depends heavily on how well the patient represents the group of other patients with the same condition.

Uniformity. The basic idea here is fairly simple. If there's a lot of natural variability in your experimental material, that variability will show up in your response and make it harder to detect and estimate the effect of the conditions you want to study.

Often it may not be possible to find material that is truly uniform; moreover, it may not even be desirable, because uniform material may not be representative. You wouldn't ordinarily want to restrict a learning experiment to only those subjects with one particular score on an IQ test, for example. Similarly, you might not want to restrict a biological study to clones of a single plant. One effective strategy that lets you have your cake and eat it too is to sort your material into groups in a way that makes each group more or less uniform, and then to regard the different groups as defining different "conditions." For example, with human or animal subjects you might sort into two groups, male and female, or into several groups based on weight or age. The complete block design (see Chapter 7) is one that makes use of this idea.

Exercise Set D

1. *Random sampling.* Suppose you wanted to do a small study of student attitudes on your campus. You could take a random sample, but it would be much easier to use a group of your friends as your sample.

   a. Which sample would give you a more uniform set of subjects—the random sample or the group of friends?

   b. Which sample would be more representative?

   c. For this study, if you have to choose between a uniform sample and a representative sample, which should you choose? Why?

2. If you have two sets of subjects available for a study, with the first set more uniform than the second, then (choose one)

   a. Chance errors are likely to be smaller

      i. using the first set of subjects

      ii. using the second set of subjects

      iii. No way to tell

   b. Bias is likely to be smaller

      i. using the first set of subjects

      ii. using the second set of subjects

      iii. No way to tell

3. If you have two sets of subjects available for a study, with the first set more representative than the second, then . . . (repeat parts [a] and [b] from # 2)

Sampling in ecology: Density of marine plankton

In Exercise 1 it is possible, at least in principle, to list the individuals in your population of interest. For Hulburt's study of plankton (Example 4.11), no list is possible for the population because the population (seawater) is not a collection of discrete items.

The data in Table 4.1 give the density (cells/ml) of five species of plankton collected at two depths and four times.

TABLE 4.1 Density of five plankton species collected at different depths and times

Date	1 Aug. 75		14 Aug. 75		23 Nov. 73		24 Nov. 73	
Depth (ft)	25	50	25	50	25	50	25	50
Species								
Nitzschia delicatissima	<.03	12.90	<.06	1.71	<.15	<.10	<.06	3.48
Emeliania huxleyi	<.03	11.70	<.06	.42	<.15	.10	8.37	5.28
Guinardia flaccida	<.03	<.03	<.06	<.03	4.80	12.50	<.03	<.03
Eutreptia marina	8.80	5.40	3.60	43.20	<.15	<.10	.03	<.03
Leptocylindrus danicus	<.03	17.70	<.03	<.03	.60	<.10	<.03	<.03

4. Suppose Hulburt wanted to know which of the five species were present at high densities and whether there was one predominant species or several present in roughly equal numbers. Describe the patterns he would have found

   a. if he had sampled only at 50 feet on 1 August 1975;

   b. if he had sampled only at 25 feet on 1 August 1975 and at 50 feet on 14 August;

   c. if he had sampled at both depths and on both dates in August 1975.

   d. Which of the patterns in a–c generalize to the whole data set?

5. Suppose Hulburt had wanted to study the relationship between depth and the abundance of the various species. Summarize the patterns of this kind

   a. if you look only at the November half of the data;

   b. if you look only at the August half of the data,

   c. How would you revise your summaries for a and b if you consider the entire data set?

6. Describe briefly what Hulburt's data suggest to you about problems of representativeness in studies of air or water pollution. Of location, depth, and time, which are relevant? Can you think of other variables that might be relevant in much the same way? What are the major difficulties in finding a good list from which to draw your sample?

SUMMARY

1. *Response*

   a. The **response** is the measurement you use to judge the effect of the conditions. Almost all the methods in this book use only one response at a time.

   Nominal: The measurement is the name of a category, that is, the measurement tells which of several groups an item belongs to; the groups have no meaningful order.
   Ordinal: The measurement sorts items into categories that are ordered, but there is no meaningful way to measure the distance between categories.
   Interval: The measurement is a number and the distance (interval) between numbers has meaning, but there is no natural zero (no lowest possible value), so that "twice as big" doesn't make sense.
   Ratio: The measurement is a number on a scale that has both a meaningful distance and a natural lowest value, so that "twice as big" (a ratio) makes sense.

Analysis of variance is impossible for nominal data, sometimes appropriate for ordinal data but at the risk of losing information, and generally appropriate for interval and ratio data.

Additional options for choice of response:

Interval and ratio data: **change** in the measurement.
Ratio data: **percent change** in the measurement.

   b. **Reliability** and **validity** both refer to the soundness of the connections between the goal of an experiment and your choice of response, that is, between the kinds of conclusions you hope to draw and the kind of evidence you plan to gather. Reliability is concerned with repeatability. Validity is concerned with relevance.

2. *Conditions*

   a. In an **experiment**, you start with one set of subjects or material to which you **assign treatments** (the conditions you want to compare). In an **observational study**, you start with several **populations** of subjects or objects (with the conditions already built in). You take samples from populations. (An experiment compares treatments; an observational study compares populations.) It is much easier to isolate the effects of interest if you can assign conditions. In an observational study the conditions you want to study will almost never be the only thing that makes one population different from another. This makes it hard to identify the effects responsible for observed differences.
   b. Two influences on the response are **confounded** if the design makes it impossible to isolate the effects of one from the effects of the other. **Selection bias** occurs in observational studies when the process of selecting groups to be compared confounds the effects of interest with other effects.

3. *Material*

   a. In an experiment, the **units** are the batches of material that receive the treatments. In an observational study, the **units** are the items or individuals that make up the population. The amount of information you get from a study depends much more on how many units you use than on how many response values you have.
   b. Material should be representative and uniform.

Exercise Set E: Review Exercises

1. *Chick thyroids.* There is a standard experiment used in endocrinology to demonstrate that if you feed chickens a chemical [propyl-thiouracil (PTU)] that prevents them from using iodine, their thyroid glands get bigger. (The idea is that without iodine, the gland can't do its job, which is to produce the hormone thyroxin. When the level of thyroxin gets low, a regulatory mechanism tells the thyroid to get bigger.)

In this version of the demonstration, young chicks are weighed and then divided into two groups. The control group gets regular food, and the treatment group gets regular food with 0.1% PTU. After 2 weeks the chicks are weighed again and killed with chloroform so their thyroid glands can be dissected out and weighed.

a. In all there are three measurements: initial weight, final weight, and weight of the thyroid. Which of Stevens' types is each?

b. Name three additional measurements that you could get by combining pairs of the original measurements. Which of Stevens' types is each?

c. What variable would you judge to be the best choice for the response, considering the purpose of the experiment?

d. What are the conditions for this study?

e. Is the study observational or experimental?

2. *Mental rotation.* Most people recognize familiar shapes like the letter R so quickly and automatically that it is easy to overlook how much we depend on more than shape alone. Context and position are also important. Suppose, for example, you want to study the effect of position. Psychologists have invented a standard way to do this, and while it's not part of the problem, you might want to stop reading long enough to try to design your own experiment: how could you arrange to measure the relationship between the time it takes to recognize a familiar shape and how far the shape has been rotated out of its usual position?

Part of the challenge in designing an experiment of this sort is to make sure that your subjects really do recognize the familiar shape. It's not enough just to show them shapes and tell them to say "now" as soon as they think they have recognized what it is. Cooper and Shepard got around this problem in a way that has become standard: give your subjects a decision to make—one that is quick and simple but can't be done without recognizing the shape. They showed subjects a series of rotated shapes, half of which were ordinary Rs, rotated; the other half were backwards (mirror-image) Rs. The task was to decide whether the R was ordinary or mirror image. The 12 stimuli they used are shown in Fig. 4.3.

Suppose each of the 12 stimuli was presented 10 times, so that each subject got a total of 120 presentations in a random order. One way to analyze the results would be to consider each of the 120 presentations separately, that is, to have one value of the response for each presentation. A second way would be to lump together the 10 repetitions of each kind of stimulus and use some summary measure to get one value of the response for each set of 10 presentations.

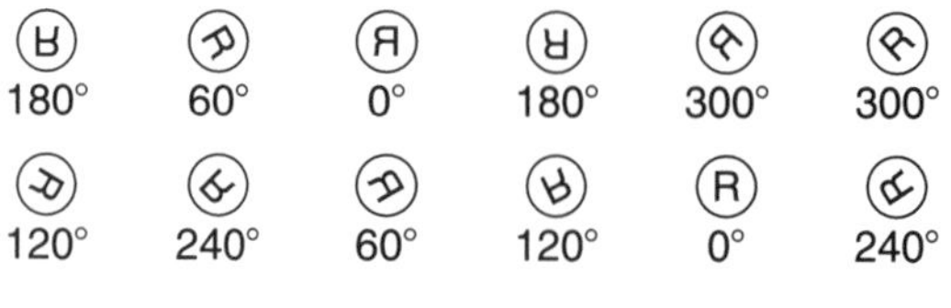

180° 60° 0° 180° 300° 300°

120° 240° 60° 120° 0° 240°

Figure 4.3 Cooper and Shepard's series of shapes.

a. What are the conditions this study was designed to compare?

b. What feature of this study makes it experimental rather than observational?

c. What are the two natural measurements to make in this experiment if you treat each presentation separately? Which of Stevens' types is each?

d. Which of the two would lend itself better to analysis of variance? Which of the two would you judge to be more directly related to the purpose of the study?

e. Now consider the second approach, where you lump the stimuli together in sets of 10 and use summaries. What are the two natural summaries that correspond to the two measurements in part c?

f. In this particular experiment, the task of deciding between R and its mirror image was one that subjects got right almost all the time. Why do you think the experiment was set up to be that way?

3. *Hamlet*, act II, scene 2, line 259: "There is nothing either good or bad, but thinking makes it so."

Robert Rosenthal, who teaches psychology and social relations at Harvard, has done a large number of experiments that show how much our expectations influence what actually happens. One experiment looked at the influence of first-grade teachers' expectations on how well their students performed. All the first graders in the study were given a test, which their teachers were told was designed to predict which children would be likely to show a sudden spurt in how well they did in school. After the test had been scored, the teachers were given three lists of names, telling which ones of their students were likely to spurt ahead, which ones were likely to make average progress, and which ones probably would not show much intellectual growth.

What the teachers did not know was that the test was not a special predictor but just a standard IQ test, and the three groups had nothing to do with the test: they were chosen using a chance device and thus were pretty much the same.

After a year the children were given another standard IQ test, and the differences in scores (second test minus first) were compared for the three groups. The results showed that how well the students did fit with what the teachers had been led to expect: IQ scores for the first group improved the most, and those for the third group improved the least.

a. List the response and tell which of Stevens' types it is.

b. List the conditions to be compared.

c. Is the study experimental or observational?

4. *A comparison of the three studies*

a. Think about the issues of uniformity and representativeness of the subjects (chicks and humans) in the last three exercises.

i. For which study would uniformity and representativeness be easiest to achieve? Why?

ii. For which study would they probably hardest to achieve? Why?

b. Now consider the reliability of the measurement process in the three studies. Imagine repeating the measurement process a second time with the same subject with the conditions the same as the first time, and judge how different the two measurements are likely to be. (Even though you can't actually dissect out the same thyroid gland twice, you can imagine how things might go if it were possible.)

i. For each study, list what you think is the most important influence that would tend to make the two measurements different.

ii. Which study has the most reliable response?

c. For which two of the studies is the validity of the response pretty much assured, given the purpose of the study? Why is validity somewhat more uncertain for the other study?

5. *Selection bias: The Salk polio vaccine.* Polio is now a very rare disease, thanks to medical research, which developed a vaccine, and thanks to statistical work, which proved the vaccine to be effective. When I was growing up in the 1950s, however, polio was still quite common; one of my classmates in sixth grade was unable to walk because polio had paralyzed his legs. When I was seven, I was one of nearly two million children who served as subjects in a study to test a vaccine that had been developed by Dr. Jonas Salk. I remember taking a form home to my parents to get their permission for me to participate; not everyone got permission. I also remember standing in line in the school cafeteria, watching the line grow shorter, knowing it would soon be my turn to get a shot. Although I was too young to understand at the time, that shot might have been the vaccine, but it might have been a placebo shot. (Once the results were in, the vaccine was made available to everyone who wanted it.)

The testing of the vaccine actually involved two different studies. One was a true experiment, with the treatment and placebo assigned by a chance device, but the other study was observational. It used the children who didn't get permission as a control group and gave the vaccine to everyone who got permission. In what ways do you think the treatment and control groups are likely to be different? How might that bias the results of the study? (The answer here might surprise you.)

6. *Portacaval shunt.* Reread Example 4.6 on the portacaval shunt.

a. If you read carefully, you should be able to construct a table that gives the information from the article in *Gastroenterology* that summarized the 51 previous studies. The purpose of your table should be to show the relationship between the soundness of a clinical study (the conditions) and the degree of enthusiasm for the operation (the response). Use as rows of your table the three kinds of controls (good, poor, and none); use as columns the degree of enthusiasm.

b. Describe the ordinal response variable by telling how you would assign the values 1, 2, and 3 to the three possible outcomes described in the example.

c. For this way of looking at the summary study, what are the three conditions to be compared?

d. For each of the three conditions, compute the average response. How good a summary do you think these numbers give?

e. Comment on any parallel you see between the conclusion here and the conclusion in Robert Rosenthal's experiment with first graders and their teachers.

7–9. *Stories mothers tell.* [These exercises assume that you have read about the Thematic Apperception Test (TAT) in the Supplementary Examples, S4.6.]

Mental health professionals are still a long way from knowing what causes mental illness. Inheritance has something to do with it and so does early family life, but only a few of the connections are well understood. One study designed to work on a small corner of this issue used the TAT to compare 20 mothers of schizophrenic children with 20 control mothers who had normal children of the same sex and approximately the same age. Each mother was shown the same set of 10 TAT pictures, and the 10 stories she told were classified into categories by blinded raters. The pictures were ones that were likely to lead to stories about parents. There were five categories, which had been developed in a similar but earlier study:

A. Personally involved, child-centered, flexible interactions

B. Impersonally involved, superficial interactions

C. Overinvolved, parent-centered interactions

D. One of A, B, or C, but can't tell which

IR. Irrelevant: none of A, B, and C applies

The results are listed in Tables 4.2 and 4.3.

TABLE 4.2 Mothers of schizophrenics

Subject	A	B	C	D	IR
1	2	2	4	1	1
2	1	0	2	1	6
3	1	1	4	1	3
4	3	1	1	1	4
5	2	1	0	1	6
6	7	0	0	0	3
7	2	1	2	2	3
8	1	0	1	0	7
9	3	2	0	1	4
10	1	1	3	0	5
11	0	1	4	0	5
12	2	1	3	0	4
13	4	1	1	0	4
14	2	2	1	0	5
15	3	0	3	1	3
16	3	0	1	0	6
17	0	0	3	1	6
18	1	1	3	0	5
19	2	0	4	0	4
20	2	1	2	2	3
Totals	42	16	42	12	87

TABLE 4.3 Control mothers

Subject	A	B	C	D	IR
1	8	0	1	0	1
2	4	0	0	1	5
3	6	0	1	0	3
4	3	0	1	1	5
5	1	0	2	1	6
6	4	0	2	1	3
7	4	1	2	1	2
8	6	0	0	0	4
9	4	0	1	0	5
10	2	1	2	0	5
11	2	1	2	1	4
12	1	1	3	2	3
13	1	1	3	0	5
14	4	0	1	1	4
15	3	0	2	1	4
16	3	0	2	0	5
17	2	1	0	2	5
18	6	0	0	0	4
19	3	1	1	0	5
20	4	0	2	0	4
Totals	71	7	28	12	82

7. *Choices for a response*

a. There are two ways to think about the data, according to whether or not you lump together the 10 stories for each mother. Suppose you think of having one value of the response for each story: either A, B, C, D, or IR. Which of Stevens' types is this?

b. Now suppose you want to have one value of the response for each mother. To do this, you have to find a number to summarize each set of 10 stories. Why can't you use an average here?

c. One set of choices for the response in b is to pick any one of the five categories and to use as the response the number of stories (out of 10) that the judges put in that category. Look over the entire set of five choices and decide which category (or categories) you would pick if you were willing to be a bit sleazy and you wanted to claim the two groups of mothers told different kinds of stories. Which would you pick if you wanted to show the opposite: that the two sets of stories weren't very different? The point here is important. Just as Morton reached a wrong conclusion (Example 4.4) because he chose a response that gave the kind of pattern he expected to find, here it is possible to do the same sort of thing. The moral is that you need to be very careful in situations where you (or the person whose study you're reading about) have several possible choices for the response.

d. One choice for a response that I like, even though the authors did not use it in their own analysis, is the number of As minus the number of Cs. My reason for liking it comes from theory: children whose actual interactions with their parents are like those in the group A stories are thought to grow up better adjusted than children whose actual interactions follow the pattern of group C stories. So theory would suggest (but be careful here: see Exercise 9!) that mothers of normal children would score much higher on this response than mothers of schizophrenics. Find the average value of this response for each set of 20 mothers. (Look for a shortcut.)

8. Scatterplot y = no. type C stories versus x = no. type A stories, with each mother as a point. Make two plots, one for each set of 20 mothers. Then write a short paragraph comparing the patterns in the two plots and telling what the differences suggest about mothers of schizophrenics.

9. The investigators did separate comparisons for four responses: number of As, number of Bs, number of Cs, and number of B-or-Cs. Using standard statistical tests, they found that the difference in number of As was "real," that is, too big to be due just to chance. The differences in numbers of Bs and numbers of Cs taken separately might reasonably be due just to chance, but lumping the Bs and Cs together to count B-or-Cs gave an average difference that was too big to be just chance variation.

a. Why is it not a surprise that if the difference in the number of As is real, the difference in the number of B-or-Cs is also real?

b. Here are two possible conclusions from the results:

(Cause) Mothers of schizophrenics tend to have fewer type A interactions with their children, and that tends to make the children grow up withdrawn into themselves.

(Effect) Schizophrenic children grow up withdrawn into themselves, and that tends to make mothers of schizophrenics have fewer type A interactions with their children.

Choose one:

The data tend to support the first conclusion.

The data tend to support the second conclusion.

One of the two must be right, but the data don't give evidence for one over the other.

None of the above.

Give reasons for your answer. (Be alert to whether you assume that the kind of stories told reflect the actual situation.)

NOTES AND REFERENCES

Sources for data sets are listed following the index to data sets at the back of the book.

Appendix: Supplementary Examples

Reliability and Validity

Example S4.1 Freud's Stages of Psychosexual Development

One of the oldest and most famous arguments in psychology concerns the validity of much of Sigmund Freud's work. Freud used as evidence things his patients had told him in their psychoanalytic sessions, and concluded (among other things) that all infants pass through three stages of psychosexual development: oral, anal, and phallic. Critics later attacked Freud's work, questioning the validity of his evidence.

Material: Two problems of representativeness: (a) Freud drew conclusions about the development of *infants*, but his subjects were all *adults*. A direct study of infants would have been more convincing. (b) Freud claimed that his three stages were a universal fact of *normal* development, but his studies were based on people who came to him with psychological problems.

Conditions: Critics have claimed that Freud's patients were influenced by their sense of what he wanted to hear, and that indeed no research based on therapy sessions can be free of this kind of placebo effect.

Response: (a) Freud's own perception of the evidence was undoubtedly influenced by his expectations: there were no blind evaluators. (b) Freud did not use a clearly defined and objective response, analogous to Jung's choice of reaction time. To do so would have narrowed the focus of his studies much more than he wanted; he was as much interested in formulating hypotheses as in testing them. Nevertheless, his decision not to do so raised serious questions about the validity of his evidence.

These and other criticisms of Freud's work led later investigators to pay more attention to validity, and while the controversy is far from resolved, many studies in the past few decades suggest that, at the very least, Freud was often on the right track. ■

Isolate the Effects of Interest

EXAMPLE S4.2 SPERLING'S PARTIAL REPORT PROCEDURE

Psychologists who studied visual memory in the 1950s often used some variation of the following plan. While subjects were sitting in front of a screen, a display of letters would appear for a small fraction of a second and then disappear; the subjects would then be asked to report as many letters from the display as they could.

U	G	M	R
A	N	P	K
Z	J	E	Q

These studies showed that subjects could usually report only four or five of the twelve letters, and so would seem to suggest some sort of upper limit of about five items for visual memory in situations of this sort.

However, in 1960 G. A. Sperling demonstrated that there was more than the effect of visual memory involved, and that the upper limit seemed to be a lot higher than five items. He showed his subjects the same sort of display, but after it was removed, instead of asking subjects to report letters from the whole display (whole report procedure), he asked for letters from just one of the three rows, which he chose each time using a chance device. He found that whichever row the chance device happened to choose, the subjects on average could report more than three of the four letters in the row. Since they could remember more than three letters from any one of the three rows, without knowing in advance which one to focus on, Sperling concluded that earlier experiments had found a low rate of recall because they had failed to isolate the effects of visual memory from the effects of the report procedure. His partial report procedure suggested that subjects could actually remember—for a short time at least—many more letters than they could report. ■

Experimental Versus Observational Studies

EXAMPLE S4.3 SMOKING AND LUNG CANCER

Back in the 1950s and 60s, when evidence began to pile up linking cigarette smoking with lung cancer, some people tried to dismiss the evidence as "merely statistical." The more thoughtful of these skeptics were not objecting to statistical evidence per se; they were concerned about the difference between a true experiment and an observational study, and the special problems that arise when you try to interpret the results of an observational study. Part of the evidence against smoking came from comparing the rates of lung cancer among smokers and nonsmokers; the cancer rate was higher among smokers.

Quite clearly, it is not possible to do a genuine experiment to compare smokers and nonsmokers. (Imagine telling one of your subjects, "Sorry, Jack. Your coin toss came up tails. For the next twenty years, please smoke two packs a day.") The subjects in the observational studies came with the conditions built in: some were smokers, some were not. The skeptics argued that smoking might not be the *cause* of cancer. "For all we can tell," they said, "there might be some people who are more likely to develop cancer than others, for underlying biological reasons we don't yet understand. And the underlying biology might make these same people more likely to enjoy smoking, or to become addicted. Imagine a nervous, hard-driven type. People like that might be more inclined to smoke, and more likely to get cancer as well, but just because the two go hand in hand doesn't mean one causes the other." ■

There are three points I want to make about this example. First, hardly anyone still thinks that smoking is safe. Second, some tobacco lobbyists no doubt used the argument cynically, to protect economic interests, for example by weakening the warnings printed on cigarette packs. But third, the issues raised by the argument are real and important, and you should think about such issues in connection with any observational study.

EXAMPLE S4.4 GENDER AND SPATIAL ABILITY

The ability to visualize 3-dimensional images is part of what makes a person good at mathematics. One way to measure this ability is to present subjects with pairs of pictures like the pair below, and ask whether they are pictures of the same object. To answer the question, a subject has to rotate one object mentally, to see whether it can be made to look like the other one. Experiments of this sort usually use the time it takes to reach a decision as the response variable.

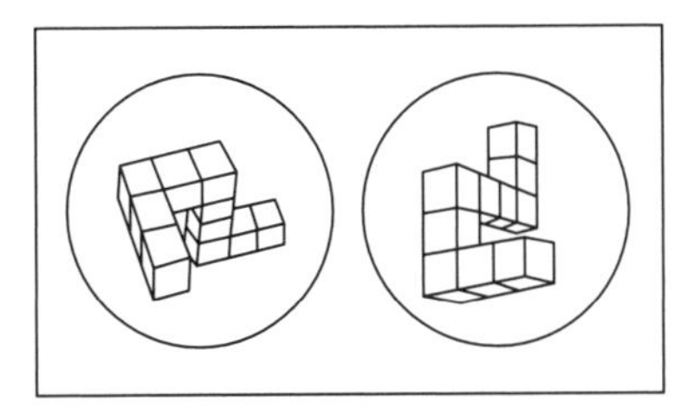

Many people believe that males are better than females at mathematics, and, in particular, some studies have shown that males are better at tasks that involve visualizing in three dimensions. However, it is almost impossible to avoid selection bias in studies of this sort.

Here's a hypothetical but realistic example. The response is reaction time, the time it takes to decide whether two pictures show the same object; the conditions are male or female, for human subjects.

Suppose your results show that males are distinctly quicker than females at mental rotation. What can you conclude? Are males naturally better at this kind of task?

Not necessarily. Being male or female involves more than just genes and hormones. In our culture, boys and girls tend to grow up with different sets of expectations and somewhat different environments and experiences. I teach at a college for women, and over the years some of my students have told me how various teachers in grade school have let them know, in one way or another, that they weren't expected to do well at mathematics.

In choosing the groups to be compared, it is impossible to get groups that are alike apart from sex; the selection process automatically brings with it effects due to differences in background, and these differences almost surely bias the results. (For a summary of studies of male/female differences, see Macoby and Jacklin's *The Psychology of Sex Differences*, and for a bio-logist's critique of recent work in this area, see Anne Fausto-Sterling's *Myths of Gender*. ■

The Response in Psychology Research: Special Considerations

The physicist Lord Kelvin is often quoted as having said, "If you cannot measure, your knowledge is meager and unsatisfactory." Indeed, those words are engraved on the facade of the Social Science Research Building at the University of Chicago. It has also been said that to social scientists, the practical meaning of Kelvin's words tends to be "if you cannot measure, measure anyway."

The pair of quotations together capture much of what makes experimental work in psychology particularly challenging: some of the things that seem most important to understand often seem impossible to measure adequately. The challenge is there in trying to understand how we think: How do people find chess strategies, invent geometry proofs, recognize familiar objects, remember their friends' phone numbers? The challenge is if anything even greater in trying to understand personality and emotional life. What really happens in psychotherapy? We know the behavioral surface: patient and therapist take turns vibrating their own their vocal chords and each other's eardrums; if this happens often enough, and goes on long enough, many patients end up saying they feel happier and more confident. What really goes on beneath the surface? Questions like these are important, but hard to study: How can you find measurements that are both reasonably objective and reliable (reproducible under the same conditions), and which at the same time capture more than a tiny piece of the thing you really want to know about?

Although I don't have a satisfactory answer, I hope you can get a partial answer by reminding yourself of the question each time you read one of the psychology examples in this book. I've used real examples that psychologists regard as important, so you can see how research scientists have chosen their response variables in a variety of specific situations.

Many experiments use as a response some quantity that can be measured directly, like reaction time. In experiments of this sort, the main challenge usually comes in inventing a set of conditions which insure that the reaction time can be used as intended, to study complexes, or memory capacity, or whatever. For other experiments, like the examples that follow, there is no directly measurable response. In such cases, you either have to construct it, by combining many pieces of information according to a set of rules that define the response, or else you have to use **blind evaluators** who act like measuring instruments, assigning numbers by subjectively combining bits of information. Or you may need to use some combination of both.

EXAMPLE S4.5 TARDIVE DYSKINESIA

A few years ago I helped a graduate student with a study of different therapies for people with tardive dyskinesia, a particular kind of nervous twitching that often afflicts long-time users of antipsychotic drugs.

The student's choices for subjects and conditions were fairly straightforward. She wanted to compare two kinds of therapy with a control; one therapy was based on relaxation methods, and the other attempted to teach conscious control of the twitching. She had little choice about her subjects: she took everyone from a particular treatment center in Boston who had tardive dyskinesia. She divided her subjects into three groups, one for each therapy condition, plus a control group who got placebo therapy in the form of regular group meetings.

The goal of the study was to see whether either therapy helped the subjects with their symptoms, so it was clear that the response should be some measure of how severe or how frequent those symptoms were. The graduate student made videotapes of the subjects through a one-way mirror, both before and after the treatment period; her challenge was to find a meaningful way to summarize the videotapes using numbers.

She decided to divide each videotape into five-minute segments, and to train two assistants to classify and count the number of twitches of various kinds in each segment. Thus she used raters to convert the videotape information into numbers, and she then used those numbers to construct response variables. One such response was the average number of facial twitches per segment.

Counting twitches is not as easy as it might seem, even with practice, because it's not always clear what to count as a twitch. So the experimenter took two precautions: she kept her assistants in the dark about which treatment group the subjects belonged to (blind evaluators), and she used two raters, not one. Using two raters allowed her to compare their counts for the same videotapes, to get a measure of reliability. ■

EXAMPLE S4.6 PROJECTIVE TESTS: RORSCHACH, TAT, DRAW-A-PERSON

In the last example, where the raters counted twitches, their task was comparatively straight-forward, and would not be a source of controversy. Projective tests are often quite controversial.

In the Rorschach Inkblot Test, which first appeared in 1921, subjects are shown a standard series of ten symmetric inkblots one at a time and asked what they see in them. The two premises underlying the test are (1) that subjects project various characteristics of their personality into their responses, and (2) that trained experts can use the responses to assign meaningful scores related to personality. The test is famous, frequently used, and frequently criticized.

The scoring of responses is far from purely subjective. Learning the scoring system typically takes three semesters of intensive work. Nevertheless, despite the existence of a large Rorschach industry, many psychologists dismiss the test as both unreliable (raters given the same set of responses don't come up with the same scores) and invalid (according to the critics, the scores don't measure what the testers claim they do). For more on this test, see A. R. Jensen's "Review of the Rorschach."

The Thematic Apperception Test (TAT) was developed by Harvard psychologist Henry Murray (1938) as an alternative to the Rorschach. Subjects are shown not inkblots, but a series of ambiguous pictures, and are asked to respond to each with an imaginative story. (For example, one picture shows a young boy looking at a violin that sits on a table in front of him.) The tester then scores these spontaneous stories using some set of well-defined rules.

The Draw-a-Person Test (Machover, 1949) is based on subjects' drawings of people, which are scored according to size, position, parts of the body, and so on. ■

These three projective tests are used mainly by some clinical and social psychologists. In an article "Test results are what you think they are," Chapman and Chapman report a series of experiments that raise doubts about the clinical use of the Draw-a-Person test; the skeptical attitude of others is nicely summarized by the last line of Jensen's review article: "The rate of scientific progress in clinical psychology might well be measured by the speed and thoroughness with which it gets over the Rorschach." Although Jensen makes a strong case against the test, I think it would be a shame to reject the goals of projective testing just because of the flaws of particular tests.

Compare the counting of twitches in Example S4.5 with the projective tests. The first is reasonably reliable, valid as one measure of symptom severity, and uncontroversial, while the projective tests have many weaknesses. At the same time, though, the projective tests are far more ambitious: they seek to measure aspects of personality and motivation, things which are universal, of tremendous importance, and devilishly hard to understand.

Moreover, here as elsewhere, validity depends not only on the response, but also on its purpose. I see a big difference between using the tests for psychiatric diagnosis and using them for less ambitious purposes. Thus I would be quite skeptical of anyone who claimed to be able to tell whether I was neurotic, and in what ways, just on the basis of stories I told about ten inkblots. However, in the next exercise set I give an example (6.F.3) where the Thematic Apperception Test was used in a way that strikes me as fundamentally sound.

One of the biggest difficulties with tests like the Rorschach and TAT is that they are open-ended: there are no restrictions on how the subjects respond, which means that someone has to turn their sentences and paragraphs into numbers, and so far no one has found an uncontroversial way to do this, except for certain specific and narrow purposes.

A natural alternative is to restrict the way subjects can respond, usually by asking multiple choice questions. This makes it easy to write out unambiguous rules for turning a set of answers into scores, and thus makes scoring the test more objective. Unfortunately, it doesn't get around the problem of deciding what the scores mean; interpretation remains a subjective and tricky matter.

Example S4.7 The MMPI Test

When I went to Dartmouth College as a freshman in 1964, I and my 800 classmates took the Minnesota Multiphasic Personality Inventory (MMPI). Twenty years later, researchers from Dartmouth asked us all to take the test again; they wanted to see whether there was any pattern relating changes in the test scores with the ways the Vietnam War had affected our lives. I and my classmates were the subjects, and the conditions were various categories of experience related to the war: served in Vietnam, served in the military but not in Vietnam, did alternate service as a conscientious objector, went to Canada, and so on.

The response variables came from the MMPI test, a written true/false test with 566 items. Each item is a statement about a person, like "I like mechanics magazines" and "My sleep is fitful and disturbed." A person taking the test is asked to decide for each statement whether it is more nearly true or false as a description of himself. The entire set of 566 true/false answers can then be used to construct a variety of scores. For example, there are standard rules for getting scores named for psychiatric categories like depression, schizophrenia, etc. Critics continue to raise questions about what the scores really mean. ■

Example S4.8 Mental Health, Memory, and Interference

Combining a bunch of yes/no answers to get a summary score isn't always controversial. For example, I worked briefly as a statistical consultant for a project to study the effect of interference on memory. The theory behind the experiment was that people with severe psychiatric problems find it harder to focus their attention; this theory predicts that psychiatrically normal subjects would be less affected than others by distracting sounds while they tried to remember things. (The results of this experiment supported the theory.)

In this experiment the conditions to be compared were defined in part by the subjects, who were classified as (1) schizophrenic, (2) schizotypal, (3) borderline, and (4) psychiatrically normal. For each subject, the experiment was divided into a series of trials; on each trial the subject heard a female voice read a string of digits, like "one eight five nine three," and was then asked to repeat them back. Each trial was scored right or wrong. There were lots of trials, presented in a random order created by a chance device, but they sorted into a small number of categories based on the number of digits in the string, and on whether or not the subjects heard a distracting male voice reading digits in between the test digits read by the female voice.

In the analysis of this study, the conditions were defined by the various combinations of psychiatric category, the number of digits in the string, and the interference condition, present or absent. One response was defined by lumping together all the trials of a given type, such as four digits/no interference, and adding up the number of times the subject remembered the string correctly. In this example, constructing the response is straightforward, largely because all the items lumped together to get a score are clearly similar. ■

EXAMPLE S4.9 SEMANTIC DIFFERENTIAL

The MMPI test is based on a very large number of true/false items. An alternative that uses fewer items but offers more than two possible choices for each is Osgood's (1957) Semantic Differential. A subject is presented with a set of pairs of words with opposite meanings, separated by a ten-centimeter line,

RELAXED ____ TENSE

and asked to mark the point between each pair of opposites that most accurately describes a particular person, usually the subject herself.

Sometimes the scores themselves serve as response variables, and sometimes the scores from several pairs of similar items are combined. For example, in a study of the effects of vitamin B_6 on premenstrual syndrome, the Ph.D. candidate running the experiment used three clusters of adjective pairs, one cluster for each of anxiety, depression, and perceived stress. ■

EXAMPLE S4.10 THE DEFENSE OF UNDOING

During the last three decades, the Psychotherapy Research Group led by Joseph Weiss and Harold Sampson at Mount Zion Hospital in San Francisco has developed a variety of new ways to measure aspects of psychoanalytic treatment. One study from this group concerns a patient's use of a particular pattern of verbal behavior known as *undoing*. For example, a patient might angrily criticize his therapist for starting a session late, only to shift from anger to add, "But, of course, whenever you start late, you always make up the time, so I get the full fifty minutes just the same. You've always been fair about that." The shift from one set of ideas and feelings to another, more or less opposite set (in this case from the angry accusation to the conciliatory afterthought), is an instance of undoing. Psychoanalytic theory regards the patient's shift as an attempt to protect himself from painful feelings associated with his anger and criticism; in this instance the threatening feelings might include guilt and fear of retaliation. In the metaphorical language of analysis, the reason the patient shifts—without being aware of it—from one set of feelings to an opposite set, is to cancel his unacceptable feelings, to make them go away by magic, as if they had never existed.

The investigators in this study had a particular hypothesis to test: that a patient's use of undoing is a symptom of his neurosis, and that if his therapist behaves in appropriate ways, then over the long term the patient will come to recognize his use of undoing as a defense, and will eventually gain control over it.

Here, as in the other examples of this section, the challenge comes from the choice of response. The choice of material wasn't hard: undoing is regarded as characteristic of patients with a particular diagnosis (obsessional disorder), and there aren't many detailed records of long-term therapy available to choose from, so there wasn't much freedom to choose. The investigators used detailed notes of 108 sessions from the therapy of a 35-year-old man, whom they called Mr. A. The conditions followed from the hypothesis that Mr. A's use of undoing would show changes over the course of his therapy. So the investigators divided the 108 sessions into 6 sets of 18 sessions each to use as conditions: Sessions 1–18, 19–36, 37–54, 55–72, 73–90, and 91–108.

Sampson and his colleagues constructed their response by first defining four ordered categories they called stages:

In the first stage, the patient was aware that he could have only *one* attitude toward any situation; in the second stage he became aware that he actually held *two or more contradictory attitudes*; in the third stage he became aware that he *shifted between* the opposing attitudes, and had assumed that each attitude canceled the preceding one; in the fourth stage he *acquired control of shifting*—that is, he could actively turn from one attitude to another, or could hold on to one attitude if he wished to do so. (p. 527)

Next, they trained three judges to recognize these stages. Finally, they showed the trained judges the records of the therapy sessions, but in a *randomly scrambled* order. Each judge assigned a stage (1 to 4) to each instance of undoing, and the three scores were combined to get the response. In all there were 44 sessions where undoing was judged to have occurred. The results are given in Table S1. ■

TABLE S1

Condition		Scores for sessions with undoing[a]	Average score
Sessions	1–18	1, 1, 1, 2	1.25
	19–36	1, 1, 1, 1, 1, 1, 2, 2, 2, 2, 2	1.45
	37–54	2, 2	2.00
	55–72	2, 2, 2, 2, 3, 4	2.50
	73–90	1, 1, 3, 3, 3, 3	2.33
	91–108	2, 2, 3, 3, 3, 3, 3, 3, 3, 3, 4, 4, 4, 4, 4	3.20

[a]For the many instances in which all three judges agreed, the score shown is their unanimous choice. For the other instances, the three judges discussed their judgments, and the score shown is their consensus rating.

EXAMPLE S4.11 Q-SORT

In the early 1950s the University of Chicago clinician Carl Rogers led a four-year research project which sought to measure personality change resulting from a particular form of therapy, which Rogers called "client centered" but which others called "Rogerian." Rogers' research design was too complicated to describe here, but his basic strategy was to compare subjects who got therapy with a control group at four times, spread over about a year and a half. The response variables he constructed were all based on a technique known as the *Q-sort*, which had just been developed. Since then, various versions of the Q-sort have come to be used quite widely in studies of personality.

In the version Rogers used, a subject would get 100 cards to sort into nine piles. Each card had a description printed on it, such as "I am a submissive person," "I don't trust my emotions," "I usually like people," "I am afraid of sex." The subject was to put in the first pile the one card "most like me"; to put in the second pile the 4 descriptions "next most like me"; to put 11 cards in the third pile, 21 in the fourth, 26 in the fifth, 21 in the sixth, 11 in the seventh, 4 in the eighth; and in the ninth pile put the one card "least like me." (The sizes of the piles—the 1, 4, 11, 21, 26, 21, 11, 4, and 1—come from the bell-shaped curve.) Once the subject had sorted the cards a first time to describe himself, he was asked to sort them again, this time to describe his ideal self, the kind of person he would most like to be.

To construct a response variable, Rogers had to find a way to reduce the huge mass of raw data. Each Q-sort in effect produces 100 numbers: each of the 100 descriptions gets assigned a score from 1 (most like me) to 9 (least like me). And each of Rogers' subjects did eight Q-sorts: at each of four times during the study, the subject sorted once for "who I am" and once for "who I'd like to be."

Rogers took the 200 numbers from each pair of Q-sorts and reduced them to a single number called a *correlation coefficient*. This number, which is always between -1 and $+1$, is one common way to measure how closely two sets of scores agree. (For more on correlation, see Chapter 14.) In this case, Rogers wanted to use the correlation coefficient to measure agreement between the Q-sort for "who I am" and for "who I'd like to be." A score near $+1$ meant the subject saw himself as very much like his ideal self; a score near -1 meant the subject saw himself as very nearly opposite from his ideal self.

Rogers found that over the course of time, the correlation for subjects in the therapy group increased, from an average near 0 at the beginning of the study to an average near 0.3 by the end of the year and a half. Correlations for the control group did not change. (However, it is important to note that the control group began with an average correlation of almost 0.6—these were people who felt no need for therapy—and because they started out high on the scale, there wasn't a lot of room for their scores to move up.) ■

The examples in this section show a lot of variety. Some of the responses are comparatively objective and free from controversy: the number of twitches per 5 minutes (S6.5) or the number of strings of digits remembered correctly (S6.8). Other responses depend heavily on subjective judgments: the Rorschach (S6.6), or the stages of undoing (S6.10). There are also some in between: the MMPI test (S6.7) is quite objective in that it can be scored by a machine, but the meaning of the scores is another matter altogether. Some of the response variables are based on commonly used tests that have become standard (Rorschach, TAT, Draw-a-Person, Semantic Differential, MMPI, Q-sort); others were custom designed for a particular study (twitches per 5 minutes, stages of undoing).

With each choice of response, it is worth asking two sets of questions, about reliability and validity, about repeatability and relevance. I've already said a bit about validity in the context of particular examples, so I'll concentrate here on reliability. (The exercises will give you a chance to practice applying both concepts.) If you focus on the process of turning the raw results into numerical scores, and ask how likely it is to get the same scores if you (or someone else) were to repeat the scoring process on the same raw results, you would find that some of the response variables in the last set of examples are much more reliable than others. The MMPI test, for example, has near-perfect reliability in this sense. There is a set of rules for turning each set of true/false answers into scores, and the rules are so unambiguous that the scoring is usually done by machines. When projective tests like the Rorschach and Draw-a-Person are used for psychiatric diagnosis, they are near the opposite extreme. Even though there are elaborate sets of rules for scoring what a subject says about the inkblots, and for scoring the kinds of drawings a subject makes, it is likely that two different raters would come up with different scores.

Reliability refers to agreement; **inter-rater reliability** refers to how closely two raters given the same raw results would agree in their scores. You may remember from the example of the Q-sort (S6.11) that statisticians have a standard way to measure how closely two sets of scores agree: the correlation coefficient. In that example, Rogers computed correlations as one way to summarize how closely a person's Q-sort scores for "who I am" matched the Q-sort scores for "who I'd like to be." You can use the same idea to measure reliability. For example, the psychiatrists who studied the defense of undoing used three trained judges to score the instances of undoing. The judges worked independently, and after the scores were in, the investigators computed correlations to see how closely they agreed. Each judge scored 44 instances of undoing, and for each pair of judges the investigators computed a correlation. Two judges who agreed perfectly would have had a correlation of 1.00, whereas if one or both judges had assigned scores purely at random, the correlation would have been very close to 0. The actual correlations were 0.76 (A with B), 0.82 (A with C), and 0.97 (B with C), which indicates that the scoring system was pretty reliable.

Supplementary Exercise Set S.A

1. *Sperling's partial report procedure*. Often the instructions you give your subjects have a lot to do with how easy it is to score the results. Reread Example S4.2 on Sperling's experiment, and remember that the goal of the study was to measure the capacity of a person's short term visual memory. Imagine that you have shown three subjects the display of twelve letters in the example, and you have told each subject, "Please repeat back as many letters as you can from the second row." Suppose these are their answers:

Subject #1: "A, N, K, Q."

Subject #2: "A, P, R, and either K or Q, I'm not sure which."

Subject #3: "A, X, P, K, R, Z, N."

Think about how you would score these answers. Then tell how you would change the instructions to make the scoring easier.

2–4. *The defense of undoing*. (Reread Example S4.10.)

2. The results show a clear pattern: early instances of undoing tend to get classified as stage 1 or 2, later instances tend to get classified as stage 3 or stage 4.

   a. Why was it important that the order of the sessions was randomly scrambled before the sessions were shown to the judges?

   b. The correlations that measured how closely the three judges agreed with each other were quite high, suggesting that the response has high reliability. Remember, though, that the judges were rating the same material that the investigators had used to develop the rules for deciding the four stages. Which of the following situations would you expect to show the highest reliability (closest agreement among judges)? Next highest? Next? The lowest? Give reasons for your answers.

      i. The actual situation: Judges rated the same material used by the investigators to develop the scale.

      ii. New material, same patient: The same three judges rate new material from additional sessions with the same patient. (Careful, this one is tricky: "Additional sessions" can only mean sessions # 109 or later, when most instances would tend to be stage 3 or 4.)

      iii. New material, another patient, same diagnosis: The judges rate sessions for another patient with obsessional disorder.

      iv. New material, another patient, different diagnosis.

3. *Discussion question*. It's not hard to imagine an argument between two extreme points of view:

A: "The reliability of the response is very high for work of this sort, and the pattern of averages is very strong: over the course of his therapy, the patient moved from near 1 to well above 3. Clearly, he got a lot better, and so the therapy must have been helpful."

B: "Baloney. Anyone could get results like that. All you need to do is to look through the patient records until you find some way that the late sessions are different from the earlier ones. There's got to be some way the sessions differ, even if it means absolutely nothing. Once you've found your difference, just make sure you describe it carefully enough that other people can see it, too. Then you'll get nice high correlations, which prove only that you did a careful job describing what you saw. Besides, even if the patient did get better, so what? That doesn't prove it was the therapy that helped him."

The authors of the paper reached a conclusion that fits somewhere between these two extremes. What is your own conclusion based on what I've told you so far?

4. There's actually more to the study than I've told you, and the part I've not told you yet illustrates one common way to investigate the validity of your response: show that it is related to other measures of the same thing.

Clinical theory makes two predictions that are relevant here. One: A healthier patient is better able and more likely to be aware of intense unpleasant feelings like shame, guilt, anxiety and depression. (This is not the same as saying that the healthier we are, the worse we feel. Rather, the theory says we all have such feelings, but that the healthier we are, the more easily we recognize the feelings and deal with them appropriately.) Two: A healthier patient is better able and more likely to experience opposite feelings simultaneously, things like depression and enthusiasm at the same time.

The authors developed scales to measure (1) the patient's awareness of intense feelings and (2) his simultaneous experience of opposite feelings. They got judges to rate the sessions on these scales, and found that both measures had high correlations with the undoing scale.

How does this additional information change your conclusions?

CHAPTER 5

RANDOMIZATION AND THE BASIC FACTORIAL DESIGN

Review and Preview

To help organize the various methods you are learning, you might think of the material of this book divided into design of experiments and analysis of the results, with the design part divided into content and structure and the analysis part divided into description and inference.

DESIGN

Content	*Structure*
Response	Random assignment
Conditions	Blocking
Material	Factorial crossing etc.

ANALYSIS

Description	*Inference*
Averages	Hypothesis testing
Dot diagrams	Confidence intervals
Scatterplots	
Interaction graphs etc.	

This chapter and the next two present three principles for choosing a design structure. Two of these principles, random assignment and blocking, give complementary strategies for assigning the treatments to units. We will use the two principles to create four basic designs: the basic factorial (BF), the complete block (CB), the Latin square (LS), and the split plot/repeated measures (SP/RM). Each of these four basic designs can be extended to form an entire family of designs, using the third of the three principles, factorial crossing, which deals with treatment structure.

Two principles for assigning treatments to units:

- Random assignment (Chapter 5)
- Blocking (Chapter 7)

Four design structures based on these principles:

- BF: basic factorial (Chapter 5)
- CB: complete block
- LS: Latin square
- SP/RM: split plot/repeated measures

(CB, LS, SP/RM: Chapter 7)

One principle for extending the basic designs:

- Factorial crossing (Chapters 6 and 9)

Chapters 5 through 7 each include a section on linear decomposition and analysis of variance, so you can choose to learn the formal analysis as you go along, one design at a time. However, it is also possible to skip these sections at first, in order to learn the ideas of design and informal analysis before turning to formal ANOVA.

1. The Basic Factorial Design ("What You Do")

Once you have pinned down the content of your study—chosen the response, conditions, and material—your next step is to choose a design structure. For a true experiment, since the conditions are treatments you apply, choosing a structure means deciding how to assign treatments to units. For an observational study, the conditions correspond to populations, and choosing a structure means deciding how to get a sample of units from each population.

The single most important idea to guide your choices is **randomization**:

> **Randomization Principle**
>
> **Whenever possible, any assigning or sampling should be done using a chance device.**

This principle always applies: no matter what design structure you use, you should randomize wherever you can. The randomization principle leads directly to the simplest design, the **randomized basic factorial (RBF) design**, whose experimental version is usually called the **completely randomized (CR) design**.

Experimental Version: The Completely Randomized (CR) Design

The *completely randomized* (CR) or *randomized basic factorial* (RBF) design uses a chance device to assign one treatment to each experimental unit. For a balanced design, each treatment gets assigned to the same number of units.

The two genuine experiments in Example 4.6, about the operation for chest pain, are both examples of completely randomized designs: a chance device was used to assign each patient either to a treatment group (ligation) or to a control group (placebo operation). Most experimental designs are more complicated than this, the simplest of all designs, but all designs for randomized experiments rely on the strategy of random assignment. For that reason, the CR is the natural choice for the design to study first.

EXAMPLE 5.1 WALKING BABIES

The goal of this experiment was to compare four seven-week training programs for infants, to see whether special exercises could speed up the process of learning to walk. One of the four training programs was of particular interest; it involved a daily 12-minute program of special walking and placing exercises. The other three programs were different forms of controls. The second program, "exercise/control," involved daily exercise for 12 minutes but without the special exercises. The third and fourth programs both involved no exercise (parents in these groups were given no instructions about exercise) but differed in follow-up; infants in the third program were checked every week, while those in the fourth group were checked only at the end of the study. (The data are in Table 5.1)

Response: Age (in months) when the infant first walked alone

Treatments: The four programs

Units: 23 one-week-old white male infants

CR design: Each infant was randomly assigned to get one of the four training programs. Three of the four programs were assigned six infants each, and one got five. ■

TABLE 5.1 Age in months when babies first walked

Group	Data					
Special exercises	9	9.5	9.75	10	13	9.5
Control 1: exercise	11	10	10	11.75	10.5	15
Control 2: weekly report	11	12	9	11.5	13.25	13
Control 3: single report	13.25	11.5	12	13.5	11.5	—

This experiment is typical of many experiments whose purpose is to evaluate different treatments. The treatments could be almost anything: ways to prevent disease (see Chapter 4, Review Exercise 5 about the vaccine for polio); attempts to cure a disease or at least make it easier to live with (Example 4.5 on mammary ligation, Example 4.6 on the portacaval shunt, Example S4.5 on tardive dyskinesia); kinds of diets to fatten pigs or get cows to give more milk; or methods for helping people learn something. (See also Example S5.1: Estradiol and Uterine Weight.)

Balance. A basic factorial design is called balanced if the treatment groups are equal in size. The walking babies data set (Example 5.1) is unbalanced: there are six babies in three of the groups and only five in the fourth group. Adding one more baby to that last group would have made the design balanced.

Balanced designs offer several advantages over unbalanced designs. For one thing they are easier to analyze. (The computations are easier to do and easier to understand and interpret.) Moreover, because balanced designs provide equal amounts of information about the conditions you want to compare, as a rule they give more precise comparisons than unbalanced designs with the same total number of observations. Finally, although analysis of variance will sometimes give misleading results if one or more of certain key assumptions do not fit the data, the fit of the assumptions tends to matter somewhat less if the design is balanced.

I think it's very likely that the plan for the walking babies experiment started out balanced, with six babies in each group, and then for some reason one family dropped out of the study. Unfortunately, this sort of thing often happens, so that a design ends up not quite balanced. Fortunately, however, designs that are only one or two observations short of being balanced are almost as good as designs that are perfectly balanced. (Chapter 12, Section 5 tells how you can estimate a replacement value for a missing observation in order to analyze almost-balanced data using methods for balanced data.)

Why should you use a chance device? There are two main reasons.

1. **Randomizing reduces the risk of bias.** For just about any study, it will be impossible to make your experimental material uniform. Some chunks of material will just naturally tend to give higher values for the response than others. If you assign treatments haphazardly, or even if you rely on expert judgment, you risk bias of the sort I described back in Chapter 1, Section 3. Random assignment tends to even out differences in experimental material in much the same way that a set of coin tosses tends to show roughly half heads, half tails.

2. **Randomizing makes the variability in your material behave in a chancelike way.** Random assignment makes it a matter of chance which units go with each treatment, so that differences get randomly associated with the treatments. Every measurement contains pieces due to the effects of the conditions of interest but also contains a leftover piece, the residual error, which is due in part to measurement error and in part to variability in your material. To judge the effects of

the conditions, we will need to know how the residual errors influence our observed values. Specifically, we want these errors to be chancelike so we can rely on the predictable regularities of chancelike behavior. It is nearly always reasonable to assume that measurement errors are chancelike, but it is not always reasonable to make the same assumption about differences between units. Randomizing the assignment of conditions makes it reasonable to assume that residual errors will be chancelike.

Observational Versions of the Basic Factorial Design

Sampling instead of assigning. As you'll see in Section 3, the observational version of the basic factorial design has the same factor structure as the experimental version: The meaningful ways to group the observations and compare group averages will be the same for both versions of the basic factorial design. In this sense of "what you get," the two versions are the same. In terms of "what you do," however, they are quite different: in an experiment you assign treatments to units; in an observational study you sample units from populations. (Example 4.9, on SAT scores, showed how important that difference can be.)

Simple random samples. There are many different ways to get your samples; some are much sounder than others. I'll describe two of these ways, one that randomizes and one that doesn't.

> **For a simple random sample (SRS) you use a chance device to make sure that (1) every member of the population has the same chance of getting in the sample and (2) all possible samples of the size you want are equally likely.**

Using SRSs gives you the observational twin of the CR experiment:

> **For the *observational* version of the randomized basic factorial (RBF) design you take a simple random sample from each population. (The design is balanced if the samples are the same size.)**

The reasons for randomizing are pretty much the same as before: randomizing protects against bias, and helps ensure that residual errors will be chancelike.

Convenience samples and compromises. Often, unfortunately, it may not be practical or even possible to get simple random samples. A **convenience sample** is one you get (without using a chance device) by taking whatever units are easily available. With convenience samples, bias is very often a big problem because the units that are handy differ in some important way from those that are not. For example, people who fill out and return a questionnaire may have different opinions from the people who just throw it in the trash.

Many observational studies use sampling plans that are compromises, not as good as SRSs but much better than convenience samples.

EXAMPLE 5.2 INTRAVENOUS FLUIDS

In 1973 Turco and Davis published a study that compared the purity of intravenous fluids manufactured by three different drug companies.

Conditions: Three drug companies—Cutter, Abbot, and McGaw

Material: Six different samples of intravenous fluid from each drug company

Response: Number of contaminant particles of a certain size (5 or more microns in diameter) per unit of fluid (Table 5.2). Here, and throughout the book, observations marked with a * were changed by ±1 in the last decimal place to simplify certain exercises.

TABLE 5.2 Number of particles per IV unit

Cutter	Abbot	McGaw
255	106*	578
265*	289*	516
343*	99*	214
332*	275	413
232*	221	401
217	240	260

■

Notice that this design is balanced: there are six observations for each condition. Notice, also, that for this example there's no way you can assign conditions: you can't take 18 samples of fluid, and assign 6 of the samples to be manufactured by Cutter, 6 by Abbot, and 6 by McGaw. Each of the conditions corresponds to a population—the set of all bags of fluid manufactured by a particular company. The six bags from each population were obtained by random sampling from those available at the time.

Exercise Set A

1. *Balance.*
   a. The design for Example S4.10 (The Defense of Undoing) is not balanced. Why?
   b. In general, would you expect it to be easier to make your design balanced for an experiment or for an observational study? Why?

2–7. *Using random number tables.* For a CR experiment you need to use a chance device to assign conditions to your subjects or other material. It is possible to do this by writing numbers or names on slips of paper, mixing them in a box, and drawing them out at random. Experience has shown, however, that the mixing is rarely thorough enough. It is generally better to use a table of random numbers generated by a computer. The purpose of this exercise is to let you practice using such a table. The first few parts are just warm-up; the last part asks you to use the table to construct a CR design in various ways.

Turn to Table 1 at the back of your book, and look at how it is arranged: there are 50 rows and 25 columns, with a two-digit number where each row and column come together; the table has spaces that divide things into five-by-five squares. Although the two-digit numbers were generated by computer, they behave as if you had written the numbers 00, 01, 02, . . . , 98, 99 on 100 slips of paper and put them in a box, then mixed them thoroughly, drawn out a number for the table, put it back, mixed again, drawn out another number, and so on.

2–3. *Tossing coins*

Here are two different ways you can use the table to get results like 10 tosses of a fair coin.

2. *Odd/even*. First, pick a place to start. One way to do this is to look at a clock, and take the number of minutes past the hour as your starting row number; take the day of the month as your starting column number. For this exercise, however, start at row 6, column 6. Next, pick a path, like across the rows or down the columns. For this exercise go across the row. Use the digits one at a time, ignoring the fact that they come in pairs. If the digit is odd, record H for heads; if it's even record T. Continue across until you have 10 tosses. Then count up how many Hs you got.

3. *High/low*. This time start at row 8, column 3, and move down the column, using the numbers in pairs. Record H if the number is between 00 and 49; T if between 50 and 99. Find how many Hs you got in 10 "tosses."

4. *Rolling one die 12 times*. Start at row 11, column 1, and move across, using the digits one at a time. If the number is a 1, 2, 3, 4, 5, or 6, write it down; if it is a 0, 7, 8, or 9, skip to the next number. Keep going until you have written down 12 rolls. Then total up how many 1s, 2s, etc., you got.

5–7. *Assigning conditions for the walking babies experiment*. Remember that for this experiment there were 23 one-week old infants and four conditions. The design called for six children in each of the first three groups, with five in the last. Here are two equally good ways to assign conditions to children. Suppose you have assigned the numbers 1, 2, . . . , 23 to your children. It doesn't matter how you do this, as long as you know which child is 1, which is 2, and so on.

5. For this method, you'll first choose the six children for group 1, then the six children for group 2, and so on. Start at row 2, column 11, and use the digits in pairs, reading down the column. Divide each two-digit number by 25, and take the remainder. If the remainder is between 1 and 23, that subject goes in group 1; if the remainder is 0 or 24, ignore it and drop down to the next pair of digits. (For example, suppose the first digit pair is 83. Since 25 goes into 83 three times, with a remainder of 8, child 8 would get chosen for the first group.) Keep going until you have assigned six subjects to group 1; then do the other groups in the same way.

There are two additional things to notice. First, you can't use a child more than once, so if you have already assigned a child to some group and you get his number again, skip him and go on to the next number. Keeping track of this requires care, but it's not hard. Second, as soon as you've filled the first three groups, you can quit: the five children left over are the ones who go in group 4.

6. For this method, you'll take the children in order and use random numbers to assign conditions. Start at row 21, column 11, and use the digits one at a time, reading across. Divide each one-digit number by 5, and use the remainder. If the remainder is a 1, 2, 3, or 4, that tells you the group for child 1; if you get a 0, skip to the next digit. Then move on to the next child, and so on through the list. Using this method, you have to keep track of how big your groups are getting. As soon as group 1 is full (six children), skip over any 1s that come up in the table; do the same for the other groups as they fill up.

7. The following method is biased, because two of the four treatment groups are more likely to be chosen than the other two. Which two are more likely? The method: same as in 6, except divide each number by 4 instead of 5. Then add 1 to get the group number: a remainder of 0 means group 1, a remainder of 1 means group 2, and so on.

2. INFORMAL ANALYSIS

This section is about two tools, averages and dot graphs, about their relationship (a dot graph balances at the average), and about using the tools for three purposes: to assess treatment differences, to detect outliers, and to assess variability.

Averages and Outliers

The average of a set of numbers equals their sum divided by how many there are. The average itself tells a typical value for the set of observations, and the deviations from the average tell about variability in the data. The deviations from an average always add to zero.

Averages summarize groups of observations. There's an important fact about averages that we'll use over and over throughout this book: when you compute an average, you treat all the numbers that go into the average in the same way, as interchangeable. The order of the numbers doesn't matter. For example, the average of 1, 2, and 3 equals 2, and you get the same average if you take the numbers in reverse order: $(3 + 2 + 1)/3 = 2$. What matters is the group of numbers that goes into the average.

For almost all data sets, several averages will be worth looking at. Each set of averages corresponds to a way of sorting the data into groups, one group for each average, and each kind of experimental design has a characteristic set of meaningful averages. Learning to recognize the sets of averages is one good way to tell the designs apart.

EXAMPLE 5.3 AVERAGES FOR THE WALKING BABIES

For this data set it makes sense to compute an average for each treatment group (Fig. 5.1).

Treatment groups							Averages
Special exercises	9	9.5	9.75	10	13	9.5	10.125
Control 1: exercise	11	10	10	11.8	10.5	15	11.375
Control 2: weekly report	11	12	9	11.5	13.3	13	11.675
Control 3: single report	13.3	11.5	12	13.5	11.5		12.350

FIGURE 5.1 Treatment averages for the walking babies data. Each average on the right is based on the group of observed values in the same row (on the left). This set of averages corresponds to sorting the data set into rows.

On average, the babies getting special exercises walked soonest (a little over 10 months), those in the first two control groups took about a month longer to walk, and those in the third control group took the longest (more than 12 months). ■

Outliers

An *outlier* is an observation that is far from the rest of the observations in its group (except, possibly, for other outliers).

Averages often provide good summaries, and that is their main advantage: they make it easier to compare groups because they give us just one number per group. This advantage has a corresponding disadvantage: an average doesn't tell us about the individual observed values, and sometimes that information can be very important. Consider, for example, the first row of the data in Fig. 5.1, for the children who got the special exercises. Five of the six ages are bunched between 9 and 10 months, but one child took 13 months. Observations like these, which stand away from the rest of their group, are called **outliers** and deserve special attention. (Different statisticians use the word "outlier" with different meanings. Many use it as I have defined it, to refer to any "outlying" value. Others, however, reserve it for those values that have been judged "bad," that is, too deviant to be included in the data set as part of a formal analysis.)

The influence of an outlier. Outliers deserve special attention because they exert more than their share of influence on the value of the average. A good way to understand this comes from looking at the relationship between the average and a dot graph of the observations.

A dot graph represents observed values as points plotted on a number line.

Suppose you have just two observations, say 9 months and 11 months. The average is 10 months, and we can show the relation between the dot graph and the average like this:

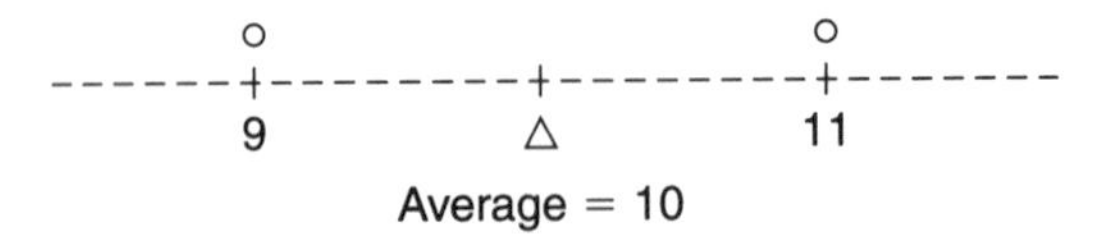

A dot graph balances at the average.

Here's a dot graph for group D (Control 3), which balances at its average.

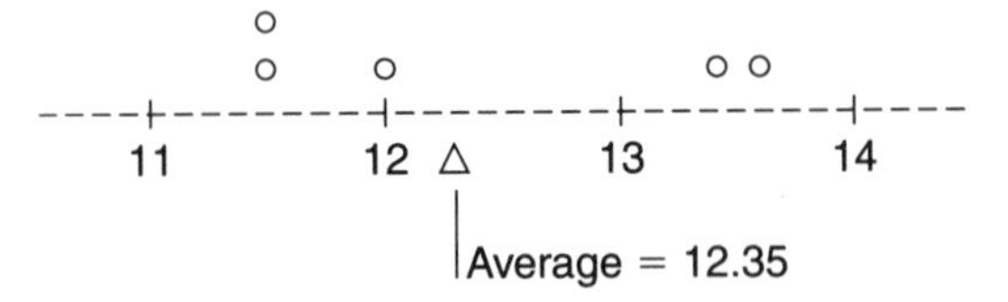

Now for the influence of an outlier. Figure 5.2 shows two graphs for group A, special exercises. The top one leaves out the outlier, the slow child who took 13 months; the other five children have an average of 9.55 months. The bottom graph is for all six children; putting the outlier back in raises the average from 9.55 to 10.125.

```
       o
o      o  o  o
--+-------+--
9    Δ   10
     |
     |Average = 9.55 (without the outlier)
```

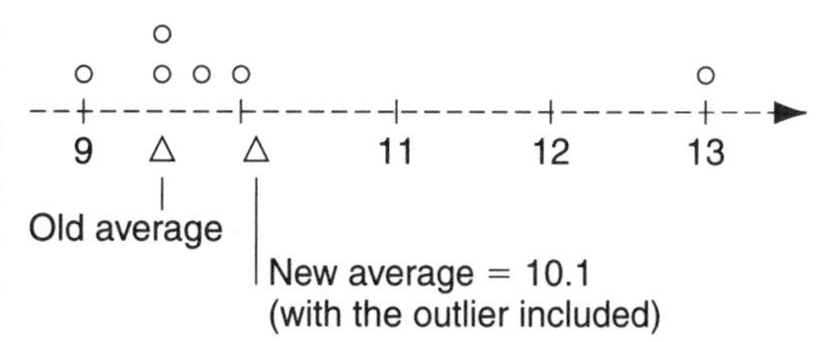

FIGURE 5.2 Influence of an outlier on the average. Taking out the outlier gives an average around 9.5; putting the outlier back in raises the average up to 10.1. Notice that if you tried to balance the bottom dot graph at the old average, the graph would tip down and to the right because the outlier has so much influence.

Remedies. Because outliers exert so much influence on an average, it is important to think about how to deal with them. Some outliers can be traced back to mistakes in measuring or recording. If you can correct the mistake, then, of course, that's what you should do. Much of the time though, you can't figure out why an outlier is so different, and it's not obvious what to do about it. If you knew the observation was truly unrepresentative, you'd do best to throw it away. However, it could happen that the nature of your experiment just naturally produces extreme values every so often. If that's the case, and you threw out the outlier, you'd be throwing out information about the experiment. Statisticians have developed special methods to deal with outliers, and this book will tell you a few of them. For example, you might use a summary number other than the usual average, one that is not so sensitive to outliers. The median (Exercise B.3) is one such summary. One general strategy that doesn't depend on special methods is to do one analysis including the outlier and another analysis without it. If you reach the same conclusions both ways, then you don't need to be concerned; but if you get different conclusions, you'll know that what you conclude from the data is sensitive to what you assume about the outlier.

Parallel Dot Graphs

We can combine dot graphs for each of the four treatment groups to get a parallel dot graph.

A parallel dot graph for comparing groups of observed values represents each observed value as a dot or an X (or some other symbol). The height (y-value) of each dot tells the observed value, and the dots go in columns, one column for each group.

EXAMPLE 5.4 PARALLEL DOT DIAGRAM FOR THE WALKING BABIES EXPERIMENT

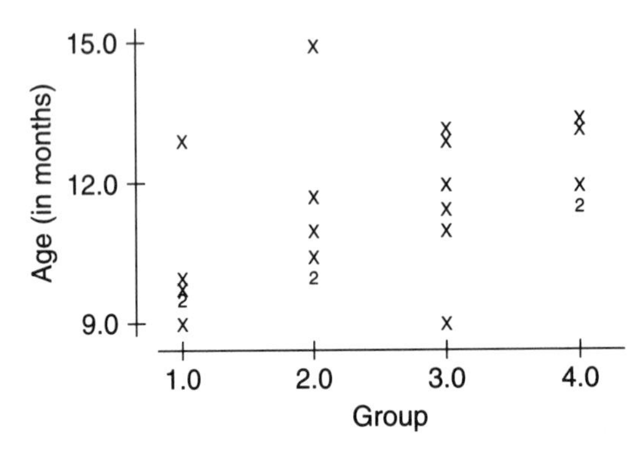

FIGURE 5.3 Parallel dot diagram for the walking babies data. There are four columns of Xs, one for each condition (1 = special exercises, 2 = exercise control, 3 = weekly report, 4 = single report). (A 2 means that two values coincided.) The graph shows at a glance that the Xs for the first group (special exercises) form a cluster that is lower than the other three clusters and that the cluster of Xs for the fourth group (Control 3) is higher than the other clusters. Moreover, the graph shows features that the averages hide: one of the six children who got special exercises took 13 months to walk, a lot longer than the other children in his group. One of the children in the first control group took 15 months, a lot longer than the rest of his group; and one zippy child in the second control group walked after only nine months, way ahead of his peers. ■

Drawing a dot graph. The graphs in this book were made by computer, and Section 5 shows one way to get parallel dot graphs using the Minitab computer package. However, it's also useful to be able to draw dot graphs by hand. The next example tells how, in three steps: x-axis, y-axis, and points. Only the second step, choosing a scale for the y-axis, sometimes calls for judgment.

EXAMPLE 5.5 DOT GRAPH FOR THE INTRAVENOUS FLUIDS (EXAMPLE 5.2)

Step 1: *x-axis*. First, draw a horizontal line, and label three columns, one for each drug manufacturer.

Step 2: *y-axis*. Next, go through the data to find the smallest value (99) and the largest value (578). Then draw a vertical line and mark it off with numbers ranging from below the smallest to above the largest. If I were plotting these particular numbers by hand, I would start the scale at 0, but 0 won't always be a reasonable starting value. (Look back at the dot graph for the walking babies. There, the numbers range from 9 to 15. A scale that starts at 0 would force all the dots into a bunch at the top of the graph, so the computer started the scale at 9. Note the break between the x-axis and the y-axis to show that the scale doesn't necessarily start at zero.)

Step 3: *points*. Finally, plot points for the observed values, one at a time, putting each point in the column for its group, with height equal to the observed value (Fig. 5.4).

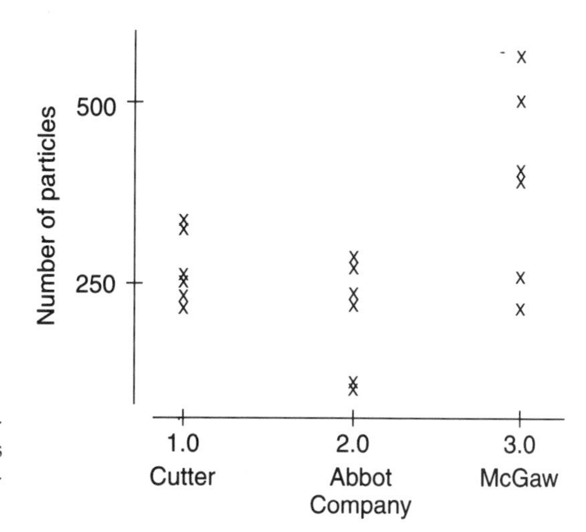

FIGURE 5.4 Parallel dot graph for the intravenous fluids data. Each observed value is the number of contaminant particles 5 microns or more in diameter. ■

Reading a dot graph. Each data set will have its own features, and so with each dot graph you should look for any striking patterns. Nevertheless, particularly at first, it can help to have a checklist of things to look for.

Four Things to Look for in a Parallel Dot Graph [G O S T]

Groups: How do the groups compare?

Outliers: Are there any?

Spread: How spread out are the dots in each group?

Transforming: If the groups have different spreads, and spread seems related to average, it is worth considering a change of scale?

EXAMPLE 5.6 INTRAVENOUS FLUIDS: READING THE DOT GRAPH

Discuss the patterns in the dot graph of the fluids data.

SOLUTION

Groups: The clusters of points show that Abbot's fluids tend to have smaller numbers of impurities, Cutter is next, and then McGaw.

Outliers: There are no outliers.

Spread: The pattern of points in each cluster gives you a picture of how much variability there is. All the values for Cutter are between 200 and 350; the values for Abbot are a bit more spread out; and the values for McGaw are still more spread out.

Transforming: The picture suggests that average and spread may be related: Cutter and Abbot have lower averages than McGaw, and their clusters are less spread out. This pattern doesn't exactly hit you in the face, but it's worth noticing because it suggests that changing the scale of measurement might make the spreads more nearly the same. (See Note 1 at the end of the chapter for more on transforming.) ■

Exercise Set B

1. Bird calcium. For many animals, the body's ability to use calcium depends on the level of certain sex-related hormones in the blood. The following data set looks at the relationship between hormone supplement (present or absent) and level of calcium in the blood. The subjects were 20 birds, 10 female and 10 male. Half the birds of each sex got a hormone supplement; the others served as controls. The response is the level of plasma calcium in mg/100 ml.

Female controls:	16.5, 18.4, 12.7, 14.0, 12.8
Male controls:	14.5, 11.0, 10.8, 14.3, 10.0
Females with hormone supplement:	31.9, 26.2, 21.3, 35.8, 40.2
Males with hormone supplement:	32.0, 23.8, 28.8, 25.0, 29.3

Use the parallel dot diagram in Fig. 5.5 for this exercise.

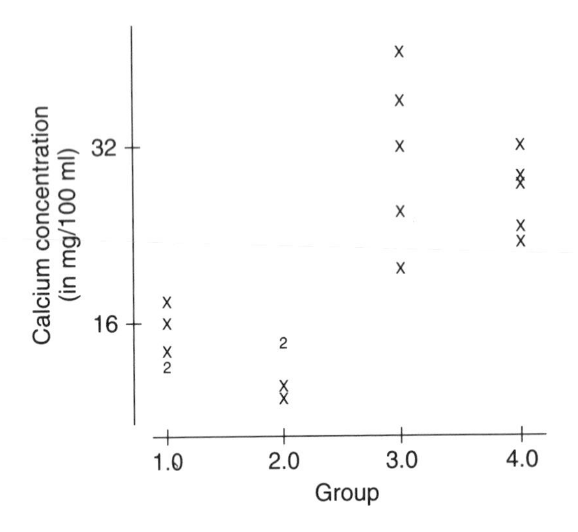

FIGURE 5.5 Parallel dot diagram for the bird calcium data (1 = male, no hormone; 2 = female, no hormone; 3 = male, hormone added; 4 = female, hormone added).

a. Groups: Use the balance point interpretation to estimate the average for each group: turn the graph on its side, locate the approximate balance point, and read off the estimated average from the scale on the y-axis.

b. Outliers: Are there any?

c. Spreads: Find the range (largest − smallest) for each group of observations. (The range is quick to compute, and so we sometimes use it even though the standard deviation—Section 4—is more reliable as a measure of spread.)

d. Transforming: When the spreads of a data set are unequal and the group averages and group spreads seem related, then transforming the data to a new scale is likely to make the analysis more sound. (See Note 1 at the end of the chapter.) The four group averages are 12.12, 14.88, 27.78 and 31.08. Use your estimates from part (a) to decide which average goes with which group. Then list the four groups with their averages and their ranges in a table. Do average and spread seem related?

2. To get a feel for the effect of an outlier, consider the following sets of observations.

a. 1, 2, 3, 4, 5 b. 1, 2, 3, 4, 10 c. 1, 2, 3, 4, 20

Draw a dot diagram for each, compute the average, and mark the balance point on the dot diagram with an arrow.

3–4. *The median*. As Exercise 2 illustrates, the usual average is sensitive to outliers: a single outlier can have a pretty big effect on the value of the average. For this reason statisticians often use the **median** of a set of observations instead of the usual average. The median, like the average, is a one-number summary of the observations, but unlike the average, it is not sensitive to outliers. (There is a difference between casual and official language here. Statisticians call the usual average the mean and use the word "average" for the mean, the median, and some other summary numbers as well. I'll stick to the everyday meaning of "average" in this book.)

The *median* of a set of numbers splits their dot diagram in half: half the observations are to the left of the median and half are to the right (Fig. 5.6).

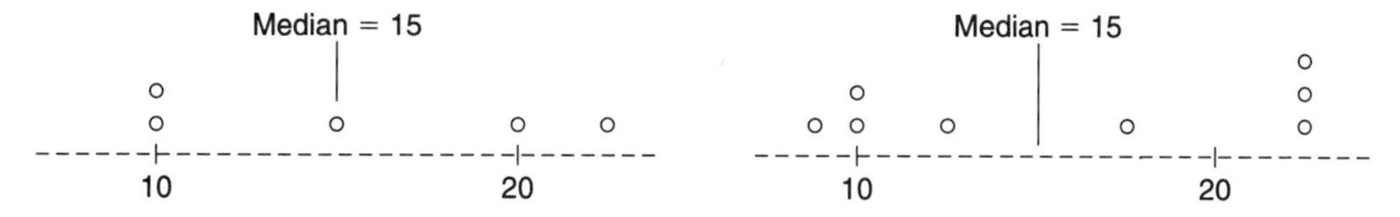

FIGURE 5.6 The median splits a dot diagram down the middle. If you have an odd number of dots, the median will be at the middle dot; if you have an even number of dots, there will be two middle dots and the median will lie halfway between them.

3. Find the median for each set of observed values in Exercise 2. (What is the point of this exercise?)

4. Use the dot graph in Exercise 1 to locate the median for each of the four groups of birds. Which group has its average and median farthest apart? Tell what makes the distribution of dots for that group different from the others.

5–7. *Bimodality: The Milgram compliance study*. Imagine yourself as a subject in one of Stanley Milgram's experiments: After reading Milgram's ad in the New Haven paper, you volunteer to be part of a Yale University study of memory and learning. When you show up for your appointment at the impressive-looking lab on the Yale campus, a stern man in his early 30s wearing a gray lab coat introduces you to another volunteer, a likable-seeming, Irish-looking man in his late 50s. Mr. Lab Coat tells you that you have already earned your fee just for showing up; it's yours regardless of how things go from then on. Then he explains the experiment:

> We know very little about the effect of punishment on learning because almost no truly scientific studies have been made of it in human beings.
>
> For instance, we don't know how much punishment is best for learning—and we don't know how much difference it makes as to who is giving the punishment, whether an adult learns best from a younger or an older person than himself—or many things of that sort.
>
> So in this study we are bringing together a number of adults of different occupations and ages. And we're asking some of them to be teachers and some of them to be learners.
>
> We want to find out just what effect different people have on each other as teachers and learners and also what effect punishment will have on learning in this situation.
>
> Therefore, I'm going to ask one of you to be the teacher here tonight and the other one to be the learner.
>
> Does either of you have a preference?

When you and the other volunteer draw slips of paper from a hat, yours says "Teacher." The experimenter takes you into the next room to watch while he straps the learner into a sort of electric chair, smears electrode paste on his wrist, and attaches the electrode. Then you are taken to another room and seated at the control panel of a machine whose label reads "Shock Generator, Type ZLB, Dyson Instrument Company, Waltham, Mass. Output 15 Volts–450 Volts." The control panel has 30 switches in a row, running from 15 volts on the left to 450 volts on the right. Each switch is labeled with the number of volts, and groups of switches are labeled "Slight Shock," "Moderate Shock," and so on, across to "Danger: Severe Shock" for the next-to-last group, and simply "XXX" for the last group.

Mr. Lab Coat then gives you your instructions: if the learner gives a wrong answer, announce the voltage and give him a shock; start with the lowest shock, and each time he makes a mistake, move up to the next-higher shock.

You begin. Each time you deliver a shock, you see a red light come on above the switch, you hear a buzz, a blue voltage energizer light flashes, and the dial on the voltage meter swings to the right. Before long, the learner has made so many mistakes that your voltages are no longer in the range of slight or even moderate shock. Do you continue? At what point would you refuse to go on?

That question—how far will you go before you refuse to do what you were told?—was the real focus of Milgram's study. The experiment was not about learning and memory: the machine was a fake, there were no shocks, and the learner was in on the secret from the beginning.

You might think you would probably refuse to go on as soon as the shocks got beyond moderate, but take a look at the dot diagram in Fig. 5.7, which gives the results for Milgram's first experiment, with 40 adult male subjects.

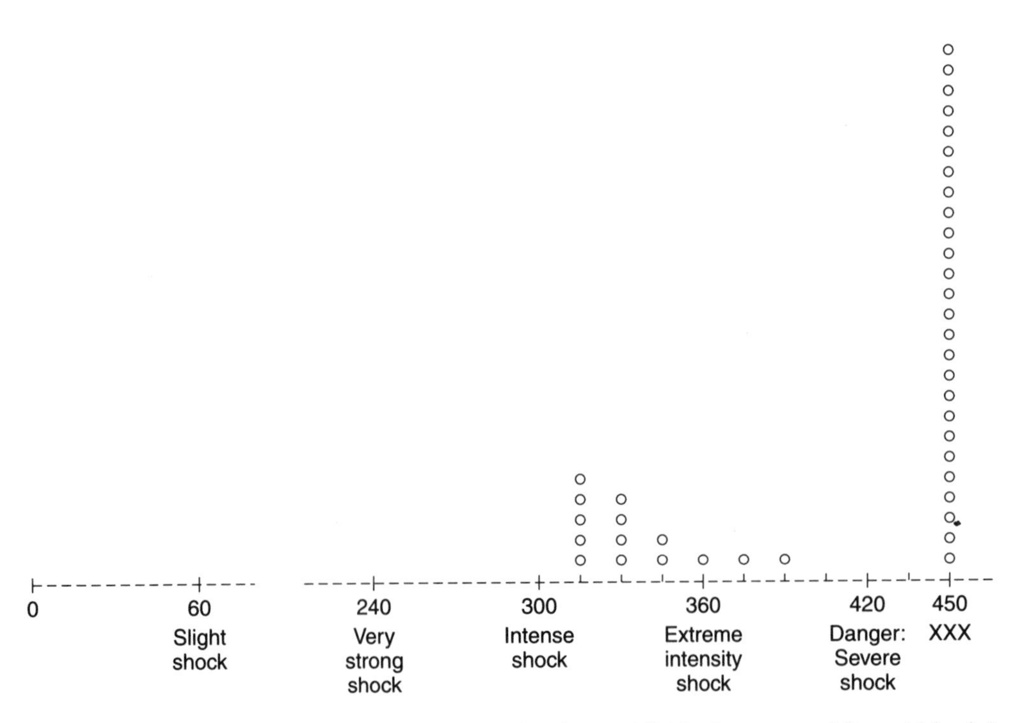

FIGURE 5.7 Highest "shock" delivered, for 40 male subjects. All 40 subjects were delivered "shocks" above 300 volts; 26 of the subjects were delivered the highest possible shock, in the range marked XXX.

5. Notice that the dot graph in Fig. 5.7 has two peaks, with a valley in between. Shapes like this are called **bimodal**. (The values corresponding to the peaks are called *modes*; the value for the highest peak is called *the* mode.) It is usually the case that when a dot graph has two peaks, there are two stories to tell, one for each peak. What are the stories for the left and right peaks in this example?

6. Guess the average by thinking of it as the balance point. Then compute the average, but use this shortcut: instead of adding up all the voltages, which would be a pain, find the average number of steps above 300. (For example, 315 volts is 1 step above 300; 450 volts is 10 steps above 300; and so on.)

7. Notice that there are no dots near the average, which falls in the valley between the two peaks. In what way does this make the average a misleading summary for these numbers?

Warning: analysis of variance is based on averages. If the shape of your data makes averages misleading as summaries, ANOVA will be misleading, too.

8–11. Rating the Milgram compliance study

The results of the Milgram study were profoundly disturbing. Many people were reluctant to acknowledge what the results had demonstrated—that if someone in authority gave the orders, most people would obey instructions to do things that under other circumstances they would regard as highly immoral. Disturbing as it may be to read that *other* people would deliver shocks at the highest level, it was far worse for many of Milgram's subjects, who not only had been deceived, but also had discovered first-hand a rather ugly fact about themselves.

Just as the experiments that compared mammary ligation to a placebo had raised questions about whether it was ethical to give people fake operations, the Milgram study raised questions about what limits ought to be on what you could ask subjects to do. The data below are from a study designed to determine whether it was the *design* of the Milgram experiment or the *results* that people found so objectionable. After all, if no one had gone along when asked to give the shocks, it seems likely that there would have been no controversy.

The subjects for this study were high school teachers, who were randomly divided into three groups. One group of teachers read the actual results of the Milgram experiment, that every one of the subjects complied with instructions and gave shocks at more than 300 volts; the second group got a fake description that said many but not all of the subjects complied; the third group got a fake description that said most of the subjects refused to comply. After reading the report, each teacher rated the ethics of the study on a scale from 1 to 7; the lower the score, the more unethical the study was thought to be.

Here are the results:

I	(actual results; all complied):	6, 1, 7, 2, 7, 1, 7, 3, 4, 1, 1, 1, 6
II	(fake results; many complied):	3, 1, 3, 7, 6, 7, 4, 3, 1, 1, 2, 5, 5
III	(fake results; most refused):	5, 7, 7, 6, 6, 6, 7, 2, 6, 3, 6

8. Construct a parallel dot diagram for the data.

9. Which of the three groups is the most bimodal? Make up a believable explanation. Notice that the bimodality is not nearly as extreme as for the Milgram data.

10. Notice the bumpiness in the dot diagram for group II (an up-down–up-down pattern). Make up a believable explanation.

11. Write a paragraph discussing what you see in the dot graph. Cover the usual four: G O S T, plus anything else that you think is worth mentioning.

3. Factor Structure ("What You Get")

Factors

Both the babies data and the fluids data come from **one-way** basic factorial, or BF[1], designs. Intuitively, you can think of a one-way BF as a design that gives just one set of treatment or condition averages to compare. (In the next chapter, which discusses two-way structure, you'll be able to use the contrast between one-way and two-way to understand the idea better.)

This section presents the **factor structure** of BF[1] designs. The factor structure gives a systematic way to organize both the informal and formal analysis of a data set, gives a way to classify and compare different designs, and gives a useful framework for thinking about how to plan an experiment or observational study. In short, learning to recognize the factor structure of a design is fundamental.

A partition of the observations is a way of sorting them into groups; each observation belongs to a group, and no observation belongs to more than one group.

A factor is a meaningful partition of the observations; a partition is meaningful if it corresponds to a set of conditions built into the design of the experiment.

Turn back to Example 5.3, which gives treatment averages for the walking babies data. Notice that these four averages came from sorting the observations into four groups, one group for each average. The grand average for a data set comes from another way of sorting: all the observations together in a single group. Each of the two ways of sorting is a partition.

Any set of observations will have lots of possible partitions. However, only a few of these will be meaningful in the sense that the observations in each group have some feature in common, the way the walking babies data set divides naturally into four groups, one for each treatment. These meaningful partitions are the factors.

Recognizing factors. The next example illustrates three ways to tell whether a particular partition is a factor.

Three Tests for Factors

1. **Do the observations in each group have some feature in common that is not common to observations in any other group?**
2. **Does it make sense to compute an average for each group and compare averages?**
3. **If you interchange two groups, putting one in place of another, do you change the meaning of the data?**

If the partition is a factor, all three answers should be yes. If not, all three answers should be no.

EXAMPLE 5.7 THE TREATMENT FACTOR FOR THE WALKING BABIES

Turn back to the data set from Example 5.1 so you can picture it in your mind as you read this example. There are lots of ways to partition the observations into groups, but the way I have arranged the numbers in a rectangle makes two partitions stand out: we could divide the data into either four rows or six columns. Dividing into rows gives a factor because each row corresponds to a treatment. All the observations in row 1 have something in common that no other observations have: they all come from children who got the special exercises. Moreover, using rows as groups, it makes sense to compare one group with another, perhaps by computing an average for each row. These averages tell us something meaningful; for instance, that on average the children who got the special exercises (row 1) walked sooner than the children in the exercise/control group (row 2).

Now consider columns. If you go by shape and forget what the numbers stand for, you might think the partition into columns is a factor. After all, the numbers in column 1 do have one thing in common: they all sit in that column. However, the columns have no meaning as far as the experiment is concerned (for example, comparing column averages wouldn't tell you anything meaningful about the experiment), and so the partition into columns is not a factor for the design.

The interchange test for a factor. There's a little test you can do in your head to help decide which partitions are meaningful; it involves switching around groups of numbers in the table and asking whether that changes the meaning of the results. For a factor, the answer will be yes.

First apply the test to rows. Switch the entire first row with the second row: have you changed the message from the data? If you just switched rows and leave the labels alone, the answer is "Yes": the new arrangement would appear to say that the children who got special exercises were slower than the children who got exercise/control. Switching rows has changed the meaning, so the partition into rows is a factor.

Now think about switching the first and second columns: doing this doesn't change the message from the data at all. The partition into columns isn't a factor.

Levels

The groups for a factor are called levels.

Each factor divides the observed values into one or more groups; each group is called a *level* of the factor. For some studies this language is fairly natural: a drug study might have levels of a factor correspond to doses of a drug, for example. For other studies, "level" may seem like a strange word to use, until you get used to it. For example, for the walking babies data "special exercises" is a level of the treatment factor; in the fluids example Abbot, Cutter, and McGaw are levels of the factor for conditions.

Factor Structure. The complete list of factors for a design is called its **factor structure**. Every design has two **universal factors:**

1. The **benchmark** or **grand mean** corresponds to the average of all the observations (there is just one group containing all the observations).
2. The **residual error** corresponds to differences between individual observations (each observation is by itself in a separate group).

Since every design has the same two universal factors, what matters in comparing designs is the set of other, nonuniversal or **structural** factors. For one-way designs there is only one structural factor.

The one-way basic factorial plan (BF[1]) has only one structural factor, which corresponds to treatments or conditions.

EXAMPLE 5.8 STRUCTURAL FACTOR FOR THE INTRAVENOUS FLUIDS

In an abstract sense, the fluids data set has the same factor structure as the walking babies data set, even though this study is observational, whereas the last was a true experiment: factor structure doesn't depend on whether or not conditions can be assigned. There is one surface difference between the two data sets, however, that I want to point out just to make sure you don't get sidetracked by it: for the babies data the conditions of interest go with rows of numbers; for the fluids data they go with columns. (Look back at the data in Example 5.2, and use whatever tests you like to check that for the fluids data the partition of the data into columns is a factor, but the partition into rows is not.) ■

Factor diagrams. In this book I always organize the data from an experiment in a rectangle in a way that lets you picture each partition as a way of carving up the rectangle. If you represent the data as a rectangular box, you can represent each factor by drawing in lines to show how the factor carves up the rectangle. Doing this for each factor gives the **factor diagram**.

EXAMPLE 5.9 FACTOR DIAGRAM FOR THE WALKING BABIES DATA

For this data set, as for any one-way basic factorial, there are three factors; the two universal ones, and one structural factor, which in this case partitions the data into rows (Fig. 5.8).

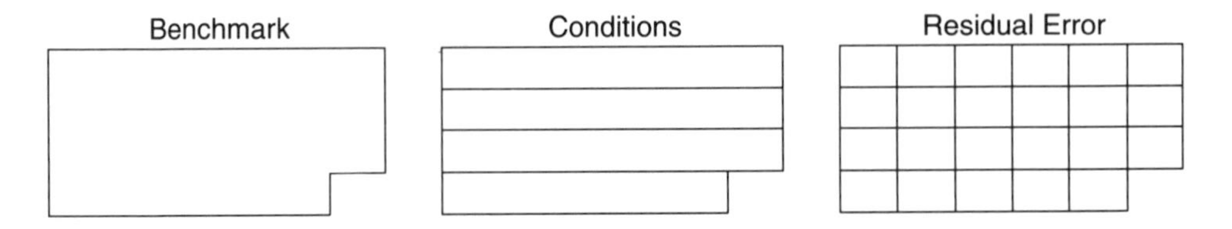

FIGURE 5.8 Factor diagram for the walking babies data.

EXAMPLE 5.10 FACTOR DIAGRAM FOR THE INTRAVENOUS FLUIDS DATA

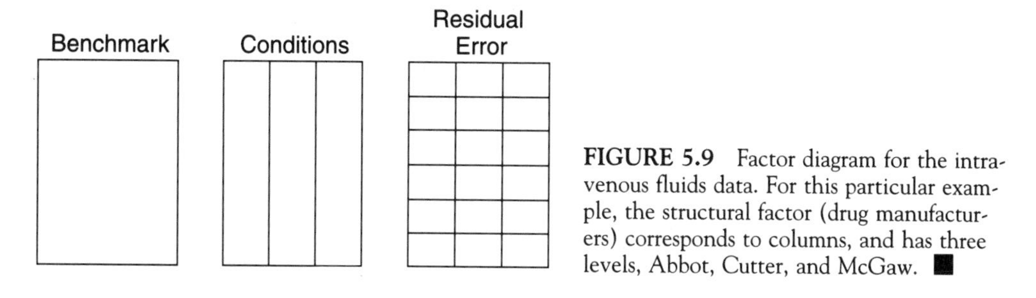

FIGURE 5.9 Factor diagram for the intravenous fluids data. For this particular example, the structural factor (drug manufacturers) corresponds to columns, and has three levels, Abbot, Cutter, and McGaw. ■

Recognizing the factors is sometimes tricky. See, for example, S5.2: Fisher's Iris data.

Exercise Set C

Factor diagrams.

1. Leafhopper survival data. If you have studied organic chemistry, you know that the word *sugar* refers not just to one chemical compound, but to a whole class of similar compounds. Sucrose, glucose, and fructose are three common ones. Glucose and fructose are simple sugars, with six carbon atoms per molecule, and sucrose (ordinary cane sugar) is a twelve-carbon sugar. It is reasonable to suppose that the structure of a sugar molecule has something to do with its food value, and it was exactly that thinking which led to this experiment with potato leafhoppers.

There were four diets to be compared: a control (2% agar), plus control with each of the three kinds of sugar. The experimenter prepared two dishes with each diet, divided his leafhoppers into eight groups of equal size, and randomly assigned them to dishes. The response was the number of days until half the insects in a dish had died: control (2.3, 1.7), control with sucrose (3.6, 4.0), control with glucose (3.0, 2.8), and control with fructose (2.1, 2.3).

Arrange the data in a rectangle and draw the factor diagram.

2. Reread Review Exercise 3, Chapter 4, on Rosenthal's experiment to study the effect of teacher expectations. Imagine you had the data from the experiment; draw and label a rectangular diagram showing conditions as columns; be sure to label the columns. (Your diagram should be similar to the one in Example 5.2 for the intravenous fluids data, except that you won't have numbers in it.) Then draw and label the complete factor diagram.

3. Draw and label a rectangular diagram, with groups as rows, for the chick thyroid experiment in Review Exercise 1, Chapter 4. Then draw and label the complete factor diagram.

Recognizing factors. Reread Exercise B.1 on the bird calcium study, and notice that there are not one, but three structural factors, that is, three meaningful sets of averages related to the conditions: (i) You can compare the four group averages; (ii) You can compare the overall average for the females (rows 1 and 3 of the data) with the overall average for the males (rows 2 and 4); (iii) You can compare the overall average for the control birds (rows 1 and 2) with the overall average for the birds that got the hormone (rows 3 and 4). For each of Exercises 4–6 below, decide whether there is just one structural factor, as in Exercises 2 and 3, or more than one, as for the bird study.

4. Exercise B.8-11: Rating Milgram
5. Example 4.10: Sponge Cells
6. Example 4.14: Morton's Skulls
7. Draw the factor diagram for the data in Example S4.10.
8. Draw the factor diagram for the data in Review Exercises 7–9, Chapter 4. Be careful: there is a trap to avoid. It might help to list factors and count levels before you draw the diagram.

4. Decomposition and Analysis of Variance for One-Way BF Designs

In the overview at the beginning of this chapter, I divided the analysis of a data set into description and inference. The kinds of informal analysis we've done so far in this chapter are part of description: the idea is to find ways to summarize and present patterns in the data without relying on any assumptions about the way residual errors behave. Descriptive methods like dot diagrams, averages, and so on basically say, "Here is a pattern." They take the data at face value, and don't try to generalize beyond what you can see in the numbers you've actually got. The methods of inference are more ambitious.

Usually when you do an experiment, you hope to be able to generalize from your results, to say something about the underlying process that gave rise to the particular set of numbers you happened to get. You hope that your experimental material—your subjects or hamsters or buttercups—will be typical of the much larger set of subjects, hamsters, or buttercups you might have used and that the particular conditions of your experiment represent the general conditions you wanted to study. If your conditions and material are in fact representative of some more general situation, then it makes sense to ask, "To what extent are the patterns in my data typical of the underlying process that created my observations? If I (or someone else) were to repeat the experiment with similar material and conditions, what parts of the new results would be different, and what parts would be the same as before?" The methods of statistical inference try to answer questions of this sort.

These methods depend on a set of assumptions about the process that created your data set. In particular, we assume that residual errors in fact behave in a chance-like way. Unless this and certain other assumptions fit your data, at least approximately, the conclusions you draw using the methods will not be sound. If you want to use the methods in a responsible way, it is essential that you come to understand the assumptions, how they are related to the methods, and how to check whether they fit your data. However, I find that when you are starting out, it is easier to concentrate on just one thing at a time. So this section will deal mainly with what to compute and how to interpret the results. Chapter 12 will deal with the assumptions in more detail.

The rest of this section comes in three parts: decomposition; degrees of freedom; and mean squares, the standard deviation, and the F-test for treatments. I'll illustrate these using the leafhopper experiment of Exercise C.1 as an example. If you haven't already read the description of that experiment, you should do that before going on.

EXAMPLE 5.11 A MODEL FOR THE LEAFHOPPER DATA

Exercise C.1 describes a one-way RBF experiment to compare the nutritional value of a control diet and three kinds of sugar diets (sucrose, glucose, and fructose) for potato leafhoppers. The diets are the four treatments, and the experimental units are batches of leafhoppers. The response is the time in days until half the leafhoppers in a batch died. (See Fig. 5.9).

Diet	Survival Times	
Control	2.3	1.7
Sucrose	4.0	3.6
Glucose	2.9*	2.7*
Fructose	2.1	2.3

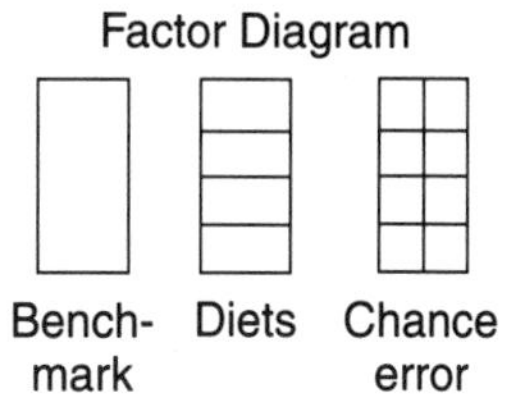

Factor Diagram

Bench-mark Diets Chance error

FIGURE 5.9 Data and factor diagram for leafhopper example.

So far, we've used the factor diagram to record which groups of averages lead to meaningful comparisons. When we decompose a data set, our factor diagram takes on a new meaning as shorthand for a set of assumptions about the data—a **model** for the data.

First, we assume that each observation is a *sum* of three pieces, one piece for each factor, added together:

$$\text{Obs} = \text{Benchmark} + \text{Diet effect} + \text{Chance error}.$$

Second, we assume the pieces that get added together are *constant* in the following sense based on the factor diagram:

1. There is one benchmark value (Fig. 5.9 shows only one group for this factor). Each of the eight observations gets that same one benchmark value.
2. There are four diet effects, one for each diet, just as there are four groups for the diet factor in Fig. 5.9. If the factor for diets puts two observations together in the same group, we assume those two observations get the same diet effect.
3. There are eight chance errors, one for each "group" shown in the chance error factor in Fig. 5.9. ■

Taken together, the two assumptions of **additivity** and **constant effects** allow us to think of the observations as created by a kind of assembly line process. (If you haven't read Chapter 2, or for a quick review, see Note 2 at the end of this chapter.) There are four more assumptions—about the way the chance errors behave—that are also part of our model for the data. (If you haven't read Chapter 2, or for review, see Note 3.)

The benchmark and diet effects are parameters, unknown "true" values we must estimate. The "true" standard deviation for the chance errors is also a parameter. If we subtract our estimated benchmark plus diet effect from each observed value, the leftover pieces will be our estimates of the chance errors, and we can use them to estimate the standard deviation.

Before you read on, look over the data, and think about the following questions:

1. Chance errors. Is the typical size of chance error closest to 0.1, 0.25, 0.5, or 1.0?
2. Diet effects. Which diet is best? Is there a pattern relating survival times and the kind of diet? (Sucrose is the 12-carbon sugar.)
3. How big are the diet differences, in comparison to chance error?

Decomposition of a Balanced One-Way Design

The one-way design has such a simple structure that you can decompose the data "by common sense": the factor structure tells us to split each observed value into a sum of three pieces: an estimate for the benchmark, plus an estimated condition effect, plus an estimated chance error. 1 Since the benchmark is the piece common to all the observations, we estimate it using the grand average. 2 To estimate each condition effect, we ask, "How far is the condition average from the grand average?" 3 To estimate each chance error, we ask, "How far is the observed value from its condition average?"

Decomposition of a Balanced One-Way Design

Estimated benchmark = Grand average.
Estimated condition effect = Condition average − Grand average.
Residual = Observed value − Condition average.

EXAMPLE 5.12 DECOMPOSITION OF THE LEAFHOPPER DATA

1. The benchmark value is common to all eight observations. We estimate it using the average of those eight numbers. For a balanced design, the grand average also equals the average of the condition averages, which in this case works out to 2.7 days.
2. Each estimated diet effect tells us the distance from the diet average to the grand average (see Fig. 5.10).

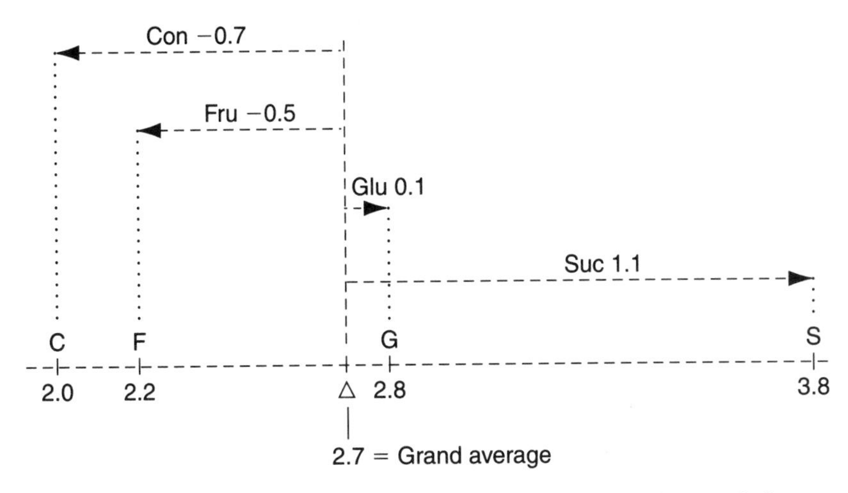

FIGURE 5.10 Diet averages and estimated diet effects. Each estimated effect equals diet average minus grand average. The control diet (no sugar) is worst: leafhoppers survived, on average, 0.7 days less than the estimated benchmark survival time. The more complex (12-carbon) sugar, sucrose, is best: leafhoppers survived 1.1 days longer than the estimated benchmark.

3. You can estimate the chance errors one diet at a time. Suppose we had only the control group, with its two values, 2.3 and 1.7. It would be natural to take the average of these two numbers as an estimate of the piece common to both, and use the distance from each number to the average as an estimate of chance error (Fig. 5.11).

   In the same way, we get estimates of the chance errors for the other three diets: sucrose (±0.2), glucose (±0.1), and fructose (±0.1).

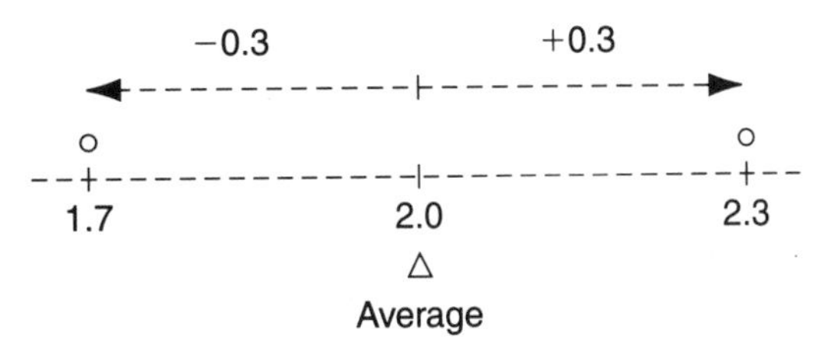

FIGURE 5.11 Deviations from the average estimate of the chance errors. For the control group, the average is 2.0, and the deviations (i.e., the estimated chance errors) are 1.7 − 2.0 = −0.3 and 2.3 − 2.0 = +0.3.

Now that we have all the pieces, we can put them together to get the complete decomposition. For example, the first observed value of 2.3 for the control group equals the grand average (2.7) plus control effect (−0.7) plus estimated chance error (0.3): 2.3 = (2.7) + (−0.7) + (0.3). We can show this for all the observations at once by writing the box of observed values as a sum of three boxes (Fig. 5.12). ■

A general decomposition rule. Although BF[1] designs are simple enough to decompose "by common sense," all other designs are more complicated. It helps to learn to rely on one general approach that works for all balanced designs. This approach simply extends the logic we just used to decompose the leafhopper data: to estimate

	Observed value			Grand average			Diet effect			Residual	
Control	2.3	1.7		2.7	2.7		−0.7	−0.7		0.3	−0.3
Sucrose	4.0	3.6	=	2.7	2.7	+	1.1	1.1	+	0.2	−0.2
Glucose	2.9	2.7		2.7	2.7		0.1	0.1		0.1	−0.1
Fructose	2.1	2.3		2.7	2.7		−0.5	−0.5		−0.1	0.1

FIGURE 5.12 Decomposition of the leafhopper data. The numbers add up according to where they sit in the boxes. For example, picking the numbers in the upper left corners gives 2.3 = 2.7 + (−0.7) + 0.3. In terms of the assembly line metaphor (Note 2), this tells us that the survival time of 2.3 days started out at a benchmark value we estimate to be 2.7 days (the same as all the other observations), then got adjusted downward by 0.7 days (the estimated effect of the control diet), and finally got increased by 0.3 days (the estimated chance error).
Notice the patterns of repetitions and adding to zero within the boxes. The box for the grand average has one number repeated eight times. The box for diets has four numbers, one per row, each appearing twice; these numbers, which are deviations from the grand average, add to zero over the whole box. In the box for residuals, each row contains deviations from the row average, and the numbers add to zero in each row.

the effect for a (meaningful) group, start with the group average. Then subtract any estimated effects that are the same for all the observations in the group. To turn this logic into a rule, we need a way to tell, from a factor diagram, which effects are the same for all the observations in a group.

Inside and outside factors. How can we tell, for example, that for the leafhopper data our model assumes the benchmark is the same for all the observations in the control group? The factor diagram tells us: the group of control observations (first row) fits completely inside the only group for the benchmark factor.

> **Definition**
> **One factor is *inside* another if each group of the first (inside) factor fits completely inside some group of the second (*outside*) factor.**

EXAMPLE 5.13 INSIDE AND OUTSIDE FACTORS FOR THE LEAFHOPPERS

For each of the three factors for the leafhopper data, list all outside factors.

SOLUTION

Benchmark: No outside factors

Diets: Benchmark is outside

Chance error: Both diets and benchmark are outside ■

Using inside and outside factors, we can now state the general decomposition rule:

> **Estimated effect for a factor = Average for the factor − Sum of estimated effects for all outside factors.**

The sum of the estimated effects for the outside factors is called the **partial fit**. (For another description of the rule, in the context of the hamster experiment, see Appendix 1 at the end of Chapter 3.)

EXAMPLE 5.14 DECOMPOSING THE LEAFHOPPER DATA USING THE GENERAL DECOMPOSITION RULE

According to the rule, each estimated effect for a factor equals the average for that factor minus the partial fit (= sum of the estimated effects for all outside factors). Show that this rule gives the same decomposition as in Example 5.12.

SOLUTION

Benchmark. There are no outside factors, so the partial fit equals 0. The estimated benchmark equals the grand average.

Diets. The benchmark is outside, so the partial fit equals the grand average. Each estimated diet effect equals the diet average minus the grand average.

Chance error. Both the benchmark and diets are outside. So,

Partial fit = [Grand average] + [Diet effect]

= [Grand average] + [Diet average − Grand average]

= Diet average.

Thus, each residual (estimated chance error) = observed value − diet average. ■

Degrees of Freedom for a Balanced BF[1] Design

Now that we've split the data into pieces corresponding to factors, we want to compare (1) the size of the differences due to diets with (2) the size of the differences due to chance error. It won't work, however, just to ask which box shows the bigger differences. The differences in the diet box are based on averages, and the chance part of an average tends to show less variation than the chance part of an observed value. To make a valid comparison, we first need to know "How did the chance errors in the original eight observations get distributed when we split them into pieces?" To answer the question, we think in terms of units of information. Each observed value provides one unit of information about chance error. When you decompose a data set into grand average plus treatment effects plus residuals, you redistribute the units of information, so many to each box of the decomposition. Each number in a box contributes one unit *unless the patterns of repetition and adding to zero tell us what that number has to be.*

Counting free numbers

Definition

The degrees of freedom (df) for a table equals the number of free numbers, the number of slots in the table you can fill in before the patterns of repetition and adding to zero tell you what the remaining numbers have to be. The df tells the number of units of information about chance error in that box of the decomposition.

The patterns of repetition are easy: within each level (group) of a factor, all the numbers in the box for that factor are the same. The patterns of adding to zero come from the averages you subtract to get the estimated effects: whenever you subtract the average from a set of numbers, the deviations (leftovers) add to zero.

EXAMPLE 5.15 DEGREES OF FREEDOM FOR THE LEAFHOPPER DATA: COUNTING FREE NUMBERS

For the leafhoppers, the general rule will give $df_{Grand} = 1$, $df_{Diets} = 4 - 1 = 3$, and $df_{Res} = 8 - (1 + 3) = 4$. Show that you get the same df if you count free numbers directly.

SOLUTION

1	R
R	R
R	R
R	R

Grand: df = 1. This factor has only one group, so the box for the grand average contains one number repeated eight times, as you can see in the decomposition in Example 5.12. (There are no outside factors, no averages to subtract, and no patterns of adding to zero.) There is one free number: one unit of information about chance error. The box at the right shows all this schematically: the 1 stands for the 1 free number, and the Rs stand for repetitions.

1	R
2	R
3	R
+	R

Diets: df = 3. Repetitions: The factor has four groups (the four rows); the repetitions, which you can see in the decomposition, are shown with Rs in the box to the right. Adding to zero: Each diet effect equals the diet average minus the grand average, so the four diet effects add to zero. As soon as you know any three, you can figure out the fourth. Thus, there are three free numbers in the box for diets. The box above uses a 1, 2, and 3 to show the free numbers and a + to show the pattern of adding to zero. Note that the R in the lower right corner could also have been a +, since the column adds to zero. (Whenever a number is both repeated (R) and forced (+) by the pattern of adding to zero, I put an R, but either one will work.)

1	+
2	+
3	+
4	+

Res: df = 4. Repetitions: None, because each number is in a group by itself. Adding to zero: Each residual is a deviation from a diet average (= row average), so the residuals add to zero separately within each row, as you can see in the decomposition. Thus, in each of the four rows, one number is free and the other is fixed by the pattern of adding to zero. There are four free numbers in all. The numbers and plus signs in the box above show these patterns. ■

A general rule for degrees of freedom, and a shortcut summary for the BF[1] design. The general rule for decomposing a data set leads to a parallel rule for degrees of freedom.

df for a factor	=	Number of levels for the factor	−	Sum of df for all outside factors.

To find the degrees of freedom, start with the number of levels, which tells the largest possible number of different numbers in the box of decomposition. (For each level, the estimated effect is the same for all observations in that group, so starting with the number of levels takes care of the repetitions.) Now we have to adjust down to account for any patterns of adding to zero. Each time you subtract the effect of an outside factor, you create patterns of adding to zero, and (I'm skipping a bit here) the df for the outside factor tells how many patterns of adding to zero you need to adjust for.

The general rule leads directly to the following summary for the one-way design:

For a balanced one-way design:

$\mathrm{df}_{\text{Grand average}} = 1;$
$\mathrm{df}_{\text{Conditions}} = (\#\text{ conditions} - 1);$
$\mathrm{df}_{\text{Residual}} = (\#\text{ observations} - \#\text{ conditions}).$

Although the rule is quick and easy, I urge you to learn to count free numbers as well, because the rule is just a way to get the answer without thinking; counting free numbers brings you closer to the logic of analysis of variance.

Mean Squares, the Standard Deviation, and the *F*-test for Treatments

Now that we've counted units of information, we can use them to compute a measure of average variability for diet effects and for residuals, then compare the sizes. We start by squaring the numbers in each box of the decomposition and adding, to get the **sum of squares (SS)** for each box.

The sum of squares (SS) for a set of deviations measures their overall variability: If the deviations are large (spread out), their SS will be large; if the deviations are small (close together), their SS will be small.

For balanced designs it turns out that just as the boxes of the decomposition add to give the original observations, and the df for the boxes add to give the total number of observations, the SSs for the boxes will also add to give the SS for the observations.

The decomposition splits three things in parallel:

- **The pieces of the observed values add up (linear decomposition).**
- **The degrees of freedom add up.**
- **The sums of squares add up (quadratic decomposition).**

Mean squares and F-ratios. Since the SS for a box in the decomposition measures overall variability in the box, and the df counts units of information about chance error, the ratio SS/df, called the **mean square (MS)**, tells the *variability per unit of information*, a kind of average variability for that box of the decomposition. Thus, the mean square for diets, MS_{Diets} tells the average variability among the diet effects, and the residual mean square, MS_{Res}, tells the average variability among the residuals.

Definition

The mean square for a set of deviations equals their sum of squares divided by the degrees of freedom: MS = SS/df. The MS measures a kind of average variability per unit of information.

We summarize all this information in an **analysis of variance table**, with one row for each box of our decomposition, and columns for SS, df, and MS. To compare the size of the diet effects with the size of the chance errors, we compute the ratio $F = MS_{Diets}/MS_{Res}$, which equals 17.42 for the leafhoppers: the average variability in the diet effects is more than 17 times as big as the average variability in the residuals.

The F-ratio for the conditions of interest in a basic factorial design equals the mean square for conditions divided by the residual mean square:

$$F = MS_{Cond}/MS_{Res}.$$

It tells us that the average variability in the condition effects is *F* times as big as the average variability in the residuals.

EXAMPLE 5.16 ANOVA TABLE FOR THE LEAFHOPPERS

Here are the sums of squares, mean squares, estimated SD, and *F*-ratio for the leafhoppers.

Source	SS	df	MS	*F*-ratio
Grand average	58.32	1		
Diets	3.92	3	1.307	17.42
Residual	0.30	4	0.075	
Total	62.54	8		

■

The F-test, significance levels, and critical values. To test whether the observed diet effects are too big to be due just to chance error, we need to know how likely it is that we'd get an observed ratio as large as 17.42 (or larger) *if the true diet effects were zero.*

Definition

The observed significance level (or *p*-value) for an F-ratio is the probability of getting a value that high or higher just by chance if in fact there were no differences in the underlying true values.

Today computer packages that do ANOVA typically compute *p*-values for you. The package I used (Section 5) does this and found a *p*-value of 0.009 for the leafhoppers: the chance is only 9 in 1000 (less than 1%) that we'd get an *F*-ratio as big as

17.42 if in fact there were no true diet differences. We can safely conclude that the observed differences are "real," that is, too big to be just chance error.

If you don't use a computer, you can rely instead on tables of **critical values** for the F-ratios. If the F-ratio in your ANOVA table is bigger than the .05 critical value, that means the p-value is less than .05. If your F-ratio is also bigger than the .01 critical value, that means your p-value is less than .01. As a rule, scientists regard a set of differences as "real" if the p-value is less than .05, that is, if the F-ratio is bigger than the .05 critical value. (By convention, the differences and the F-ratio are called **significant**, but in this specialized sense, the word means "detectable," which is not at all the same as "important.") To find a .05 critical value, turn to the tables at the back of your book, and find the one headed "5% F-values." The rows of the table correspond to the df for the denominator MS of the F-ratio you want to test; the columns correspond to the df for the numerator MS.

Notation

$F_{.05;3,4}$ means the 5% critical value with numerator df = 3, denominator df = 4.

To test diets for the leafhopper data, find row 4 ($df_{Res} = 4$) and column 3 ($df_{Diets} = 3$). The number in row 4, column 3 is 6.59, the 5% critical value for this test. Since our F-ratio of 17.42 is bigger than 6.59, we know the p-value is less than .05, and we conclude that the differences are "real."

The estimated standard deviation. We'd like to use our decomposition to measure the typical size of the chance errors for the leafhopper experiment. Since the mean square for residuals measures their average variability, I hope you find it reasonable to work with MS_{Res}. In fact, we just need to make one adjustment: since MS_{Res} is an average of the *squared* residuals, we take a square root to get back to the same scale as the observations themselves.

Definition

The (estimated) standard deviation **for a data set equals the** root mean square of the residuals:

$$SD = \sqrt{(MS_{Res})}.$$

The SD tells the typical size of a chance error.

The estimated SD $= \sqrt{MS_{Res}} = \sqrt{0.075} \approx 0.274$: each observed value equals a true value, give or take about .27 days, or roughly 6 hours. If the chance errors follow a normal curve (see Note 3), then roughly two-thirds of the observed values will be within 6 hours of their corresponding true values, and roughly 95% will be within 12 hours.

If we were to continue the formal analysis of the leafhopper data, it would be natural at this point to construct confidence intervals to estimate the likely size of the true diet differences. Chapter 11 shows how to construct these intervals.

Exercise Set D

Many of these exercises use fake data in order to keep the arithmetic simple while you learn the mechanics.

Decomposition

1. Decompose each of the following into a box containing the average plus a box of deviations from the average (leftovers).

a. 10 20 b. 5 5 20

2. A small version of the walking babies data. (To make the arithmetic simple, I made the babies learn fast.) Decompose the little CR into a box for the grand average plus a box of estimated condition effects plus a box of residuals.

Exercise	2	6
Control 1	3	5
Control 2	4	4

3–10. *Patterns of repetitions and adding to zero: Counting degrees of freedom.* Fill in the missing numbers in the decompositions below. Then write the df under each box.

3.

Avg.		Res.
10		2
?	+	−1
?		?

4.

Avg.		Res.
?		−3
4	+	?
?		2

5.

Cond. avg.			Res.	
3	5		1	−1
?	?	+	0	2
?	?		?	?

6.

Cond. avg.				Res.		
1	6	3		2	?	−2
?	?	?	+	−1	3	?
?	?	?		?	4	3

7.

Cond. avg.			Res.	
4	?		−1	2
?	?	+	−2	−1
?	6		?	0
?	?		4	?

8.

Gr. avg.			Cond. eff.			Res.	
10	?	+	2	?	+	1	?
?	?		?	?		?	−2

9.

Gr. avg.			Cond. eff.			Res.	
10	?		3	?		2	0
?	?	+	?	?	+	−1	?
?	?		?	?		?	3

10.

Gr. avg.				Cond. avg.				Res.		
?	?	?		?	?	?		?	1	?
?	7	?	+	?	−3	?	+	3	?	−5
?	?	?		4	?	?		−1	2	1

11–15. *Standard deviation.*

11. *Effect of an outlier.* The batch of observed values whose decomposition appears below contains an outlier.

Observations		Average		Residuals
20 8 5 9 6 12	=	10 10 10 10 10 10	+	10 −2 −5 −1 −4 2

a. The typical size of chance error is closest to ______________. (Choose from 1, 3, 5, or 9.)

b. Omit the outlier, and decompose the remaining set of numbers.

c. Without the outlier, the typical size of chance error is closest to __________. (Choose from 1, 3, 5, or 9.)

d. Which of (i) or (ii) below more accurately completes the following sentence?

For this data set, omitting the outlier

i. leaves the estimated size of chance error about the same.

ii. makes the estimated size of chance error roughly half as big as before.

e. Now compute $SD = \sqrt{MS_{Res}}$, first for the original set of 6 numbers, and then for the set of 5 you decomposed in (b).

12. *Bird calcium: Effect of the "wrong" scale*. Exercise B.1 described a study of the effect of hormone supplements on the level of plasma calcium in male and female birds. (Look back to the parallel dot graph in that exercise, which showed that larger spreads went with larger averages, suggesting that the data should be transformed.)

Here is a decomposition of the (untransformed) concentrations into condition averages plus residuals. Look over the table of residuals, and guess a typical size of chance error separately for each of the four conditions; choose from 1/2, 2, 3, 7, 10 and 13. Then use your estimates to estimate SD_{Max}/SD_{Min}. As a rough rule, you should look for a transformation if this ratio is bigger than 3 and the group averages and SDs are related. Should these data be transformed? (The two observations marked * were each lowered by 0.1 to make the arithmetic less messy.)

Control		Hormone			Condition Averages					Residuals			
F	M	F	M										
16.5	14.5	31.9	32.0		14.88	12.12	31.06	27.78		1.62	2.38	0.84	4.22
18.4	11.0	26.2	23.8		14.88	12.12	31.06	27.78		3.52	−1.12	−4.86	−3.98
12.7	10.8	21.3	28.8	=	14.88	12.12	31.06	27.78	+	−2.18	−1.32	−9.76	1.02
14.0	14.3	35.7*	25.0		14.88	12.12	31.06	27.78		−0.88	2.18	4.64	−2.78
12.8	10.0	40.1*	29.3		14.88	12.12	31.06	27.78		−2.08	−2.12	9.04	1.52

13. *Fake data: The effect of adding constants or multiplying by constants*. Decompose each of the following fake data sets into average plus residuals. Then compute sums of squares and count degrees of freedom. Write your SSs and dfs under the boxes of your decomposition. Then compute $SD = \sqrt{MS_{Res}}$. For these fake data sets, all the arithmetic should involve only whole numbers. Parts (a)–(e) are related to each other and to the next two questions: look for patterns as you do the arithmetic.

a.	b.	c.	d.	e.
2	12	20	102	120
−2	8	−20	98	80

14. Mark the following statements true or false. In trying to decide, you might refer to your numerical results from Exercise 13, and you might rely also on any intuitive ideas you have about what ought to be true of a number that measures how spread out the data are.

a. If you add 100 to every number in a data set, the SD goes up by 100.

b. If you add 100 to the largest number in a data set and leave the rest of the data unchanged, the SD goes up, but by less than 100.

c. If you add the same number c to every number in a data set, you don't change how spread out the numbers are, and the SD stays the same.

d. If you multiply every number in a data set by 10, the new SD will be 10 times as big as the old one.

e. If you multiply every number in a data set by 10, the new SD will equal 10 plus the old SD.

f. If you multiply every number in a data set by -10, the new SD will be (-10) times the old SD.

15. Fill in the blanks.

a. If you multiply every number in a data set by the same number c, the new SD will equal the old SD times ________.

b. If the SD for a data set equals 5, and you multiply every number in the data set by 3, and then add 10 to every number, the new SD will equal ________.

c. If the SD for a data set equals 5, and you add 10 to every number in the data set, and then multiply every number by 3, the new SD will equal ________.

Real data sets: Decomposition and ANOVA

16–20. *Intravenous fluids*

16. Decompose the data of Example 5.2 into condition averages plus residuals. (The arithmetic should involve only whole numbers.)

17. Estimate the typical size of chance error for each group; choose your estimates from 20, 50, 80, 110, 140, and 200. Do your estimates suggest that a transformation might be worthwhile?

18. Compute a separate SD for each group: get SS_{Res} for the group, divide by $5 = df_{Res}$, and take the square root. Then compute the ratio of largest to smallest SD. Do you need to transform?

19. Complete the decomposition of these data by splitting each condition average into the grand average plus a condition effect. Then fill in the blanks in the following sentence three times, once for each manufacturer: For fluids manufactured by ______ laboratory, the number of contaminant particles is on average ______ (higher/lower) than the grand average by ______ particles per unit of fluid.

20. Give the ANOVA table, and state your conclusion based on the F-ratio.

21–23. *Walking babies*

21. Here is the ANOVA table for the data of Example 5.1. Fill in any missing numbers, and state your conclusion: is the effect of the conditions big enough to be detected?

Source	*SS*	*df*	*MS*	F	*Critical value*
Grand Average			2950.45		
Conditions	14.45				
Residual		19	2.321		
Total	3009.00				

22. The residuals from the four condition averages are given below. Use these to compute a separate SD for each group, and find SD_{Max}/SD_{Min}. Should these data be transformed?

Special exercise	−1.13	−0.63	−0.38	−0.13	2.88	−0.63
Exercise control	−0.38	−1.38	−1.38	0.38	−0.88	3.63
Weekly report	−0.63	0.38	−2.63	−0.13	1.63	1.38
Single report	0.90	−0.85	−0.35	1.15	−0.85	

23. Draw a dot diagram, using solid circles for the residuals that correspond to the outliers and using open circles for the others. Then mark off SDs underneath your graph, and check the rule based on the normal curve, as in Example 4.9. What percent of the residuals are within 1 SD of 0? Within 2 SDs?

24. *Mothers of schizophrenics*. (See Review Exercises 7–9, Chapter 4.)

The data set below gives (the number of type A stories minus the number of type C) stories for the 20 mothers of schizophrenics (first row) and the 20 control mothers (second row).

The total sum of squares is 347. Complete an ANOVA table, and state your conclusions. (Hint: You don't have to do a complete decomposition. Find the grand average and the two condition effects; then use these to get the corresponding SSs; then subtract from the total SS to get the residual SS.)

																					Avg.
Schizo.	−2	−1	−3	2	2	7	0	0	3	−2	−4	−1	3	1	0	2	−3	−2	−2	0	0.00
Normal	7	4	5	2	−1	2	2	6	3	0	0	−2	−2	3	1	1	2	6	2	2	2.15

5. Using a Computer [Optional]

Computers have become so important in statistical work that they can no longer be considered optional. Thus the "Optional" in the section heading refers not to the computer, but to this section of the book. Some of what is contained in this section and the others like it in later chapters is true of pretty much any computer package for statistical analysis, but if you are using software other than Minitab, which I've chosen for illustration here, you may prefer to skip the sections dealing with the computer.

If you're new to computers. Before you can use Minitab to analyze data sets, you first need a general introduction to your computer: how to use the keyboard and mouse, how to use menus and dialog boxes, how to save and print results, and so on. You'll also need a basic introduction to Minitab, which you can get from the beginning part of the *Minitab Reference Manual*, the *Minitab Handbook*, or *The Student Edition of Minitab*.

Entering data from the keyboard. When you first start Minitab, the bottom half of your screen should look like an empty spreadsheet, a table with numbered rows and columns, with nothing in the body of the table. (If your screen shows something else, click on the word Window at the top of the screen to open the Window menu, and holding down the mouse button, move it down to 2. Data, then release.) To enter data, you use the mouse to move the cursor to a location in the table, click, and type a number. To enter a new number in the cell just below in the same column, use Enter or the down arrow key or the mouse. To move right, use the right arrow key or the mouse.

Data in Minitab. Minitab is column-oriented. This means that to analyze a data set using Minitab, all your response values must be in a single column. For a BF[1] design you must also have a second column whose numbers tell the levels of the factor of interest. The right half of Figure 5.13 shows a Minitab listing of the Leafhopper Survival data from Example 5.11. (The more recent versions of Minitab allow you to use names for factor levels, so instead of 1, 2, 3 and 4, you could type Control, Sucrose, Glucose and Fructose.)

As shown in Example 5.11

Diet	Survival Times	
Control	2.3	1.7
Sucrose	4.0	3.6
Glucose	2.9	2.7
Fructose	2.1	2.3

Minitab Column Format

Row	Survival	Diet
1	2.3	1
2	4.0	2
3	2.9	3
4	2.1	4
5	1.7	1
6	3.6	2
7	2.7	3
8	2.3	4

FIGURE 5.13 Leafhopper survival data. On the left are four rows of two observations each, one row per diet. On the right, the same data have been rearranged in the form required by Minitab. The two side-by-side columns of response values on the left ("Survival Times") have been stacked on top of the other to form a single column of response values on the right ("Survival"). Another column ("Diet") tells the level of the treatment factor, with 1 = control, 2 = sucrose, and so on.

Naming variables. Minitab lets you refer to a variable either by column number or by name. For example, in Figure 5.13 the survival times are in column 1, so c1 refers to the response. The second column, c2, contains the levels of the structural factor. Instead of using column numbers, you can give each variable a name (up to eight letters long) and refer to variables by name. The data window shows each variable as a column, with a cell for the column name at the very top. Move the cursor to that cell, click, and type in the variable name. Using names has two advantages: (1) It may be easier to remember the name "DIET" than to remember the number of the column that holds the structural factor. (2) Your output will be easier to read, because Minitab will use names instead of numbers to label graphs and tables.

Using stored data. Instead of typing in data from the keyboard, you can use data stored on disk. First open the `File` menu: move the cursor to the word `File` in the upper left corner of the screen, press the mouse button, and hold it down. Now select `Open Worksheet`: move the cursor down until the words `Open Worksheet` are highlighted, then release the mouse button. These two steps are abbreviated as follows:

```
File > Open Worksheet ...
```

The three dots after "Open Worksheet" are Minitab's way to let you know that you will have to fill in a **dialog box**, that is, you will have to give the computer additional information by filling in your choices as requested in the next screen. To load data stored on a disk, you'll need to identify the disk drive and also the name of the Minitab worksheet where the data set is stored.

Group summaries and parallel plots. From the `Stat` menu, choose `Basic Statistics`, and from that menu choose `Descriptive Statistics`:

```
Stat > Basic Statistics > Descriptive Statistics ...
```

The resulting dialog box will ask you to supply the name of the variable you want summarized, and will give you the option of asking for separate summaries for each level of the structural factor. Here's how you would get group summaries (one set for each diet) and a parallel dot plot of survival times for the Leafhopper data. First make sure the insertion point—the vertical bar—is in the box labeled `Variables`. Move the mouse to the list of variables at the left of the screen until the arrow is over the variable you want (`Survival` in our case), and either click twice, or click once and then click `Select`. Now move the arrow to the little square to the left of the words `By Variable`, and click: an X should appear in the box. Move the arrow to the larger box to the right of the words, click to position the insertion point, and go back to the list of variables on the left to choose the structural factor (Diets). We can abbreviate all this as follows:

```
Variables:     Survival
By variable:   Diet
```

To request a parallel dot plot, click on the button labeled Graphs at the bottom right of the dialog box. A new dialog box appears, and gives you the chance to select any of a number of plots. Choose whatever you'd like to see, then click OK. That takes you back to the previous dialog box. Click OK once again, and Minitab will show you group means and standard deviations along with various other summaries, followed by whichever plots you chose.

ANOVA using Minitab's `Oneway` command. Your data must be in the two-column format shown in the bottom part of Figure 5-13, with one column for response values and the other for levels of the factor of interest.

From the `Stat` menu choose `ANOVA`, and from the `ANOVA` menu choose `Oneway`:

```
Stat > ANOVA > Oneway ...
```

Fill in the dialog box with your response and structural factor:

```
Response: Survival
Factor:   Diet
```

Clicking on the `Graphs` button brings up a dialog box that gives you a different way to get a parallel dot plot. There are other choices for plots as well, useful for checking the Fisher assumptions. (See Chapters 2 and 12.) To request these you will have to go back to the previous dialog box and click the two little squares for `Store residuals` and `Store fits`. Then click the `Graphs` button again to choose your graphs.

Reading Minitab's ANOVA Table. Minitab's printout comes in two parts, an ANOVA table followed by a set of treatment averages. First compare Minitab's ANOVA table with the one in Example 5.16, and notice the following features:

1. Minitab's table does not include a row for the grand average.
2. Minitab's rows for Diet and Error are the same as the corresponding rows for Diets and Residual in Example 5.16.
3. Minitab's row for Total tells what you get by adding the two rows for Diet and Error. In Example 5.16 the row for the grand average also gets included in the totals.

Analysis of Variance on Survival

Source	DF	SS	MS	F	p
Diet	3	3.9200	1.3067	17.42	0.009
Error	4	0.3000	0.0750		
Total	7	4.2200			

The differences boil down to what gets done with the grand average. Giving it a row in the table and counting its SS and df in the totals as in Example 5.16 emphasizes that ANOVA splits apart the observations themselves. The total df equals the number of observed values, and the total SS comes from squaring and adding those values. The alternative approach, used by Minitab and most other computer packages, emphasizes *variability*. Because there is no variability in the decomposition table for the grand average, these computer packages regard the grand average as irrelevant. They work with the numbers you get by subtracting the grand average from the observed values. Their total SS comes from squaring and adding those numbers and equals the SS for the observations minus the SS for the grand average. Their total df is the number of observed values − 1.

Other output from **`Oneway`.** Minitab does not show the linear decomposition of a data set. Instead, for a BF[1] design, you automatically get the average for each treatment group, along with the group SD, and pictures showing confidence intervals for the underlying true values for each group. (For now, you may want to ignore confidence intervals, which this book treats in Chapter 11.) Minitab also prints out the overall SD for the experiment ('POOLED STDEV'), computed, in the usual way, as the root mean square for residuals.

```
                                   Individual 95% CIs for Mean
                                   Based on Pooled STDEV
Level   N     Mean     STDEV   --+---------+---------+---------+----
  1     2   2.0000    0.4243   (------*------)
  2     2   3.8000    0.2828                            (------*------)
  3     2   2.8000    0.1414                 (------*------)
  4     2   2.2000    0.1414        (------*------)
    Pooled                     --+---------+---------+---------+----
    STDEV = 0.2739             1.60      2.40      3.20      4.00
```

6. Algebraic Notation for Factor Structure [Optional]

The purpose of this optional section, and others like it in the chapters that follow, is to show you the way statisticians usually write out the factor structure of an experimental design. Although you don't need to learn the notation to read this book, you will need it if you want to read almost any other book on analysis of variance. Moreover, the notation is particularly useful if you want to understand the relationship between underlying true values (parameters) and their estimates from the data.

Notation for the Model

The basic idea comes from the assembly line metaphor: each observed value is written as a sum of pieces, one piece for each factor. Statisticians use letters to stand for the pieces, one letter for each factor.

For a one-way design there are three factors: benchmark, treatments (or conditions), and residual error. Statisticians often use μ for the benchmark piece, α for the condition effect, and e for the error. They usually use y to stand for the observed value.

I'll use as an example a data set with four rows and two columns, with conditions for rows, just like the leafhopper data of Example 5.11.

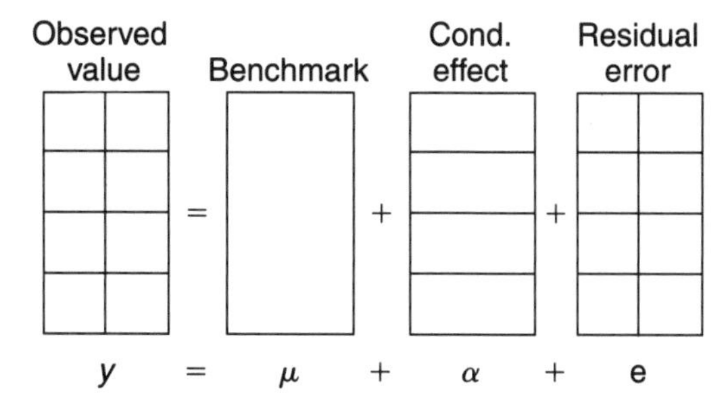

All three symbols to the right of the equals sign stand for unknown numbers. The Greek letters μ and α stand for parameters. The true standard deviation for the errors is also a parameter, usually represented by the Greek letter σ.

As it stands, the notation tells that each observed value is the sum of three pieces, but it doesn't yet tell the number of choices at each work station or even the number of observed values. To show these things, statisticians add subscripts.

Here's how they'd write the set of observed values:

y_{11}	y_{12}
y_{21}	y_{22}
y_{31}	y_{32}
y_{41}	y_{42}

Typical observed value: y_{ij}

Each y has two subscripts. The first tells the row number, and the second tells the column number: y_{21} stands for the observed value in row 2, column 1; the observed value in row 3, column 2 is written y_{32}.

For the benchmark, everybody gets the same amount; there's only one choice, μ, without subscripts:

μ

For condition effects, there are four choices, written α_1, α_2, and so on.

α_1
α_2
α_3
α_4

Typical condition effect: α_i

According to the factor diagram, the factor for conditions has four levels, one for each row. The subscript for each α tells the row number. So α_1 is the piece added on for the observations y_{11} and y_{12}, which sit in the first row. In general, the subscript for α matches the first subscript for the ys in the same row. We assume that the true treatment effects α_i are deviations from the true benchmark μ, and so we assume the α_i sum to zero. (See Exercise E.6.)

There's one residual error for each observation.

e_{11}	e_{12}
e_{21}	e_{22}
e_{31}	e_{32}
e_{41}	e_{42}

Typical residual error: e_{ij}

The subscripts for the errors follow the same pattern as for the ys: the first subscript gives the row and the second subscript gives the column.

Putting this all together gives the following:

$$\begin{array}{c}\text{Observed value}\\ \begin{array}{|c|c|}\hline y_{11} & y_{12}\\ \hline y_{21} & y_{22}\\ \hline y_{31} & y_{32}\\ \hline y_{41} & y_{42}\\ \hline\end{array}\end{array} = \begin{array}{c}\text{Benchmark}\\ \begin{array}{|c|}\hline \\ \\ \mu \\ \\ \\ \hline\end{array}\end{array} + \begin{array}{c}\text{Cond. effect}\\ \begin{array}{|c|}\hline \alpha_1\\ \hline \alpha_2\\ \hline \alpha_3\\ \hline \alpha_4\\ \hline\end{array}\end{array} + \begin{array}{c}\text{Residual error}\\ \begin{array}{|c|c|}\hline e_{11} & e_{12}\\ \hline e_{21} & e_{22}\\ \hline e_{31} & e_{32}\\ \hline e_{41} & e_{42}\\ \hline\end{array}\end{array}$$

Notice that the way the subscripts go together tells how the pieces are added together:

Row subscripts match

$$\overset{i}{y}_{1\underset{j}{1}} = \mu + \overset{i}{\alpha}_1 + \overset{i}{e}_{1\underset{j}{1}} \qquad \overset{i}{y}_{1\underset{j}{2}} = \mu + \overset{i}{\alpha}_1 + \overset{i}{e}_{1\underset{j}{2}}$$

Column subscripts match

$$y_{21} = \mu + \alpha_2 + e_{21} \qquad y_{22} = \mu + \alpha_2 + e_{22}$$

Finally, to get the statistician's shorthand version, we just write the equation for one typical observed value:

$$y_{ij} = \mu + \alpha_i + e_{ij},$$

where the α_i sum to zero.

Using this notation, we can rewrite the six Fisher assumptions, which are also part of the model:

A: *Additivity*. This assumption is already built into the notation—it shows up in the plus signs to the right of the equals sign.

C: *Constant effects*. This assumption is also built in, by way of the subscripts. The μ has no subscripts, which indicates that it is the same regardless of i and j. Each α_i has only an i subscript, to record that the treatment effect depends only on the condition i but is the same regardless of j.

Z, S, I, N: *The assumptions about the residual errors*. We assume the e_{ij} behave like independent draws from the same box of numbered tickets (same standard deviation), with the average of box equal to zero, and that a dot graph for the tickets follows a normal curve.

Notation for the Estimates

We can extend the y notation to describe averages and estimated effects. This extended notation is based on two conventions: a y with a bar over it ($\bar{y}$) stands for an average of observed values, and a dot in place of a subscript tells to compute the average by adding up all the observations corresponding to the missing subscript. For example, the grand average is abbreviated as

$$\bar{y}_{..} = \text{average of all the observations.}$$

1. The bar tells you it's an average.
2. The y tells you it's an average of observed values.
3. Both subscripts are missing, so you add up *all* the ys to get the average.

The row averages are written $\bar{y}_{1.}$ (first row average), $\bar{y}_{2.}$ (second row average), and so forth:

$$\bar{y}_{1.} = \text{average of the observations in row 1.}$$

1. The $\bar{y}$ part tells you it's an average of observed values.
2. The first subscript is a 1: this tells you the average is for row 1.
3. The second subscript is missing: this tells you to get the average by adding all the observations corresponding to that subscript: $y_{11} + y_{12}$.

There's quite a bit more to the notation—I've just given you the basics here; I've saved a lot for the exercises. If you want to become fluent with the algebraic language, you've got to practice, as with any language.

Exercise Set E

1. Refer to the walking babies data set (Example 5:1) for this exercise.
   a. What is the largest value of i? of j?
   b. Tell in words what α_3 stands for.
   c. What are the subscripts for
      i. the first child who got the special exercises?
      ii. the third child who got no exercise instructions, with weekly reporting?
      iii. each of the children in the last control group ? (List them all, in order.)
   d. Compute $\bar{y}_{1.} - \bar{y}_{..}$, and tell in words what the number tells you.

2. Turn the intravenous fluids data set (Example 5.2) on its side so that rows correspond to drug companies.
   a. What is the largest value of i? of j?
   b. Write out the symbols for the effect of Cutter; of McGaw; of Abbot.
   c. Write the symbols (y with subscript pairs) for the observations whose values are 99; 343; 401; 214.

3. *Degrees of freedom add up*. Many statisticians use capital letters to stand for the largest values of subscripts: I = largest i subscript = number of treatments, and J = largest j sub-

script = number of units per treatment. (For the leafhoppers, $I = 4$ and $J = 2$.) Write the following in terms of I and J:

a. df_{Total}

b. df_{Cond}

c. df_{Res}

d. Write the sum $df_{Grand} + df_{Res} + df_{Cond}$ in terms of I and J, and show that the sum equals df_{Total}.

4. *Pieces of the decomposition add up*. Using the $\bar{y}$ notation, write expressions for (a) the estimated benchmark, (b) the estimated effect for condition i, and (c) the residual for the observation in row i, column j. (d) Then write the sum of (a), (b), and (c), and show that this sum equals the observed value y_{ij}.

5. *Deviations from an average add to zero*. Suppose you have a set of n numbers represented by $y_1, y_2, \ldots, y_n$. Let $\bar{y}_. = (y_1 + y_2 + \cdots + y_n)/n$ be their average. Then the deviations are $y_1 - \bar{y}_.$, $y_2 - \bar{y}_.$, ..., $y_n - \bar{y}_.$. Write out the sum of the deviations, and show that this sum equals zero.

6. *Why we assume the α_i sum to zero*. We assume that all the observations made under the same conditions have the same "true" value and that each observation equals that true value plus a residual error. If we let μ_i stand for the true value for condition i, then $y_{ij} = \mu_i + e_{ij}$. Now define the benchmark value $\bar{\mu}_.$ to be the average of the μ_i, and let α_i be the deviation $\mu_i - \bar{\mu}_.$. Using Exercise 5 as a guide, show that the α_i add to zero.

7. *Relationship between estimates and parameters*

a. For the leafhopper example, show that the ith treatment average $\bar{y}_{i.} = \mu + \alpha_i + \bar{e}_{i.}$, where $\bar{e}_{i.} = (e_{i1} + e_{i2})/2$.

b. Starting from Exercise 4c, show that the residual for the observation in row i, column j equals $e_{ij} - \bar{e}_{i.}$; notice that this expression does not involve either μ or the α_i.

c. Starting from Exercise 4a, and using the assumption that the α_i add to zero, show that the grand average $\bar{y}_{..} = \mu + \bar{e}_{..}$ (= true benchmark plus the average of all the errors).

d. Show that the estimated effect for treatment i equals $\alpha_i + \bar{e}_{i.} - \bar{e}_{..}$ (= true treatment effect plus an error part).

8. *Sums of squares*. Show that

a. $SS_{Grand} = IJ\,[\bar{y}^2_{..}]$.

b. $SS_{Cond} = J[(\bar{y}_{1.} - \bar{y}_{..})^2 + (\bar{y}_{2.} - \bar{y}_{..})^2 + \cdots + (\bar{y}_{I.} - \bar{y}_{..})^2]$.

9. *Sums of squares add up*. Suppose you have n observations, all made under the same conditions, and you decompose them each into average plus deviation:

$$\begin{aligned} y_1 &= \bar{y}_. + (y_1 - \bar{y}_.) \\ y_2 &= \bar{y}_. + (y_2 - y_.) \\ &\vdots \\ y_n &= \bar{y}_. + (y_n - \bar{y}_.) \end{aligned}$$

Use the definition of $\bar{y}_.$ to show that the SSs for the two right-hand columns add up to give the SS for the column on the left.

10. *Shortcut formulas for sums of squares*

    a. Show $SS_{Cond} = J[\bar{y}_{1.}^2 + \bar{y}_{2.}^2 + \cdots + \bar{y}_{I.}^2] - SS_{Grand}$.

    b. Starting from the fact that each residual equals $y_{ij} - \bar{y}_{i.}$, show that

$$SS_{Grand} + SS_{Cond} + SS_{Res} = SS_{Total},$$

and so

$$SS_{Res} = SS_{Total} - J[\bar{y}_{1.}^2 + \cdots + \bar{y}_{I.}^2].$$

SUMMARY

1. *The basic factorial design*

   a. The **completely randomized (CR)** or **randomized basic factorial (RBF)** design uses a chance device to assign one **treatment** to each **experimental unit**. (For a **balanced** design, each treatment gets assigned to the same number of units.)

   b. For the *observational* version of the randomized basic factorial (RBF) design, you take a **simple random sample (SRS)** from each population. (For an SRS you use a chance device to make sure that every member of the **population** has the same chance of getting in the sample.)

2. *Informal analysis*

   a. Two tools, averages and dot graphs, can be used to assess group differences (G), to detect outliers (O), to assess spread (S), and see whether a transformation is needed (T).

   b. The **average** of a set of numbers equals their sum divided by their number. The average itself tells a typical value for the set of observations, and the **deviations** from the averages tell how far each number is above (+) or below (−) its average. The deviations from an average add to zero.

   c. A **parallel dot diagram** for comparing groups of observations represents each observed value as a dot. The height of each dot tells the observed value, and the dots go in columns, one column for each group. If you turn a dot graph for a group on its side, it will balance at the average for the group.

   d. An **outlier** is an observation that lies far away from the rest of the numbers in its group. Because outliers exert so much influence on an average, it is often a good idea to do one analysis including the outlier and another without it, and compare the results.

   e. If the **spreads** for the groups in a dot graph are very unequal and the sizes of the spreads seem related to the sizes of the averages, it is worth considering a **transformation** to a new scale.

3. *Factor structure*

   a. A **factor** is a meaningful partition of the observations. A **partition** of the observations is a way of sorting them into groups; each observation belongs to a group, and no observation belongs to more than one group. A partition is meaningful, and hence a factor, if it corresponds to a set of conditions built into the design of the experiment.

   b. Three tests for factors:

      i. Do the observations in each group have some feature in common that is not common to observations in any other group?

      ii. Does it make sense to compute an average for each group and compare averages?

      iii. If you switch groups, putting one in place of another, do you change the message from the data?

If the partition is a factor, all three answers should be yes. If it's not a factor, all three answers should be no.

c. Every design has two **universal factors**, the factor for benchmark (one group containing all the observations) and the factor for residual error (each observation goes by itself into a separate group).

d. A one-way design has only one structural factor, corresponding to treatments or to conditions.

e. The **factor diagram** for a data set is a series of boxes, one box per factor, showing each of the meaningful ways to partition the data.

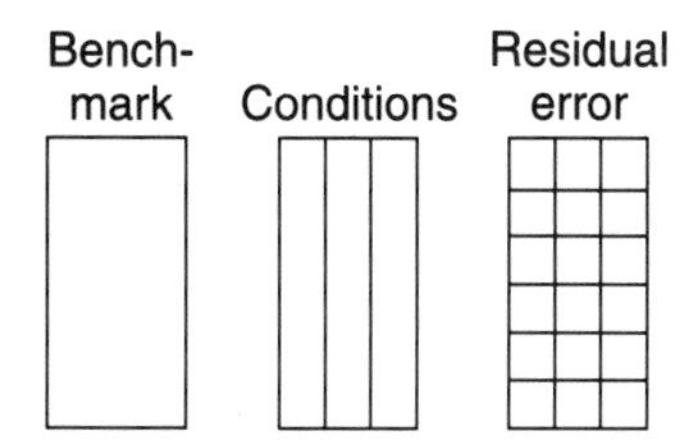

4. *Decomposition and analysis of variance.*

a. **General decomposition rule.** One factor is **inside** another if each group of the first (inside) factor fits completely inside some group of the second (**outside**) factor. The estimated effect for a factor equals the average for that factor minus the partial fit (= sum of the estimated effects for all outside factors).

b. Decomposition for one-way designs. Each observed value is split into three pieces:

i. grand average (= estimated benchmark), plus

ii. condition average minus grand average (= estimated condition effect), plus

iii. observation minus condition average (= residual).

c. The **degrees of freedom (df)** for a table equals the number of *free* numbers—the number of slots in the table you can fill in before the patterns of repetitions and adding to zero tell you what the remaining numbers have to be. The df tells the number of units of information about chance error in that box of the decomposition.

d. **General rule for df:** The df for a factor equals the number of levels minus the sum of the df for all outside factors. For the BF[1] design: $df_{Grand} = 1$; $df_{Cond} =$ (number of conditions $-$ 1); and $df_{Res} =$ (number of observations $-$ number of conditions).

e. The **sum of squares (SS)** for a set of deviations measures their overall variability: if the deviations are large (spread out), their SS will be large; if the deviations are small (close together), their SS will be small.

f. The **mean square** for a set of deviations equals their sum of squares divided by the degrees of freedom. The MS measures a kind of average variability per unit of information. MS = SS/df.

g. The **F-ratio** for the conditions of interest in a basic factorial design equals the mean square for conditions divided by the residual mean square. It tells us that the average variability in the condition effects is *F* times as big as the average variability in the residuals. $F = MS_{Cond}/MS_{Res}$.

h. The **observed significance level** (or ***p*-value**) for an *F*-ratio is the probability of getting a value that high or higher just by chance if in fact there were no differences in the underlying true values.

i. The (estimated) **standard deviation** for a data set equals the **root mean square of the residuals.** The SD tells the typical size of a chance error. $SD = \sqrt{MS_{Res}}$.

Exercise Set F: Review Exercises

1. *Mothers' stories*. The dot graph below shows residuals from a decomposition of the data given in Exercise D.24. Guess the SD by using the rule based on the normal curve: find a distance such that two-thirds of the dots fall within that distance of zero.

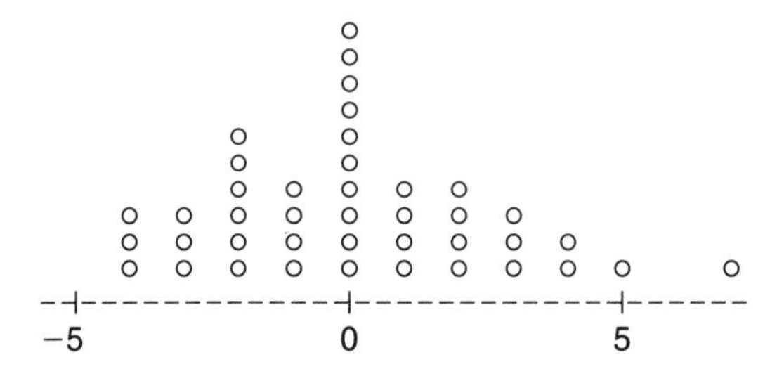

2–4. *Lung cancer*. The numbers below come from a study of lung cancer patients. Nine patients who survived 19 or fewer months were compared with nine others who survived longer than 19 months. The numbers give the morphometric index, a characteristic of tumor tissue.

Short-term:	123	114	99	103	83	106	113	109	95
Long term:	142	117	106	170	116	148	226	190	165

2. Tell the response. Then list all structural factors, give the number of levels, and tell whether each is observational or experimental. Tell what the unit is.

3. Draw a parallel dot graph, and use it to discuss the data.

4. Use the balance point interpretation to estimate the average for each group; then estimate (do not compute) the SD separately for each group, choosing among 1, 5, 10, 20, 40, 60, and 100. What do your estimates tell you?

5–8. *Walking babies*. Turn back to the parallel dot graph for the walking babies data (Example 5.4). Identify three outliers, and notice that if you simply leave out all three outliers, you get a balanced design. Consider a new analysis that leaves out these three observations.

5. True or false: if you leave out all three outliers, then
   a. SS_{Cond} will be larger than before.
   b. df_{Res} will go from 19 down to 16.
   c. MS_{Res} will get bigger because df_{Res} gets smaller.

6. Compute new condition averages and condition effects. Then fill in the blanks in the following sentence four times, once for each condition: on the average, babies in the __________ group walked __________ months __________ (sooner/later) than the overall average time.

7. Now find MS_{Cond}, MS_{Res}, and the F-ratio. Taking the F-ratio at its face value, what do you conclude about the effects of the treatments?

8. The two analyses, with (Exercise D.21) and without the outliers, give very different results. Overall, what is your conclusion from the two analyses taken together? What would you say about a scientist who never pays attention to outliers, and who always analyzes only the complete data set? What would you say about a scientist who throws away 3 observations when there were only 23 to begin with?

9–13. *Bird calcium*. In your work with this data set, you should have discovered that these data should be transformed. In Chapter 12, Section 1 I describe how to use a special kind of scatterplot to choose a transformation to equalize the group SDs in order to make the assumption of same SDs more appropriate. The plot for the bird calcium data suggests transforming to logarithms. (You already know enough to read that section if you want to see how the plot works.)

I used the logarithm button on my calculator to replace each observed value by 100 × log of observed value. Here are the data in the new scale, with a few observations (marked *) changed by ±1 to simplify the arithmetic.

I. Female control:	120*	124*	111*	115	110*
II. Male control:	117*	104	103	116	100
III. Female/hormone:	150	142	133	155	160
IV. Male/hormone:	150*	138	145*	140	147

9. Draw a parallel dot diagram, and write a short paragraph about what you see.

10. Decompose the data into condition averages plus residuals. Compute a separate SD for each condition, and find the ratio SD_{Max}/SD_{Min}. Compare this ratio with the one based on the untransformed data: has the transformation done what it was supposed to?

11. Write out the ANOVA table, and tell what you conclude from the *F*-ratio. Is your conclusion the same as the one based on the untransformed data?

12. Using $\sqrt{MS_{Res}}$ as your SD, check the rule based on the normal curve: what percent of the residuals fall within 1 SD of 0? 2 SDs? 3 SDs?

13. For the data in the original scale, $\sqrt{MS_{Res}}$ = _____. Using the residuals given in Exercise D.12, check the rule for the raw concentrations.

14. a. Use your work in Exercises 12 and 13 to fill in the following table:

	Percent of residuals		
	Within 1 SD	**More than 1 SD, less than 2 SDs**	**More than 2 SDs**
Original data			
Transformed data			
Normal curve	68%	27% (95% − 68%)	5%

b. Fill in the blanks, choosing from higher/lower: The bird calcium data suggest that when you have chance errors of two different sizes, the percentage of residuals within ±1 SD tends to be ___________ than what the 1-2-3-SD rule predicts: the percentage of residuals more than 1 SD and less than 2 SDs from zero tends to be _________ than what the rule predicts; and the percentage of residuals more than 2 SDs from zero tends to be _________ than predicted. Why should this be the case?

15. *Bivariate CR[1] model for Kelly's hamsters*. Construct two ANOVA tables for the untransformed enzyme concentrations from Kelly's hamster experiment (Chapter 1), one ANOVA using heart concentration as your response and the other using brain concentration. What do you conclude about the effects of day length?

16. *Rating Milgram*. Analyze the data given in Exercise B.8: decompose the data; use the residuals to check the assumption of same SDs and the normality assumption; and then do the ANOVA. Were the teachers influenced by the results they thought Milgram got when they judged how ethical his study was?

NOTES AND REFERENCES

1. ***Transforming to equalize spreads.*** The IV fluids data (Example 5.6) and the bird calcium data (Exercise B.1) both show two features that suggest thinking about a new scale: (1) The groups have different spreads, and (2) Group average and group spread seem related.

Why change scales? One of the goals of analysis of variance is to judge whether differences in group averages are too big to be due just to chance error. The judgment depends on comparing (1) the size of the differences between group averages—the *between-group variation*—to (2) the size of the differences between individual observations within the groups—the *within-group variation*. For the method to work, the spreads of the groups need to be roughly equal. If they're not, changing to a new scale can sometimes help equalize spreads.

It's in fact quite common in science for variability in a set of measurements to depend on the typical size for the measurement. For example, differences in the weights (kg), heights (cm), and running speeds (m/sec) would be a lot bigger for a group of adult elephants than for a group of adult mice. However, suppose you reexpress the differences as a percent of the average value for the group. You'd find that in your new scale the variation in elephant weights (as a percent of the average weight of the elephants) is much more nearly the same as the variation in mouse weights (as a percent of the average weight of the mice).

You can read more about transforming in Chapter 2, Section 3 and in Chapter 12.

2. ***The assembly line metaphor.*** (This metaphor is an invention, of course. Its purpose is to describe the way statisticians assume our observed values behave.) We can think of each of the eight leafhopper observations as built in three stages, at three work stations on an assembly line:

1. Each observation starts out at the Benchmark station, where the worker puts a certain fixed amount—always the same amount for all the observations—on the conveyer belt.
2. Our observation then moves to the next station, Diet Effects, where the worker adds or subtracts a second piece. The pieces for the diet factor are of different sizes depending on the diet group an observation belongs to, but for each group there is just one constant size.
3. Finally, the observation moves to the last work station, where a residual error is added or subtracted, and the completed observation rolls off the assembly line, ready to be measured.

3. ***Four assumptions about chance errors.*** To describe the assumptions, I'll rely on a metaphor—the box model—from a book by three Berkeley statisticians. Imagine a box of tickets with one number printed on each ticket. To get the chance error for an observation, you reach into the box, mix the tickets, and randomly draw one out. The number on the ticket tells the chance error. Here are the four assumptions about the way these errors behave:

- Z The average of the tickets in the box is *zero*.
- S The chance errors are all drawn from the *same* box. In particular, there is one typical *size*, as measured by the **standard deviation**, for the tickets in the box.
- I The draws are **independent**: the outcome of one draw doesn't depend on or influence the outcomes of any other draws.
- N The set of tickets in the box follows a bell-shaped **normal** curve.

In practical terms, the assumptions Z, S, and N tell us that

Roughly	2/3 of our chance errors will fall within ± 1 SD of zero.
	95% of our chance errors will fall within ± 2 SDs of zero.
Essentially	all of our chance errors will fall within ± 3 SDs of zero.

4. ***Outliers.*** One rough rule for judging outliers is to measure the distance from a suspicious value to its group average, and throw out the value if the distance is 3 standard deviations or

more. However, since the value of the SD is itself sensitive to outliers, statisticians have developed other, less sensitive rules.

Rather than rely on an "all or nothing" rule to keep or throw out an observation, you can instead do two analyses, one including the questionable value and one without it, and see how much your conclusions depend on how you handle the one observation. (One way to reanalyze a data set that contains an outlier is to use the rest of the data to estimate a replacement value, using the methods of Chapter 12, Section 5.) Few scientists would put much confidence in any conclusion that depends on a single observation.

Appendix: Supplementary Examples

The Completely Randomized Design

EXAMPLE S5.1 ESTRADIOL AND UTERINE WEIGHT

The CR design is very common in experimental endocrinology, where many experiments have essentially the same form. Each experiment is designed to study a particular hormone, like thyrotropin (TSH) or parathormone (PTH) or insulin; each hormone is produced by a particular gland (pituitary, parathyroid, and pancreas for the hormones I just listed); and each hormone has a particular set of jobs to do: thyrotropin stimulates the thyroid and helps regulate its use of iodine, PTH stimulates the kidneys to excrete phosphorus and helps the bones to use calcium, insulin promotes the uptake of glucose by the liver and muscles. A typical CR experiment divides rats, mice, chicks, or other lab animals into groups; one or more groups serve as controls, and the rest have the gland de jour surgically removed; some of these groups then get a variety of treatments, like hormone injections, special diets, etc. The response is chosen to be some measurement related to the job the hormone is thought to do. For example, a PTH experiment might measure the amount of phosphorus in the urine.

In females the hormone estradiol is produced by the ovaries. If the ovaries are surgically removed, the rest of the reproductive system will lose its capacity to function, and the uterus will get smaller. One way to prevent this is to give estradiol supplements. In this experiment, the purpose is to measure the relationship between the dose of estradiol and the weight of the uterus. Sixty mice are randomly assigned to one of six groups, ten mice per group, as in Table S2.

TABLE S2

Group	Operation	Daily dose of estradiol
I. Control 1	None	None
II. Control 2	Sham	None
III. Treatment 1	Remove ovaries	.001 micrograms
IV. Treatment 2	Remove ovaries	.005 micrograms
V. Treatment 3	Remove ovaries	.010 micrograms
VI. Treatment 4	Remove ovaries	.020 micrograms

After one week of hormone supplements, each animals is killed and its uterus is weighed.

Factor Structure

EXAMPLE S5.2 Fisher's Iris Data: Several Response Variables

You might think it will always be easy to tell the response and the factor for conditions apart. After all, the conditions are the things you compare, and the response is the thing you use to measure differences, so in terms of the structure of the experiment, the response and the conditions have very different functions. Unfortunately, though, when there is more than one measurement, things can get tricky.

Table S3 is a small portion of a large data set collected in the 1930s by Dr. Edgar Anderson, and made famous by R. A. Fisher, who used the data to illustrate a statistical method called *discriminant analysis*.

TABLE S3

Iris setosa				*Iris versicolor*				*Iris virginica*			
Sepal		Petal		Sepal		Petal		Sepal		Petal	
Length	Width	Length	Width	Length	Width	Length	Width	Length	Width	Length	Width
5.1	3.5	1.4	0.2	7.0	3.2	4.7	1.4	6.3	3.3	6.0	2.5
4.9	3.0	1.4	0.2	6.4	3.2	4.5	1.5	5.8	2.7	5.1	1.9
4.7	3.2	1.3	0.2	6.9	3.1	4.9	1.5	7.1	3.0	5.9	2.1
4.6	3.1	1.5	0.2	5.5	2.3	4.0	1.3	6.3	2.9	5.6	1.8
⋮	⋮	⋮	⋮	⋮	⋮	⋮	⋮	⋮	⋮	⋮	⋮
5.3	3.7	1.5	0.2	5.1	2.5	3.0	1.1	6.2	3.4	5.4	2.3
5.0	3.3	1.4	0.2	5.7	2.8	4.1	1.3	5.9	3.0	5.1	1.8

What is the response, and what are the structural factors?

Here's one way to think about the data: The data sit nicely in a rectangle, and we know that each meaningful way to divide up the data corresponds to a factor, so there must be several, including the three in Figure S1.

Species

setosa	*versicolor*	*virginica*
setosa	*versicolor*	*virginica*
setosa	*versicolor*	*virginica*
setosa	*versicolor*	*virginica*
setosa	*versicolor*	*virginica*
setosa	*versicolor*	*virginica*
setosa	*versicolor*	*virginica*
setosa	*versicolor*	*virginica*

Three groups

Sepal vs. Petal

Sepal	Petal	Sepal	Petal	Sepal	Petal
Sepal	Petal	Sepal	Petal	Sepal	Petal
Sepal	Petal	Sepal	Petal	Sepal	Petal
Sepal	Petal	Sepal	Petal	Sepal	Petal
Sepal	Petal	Sepal	Petal	Sepal	Petal
Sepal	Petal	Sepal	Petal	Sepal	Petal
Sepal	Petal	Sepal	Petal	Sepal	Petal
Sepal	Petal	Sepal	Petal	Sepal	Petal

Two groups

Length vs. Width

L	W	L	W	L	W	L	W	L	W	L	W
L	W	L	W	L	W	L	W	L	W	L	W
L	W	L	W	L	W	L	W	L	W	L	W
L	W	L	W	L	W	L	W	L	W	L	W
L	W	L	W	L	W	L	W	L	W	L	W
L	W	L	W	L	W	L	W	L	W	L	W
L	W	L	W	L	W	L	W	L	W	L	W
L	W	L	W	L	W	L	W	L	W	L	W

Two groups

FIGURE S1

This way of looking at the data may seem a lot like what I did with the other data sets in Chapter 5, but for some purposes, it would be wrong to think of all three of the partitions above as factors. Instead, we could regard the data as having four measurements (sepal length, sepal width, petal length, petal width), any one of which could serve as the response, but with only one structural factor, species.

Which way is right, and how can you tell? This is an example of the kind of data set that multivariate methods were invented for. If you had a data set like this one, with more than one response, and you wanted to analyze the responses simultaneously, and you knew multivariate methods, that would be a sound strategy: one structural factor, and four responses. If you're reading this, however, you probably don't have that option. You've got to choose between four separate analyses—one for each response, with species as the only structural factor—or something along the lines suggested by the set of partitions above. There are two things to think about in deciding which way to go: What do you want your analysis to tell you about, and how well does your data set fit the assumptions of the method you plan to use? I would argue that for this data set, you could learn something useful from either approach—multivariate, or single response with sepal vs. petal and length vs. width as factors. I'll return to these issues in Chapter 8.

CHAPTER 6

INTERACTION AND THE PRINCIPLE OF FACTORIAL CROSSING

OVERVIEW

Two Reasons for Choosing Factorial Treatment Structure

- **To study the effects of two or more sets of conditions (treatments) in a single experiment**
- **To study how sets of conditions interact**

In the examples of the previous chapter, I took the set of treatments more or less for granted, in order to focus on how those treatments got assigned to units. This chapter looks more closely at the treatments themselves.

In many situations you may be interested in two or more sets of conditions that influence the response. Although it would be possible to study each set of conditions separately, using a different experiment for each set, it is often more efficient to run a single experiment that lets you study all your sets of conditions at once. For example, if you are interested in plant nutrition, you might want to design a single experiment to study both the effects of nitrogen and the effects of phosphorus on the size of your plants. Section 1 shows how to design such experiments, using the idea of factorial treatment structure.

Designs with factorial treatment structure give the information you need to measure interaction between two sets of conditions that influence your response. (As you may remember if you read Chapters 1 to 3, it was the presence of interaction that I found most interesting about the results of Kelly's hamster experiment: the effect of day length on enzyme concentration was not the same for hearts and brains.)

The concept of interaction is one of the most important general concepts in science because interaction is so much a part of the way things work. If your doctor prescribes penicillin, you should be careful to take it on an empty stomach because penicillin interacts with many foods: the amount of penicillin that reaches your bloodstream depends on how much of those foods you have in your stomach. Symbiosis is another instance of interaction: two organisms interact in a way that makes them better off living together than living apart. Debates about heredity versus environment are often thorny because the interactions between heredity and environment are both subtle and complex. (Educational testing is made more difficult because of the way cultural background, academic achievement, and natural ability interact.) Virtually everything studied by social psychologists involves interaction of one sort or another. In cognitive psychology, what we remember and how we perceive things depends not on the stimulus alone, but also on context: context and stimulus interact.

The word "treatment" in "factorial treatment structure" can remind you that the idea deals with choosing a set of treatments but says nothing about how you assign those treatments to units. Indeed, you can use any of a number of strategies. You can use complete randomization with factorial treatment structure, but you can also rely on various alternative strategies involving blocking (Chapter 7). To keep things simple at first, however, this chapter will look only at the simplest factorial plans, and leave other possibilities for later.

1. Factorial Crossing and the Two-Way Basic Factorial Design
2. Interaction and the Interaction Graph
3. Decomposition and ANOVA for the Two-Way Design

1. Factorial Crossing and the Two-Way Basic Factorial Design (BF[2])

Factorial Crossing

The main idea of this section is that if you want to study two or more sets of conditions in the same experiment, and/or you want to study how they interact, your design should include all possible combinations of the conditions.

> **Two sets of treatments are crossed if all possible combinations of treatments occur in the design. The design is called a** two-way factorial **and has** factorial treatment structure.

Example 6.1 Pig Out

Much of the theory of experimental design grew out of a search for more efficient farming methods: ways to improve crop yields and control insect damage, ways to raise healthier cows and fatter pigs. This is a pig-fattening example.

It seems natural to think that adding the right vitamins to a pig's diet might produce fatter pigs faster. Suppose you've decided to study the effect of vitamin B12. You could take B12 as your treatment, with two levels: 0 mg (control) and 5 mg.

So far, so good, but there can be complications. Pigs have bacteria living in their intestines; some of these bacteria can prevent a pig from using the vitamins. To control the bacteria, you might decide to add antibiotics to the pigs' diets.

Consider two possible plans. For the simpler plan, you'd give antibiotics to all the pigs, regardless of whether they got B12. This plan would have just one set of two treatments: (1) antibiotics plus 0 mg B12 (control), and (2) antibiotics plus 5 mg B12.

As an alternative, you could regard antibiotics as a second set of treatments: 0 mg (control) and 40 mg. If you take both treatments together, B12 and antibiotics, there are four possible treatment combinations:

(1) standard diet (control)
(2) standard with 5 mg vitamin B12 added
(3) standard with 40 mg antibiotics added
(4) standard with 5 mg B12 and 40 mg antibiotics added

This set of four diets has a two-way factorial treatment structure (Fig. 6.1). Each diet is a treatment combination that comes from choosing 0 mg or 5 mg of B12 (first treatment = first basic factor) and 0 mg or 40 mg of the antibiotic (second treatment = second basic factor).

		Antibiotics	
		0 mg	40 mg
B12	0 mg	1: Control	3: Antibiotics
	5 mg	2: B12	4: Both

FIGURE 6.1 The four pig diets have a two-way factorial structure.
First basic factor = rows = B12, with 2 levels (0 or 5 mg)
Second basic factor = columns = antibiotics, with 2 levels (0 or 40 mg)
Compound factor = diets, with 4 levels (control, B12, antibiotics, or both) ■

In a two-way factorial design like the pig example, each little box, called a **cell**, represents one of the possible combinations of a row treatment and a column treatment. Each experimental unit is assigned to a cell, and there are usually two or more units (and so two or more observations) per cell. The row factors and the column factors are *crossed*: every possible combination is present. (Notice that crossing has a visual meaning here: every row extends all the way from left to right and crosses each of the columns; every column goes from top to bottom and crosses all the rows.)

EXAMPLE 6.2 SUBMARINE MEMORY

Cognitive psychologists who study memory have known for some time that we tend to remember more if the conditions when we are tested match the conditions when we first did the learning. In other words, there is an interaction between learning conditions and recall conditions.

One striking demonstration of this interaction comes from an experiment by Godden and Baddeley (1975), who had some of their subjects learn lists of words while sitting under 20 feet of water. The subjects for their experiment were 16 members of a diving club; the response was the number of words remembered correctly from a list of 36; and there were four conditions to be compared (Fig. 6.2). These conditions were treatment combinations that came from crossing two treatments—learning environment (dry or wet) and recall environment (dry or wet). For the dry environment the diver sat on the shore; for the wet environment the diver wore weights and sat under 20 feet of water.

		Recall environment	
		Dry	Wet
Learning	Dry		
environment	Wet		

FIGURE 6.2 Four treatment combinations for the submarine memory experiment. The four cells represent all possible combinations of two treatments (learning environment and recall environment) each with two levels (dry and wet). The two treatments are crossed. ■

Basic and compound factors. As usual, we reserve the word "treatments" for conditions you can assign. Factorial crossing, however, can occur with conditions that are not necessarily experimental. To describe the structure of the conditions in a way that includes observational studies, we describe crossing in terms of *basic* and *compound* factors (Fig. 6.3).

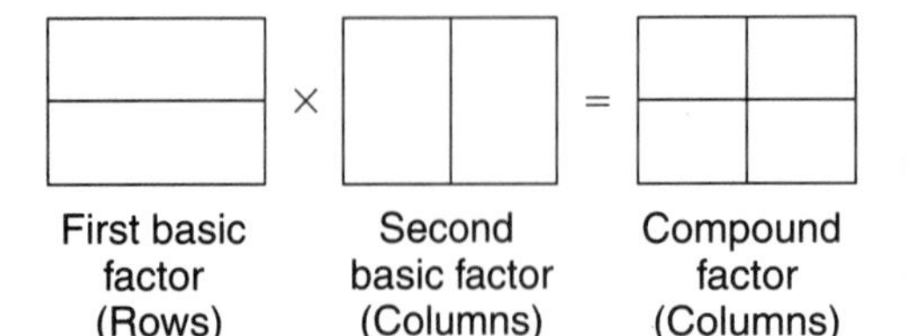

FIGURE 6.3 Basic factors and compound factors. A factor is **compound** if its levels come from crossing two (or more) other factors. A factor is **basic** if it is not compound.

Two-Way Basic Factorial Designs: "What You Do"

If your conditions come from factorial crossing and you can assign them using a chance device, you get the two-way completely randomized, or CR[2], design. There are also versions of the design with observational conditions. Taken together, the CR[2] and its observational twins are called two-way basic factorial, or BF[2], designs. If both factors are randomized, we can write RBF[2].

The two-way completely randomized experiment (CR[2]). When you have two experimental factors, there are several strategies for assigning treatment combinations to experimental units, and Chapter 7 will present two of them. This chapter presents the simplest strategy, complete randomization:

The Two-Way Completely Randomized Experiment (CR[2])

- **The treatment combinations come from crossing two basic treatment factors.**
- **Treatment combinations are assigned to units completely at random, as in any CR experiment.**
- **For balance each treatment combination is assigned to the same number of units.**
- **If you want to measure interaction in a two-way design, you have to have more than one observation per cell.**

EXAMPLE 6.3 CR[2] DESIGNS

Tell how to run the pig and submarine memory studies as CR[2] designs.

SOLUTION

a. *Pigs*. Once you've chosen your set of treatment combinations, you need to decide how to assign them to experimental units. In principle, various strategies are possible, but for this particular experiment the most suitable strategy is also the simplest: complete randomization.

Response:	Average daily weight gain, in pounds
Treatment combinations:	4 diets
Experimental units:	12 piglets
CR design:	Randomly assign a diet to each piglet, with each diet going to 3 piglets

b. *Memory*. There are various strategies for assigning the four treatment combinations. The simplest, CR strategy, would take the 16 divers as the experimental units and would assign each diver to one of the four cells, four divers to a cell. This is a situation, however, where it makes sense to reuse subjects. The actual design for this experiment is one I'll describe in a later chapter. ■

Notice that for both these experiments, each treatment combination gets assigned to more than one unit. In other words, these designs give you more than one observation per cell. Although you can have a balanced CR[2] design with just one observation per cell, such designs do not give enough information to measure interaction. You need two or more observations under the same conditions in order to estimate the size of chance error. If you have only one observation per cell, there's no way to tell how much of any differences you observe is due to chance error and how much is due to interaction.

Variations on the CR[2]: observational factors of interest. The two examples so far, pigs and divers, were both examples of genuine experiments: conditions were assigned using a chance device. The next examples have essentially the same structure ("what you get"), but one or more sets of conditions will be built into the material, not something you can assign.

EXAMPLE 6.4 MORTON'S SKULLS, REVISITED (EXAMPLE 4.4)

One of Morton's data sets had the structure of a two-way factorial plan, with race and sex as the two basic factors of interest, both observational. Morton couldn't *assign* race or sex to the skulls; instead, he used built-in characteristics to *define* the populations he wanted to compare and then took convenience samples from the populations. ■

EXAMPLE 6.5 BUTTERCUPS

This study was done by a colleague of mine, a plant ecologist. She wanted to study whether buttercups adjust to their environment in ways that make them less healthy if they are transplanted to a new environment. Her study had two factors of interest, original environment and new environment; each had two levels, sunny field and damp woods (Fig. 6.4). She dug up a bunch of plants from each environment and then randomly assigned half the plants from each environment to go back to the same environment; the rest were assigned to the other environment.

		New environment	
		Field	Woods
Original	Field		
environment	Woods		

FIGURE 6.4 Buttercups: One observational and one experimental factor of interest. Original environment is observational: it was essential to the goal of the experiment that the plants come with the environment already determined. This factor defines two populations (plants growing in the field, plants growing in the woods). New environment is experimental: levels of this factor (field or woods) were assigned at random to samples of plants from each population. ■

Here's a summary of the three kinds of BF[2] designs:

1. If both factors of interest are experimental, you use a chance device to assign a treatment combination to each experimental unit.
2. If both factors of interest are observational, you cross-classify your populations into cells and get a sample from each population.
3. If there's one factor of each kind, you obtain a sample of units from each of your populations, then use a chance device to assign levels of the experimental factor (treatments), separately within each sample.

Factor Structure of the BF[2] Design: "What You Get"

Every two-way BF design has three structural factors:

1. **First treatment factor**
2. **Second treatment factor**
3. **Interaction**

All examples in this chapter have the same factor structure. You can find that structure in two stages—a first stage that ignores the factorial crossing, and a second stage that takes the treatment structure into account.

Stage 1 (Reduced form): If you ignore the special structure of the treatments and regard the treatment combinations as just a single set of treatments, you get the same three factors as for the one-way designs in Chapter 5: benchmark, conditions, and residual error.

Stage 2 (Expanded form): If you then ask, "Which sets of treatment averages are meaningful?" you find that the factor for conditions can be "expanded" to give three factors: first basic factor, second basic factor, and their interaction.

Learning to switch back and forth between the two ways of thinking about the factor structure is quite useful because once you can do that, you'll be able to take any design in this book and create an entire family of designs just by building in new factors of interest.

Example 6.6 Pig Out, Again

The data for the pig example (Example 6.1), arranged into cells, appear in Table 6.1. Find the factor diagram.

TABLE 6.1 Weight gains (lbs/wk) of pigs

		Antibiotics	
		0 mg	40 mg
B12	0 mg	1.30 1.19 1.08	1.05 1.00 1.04*
	5 mg	1.26 1.21 1.19	1.52 1.56 1.54*

Grand average:
1.245

Diet Averages

		Antibiotic:	
		0 mg	40 mg
B12	0 mg	1.19	1.03
	5 mg	1.22	1.54

Solution

Stage 1 (Reduced form)

If we regard the data as coming from a one-way BF with four conditions, then there are three factors.

1. Benchmark: it makes sense to compute a grand average, with all the observations lumped together in one group.
2. Diets: It also makes sense to compute a separate average for each of the four diet groups.
3. Residual error: Finally, it makes sense to compare individual observed values with each other in order to estimate the typical size of chance error.

Looking at the data in this way gives a factor diagram with three boxes (Fig. 6.5).

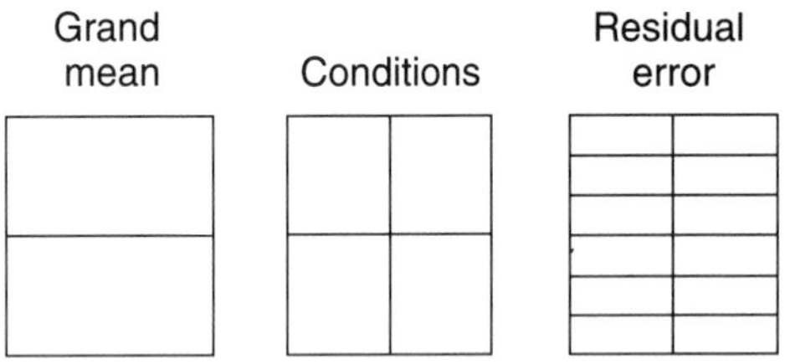

Grand mean | Conditions | Residual error

FIGURE 6.5 Factor diagram for the pig data, regarded as a one-way BF design. You can always begin the factor diagram for a BF[2] design in this way, starting with grand mean, conditions, and residual error. Here, as always, each factor corresponds to a meaningful partition of the data, and these three partitions are meaningful in the same sense as for any BF[1].

Stage 2 (Expanded form)

Now consider the factorial treatment structure: each diet is a combination of two treatments, B12 (0 mg or 5 mg) and antibiotics (0 mg or 40 mg). Such a structure gives three sets of meaningful averages.

1. *B12*. To summarize the overall effect of B12, we can find two average weight gains, one for the six pigs who got 5 mg of B12 and a second average for the six little piggies who got none. (Sad to say, all 12 little piggies went to market.) Each average lumps together three piglets who got antibiotics with three who got none. (See Fig. 6.6.)
2. *Antibiotics*. In the same spirit, we can summarize the overall effect of antibiotics by computing an average for the six pigs who got 0 mg (first column of the data) and another average for the six who got 40 mg (second column of the data). (See Fig. 6.7.)

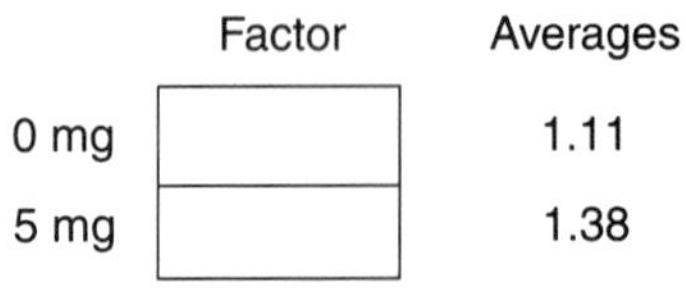

	Factor	Averages
0 mg		1.11
5 mg		1.38

FIGURE 6.6 Factor for the overall effect of B12.

Factor 0 mg	Factor 40 mg	Averages (0 mg)	Averages (40 mg)
		1.205	1.285

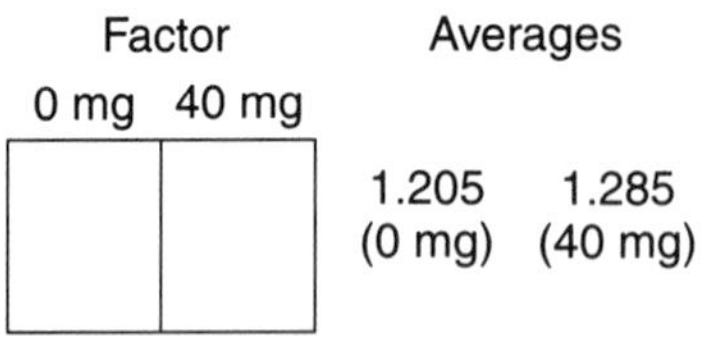

FIGURE 6.7 Factor for the overall effect of antibiotics.

3. *B12 × Antibiotics*. Finally, we can ask about interaction: to see whether the effect of B12 is the same for pigs who got no antibiotics as for those who got 40 mg, we need all four of the diet averages we got in connection with the first fac-

tor diagram. In all, then, we have three factors that describe the factorial treatment structure: B12 (treatment 1), antibiotics (treatment 2), and their interaction. Putting these three in place of the factor for diets in the first factor diagram gives the complete set of factors for the two-way CR experiment (Fig. 6.8).

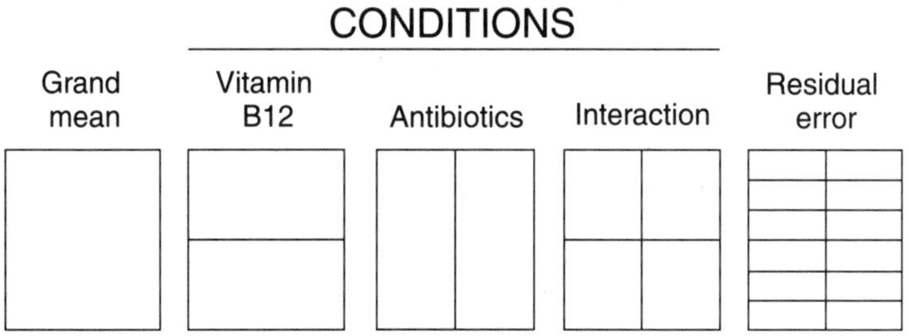

FIGURE 6.8 Factor diagram for the pig data, a two-way factorial CR design. ■

Looking ahead. For any factor diagram, there are two closely related ways to think about what it tells you. The first is in terms of meaningful ways to group and compare observations, essentially in terms of sets of averages. A second way is in terms of the assembly line metaphor. The diagram says that we regard each observed value as a sum of five pieces. Starting with a benchmark value, the observation next gets an amount added for the overall effect of B12, then a piece for overall effect of antibiotics, then a piece for interaction, and finally a residual error.

Observed value = Benchmark amount
+ Overall effect of B12
+ Overall effect of antibiotics
+ Interaction effect
+ Residual error.

This structure is what we will use to guide our formal analysis in Section 3: we'll split apart the observed values in order to estimate each of the terms in the model.

Exercise Set A

1. *Hornworms: What's for lunch? Sawdust.*

The following experiment has a two-way factorial CR design. Draw and label a two-way table showing the two treatments and the combinations of their levels. To label your table, be sure to do all of these five things:

a. Label the table as a whole.

b. Label the treatment for rows.

c. Label each level of the row treatment.

d. Label the treatment for columns.

e. Label each level of the column treatment.

Twenty-four tobacco hornworms served as subjects for this experiment, which was designed to see how worms raised on low-quality food would compare with those on a normal diet. Hornworms grow through stages called instars, shedding their skin between instars. Thus, counting instars is a natural way to keep track of a hornworm's age. The two dozen hornworms were randomly divided in half, and the lucky half were raised on regular rearing food for the first three instars. The unlucky half got a mixture of 20% regular food, 80% cellulose. Cellulose has no more food value for a hornworm than it has for you or for me: neither you nor a hornworm can digest the stuff. When the two groups of hornworms reached the fourth instar, each group was randomly divided in half for testing. One half of each group got regular food during instar four, and the other half got the mixture with cellulose. The experimenter kept track of how much each hornworm ate, and computed a response value based on total food eaten in relation to body weight.

2. Reread Exercises 5.C.4–6, which describe four studies. For exactly two of these, the conditions have a factorial treatment structure. Which ones are they? For each, tell the names of the factors of interest, tell how many levels each factor has, and tell whether each is experimental or observational. Then draw and label a two-way table showing the basic factors and their combinations.

3. Exercise 5.A.1 described three post-surgical treatments for breast cancer: chemotherapy, radiation therapy, and both. You can think of these as three out of four possible treatment combinations that come from crossing two sets of treatments. Show this by drawing and labeling a two-way table; put an X in the cell for the missing treatment combination. Why wouldn't you use a complete factorial design (all four combinations) here?

4. *Aggressive mice*. Leshner and Moyer investigated the relationship between the male hormone testosterone and two aspects of mouse behavior: aggressiveness and how long it took the mice to learn to avoid attack. They compared three groups of male mice: castrated, sham operated (placebo), and castrated with testosterone supplements. (Their results showed that the hormone had a dramatic effect on aggressiveness and essentially no effect on learning to avoid attack.)

This is another example of a two-way factorial plan with a missing cell. Show the factorial structure and missing cell with a table, just as in Exercise 3. Why were there no mice for the missing cell?

5. *Warm-blooded, cold-blooded*. The BF[1] design of Chapter 5 has served as the structure for many old-style studies in comparative physiology, which picked a bunch of animals (the conditions) and some physiological measure (the response) as a basis for comparing them. This idea can be extended to compare the way different species react to changes in their environment. Here's a simple example.

Suppose you want to use a two-way factorial design to compare the effects of temperature on warm-blooded and cold-blooded animals. Think how you would do this, then answer the questions. (Be sure to include spiny anteaters among your subjects: these animals are in between warm- and cold-blooded.)

a. Draw and label a two-way table showing your factors of interest and the combinations of their levels.

b. Which of your factors of interest are observational and which are experimental?

c. Consider the choice between the two-way BF plan and a different plan that reuses subjects. Under what circumstances would you prefer each one?

6. *Bird calcium*. Copy the (transformed) data from Exercise 5.F.9 into the same form as the pig data of Example 6.6, and draw the factor diagram.

7. *Hornworms and cellulose*. The response used for the hornworm experiment of Exercise A.1 was a measure of total food eaten during the test period in relation to body weight:

$$\text{Consumption index} = 10 \times \frac{\text{Total grams eaten during 4th instar}}{\text{Average body weight during 4th instar}}.$$

Here are the results:

RR:	43*	61	45*	35	31	31*
RC:	105	90	104*	107	123*	113
CR:	41*	39	38*	36	41	27*
CC:	103*	109	100	110	118	90

R means regular rearing food, C means cellulose mixture; the first letter tells what diet the worms were reared on; the second letter tells the test diet.

Draw and label a rectangular format that shows the crossing, and write out the factor diagram.

2. Interaction and the Interaction Graph

Interaction as a Difference of Differences

Because interaction is so common in science, the word itself is used with a variety of meanings, not all of them precise. The purpose of this section is to give a precise definition of what statisticians mean by interaction—interaction is a difference of differences—and to show how to draw and read interaction graphs.

I'll start with an informal, unstructured description of the interaction that is present in the pig data, so you can see how the formal definition gives an organized structure to the comparisons that interaction involves.

Example 6.7 Interaction in the Pig Data

Turn back to the table of four diet averages at the beginning of Example 6.6, and think about the patterns. If you are a pig with a choice of diet and you want to get fat fast, you'll do best with B12 and antibiotics together. That's only part of the message, however. Suppose you can't have B12; should you want antibiotics? No: the pigs with neither (control diet) did better than the pigs with antibiotics only, by .16 pounds a day, which amounts to an extra pound a week. It seems reasonable to speculate that the antibiotics might be killing off microorganisms that actually help a pig's digestion.

Now suppose you're going to get B12, like it or not; do you want antibiotics? This time the answer is yes: pigs who got antibiotics along with their B12 gained about a third of a pound more per day than those who only got B12. The effect of antibiotics depends on whether or not you get B12 along with it.

Now notice that we can structure the comparisons of the last paragraph by computing and comparing differences. One way to measure the effect of antibiotics is to subtract: average with antibiotics − average without (Fig. 6.9).

		Antibiotics		Difference
		0 mg	40 mg	With − Without
B12	0 mg	1.19	1.03	−0.16
	5 mg	1.22	1.54	0.32

FIGURE 6.9 Interaction as a difference of differences. The effect of antibiotics, as measured by the difference, is different for the two levels of the B12 factor. For the pigs who got no B12, adding antibiotics cut back their weight gain by 0.16 pounds per day. For those who did get B12, adding antibiotics increased their weight gain by 0.32 pounds per day.

If we prefer, we can focus instead on the effect of B12 (Fig. 6.10).

		Antibiotics		
		0 mg	40 mg	
B12	0 mg	1.19	1.03	Each difference equals
	5 mg	1.22	1.54	average with B12 −
Difference		0.03	0.51	average without

FIGURE 6.10 Interaction is present: The differences are quite different. Each difference equals the average with B12 − average without. For pigs who got no antibiotics, adding B12 had almost no effect. For pigs who did get antibiotics, adding B12 was worth an extra half-pound per day. ■

We can generalize from the last example to give a definition of interaction:

There are three essential pieces to the structure of interaction: two crossed factors, and a response. Interaction is present if the effect of one factor, as measured by differences in the response averages, is different for different levels of the other factor.

EXAMPLE 6.8 THE STRUCTURE OF INTERACTIONS

a. Pigs, again (Example 6.1)

Factor 1: B12, at two levels (0 mg and 5 mg)

Factor 2: Antibiotics, at two levels (0 mg and 40 mg)

Response: Average daily weight gain, in pounds

Interaction: The difference in average weight gain due to antibiotics is different depending on whether or not B12 is present.

b. Submarine memory (Example 6.2)

Factor #1:	Learning environment, two levels (dry and wet)
Factor #2:	Recall environment, two levels (dry and wet)
Response:	Number of words recalled correctly, from a list of 36
Interaction:	The difference in average numbers of words remembered in the wet and dry recall environments is different for the two learning environments.

c. Buttercups (Example 6.5)

Factor #1:	Original environment, two levels (woods and field)
Factor #2:	New environment, two levels (woods and field)
Response:	Total dry weight of the plant
Interaction:	The effect of the new environment on the plant's size (as measured by the difference in average response) depends on which environment the plant had originally adjusted to. ■

Notice that if any of the three essential pieces are missing, it doesn't make sense to talk about interaction. If you don't have a response, there's no way to measure the effects of the conditions. If you have only one factor, not two, there's nothing for that factor to interact with. One way to see this is to think of interaction as a kind of two-layer comparison: for each level of one basic factor, you compare the effects of the other factor, and then you compare the comparisons to see whether the effects are the same in each case.

The Interaction Graph

The interaction graph plots cell averages as points and joins the points by line segments in a way that lets you compare differences by comparing slopes of the line segments. If the differences are equal, the line segments will be parallel. If the lines are not close to parallel, interaction is present. Figure 6.11 shows an interaction graph for the pig data.

Section 4 shows how to get interaction graphs using the Minitab computer package. However such graphs are not hard to do by hand. Here's a step-by-step method to follow:

1. Start with the table of cell averages, with the levels of one basic factor as rows and the levels of the other basic factor as columns.
2. Draw a set of axes: (a) label the y-axis so that it starts lower than the lowest cell average and ends higher than the highest average; (b) put tick marks along the x-axis, one for each column, and label these with the levels of the column factor.
3. For each row of the table, plot each average as a point. (a) Each point should lie above and in line with the tick mark for its column in the table; (b) the height of the point (on the vertical scale) should equal the value of the average.

4. Join all the points from the same row in the table with line segments, and label each set of line segments with its level of the row factor.

EXAMPLE 6.9 INTERACTION GRAPH FOR THE PIG DATA

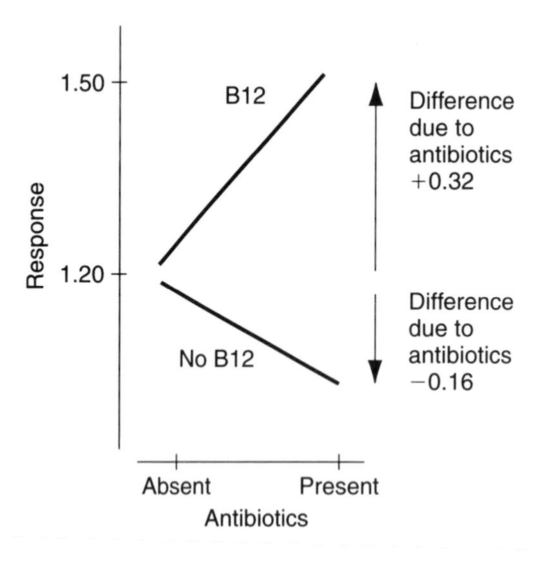

FIGURE 6.11 Interaction graph for the pig data. The top line, for the two averages with B12 present, goes up, from 1.22 on the left (no antibiotics) to 1.54 on the right (antibiotics). The change in y-values equals the difference of +0.32 in the averages. The bottom line (no B12) goes down, from 1.19 (no antibiotics) to 1.03 (antibiotics). The change in y-values equals the difference of −0.16. The two lines have opposite slopes because the two differences have opposite signs. ■

The following example is just one of the many, many instances in science where the entire hypothesis of interest boils down to a statement that a certain kind of interaction exists. Drawing an interaction graph gives a good visual summary of the evidence for the hypothesis.

EXAMPLE 6.10 IMAGERY AND KNOWLEDGE REPRESENTATION

Suppose you are a subject in a cognitive psychology experiment in which you are asked to respond "true" or "false" to statements like "A cat has claws." Do you call up an image of a cat in your mind, and check to see whether the cat in your mental picture has claws, or do you rely more directly on the *meaning* of the statement, without referring to an image? One major hypothesis in developmental psychology is that young children tend to rely more on images, whereas adults tend to rely more on meaning.

An experiment by S. M. Kosslyn was designed to measure differences related to this hypothesis. There were two basic factors, one observational and one experimental. The observational factor was age, with three levels: first-graders, fourth-graders, and adults. The experimental factor was instructions, with two levels: subjects who got "imagery instructions" were told to inspect a mental image to decide whether a statement was true; subjects who got "no imagery instructions" were simply told to respond as fast as they could. The response variable was the time it took to respond. If the hypothesis is correct, then the response time for children should be pretty much the same for both sets of instructions, because they will tend to rely on images even when they aren't told to. In contrast, the adults will be slower when they have been told to rely on images than when they are free to respond however they choose. Thus the hypothesis predicts an interaction effect.

To construct an interaction graph, we follow the four steps.

Step 1: Start with the condition averages:

	First	Fourth	Adult
Imagery instructions	2000	1650	1600
No imagery instructions	1850	1400	800

FIGURE 6.12 Response times in milliseconds for Kosslyn's experiment

Step 2: Draw a set of axes. We want a vertical scale that starts below 800 milliseconds and extends to above 2000 milliseconds. On the horizontal axis, we mark the three levels of the basic factor for columns.

Steps 3 and 4: For each of the two rows in the table, we plot the average values as points and then join the points by lines. The completed graph is shown in Fig. 6.13.

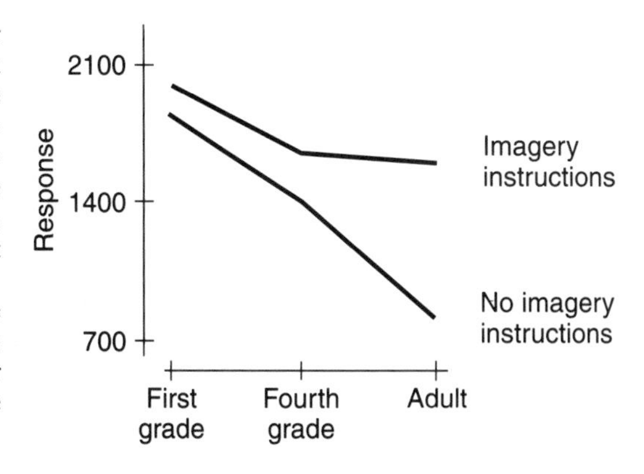

FIGURE 6.13 Interaction graph for Kosslyn's experiment. Notice that if you look only at first- and fourth-graders, covering up the right half of the graph, the lines are nearly parallel: there's not much interaction. Fourth-graders are faster than first-graders, and response times are slower for the imagery instructions, but the size of the effect of instructions is about the same for the two younger age groups.

For the right half of the graph, on the other hand, there is considerable interaction: the instructions have a much bigger effect for adults than for fourth-graders; the lines here are not even close to parallel.

Exercise Set B

1. *Kosslyn's imagery experiment*

   a. Follow the format of Example 6.8 to describe the interaction structure for the experiment in Example 6.10.

   b. For each level of the age factor, compute a difference of averages to measure the effect of instructions. Then write a sentence comparing the three differences.

   c. Use Kosslyn's data to decide whether the following statement is true or false; then write a sentence or two in support of your position. True or false: whether or not you find interaction between two factors can depend on the levels you choose for those factors.

2. *Kosslyn, continued: The other interaction graph.* For every BF[2], there are two interaction graphs, depending on which factor you use for rows and which for columns.

a. Rewrite the table of cell averages in Example 6.10, this time using age for rows and instructions for columns.

b. Draw the corresponding interaction graph.

c. Which of the two interaction graphs do you find easier to read? Tell why.

3. *Personality and selective attention*. The popular press has given a lot of attention to the type A personality: hard-driven, in a hurry, competitive, aggressive, and hostile. Such people are thought to be more inclined than others to judge themselves by their failures and accomplishments, and they are thought to be more likely than others to have heart attacks.

Although I don't disagree with these findings, I do urge you to be cautious: ask yourself how we know the type A personality really exists. Acorns exist, and it's easy to tell that I'm not an acorn: no one would argue about that except perhaps a philosopher. But how would you decide whether I'm a type A? How would you persuade the world you were right?

In the study I'm about to describe, "type A" refers to those subjects who answered a set of questions one way instead of another (Type B). The point here is that anyone could do that sort of thing with any set of subjects: make up a set of questions, collect the answers, sort people into two piles, and give names to the piles. If you set out to get two piles, you'll get them, whether they mean anything or not. The key issue, as I hope you recognize, is one of validity. I don't want to go into detail about how you would try to establish the validity of the two categories, but if you're interested, you could start with the references I've given in the data sources.

This study compared two groups of subjects, those who scored in the top quarter (type A) and those who scored in the bottom quarter (Type B) of the Jenkins Activity Survey, Form T. The goal was to see whether the two groups would differ in the amount of time they spent paying attention to what they thought were negative facts about themselves.

Imagine yourself as a subject. The experimenter tells you that the focus of the study is on the relationship between personality and learning. In the middle of what you think is the experiment, she tells you she forgot some cards and that she needs to go get them. She leaves you alone in the room with two stacks of cards and a code book of adjectives. The two stacks are labeled Data cards—Positive Assets and Data cards—Negative Liabilities. The experimenter has explained that the numbers on the cards are coded descriptions of you, chosen on the basis of the test you took earlier, and that while she is gone you can use the code book to find out the adjectives that go with the numbers.

In fact, the numbers were chosen by a chance device, and the purpose of the experiment is to see how much time you spend with the cards in the negative pile. Type A individuals are generally thought to be highly critical of themselves. The hypothesis behind the study is that the pattern of what type As do and don't pay attention to tends to confirm their negative view of themselves.

The cell averages are given in Table 6.2.

a. Draw both interaction graphs. Which do you find easier to understand?

b. Summarize the interaction using the format of Example 6.8.

TABLE 6.2 Time (seconds) spent attending to negative information.

	Type A	Type B
Male	170	72
Female	159	72

c. Discuss whether the results tend to support the hypothesis of the study. Is the hypothesis one that necessarily involves interaction?

4. *Classifying habitats*. Ecologists use the relationship between size and reproductive capacity to classify habitats as size-beneficial or size-neutral. For example, in a relatively dry habitat, larger seeds will be less likely to dry out than smaller seeds and so will be more likely to germinate: for this habitat, size has an advantage (which compensates for the extra energy it takes to produce larger seeds). In a relatively moist habitat, on the other hand, seeds don't need the same protection from low humidity, and seed size carries no advantage to offset the cost. The first habitat is size-beneficial, the second is size-neutral. (To be more precise, the first is offspring-size-beneficial and the second is offspring-size-neutral, because it is the seeds rather than the plants whose size matters.)

Consider an experiment to compare two such habitats. Pick a genus of plants, for example, and pick three species whose seeds are of different sizes, small, medium, and large. Then put 100 seeds of each species in each of two lab environments with controlled humidity, one dry and one moist. Take as your response the number of seeds that germinate.

a. Suppose that in fact the dry environment is size-beneficial and the moist environment is size-neutral. Make up, draw, and label an interaction graph to show this.

b. Describe the interaction structure using the format of Example 6.8.

5. *Dandelions*. A series of papers reported on three habitats for dandelions. Habitat 1 was a footpath, where mortality was high and size brought no advantage. Habitat 3 was an old pasture, where there was a lot of competition among adult plants, and there was an advantage to being big. Habitat 2 was in between.

There were four types of dandelions (A, B, C, D) growing in these habitats. For each habitat, the number of plants of each type was recorded. (The numbers in Table 6.3 are approximations I took from a graph.)

a. Draw the interaction graph. (Turn the table on its side first.)

b. Summarize the interaction structure using the format of Example 6.8.

c. According to the ecological theory, the largest of the four types would be likely to predominate in which habitat? Which do the data suggest is the largest type? The smallest?

6. (*Continuation of* 5.) Bigger plants need more energy and nutrients, which are in finite supply. This is the cost of being big: you need more of things that everybody needs. Theory predicts that because of this cost, larger plants will have fewer resources left over, and so will tend to produce fewer flowers or seeds. The average numbers of flower heads per plant for the same set of dandelions are listed in Table 6.4.

TABLE 6.3 Average number of plants of each type.

	Type of dandelion			
	A	B	C	D
Habitat 1	55	20	30	1
Habitat 2	30	50	25	10
Habitat 3	10	15	20	70

TABLE 6.4 Average number of flower heads per plant.

	Type of dandelion			
	A	B	C	D
Habitat 1	3.6	2.2	1.2	0.0
Habitat 2	2.6	1.5	2.0	0.0
Habitat 3	3.4	1.6	0.5	0.6

a. Draw and label an interaction graph. (Turn the table on its side first.)

b. Describe the pattern of interaction in words, and discuss its relationship to the theoretical prediction. Taking the theory and the data together, which plants do you think were the largest? The smallest? Do your inferences about size here agree with the ones in Exercise 5?

7. *Warm-blooded/cold-blooded.* Reread Exercise A.5, and suppose you had done the experiment using cats and lizards as your subjects (forget the spiny anteaters for this exercise), using temperatures of 5°C (41°F), 15°C (59°F), and 25°C (77°F).

a. Make up a table of cell averages, assuming the lizards will have body temperatures close to the temperature of their environment and cats will have body temperatures that are pretty much the same regardless of air temperature.

b. Visualize two interaction graphs, one for the table as you wrote it and another with the table sideways. Which graph would be easier to understand? Draw that graph.

c. Write a summary of the interaction structure, following the format of Example 6.8.

8–9. *The sizes of rectangles: Interaction depends on the scale.* Consider an experiment to study the effect of two factors, length and width, on the areas of rectangles. Suppose you choose three levels for the length factor (1", 4", 10") and two levels for the width factor (1", 4"). Take as your response the area of the rectangle, in square inches.

8. a. Write out a table showing the factorial crossing. Put a suitable average response value in each of the six cells of your table.

b. For each level of the length factor, measure the effect of width by computing the difference in areas for the two widths. Are these differences the same for all levels of the length factor?

c. Draw and label an interaction graph.

9. Now transform your data to logarithms, and repeat parts (a) to (c) in this new scale. Then (d) compare your two sets of results.

10. *Informal analysis of the pig data.* Because you can always ignore factorial treatment structure and regard any two-way BF design as having just one set of conditions, you can use averages and parallel dot graphs in exactly the same way as you would for a one-way BF design: to compare conditions, to check for outliers, and to compare the variability in the response. Do such an analysis for the pig data.

3. Decomposition and ANOVA for the Two-Way Design

Just like Section 4 of the last chapter, this section comes in three parts: first, decomposition; then degrees of freedom; finally, mean squares, the standard deviation, and F-tests. Because the F-tests depend on the assumption that residual errors are chancelike, I'll refer to them in this section as chance errors.

I'll use the pig data from Example 6.1 for illustration, but first, to make the arithmetic easier to follow, I'll use what statisticians call **coding** to change the scale for the response.

EXAMPLE 6.11 THE PIG DATA IN CODED FORM

For Example 6.1, the CR experiment to compare the effects of Vitamin B12 and antibiotics on how fast pigs gained weight, the response was measured in pounds per day. Because every observation equals "1.something," I've left off all the 1s. Then, to get rid of the decimals, I multiplied every number by 100. Thus, 1.30 becomes .30 and then 30; 1.19 becomes .19 and then 19, and so forth. In effect, I've changed from measuring in pounds to measuring in hundredths of a pound above 1 pound. (See Fig. 6.14.) ■

	Original Data Antibiotics 0 mg	 40 mg		Coded Data Antibiotics 0 mg	 40 mg	Cell Averages	
0 mg	1.30	1.05	0 mg	30	5		
B12	1.19	1.00	B12	19	0	19	3
	1.08	1.04		8	4		
5 mg	1.26	1.52	5 mg	26	52		
B12	1.21	1.56	B12	21	56	22	54
	1.21	1.54		19	54		

FIGURE 6.14 Coding the pig data makes the patterns easier to see. The numbers on the left are in pounds. The numbers on the right give the same information measured in hundredths of a pound above 1 pound. Before reading ahead, you might try to anticipate the formal analysis to come: guess whether the SD for the coded data is closest to 1, 3, 5 or 10; whether the overall effect of B12 is large or small in comparison with this SD; and likewise for the overall effect of antibiotics and the interaction effect.

Decomposition

As always, the decomposition of a data set is based on its factor structure. (Fig. 6.15).

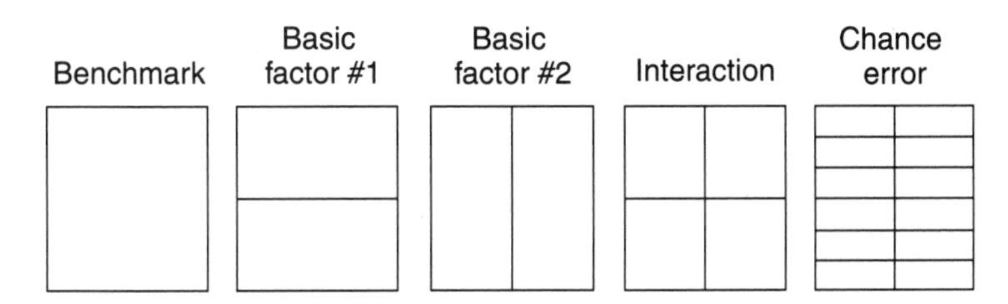

FIGURE 6.15 Factor diagram for a CR[2] design with two levels for each basic factor. The factor structure for every BF[2] design tells us to split each observed value into a sum of five pieces: estimated benchmark, plus estimated effects for each of the two basic factors (also called estimated **main effects**), plus estimated interaction effect, plus estimated chance error.

To split the set of observations into five pieces, one for each factor, we apply the general decomposition rule: The estimated effect for a factor equals the group average minus the partial fit (= sum of estimated effects for all outside factors). To prepare for the rule, first look over the factor diagram, and find the inside/outside relationships: the benchmark, as always, is outside all other factors; each of the two basic factors is inside the benchmark and outside both the interaction and the residual error; interaction is inside both basic factors and the benchmark and outside only residual error; residual error, as always, is inside all other factors.

1. *Benchmark.* There are no outside factors; the partial fit is zero; the estimate equals the grand average.
2. *Main effects.* For each basic factor, only the benchmark is outside, so the partial fit equals the grand average, and each estimated main effect equals the average for a level of the basic factor minus the grand average. Notice that these estimates follow the same pattern as for condition effects in the BF[1]: factor average − grand average.
3. *Interaction.* There are three outside factors: the benchmark plus both basic factors. The partial fit estimates what you would get for cell averages if there were no interaction: just the grand average plus the two estimated main effects. (See Exercises C.22–24.) Each interaction effect equals a cell average minus this partial fit.
4. *Chance error.* The factor for chance error is always inside all other factors, so each fitted value always equals the sum of the effects for the other factors. Exercise E.3 asks you to show that for every BF[2], the sum of these effects equals the cell average, which makes each residual equal to observed value minus cell average (just as for any BF[1]).

You can always compute all the averages you need directly from the observed values, but if you're doing the arithmetic by hand, you can save a little time if you compute the cell averages first, and use them to get the other averages. The box below gives one reasonably efficient way to do this, and Example 6.12 illustrates the steps. (However, it would be a waste of effort to memorize the steps. All you need to remember is the one general rule for all balanced designs.)

Decomposition for the Balanced Two-Way Design

1. **Decompose each observed value into cell average plus residual.**
2. **Decompose the cell averages into four pieces, in two steps:**
   a. **Use the table of cell averages to compute the averages for the two basic factors, then the grand average, and then the estimated main effects (= factor average − grand average).**
   b. **Interaction. Compute partial fit = grand average plus effect for basic factor 1 plus effect for basic factor 2; then interaction effect = cell average − partial fit.**

Example 6.12 Decomposition for the Coded Pig Data

Step 1: Decompose each observed value into a cell average plus residual (see Fig. 6.16).

Obs		Cell avg		Residual	
30	5	19	3	11	2
19	0	19	3	0	−3
8	4	19	3	−11	1
26	52	22	54	4	−2
21	56	22	54	−1	2
19	54	22	54	−3	0

FIGURE 6.16 Observation = Cell average + Residual. Here, as always, the numbers go together according to where they sit in the boxes, and the deviations from each average add to zero.

Step 2a: Use the cell averages to compute first the averages for the basic factors, then the grand average, and then the estimated main effects (Fig. 6.17).

Step 2b: Interaction effects.

		Antibiotics 0 mg	Antibiotics 40 mg	Rows Tot	Rows Avg	Rows Eff
B12	0 mg	19	3	22	11	−13.5
B12	5 mg	22	54	76	38	13.5
COL	Tot	41	57	98		
COL	Avg	20.5	28.5		24.5	Grand avg
COL	Eff	−4.0	4.0			

FIGURE 6.17 Using the cell averages to estimate the benchmark and main effects. First compute row and column totals, row and column averages, and the grand average. Then subtract the grand average from the row and column averages to estimate the main effects.

Partial fit = grand average plus sum of the two main effects.

Grand avg			B12 effect			Antibiotic effect			Partial fit	
24.5	24.5	+	−13.5	−13.5	+	−4.0	4.0	=	6.5	15.5
24.5	24.5		13.5	13.5		−4.0	4.0		34.5	43.5

Estimated interaction effect = Cell average − partial fit

Cell avg			Partial fit			Inter. effect	
19	3	−	6.5	15.5	=	12.5	−12.5
22	54		34.5	43.5		−12.5	12.5

Putting the pieces together gives the complete decomposition:

Obs			Grand avg			B12			Antibiotics			Interaction			Residuals	
30	5		24.5	24.5		−13.5	−13.5		−4	4		12.5	−12.5		11	2
19	0		24.5	24.5		−13.5	−13.5		−4	4		12.5	−12.5		0	−3
8	4	=	24.5	24.5	+	−13.5	−13.5	+	−4	4	+	12.5	−12.5	+	−1.1	1
26	52		24.5	24.5		13.5	13.5		−4	4		−12.5	12.5		4	−2
21	56		24.5	24.5		13.5	13.5		−4	4		−12.5	12.5		−1	2
19	54		24.5	24.5		13.5	13.5		−4	4		−12.5	12.5		−3	0

First, read the decomposition one observation at a time. The numbers tell you, for example, that for the first control pig in the upper left corner, the observed daily gain of 30 (hundredths of a pound above 1 pound) equals the grand average of 24.5, plus an estimated effect of −13.5 (no B12), plus an estimated effect of −4.0 (no antibiotics), plus an estimated interaction effect of +12.5, plus an estimated chance error of 11.

Now read the decomposition one box at a time, starting with the residuals. The SD will turn out to be about 6, which seems a reasonable number for the "average-sized" residual. The overall effect of antibiotics is roughly the same size as this; the overall effect of B12 and the interaction effect are roughly twice this big. Finally, notice the patterns of repetitions and adding to zero—patterns you will need to use in order to count free numbers. ■

Degrees of Freedom

There are three ways you can find the degrees of freedom: by counting free numbers; by the general rule based on inside and outside factors; or by a shortcut formulas for the BF[2] design. Exercise E.4 will ask you to show that if you apply the general rule for degrees of freedom, you get the results summarized in the following box.

Degrees of Freedom for the Balanced Two-Way Design

Factor	Degrees of Freedom
Benchmark	1
Each basic factor	number of levels − 1
Interaction	(df for 1st factor) × (df for 2nd factor)
Chance error	(number of cells) × (number of observation per cell − 1)

Although I've included these shortcuts (here and elsewhere) because almost every book on ANOVA does, I urge you not to memorize them or overuse them. Try to rely instead on the other two methods, which work for all balanced designs, not just the BF[2], and which are more closely tied to the meaning of df as units of information about chance error.

Example 6.13 Degrees of Freedom for the Pig Data: Counting Free Numbers

According to the boxed summary, the df for the pig data are

$$\begin{aligned} df_{Grand} &= 1, \\ df_{B12} &= (2-1) = 1, \\ df_{Anti} &= (2-1) = 1, \\ df_{Inter} &= (2-1)(2-1) = 1, \\ df_{Res} &= 4(3-1) = 8. \end{aligned}$$

Using boxes with Rs and pluses to show repetitions and patterns of adding to zero, show by counting free numbers that the degrees of freedom are correct.

SOLUTION. See Figure 6.18.

Grand
df = 1

1	R
R	R
R	R
R	R
R	R
R	R

Entire box is constant. No patterns of adding to zero.

B12
df = 1

1	R
R	R
R	R
+	R
R	R
R	R

Top half is constant. Bottom half is constant. Halves add to zero.

Antibiotics
df = 1

1	+
R	R
R	R
R	R
R	R
R	R

Columns are constant. Column effects add to zero.

Interaction
df = 1

1	+
R	R
R	R
+	+
R	R
R	R

Effects constant in each cell. Interaction effects add to zero, both across, and down.

Residuals
df = 8

1	5
2	6
+	+
3	7
4	8
+	+

No repetitions. Residuals add to zero within each cell.

FIGURE 6.18 Degrees of freedom for the pig data.

Mean Squares, Standard Deviation, and F-tests

Once you have the decomposition and degrees of freedom, constructing an ANOVA table is straightforward. You follow the same pattern as for a one-way design, except that instead of just one row for conditions you have three: one for each basic factor, and one for interaction. In each case, you get the *F*-ratio using MS_{Res} for the denominator.

Before carrying out the *F*-tests, however, it is a good idea to estimate the SD and use it for a rough check of the normality assumption. (See Note 3 at the end of Chapter 5.)

EXAMPLE 6.14 THE SD AND NORMALITY ASSUMPTION FOR THE PIG DATA

Compute an estimate for the SD of the chance errors, and check the normality assumption by comparing the residuals with the estimated SD.

SOLUTION. To estimate the SD, we take the square root of MS_{Res}. From the residual box of the decomposition (Example 6.12), we first compute the residual sum of squares, which equals 290. Then

$$MS_{Res} = \frac{SS_{Res}}{df_{Res}} = \frac{290}{8} = 36.25,$$

$$SD = \sqrt{MS_{Res}} = \sqrt{36.25} = 6.02.$$

Then, to check the normality assumption, we mark off ± 1 SD, ± 2 SDs, ± 3 SDs under a dot diagram for the residuals, and find the number of residuals in each interval (Fig. 6.19).

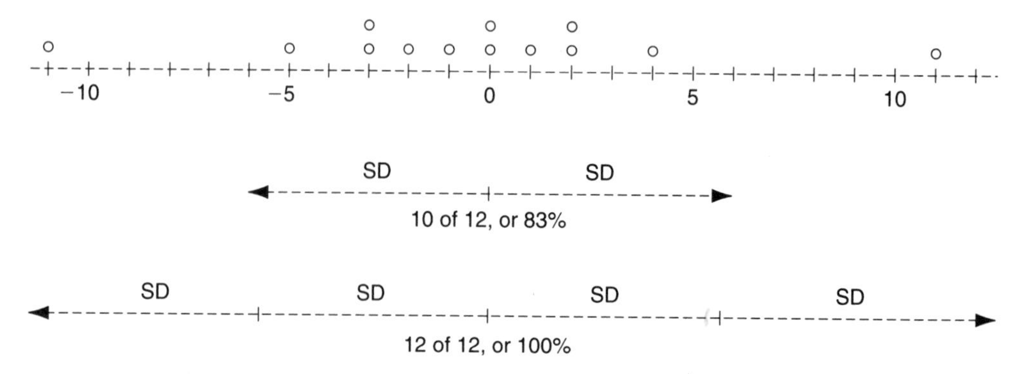

FIGURE 6.19 Checking the 1-2-3-SD rule for the pig residuals.

- 83% of the residuals are within 1 SD of zero. The rule predicts 2/3 or about 67%.
- 100% of the residuals are within 2 SDs of zero. The rule predicts 95%.

The fit is good enough to go ahead with ANOVA and the *F*-test, although two points deserve comment:

1. Note that the residuals tend to be somewhat closer to zero than the rule predicts. This is quite generally true in ANOVA: residuals always tend to underestimate the actual chance errors. See Note 1 at the end of the chapter for more on this.
2. The twin residuals at ± 11 are isolated from the other 10 residuals. Should we regard the corresponding observations with suspicion? Although they are different enough to be worth thinking about, on balance I don't regard them as deviant. Here's why: the two extreme residuals correspond to observed values of 8 and 30 for the control group, whose other observation is 19. Because the three values in the group are evenly spaced (8 19 30), I can't reasonably consider one of the 8 and 30 "bad" unless I'm willing to consider the other one "bad" as well. Although it is possible to get two such deviant observations in a set of 12, both in the same group of three, I judge it unlikely. Nevertheless, it's worth thinking about what happens to the analysis if you replace either of the extreme observations with the average of the other two. (See Exercise C.25.) ■

EXAMPLE 6.15 ANALYSIS OF VARIANCE TABLE FOR THE CODED PIG DATA

Table 6.5 is the ANOVA table for the coded pig data.

Before you think about the *F*-tests, notice how the SSs are related to the tables whose variability they summarize. The total sum of squares is about 11,750, which gets divided among the five tables of the decomposition in Example 6.12. Most of it goes to the grand average, about a third goes to the three boxes for main effects and interaction, and only a small piece goes to residuals. Notice that SS_{B12} and SS_{Inter} are both much larger than SS_{Anti}. You can see this in the tables themselves: the main effects for B12 (± 13.5) and interaction effects (± 12.5) are roughly equal in size and are two to three times as big as the main effect for antibiotics (± 4). Finally, look back at the residual table, and notice that most of the variability comes from the two

TABLE 6.5 ANOVA table for the coded pig data.

Source	SS	df	MS	*F*-ratio	Table value
Grand average	7,203	1	7,203		
B12	2,187	1	2,187	60.33*	5.32
Antibiotics	192	1	192	5.30	5.32
Interaction	1,875	1	1,875	51.72*	5.32
Residual	290	8	36.25		
Total	11,747	12			

A * indicates that the true diet differences are "statistically significant," that is, big enough to be detected by the experiment.

big residuals: $(11)^2 + (-11)^2 = 242$. The remaining 10 residuals together contribute only 48 to $SS_{Res} = 290$.

Each of the three *F*-ratios compares the mean square for a main effect or interaction with the mean square for chance error. For example, the *F*-ratio for B12 tells us that MS_{B12} is 60.33 times as big as MS_{Res}. The average variability per unit of information about chance error in the B12 box is almost 65 times the average variability in the residual box. Informally, there appears to be much more variability in the B12 box than would be due to chance error alone. Since the *F*-ratio is much bigger than the critical value (5.32), we reject the null hypothesis, which says the true main effects for B12 are zero, and we call B12 statistically significant: the "true" overall differences due to B12 are big enough to be detected. Similarly, we find that the difference due to interaction is also "real." The overall difference due to antibiotics (averaged over B12 present and absent) is right on the border, too close to call, but because the interaction effect is real, we know that the difference due to B12 is different depending on whether or not antibiotics are present in the diet. ■

Exercise Set C

1. Decompose the following fake version of the hornworm data.

   RR: 5, 9 RC: 2, 4
   CR: 10, 12 CC: 16, 22

2. *Degrees of freedom*. A decomposition of a fake BF[2] data set is given below. Use the patterns of repetition and adding to zero to fill in the missing numbers. Then write the df below each box.

Grand average

12	?	?	?	?	?
?	?	?	?	?	?

\+ Row effects

?	?	?	?	?	12
?	?	?	?	?	?

\+ Column effects

?	?	?	?	?	1
?	?	?	−4	?	?

\+ Interaction effects

−2	?	?	?	?	5
?	?	?	?	?	?

\+ Residuals

−2	?	0	?	?	1
?	3	?	−4	?	3

3–9. *Puzzled children: Decomposing a data set.* The data set below comes from a much larger study done in the Department of Psychology and Education at Mount Holyoke College. The response is the number of simple puzzles solved by a child in a fixed length of time. The data are for 12 children, cross-classified by age and sex.

	Age (years)					
Sex	3		5		7	
Male	24	20*	18	16	31	53
Female	20	8	50	36	54	66

3. Of the two basic factors, which are observational, which are experimental?

4. Construct the table of cell averages and draw the interaction graph. Which cell average seems most "out of line" with the others? What is your reason for judging it "out of line"? Fill in the blanks with either "boys" or "girls": The numbers suggest that at age 3, ________ are better than ________ at solving the kind of puzzles used in the experiment but that within a few years, the ________ have overtaken the ________.

5. *Estimated effects.*

   a. Find the averages for the two basic factors.

   b. Then complete the following sentence, filling in the blanks with appropriate numbers or "boys" or "girls": Overall, the ________ solve ________ puzzles fewer than the grand average; the ________ solve ________ puzzles more.

   c. Write a sentence interpreting the estimated effects for age, following the pattern in (b).

6. *Interaction.*

   a. Find the table of partial fits for estimating interaction.

   b. If there were no interaction, then we would expect 3-year-old boys to solve ________ puzzles, and would expect 7-year-old girls to solve ________ puzzles.

   c. Estimate the interaction effects.

7. Write the complete decomposition of the data set.

8. *Degrees of freedom.* For each factor in your decomposition of the data in Exercise 1, write out a box of Rs, pluses, and numbers to show the patterns of repetitions, patterns of adding to zero, and numbers of free numbers. (Follow Example 6.13 as a model.)

9. ANOVA. Use your results from Exercises 7 and 8 to write the ANOVA table for the puzzled children data. What do you conclude from the *F*-ratios?

10. *Fake data for quick practice.* Construct the complete ANOVA table for data from a BF[2] if there are 5 observations per cell, the estimated SD is 2, and the condition averages are as follows:

15	5	25
15	25	35

11–12. *Bird calcium*

11. Construct an interaction graph for the transformed bird calcium data given in Exercise 5.C.9. Is the effect of the hormone supplement on the log concentration of calcium in the plasma about the same for male and female birds?

12. Decompose the data, give the ANOVA table, and tell what you conclude from the F-ratios.

13–15. *Hornworms.*

13. Draw a parallel dot graph for the data given in Exercise A.7. Discuss the usual four features (G O S T).

14. The purpose of the experiment was to see whether worms raised on the cellulose mixture would get so accustomed to eating a lot (in order to compensate for the low nutritional value) that they would continue to eat a lot even after they were switched over to regular food during the fourth instar. Thus, there are two alternative outcomes: either the worms in the CR group would continue to eat a lot, or they would not.

   a. Which outcome corresponds to "no interaction"?

   b. Draw the interaction graph, and describe what it tells you in light of the purpose of the experiment.

15. An ANOVA will yield three F-ratios: one for rearing food, one for test food, and one for interaction.

   a. Which of these do you expect to be larger than their corresponding critical values?

   b. Do the decomposition and ANOVA; then write a short summary stating your conclusions.

16–18. *Hornworms (continued): Informal checks of assumptions*

16. Same SDs. Use your decomposition from Exercise 15(b) to compute a separate SD for each of the four cells. Compute the ratio SD_{Max}/SD_{Min}. Are the SDs roughly equal? If not, does the size of the cell SD seem related to the size of the cell average?

17. *Normality*. Use a dot graph like the one in Example 6.14 to do a rough check of the normality assumption. What do you conclude?

18. Based on Exercises 16 and 17, do you think your conclusions based on the F-tests in Exercise 15(b) should be revised? Why or why not?

19–21. *Hornworms (continued): Estimating replacement values for outliers*

19. Notice from your dot graph that the value of 61 for the second worm in the RR group is a clear outlier. One way to handle an outlier (Chapter 12, Section 5) is to regard it as a "missing" observation, estimate a replacement value, and analyze the data set with this estimate in place of the outlier. If you do this, you should reduce your df for residuals by 1. For any BF design, you get the replacement value by taking the average of all the observations except the outlier in the group that contains the outlier. Do this to estimate a replacement value for the 61. Does replacing the outlier have much effect on the cell average for RR? On the interaction pattern? Will replacing the outlier change: the grand average? any of the cell averages for RC, CR, or CC? any of the residuals for these three cells?

20. The assumption of same SDs. In Exercise 16 you computed and compared SDs for the four conditions. Recompute the SD for RR, and notice that replacing the outlier changes the cell SD a lot more than it changed the cell average. Does the ratio of largest-to-smallest SD change?

21. Notice that with the outlier replaced, there is now a clear pattern of large SD with large average, small with small, suggesting that a transformation would be able to make the cell SDs more nearly equal. It would seem that we ought to transform and then redo the whole analysis. What evidence can you give to show that all this additional work would not change the conclusions?

22–24. "No interaction means the effects of the basic factors are additive."

The purpose of these exercises is to illustrate the relationship between interaction and the terms in a linear decomposition.

22. In part (a) of Exercise B.8, you made up a set of cell averages for a "study" of factors affecting the areas of rectangles.

a. Use that table of cell averages to find the grand average and estimated effects for length and width.

b. Now find the partial fit for interaction. (These numbers tell you what the areas would be if you could find areas simply by adding the effects of length and width.) Draw two interaction graphs, one using the actual cell averages and a second one using the partial fits.

23. Now repeat Exercise 22, this time using the logs of the areas, from part (d) of Exercise B.9.

24. a. For the puzzled children: are the effects of age and sex additive?

b. For the birds: are the effects of sex and the hormone treatment additive?

c. For the hornworms: are the effects of test food and rearing food additive?

25. *Pigs, revisited.* In Example 6.12 we found a pair of extreme residuals, at ±11, corresponding to weight gains of 8 and 30 (hundredths of a pound) for two of the pigs on the control diet. To judge the effect of one of these observations on the analysis, you can replace it with the average of the two other observations in the group, and redo the analysis. (For the reanalysis using the artificial observation, you should reduce df_{Res} by 1, from 8 to 7.

a. Replace the 8 by the average of 19 and 30, and redraw the interaction graph. Then put back the 8, replace the 30 by the average of 8 and 19, and draw another interaction graph. Write a short paragraph on the sensitivity of the interaction pattern to these changes.

b. Suppose you replaced one of the two questionable observations as described in (a) and were then to redo the decomposition and ANOVA. The residual sum of squares for the new analysis will be a lot less than for the analysis based on all 12 actual observations. Will SS_{Res} be lower by roughly 10, 25, 50, or 100?

4. Using a Computer [Optional]

This section shows how to use Minitab (Release 11) to analyze the data from Example 6.1

Required form for data. The example in Fig. 6.20 shows how to put the data from the Pigs and Vitamins experiment into the form required by most computer packages. I've used numbers for the levels of the three structural factors (B12, Anti, and Inter), as required by older packages. Newer packages, including Release 11 of

	Antibiotics 0 mg	40 mg	Gain	B12	Anti	Inter
			30	1	1	1
0 mg	30	5	19	1	1	1
B12	19	0	8	1	1	1
	8	4	26	2	1	2
5 mg	26	52	21	2	1	2
B12	21	56	19	2	1	2
	19	54	5	1	2	3
			0	1	2	3
			4	1	2	3
			52	2	2	4
			56	2	2	4
			54	2	2	4

FIGURE 6.20 Coded pig data in rectangular and columnar forms. The listing on the left shows the data in the rectangular layout used throughout this chapter. Levels of the two treatment factors are shown by an observed value's position in the table. The listing on the right shows the column-oriented format required by Minitab and most other computer packages. The first column ("Gain") holds the response values and comes from stacking the two side-by-side columns of the rectangular layout on top of each other. The next two columns on the right ("B12" and "Anti") tell the levels of the two treatment factors. The last column ("Inter"), which tells levels of the factor for interaction, would not be needed for many computer programs.

Minitab, allow you to use names instead of numbers. So, for example, in the column for the factor B12, I could have typed labels 0 mg and 5 mg instead of the numbers 1 and 2. (Using names will make some displays, such as interaction graphs, a bit easier to read.) Most computer packages, Minitab included, do not require you to include a separate column for interaction. However, you will need this column if you want Minitab to compute separate SDs for the cells of your two-way design as a check on assumption S of same SDs.

Interaction graphs. Newer versions of Minitab will now do these for you. From the `Stat` menu, choose `ANOVA`, then `Interactions Plot`:

```
Stat > ANOVA > Interactions Plot ...
```

In the dialog box that results, tell the computer the names of your two factors, and of the response:

```
Factors:                 B12 Anti
Raw response data in:    Gain
```

To get the second version of the interaction graphs, repeat the sequence of commands, but reverse the order of the two factors:

```
Stat > ANOVA > Interactions plot ...
Factors:                 Anti B12
Raw response data in:    Gain
```

ANOVA using Minitab. You'll need your response values in one column, with levels of your two treatment factors in two more columns. (The first three columns in the right-hand listing of Fig. 6.20 illustrate one acceptable format.) In Minitab you can do a balanced two-way ANOVA with either of two commands: `Twoway`, which only works for balanced two-way designs, and `ANOVA`, which works for any balanced design. Although `Twoway` is a bit simpler to use, it's worth the effort to learn the more flexible `ANOVA` command instead because you can use it for many other designs besides the BF[2].

From the `Stat` menu, choose `ANOVA`, then `Balanced ANOVA ...`:

```
Stat > ANOVA > Balanced ANOVA ...
```

The resulting dialog box asks for the response and your model. In the pig example, these are as follows:

```
Response:   Gain
Model:      B12 Anti B12*Anti
```

The model is simply a list of all the structural factors, separated by spaces, with an asterisk used to specify interaction. (Minitab offers a shortcut when you tell the model. A vertical line or an exclamation point between two factors tells Minitab to include the interaction. Thus, instead of listing all three factors to the right of the `=` you could have typed `B12|Anti` to say the same thing.)

Averages. Minitab won't show you the decomposition of the data, but you can ask to see the averages for the levels of any or all factors. Simply click the `Options` button toward the bottom of the dialog box for `Balanced ANOVA`, and list the structural factors in the resulting box:

```
Display means corresponding to the terms:   Anti B12 Anti*B12
```

Here, as before, you can use the shortcut `B12!Anti` to abbreviate the two factors and their interaction.

Fig. 6.21 shows the resulting computer output.

Patterned data: a shortcut for typing in factor levels. Look back at the listing of the Pig data at the beginning of this section, and focus on the three structural factors. Although it doesn't take very long to type in the factor levels for a data set as small as this one, there are times when you may wish you had a shortcut. Minitab has one.

Here's how you would use the shortcut to enter the levels of the factor B12. From the `Calc` menu, choose `Make Patterned Data` and then `Simple Set of Numbers ...`

```
Calc > Make Patterned Data > Simple Set of Numbers ...
```

Then fill in the dialog box as follows:

```
Store patterned data in:        B12
From first value:               1
To last value:                  2
In steps of:                    1
List each value:                3      times
Repeat the whole sequence:      2      times
```

This tells Minitab to start with the sequence 1 2, then list each value 3 times to get 1 1 1 2 2 2, and finally repeat that sequence two times to get 1 1 1 2 2 2 1 1 1 2 2 2.

```
        Analysis of Variance (Balanced Designs)
Factor      Type    Levels   Values
B12         fixed   2 0 mg    5 mg
Antibiot    fixed   2 0 mg   40 mg

Analysis of Variance for Gain

Source          DF      SS        MS        F       P
B12              1   2187.0    2187.0    60.33   0.000
Antibiot         1    192.0     192.0     5.30   0.050
B12*Antibiot     1   1728.0    1728.0    47.67   0.000
Error            8    290.0      36.3
Total           11   4397.0

F-test with denominator: Error
Denominator MS = 36.250 with 8 degrees of freedom

Numerator       DF      MS        F        P
B12              1   2187.0    60.33    0.000
Antibiot         1    192.0     5.30    0.050
B12*Antibiot     1   1728.0    47.67    0.000

Means

B12            N       Gain
0 mg           6     11.000
5 mg           6     38.000

Antibiot       N       Gain
0 mg           6     20.500
40 mg          6     28.500

B12      Antibiot      N       Gain
0 mg        0 mg       3     19.000
0 mg       40 mg       3      3.000
5 mg        0 mg       3     22.000
5 mg       40 mg       3     54.000
```

FIGURE 6.21 Minitab output for ANOVA of the Pigs and Vitamins data (Example 6.1).

5. Algebraic Notation for the Two-Way BF Design [Optional]

In this section I'll use the pig weight example one last time, to introduce algebraic notation first for the BF[2] model and then for the estimated effects. Here's the factor structure we want to represent using symbols:

Grand mean | Vitamin B12 | Antibiotics | Interaction | Residual error

The notation and basic ideas here are the same as for the BF[1] example in Section 6 of Chapter 5: y stands for the observed value; each y is written as a sum of terms, one for each factor; μ stands for the benchmark amount; e stands for the chance errors; and we use subscripts to tell how many levels there are for each factor and how the terms get added together.

Here are the standard symbols, without their subscripts:

Observed value =	y
Benchmark	μ
+ B12 effect	α
+ Antibiotics effect	β
+ Interaction effect	$(\alpha\beta)$
+ Residual error	e

This is usually written horizontally: $y = \mu + \alpha + \beta + (\alpha\beta) + e$.

Now for the subscripts. A two-way design with more than one observation per cell has three subscripts:

First subscript	i	tells the level of the first basic factor.
Second subscript	j	tells the level of the second basic factor.
Third subscript	k	counts observations in each cell.

The first two subscripts together tell you which cell an observed value belongs to. For the pig data, there are four cells in a two-by-two rectangle.

		No antibiotics $j = 1$	Antibiotics $j = 2$
No B12	$i = 1$	$ij = 11$	$ij = 12$
B12	$i = 2$	$ij = 21$	$ij = 22$

For our example, there are three observations in each cell, so the third subscript will be 1, 2, or 3.

		Antibiotics	
		$j = 1$	$j = 2$
B12	$i = 1$	y_{111} y_{112} y_{113}	y_{121} y_{122} y_{123}
	$i = 2$	y_{211} y_{212} y_{213}	y_{221} y_{222} y_{223}

Part of what makes algebraic notation hard to get used to is its efficiency. We're used to reading English, which has a lot of redundancy built into it. You don't have to pay attention to every little squiggle to get the whole message. In fact, you don't even have to see everything. I bet you can read this: Mry hd a lttl lmb. Algebraic notation makes every little squiggle carry a message, which means that to read it, you have to train yourself to pay attention to small differences that you could ignore in ordinary English.

For example, y_{112} and y_{121} look almost alike, but y_{112} stands for cell 11, pig 2, the second pig who got no B12 and no antibiotics; and y_{121} stands for cell 12, pig 1, the first pig who got no B12 but 40 mg antibiotics.

To finish the shorthand for the factor structure we add subscripts. The benchmark value is the same for all the observations, so μ gets no subscripts. There are two levels for the first basic factor, B12, so we need an α_1 and an α_2:

Grand mean
μ

Vitamin B12
α_1
α_2

Antibiotics

Interaction

Residual error

There are also two levels for the second basic factor, antibiotics, so we have a β_1 and a β_2 :

Grand mean
μ

Vitamin B12
α_1
α_2

Antibiotics
β_1 β_2

Interaction

Residual error

Notice that there's a new wrinkle in the notation here. According to the code we've set up, y_{213} means level 2 of the first factor, level 1 of the second factor, pig 3 in that cell. The first subscript tells you the level of the first factor, the second subscript tells the level of the second factor. The same convention holds for α_1 and α_2: the first (and only) subscript tells you the level of the first factor. For β, however, which also has only a first subscript, we can't rely on the same convention. The β subscript tells you the level not of the first factor, but of the second. The only way you know which basic factor the notation refers to is from the Greek letters: α is for the first factor, β is for the second.

The interaction terms are $(\alpha\beta)_{11}$, $(\alpha\beta)_{12}$, etc., and the two subscripts tell you the cell:

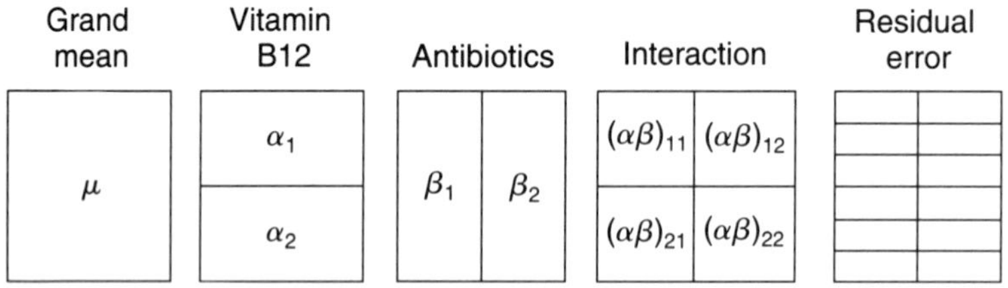

Grand mean	Vitamin B12	Antibiotics		Interaction		Residual error
μ	α_1	β_1	β_2	$(\alpha\beta)_{11}$	$(\alpha\beta)_{12}$	
	α_2			$(\alpha\beta)_{21}$	$(\alpha\beta)_{22}$	

Finally, the residual errors have three subscripts that follow the same pattern as the observed values.

		Antibiotics	
		$j = 1$	$j = 2$
B12	$i = 1$	e_{111} e_{112} e_{113}	e_{121} e_{122} e_{123}
	$i = 2$	e_{211} e_{212} e_{213}	e_{221} e_{222} e_{223}

Here, then, is how most statistics books abbreviate the notation for the factor structure:

$$y_{ijk} = \mu + \alpha_i + \beta_j + (\alpha\beta)_{ij} + e_{ijk}.$$

The way I read this is first to ignore the subscripts and then to translate:

$y = \mu + \alpha + \beta + (\alpha\beta) + e$, or
"Observed value (y) = grand mean (μ) + effect of first basic factor (α) + effect of second basic factor (β) + interaction adjustment ($(\alpha\beta)$) + residual error (e)."

Next, I put the subscripts back, one at a time, thinking about one particular observed value, say y_{121}. The subscripts 121 tell you $i = 1$, $j = 2$, $k = 1$, so every time an i appears in the equation for the model, I put a 1:

$$y_{1jk} = \mu + \alpha_1 + \beta_j + (\alpha\beta)_{1j} + e_{1jk}$$

$$i = 1$$

I go back and put a 2 for every j:

$$y_{12k} = \mu + \alpha_1 + \beta_2 + (\alpha\beta)_{12} + e_{12k}$$

$$j = 2$$

Finally, I put a 1 for each k:

$$y_{121} = \mu + \alpha_1 + \beta_2 + t_{12} + e_{121}$$

$$k = 1$$

The notation tells me that the observation in cell 12, for pig 1, equals the benchmark (μ), plus the effect for level 1 of factor 1 (α_1), plus the effect for level 2 of factor 2 (β_2), plus the interaction effect for cell 12 [$(\alpha\beta)_{12}$]; plus the chance error for cell 12, pig 1 (e_{121}).

The complete model for a BF[2] design has two more parts, a set of restrictions on the parameters, and the usual four Fisher assumptions about the residual errors: they behave like independent draws (I) from the same box (S) whose numbered tickets follow a normal curve (N) centered at zero (Z). The restrictions on the parameters are that they add to zero following the same patterns as the estimated effects: the α_i add to zero; the β_j add to zero; and the $(\alpha\beta)_{ij}$ add to zero across each row and down each column (that is, if you fix either subscript and add over the other you get zero). Exercise D.5 deals with the logic of these restrictions.

We can extend the convention for averages from the Section 6 of Chapter 5 to get a shorthand notation for averages and estimated effects for the BF[2] design. A $\bar{y}$ stands for an average of observed values, and dots in place of subscripts tell you which observations get added together:

$\bar{y}_{...}$ = grand average; add together $y_{111} + y_{112} + \cdots + y_{223}$ (then divide).

$\bar{y}_{2..}$ = average of all the observations that have a 2 as the first subscript, that is, all the observations for pigs who got B12.

$\bar{y}_{.1.}$ = average of all the observations that have a 1 as the second subscript, that is, all observations for pigs who got no antibiotics.

$\bar{y}_{21.}$ = average of all the observations that have a 2 as the first subscript and a 1 as the second subscript, that is, all the observations in cell 21 (5 mg B12, 0 mg antibiotics).

Two exercises, D.4 and D.6, give you a chance to put this notation to work; first, to get shorthand expressions for the estimated effects in the decomposition of a BF[2]; then to use the model to rewrite these expressions in terms of the parameters and residual errors, in order to see the relationship between the estimates and the true values.

Exercise Set D

1. Refer to Exercise A.6, where you wrote out a factor diagram for the bird calcium data.
   a. What is the largest value of subscript i? of j? of k?
   b. Tell in words what α_2 stands for; what β_1 stands for; what $(\alpha\beta)_{12}$ stands for; what e_{112} stands for.
2. Refer to Exercise A.7, where you wrote out the factor diagram for the tobacco hornworm data.
   a. What is the largest value of subscript i? of j? of k?
   b. Translate each of the following into algebraic notation:
      i. The overall effect for regular test food
      ii. The observed value that sits in the second row, sixth column of the rectangle
      iii. The observed value that sits in the first row, third column of the rectangle
      iv. The interaction effect for rearing food = cellulose mixture, test food = regular.
3. *Degrees of freedom*
   a. Use the capital letter convention for the largest values of subscripts (I stands for the largest value of i, J for the largest value of j, etc.) to abbreviate the expressions for df given in the box on page 220 in Section 3 of this chapter.
   b. Show algebraically that the df add up to give df_{Total}.
4. *Notation for estimated effects*. Write expressions for the following estimates using the y-bar notation:
   a. The estimated benchmark
   b. The estimated effect for level i of the first basic factor
   c. The estimated effect for level j of the second basic factor
   d. The estimated interaction effect for cell i,j
   e. The estimated residual error for the kth observation in cell i,j.
   f. Write the sum of these five terms, and show that the sum simplifies to give you y_{ijk}.
5. *Restrictions on the parameters: summing to zero*. The model for the BF[2] design includes a set of restrictions on the α, β, and $(\alpha\beta)$ terms. The goal of this exercise is to show you where these restrictions come from. First, a bit more notation. A handy shorthand for "the α_i add to zero" is $\alpha_+ = 0$. Putting a $+$ in place of a subscript means "add up all the terms with that subscript," just as putting a dot in place of a subscript tells you to take an average. In this notation, the restrictions on the parameters of the BF[2] become

$$\alpha_+ = 0, \qquad \beta_+ = 0,$$
$$(\alpha\beta)_{i+} = 0 \quad \text{for every } i,$$
$$(\alpha\beta)_{+j} = 0 \quad \text{for every } j.$$

Here's where the restrictions come from. Assume that all the observations made under the same conditions—all the observations in the same cell—have the same true value, say μ_{ij} for cell i,j (so that each $y_{ij} = \mu_{ij} + e_{ijk}$). Now define the parameters of the model in terms of these true cell means:

$$\mu = \overline{\mu}_{..},$$
$$\alpha_i = \overline{\mu}_{i.} - \overline{\mu}_{..},$$
$$\beta_j = \overline{\mu}_{.j} - \overline{\mu}_{..},$$
$$(\alpha\beta)_{ij} = \overline{\mu}_{ij} - [\mu + \alpha_i + \beta_j].$$

Show that with these definitions, the α, β, and $(\alpha\beta)$ terms add to zero in the way I have claimed.

6. *Relation between estimates and true values*. The model for the BF[2] says that each observed value equals a sum of unknown parameters plus a residual error. Thus, any average of the observed values equals some combination of parameters and errors; likewise for the estimated effects. By writing each estimate in terms of parameters and errors, you can see how the estimates are related to the parameters they are supposed to estimate. (You might want to review Exercise 5.E.7 before going on.)

a. Using the model to rewrite each observed value as a sum of parameters plus a residual error, express each of the averages $\overline{y}_{...}$, $\overline{y}_{i..}$, $\overline{y}_{.j.}$, and $\overline{y}_{ij.}$ in terms of parameters and averages of the errors. Use the restrictions on the parameters from Exercise 5 to simplify your expressions. You should find, for example, that $\overline{y}_{...} = \mu + \overline{e}_{...}$ (= true benchmark plus error part).

b. Now use your results from part (a) to rewrite the estimated effects (Exercise 4) in terms of parameters and averages of errors. Is it true in every case that the estimated effect equals the true value plus an error part? Does your expression for the estimated error involve any parameters?

SUMMARY

1. *Factorial crossing*

a. Two factors in a design are **crossed** if every combination of levels of the two factors occurs in the design. (A factor is **compound** if its levels come from crossing two (or more) other factors. A factor is **basic** if it is not compound.)

b. Experiments with crossed factors of interest let you study the effects of two or more sets of conditions (treatments) in a single experiment and study the way the sets of conditions **interact**.

2. *The two-way completely randomized experiment, or CR[2]*

a. The treatment combinations come from crossing two sets of treatments: all possible combinations are present. Treatment combinations get assigned to units completely at random, as in any CR experiment. A two-way CR is balanced if each treatment combination gets assigned to the same number of units.

b. If you want to measure either interaction or chance-error size in a two-way design, you have to have more than one observation per cell.

c. There are two observational variants of the CR[2] design: if both factors of interest are observational, you cross-classify your populations into cells and get a sample from each population. If there's one factor of each kind, you obtain a sample of units from each of your populations, then use a chance device to assign levels of the experimental factor (treatments), separately within each sample.

3. *Factor structure*. All three of the designs in 2b above are called two-way basic factorial (BF[2]) designs and have the same factor structure. You can find the factor structure of a BF[2] in two stages.

Stage 1 (*Reduced form*): Ignore the treatment structure. The design has the same factors as the BF[1]: benchmark, conditions, and residual error.
Stage 2 (*Expanded form*): Replace the factor for conditions with three factors, as in the diagram below, to get five factors in all.

Benchmark	Basic factor #1	Basic factor #2	Interaction	Chance error

4. *Interaction*

a. There are three essential pieces to the structure of interaction: two crossed factors and a response. Interaction is present if the effect of one factor, as measured by differences in the response averages, is different for different levels of the other factor.

b. An interaction graph plots cell averages as points, with one column of points for each level of the column factor. Line segments join the points for each level of the row factor. If the line segments are approximately parallel, there is no interaction.

5. *Decomposition and ANOVA*. The decomposition and ANOVA follow the general rule based on inside and outside factors. Each estimated effect equals the average for one level of a factor minus the sum of effects for all outside factors. The df for a factor equals the number of levels for the factor minus the sum of df for all outside factors. To get the ANOVA table, you follow the same pattern as for a one-way design, except that instead of having just one row in the table for conditions, you have three rows, one for each basic factor and one for interaction. To get *F*-ratios, you use MS_{Res} in the denominator.

REVIEW EXERCISES: EXERCISE SET E

1. *Behavior therapy for alcoholics*. Sometimes alcoholics are treated by aversion therapy: they are taught to associate alcohol with electric shocks. The hope is that their "conditioned fear response" will lead them to avoid alcohol. However, behavioral psychologists have reported that punishments seem to be generally less effective with alcoholics in comparison to a control group. This study was designed to test that hypothesis.

The design was a two-way factorial. One factor was observational: 30 subjects were alcoholics recruited from the Alcoholic Treatment Center at Madison (Wisconsin) General Hospital; another 30 (nonalcoholic) subjects from a local VA hospital served as controls. Each group of 30 was randomly divided into groups of 10 and assigned to one of three levels of the experimental factor: nonreward, punishment, and reward–punishment. Each condition had two phases that I'll call "learning" and "extinction." During the learning phase, which was the same for all three

	NonR	P	R/P
Alcoholics	25.5	36.8	87.5
Controls	17.0	8.5	65.8

NonR = Nonreward
P = Punishment
R/P = Reward–punishment

FIGURE 6.25 Average number of target sequences before quitting.

conditions, subjects were told they would get a nickel each time they pushed a set of buttons in the right order, which they could discover by trial and error. They were also told they might receive small shocks during the experiment and that they could quit at any time just by pushing button 1 four times in a row. The learning phase continued until the subject had won 20 nickels; then the extinction phase began although the subject was not told of the change. Subjects in the nonreward condition simply stopped getting nickels. Subjects in the punishment condition not only stopped getting nickels but now got a shock each time they pressed the target sequence. Subjects in the reward–punishment condition continued to get nickels but got a shock along with each nickel. The response was the number of times the subject pushed the target sequence before quitting. The results appear in Fig. 6.25.

a. Draw both interaction graphs. Which do you find easier to read? Why?

b. Describe the pattern of the graph (whichever of the two you like better) in words, and discuss whether the pattern tends to support the hypothesis that punishment is less effective with alcoholics.

2. *Compensating for parsnip webworms*. Under the control conditions of this study, wild parsnip plants averaged about a thousand seeds from their first set of flowers (primary umbels), about twice that many from the second set of flowers, but only about 250 from the third set because the flowers in the third set tend to abort before the seeds are fully formed. However, for plants attacked by the parsnip webworm, which destroyed most of the primary umbels, the pattern was quite different: the seed production for primary, secondary, and tertiary umbels averaged about 200, 2400, and 1300.

a. Put the cell averages in a two-way table, and label it.

b. Draw both interaction graphs. Which one do you find easier to read? Why?

c. Describe the interaction pattern in words, and discuss how the pattern illustrates the way the wild parsnips attacked by webworms compensated for the damage.

3. *For any BF design, fitted value equals cell average*. The fitted value is the same as the partial fit for chance error, that is, fitted value = sum of estimated effects for all factors other than chance error. Show that for the BF[2] the fitted value equals the cell average: First write out expressions for the estimated effects for benchmark, the two main effects, and interaction. (To make things concrete, use the pigs and vitamins study to name the factors. For example, "B12 effect = B12 average − grand average.") Then add the four expressions together and simplify, to show that they sum to give you the cell average.

4. *The general rule for degrees of freedom*. Show that if you compute df for the BF[2] design using the rule "df for a factor = (number of levels) − (sum of df for all outside factors)" you get the df given in the boxed summary on page 220 in Section 3 of this chapter. (You might want to abbreviate: I = number of levels for 1st basic factor, J = number of levels for 2nd basic factor, K = number of observations per cell.)

5. *Sugar metabolism*. The data below come from a CR[2] experiment designed to study the effects of oxygen concentration on the amount of ethanol produced from two sugars, galactose

and glucose, by Streptococcus bacteria. The response is the concentration of ethanol in micromoles per 0.1 micrograms of sugar.

	Oxygen Conc. (micromoles)							
Sugar	0		46		92		138	
Galactose	59	30	44	18	22	23	12	13
Glucose	25	3	13	2	7	0	0	1

Do as complete an analysis as you can. Include parallel dot and interaction graphs, decomposition and ANOVA, and informal checks of the assumptions. If you decide the data should be transformed, your best bet is probably the log transformation, which is the one most often used for concentrations. (Note that because some of the concentrations are 0, you can't take their logs. In situations like this, the standard practice is to add 1 to all the numbers first, then take logs.)

See also: Example S6.1 Mondays Are Tough (a data set that appears to have a BF[2] design but doesn't), and Exercise Set S.B.: Additional Exercises on Interaction.

NOTES AND REFERENCES

1. ***Residuals underestimate chance errors.*** In Example 6.14 we found that the residuals tended to fall closer to zero than predicted by the 1-2-3 SD rule based on the normal curve. Here's part of the reason why.

Each of the 12 observed weight gains contains one chance error. When we decompose the data, we redistribute those 12 chance errors: The grand average ends up with a chance part, and each of the four estimated diet effects ends up with a chance part. This leaves fewer than the full 12 units of chance error for the residual box of the decomposition, to be shared among 12 residuals. (The number $df_{Res} = 8$ tells how many units there are.) In a sense, taking out the units that go with the other two boxes of the decomposition "shrinks" the residuals, leaving them closer to zero than the chance errors themselves.

The reason we divide by $df_{Res} = 8$ instead of number of Res = 12 when we estimate the SD is to compensate for the "shrinking" so that the number we compute won't underestimate the true SD. In the same way, we can use the 12/8 ratio to adjust the residuals to make them closer in size to the actual chance errors: we multiply each residual by $\sqrt{12/8}$, or, in general, by $\sqrt{(\#obs)/df_{Res}}$.

Appendix: Supplementary Examples

Factor Structure of the BF[2] Design

Example S6.1 Mondays Are Tough: It May Look Like a BF[2], But Is It?

Part of my reason for including the following example is to show you a kind of data set—one where the numbers are counts—which *could* be analyzed by analysis of variance, but which would better be analyzed by methods designed especially for counted data. Each analysis is based on a particular model, and the choice between the analyses comes down to a choice between models. I won't go into details right now, but I do want to remind you that a good analysis depends on finding a good model.

About fifteen years ago, the infirmary at Smith College in Northampton, Massachusetts decided to analyze their records of visits that took place after regular office hours. Part of what made the study interesting was that at the time Smith had an unusual academic schedule. Instead of spreading classes over an entire week, like Mon-Wed-Fri at 10:00, many courses packed them together at the beginning of the week, like Mon-Tue-Wed at 10:00. This made it fairly easy for students to schedule all their classes for the first three days of the week. Administrators at the infirmary wondered whether the unusual schedule might have an effect that would show up in the pattern of visits.

For this data set, visits were classified two ways—according to day of the week, and to the seriousness of the complaint. Day of the week had five levels (Mon, . . . , Fri), and seriousness had four: visits were classified as unnecessary (least serious), requiring a nurse's attention, requiring a doctor's attention, and requiring admission, either to the infirmary or to the local hospital. The Table S4 gives the number of visits of each kind, over a two-year period.

TABLE S4 Numbers of visits to the Smith College infirmary.

	M	T	W	Th	F
Unnecessary	77	67	77	8	70
Nurse	80	66	53	73	62
Doctor	90	71	76	95	75
Admitted	61	49	42	52	28

You *can* think of this as a two-way observational study, with visits classified by seriousness of the complaint (rows) and day of the week (columns). However, there is another, better way to think about the structure of this data set.

This study is observational, not experimental: no conditions got assigned. Notice, too, that if you think of the data set as having a two-way design, there is only one observation per cell, so that it is not possible to measure interaction, at least not in the usual way. Finally, notice that this data set is quite different from the other examples so far, in that nothing gets measured. For Morton's data (Example 6.4), each skull first got classified, and then measured. The response was the volume of the skull, in cubic centimeters. For these data, the numbers in the table are counts: they tell how many visits there were of each kind. Each visit gets classified, but not measured. Many statisticians would analyze this data set as a one-way observational study with a nominal (or ordinal) response: the factor of interest is day of the week, each visit is an observational unit, and the response is the degree of seriousness. ■

Supplementary Exercise Set 2: Additional Exercises on Interaction

1. *Whose fault: attributions in troubled marriages*. Attribution theory in psychology deals with the ways we think about what causes us and others to behave as we do. Frank Fincham, at the University of Illinois, used ideas from attribution theory to compare 18 couples who had just begun marital therapy with a control group of 19 couples who volunteered after seeing an ad in their local paper. Fincham asked each person to list what s/he regarded as the two biggest problems in her/his marriage, things like making decisions, disagreements about children, etc. Then he asked his subjects to list what they saw as the causes of the problems. Finally, he asked a set of eight questions, paraphrased below, about the causes.

1–4: To what extent does the cause of the problem rest in

1. yourself?
2. your spouse?
3. the relationship?
4. outside circumstances?

5–7: To what extent is the cause of this problem

5. something that affects other areas of your marriage, as opposed to just this one problem?

6. likely to be a cause of problems in the future as well as now?

7. due to your spouse's negative attitude toward you?

8: To what extent do you blame your spouse for the problem?

Each subject answered all eight questions twice, once for each of the two problems, using a seven point scale with low numbers meaning "a little" and high numbers meaning "a lot." The two scores for each of the eight questions were then added together.

Figure S2 shows eight interaction graphs, one for each question. Use the graphs to answer the following questions:

a. Male versus female. Which two graphs show the largest overall effect of sex?

b. Troubled versus control. Which two graphs show the smallest overall difference between the therapy and control groups? Which two graphs show the largest differences?

c. Interaction. Which graph shows the biggest interaction? The smallest?

2. (*Whose fault, continued*). Here is another summary that Fincham might have looked at. I think it does a particularly nice job of pointing out an important pattern that makes the couples with troubled marriages different from the control couples.

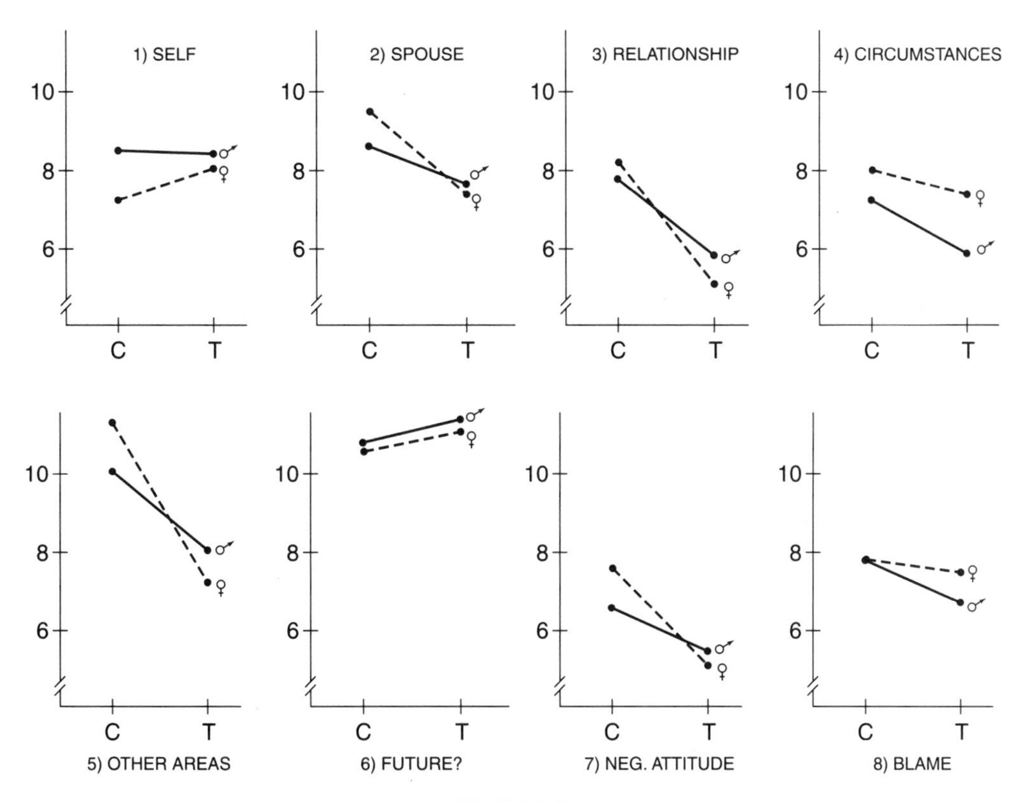

FIGURE S2

Look again at the first two of Fincham's questions, and notice that each person gets rated twice: if I'm in the study, I rate myself in Question 1 of my own questionnaire, and then my wife rates me in Question 2 of her questionnaire. So it is possible to reorganize Fincham's data to compare the two ratings of the husband (his own rating, and his wife's rating of him), and the two ratings of the wife (Table S5):

TABLE S5 Ratings of the husband and wife

	Husband rated by		Wife rated by	
	Self	Spouse	Self	Spouse
Couples with troubled marriages	8.6	9.6	7.3	8.7
Control couples	8.4	7.6	8.2	7.7

The higher the rating, the more the rater thinks the cause of the problem lies with: the husband (first table) or the wife (second table).

a. Draw interaction graphs for both tables.

b. Fill in the blanks, choosing from: practically zero, moderate, and large. For the first table (husband), the overall effect for rater (self versus spouse) is ________________; the overall effect for troubled versus control is ______________; and the interaction effect is ________________.

c. Repeat for the second table (wife), and comment on the similarities and differences in your choices for (b) and (c).

d. Complete the following summary, which I particularly like:

Whether you look at husband or wife, troubled couples or controls, there is in every case a difference between the rating a person gives him/herself and the rating of the same person by the spouse. However, the differences go in opposite directions for the two kinds of couples: people in the troubled relationships saw themselves as ______________ (more/less) at fault than their spouses saw them; people in the control relationships saw themselves as ______________ (more/less) at fault.

CHAPTER 7

THE PRINCIPLE OF BLOCKING

Overview: Designs with Nuisance Factors

A nuisance influence **is any possible source of variability other than the conditions you want to compare; that is, anything other than effects of interest that might affect the response.**

Uncontrolled nuisance influences can bias the results of a study, distorting the effects of the conditions or making the effects impossible to isolate.

This chapter deals with **blocking**, the last of three major principles for choosing a design. Together, blocking and completely random assignment provide two alternative strategies for handling unwanted variability.

So far, all of the examples of experiments in this book have had to deal with such variability. For example, in the study of pigs' diets (Example 6.1), the 12 piglets were in various ways different from each other. This variability showed up in the response, making it harder to measure the effects of the four diets. The divers who learned lists of words (Example 6.2) differed in their ability to memorize, and these differences make it harder to measure the effects of environment on memory.

Thinking about variability of this sort—and about how to deal with it—is crucial in choosing a design. Once you have made the initial three decisions about content, as you consider possible sources of variability in your data, you should think about the various nuisance influences that might affect the response variable. Randomization and blocking provide two different ways to handle the nuisance influences.

Some nuisance influences will usually be easy to spot: differences among individual human subjects, pigs, hamsters, or sponges. Others may be less obvious. For example, in the study of premenstrual syndrome (PMS; Example 7.2) designed to compare the effects of vitamin B6 and a placebo, one not-so-obvious nuisance influence came from the calendar. Symptoms of PMS may tend to be more severe at times of stress, and certain times of the year tend to be more stressful than others. For many people major holidays are often stressful. For many students the beginning of a semester tends to be less stressful than the end, when there are exams to take and papers to write. Because of the threat from bias in the PMS study, it was important to keep the nuisance effect of the calendar in mind when choosing an experimental plan. Suppose, for example, all the subjects got the placebo during December and January and then got B6 supplements in February and March. The extra stress in December and January might raise anxiety levels when subjects were taking the placebo; this would make the B6 supplements look better than they really were.

Randomizing turns a nuisance influence into chance error. We could use a coin toss or a table of random numbers to decide which subjects got B6, and when. Here, as always, random assignment turns possible bias into chance error. Reducing the threat from bias is crucial, but turning possible bias into chance error has its price: the nuisance effects that might have shown up as bias have now been mixed in with chance error, which means that these effects have increased the size of chance error. Larger chance errors make it harder to detect and to measure the effects of interest.

Blocking turns a nuisance influence into a factor of the design. The basic idea is to sort your material into uniform batches then run a bunch of little CR experiments in parallel, one for each batch. This makes it possible to compare the conditions of interest when other things are more nearly uniform. Suppose, for example, that we had 20 volunteers who were available during December and January and another 20 who were available during February and March. We could run two experiments, one for each pair of months, and analyze the results together, with time of year as a factor. Then the design would have two structural factors: the factor of interest (B6 versus placebo) and a nuisance factor (time of year).

The simplest design based on the principle of blocking has just these two structural factors: one factor of interest and one nuisance factor. Sometimes you may have to deal with more than one major nuisance influence. There are designs with two or more nuisance factors as part of their structure.

This chapter covers the following topics:

1. Complete Block Design (CB): One Nuisance Factor, One (or More) Factor(s) of Interest
2. Latin Square Design (LS): Two Nuisance Factors, One (or More) Factor(s) of Interest
3. Split Plot/Repeated Measures Design (SP/RM): One Nuisance Factor, Two (or More) Factors of Interest

1. Blocking and the Complete Block Design (CB)

The randomized complete block experiment: what you do. Complete block designs differ from completely randomized designs in the way treatments are assigned to experimental units. Instead of regarding all units as essentially the same and interchangeable (CR principle), you first sort your units into groups called blocks and then randomly assign treatments to units separately within each block.

The Randomized Complete Block Design (RCB)

- **First sort your units into groups (= blocks) of similar units. The number of units in each block must equal the number of treatments (or treatment combinations).**
- **Then randomly assign treatments to units separately within each block. Each unit gets one treatment, and each block of units gets a complete set of treatments.**

In what follows, we'll look more carefully at what I mean by "sorting" and "similar," but first, here's an example.

Example 7.1 Varieties of Wheat

One of the earliest published examples of a complete block design appeared in Fisher's 1935 book, *The Design of Experiments.* The goal of the experiment was to compare five varieties of wheat, to see which gave the highest yield. The five varieties provided the treatments, the yield in bushels per acre was the response, and eight blocks of farm land were available for planting.

If you think about possible nuisance influences on the crop yield, you might find several. Certainly the weather would have an influence: some growing seasons are better than others. If we assume that Fisher wanted to complete the experiment by the end of the next season, there wouldn't be much he could do about the weather except to hope for a growing season that was fairly typical, neither too cold and wet nor too hot and dry. Weather aside, there are many sources of variability in the blocks of land, mainly differences in the composition of the soil: what nutrients were present, how well the soil held moisture, and so on. Since many of the possible nuisance influences were related to the soil, Fisher made the reasonable assumptions that each of his eight blocks would be more or less uniform and that the variability to be concerned about was that from one block to another, the **between-block** variability. His goal in designing the experiment was to keep this between-block variability out of chance error, by tying it to one of the factors in his design. That way he'd get more precise information about the yields of the five varieties.

Fisher's plan was to divide each of the eight blocks of land "into five plots running from end to end of the block, and lying side by side, making forty plots in all." These 40 plots were his experimental units. Fisher then used each block of five plots to run a little CR experiment, using a chance device to assign one variety to each of the five plots in the block. Figure 7.1 shows how his plan might have looked.

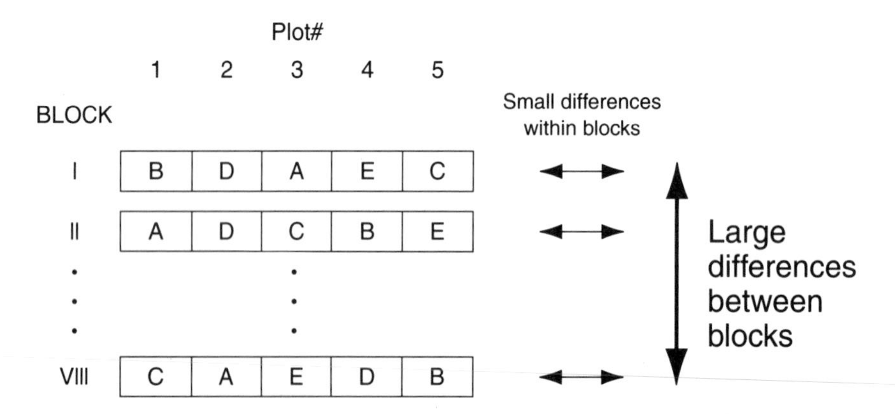

FIGURE 7.1 Fisher's complete block design. Each of the eight blocks is divided into five plots. Within each block, varieties (A, B, C, D, E) are assigned to plots at random: the $8 \times 5 = 40$ plots are the experimental units. The resulting data set would have two structural factors: varieties (the factor of interest) and blocks (a nuisance factor).

Notice how Fisher's design helps isolate the effects of interest. Each block of land is more or less uniform, and each block gets all five varieties, so it is possible to compare varieties grown under uniform conditions. Contrast this plan with a simple CR design: suppose Fisher had had 10 blocks of land instead of 8, and had randomly assigned each variety to 2 blocks. Then the block-to-block variability would have become part of chance error, the chance errors would have been much larger than for the complete block design, and differences among varieties would have been harder to detect. ■

Similar units. The effectiveness of blocking depends on the way your units are sorted into blocks. In Fisher's wheat example, it was reasonable to expect that plots from the same block of land would tend to give similar crop yields, whereas plots from different blocks would tend to give different yields. Although there are lots of ways the plots might be similar—similar in shape, for example—what matters for the experiment is that plots in a block should be similar in terms of the response, crop yield.

"Similar" units are those that would tend to give similar values for the response if they were to be assigned the same treatments. The more similar the units in a block, the more effective blocking will be.

The following example illustrates this meaning of "similar."

EXAMPLE 7.2 MATCHED PAIRS FOR THE PMS STUDY

Kim, the Ph.D. candidate who studied the effect of vitamin B6 on premenstrual syndrome, used human volunteers as experimental units and sorted her subjects into pairs to use as blocks. One woman from each pair got B6, and the other got a placebo.

Think about how she might have sorted her subjects:

1. By height: put the two tallest people together in a pair, the next two tallest together, and so on.
2. By weight: put the two heaviest together, etc.
3. By age: put the two oldest together, etc.

Each of these ways of sorting puts similar units together in pairs, but none of the ways would be very effective as a design strategy because height, weight, and age are not closely related to Kim's chosen responses, which deal with severity of symptoms. The subjects in a pair are similar in the everyday sense of the word but not in the sense that matters for an effective design.

Kim got her blocks by giving her subjects a questionnaire that asked about their symptoms. Then she grouped her subjects according to how severe those symptoms were. Two subjects were grouped together in a block if their symptoms were similar. ■

Ways to get blocks. In the last example, Kim actually sorted her units into blocks. In the wheat example, Fisher got his blocks of units by starting with the blocks of land and then subdividing each block into a set of smaller pieces to use as units. A third way to get blocks is by reusing subjects or chunks of material in each of several time slots.

Three Ways to Get Blocks:

1. ***Sort*** **units into groups (= blocks).**
2. ***Subdivide*** **larger chunks (= blocks) of material into sets of smaller pieces (= units).**
3. ***Reuse*** **subjects or chunks of material (= blocks) in each of several time slots (= units).**

The following example illustrates a design that creates blocks by reusing subjects.

EXAMPLE 7.3 SUBMARINE MEMORY: REUSING SUBJECTS TO GET BLOCKS

Recall the experiment that used 16 members of a diving club as subjects in order to study influences on learning (Example 6.2). There were four treatment combinations, which came from crossing learning environment with recall environment:

Treatment combination

1. Learn dry/Recall dry

2. Learn dry/Recall wet
3. Learn wet/Recall dry
4. Learn wet/Recall wet

Learning environment	Recall environment	
	Dry	Wet
Dry	1	2
Wet	3	4

How would you run this experiment using a complete block design?

SOLUTION: You *could* get a block design by sorting subjects: give your 16 divers a pretest to measure how well they learn lists of words. Then put the four subjects with the highest pretest scores together in a block, and so on. The 16 divers give you 4 blocks of 4 units each. Within each block, randomly assign one of the four treatment combinations to each subject.

A better strategy reuses subjects. Each subject is used four times, once for each treatment combination. Thus, each subject provides four units (= time slots); each set of four time slots for a subject is a block. The resulting data set would have subjects as a nuisance factor. To make the study a true experiment, you'd use a chance device to assign treatment combinations to time slots, separately for each subject: 16 subjects, and 16 random orders. ■

This example illustrates one of the most common uses of blocking in psychology: human subjects serve as blocks. The differences between individuals are often large when compared to the size of the effects psychologists study. Using subjects as blocks is very often the best way to isolate the effects of interest.

Observational studies in complete blocks. All three of the CB examples so far have been true experiments: treatments were randomly assigned. If, instead, your units come with conditions built in, you have an observational study in complete blocks.

EXAMPLE 7.4 THE DREAMING OF A SHREW

Scientists who study sleep have identified three kinds of sleep that can be distinguished by the kind of brain waves that occur: light slow-wave sleep (LSWS); deep slow-wave sleep (DSWS), and rapid eye movement sleep (REMS). Research on humans has established that our dreaming occurs during REM sleep.

For this study, the three kinds of sleep were the conditions of interest, and the subjects were six tree shrews. The response was heart rate, in beats per minute. Each shrew served as a block and provided three units: each shrew was measured under the three conditions of sleep (see Fig. 7.2).

Block(shrew)	Condition		
	LSWS	DSWS	REMS
I	14.0*	11.7	15.7
II	25.8*	21.1	21.5
III	20.8*	19.7	18.3
IV	19.0	18.2	17.1*
V	26.0*	23.2	22.5
VI	20.4*	20.7	18.9

FIGURE 7.2 Heart rate (beats per minute) of sleeping shrews. This study has a complete block design with shrews as blocks and three kinds of sleep as the (observational) factor of interest. ■

With any observational study, you need to pay particular attention to the possibility of confounding and bias. For the shrew example, having the factor of interest be observational does not threaten the validity of the study, but such is not always the case. For an example that has been used to justify false claims in advertising, see Example S7.1: "SAT Scores: Confounding in an Observational Study."

Factor structure of the one-way CB design. All the examples in this section are examples of the one-way complete block design or CB[1], except for Example 7.3 (submarine memory), which was a two-way complete block, or CB[2]. All one-way CB designs have the same two structural factors.

The one-way CB design has the usual two universal factors, plus two structural factors:

One factor of interest—treatments or conditions

One nuisance factor—blocks.

EXAMPLE 7.5 FACTOR STRUCTURE FOR THE SHREWS

Reread the last example, and identify the structural factors by deciding which sets of averages are meaningful.

SOLUTION. There are two structural factors, corresponding to rows = shrews = blocks, and columns = kinds of sleep = conditions of interest. There are also the usual two universal factors, corresponding to the benchmark (grand mean) and residual error. (See Fig. 7.3.)

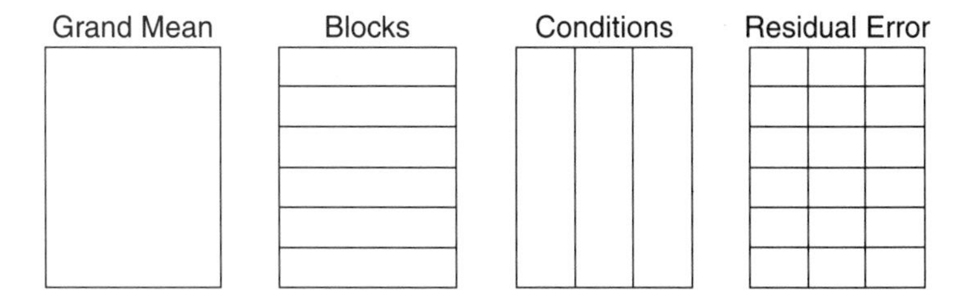

FIGURE 7.3 Factor diagram for a one-way complete block design.

Every complete block design has these four factors:

1. Benchmark: one group
2. Blocks: each row is a group
3. Conditions: each column is a group
4. Residual error: each observation has its own group.

(Some data sets may have blocks as columns and conditions as rows: you need to pay attention to which is which.) ■

See also Example S7.2: "Radioactive Twins" and Example S7.3: "Bee Stings."

Exercise Set A

Each of the examples listed below has blocks as a nuisance factor.

a. For each, name the nuisance factor. Be as specific as you can: don't say "subjects" if you know the subjects are tree shrews.

b. Tell what the unit is.

c. Tell how the units and blocks are related: do you get blocks by sorting/grouping units according to some shared property? or by subdividing a block of material into smaller chunks? or do you get units (= time slots) by reusing subjects or material?

1. Example 7.1: Varieties of wheat
2. Example S7.2: Radioactive twins
3. Example S7.3: Bee stings
4. None of the examples below uses blocks as a nuisance factor in its design. For each, I have listed a nuisance influence. Give a good reason why the influence I listed was not made a factor for the design.

    a. Example 5.1: Walking babies (parents)

    b. Example 5.2: Intravenous fluids (samples of fluid)

    c. Exercise 5.B.1: Bird calcium (birds)

    d. Exercise 5.B.8: Rating the Milgram study (subjects)

5. Lab groups as blocks

For many science labs, students divide into lab groups, and an experiment is done several times in parallel, once by each group. For example, at Mount Holyoke College where I teach, the chick thyroid experiment (Chapter 4, Review Exercise 1) is done that way. Each lab group gets two chicks; one serves as a control, and the other has PTU added to its food. After two weeks the thyroid glands get weighed. Then the data from all the groups are collected, tabulated, and handed out to the class.

a. If you think of the experiment as what it is that a single lab group does, then there is no way to measure the size of residual error, because ____________.

b. If you think of the experiment as what it is that the whole class does together, then the problem of measuring residual error goes away. Name the nuisance factor and factor of interest for this way of looking at the data.

6. Reread Exercise 6.E.2 about parsnips and webworms.

   a. If you think about how best to run this study, you should find a nuisance influence that could be turned into blocks. What is it?

   b. The study did, in fact, have a nuisance factor (blocks) as part of its structure. However, the design was not a CB. How can you tell this from the description? What is it about the study that makes it different from a CB?

7–9 *Radioactive twins* (Example S7.2)

7. Write out the factor diagram.

8. Draw a parallel dot diagram, with two columns of dots, one for each environment. (Ignore the block structure.) Discuss the usual four features.

9. Compute the seven block averages, and plot them as dots on a number line. Do any of the twin pairs stand out as unusual?

10–13 *Bee stings* (Example S7.3)

10. Write out the factor diagram.

11. Draw a parallel dot diagram, with two columns—one for stung, one for fresh—and discuss what you see.

12. Compute the block averages, and plot them on a number line. Do any of the nine occasions stand out as unusual?

13. *Replacement values for outliers.* You should have identified one outlier as part of your work in #11. One way to handle an outlier is to treat it as a missing observation, estimate a replacement value, and analyze the data set twice, once with the outlier included, and a second time using the replacement value instead. I used the standard method (Chapter 12, Section 5) to estimate a replacement value of 16 for the outlier. The column totals, including the outlier, are 252 (stung) and 144 (fresh). Compute two sets of condition averages, one set including the outlier, and the other set using the replacement value instead; use these to comment on the effect of the outlier.

2. Two Nuisance Factors: The Latin Square Design (LS)

STRUCTURAL FEATURES OF A LATIN SQUARE DESIGN

1. **There are two nuisance factors and one (possibly compound) factor of interest.**
2. **The number of levels is the same for all three of these factors.**
3. **Every pair of factors is crossed: for each choice of two factors, every combination of levels of the two factors occurs exactly once.**

In Example 7.3 I described how the submarine memory experiment could have been run in randomized complete blocks with subjects as blocks. That design makes the nuisance influence—subjects—a factor of the design. However, I mentioned that there is another nuisance influence to worry about as well. You might suppose I'm thinking of the order of the conditions and the effect of time, but I'm not. It's true that there might be some sort of practice effect, with subjects tending to do better under the later conditions; for some situations there might also be an effect due to tiredness or boredom, with subjects tending to do better early on, when they are fresher.

However, the nuisance influence I have in mind comes from the measurement process. Remember that the response was the number of words recalled correctly from a list of 36 words. Obviously, it would be a stupid plan to use the same list of 36 words four times for each subject. There ought to be four lists, so that each condition could be measured using a new list. That's the other nuisance factor: the four lists.

The **Latin square** plan allows you to control simultaneously for two nuisance influences, by making each one a structural factor of your design. As you read through this section, see if you can figure out how to apply the ideas of the Latin square to the submarine memory experiment.

The Latin square (LS) design is usually represented by a diagram like those in Fig. 7.4.

A	B	C
B	C	A
C	A	B

A	B	C	D
C	D	A	B
D	C	B	A
B	A	D	C

FIGURE 7.4 Two Latin square designs. Notice the distinguishing features of the Latin square: for each design, the number of different letters equals both the number of rows and the number of columns; the letters are arranged so that each letter appears exactly once in each row and exactly once in each column.

In applications, rows and columns usually represent levels of two nuisance factors, and letters stand for the conditions of interest.

Latin square designs are often used as an alternative to complete block designs for experiments on human or animal subjects.

A Common Situation Where a Latin Square Design Is Useful

1. ***Subjects exhibit a lot of variability*: make subjects a nuisance factor in the design, if it is possible to measure each subject under each of the conditions.**
2. ***The order of the conditions has a systematic effect*: make time periods a nuisance factor in the design, and use a Latin square to ensure balance.**
3. ***The factor of interest is experimental*: you must be able to assign treatments to fit the pattern of the Latin square.**

The following example is from agriculture, but I chose it in part because I think it illustrates the kind of reasoning that leads to LS designs in a variety of subject areas.

Example 7.6 Milk Yields

The goal of this study was to compare three diets (full grain, partial grain, and roughage) to see what effect diet has on how much milk a cow gives. The response was pounds of milk per week.

You could compare the diets using a CR design, assigning one diet to each cow (= unit), but experience has shown that you would need a large number of cows for the CR plan because the cow-to-cow variation is very large. A better plan is to use cows as blocks, and give each cow all three diets, one at a time, in a random order. (For a plan like this one, the unit would be a time slot, with each cow providing three units.) Six weeks is about the right time on each diet: you want the same diet for long enough to affect milk yield but not for too long, because the longer the time since a cow was last pregnant, the less milk she gives—you wouldn't want the pattern of declining milk yields to hide the effects of diet.

Imagine that you've decided to run the experiment as a randomized block design, using three cows as blocks, with each cow getting the three diets, six weeks on each diet. Here's a block design that would be disastrous:

	Weeks		
Cow	1–6	7–12	13–18
I	Roughage	Partial grain	Full grain
II	Roughage	Partial grain	Full grain
III	Roughage	Partial grain	Full grain

FIGURE 7.5 A nonrandomized complete block design. This (nonrandomized) design confounds the effect of diet with the effect of time, and the usual decline in milk yields could end up making roughage look better than grain.

To rescue the design, we should use a chance device to choose the order of the diets, separately for each cow. I just did that, and here's my randomized CB design:

	Weeks		
Cow	1–6	7–12	13–18
I	Full grain	Partial grain	Roughage
II	Roughage	Full grain	Partial grain
III	Full grain	Roughage	Partial grain

FIGURE 7.6 A randomized complete block design. Each cow serves as a block of three units (= time slots). For each cow, the order of the diets was chosen using a chance device.

The new design is much better than the previous one, but no doubt you can spot this design's weakness: just by the luck of the draw, full grain appears twice during the first six-week period, when milk yields are likely to be highest; partial grain appears twice during the last six-week period, when milk yields are likely to be lowest. This design might make full grain look better than it really is and make partial grain look worse. The problem is a lack of balance.

You can probably figure out how to make the design balanced. (One way is to reorder the diets for cow III, from full/roughage/partial to partial/roughage/full.) Any of the resulting balanced designs would be an example of a Latin square. Here's the one that was actually used, together with the results:

Cow	Weeks 1–6	7–12	13–18
I	A: 608	B: 716*	C: 845*
II	B: 885	C: 1086*	A: 711
III	C: 940	A: 766	B: 832

FIGURE 7.7 Latin square design for milk yields. The numbers give yields in pounds per week, and the letters stand for the diets: A = roughage; B = partial grain; and C = full grain. Notice the balance: each cow gets each diet exactly once, and each diet occurs once in each of the three periods. (The numbers with an asterisk were changed by ±1 to simplify arithmetic later on.) ■

Example 7.6 illustrates several features of the LS design. There are two nuisance factors (in this case, cows and time periods) and one factor of interest (diets). Each factor has the same number of levels (3), and for each pair of factors, each combination of levels occurs exactly once. Thus, for example, each cow gets each diet exactly once.

Randomization in the LS plan. How do you randomize the assignment of treatments to units? If your two nuisance factors are subjects and time slots, you should randomize in three steps:

Randomizing Subjects, Times Slots, and Treatments

Step 1. Randomly assign subjects to row numbers.
Step 2. Randomly assign time slots to column numbers.
Step 3. Randomly assign treatments to letters.

EXAMPLE 7.7 RABBIT INSULIN: RANDOMIZATION FOR AN LS DESIGN

The goal of this experiment was to compare four doses of insulin, A, B, C, and D; more specifically, to compare their effect on the blood sugar level of rabbits. The design is very much like the one in the last example, except this time rabbits replace cows, doses replace diets, and there were four levels of each factor: four rabbits; each rabbit got all four doses, one dose on each of four days. The four time slots were spaced apart, to let the rabbits recover from the effects of one dose before getting the next.

In Fig. 7.8 I've shown the data twice. The randomization is easy to see in the data on the left. On the right, the same data have been rearranged.

Rabbit	Date *4/23*	*4/27*	*4/26*	*4/25*	*Rabbit*	Date *4/23*	*4/25*	*4/26*	*4/27*
III	A 57	B 45	C 60	D 26	I	B 24	C 46	D 34	A 48
I	B 24	A 48	D 34	C 46	II	D 33	A 58	B 57	C 60
IV	C 46	D 47	A 61	B 34	III	A 57	D 26	C 60	B 45
II	D 33	C 60	B 57	A 58	IV	C 46	B 34	A 61	D 47

FIGURE 7.8 Response of four rabbits to four doses of insulin. The letters A to D are the doses; the response is the rabbit's blood sugar level (mg %) 50 minutes after injection with insulin. The table on the left shows the randomization more clearly; the table on the right is equivalent but arranged more conventionally. ■

For some applications of the Latin square plan, the arrangement of rows and columns will be fixed by the situation, and the three-step randomization must be done using the row and column labels as the next example shows.

EXAMPLE 7.8 LATIN SQUARE PLANS IN AGRICULTURE

Suppose you want to compare five varieties of wheat and you have one large rectangular plot of land which you divide into 25 subplots, arranged in 5 rows by 5 columns. How would you randomize the assignment of varieties to subplots, using the following 5 × 5 Latin square?

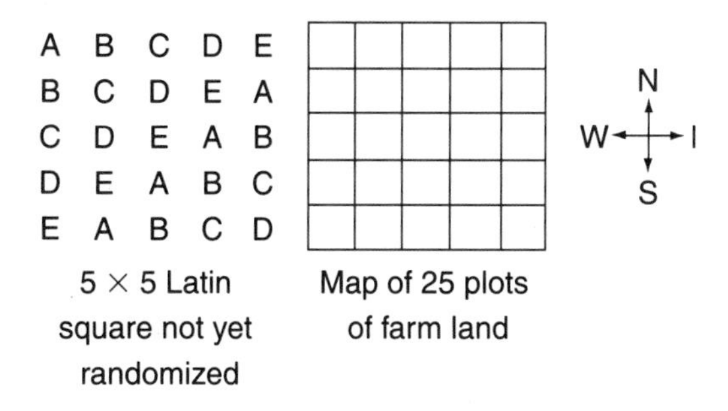

Comment. It would be a gross blunder to use this Latin square "as is." Suppose your large field happened to be more moist and fertile toward the northwest corner and sandy and less fertile toward the southeast. Then, for example, whichever variety was assigned to the D subplots would have an advantage over the variety assigned to the A subplots.

SOLUTION. Follow the three-step randomization, as follows:

Step 1. Randomize rows. Choose digits from a random number table to get a random rearrangement of the numbers 1 to 5, and use them to reorder rows. I got 2 4 3 5 1: the first row of my randomized LS will be row 2 from the original; the second row of the randomized LS will be row 4 from the original; and so on:

2	B	C	D	E	A
4	D	E	A	B	C
3	C	D	E	A	B
5	E	A	B	C	D
1	A	B	C	D	E

In the diagram the top row corresponds to the northmost row of five subplots on the map. The letters in this row come from the second row of the nonrandomized Latin square I started with.

Step 2. Randomize columns. I got a random order 1 4 2 5 3: the first column of my randomized LS is column 1 from the square in step 1; the second column is column 4 from that square, etc.:

	1	4	2	5	3
2	B	E	C	A	D
4	D	B	E	C	A
3	C	A	D	B	E
5	E	C	A	D	B
1	A	D	B	E	C

In the diagram the rightmost column corresponds to the eastmost column of five subplots on the map. The letters in column 5 came from column 3 of the row-randomized square in step 1.

Step 3. Randomize treatments. Randomly assign letters to varieties, and sow each subplot with the variety given in the Latin square from step 2. ■

Factor structure of the LS design. The factor diagram for the simplest Latin square experiment has three structural factors: rows, columns, and treatments (letters). (There can't be any interaction factors, because for any two factors you look at, each combination of levels occurs only once.)

Grand Mean	Nuisance factor 1	Nuisance factor 2	Treatments	Residual error

A	B	C
B	C	A
C	A	B

FIGURE 7.9 Factor diagram for a 3 × 3 Latin square. There are two nuisance factors—rows and columns—and one factor of interest, usually called treatments. The factor of interest partitions the observed values into three groups of three observations each: all the As go together in one group, the Bs in another, and the Cs in a third. The factor diagram for other Latin squares would have the same five factors, with the number of groups (and their size) depending on the size of the square.

Variations on the simplest Latin square plan. The basic LS plan is often extended in either of two ways: some experiments take the form of several Latin squares rather than just one; and sometimes the factor of interest is compound, not basic. (A Latin square with two crossed factors of interest would be called a **two-way LS**, or **LS[2]**.) Example S7.4: "Submarine Memory" illustrates both of these extensions: several Latin squares, with two crossed factors of interest.

Exercise Set B

1. *Milk yields*

   a. To remind yourself about the balance in a Latin square design, complete the following list of the nine observations for the milk yield example (Example 7.6). Notice that I've set up the list so that observations are grouped by diet.

Week	Diet	Cow	Yield	Average
_____	Roughage	I	_____	
_____	Roughage	II	_____	_____
_____	Roughage	III	_____	
_____	Partial grain	I	_____	
_____	Partial grain	II	_____	_____
_____	Partial grain	III	_____	
_____	Full grain	I	_____	
_____	Full grain	II	_____	_____
_____	Full grain	III	_____	

b. Compute the average milk yield for each diet.

c. Estimate the averages for the columns (periods) of the Latin square design. What do these column averages tell you? Do the data support the idea that milk yields tend to decline over the course of the experiment?

d. Which differences are bigger: between cows or between periods?

e. Is the factor of interest for this design observational or experimental? Why is it unusual for an LS design to have the factor of interest be observational?

2–4 *Finger tapping*. Many people rely on the caffeine in coffee to get them going in the morning or to keep them going at night. Wouldn't you really rather eat chocolate? Chocolate contains theobromine, an alkaloid quite similar to caffeine, both in its structure and in its effects on humans.

In 1944, Scott and Chen reported the results of a study designed to compare caffeine, theobromine, and a placebo. Their design used four subjects as blocks and assigned the three treatments to each subject in a random order, one drug on each of three different days. Subjects were trained to tap their fingers in such a way that the rate could be measured; presumably the training got rid of any practice effect. The response was the rate of tapping two hours after taking a capsule containing either a drug or the placebo.

2. a. Draw and label a rectangular table for the data from this experiment. List the structural factors, and tell how many levels each has. What is the unit for this experiment?

   b. Write the factor diagram.

3. Suppose there had been six subjects instead of four. What design, based on Latin squares, should the investigators have used? Draw and label a diagram that shows your design; tell what each row, column, and letter means. Then write the factor diagram.

4. Now suppose you had no additional subjects beyond the original four but you wanted to include a third alkaloid, nicotine, in your comparisons. Draw and label a diagram that shows what design you would use. Then write the factor diagram.

5–8 *Wireworms*. This experiment was designed to compare the effectiveness of five different soil fumigation treatments (K, M, N, O, P) for controlling wireworms. The treatments were applied to the plots of a Latin square, and the numbers of wireworms per plot were counted a year later.

P 3	O 2	N 5	K 1	M 4
M 6	K 0	O 6	N 4	P 4
O 4	M 9	K 1	P 6	N 5
N 17	P 8	M 8	O 9	K 0
K 4	N 4	P 2	M 4	O 8

5. Reorder the data, listing the observed numbers of wireworms by treatment. For example, for treatment K: 1, 0, 1, 0, 4. Draw a parallel dot graph with the five treatments as columns, and discuss what you see.

6. Compute the average number of wireworms for each group. Then compute an SD separately for each group, ignoring the LS structure and thinking of the data in Exercise 5 as coming from a BF[1] design. Are the group SDs roughly equal? Do the average and the SD seem related?

7. Replacement value for the outlier. From looking at the SDs for groups K, M, O, and P, I judge a typical chance error size to be about 2.5. The distance from 17 to the group N average is four times this typical size. Using the method from Chapter 12, Section 5, I estimated a replacement value of 7 (rounded off) for the outlier. Recompute the average and the SD for group N, using the replacement value in place of the outlier. Notice what a huge influence the outlier has on both the average and the SD for treatment group N, but notice, too, that replacing the outlier doesn't change the conclusion about which treatment is best.

8. *Wireworms (continued): Transforming instead of replacing the outlier*. We can't tell just from the data whether the 17 in Exercise 5 is a stray value that ought to be omitted from the analysis, or whether natural variability leads to occasional unusually large numbers of worms. Perhaps in a different scale the outlier would not seem so extreme. Although there is a mild relationship between average and spread—the smallest range goes with the smallest average and the largest with the largest—I don't regard that pattern as strong enough by itself to make me look for a transformation. However, three of the groups—K, N, and P—have lower values bunched together and higher values more spread out, another suggestion that bigger chance errors tend to go with bigger observed values. Since both patterns suggest transforming, I think it is worth a try. The transformation I chose to use was the square root: essentially, I replace each observed value by its square root, although I made two minor modifications: For technical reasons, when the observed values are small, the square root transformation works a bit better if you add 1 to all the numbers first; so I did that. Then, to get rid of decimal fractions, I multiplied all the square roots by 10, and then I rounded off to the nearest whole number. For example, a 3 in the original scale gets 1 added to give 4, then has a square root taken to give 2, and then gets multiplied by 10 to give 20. It would have then been rounded off had it not already been a whole number.

Here are the transformed values, already sorted by treatment group:

Treatment	$10 \times \sqrt{(1 + \text{no. of worms})}$	*Average*	*SD*
K	22, 10, 14, 14, 10	14.0	4.9
M	26, 32, 30, 22, 22	26.4	4.6
N	42, 22, 24, 22, 24	26.8	8.6
O	22, 17, 26, 32, 30	25.4	6.1
P	20, 30, 17, 26, 22	23.0	5.1

Notice that the transformation has done its job. The spreads are now much more nearly equal, with the largest SD less than twice as big as the smallest. Moreover, in the new scale the outlying value seems not quite as far out as before.

a. Construct a parallel dot graph, and compare what you find with your dot graph in Exercise 5.

b. Which analysis do you prefer, that in Exercise 5 or that in Exercise 8? If your purpose is to find the best soil treatment, does it matter which way you analyze the data?

3. THE SPLIT PLOT/REPEATED MEASURES DESIGN (SP/RM)

The Simplest Split Plot/Repeated Measures Design

- **has experimental units of two different sizes. The larger units are blocks of smaller units.**
- **has two sets of treatments or treatment combinations. One set of treatments gets assigned to larger units as in a CR design. The other set of treatments gets assigned to smaller units separately with each block, as in a CB design.**

Although there are many versions of the split plot/repeated measures design, in this section I shall present only the simplest version, which I'll call the **basic SP/RM.** (Psychologists often call this version a **mixed** design.) It is one of the most frequently used designs in psychology and is often used in biology as well. In several respects this design is quite different from the others you have seen so far.

Comparison with simpler designs. The CR, CB, and LS designs represent three different ways of assigning treatments to experimental units. For the CR, the units were not grouped or arranged in any way before treatments were randomly assigned; for the CB, the units were grouped into blocks, and treatments were randomly assigned to units separately within each block; for the LS, the units were arranged in rows and columns, a kind of two-way blocking, before treatments were randomly assigned.

For each of these three kinds of designs, the experimental units were all of just one size, there was only one set of treatment combinations to assign, and the process of random assignment handled all the treatments or treatment combinations in the same way, regarding them as interchangeable. Split plot/repeated measures designs have units of two different sizes and have two distinct sets of treatments or treatment combinations. One set is assigned to the larger units; the assignment process is the same as for a CR. Each of the larger units functions like a block of smaller units; the second set of treatments is assigned to the smaller units separately within each block, just as for a CB.

What You Do

Terminology. The standard vocabulary for describing SP/RM designs depends on how you get your blocks.

a. If the blocks come from subdividing larger chunks of material, the design would be called a **split plot**, the larger experimental units would be called **whole plots**, and the factor of interest whose levels are assigned to the larger units would be called a **whole plot factor**. The smaller units would be called **subplots**, and the associated factor of interest would be a **subplot factor**.

b. If, on the other hand, the blocks come from reusing subjects or material in each of several time slots, the design would be called a **repeated measures** design. The factor assigned to subjects (= larger units) would be called a **between-subjects factor**; the factor assigned to time slots would be a **within-subjects factor**.

c. There's also a set of terms that you can use regardless of how you get your blocks: the larger units are simply called **blocks**, and the two kinds of factors are called **between-blocks** (assigned to larger units) and **within-blocks** (assigned to smaller units).

Experimental version of the SP/RM. We can think of carrying out the experimental plan in two steps that correspond to the CR (whole plot) and CB (subplot) parts of the design:

> 1. **CR step: Randomly assign levels of the between-blocks factor to equal numbers of blocks = whole plots.**
> 2. **CB step: Randomly assign levels of the within-blocks factor to subplots, separately within each block.**

EXAMPLE 7.9 DIABETIC DOGS

(Identify the CR and CB steps in the following experiment.) The disease diabetes affects certain aspects of the way a body uses sugar. In particular, the disease affects the rate of turnover of lactic acid in a system of biochemical reactions called the Cori cycle. The purpose of this experiment was to compare two methods of using radioactive carbon-14 to measure the rate of turnover. In one method the radioactive tracer was injected all at once; in the other method it was infused continuously.

The investigators randomly sorted ten dogs into two groups. Five dogs served as controls, and each of the remaining five had its pancreas removed, to make it diabetic. The rate of turnover was then measured twice for each dog, once using the injection method and once using the infusion method. For each dog, the order of the two methods was randomly assigned.

SOLUTION. For this experiment the larger units are the dogs (= subjects = blocks), and the smaller units are time slots; the blocks come from reusing subjects (RM) rather than subdividing (SP). The between-subjects factor is the operation (control, or remove pancreas). The within-subjects factor is the method (injection or infusion).

CR step: Randomly assign an operation to each dog.

CB step: For each dog, randomly choose an order for the two methods. ■

SP/RM designs with observational and nonrandomized factors. In Example 7.9, both factors of interest were randomly assigned. In the next two examples, one factor of interest is either observational or else is experimental but the assignment is not randomized.

EXAMPLE 7.10 MEMORY AND INTERFERENCE

The purpose of this study (see Example S4.8) was to see whether there was a relationship between psychiatric disorders and the effect of distracting noises on the memory capacity of human subjects. Each subject belonged to one of four psychiatric classifications: schizophrenic, schizotypal, borderline, or normal. Each subject was tested under two sets of conditions: with and without interference. For the non-interference condition the subject heard a male voice read strings of digits and was asked to repeat them; the response was the total number of strings the subject got right. For the interference condition the subject heard the same male voice read strings of digits, but there was also a female voice reading digits in between those of the male voice. The subject was supposed to remember and repeat only the digits spoken by the male voice; here, as before, the response was the number of strings the subject got right.

For this study, subjects served as blocks. Psychiatric classification (diagnosis) was the between-blocks factor, with four levels. (To check this, notice that to compare levels of this factor—for example, to compare schizophrenics with normals—you compare *different* sets of subjects.) The interference condition was the within-blocks factor, with two levels. Each subject was tested under both conditions.

Diagnosis is an observational factor; you can't assign it to subjects by tossing a coin (cynics to the contrary notwithstanding); instead, you choose samples from the populations of interest. The interference condition is experimental: the assignment of the test conditions to time slots could be done using a chance device, and in fact it was. ■

EXAMPLE 7.11 LOSING LEAVES

This experiment was part of a student project designed to study two factors thought to affect the rate at which plants lose their leaves. The basic theory says that the chemical auxin (IAA) must be present to keep a leaf from falling off the plant and that as a plant's supply of naturally occurring auxin gets depleted, the leaves fall off. If this theory is correct, then adding auxin should slow the rate at which the leaves fall off.

There are two competing theories about the source of the auxin. One says the auxin flows to the leaf from the stem via the leaf blade; the other says each leaf gets a fixed amount of auxin when it is formed and as that amount gets used up, the leaf tends to fall off. If the first theory is correct, then removing the leaf blade (deblading) should make the leaves fall off sooner.

The design was a basic SP/RM, with 16 *Coleus* plants in separate pots as the larger units. Each plant was randomly assigned to get one of four concentrations of auxin added to the soil in its pot. Thus, the whole plot factor was auxin, with four levels. Four randomly chosen plants served as controls; another four were treated with a water and lanolin mixture; four more were treated with auxin added to the mixture at a high concentration; the last four were treated with auxin added to the mixture at a low concentration.

To make the plants as uniform as possible, the student pruned each one of all its auxiliary branches and all leaves except for three pairs, plus an apical bud (where leaves are formed). The two lowest leaf pairs left on each plant served as subplots. The lowest pair was left untreated; the response was the number of days after pruning before this bottom pair fell off. Then the new bottom pair (originally the second lowest pair) was debladed; the response was the number of days after deblading before the leaf pair fell off. Thus, the subplot factor was deblading, no or yes.

Although both factors of interest in this experiment were in fact treatments, only the whole-plot treatment was randomly assigned. ■

Variations on the basic SP/RM design. The key feature of any SP/RM design is that there are units of two sizes, with the larger units corresponding to groups of smaller units. The examples in this section have all illustrated the simplest of all possible designs with this key feature. Many variations are possible: you can have fancier treatment structures, you can use fancier strategies for assigning treatments to units, or both.

Treatment structure: Either of the factors of interest can be compound. In other words, either of the whole-plot treatments or the subplot treatments could be treatment combinations that you get by factorial crossing. For example, the study of memory and interference (Example 7.10) actually tested subjects under four conditions that came from crossing two factors. One factor was interference condition, present or absent; the other was length of the digit string, long or short. For this experiment the within-blocks factor would be compound, with $2 \times 2 = 4$ treatment combinations as its levels.

Assigning treatments to units: In the simplest SP/RM, whole-plot treatments are assigned to whole plots as in a CR; subplot treatments are assigned to subplots as in a CB. For some situations it might make sense to sort whole plots into groups of similar units and then assign whole-plot treatments separately within each group, as in a CB. For other situations you might want to arrange your whole plots in a Latin square. In the same spirit, for some situations it might be a good plan to arrange each block's worth of subplots in a Latin square.

What You Get

Factor structure of the basic SP/RM. The simplest SP/RM has four structural factors: two basic factors of interest; their interaction; and a nuisance factor (blocks). The following example shows a standard rectangular format for data from an SP/RM and uses that format to give the factor diagram.

EXAMPLE 7.12 DIABETIC DOGS: FACTOR STRUCTURE OF THE SP/RM

The lactic acid turnover rates from the experiment described in Example 7.9 appear in Fig. 7.10.

		Inject	Infuse
Control dogs	#0	44	28
	#4	33	23
	#5	38*	34
	#21	59*	19
	#23	46	26
Diabetic dogs	#16	54*	42
	#17	43	23
	#18	55	23
	#19	71	27
	#24	57*	35

FIGURE 7.10 Standard format for the basic SP/RM experiment.

- Rows correspond to blocks = whole plots.
- Groups of rows correspond to levels of the between-blocks factor.
- Columns correspond to levels of the within-blocks factor.
- Groups of rows crossed with columns correspond to cells, the combinations of levels of the two factors of interest.
- Rows crossed with columns correspond, to subplots, the levels of the factor for (subplot) residual error.

The first four sets of groups listed above are precisely the structural factors for the SP/RM design (Fig. 7.11).

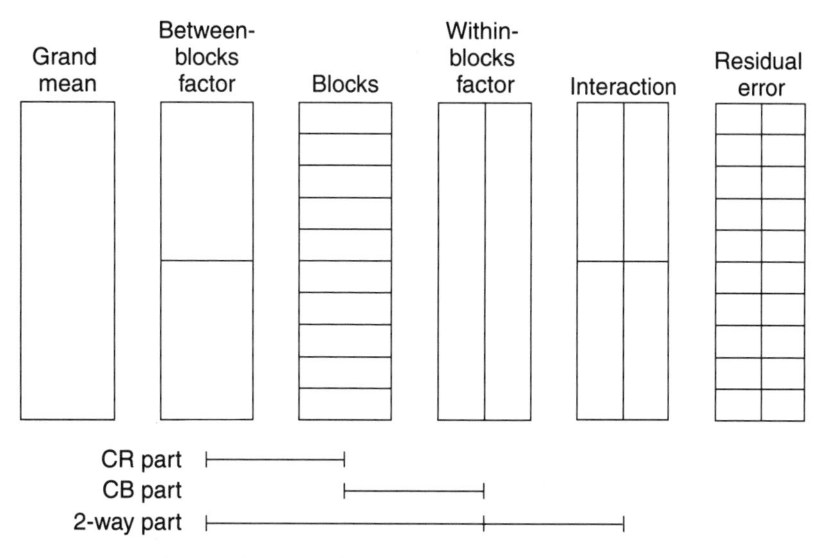

FIGURE 7.11 Factor diagram for the SP/RM design. The SP/RM design has four structural factors:
1. Between-blocks factor (groups of rows)
2. Blocks = whole plots = larger units (rows)
3. Within-blocks factor (columns)
4. Interaction (groups of rows × columns). ■

Crossing versus nesting. The SP/RM design illustrates two different ways that a pair of factors can be related: by crossing or by **nesting**. The within-blocks factor is crossed with blocks: every combination of levels of the two factors occurs in the design. In the dog experiment, for example, method (within-blocks factor) is crossed with dogs (blocks): every dog provides two turnover rates, one each for injection and for infusion. Notice that crossing is a symmetric relationship. "Method is crossed with Dogs" is equivalent to "Dogs is crossed with Method."

The other relationship, nesting, is not symmetric. For the dogs experiment, each dog gets only one operation, either control or removal of pancreas, not both. Each level of the dog factor occurs with only one level of operation. Dogs is **nested within** operation.

> Nesting. **One factor is** nested within **another if each level of the first ("inside") factor occurs with exactly one level of the second ("outside") factor.**

For the basic SP/RM design, the factor blocks is nested within the between-blocks factor. (The vocabulary is standard and logical but potentially confusing if you go by sound rather than meaning: the nesting is within the between-blocks factor.)

Check that nesting is not a symmetric relationship in that, for example, Dog is nested within Operation but not the other way around. (As an exercise, think about what it would mean if each of two factors were nested within the other.)

One way to visualize the way two factors are related is by drawing the box for one factor on top of the box for the other. The resulting box represents the **product** of the two factors.

> **For any two factors I and II, their** product **I × II is the factor represented by writing the boxes for I and II on top of each other.**

The next example shows how to use this idea.

EXAMPLE 7.13 CROSSING VERSUS NESTING

The factor diagrams in Figs. 7.12–7.14 illustrate three of the most common ways that two factors can be related: by crossing, by nesting, and by complete confounding.

Imagine a set of eight observed values, arranged in four rows by two columns. Consider various ways you might sort the eight observations into groups. In all three cases below, one factor corresponds to sorting into pairs of rows: rows 1 & 2 (level 1) and rows 3 & 4 (level 2). In each case there is also a second factor, which corresponds to sorting into columns, or into rows, or into pairs of rows.

a. **Crossing.** If the second factor corresponds to columns, A and B, then each combination of levels of the two factors occurs in the data set (Fig. 7.12).

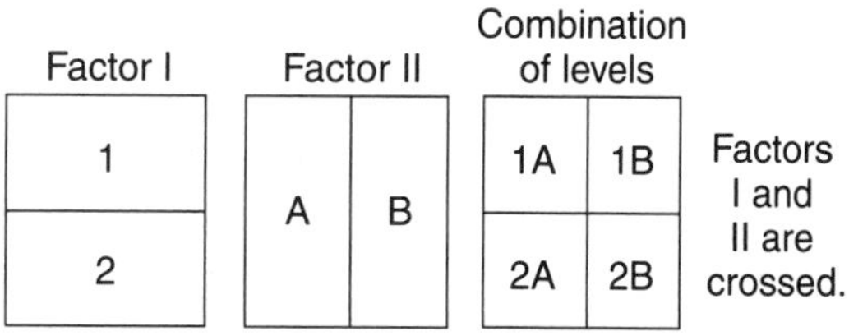

Factor I	Factor II		Combination of levels		
1	A	B	1A	1B	Factors I and II are crossed.
2			2A	2B	

FIGURE 7.12 If I and II are crossed, then their product I × II is the usual factor for the interaction of I and II. All possible combinations of levels of I and II occur as levels of I × II.

b. **Nesting.** Now suppose that the second factor corresponds not to columns but to rows: A, B, C, and D. Then each level of the second factor occurs with exactly one level of the first factor. Factor II is nested within factor I (Fig. 7.13).

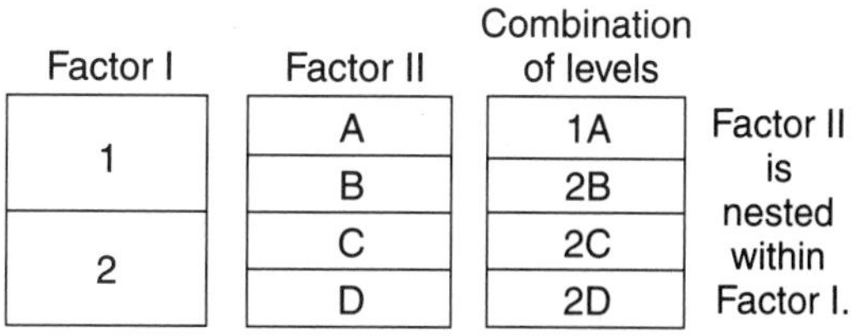

Factor I	Factor II	Combination of levels	
1	A	1A	Factor II is nested within Factor I.
	B	2B	
2	C	2C	
	D	2D	

FIGURE 7.13 If factor II is nested within factor I, then their product I × II has the same groups as factor II.

c. **Complete confounding.** This time suppose the second factor corresponds to the same groups of rows as the first factor. Then I is nested within II *and* II is nested within I. The groups for I and II coincide; the two factors are completely confounded (Fig. 7.14).

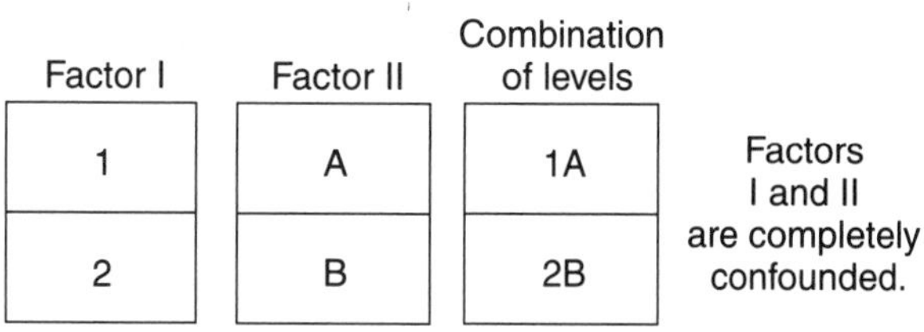

Factor I	Factor II	Combination of levels	
1	A	1A	Factors I and II are completely confounded.
2	B	2B	

FIGURE 7.14 If factors I and II are completely confounded, then the groups for I, II, and I × II are all the same. ■

In the basic SP/RM:

- **Blocks are nested within levels of the between-blocks factor.**
- **Smaller units (subplots) are nested within blocks.**
- **The between-blocks and within-blocks factors are crossed.**
- **Blocks and the within-blocks factors are crossed.**

See also Example S7.5: "Murderers' Ears."

Exercise Set C

1. Suppose the sponge study (Example 4.10) had used three sponges of each color. Suppose also that you measured several cells from each part (tip and base) of each sponge and used the average of these measurements as your response. Then your study would have the structure of an SP/RM design.

   a. What are the whole plots, that is, what is the nuisance factor?

   b. What is the between-blocks factor? Is it observational or experimental?

   c. What is the within-blocks factor? Is it observational or experimental?

   d. Draw and label a rectangular table for the data from such a study, using the description in Example 7.10 and the factor diagram in Example 7.11 as models.

   e. Write the factor diagram.

2. Read Exercise 1 on attributions and troubled marriages, in Supplementary Exercise Set S.B. Pick any one of Fincham's eight questions to think of as the (one) response. Then you can think of the data as having a repeated measures structure with ____________ as the nuisance factor, ____________ as the between-blocks factor, and ____________ as the within-blocks factor.

3. Read Exercise 6.E.2 on webworms and parsnips; there the design is an SP/RM. Identify the nuisance factor, the between-blocks factor, and the within-blocks factor. Which of the two factors of interest is experimental? Which is observational?

4. Look at Fisher's iris data (Example S5.2), and think about how you could regard the structure as an SP/RM design.

   a. Identify the nuisance factor, the between-blocks factor, and the within-blocks factor.

   b. One of the two factors of interest has itself a factorial structure that comes from crossing two more basic factors. Draw and label a two-way table to show this. (Caution: This is not necessarily a good way to think about the data.)

5. Look back at Chapter 4, Review Exercise 7, which reported scores for stories told by mothers of normal children and of schizophrenic children. What is wrong with the following statement?

   The data set has an SP/RM structure. Mothers serve as blocks, schizophrenic versus normal is the between-blocks factor, and story category (A, B, C, D, IR) is the within-blocks factor.

6–9 *Diets and dopamine*. Phenylketonuria (PKU) is a disease in which an enzyme deficiency inhibits the synthesis of dopamine, which is needed for the transmission of nerve impulses. To some extent the symptoms can be relieved by reducing the amount of the amino acid phenylalanine in the diet, because phenylalanine inhibits the enzyme responsible for the synthesis of dopamine. The following SP/RM study was designed to measure the effects of diet on the levels of dopamine in humans. There were two groups of patients with PKU, five with poor dietary control and five with good control. Each patient was measured twice, once after a week on a low-phenylalanine diet and once after a week on a normal diet. The response is the concentration of dopamine (in micrograms per milligram of creatinine) in the urine. (Several of the values were altered slightly, in ways that would not change the analysis except to simplify the arithmetic.) Note that the data set contains an outlier so gross that our first step should be to estimate a replacement value using the method of Chapter 12, Section 5. Cross out the outlier, and substitute 107, its estimated replacement.

6. What are the larger units, the smaller units, the between-blocks factor, and the within-blocks factor?

Dietary Control	Patient	Dietary Phenylalanine	
		Normal	Low
Poor	BR	87	166
	WJ	35	27
	KK	36	70
	TK	72	94
	AS	20	52
Good	DA	50	192
	MB	40	158
	MF	117	103
	MK	97	197
	TW	36	150

7. Draw a parallel dot graph with the four combinations of dietary control and dietary phenylalanine as columns. Discuss what you see.

8. Compute cell averages for the four conditions, draw an interaction graph, and describe the pattern.

9. Compute separate SDs for the four conditions, ignoring the blocks, as though the data came from a BF[2] design. Compute SD_{Max}/SD_{Min}, and check to see if the averages and the SDs are related. Should the data be transformed?

4. Formal Analysis: Decomposition and Analysis of Variance

This section comes in three parts, one on each of the designs of this chapter. By now I hope you are beginning to develop your skill at applying the two general rules to decompose data and find degrees of freedom, so that you can apply these to new designs on your own. If you have reached that point, you won't find much that is truly new in this section. The same general rules from before work for the CB, LS, and SP/RM designs. The only new wrinkle is that for the SP/RM design, there is an exception to the usual pattern for denominators in the F-ratios: to test the between-blocks factor, you use MS_{Blocks} instead of MS_{Res} for the denominator of the F-ratio.

The Complete Block Design

Summary for the Complete Block Design

Factor	*Estimated effect*	*Degrees of freedom*
Benchmark	**Grand average**	**1**
Blocks	**Bl Avg − Gr Avg**	**# blocks − 1**
Treatments	**Tr Avg − Gr Avg**	**# treatments − 1**
Residual error	**Obs − Fit**	**(# blocks − 1) × (# treatments − 1)**

Fit = Block Average + Treatment Average − Grand Average

EXAMPLE 7.14A FINGER TAPPING: PREVIEW OF THE FORMAL ANALYSIS

The data from the finger-tapping experiment (Exercise 7.B.2), together with row, column, and grand averages, are shown in Fig. 7.15.

Subject	Placebo	Caffeine	Theobromine	Average	Effect
I	11	26	20	19	−15
II	56	83	71	70	36
III	15	34	41	30	−4
IV	6	13	32	17	−17
Average	22	39	41	34 = Grand average	
Effect	−12	5	7		

FIGURE 7.15 Finger-tapping data. Rows (blocks) correspond to subjects, and columns (treatments) correspond to drugs. The response is the rate of finger tapping two hours after taking the drug. Because this CB design has more structure than a CR, it is almost impossible to tell the sizes of the residual errors without decomposing the data.

A complete decomposition of these numbers shows each observed value as a sum of four pieces:

Observed value = Grand average + Subject effect + Drug effect + Residual error

The grand average of 34 is, as usual, an estimate of the benchmark amount. The subject effects will show us the pattern of how the subjects differ, on average, from the benchmark amount and from each other. Notice that one subject tapped a lot faster than the others and that the two slowest subjects averaged pretty much the same tapping rate. The drug effects will tell us just what their name suggests, the effects of the drugs; more specifically, they tell us how far above or below the grand average each drug average is. Look at the three column averages, and notice that the caffeine and theobromine averages are roughly equal, and both are higher than the grand average; the placebo average is quite a bit lower. ■

EXAMPLE 7.14B DECOMPOSING THE FINGER-TAPPING DATA

Use the factor diagram in Fig. 7.16 to decompose the finger-tapping data; then interpret block and treatment effects.

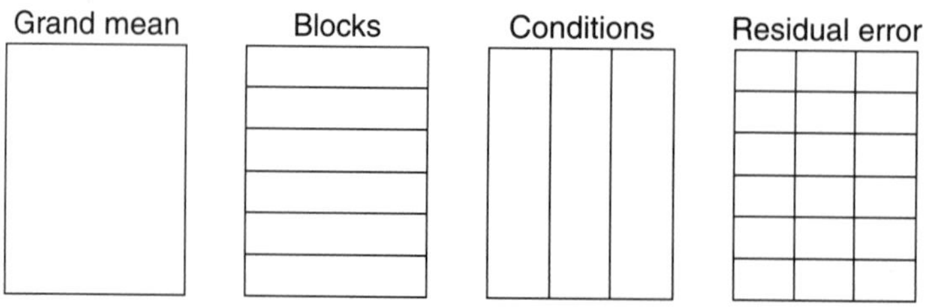

FIGURE 7.16 Factor structure for the finger-tapping data.

SOLUTION

a. Block effects. Apart from the benchmark, there are no outside factors for blocks. Each block effect = block average − grand average:

I: $19 - 34 = -15$ III: $30 - 34 = -4$
II: $70 - 34 = 36$ IV: $17 - 34 = -17$

Subjects I and IV were, respectively, 15 and 17 taps slower than the overall average, III was about average, and II was 36 taps faster than the overall average. The subject effects add to zero because they are deviations from the grand average.

b. *Drug effects*. The only outside factor is the benchmark, and so each drug effect = drug average − grand average.

Placebo:	$22 - 34 = -12$
Caffeine:	$39 - 34 = 5$
Theobromine:	$41 - 34 = 7$

The estimated drug effects show that subjects were on average 12 taps slower than the grand average after the placebo and that caffeine and theobromine raised the rate of tapping by roughly equal amounts, 5 and 7, respectively, above the benchmark rate. The drug effects add to zero.

c. *Residual error*. The outside factors are the benchmark, subjects, and drugs. Each residual equals Obs. − Fit, where

Fitted value = Grand average
+ (Block average − Grand average)
+ (Drug average − Grand average)
= Block average + Drug average
− Grand average

For example, the observed value of 11 in the upper left corner of Fig. 7.17 has a fitted value of $[19 + 22 - 34] = 7$ and a residual of $11 - 7 = 4$. You get the other

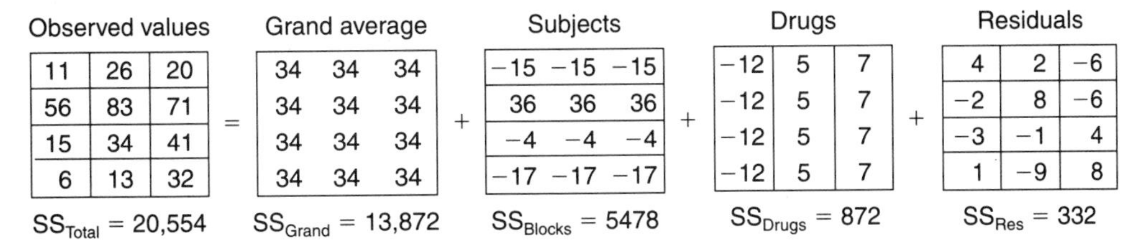

Observed values
11 26 20
56 83 71
15 34 41
6 13 32
=
Grand average
34 34 34
34 34 34
34 34 34
34 34 34
+
Subjects
−15 −15 −15
36 36 36
−4 −4 −4
−17 −17 −17
+
Drugs
−12 5 7
−12 5 7
−12 5 7
−12 5 7
+
Residuals
4 2 −6
−2 8 −6
−3 −1 4
1 −9 8
SS_{Total} = 20,554
SS_{Grand} = 13,872
SS_{Blocks} = 5478
SS_{Drugs} = 872
SS_{Res} = 332

FIGURE 7.17 Decomposition and sums of squares for the finger-tapping data. Check that the numbers add up as they should. For example, the observed value of 11 in the upper left corner does in fact equal the sum of its pieces: grand average (34) + effect for subject I (−15) + effect of placebo (−12) + residual (4) = observed value (11). Check, also, the patterns of adding to zero: The drug effects add to zero across rows; the subject effects add to zero down columns; and the residuals add to zero across rows and down columns. Finally, check the message from the sums of squares: there are huge subject differences, much larger than the drug differences. These SSs confirm that the CB design was a good choice. When the variability between subjects is so large, you would need a very large number of subjects in order to detect these drug differences using a CR design.

residuals in the same way; they are shown in the last box of the complete decomposition in Fig. 7.17. By inspection, the typical size is somewhere around 5. (The SD, which adjusts for df, is about 7.4. See Note 1 at the end of Chapter 5 for more details.) ■

EXAMPLE 7.14C DEGREES OF FREEDOM FOR THE FINGER-TAPPING DATA

Most of the dfs are straightforward:

i. df_{Total} = # observations = 12.

ii. df_{Grand} = 1, as always.

iii. df_{Blocks} = # blocks − 1 = 3. You can get this either from the general rule (df_{Blocks} = # levels for blocks − df_{Grand}) or by direct count (ignoring repetitions, the box for blocks has four numbers that add to zero, so there are three free numbers).

iv. $df_{Treatments}$ = # treatments − 1 = 2. The logic here is the same as for blocks.

v. df_{Res} = (# blocks − 1) (# treatments − 1) = (3)(2) = 6. According to the general rule, df_{Res} = # obs − [df_{Grand} + df_{Blocks} + $df_{Treatments}$] = 12 − 6 = 6. The box in Fig. 7.18 shows how to get df_{Res} = 6 by direct count.

1	2	+
3	4	+
5	6	+
+	+	+

The residuals add to zero across rows because they are deviations from row (block) averages, and they add to zero down columns because they are also deviations from column (treatment averages. You can think of the free numbers forming a rectangle of size (# rows −1) by (# columns −1). Thus, the number of free numbers equals (# blocks −1)(# treatments −1). ■

FIGURE 7.18 Residual DF for the CB design − (# blocks − 1)(# treatments − 1).

EXAMPLE 7.14D ANOVA TABLE FOR THE FINGER-TAPPING DATA

Source	SS	df	MS	F-ratio	5% crit. val.
Grand Avg.	13,872	1	13,872		
Blocks	5,478	3	1,826	33.00 *	4.76
Treatments	872	2	436	7.88 *	5.15
Residual	332	6	55.33		
Total	20,554	12			

* > 5% critical value.

Both *F*-ratios are bigger than the critical value, so we conclude that the observed differences due to subjects and to drugs are too big to be due just to chance error. These *F*-tests don't tell the whole story, however. If you look back at the decomposition, you can see that the estimated effects of caffeine and theobromine are practically the same and that almost all of the observed drug differences come from the difference between the placebo and the two stimulants taken together. It would be natural to continue the formal analysis by testing whether the observed difference between the two stimulants was "real." Chapter 11, on contrasts and confidence intervals, shows how to do this. ■

The Latin Square Design

Summary for the Latin Square Design

Factor	*Estimated effect*	*Degrees of freedom*
Benchmark	Grand average	1
Rows	Row Avg − Gr Avg	# rows − 1
Columns	Col Avg − Gr Avg	(# cols − 1) = (# rows − 1)
Treatments	Tr Avg − Gr Avg	(# treats − 1) = (# rows − 1)
Residual error	Obs − Fit	(# rows − 1) × (# rows − 2)

Fit = Row Avg + Column Avg + Treatment Avg − 2[Grand Avg]

EXAMPLE 7.15A DECOMPOSING THE MILK YIELD DATA (EXAMPLE 7.6)

The decomposition will split the numbers into five pieces, one for each factor in the factor diagram in Fig. 7.19. As always, we estimate the benchmark using the grand average. For each of the rows, columns, and treatments, the only outside factor is the benchmark, so each estimated effect equals a row, column, or treatment average minus the grand average. To get residuals, we first add the grand average plus row, column, and treatment effects to get fitted values, and then we subtract to get Res = Obs − Fit.

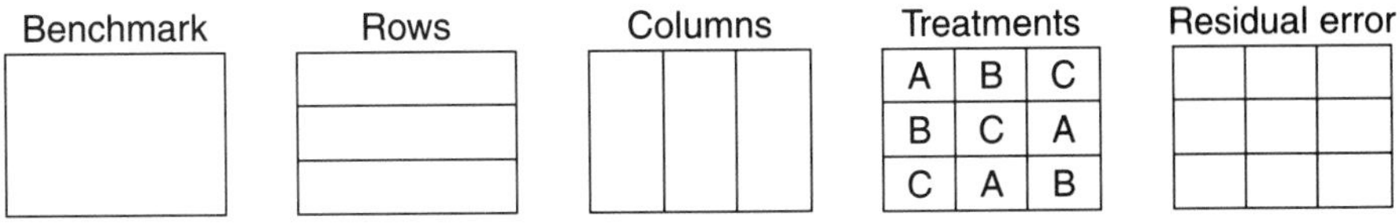

Benchmark

Rows

Columns

Treatments

A	B	C
B	C	A
C	A	B

Residual error

FIGURE 7.19 Factor structure for the milk yield experiment. For each of rows, columns, and treatments, only the benchmark is an outside factor, so each estimated effect equals a row, column, or treatment average minus the grand average.

Look over the numbers below, from Example 7.6. (I added or subtracted 1 for numbers marked * to simplify the arithmetic). Along with the observed values themselves, I've also given row, column, and treatment averages, together with the corresponding effects (= average − grand average).

	Weeks				
Cow	1–6	7–12	13–18	Average	Effect
I	A 608	B 716*	C 845*	723	−98
II	B 885	C 1086*	A 711	894	73
III	C 940	A 766	B 832	846	25
Average	811	856	796	821 = Grand average	
Effect	−10	35	−25		

Treat.	Average	Effect
A	695	−126
B	811	−10
C	957	136

A = roughage
B = partial grain
C = full grain

We can learn quite a bit from the averages and effects.

Rows: The reason for not using a CR design for this experiment was the expectation that cow-to-cow differences would be big compared to the effects of diet. Indeed, the cow averages show large differences. Cow II produced the most; cow I the least. It is not hard to check that the row effects add to zero.

Columns: Remember that experience with cows had suggested that the average yields might decline over the course of the experiment. The column averages for the three time periods, however, tell us a different story: they do not show the pattern of falling off from left (weeks 1–6) to right (weeks 13–18) that the experiment's designers worried about. Check that the column effects add to zero.

Treatments: Check first to see how I got the table of treatment effects: for A, I collected the three A observations and took their average: $(608 + 711 + 766) \div 3 = 695$. Then I subtracted the grand average to get the estimated A effect: $695 - 821 = -126$. Similarly for B $[(716 + 885 + 832) \div 3 = 811;\ 811 - 821 = -10]$ and C. The treatment effects certainly fit what you would expect from knowing what the diets were. Cows on a full-grain diet (C) averaged 136 pounds (17 gallons) per week higher yield than the estimated benchmark, and cows fed a diet of roughage (A) retaliated by giving 126 pounds per week less than the estimated benchmark. The treatment effects add to zero, as they should.

To finish the decomposition, we need the residuals. As always, each residual equals Obs. − Fit, where the fitted value equals the sum of all the other estimated effects:

$$\begin{aligned}\text{Fit} &= \text{Gr Avg} + [(\text{Row Eff}) + (\text{Col Eff}) + (\text{Tr Eff})] \\ &= \text{Gr Avg} + (\text{Row Avg} - \text{Gr Avg}) + (\text{Col Avg} - \text{Gr Avg}) + (\text{Tr Avg} - \text{Gr Avg}) \\ &= \text{Row Avg} + \text{Col Avg} + \text{Tr Avg} - 2[\text{Gr Avg}]\end{aligned}$$

For example, the fitted value for the upper left corner equals $(725 + 811 + 695 - 2 \cdot 821) = 587$, and the residual equals $608 - 587 = 21$. The remaining residuals are shown in the last box of the decomposition in Fig. 7.20. Notice that there are only three different numbers in the box for residuals: 21, 11, and −32. The reason is that the patterns of adding to zero leave few possibilities. The residuals add to zero across rows, down columns, and within each of the three treatment groups as well. (The restrictions imposed by all these patterns mean that the individual residuals tend to underestimate the actual error sizes. From looking at the box of residuals, it would seem reasonable to guess a number near 20 or so as the typical size. In fact, the standard deviation, which includes an adjustment for the restrictions, is about 50.)

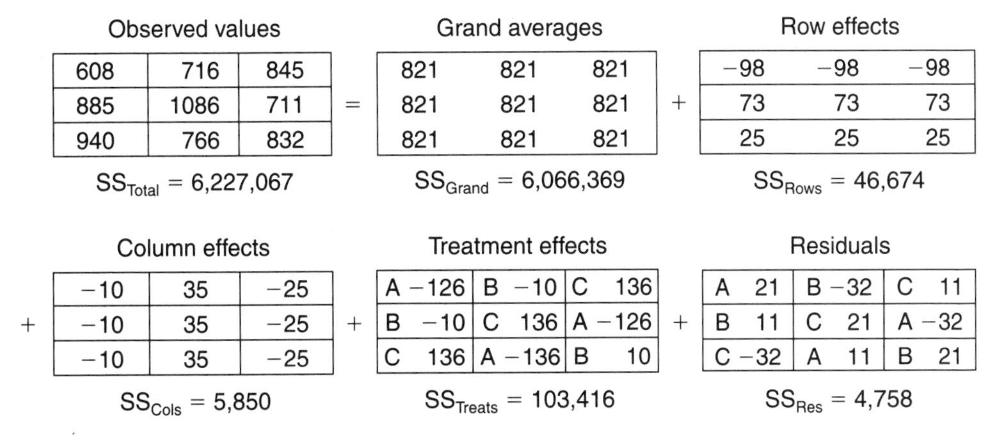

FIGURE 7.20 Decomposition and sums of squares for the milk yield data. ■

EXAMPLE 7.15B DEGREES OF FREEDOM FOR THE MILK YIELD DATA

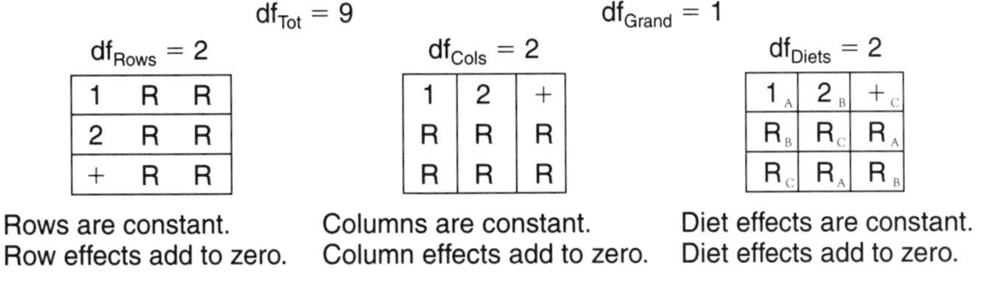

For the LS, there is a trick (Example S7.6) for finding df_{Res} by counting free numbers. You can also rely on subtraction: $df_{Res} = 9 - (1 + 2 + 2 + 2) = 2$. ■

EXAMPLE 7.15C ANOVA TABLE FOR THE MILK YIELD DATA

Source	SS	df	MS	*F*-ratio	5% crit. val.
Grand Avg	6,066,369	1	6,066,369		
Rows	46,674	2	23,337	9.81	19.0
Columns	5,850	2	2,925	1.23	19.0
Diets	103,416	2	51,708	21.74 *	19.0
Residual	4,758	2	2,379		
Total	6,227,067	12			

With so few degrees of freedom for residuals, the experiment is not very sensitive. That is, the experiment is only able to detect differences that are quite large. You can see this from the large critical value of 19.0: it takes a large *F*-ratio to declare a difference "significant." Not even the large cow-to-cow differences are "too big to be due just to chance," although the diet differences do manage to squeak by. ■

The Split Plot/Repeated Measures Design

Chapter 3 uses Kelly's hamster experiment to illustrate how to decompose data from an SP/RM design and count the degrees of freedom. If you think it would help you to review that example, now would be a good time. Example 7.16 below will go over some of the same issues a bit more quickly.

Summary for the Split Plot/Repeated Measures Design

Factor	*Estimated effect*	*Degrees of freedom*
Benchmark	Grand Average	1
Between	Between Avg − Grand Avg	# Between levels − 1
Blocks	Block Avg − Between Avg	# blocks − # Between levels
Within	Within Avg − Grand Avg	# Within levels − 1
Interaction	Cell Avg − [Between Avg + Within Avg − Grand Avg]	$df_{Between} \times df_{Within}$
Residual error	Obs − Fit	$df_{Blocks} \times df_{Within}$

Fit = Cell Avg + Block Avg − Between Avg

Exception: To get the *F*-ratio for testing the between-blocks factor of an SP/RM design, use the MS for *blocks* in the denominator.

The estimated effects and df in the summary come from the general rules based on inside and outside factors. (See **Example 7.12** for the factor diagram.)

a. **Between and within.** Both factors have no outside factors other than the benchmark. For each factor, Effect = Factor average − grand average, and df = (# levels − 1).

b. **Blocks.** The outside factors are the benchmark and the between-blocks factor. The partial fit equals (Grand Avg) − (Between Avg − Grand Avg) = Between Avg, so Block Eff = Block Avg − Between Avg. For blocks, the rule for df gives df = (# blocks − [df_{Betw} + df_{Grand}]) = (# blocks − # Between levels).

c. **Interaction.** Here the situation is the same as for a BF[2], with the between-blocks and within-blocks factors as the two basic factors. The partial fit equals [Grand Avg + (Between Avg − Grand Avg) + (Within Avg − Grand Avg)] = [Between Avg + Within Avg − Grand Avg]. Each estimated interaction effect equals a cell average minus the partial fit. To get the df, first check that # cells = (df_{Betw} + 1)(df_{Within} + 1); then check that df = (# cells − [df_{Grand} + df_{Betw} + $df_{W/in}$]) simplifies to ($df_{Betw} df_{Within}$).

d. **Residual error.** Res. = Obs. − Fit. Exercises G.17 and 18 ask you to check that the fitted value is equal to [Cell Avg + Block effect] and that the residuals and df are equal to those given in the summary table.

EXAMPLE 7.16 DIABETIC DOGS

The data set on lactic acid turnover rates was given in Example 7.12. (a) Decompose the data. (b) Then count free numbers to get the df, and give the ANOVA table. (Remember the exception: to get the F-ratio for the between-blocks factor (operations), divide by the MS for *blocks*, not by the MS for residuals.)

SOLUTION

a. Decomposition. (i) First we compute a complete set of averages. (ii) Then we'll decompose the cell averages just as for a BF[2] (Chapter 6, Section 3). (iii) Next, we get block effects (= Block Avg − Betw Avg). (iv) Finally, we compute Fit = Cell Avg + Block Eff, and Res = Obs − Fit.

i. *A complete set of averages.*

Inject	Infuse	Dogs Avg	Dogs Eff
44	28	36	1
33	23	28	−7
38*	34	36	1
59*	19	39	4
46	26	36	1
54*	42	48	5
43	23	33	−10
55	23	39	−4
71	27	49	6
57*	35	46	3

	Condition averages: Inject	Infuse	Operation Avg	Operation Eff
Cont	44	26	35	−4
Diab	56	30	43	4
Method Avg	50	28	39 (Grand average)	
Method Eff	11	−11		

ii. Decomposing the cell averages. The right half of the table of averages shows cell averages together with row averages and effects (for Operations, the between-blocks factor) and column averages and effects (for Methods, the within-blocks factor).

Now consider the interaction effects, which turn out to be ±2. Here's a sample calculation, for the upper left cell (control dogs, injection method):

$$\text{Partial Fit} = \text{Gr Avg} + \text{Row Eff} + \text{Col Eff}$$
$$(39) \quad + \quad (-4) \quad + \quad (11) \quad = 46$$
$$\text{Interaction Eff} = \text{Cell Avg} - \text{Partial Fit}$$
$$(44) \quad - \quad (46) \quad = -2$$

iii. Block effects. After the two left-most columns in the table, which show the raw data, the next two columns show the block averages and block effects. Each block effect equals the block average minus the average for all the dogs who got the same treatment.

iv. *Fitted values and residuals*. Here's a sample calculation for the lower left corner of the data set (the diabetic dog in the last row, with injection method):

Fit = Cell Avg + Block effect
(56) + (3) = 59

Res = Obs − Fit
(57) − (59) = −2

b. Degrees of freedom and ANOVA

df_{Grand} = 1

$df_{Operations}$ = 1 There are two effects, which add to zero.

df_{Dogs} = 8 Each group of five dogs has five dog effects that add to zero in the group: 4 df from each of 2 groups.

$df_{Methods}$ = 1 There are two effects, which add to zero.

df_{Inter} = 1 There are 4 effects, in a 2 × 2 rectangle. They add to zero across rows and down columns, leaving 1 × 1 free numbers.

df_{Res} = 8 By subtraction, using df_{Total} = 20.

In constructing the ANOVA table, we use MS_{Dogs} in the denominator of the *F*-ratio for Operations, the between-blocks factor, and we use MS_{Res} in the denominator of the other *F*-ratios.

Source	SS	df	MS	*F*-ratio	Crit.val.
Grand Average	3042	1	3042.0		
Operations (betw.)	320	1	320.0	4.96	5.32
Dogs (blocks)	508	8	64.5	0.76	3.44
Methods (w/in)	2420	1	2420.0	28.47*	5.32
Interaction	80	1	80.0	0.94	5.32
Residual	680	8	85.0		
Total	7050	20			

From the ANOVA table we conclude that overall differences due to operation are not quite big enough to be declared "real," but there is a detectable difference between the two methods. The other observed differences, however, for dogs and interaction, are roughly the same size as the residual errors and could easily be due just to chance variation. ■

Two kinds of units, two kinds of chance error. For the SP/RM design there is an exception to the intuitive principle that you should use the mean square for residuals in the denominator of your *F*-ratios. Chapter 14 discusses the logic of choosing these denominators and gives a single rule that works for all balanced designs. For now, however, I'll give an informal account of the logic behind the *F*-ratios for SP/RM designs:

The SP/RM has two kinds of error variability:

- **whole-plot error comes from differences among the larger units;**
- **subplot error comes from differences among the smaller units.**

The denominator MS you use for testing a factor depends on to which unit the levels of the factor are assigned.

EXAMPLE 7.17 AN ANALOGY: KIDS ON SHIPS

(This analogy asks you to visualize error variability as up-and-down motion.) Think of a fleet of ships—the larger units—riding up and down on swells of the ocean. This up-and-down motion of the ships is the variability associated with the larger units—the "whole-plot error."

Next imagine bunches of children—the smaller units—jumping up and down on the decks of the ships. Their motion is the variability associated with the smaller units—the "subplot error."

Now for the treatments. Suppose we load the ships with various amounts of cargo—we apply this (whole-plot) treatment to the ships, not to the children. But we might also outfit the jumping children with different brands of tennis shoes—we apply this (subplot) treatment to the children, not to the ships.

The virtue of this example (if there is one!) is that it reduces the logic of the F-tests to common sense: to judge the effects of the cargo (between-blocks factor), you would compare ships (blocks). Judging cargo by comparing children is obviously wrong. Similarly, to judge the effects of the shoes, you'd compare children; judging shoes by comparing ships would be ridiculous. ■

Between-blocks treatments get assigned to the larger units, blocks, or whole plots; in a sense what goes on in the subplots is irrelevant to testing the between-blocks factor. In thinking about F-ratios for between-blocks factors, it is useful to think of each block as contributing just one response value, the block average, and to ask yourself how you would analyze the resulting set of averages. If the pattern that assigns treatments to blocks is CR (as it is for all the SP/RM designs so far), then for these treatments MS_{Blocks} plays the same role as MS_{Res} for the CR design.

EXAMPLE 7.18 DIABETIC DOGS: THE BETWEEN-BLOCKS F-RATIO

I'll use the data from Example 7.16 to illustrate why you use MS_{Blocks} in the denominator of the F-ratio for the between-blocks factor. For these data the between-blocks factor—Operations—was assigned to Dogs (blocks) completely at random, as in a CR design. If you think of each block average as a single response value for its block, the resulting set of 10 block averages has the same structure as a CR[1] (Fig. 7.21).

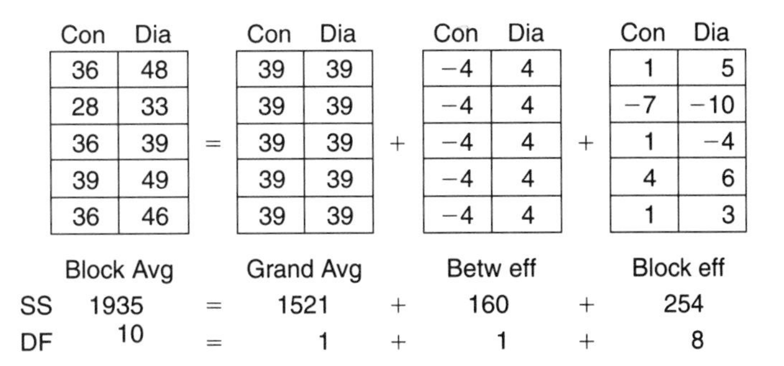

Con	Dia		Con	Dia		Con	Dia		Con	Dia
36	48		39	39		−4	4		1	5
28	33		39	39		−4	4		−7	−10
36	39	=	39	39	+	−4	4	+	1	−4
39	49		39	39		−4	4		4	6
36	46		39	39		−4	4		1	3

	Block Avg		Grand Avg		Betw eff		Block eff
SS	1935	=	1521	+	160	+	254
DF	10	=	1	+	1	+	8

FIGURE 7.21 CR decomposition of the dog (= block) averages. The decomposition is based on the CR structure used to assign operations to dogs in the SP/RM design.

If you analyze the set of block averages using the CR structure, the F-ratio, and df for testing Operations are the same as those you would get by applying the exception to the set of 20 observed values, that is, using MS_{Blocks} in the denominator of the F-ratio for operations:

The F-Test Based on Block Averages

	CR for Block Averages			
Source	**df**	**SS**	**MS**	**F**
Grand Avg	1	1521		
Operations	1	160	160	5.04
"Residual"	8	254	31.75	

	Exception, applied to the observations			
Source	**df**	**SS**	**MS**	**F**
Grand Avg	1	3042		
Operations	1	320	320	5.04
Blocks	8	508	63.5	

Both ANOVA tables have the same dfs and F-ratios. (The SSs and MSs in the top table are 2 times as big as those in the bottom table, because there are 2 times as many repetitions in the boxes of the decomposition.) ■

Exercise Set D

1–8. *Fake data sets for practice*. I've presented the two data sets below as having the same structure as Kelly's hamster data, regarded as an SP/RM. However, in the exercises that follow, I'll ask you to use each data set four times, regarding it first as a BF[1], then as a CB[1], then as a BF[2], and finally as an SP/RM.

a.		Body	Soul
Long	1	80	46
	2	70	44
Short	3	30	16
	4	36	18
Medium	5	46	32
	6	38	24

b.		Blood	Sweat	Tears
Long	1	9	7	5
	2	5	7	3
Short	3	3	0	3
	4	3	2	1

1. Think of data set (a) as a BF[1]: ignore the row labels, and think of columns as corresponding to the only factor of interest, as though each response value came from a different hamster. Decompose the data, and write the df under each box of your decomposition.

2. Now think of the data set as a CB[1]: ignore the day length labels for groups of rows, think of the individual rows as hamsters (= blocks), and think of columns as corresponding

to the factor of interest. Decompose the data, and write the df under each box of your decomposition. (If you have trouble counting df, do Exercises 9 and 10 first.)

3. Next think of the data set as a BF[2], with day length and "organ" as the two crossed factors of interest. For this part of the problem, there are no blocks: think of each response value as though it came from a different hamster. Decompose the data, and write the df under each box of your decomposition.

4. Finally, think of the data set as an SP/RM, with the same two factors of interest as in Exercise 3 and the same blocks = whole plots as in Exercise 2. Decompose the data, and write the df under each box of your decomposition. (If you have trouble counting df, do Exercise 11 first.)

5–8. Repeat Exercises 1–4, this time with data set (b).

9–11. Use the patterns of repetition and adding to zero to fill in the missing numbers in the following decompositions; then write the df under each box. Exercises 9 and 11 are both CB[1] designs with rows as blocks and columns as treatments; Exercise 11 is an SP/RM, in the standard format.

9.

Grand Avg				Block Eff				Treat Eff				Residuals		
?	?	?		?	2	?		?	?	?		4	2	?
?	?	?		1	?	?		?	?	?		1	1	?
?	17	?	+	?	−2	?	+	3	−2	?	+	3	−2	?
?	?	?		?	?	?		?	?	?		−5	1	?
?	?	?		4	?	?		?	?	?		?	?	?

10.

Grand Avg					Block Eff					Treat Eff					Residuals			
8	?	?	?		3	?	?	?		0	1	3	?		3	−2	?	?
?	?	?	?	+	−4	?	?	?	+	?	?	?	?	+	?	2	6	6
?	?	?	?		?	?	?	?		?	?	?	?		1	?	?	−3

11.

	Grand Avg			+	Between			+	Blocks			+	Within			+	Interact			+	Residuals		
	14	?	?		2	?	?		1	?	?		1	−4	?		1	−4	?		1	−4	?
L	?	?	?		?	?	?		−2	?	?		?	?	?		?	?	?		2	3	?
	?	?	?		?	?	?		?	?	?		?	?	?		?	?	?		?	?	?
				+				+				+				+				+			
	?	?	?		?	?	?		−1	?	?		?	?	?		?	?	?		5	2	?
S	?	?	?		?	?	?		?	?	?		?	?	?		?	?	?		−4	2	?
	?	?	?		?	?	?		?	?	?		?	?	?		?	?	?		?	?	?

12–13. Decompose the following fake milk yield data sets, and write the df under each box of your decomposition. (If you have trouble with the df, remember that the first three boxes of the decomposition—grand average, row effects, and column effects—are the same as for the decomposition of a CB data set.)

12.

15 A	2 B	1 C
2 B	1 C	3 A
−2 C	6 A	17 B

13.

16 A	11 B	6 C	7 D
7 C	10 D	9 A	10 B
15 D	4 A	9 B	8 C
18 B	7 C	12 D	11 A

14–16. *Bee stings* (*Example S7.3*). The bee-sting data are shown below, together with row, column, and grand averages.

Block	Stung	Fresh	Average
1	27	33	30
2	9	9	9
3	33	21	27
4	33	15	24
5	4	6	5
6	22*	16	19
7	21*	19	20
8	33	15	24
9	70	10	40
Average	28	16	22

14. Decompose the data, give the ANOVA table, and state your conclusions based on the F-ratios.

15. Now substitute the estimated replacement value of 16, from Exercise 7.A.13, for the 70 in block 9. Decompose the data a second time, give the ANOVA table, and tell what conclusions you would draw from the new pair of F-ratios.

16. Write a short paragraph comparing your two analyses and commenting on the influence of the outlier. What is your overall conclusion: are bees more likely to sting the previously stung balls?

17–19. *Sleeping shrews* (*Example 7.4*). A decomposition of the sleeping shrew data is shown below. Use this decomposition to answer the questions that follow. The rows are the six shrews, and the columns are the three kinds of sleep: deep slow wave; light slow wave; and rapid eye movement.

Observed values

14.0	11.7	15.7
25.8	21.1	21.5
20.8	19.7	18.3
19.0	18.2	17.1
26.0	23.2	22.5
20.4	20.7	18.9

=

Grand average

19.7	19.7	19.7
19.7	19.7	19.7
19.7	19.7	19.7
19.7	19.7	19.7
19.7	19.7	19.7
19.7	19.7	19.7

+

Shrews

−5.9	−5.9	−5.9
3.1	3.1	3.1
−0.1	−0.1	−0.1
−1.6	−1.6	−1.6
4.2	4.2	4.2
0.3	0.3	0.3

+

Kind of sleep

1.3	−0.6	−0.7
1.3	−0.6	−0.7
1.3	−0.6	−0.7
1.3	−0.6	−0.7
1.3	−0.6	−0.7
1.3	−0.6	−0.7

+

Residuals

−1.1	−1.5	2.6
1.7	−1.1	−0.6
−0.1	0.7	−0.6
−0.4	0.7	−0.3
−0.8	−0.1	−0.7
−0.9	1.3	−0.4

Look over the table of observed values, and think about the following questions. Do any of the shrews stand out as different from the rest? Is there greater variability between shrews or between kinds of sleep? Which kind of sleep has the fastest average heart rate? the slowest? Do any individual values stand out as unusual? Try to guess the typical size of residual error from the observed values, and notice how hard this is to do.

17. *Shrews*. The table of observed values reveals that the heart of shrew I beat more slowly than average while the hearts of shrews II and V beat faster than average. Draw a dot diagram of the shrew effects, and use the pattern to comment on whether you regard any of these three shrews as an outlier.

18. a. Is the typical size of residual error closest to .5, 1, or 2?

b. Find the residual farthest from zero. To which observed value does it belong? In what way is this observed value different from the rest? Do you consider the observed value that goes with the biggest residual to be atypical?

c. The residual mean square is 2.00. Compute the SD and use a dot graph of the residuals to check the normality assumption.

19. An incomplete ANOVA table for the sleeping shrew data is shown below. Use the information given to fill in any missing numbers. Then discuss what the numbers tell you about shrews, heart rates, and kinds of sleep.

Source	SS	df	MS	*F*	5% crit. val.
Grand avg.	6985.6				
Shrews					
Sleep				3.8	
Residual			2.00		
Total	7215.0				

20. *Rabbit insulin (Example 7.7)*. Decompose the data, give the ANOVA table, and state your conclusions.

21–24. *Losing leaves (Example 7.11)*. Here are the data from the experiment in Example 7.11:

Auxin	Plant	Deblading No	Deblading Yes
Control	1	24	20
	2	20	17
	3	21	11
	4	23	16
Lanolin only	5	13	11
	6	15	11
	7	14	17
	8	14	13
Lanolin with	9	25	35
10^{-3}M auxin	10	24	33
(high conc.)	11	15	22
	12	20	31
Lanolin with	13	32	38
10^{-5}M auxin	14	26	33
(low conc.)	15	21	26
	16	19	20

21. *Decomposition.* Estimate the effects for all factors of interest for the leaf data, and use your estimates to compute sums of squares. (Check that your SSs are the same as the ones in Exercise 23 below.)

22. *Counting df.* Draw boxes for all factors, and count degrees of freedom by filling in the boxes with numbers, +s, and Rs. Then check that your answers agree with the df in the ANOVA table.

23. $SS_{Plants} = 483.75$; $SS_{Debl} = 24.5$; $SS_{Inter} = 265.75$; $SS_{Res} = 42.75$.

Complete the ANOVA table and discuss the results.

24. *Discussion question: Confounding in the leaf experiment.* Reread the description of the experimental design, looking for nuisance influences that might be confounded with the effect of deblading. Then discuss your sense of how important this confounding is and how you would take it into account in interpreting the results of the experiment.

25. *Diets and dopamine (Exercises 7.C.6–9).* Decompose the dopamine data set, using the replacement value instead of the outlier. Give the ANOVA table, and tell what your conclusions are. (Remember to reduce the total and residual dfs by 1, since one of your "observations" is really just an estimate.)

5. Using Scatterplots to Check the Fit of Models with Blocks

Whenever you have numbers that come in pairs, you should consider drawing a scatterplot to help see how the numbers in the pairs might be related. Is there a positive relationship (pattern of high numbers with high and low numbers with low); or a negative relationship (high with low, low with high) as in Kelly's hamster data; or is there some other pattern?

Scatterplots Are Useful Whenever Your Numbers Come in Pairs:

1. Within-blocks scatterplots. **(For designs with blocks as a nuisance factor.) Pick two levels of the within-block factor to scatterplot, with blocks as points.**
2. Res vs. Fit. **(For *any* analysis that splits observations into fitted values plus residuals.) Scatterplot residuals versus fitted values.**

Within-Blocks Scatterplots

For a CB[1] design, if the model and Fisher assumptions fit the data, a within-blocks scatterplot of readings for any one condition against readings for any other should suggest a line parallel to the identity line $y = x$. (If horizontal and vertical scales are equal, this is the 45° line.)

EXAMPLE 7.19 COOL MICE: A CB OBSERVATIONAL STUDY

The *cooling constant* measures the rate at which the temperature of a warm object returns to the temperature of its environment. Each kind of object has its own cooling constant, and medical examiners rely on this fact when they use the temperature of a corpse to determine the time of death.

Imagine for a moment that you have decided to commit murder. To cook up an alibi for yourself, you plan to heat up your victim's body to mislead the coroner about the time of death. Do reheated bodies cool at the same rate as freshly killed ones?

Believe it or not, in 1951 J. S. Hart published (in the *Canadian Journal of Zoology*) the results of a study designed to answer this question. He chose for his response the cooling constant; he used 19 mice as his experimental material; and there were two conditions of interest: freshly killed and reheated. Hart used a complete block design, with each mouse as a block. Each mouse provided two observational units: each mouse was measured twice, once when freshly killed and then again after reheating.

Hart's Cooling Constants for Mice Under Two Conditions

Mouse (Block)	Freshly killed	Reheated	Mouse (Block)	Freshly killed	Reheated
1	573	481	11	445	443
2	482	343	12	383	342
3	377	383	13	391	378
4	390	380	14	410	402
5	535	454	15	433	400
6	414	425	16	405	360
7	438	393	17	340	373
8	410	435	18	328	373
9	418	422	19	400	412
10	368	346			

Blocks come from reusing subjects: each mouse provides a pair (= block) of time slots. For this study you can't assign conditions at random, because the nature of the conditions fixes their order. The factor of interest is observational, not experimental. Figure 7.22 shows the within-blocks plot. ■

The logic behind the line. How do the model and Fisher assumptions lead to a line with a 45° slope? Remember that we get such a line if (and only if), apart from the errors, each *y*-coordinate (reheated) equals the *x*-coordinate (fresh) plus a constant. Here, almost word for word, is an explanation by Professor Francis Anscombe, of Yale University:

According to our model for each block we have

$$\text{Reheated reading} = \text{Benchmark} + \text{Block effect} + \text{Reheated effect} + \text{Error} \quad (1)$$

$$\text{Fresh reading} = \text{Benchmark} + \text{Block effect} + \text{Fresh effect} + \text{Error} \quad (2)$$

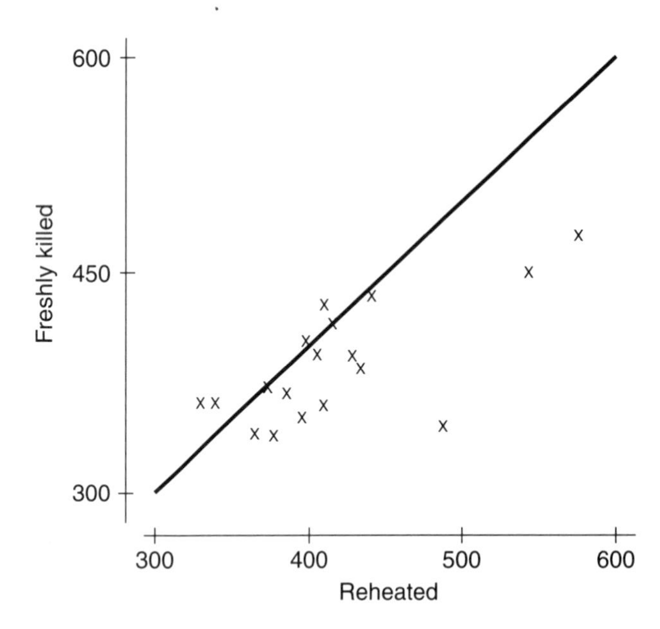

FIGURE 7.22 Scatterplot of reheated versus freshly killed for Hart's 19 mice. For this plot each point corresponds to a mouse (block), and thus to a pair of cooling constants: reheated on the y-axis, freshly killed on the x-axis. If the model and Fisher assumptions fit, it ought to be possible to draw a line with slope 45° passing through the middle of the plotted points, with the points scattered equally above and below the line. In fact, the plot suggests a line with a slope somewhat less than 45° unless you separate the three rightmost points and treat them as a second group, in which case two parallel 45° lines can be drawn satisfactorily through the two groups. (However, the published source gives no reason why there should be two groups.)

Subtracting (1) − (2) gives

$$\text{Reheated reading} - \text{fresh reading} = (\text{Difference of the two condition effects}) + (\text{Difference of the two errors})$$

Since the difference of the condition effects is a constant, and the difference of the two errors behaves like one error with a larger SD, we get

$$\text{Reheated reading } (y) = \text{Fresh reading } (x) + \text{Constant} + \text{Error}$$

Apart from the error, the equation simplifies to $y = x + \text{constant}$, which is the equation of a line with slope 45°.

Now consider a CB design where the only factor of interest has more than two levels, such as the sleeping shrew study (Example 7.4). You can plot readings for any one of the three conditions against those for another one. Are lines parallel to $y = x$ suggested? If not, the Fisher assumptions seem to be contradicted.

EXAMPLE 7.20 SLEEPING SHREWS: WITHIN-BLOCKS PLOTS

Two scatterplots for the sleeping shrews (Example 7.4) are shown in Fig. 7.23. Reread the description of the data, and then write a short paragraph discussing any patterns you see in the scatterplots. (Note in these plots, which use different scales for the x- and y-axes, the identity line $y = x$ does not have slope equal to 45°.) Are lines parallel to the identity line $y = x$ suggested?

SOLUTION. For the plot on the right, the points cluster near a line parallel to, but somewhat above, the identity line $y = x$, suggesting that LSWS = DSWS + (a positive constant). For the plot on the left, any line passing near all six points would be quite a bit less steep than the identity line $y = x$. However, if you exclude the left-

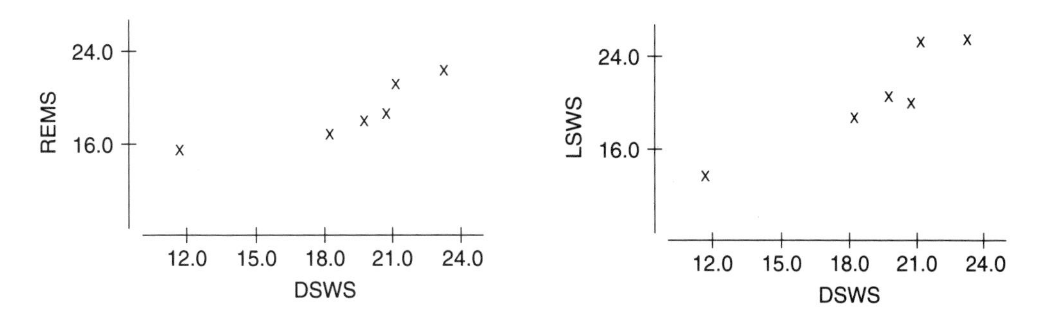

FIGURE 7.23 Within-blocks scatterplots for the sleeping shrews. Each point is a shrew. Heart rates for rapid eye movement sleep (REMS, left) and light slow wave sleep (LSWS, right) are plotted against heart rates for deep slow wave sleep (DSWS).

most point (shrew I) in the left-hand plot, the remaining five points suggest a line very near the identity line. Thus, the plot suggests shrew I as an outlier. ■

Scatterplots of Residuals Versus Fitted Values

The single, most useful way to check the fit of the model and Fisher assumptions is to scatterplot residuals versus fitted values. If the fit is good, the plot will suggest an oval balloon with the *x*-axis running lengthwise through the middle.

EXAMPLE 7.21 RESIDUAL VERSUS FIT FOR THE COOL MICE (EXAMPLE 7.19)

Take a look at the scatterplot of residuals versus fitted values for the cool mice data from Example 7.19 (Fig. 7.24).

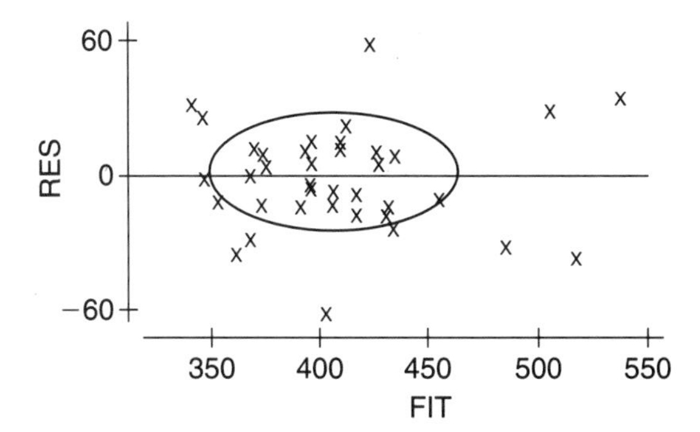

FIGURE 7.24 Scatterplot of residuals versus fitted values for the cool mice: a good fit. This plot shows roughly the sort of pattern you should expect for situations where the model and Fisher assumptions are reasonable: an oval balloon with the *x*-axis running lengthwise through the middle.

Because the horizontal and vertical scales are the same, the shape of the balloon tells roughly what fraction of the overall variability in the data is accounted for ("explained") by the model. The balloon in the picture stretches horizontally from 350 to 470, or 120 units, and vertically from −25 to 25, or 50 units. Very roughly, the residual variability is 50/120, or about 40% of the variability "explained" by the model.

Features Suggesting Poor Fit

Outliers: one or more isolated points (the balloon may be tilted).

Unequal SDs: if small residuals (+ or −) go with small fitted values, and large with large, the plot will suggest a wedge, opening to the right. Transform to equalize SDs.

Nonadditive model: a curved plot suggests nonadditivity. Transform to a new scale and/or add more factors (e.g., interactions) to the model.

EXAMPLE 7.22 SUGAR METABOLISM

Exercise 6.E.5 gave data from a CR[2] experiment designed to study the effects of oxygen concentration on the amount of ethanol produced from two sugars, galactose and glucose, by *Streptococcus* bacteria. The response was the concentration of ethanol, in micromoles per 0.1 micrograms of sugar. The version of the data given below has some of the observations altered slightly to make the arithmetic easier. Compute cell averages (= fitted values) and residuals. Then scatterplot Res. vs. Fit, and tell what you conclude from the plot.

	Oxygen conc. (micromoles)							
Sugar	0		46		92		138	
Galactose	59	30	44	18	22	23	12	13
Glucose	25	3	13	2	7	0	0	1

SOLUTION

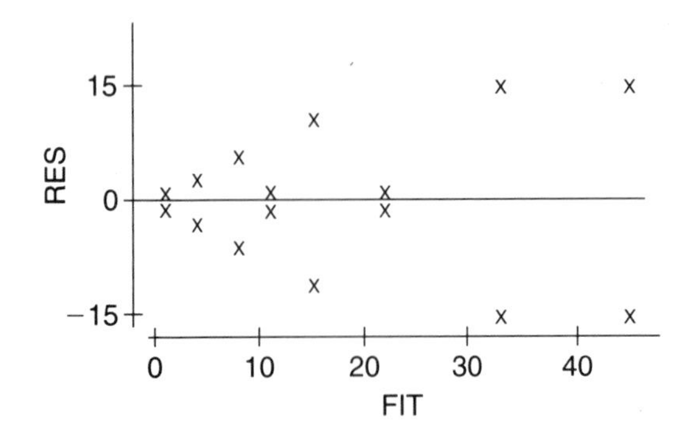

FIGURE 7.25 Scatterplot of Res. vs. Fit for the sugar data. The plot suggests a wedge, with points at the left (low fitted values) near the x-axis (small residuals, + or −) and points at the right (larger fit) farther from the axis (larger residuals). The data set should be transformed to equalize SDs. ■

EXAMPLE 7.23 SLEEPING SHREWS

The plot in Fig. 7.26 shows Res vs. Fit for the data of Example 7.4. Comment on the pattern.

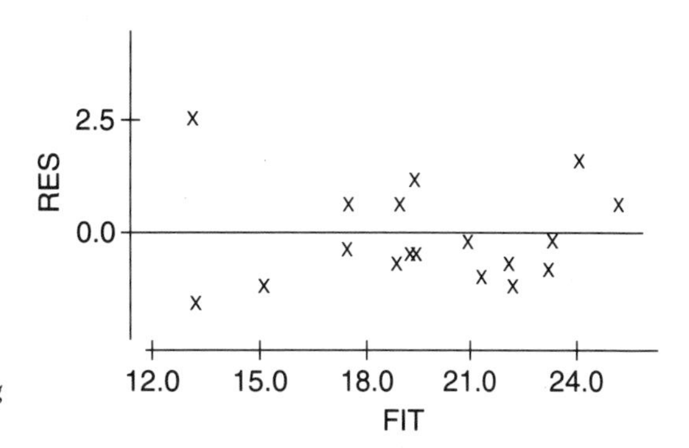

FIGURE 7.26 Residual plot for the sleeping shrews.

SOLUTION: The plot shows one isolated point, for shrew I, REM sleep (Fit = 13.1, Res = 2.6), already identified as suspicious in Exercise 7.D.18. The remaining points suggest a tilted balloon, running from lower left to upper right. ■

Exercise Set E

1. Sleeping Shrews

Exercises D.17–19 give a decomposition of the shrew data and ask about the outlying value of 15.7 for shrew I, REM. An estimated replacement value (Chapter 14, Section 5) for the 15.7 turns out to be 11.0. Locate the point for shrew I in each of the scatterplots of Example 7.20, cross it out, and plot a new point using the replacement value. Then tell how the replacement value affects the pattern in each plot. Does the new plot suggest a line parallel to the identity line $y = x$?

2–3. *Losing leaves* (Example 7.11)

2. Plot debladed versus not debladed for the leaf data, using different symbols for the four different treatment groups. Describe the patterns, and briefly discuss what those patterns might mean.

3. A scatterplot of residual versus fit for the leaf data is shown below. Briefly describe the pattern. What does it tell you about how well the model and Fisher assumptions fit?

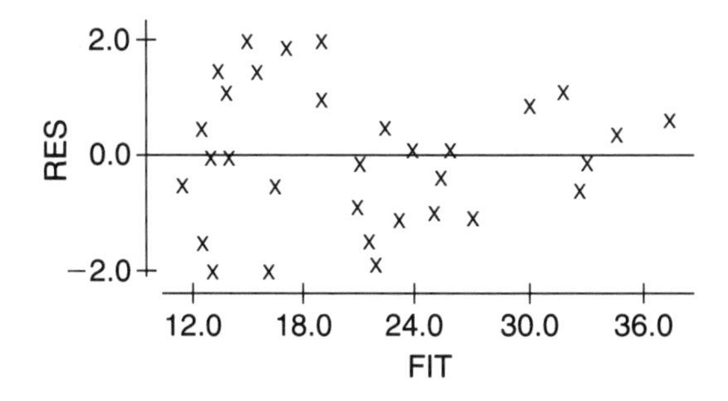

Scatterplot of Residual vs. fit for the leaf data

4. *Bee stings* (Example S7.3 and Exercises A.10–13)

a. Draw a within-blocks scatterplot, as in Example 7.19: plot y = stung versus x = fresh, with each occasion as a point, and using occasion number instead of a dot as your plotting symbol. Describe the pattern in words. What does it tell you about the bees? Do the points suggest a line parallel to the identity line $y = x$?

b. Adjust your scatterplot from part (a) by crossing out the point for occasion 9 and plotting a new point using the replacement value. Is the resulting pattern more consistent?

5. *Res vs. Fit #1*

a. Scatterplot Res vs. Fit for the complete bee sting data, including the outlier. (If you did the decomposition as part of Exericse D.14, you already have the residuals. To get fitted values, note that for a CB design, Fit = Block average + Treatment effect.) Once you have the plot, circle the two points for block 9, which contains the outlier, and draw a "balloon" around the other points, as in Example 7.21.

b. Pick one: Apart from the two outlying points: (I) The plot shows about what you would expect if the model and Fisher assumptions are appropriate. (II) The plot shows a curved relationship between residuals and fitted values, suggesting that the model does not fit the data in the present scale. (III) The plot is fatter toward the right: larger residuals tend to go with larger fitted values, suggesting that the assumption of same SDs is not appropriate for the data in the present scale. (IV) The plot shows a balloon with a slight tilt.

6. *Res vs. Fit #2*

a. Scatterplot Res. vs. Fit. using the modified data set with the replacement value for the outlier, and draw a balloon around the points. Note that this second plot has no outliers and that the balloon does not stretch as far in the vertical direction.

b. Which of (I)–(IV) from Exercise 5(b) best describes this plot?

7. *Puzzled children.* Reread Exercise 6.C.3, in which you decomposed a BF[2] data set showing the numbers of puzzles solved by 12 children, cross-classified by sex and age. Notice that for any BF design, the fitted value equals the cell average, and the residual equals the observed value minus the cell average.

a. Scatterplot Res vs. Fit for this data set, and draw a balloon around the points.

b. Describe the shape of the plot in words. Which of (I)–(IV) from Exercise 5(b) best describes the shape of your plot?

8. *Areas of rectangles: Nonadditivity.* The purpose of this exercise is to show what a plot of Res vs. Fit looks like when the model you have chosen does not give an additive fit to your data in its present scale. As you know, you can't find the area of a rectangle by adding the "effects" of length and width. In real life (as opposed to high-school geometry) you often can't tell ahead of time whether an additive model will fit, but a scatterplot of Res vs. Fit can help you decide.

Exercise 6.C.22 asked you to decompose the areas (in square inches) of a set of six rectangles. For this exercise, think of that data set as coming from a BF[2] design, with one observation per cell. Pretend that the effects of the two factors, length and width, are additive, so that the table we called "interaction" is now the table of residuals, and the table we called "partial fit for interaction" is now the table of fitted values. Scatterplot Res vs. Fit. Which of (I)–(IV) from Exercise 5(b) best describes your plot?

9. *Radioactive twins* (Example S7.2 and Exercises A.7–9)

a. Scatterplot y = the urban reading versus x = the rural reading, with one point for each twin pair. (Instead of using dots for your plot, use little numbers: 1 for the first twin pair, 2 for the second, etc.) Which twin pair stands out as unusual? In what way is that pair different from the others? Describe the pattern of the other six twin pairs.

b. The totals for the two environments are 290.6 (rural) and 318.8 (urban). Compute two sets of averages for the two environments, a first set using all seven twin pairs, and a second set omitting the outlying pair you identified in part (a).

c. How much influence does this one pair have on the message from the data?

10. *Diets and dopamine* (Exercises C.6–9)

Scatterplot y = low versus x = normal, using solid circles for the patients with poor control and open circles for the others. Discuss what you see.

6. Using a Computer [Optional]

The examples of this section show how to use Minitab to get an ANOVA table and Res. vs. Fit. plots for the CB design (Example 7.24), the LS design (Example 7.25), and the SP/RM design (Example 7.26). Example 7.27 shows how you tell Minitab to "unstack" your data in order to use Minitab commands to get within-blocks plots. ■

Example 7.24 ANOVA and Residual Plot for the Finger-Tapping Data (CB)

To use Minitab to analyze a CB[1] design, you follow almost exactly the same steps you'd use for a BF[2] design, as in Section 4 of Chapter 6. Your data must be in three columns, one for the response (named "Rate" in the printout ahead), and one each giving factor levels for treatments ('Drugs') and blocks ('Subj'). You use the same menu and dialog box as for a BF[2], with just one difference: You don't include an interaction term in your model.

If you select Store Residuals and Store Fits in the dialog box (by clicking on the little squares) Minitab will compute these two variables and store them in columns named 'RESI1' and 'FITS1'.

Here are the steps for the ANOVA:

```
Stat > ANOVA > Balanced ANOVA...
```

Fill in the dialog box as follows:

```
Response:  Rate
Model:     Drug Subj
```

You can also click on various buttons for additional choices:

`Storage:` Select Fits and Residuals if you want these stored in columns of your worksheet.

`Options:` Minitab will compute treatment and block averages for you if you enter the factor names in the box for Display Means.

`Graphs:` Residual versus fit plots, plus others.

EXAMPLE 7.25 ANOVA AND RESIDUAL PLOT FOR THE MILK YIELD DATA (LS)

For a Latin square design, you need four columns for your data—one for the response, and one each for the levels of the row, column, and treatment factors. The steps you follow for the LS design are the same as for the CB, except for two things:

1. You choose `Stat > ANOVA > Analysis of Covariance...` instead of `Balanced ANOVA...`
2. For your model, you list all three structural factors.

Here is what you get for the Milk Yield data, if you also click the Options button and ask for factor means:

```
Factor  Levels  Values
Period       3       1      2      3
  Cow        3       1      2      3
 Diet        3       1      2      3

Analysis of Variance for Yield

Source    DF       SS        MS       F       P
Period     2     5850      2925    1.23   0.449
Cow        2    46674     23337    9.81   0.093
Diet       2   103416     51708   21.74   0.044
Error      2     4758      2379
Total      8   160698

MEANS

Period    N    Yield
1         3   811.00
2         3   856.00
3         3   796.00

Cow   N   Yield
1     3  723.00
2     3  894.00
3     3  846.00

Diet   N   Yield
1      3  695.00
2      3  811.00
3      3 957.00 ■
```

EXAMPLE 7.26 ANOVA FOR THE DIABETIC DOGS (SP/RM)

For an SP/RM design like the dog experiment, your data must be in four columns—one for the response, and one each for the between-blocks factor, blocks, and the within-blocks factor. Moreover:

Important: Minitab requires that you number your blocks as though they were crossed with the between-blocks factor.

For the diabetic dogs, for example, this means we number the dogs from 1 to 5 even though there are 10 dogs in all: we renumber dog 6 as dog 1, dog 7 as dog 2, . . . , and dog 10 as dog 6. How, then, does Minitab know to analyze the data as an SP/RM with dogs nested within treatments if we've numbered the dogs as though they were crossed with treatments? *You have to show the nesting when you list the model.*

In what follows, Turnover is the response, Treatmnt is the between-blocks factor, Dog2 is the renumbered factor for blocks = dogs, and Method is the within-blocks factor. To indicate that Dog2 is nested within Treatmnt, you type the outside variable (Treatmnt) in parentheses after the inside variable (Dog2) like this: Dog2(Treatmnt). The model also includes the interaction Method*Treatmnt.

Once you have your data listed properly, start with the same choices as for a Latin square:

```
Stat > ANOVA > Analysis of Covariance...
```

Then fill in the dialog box:

```
Response:  Turnover
Model:     Treatment Dog2(Treatment) Method Method*Treatment
```

As before, you can use the Options button to ask for factor means. Here is what you get:

```
Factor        Type Levels Values
Treatmnt      fixed  2  1  2
Dog2          fixed  5  1  2  3  4  5
(Treatmnt)
Method        fixed  2  1  2

Analysis of Variance for Turnover

Source             DF       SS        MS       F       P
Treatmnt            1    320.00    320.00    3.76   0.088
Dog2(Treatmnt)      8    508.00     63.50    0.75   0.655
Method              1   2420.00   2420.00   28.47   0.000
Treatmnt*Method     1     80.00     80.00    0.94   0.36
Error               8    680.00     85.00
Total              19   4008.00

MEANS

Treatmnt    N   Turnover
1          10   35.000
2          10   43.000
```

```
Treatmnt   Dog2   N   Turnover
1            1    2    36.000
1            2    2    28.000
1            3    2    36.000
1            4    2    39.000
1            5    2    36.000
2            1    2    48.000
2            2    2    33.000
2            3    2    39.000
2            4    2    49.000
2            5    2    46.000

Method     N   Turnover
1         10    50.000
2         10    28.000

Treatmnt   Method   N   Turnover
1             1     5    44.000
1             2     5    26.000
2             1     5    56.000
2             2     5    30.000
```
■

EXAMPLE 7.27 THE UNSTACK COMMAND AND WITHIN-BLOCKS PLOTS

To get within-blocks plots using Minitab, you first have to "unstack" your response variable. Instead of one long column of response values (the "stacked" version), you need a rectangular arrangement like the standard format used throughout this book: Your response values must be unstacked into several short columns, with one row for each block, and one column for each level of the factor of interest (in the case of a CB) or within-blocks factor (SP/RM).

To get Minitab to unstack your data, you need to tell the response, tell the column for the factor of interest (CB) or within-blocks factor (SP/RM)—Minitab refers to the levels as subscripts—and then tell, one at a time, the columns where you want the unstacked data to go.

Here's how you'd unstack the finger-tapping data. From the Manip menu, select Stack/Unstack... . Then fill in the dialog box:

```
Manip > Stack/Unstack > Unstack...
Unstack the data in:           Rate
Store the unstacked data in:   c4 c5 c6
Using subscripts in:           Subj
```

Once you've unstacked your data, choose Plot from Graph menu and fill in your choices of X and Y to get your within-blocks scatterplots. ■

7. Algebraic Notation for the CB, LS, and SP/RM Designs

I assume that if you're reading this (optional) section, you're already familiar with the algebraic notation for the BF[1] and BF[2] designs, from the final sections in Chapters 5 and 6. In this section I'll rely on the ways in which the notation for the designs of this chapter fits the same pattern that you've already seen, in order to focus on what's new. To learn the notation for a new design, you need to learn two pieces of "code": (1) the symbols for the factors in the design, and (2) the subscripts.

The CB[1] Design

Symbols for the Factors		Subscripts
Factor	**Symbol**	
		i = first subscript
Benchmark	μ	= level of treatment factor
Treatments	τ	
Blocks	β	j = second subscript
Residual error	e	= level of block factor

The Model	
y_{ij}	observed value for treatment i in block j
$= \mu$	benchmark
$+ \tau_i$	effect of treatment i
$+ \beta_j$	effect of block j
$+ e_{ij}$	residual error for treatment i in block j

(Note that j is the first and only subscript for the block effects.)

The LS[1] Design

Symbols for the Factors		
Factor	**Symbol**	**Subscripts**
		i = first subscript
Benchmark	μ	= level of row factor
Nuisance factor for rows	α	
Nuisance factor	β	j = second subscript
		= level of column factor
Treatments	τ	
Residual error	e	k = third subscript
		= level of treatment factor

The Model	
y_{ijk}	observed value for row i, column j, treatment k
$= \mu$	benchmark
$+ \alpha_i$	effect for row i
$+ \beta_j$	effect for column j
$+ \tau_k$	effect for treatment k
$+ e_{ijk}$	residual error for row i, column j, treatment k

(Note that j is the first and only subscript for the column effects, and k is the first and only subscript for the treatment effects.)

The notation for the LS[1] design includes a new wrinkle: not all combinations of the subscripts i, j, k occur. For each combination of row (i) and column (j) there will be only one treatment level (k). Figure 7.27 shows how the subscripts go for the milk experiment of Example 7.9:

FIGURE 7.27 Subscripts for the milk yield data. The first subscript tells the row, the second tells the column, and the third tells the treatment (diet): 1 = A = roughage, 2 = B = partial grain, 3 = C = full grain. Notice that the third subscripts themselves follow the same Latin square pattern as the letters in the left-hand square.

Treatments

A	B	C
B	C	A
C	A	B

Observed values

y_{111}	y_{122}	y_{133}
y_{212}	y_{233}	y_{231}
y_{313}	y_{321}	y_{332}

The SP/RM[1;1] Design

Symbols for the Factors		Subscripts
Factor	**Symbol**	
Benchmark	μ	i = first subscript
Between-blocks factor	α	= level of between-blocks factor
Blocks	β	j = second subscript
Within-blocks factor	γ	= level of the block factor
Interaction	$(\alpha\gamma)$	k = third subscript
Residual error	e	= level of within-blocks factor

The Model	
y_{ijk}	observed value for level i of the between-blocks factor, in block j, with level k of the within-blocks factor
$= \mu$	benchmark
$+ \alpha_i$	effect of level i of the between-blocks factor
$+ \beta_{ij}$	effect of block j (for level i of the between-blocks factor)
$+ \gamma_k$	effect of level k of the within-blocks factor
$+ (\alpha\gamma)_{ik}$	interaction effect for level i of the between-blocks factor with level k of the within-blocks factor
$+ e_{ijk}$	residual error for the observed value y_{ijk}

Two features of this notation deserve special attention. First, notice that the interaction terms $(\alpha\gamma)_{ik}$ have two subscripts, i and k, which together tell the cell (= treatment combination). The first (i) tells the level of the between-blocks factor (= group of rows); the second (k) tells the level of the within-blocks factor (= column). We use i for the first subscript to show that it will always be the same as the first subscript of the observed value (y_{ijk}) and the (first and only) subscript of the term α_i for the between-blocks factor. We use k for the second subscript to show that it will be the same as the third subscript of the observed value (y_{ijk}) and the (first and only) subscript for the term γ_k for the within-blocks factor.

The second feature: this notation uses two subscripts to identify a block or whole plot. The first subscript (i) tells the level of the between-blocks factor (= group of rows); the second subscript (j) tells which block (= row) for that level of the between-blocks factor. For example, β_{23} would be for the third of the blocks in the second group of blocks.

Both the terms for block effects and the terms for interaction effects have two subscripts, but there is a crucial difference. The interaction terms $(\alpha\gamma)_{ik}$ come from crossing the two factors that correspond to the subscripts i and k. This means that $(\alpha\gamma)_{13}$ and $(\alpha\gamma)_{23}$ have something in common: the two 3s refer to the same thing, the third level of the within-blocks factor. The blocks, on the other hand, are nested within the between-blocks factor. This means that β_{13} and β_{23} refer to different blocks; the fact that both happen to have a 3 for the second subscript is purely a matter of chance. Statisticians often show this by writing $\beta_{j(i)}$ instead of β_{ij}. In this notation the first subscript j is for the inside factor (blocks) and the second subscript i, in parentheses, is for the outside factor. (Using this convention, we'd write $e_{jk(i)}$ instead of e_{ijk} for the error term.)

Exercise Set F

CB[1] Design

1. In each of the equations below, exactly one term has a wrong subscript. Rewrite the term using the correct subscript.

   a. $y_{12} = \mu + \tau_2 + \beta_2 + e_{12}$

   b. $y_{42} = \mu + \tau_4 + \beta_1 + e_{41}$

   c. $y_{22} = \mu_2 + \tau_2 + \beta_2 + e_{22}$

2. a. Use the convention of dots and bars to write out abbreviations for the grand average, the average for treatment 1, and the average for block 2.

   b. Now write abbreviations for the estimated effect for treatment i, the estimated effect for block j, and the residual for treatment i, block j.

   c. Add your three expressions in (b) together, and show that this sum plus the grand average simplifies to give y_{ij}.

3. If a CB[1] design has I treatments and J blocks, write expressions for $df_{\text{Treatments}}$, df_{Blocks},

and df_{Res} in terms of I and J. Then show that the sum of these plus df_{Grand} $(= 1)$ equals df_{Total} $(= IJ)$.

LS[1] Design

4. Here is an LS plan with 4 rows and columns, together with a list of 10 possible observed values. List those whose subscripts do not fit the plan.

A	B	C	D	y_{441}, y_{433}, y_{342}, y_{111}, y_{224},
B	D	A	C	
C	A	D	B	y_{244}, y_{333}, y_{212}, y_{321}, y_{134},
D	C	B	A	

5. a. Use the convention of dots and bars to write out abbreviations for the grand average, the average for row 1, the average for column 2, and the average for treatment 3.

   b. Now write abbreviations for the estimated effects for row i, column j, and treatment k, and the residual for row i, column j, and treatment k.

   c. Add your four expressions in (b) together, and show that this sum plus the grand average simplifies to give y_{ijk}.

6. If an LS[1] design has I rows, columns, and treatments, write expressions for df_{Rows}, df_{Cols}, $df_{Treatments}$, and df_{Res} in terms of I. Then show that the sum of these plus df_{Grand} $(= 1)$ equals df_{Total} $(= I^2)$.

SP/RM Design

7. Fill in the missing subscripts:

   a. $y_{\_\_\_} = \mu + \alpha_1 + \beta_{\_2} + \gamma_3 + (\alpha\gamma)_{\_\_} + e_{\_\_\_}$

   b. $y_{321} = \mu + \alpha_{\_} + \beta_{\_\_} + \gamma_{\_} + (\alpha\gamma)_{\_\_} + e_{\_\_\_}$

   c. $y_{3\_\_} = \mu + \alpha_{\_} + \beta_{\_2} + \gamma_{\_} + (\alpha\gamma)_{\_4} + e_{\_\_\_}$

8. a. Use the convention of dots and bars to write out abbreviations for the grand average and the average for the first level of each of the structural factors of the SP/RM design.

   b. Now write abbreviations for the estimated effects for level i of the between-blocks treatment, for block ij, for level k of the within-blocks treatment, and for the interaction in cell ik; write the abbreviation for the residual corresponding to y_{ijk}.

   c. Add all your expressions from (b) together, and show that this sum plus the grand average simplifies to give y_{ijk}.

9. If a balanced SP/RM design has I levels of the between-blocks treatment, with J blocks per level and K levels of the within-blocks treatment, write expressions for $df_{Between}$, df_{Blocks}, df_{Within}, df_{Inter}, and df_{Res} in terms of I, J, and K. Then show that the sum of these plus df_{Grand} $(= 1)$ equals df_{Total} $(= IJK)$.

SUMMARY

All three designs of this chapter (CB, LS, SP/RM) are based on blocking, a strategy that turns a nuisance influence into a factor of your design. A **block** is a group of experimental units that would tend to give similar values for the response. Three of the most common ways to get blocks are by **sorting**, by **subdividing**, or by **reusing** your experimental material.

1. **The randomized complete block design (CB)**

   a. For a CB, sort your units into blocks of equal size; then randomly assign treatments to units separately within each block so that each block gets a complete set of treatments.

b. The one-way CB design has two structural factors: one factor of interest (treatments or conditions), and one nuisance factor (blocks).

Grand mean Blocks Conditions Residual error

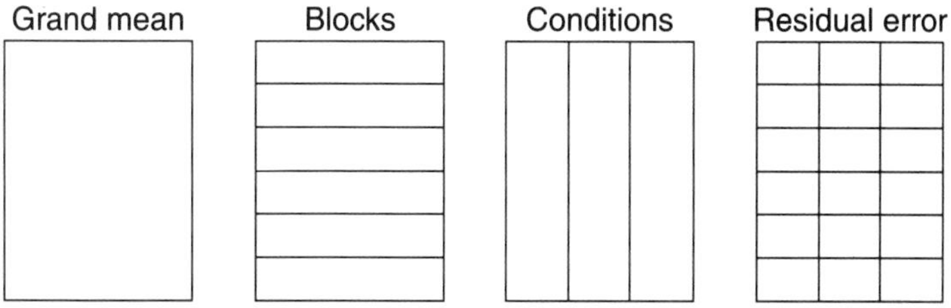

2. **The Latin square design (LS)**

a. Structural features: there are two nuisance factors and one factor of interest. The number of levels is the same for all three of these factors. Every combination of levels of the two factors occurs exactly once.

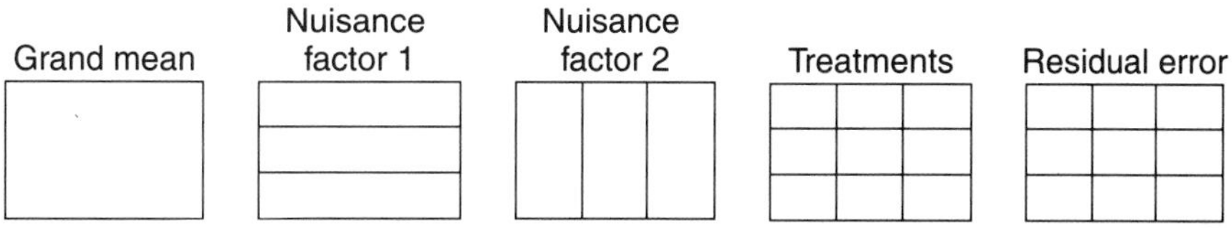

Grand mean Nuisance factor 1 Nuisance factor 2 Treatments Residual error

b. The LS is often used with subjects and time periods (order) as the two nuisance factors. You should consider using an LS design whenever subjects exhibit a lot of variability, the order of the conditions has a systematic effect, and the factor of interest is experimental. The assignments of rows, columns, and treatments should all be randomized.

3. **The simplest split plot/repeated measures design (SP/RM)**

a. There are experimental units of two different sizes; the larger units are blocks of smaller units. There are two sets of treatments or treatment combinations. One set is assigned to larger units as in a CR; the other is assigned to smaller units separately with each block, as in a CB.

b. The basic SP/RM has four structural factors: two factors of interest (one between-blocks, one within-blocks); their interaction; and one nuisance factor (blocks).

Grand mean Between-blocks factor Blocks Within-blocks factor Interaction Residual error

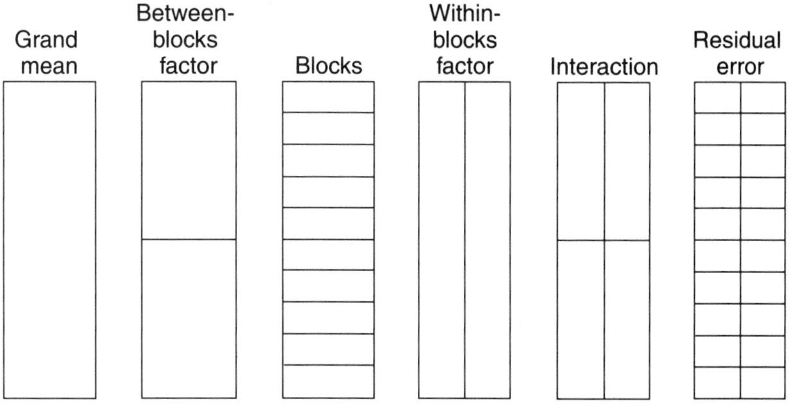

c. **Nesting**. One factor is nested within another if each level of the first ("inside") factor occurs with exactly one level of the second ("outside") factor. In the basic SP/RM:

i. Blocks are nested within levels of the between-blocks factor.

ii. Smaller units (subplots) are nested within blocks (whole plots).

iii. The between-blocks and within-blocks factors are crossed.

iv. Blocks and the within-blocks factors are crossed.

4. Formal analysis. The decomposition and degrees of freedom for CB, LS, and SP/RM designs follow the general rules based on inside and outside factors; SSs and MSs follow the same pattern as for BF designs. All *F*-ratios have the usual form, $MS_{Factor}/MS_{Residual}$, with one exception: the *F*-ratio for the between-blocks factor of the SP/RM design uses MS_{Blocks} instead of MS_{Res} in the denominator.

5. a. For a CB[1] design, if the model and Fisher assumptions fit the data, a within-blocks scatterplot of readings for any one condition against readings for any other should suggest a line with slope 45°.

b. The most useful way to check the fit of the model and Fisher assumptions is to scatterplot residuals versus fitted values. If the fit is good, the plot will suggest an oval balloon with the *x*-axis running lengthwise through the middle. Features suggesting poor fit:

Outliers: one or more isolated points; the balloon may be tilted.

Unequal SDs: the plot will suggest a wedge, not an oval.

Nonadditive model: the plot will suggest a curve.

EXERCISE SET G: REVIEW EXERCISES

1. *Needle threading*

a. Reread Review Exercise 15 in Chapter 1. For each of the three designs in that exercise (CR, CB, SP/RM), tell what the units are.

b. Now suppose all your subjects are of the same sex, so you have only the six combinations of thread color and background color to compare. Tell how to run the experiment as a Latin square. In what way is this design better than the CR and CB designs?

c. Tell how you could use Latin squares to improve the SP/RM design.

2. *Radioactive twins*. Read Example S7.2. Then decompose the data twice and give the ANOVA tables, first using all seven twin pairs, and then a second time, omitting the outlying twin pair. Compare your two ANOVAs: how much influence does pair #2 have on the results and conclusions? On the whole, what do you conclude about the effect of environment on lungs?

3–5. *Wireworms*

3. *Analysis 1: Original scale, outlier replaced* (see Exercises 7.B.5–8)

In the original scale (no. of worms), with the replacement value of 7 substituted for the outlying 17, you get the following averages:

						Total
Rows:	3.2,	4.0,	5.0,	6.4,	4.4	23
Columns:	5.0,	4.6,	4.4,	4.8,	4.2	22.8
Treatments:	K 1.2,	M 6.2,	N 5.0,	O 5.8,	P 4.8	25

a. The row and column averages suggest that soil conditions within the square of 25 plots varied in a systematic way that made some parts of the square more attractive to the worms than other parts. Look over the averages; then describe in words this pattern of attractiveness to the worms.

b. Use the averages above to find the grand average and the estimated effects for rows, columns, and treatments.

c. The total SS is 699. Use your results from (b) to give the complete ANOVA table. What do you conclude? Note that you can get the residual SS by subtraction, so you don't have to decompose the data to do this problem. Be sure to reduce the total df by 1 (to 24), and also reduce the residual df by 1, since your analysis uses only 24 actual observations.

4. *Effect of the outlier.* Consider how the analysis would be different if you used the actual number of 17 worms (for treatment N, row 4, column 1) rather than the replacement value of 7. With the outlier included, the total SS is 939. Recompute any of the averages and effects that would be different. Then redo the ANOVA table, and describe any major differences between it and your table in Review Exercise 3.

5. *Analysis 2: Transformed data, outlier included* (see Exercise 7.B.8)

In the new scale [$10 \times \sqrt{(1 + \text{no. of worms})}$] you get the following averages:

Rows:	19.4,	21.2,	23.6,	28.8,	22.6
Columns:	26.4,	22.2,	22.2,	23.2,	21.6
Treatments:	K 23.0,	M 26.4,	N 25.4,	O 26.8,	P 14.0

The total SS is 14,650. Find the ANOVA table, and compare it with the one from part (c) of Review Exercise 3.

6–8. *Hypnosis and learning*

The purpose of this study was to compare the effects of two kinds of hypnosis (traditional and "alert" hypnosis) on learning. Fifteen undergraduates served as subjects. Each subject learned two lists of 16 word–number pairs—one list under normal waking conditions, and a second list under one of three randomly assigned conditions: waking; traditional hypnosis; or alert hypnosis. The design was balanced, so that for the second list five subjects were under each of the three conditions. The two lists were of a similar degree of difficulty.

The lists were learned as follows: a subject heard a tape recording of the list and was then asked to repeat the list. If the subject made any errors, the process was repeated, up to a maximum of 15 times, until the subject gave back the list without errors. The numbers below give the total number of errors for each list, for each subject.

Subject	1	2	3	4	5
Test I	67	65	12	11	3
Test II	8	85	18	28	3

Condition: Normal waking state

Subject	6	7	8	9	10
Test I	106	30	22	17	15
Test II	105	31	29	38	49

Condition: Traditional hypnosis

Subject	11	12	13	14	15
Test I	71	65	39	16	4
Test II	36	37	26	29	2

Condition: Alert hypnosis

6. Design structure:

a. Tell the response.

b. Then list all structural factors for the design.

c. For each factor tell whether it is a nuisance factor or factor of interest, and

d. Give the number of levels.

e. For each factor of interest, tell whether it is observational or experimental.

f. Tell what the units are.

g. Then name the design.

7. Do as complete an analysis as you can, both informally (graphs, etc.) and formally (decomposition and ANOVA).

8. Tell how you could run the experiment in the last problem with the same response, the same set of subjects, and the same set of conditions to compare, but using a randomized complete block design: tell what the units and blocks are, and tell how many units per block you would have.

9. Consider how you could use Latin squares to improve on the RCB plan in Review Exercise 8:

a. Draw (and label!) a diagram showing how you would assign conditions to units.

b. Name the two nuisance factors in your design.

c. There is a third nuisance influence, one that is completely confounded with one of the structural factors in the design actually used in Review Exercise 6. What is it?

10–11. *Premature infants*. Interruptions in breathing are particularly common among premature infants. In the past, hospitals have kept babies on ordinary bassinet mattresses, but someone thought the babies might do better on waterbeds because of their gentle rocking motion. To study the effect of waterbeds on breathing, investigators attached an alarm to each of 8 premature babies; the alarm would sound whenever the baby's breathing stopped for more than 20 seconds. Each baby was monitored for two six-hour periods; during one (randomly chosen) period, the baby slept on a waterbed; during the other, the control period, the baby slept on a regular mattress. During each six-hour period, the research team counted the number of times the alarm went off, and they measured the length of time the baby was asleep. The following numbers give the number of interruptions per hour of sleep.

Waterbed	0.89	0.77	0.00	0.65	0.88	1.36	1.22	0.30
Control	1.36	1.66	0.11	1.44	1.63	1.52	1.53	0.48

10. Design structure:

a. Tell the response.

b. Then list all structural factors for the design.

c. For each factor, tell whether it is a nuisance factor or factor of interest, and

d. Give the number of levels.

e. For each factor of interest, tell whether it is observational or experimental.

f. Tell what the units are.

g. Then name the design.

11. Do as complete an analysis as you can, both informally (graphs, etc.) and formally (decomposition and ANOVA).

12. *A fake data set for quick ANOVA practice.* Suppose a CB design has 5 blocks crossed with 4 conditions and that the block averages are 10, 15, 20, 25, 30; the condition averages are 10, 20, 20, 30; and the SD equals 5. Construct a complete ANOVA table.

13–16. *Averages for a factor as a sum of effects.* You can use the decomposition rule in reverse, to express a set of averages as the grand average plus a sum of effects. For example, for a CR design:

$$\text{Cond Avg} = \text{Gr Avg} + \text{Cond Eff}$$
$$\text{Obs} = \text{Gr Avg} + \text{Cond Eff} + \text{Res}$$

13. For the CB design, express each of Cond Avg, Block Avg, and Obs as Gr Avg plus a sum of effects.

14. For the LS design, express each of Row Avg, Col Avg, Tr Avg and Obs as Gr Avg plus a sum of effects.

15. For the SP/RM, do the same for Between, Blocks, Within, Cells, and Obs.

16. For the two-way factorial BF, do the same for the averages for First basic factor, Second basic factor, Cells, and Obs.

17–18. Fitted Values and df_{Res} for the SP/RM Design

17. The fitted value for the SP/RM equals

$$\text{Fit} = \text{Grand Avg} + \text{Between Eff} + \text{Block Eff} + \text{Within Eff} + \text{Interaction Eff}.$$

Use the boxed summary for the SP/RM to rewrite each of the four effects in this expression in terms of averages. Simplify, and show that you get

$$\text{Fit} = \text{Cell Avg} + \text{Block Avg} - \text{Between Avg}.$$

18. If you put the SP/RM in our usual rectangular format, then

$$\text{\# rows} = \text{\# blocks} = (df_{Blocks} + df_{Between} + 1)$$

and

$$\text{\# columns} = df_{Within} + 1.$$

Use these facts, together with # obs = (# rows)(# columns) to show that if you compute df_{Res} using the general rule (df_{Res} = # obs − sum of df for all outside factors), you get the same result as from the shortcut in the boxed summary ($df_{Res} = df_{Blocks} \times df_{Within}$).

19. Degrees of Freedom: *True* or *false*

a. For every CB design, $(1 + df_{Blocks})$ = # blocks.

b. For every CB design, $df_{Cond} = df_{Blocks}$.

c. For every LS design, $df_{Rows} = df_{Cols} = df_{Treatments}$.

d. For every CB design, $df_{Res} = (df_{Blocks})(df_{Cond})$.

e. For every LS design, $df_{Res} = df_{Rows}$.

f. For every CR design, $df_{Res} = (df_{Cond})^2$.

g. For every SP/RM design, # blocks = $df_{Blocks} + df_{Between} + 1$.

20–22. *Shorthand expressions for df*

20. CB: Suppose you have I blocks and J treatments. Express df_{Treats}, df_{Blocks}, and df_{Res} in terms of I and J.

21. LS: Suppose your square has I rows and I columns. Find the df for rows, columns, treatments, and residuals in terms of I.

22. SP/RM: Suppose your between-blocks factor has I levels, that each level is assigned to J blocks, and that your within-blocks factor has K levels. Find the df for each of the six factors for this design, in terms of I, J, and K. Check that your df add to $IJK = df_{Total}$.

NOTES AND SOURCES

1. ***Transforming the wireworm data*** (Exercise 7.B.5–8)

Why did I use square roots instead of logs or some other transformation? I had three reasons. First, I knew that square roots often work well for counted data. (If it is reasonable to think the data fit a particular model, the Poisson, the typical size of chance error, as measured by the SD, equals the square root of the average, and a theoretical argument shows that taking square roots makes the SDs roughly equal, independent of the averages.) Second, I knew that taking square roots is something of a compromise, in a certain sense halfway between taking logs and leaving the data in the original scale. I was prepared to try logs if square roots didn't even out the spreads enough. Third, I cheated: the wireworm data set is well known to many statisticians and is, in fact, one of the data sets given to me for practice when I was first learning about transformations, so I had known for a long time that square roots would work here. (Section 12.1 shows a systematic way to choose a transformation.)

2. ***Diabetic dogs*** (Example 7.9)

In the actual experiment, there were only nine dogs. I invented a tenth dog to balance the design. In the listing of the data, dog 0 is a convenient fiction.

3. ***The SS and MS for blocks in the SP/RM design***

What I call MS_{Blocks} is computed from block effects that you get by first subtracting averages for all between-blocks factors; this subtraction is crucial. To be explicit that you have done the subtraction, you can say "the mean square for Blocks, nested within the between-blocks factor(s)," and write $MS_{Blocks(Between)}$ instead of just MS_{Blocks}. I'll continue to use the shorter MS_{Blocks} because for the SP/RM I never use MS_{Blocks} to mean anything other than $MS_{Blocks(Between)}$.

4. ***Losing leaves*** (Example 7.11 and Exercises D.21–24)

Notice how little the bare-bones ANOVA tells us by itself. We know only that the chemical treatment and the deblading have detectable effects and that there is a real interaction as well. The analysis so far tells us nothing about the pattern of these effects. Does auxin cause the leaves to stay on the plant longer? Is there a difference in the effects of the two concentrations of auxin? If so, which is more effective? What is the effect of deblading, and what is the pattern of the interaction? And what do the answers to these questions tell us about the theories that led to the experiment?

Informal answers to these questions could come from an informal analysis, most particularly from condition averages and an interaction graph. But more formal answers are also possible using the methods of Chapter 11, on contrasts and confidence intervals.

Appendix: Supplementary Examples

Blocking and the Complete Block Design

Example S7.1 SAT Scores: Confounding in a CB Observational Study

The goal of this study is to see whether taking a special course can raise your SAT scores. The subjects for the study are high school seniors who agree to take the SAT tests twice, and to take the special course between their first and second SAT tests.

The factor of interest has two levels: before versus after the special course. Each student serves as a block of two time slots (= units): each student provides a response (SAT score) under each condition. A standard format for recording the data would look like Figure S3.

Students (Blocks)	Conditions (Factor of interest)	
	Before	After
1		
2		
.		

FIGURE S3 Disastrous confounding in a CB observational study. The factor of interest (before or after the special course) is observational, and its effects are confounded with effects of order.

The problem with this design is that the factor of interest is not experimental. As a result, the effect of the special course gets confounded with another important effect, that of practice. Suppose a study based on this design finds that on average, students score about 30 points higher after they've taken the course. Would that convince you that the course is worth taking?

By the time students take the SATs a second time, they have already had the practice of taking the test once, and besides they have gotten a little older, and perhaps a little wiser. There's no way to tell (from a study like this one) how much of the 30 point increase is due to the course, how much is due to practice, and how much is due to getting a little older.

You might want to think about ways to design a study that *would* permit you to isolate and measure the effect of the course. ■

Other ways to get blocks. The principle of blocking is extremely powerful, in part because there are so many different ways to create blocks. The examples of Chapter 7 have shown you blocks of land and subjects as blocks; the next two examples illustrate other possibilities. As you read each one, try to identify the design structure: What is the factor of interest? Is it experimental or observational? What is the nuisance factor (what serves as a block), and what are the units? How would the data look in a rectangular layout?

Example S7.2 Radioactive Twins: Blocks from Matched Pairs

Most people believe that country air is better to breathe than city air, but how would you prove it? You might start by choosing a response that narrows down what you mean by "better." One feature of healthy lungs is that they are quick to get rid of nasty stuff they breathe in, things like particles of dust, smoke, and so on. The following study used as its response the rate of tracheobronchial clearance, that is, how quickly the lungs got rid of nasty stuff.

A problem in planning the study was what to do about individual differences. There are lots of people living in the country, and lots more living in cities, but it is likely that heredity could have a big influence on how well a person's lungs work, so the variability between individuals would be large. This is the sort of situation where you'd like to be able to use subjects as blocks, but, unfortunately, for a study like this one you can't, partly because few people would be willing to move to a new environment just for the sake of your research, and partly because the effects of environment are likely to be such long-term effects that neither you nor your subjects would live long enough for each of them to be used twice.

These problems are not trivial, but the investigators found a way around them. Incredibly, they managed to find seven pairs of identical twins that satisfied two requirements: first, one twin from each pair lived in the country and the other lived in a city; second, both twins in the pair were willing to inhale an aerosol of radioactive Teflon particles! That was how the investigators measured tracheobronchial clearance. The level of radioactivity was measured twice for each person: right after inhaling, and then again an hour later. The response was the percent of original radioactivity still remaining one hour after inhaling.

The (observational) factor of interest is the environment, with two levels, rural versus urban. The nuisance factor is twin pairs (blocks), with seven levels; the unit is a person. The data are given in Table S6. ■

TABLE S6 Effect of environment on rate of lung clearance.

	Environment	
Twin pair	Rural	Urban
1	10.1	28.1
2	51.8	36.2
3	33.5	40.7
4	32.8	38.8
5	69.0	71.0
6	38.8*	47.0
7	54.6	57.0

Response is percent of radioactivity remaining one hour after inhaling an aerosol of radioactive Teflon. Does the rural environment seem healthier?

EXAMPLE S7.3 BEE STINGS: TIME SLOTS AS BLOCKS

Beekeepers sometimes use smoke from burning cardboard to reduce their risk of getting stung. It seems to work, and you might suppose the smoke acts like an insect repellent, but that's not the case. When J. B. Free, at Rothamsted Experimental Station, jerked a set of muslin-wrapped cotton balls in front of a hive of angry bees, the numbers of stingers left by the bees made it quite clear they were just as ready to sting smoke-treated balls as they were to sting the untreated controls. Yet in tests using control balls and others treated with the repellent citronellol, the bees avoided the repellent: on 70 trials out of 80, the control balls gathered more stings.

What, then, is the effect of the smoke? One hypothesis is that smoke masks some other odor that induces bees to sting; in particular, smoke might mask some odor a bee leaves behind along with his stinger when he drills his target, an odor that tells other bees, "Sting here." To test this hypothesis, Free suspended 16 cotton balls on threads from a square wooden board, in a 4-by-4 arrangement. Eight had been freshly stung, the other eight were pristine and served as controls. Free jerked the square up and down over a hive that was open at the top, then counted the number of new stingers left behind in the treated and control balls. He repeated this whole procedure eight more times, each time counting the numbers of new stingers left. For each of the nine occasions, he lumped together the eight balls of each kind, and took as his response the total number of new stingers.

The resulting data have a CB structure. What are the two factors, and what are their levels? What would be a suitable rectangular arrangement for Free's data?

SOLUTION. The nuisance factor is occasions, with nine levels. (It is in fact quite common for a design to have chunks of time as levels of a nuisance factor.) The factor of interest has two levels: control and previously stung. Here are the data. I've arranged them in a rectangle with occasions as rows and treatments as columns (Table S7).

TABLE S7 Numbers of stings left in previously stung and fresh cotton balls.

Occasion	Stung	Fresh
1	27	33
2	9	9
3	33	21
4	33	15
5	4	6
6	22*	16
7	21*	19
8	33	15
9	70	10

Do bees show a predisposition to sting the ones that were previously stung?

Notice that in this last example, there are in fact two ways to think about the occasions: as a nuisance factor (blocks), or as an observational factor of interest. The first way regards the nine occasions as a sample whose purpose is to represent some hypothetical population of possible occasions that might have been used. The second way takes the occasions as being of interest in their own right. For example, the nine occasions come in a natural order, and it might be worth looking at the data for evidence of a trend over time. As Francis J. Anscombe, Professor of Statistics at Yale, puts it, nuisance factors

> are not only a nuisance (= not wanted), but not some other things—not known, not identified, anonymous. As soon as something is known about them they aren't just blocks but observational conditions that are of interest. The subjects in a psychological experiment may really have names and identifications and known characteristics, but only to the extent that they can be thought of as a sample of some kind of population can the experiment be published. In an agricultural field experiment the randomized blocks recognize the similarity of neighboring plots in what seems to be a homogeneous piece of ground. But in a series of similar experiments carried out at different locations in different years, there are known (uncontrolled) properties of the locations and of the weather that will be invoked when attempts are made to explain why the treatments have different effects in the different experiments. ■

In Chapter 8, I give an example (8.9) where the difference between the two points of view is crucial: a design that some authors presented and analyzed as a CB with months as a nuisance factor (blocks) can be better understood by thinking of the blocks as an observational factor of interest. The shift in point of view leads to a big change in the analysis.

Two Nuisance Factors: the Latin Square Design

EXAMPLE S7.4 SUBMARINE MEMORY: A PLAN BUILT FROM LATIN SQUARES

The experiment by Godden and Baddeley was designed to look for interaction between learning environment (dry or wet) and recall environment. Crossing the two factors of interest gave a total of four treatment combinations to be compared (dry/dry, dry/wet, wet/dry, and wet/wet). Sixteen members of a diving club were available as subjects.

The experiment could have been run as a CR: randomly assign four subjects to each of the four conditions. However, because memory capacity varies a lot from one person to another, it would be better to use subjects as blocks, with each subject tested under all four conditions. For each subject, you should randomize both the assignment of word lists to conditions, and the order (assignment of conditions to time slots). A randomized CB design would certainly be a reasonable plan, particularly because with as many as 16 subjects you would not be very likely to end up with a plan that was drastically unbalanced.

However, you can use Latin squares to insure that the matching of lists with treatment combinations is exactly balanced. Figure S4 is a 4-by-4 Latin square plan for 4 of the 16 subjects. (I haven't bothered to do the 3-step randomization, but I would have if I were using the plan for my own experiment.)

FIGURE S4 A (not yet randomized) Latin square plan for four divers. For this plan, the order of the four conditions would be chosen separately for each diver using a chance device. One way to use this LS plan would be to divide the 16 divers into 4 sets of 4 divers, and run 4 parallel Latin squares, one for each set of divers. That is very close to the plan that Godden and Baddeley used. ■

	List of words				
Subject	**1**	**2**	**3**	**4**	**Conditions:**
1	A	B	C	D	A: dry/dry
2	B	A	D	C	B: dry/wet
3	C	D	A	B	C: wet/dry
4	D	C	B	A	D: wet/wet

The Split Plot/Repeated Measures Design

EXAMPLE S7.5 MURDERER'S EARS

A century ago, there was a small army of scientists trying to find patterns that related people's shapes to aspects of their character. Was there something about the shape of a person's head, for example, that you could use to predict whether he had criminal tendencies?

The following study was published as having been done at Parkhurst Prison in Britain. The data set consists of 120 ear lengths, in millimeters: one right ear length and one left ear length for each of 60 men. Twenty of these men had been classified as "lunatic criminals," another twenty were "ordinary murderers," and the remaining twenty were "other criminals."

What are the factors? Which pairs of factors are crossed? Which are nested?

SOLUTION. The sixty criminals are blocks; criminal type is the between-blocks factor, with three levels; left versus right is the within-block factor. Figure S5 shows an abbreviated format for the data, which might make it easier to see the crossing and nesting.

Criminal type (Between-blocks factor)	Subject # (Block)	Side of head (Within-block factor) Left	Right
Lunatic criminals	1		
	2		
	.		
	.		
	20		
Ordinary murderers	21		
	22		
	.		
	.		
	40		
Other criminals	41		
	42		
	.		
	.		
	60		

FIGURE S5

Consider relationships among the three factors: Criminal Type (whole plot factor), Prisoners (blocks), and Side of Head (subplot factor). Prisoner is nested within Type; Prisoner and Side of Head are crossed; so are Side of Head and Type. Notice also that Ear (subplot) is nested within Prisoner. ■

Decomposition and ANOVA for the Latin Square Design

EXAMPLE S7.6 RESIDUAL DF FOR THE LATIN SQUARE

The purpose of this example is to show you that the number of free numbers in the residual table of a Latin square design equals (# rows − 1)(# rows − 2). For concreteness, we'll work with a 5×5 design. The number of free numbers will turn out to correspond to the 3×4 rectangle you get by crossing out two rows and one column of the 5×5 square in Figure S6.

+	+	+	+	+
1	2	3	4	+
5	6	7	8	+
9	10	11	12	+
+	+	+	+	+

FIGURE S6

Remind yourself that in the residual table of a decomposition, each number is in a group by itself, so there are no forced repetitions. All redundancies come from patterns of adding to zero. For a Latin square you get residuals by taking out three sets of averages—rows, columns, and treatments—so there will be three patterns of adding to zero. Rows and columns will be easy, but there is a trick to seeing the right way to handle the pattern for treatments. Think first about treatment A. There are five residuals, corresponding to the five occurrences of Treatment A in the square, and these five residuals must add to 0, which we can indicate by writing a + in one of those five squares. In the same way, the five residuals for B add to 0, and we should show this by writing a + in one of the five B squares. Likewise for each of the other three treatments. Question: Which squares should we choose for the +s? Answer: The Latin square design guarantees that each row has exactly one residual corresponding to each treatment. So we can choose to think of the first row as made up entirely of +s, that is, of residuals whose values are determined by the pattern of adding to 0 within treatment groups.

With the first row filled in, we're left with a 4×5 set of residuals which must add to 0 across rows and down columns. We can represent the adding to 0 across rows by writing a + in the last position of each row; doing this fills in the last column with +s, leaving us with a 4×4 square of residuals. In a similar way, adding to 0 down each column fills the entire bottom row with +s, which leaves us with a 3×4 rectangle of free numbers, as promised. ■

CHAPTER 8

WORKING WITH THE FOUR BASIC DESIGNS

OVERVIEW

In one sense, this chapter won't introduce you to much that you haven't seen before. By now you've seen four basic designs, their factor structures, and a set of tools for analyzing data from these designs. However, each design had a chapter or section to itself, and the exercises pretty much told you which set of ideas to work with. Real life rarely comes to you so neatly packaged.

This chapter is largely one of review and integration, taking the ideas that earlier chapters presented one at a time and looking at how they are related and at how to choose the right ideas to use when a new situation comes along. Section 1 compares the structures of the BF, CB, LS and SP/RM designs, and uses examples to illustrate how to tell which is which. Then Section 2 gives examples that take you through the planning of actual experiments. It may seem that the natural order ought to go the other way—planning first, recognizing the designs next—but I think the planning goes better if you have a clear sense of what your options are.

Section 3 gives two examples of informal analyses. Then Section 4 sketches some alternatives to ANOVA, and illustrates ways to recognize the kinds of data sets that might better be analyzed using one of these alternatives.

3. Informal Analysis: Examples
4. Alternatives to ANOVA

With all that behind you, I hope you'll feel ready, in Chapter 9, to extend the four basic designs using factorial crossing.

1. Comparing and Recognizing Design Structures

The goal of this section is to show you ways to recognize and summarize the structure of the four basic designs and ways to tell them apart. I hope that in learning these things, you will be moving toward a more ambitious goal: to be able to recognize the same sets of ideas when they apply to design structures you have not met before.

Preliminary Steps: Recognizing the Units

Preliminary Steps in Analyzing the Design

Are the conditions:

- **Treatments, which get *assigned*?**
- **Populations, from which *samples* are taken?**
- **Characteristics (*multiple measurements*)?**

Are the units:

- **Individual subjects or objects?**
- **Groups of subjects or objects?**
- **Parts or time slots for subjects or objects?**

Are there any grouping factors (blocks):

- **Groups of subjects or objects (unit = subject or object)?**
- **Individual subjects or objects (unit = part or time slot)?**

Analyzing the structure of a design usually boils down to recognizing the units, then telling how the units are related to the conditions of interest and any grouping factors. It's usually easiest to recognize the units when they correspond to distinct physical entities, individual subjects or objects. Sometimes, however, the units correspond to groups, and quite often they correspond to parts of individuals or time slots, with the individuals serving as blocks.

It's often easier to identify the units if you first classify the kinds of conditions being compared: are they treatments, or populations, or characteristics which correspond to multiple measurements on the same subject or object? If the conditions are treatments, you can find the units by asking "What do the treatments get assigned to?" If the conditions correspond to populations, you ask "What are the things you select when you sample?" If the conditions correspond to characteristics, ask "Characteristics of what?" The answer (hamsters, leaves, etc.) will ordinarily correspond to blocks of units.

When your conditions correspond to treatments or populations, finding the units and the design family tends to be fairly straightforward, although there are a few pitfalls to avoid. If your conditions correspond to multiple measurements, however, it may not be clear what model to use. The ANOVA model regards the characteristics as levels of a factor, and whether this is appropriate will depend both on the purpose of the analysis and on whether the residual errors behave properly. (There's more on this in Section 4 and in Chapter 12.)

Identifying the units

A. Suppose first that your conditions are *treatments*.

1. In the simplest case, your units will be distinct physical entities, individual subjects or objects. For example, in the walking babies experiment (Example 5.1), the treatments were assigned to infants; in the experiment on pigs and vitamins (Example 6.1), the diets were assigned to piglets.

2. If, however, the treatments are applied to groups of individuals, then the unit is the group. When you get just one measurement from the group, as in the leafhopper experiment (Exercise 5.C.1 and Example 5.11), where there was one measurement for each dish of eight leafhoppers, it won't be hard to identify the group as the unit. But if each individual in a group provides a measurement, it's easy to get the units wrong. Suppose you were comparing teaching methods, using test scores as your response. There would be one score for each student, but if the methods got assigned to classes, the unit would be a class, not a student.

3. You should always ask what gets done to the individual subjects or objects. If two treatments for glaucoma are randomly assigned to right and left eyes, the unit is an eye, not a person. If, as in the finger tapping experiment (Exercise 7.B.2–4), different drugs are given to the same person at different times, the unit is a time slot, not a person.

B. Now suppose your conditions correspond to *populations*.

1. The units will usually be individual subjects or objects, which you can recognize by asking "What are the items being selected when you sample?" For example, in the study of intravenous fluids (Example 5.2), the investigators chose samples of bags of fluid; the unit is a bag.

2. With observational studies, just as with experiments, one of the main traps to look out for is assuming the unit is an individual when in fact the unit is a group. In the study of sponges (Example 4.10), the fact that there was a measurement for each individual cell misled the graduate student, who assumed the unit was a cell. If you think carefully about the populations—green sponges and white sponges—you can tell that the unit is actually a sponge. Usually, as with

the sponges, the units will be the same as the things you choose when you sample, but there are exceptions. In the study of rural and urban environments (Example S7.2), the investigators selected twin pairs, but the unit was the individual person.

3. For experimental conditions, it's quite common to have parts of individuals or time slots for units. But for built-in conditions tied to parts or time slots, it may be more natural to think of the conditions as characteristics which correspond to multiple measurements. For example, consider the sleeping shrews (Example 2.4). The three kinds of sleep do correspond to different time slots, but you can also think of the heart rates for the kinds of sleep as characteristics of the shrews.

C. When your conditions correspond to *multiple measurements*, there won't really be any units in the usual sense. True, you can stretch things and say the kinds of sleep for the shrews have time slots as units, and the hearts and brains in Kelly's study have chunks of hamsters as units, but the only practical value is that it may help you see how the "units" are grouped. When your conditions correspond to multiple characteristics, and you decide to use an ANOVA model, the individuals will always function like blocks of "units."

Starting from individual subjects or objects. For most studies, there will be individuals—people, animals, plants, or objects—that stand out clearly, even though they may not be units. All the same, they make a natural starting point for trying to identify units. In general, the units will be the individuals themselves, or groups of individuals, or else parts or times slots for individuals. Here's a set of most common possibilities:

A. The individuals are grouped.

1. One response per group. The unit is a group of individuals. There are no blocks. For example, in the leafhopper experiment, individual leafhoppers are grouped 8 to a dish. There is one response per dish, and the unit is a dish.

2. One response per individual. Here you need to ask: Are the treatments applied to entire groups, or separately to individuals? (For observational conditions, are they properties of entire groups, or of separate individuals?)

a. Entire groups: The unit is a group. For example, in the study of white and green sponges, you might think of individual cells, grouped into sponges. There is one response per cell, but the conditions (green and white) are properties of entire sponges. The unit is a sponge. In the same spirit, a study to compare soaps might replace all soap in randomly chosen college dormitories with antibacterial soap, leaving a control soap in the other dormitories. It would be natural to record the incidence of flu and colds for individual students, but unit would be a dormitory, not a student.

b. Individuals: The unit is an individual, and the groups are blocks. The matched pairs design of the PMS study provides an example of this sort.

B. The individuals are not grouped.

1. One response per individual. The unit is an individual. There are no blocks. Typical BF examples include the walking babies, intravenous fluids, and pigs and vitamins data sets.

2. More than one response per individual. The unit is a part or time slot. Each individual is a block. The finger tapping, milk yields, and diabetic dog data sets of Chapter 7 are examples of this situation.

A Checklist for Analyzing Designs

I find that it helps organize your thinking about the various designs if you follow a standard format when you write out your analysis. The format I like comes in three parts: (A) Purpose and response(s), (B) Units and design structure, and (C) Treatment structure.

A. Purpose and response

Usually the general purpose of a study will be clear from its description, but what I have in mind for "Purpose and response" is something narrower: a summary in the form "The goal of this study was to compare the effect of (conditions) on (material), as measured by (response)." If there's only one response, this summary will usually be straightforward; but, as I'll illustrate at the end of the section, sometimes when there is more than one possible response, things can get tricky.

B. Design structure

1. *Units.* How many kinds (sizes) of units are there?

a. Just one. Each of the basic BF, CB, and LS designs has units of just one size. (The same is true of some other designs, too; for example, the generalized complete block (GCB) designs treated in Chapter 13.)

b. Two. The SP/RM design has units of two different sizes, whole plots (blocks) for the between-blocks treatments, and subplots for the within-blocks treatments. However, not every design with units of two sizes has the same SP/RM structure as in Chapter 7. (See Example S8.2)

c. More than two. (See Chapter 13).

2. *Grouping.* For each kind of unit, what is the structure that relates treatments (or conditions) to units?

a. BF: Units are not grouped or arranged before they are assigned to treatments.

b. CB: Units are arranged in blocks of equal size; blocks are crossed with treatments.

c. LS: Units are arranged in rows and columns; rows and columns are crossed, rows and treatments are crossed, columns and treatments are crossed.

d. Other: Various other structures are possible, for example the GCB designs mentioned in (1a) above.

3. Design family. Your answers to (1) and (2) tell the design family. For example, the SP/RM family from Chapter 7 uses a CR structure to assign between-blocks treatments to the larger units, and uses a CB structure to assign within-blocks treatments to the smaller units.

C. **Treatment structure**
Separately for each kind (size) of unit, list the treatments or factor(s) of interest. For any compound factors of in-terest, list the basic factors and interactions.

In what follows, I will illustrate these ideas using several examples. For many of these examples, I will not show you the data, and the description I give may seem a bit compact. All this is deliberate: I want to give you examples that read something like what you might find in the abstract at the front of a journal article, so you can practice recognizing design structure from a brief description.

EXAMPLE 8.1 EMOTIONS AND SKIN POTENTIAL

In this experiment, eight subjects were hypnotized, and while under hypnosis, were asked to feel each one of four emotions (fear, happiness, depression, and calmness), one emotion at a time. One of the physiological measurements taken was the skin potential, in millivolts (mV). Summarize the design.

A. *Purpose and Response(s)*: to compare the effects of four requested emotions on hypnotized subjects, using skin potential as the response.

B. *Design structure*
   1. Units: There is only one size of unit; the unit is a time slot for a subject.
   2. Grouping: The units come grouped in blocks, four time slots per subject.
   3. Design family: CB.

C. *Treatment structure*
   There is only one factor of interest, the four emotions, which get assigned to time slots. You can't tell from my brief description whether the order of the emotions was randomized, but it could (and should) have been.

Table 8.1 summarizes the design and treatment structure:

TABLE 8.1 Summary table of factors: Emotions and skin potential—a CB[1]

Unit	Grouping (Nuisance factors)	Treatments
Time slot	Subjects (8)	Emotions (4)

A standard rectangular format for the data would have 8 rows, one per subject, and 4 columns, one per emotion. We can write the factor diagram in the standard form for a CB[1] design, with rows as blocks (Figure 8.1):

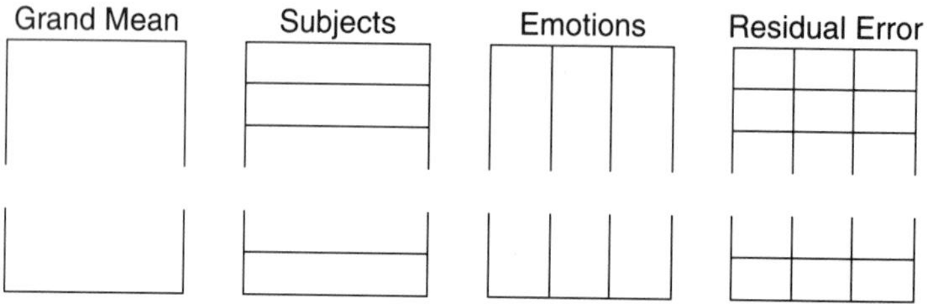

Grand Mean Subjects Emotions Residual Error

FIGURE 8.1 ■

EXAMPLE 8.2 WIREWORMS

Exercise 7.B.3 described an experiment in which five different soil fumigation treatments were used to control wireworms. The treatments were applied to the plots of a Latin square, and the numbers of wireworms per plot were counted a year later. Summarize the design.

A. *Purpose and Response(s)*: to compare the effects of fumigation on the numbers of wireworms in soil.

B. *Design structure*

1. Units: There is only one size of unit; the unit is a plot of soil.

2. Grouping: The units are arranged in rows and columns before treatments are assigned. (Each of the three pairs of factors is crossed, with exactly one observation for each pair of levels (each row and column, each row and treatment, each column and treatment).

3. Design family: LS. Here, as nearly always, it is easy to recognize someone else's use of a Latin square. With these designs, the main place to put your effort is in learning to spot situations where you ought to use one yourself.

C. *Treatment structure*

There is one factor of interest, the 5 kinds of fumigation. These treatments were randomly assigned to units.

Table 8.2 lists the factors to show the design and treatment structure:

TABLE 8.2 Summary table of factors: Wireworms—an LS[1].

Unit	Grouping (Nuisance factors)	Treatments
Plots of land (25 in all)	Rows (5) Columns (5)	Fumigations (5)

The factor diagram would be essentially the same as for other LS[1] designs, with five factors: benchmark, rows, columns, treatments, and residual error. I won't take space to show the diagram, but take a minute to visualize it for yourself. ■

I've chosen the next example mainly because it shows statistical ideas at work in an interesting setting, but note that it also raises an ethical issue: The subjects for the experiment are prison inmates.

EXAMPLE 8.3 SOLITARY CONFINEMENT

The purpose of this experiment was to study the effects of sensory deprivation on the arousal level of human subjects. The measure of arousal level used was the frequency of alpha waves in the brain, as recorded by an electroencephalogram (EEG). Twenty subjects were chosen from an initial pool of 82 inmates of the Kingston (Ontario) Maximum Security Penitentiary. Ten subjects were randomly assigned to solitary confinement for a week; the remaining ten served as controls. Each subject was measured three times, once each on days 1, 4, and 7 of the experiment. Summarize the design.

A. *Purpose and response(s):* To compare the effects of solitary confinement versus control conditions on the arousal level of prison inmates, as measured by the frequency of alpha waves.

B. *Design structure*

   1. Units: There are units of two sizes. The larger unit is a subject; the smaller unit is a time slot (day) for a subject.

   2. Grouping:

      a. The subjects were not arranged or grouped before the treatments (solitary or control) were randomly assigned; the structure is CR.

      b. The time slots come grouped, 3 per subject, as in a CB.

   3. Design family: SP/RM.

C. *Treatment structure*

   Both the between-blocks factor (solitary or control) and the within-blocks factor (time) are basic. The assignment of conditions to subjects was randomized. For times, however, although conditions are associated with the smaller units in a CB pattern, the conditions are defined rather than assigned or selected, and no randomization is possible.

TABLE 8.3 Summary table of factors: Solitary confinement—an SP/RM[1;1].

Unit	Grouping (Nuisance factors)	Treatment Combinations: Main effects	Treatment Combinations: 2-Factor interaction
Subjects (20 in all)	(none)	Deprivation (2)	
Time slots (3 per subj)	Subjects	Days (3)	Deprivation × Days

A standard rectangular format for SP/RM designs has one row per block, one column per level of the within-blocks factor. We could arrange these data in 20 rows by 3 columns. The factor diagram would have 6 boxes, one each for benchmark, deprivation, subjects, days, interaction, and residual error. ■

The following experiment is easy to misclassify. When I first read its description in a journal article, I didn't think carefully enough about the units, and my first guess for the design structure was wrong. I'll present my faulty analysis first: see if you can spot what I overlooked.

EXAMPLE 8.4 FEEDING FROGS: A FAULTY ANALYSIS

Twelve South African clawed frogs (*Xenopus laevis*) were kept in one-gallon jars except for feeding. The feeding behavior was observed under three conditions: isolated, paired with another frog, and in a group of twelve. Each frog was observed for 200 minutes in each condition, and the amount of food it ate was recorded.

A. *Purpose and response(s)*: To compare the effect of grouping (isolated, paired, in a group of twelve) on the feeding behavior of frogs, as measured by the amount of food eaten in 200 minutes.

B. *Design structure*

1. Units: There is one kind of unit, a time slot for a frog.

2. Grouping: The units come grouped in blocks, three time slots per frog. Treatments get assigned to time slots separately for each frog.

3. Design family: CB.

C. *Treatment structure*

There is one factor of interest, so the design is a CB[1].

A rectangular format and factor diagram would be much the same as for the CB[1] design of Example 8.1 (Figure 8.2.):

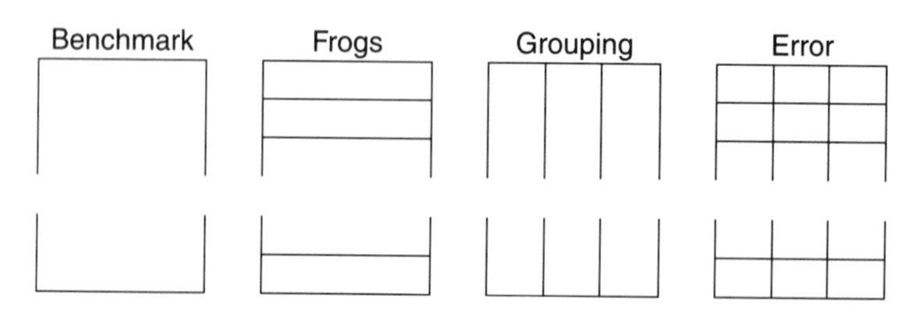

FIGURE 8.2

What I overlooked in my first, faulty analysis is that two of the treatment conditions almost surely got assigned to *groups* of frogs, not to individual frogs. For the paired condition, the 12 frogs were probably sorted into 6 pairs: the unit here is a pair. For the grouped condition, all 12 frogs received the treatment as a group: the unit is the entire group of 12. The design doesn't fit into any standard family. ■

See also "Examples S8.1 Perch Eggs," and S8.2 "Cats and Tetanus."

Exercise Set A

1–10. For each of the following studies, carry out the three preliminary steps from Section 1: (a) Do the conditions correspond to treatments, populations, or characteristics? (b) Tell the units; then write Ind, Gr, or P/TS to indicate whether the units correspond to individual subjects or objects, groups of individuals, or parts/time slots for individuals. (c) Are there blocks? If so, tell what they are, and write GR or Ind to indicate whether they correspond to groups or individuals.

1. Milk yields (Example 7.6)
2. Diabetic dogs (Example 7.9)
3. Bird calcium (Exercise 5.B.1)
4. Wireworms (Exercise 7.B.5–8)
5. Beestings (Example S7.3)
6. Rating Milgram (Exercise 5.B.8–11)
7. Cooling mice (Example 7.19)
8. Losing leaves (Example 7.11)
9. Dopamine and diets (Exercise 7.C.6–9)
10. Premature infants (Exercise 7.G.10–11)

11–14. Each of the following data sets is one whose design structure has already been given; each is paired with an example (from Section 1) with the same structure. For each, use the checklist to review the design, following the example from the text as a guide. Then write out a summary table of factors and a factor diagram.

11. Exercise A.1: Tobacco hornworms (CR, like Example S8.1a)
12. Example S7.2: Radioactive twins (CB, like Example 8.1)
13. Example 2.9: Rabbit insulin (LS, like Example 8.2)
14. Kelly's hamsters (SP/RM, like Example 8.3)

15–18. Write out a summary of the design structure for each of the following experiments. Follow the format of the examples.

15. *Trout.* In this experiment 32 rainbow trout were maintained in one of three conditions: sea water (32% salinity), brackish water (18% salinity), or freshwater (0.5% salinity). One measure of the effect of salinity was the weight gained. Be careful about units: Tell how this experiment should be done if the unit is to be a trout. Under what circumstances would the unit be a group of trout?

16. *The Transylvania effect.* The word "lunatic" comes from the Latin word for moon, and reflects a centuries-old belief that the moon affects mental health. Between August 1971 and July 1972 there were twelve full moons, one per month; researchers used the records of admissions to a Virginia mental health clinic to see if they could detect an effect of the moon.

They divided each month into three periods: before, during, and after the full moon. Then they computed the admission rate (patients per day) for each period of each month.

17. *Pine pruning.* Monterey pine trees were used in this experiment to compare the effects of five pruning strategies. Each strategy had the same form: prune away all branches from the main trunk, up to a particular height; the heights used were (in feet) 10, 8, 6, 4, and 0 (control). The experiment was carried out on trees growing in a 5 × 5 arrangement of plots on a pine plantation. The effect of pruning was measured by the annual increase in square inches of basal area; these numbers are given below. The number in parentheses is the pruning height.

43 (10)	43 (4)	53 (0)	51 (6)	48 (8)
43 (0)	49 (6)	45 (8)	46 (4)	41 (10)
40 (8)	44 (0)	47 (6)	41 (10)	46 (4)
43 (6)	36 (10)	48 (4)	44 (8)	40 (0)
42 (4)	38 (8)	38 (10)	42 (0)	42 (6)

18. *Elaboration and memory.* The theory behind this experiment is that "deeper and more elaborate processing" is an aid to memory. To see what I mean, think about the words "pie," "pig," and "pay" as you compare two tasks: (1) to check whether each word has an "e" or a "g" in it, and (2) to rate the pleasantness of each word. Hyde and Jenkins, who ran this experiment, hypothesized that the letter-checking task was not as deep or elaborate as rating the pleasantness of the words, and would not be as great an aid to memory.

The subjects in this experiment saw a set of 24 words presented at a rate of three seconds per word. One group of subjects was asked to check for *e* or *g*; the other group rated pleasantness. Each of these groups had been subdivided: half of each group thought the letter-checking or rating was the main task; the other half had been told that they would be asked to recall as many words as they could. Once the initial task was over, all subjects were tested for recall. (Surprisingly, it made almost no difference whether subjects had been told they would be tested for recall. As predicted, subjects rating pleasantness could recall a lot more words than the letter-checkers—roughly 70% recall versus 40%.)

2. Choosing a Design Structure: Deciding About Blocking

Thinking about factors. In planning an experiment, once you have made your three decisions about content, your next step is to choose a design structure. In looking ahead to choices about structure, however, it is usually worth reviewing your decisions about content. Could you get more information, with only a little extra effort or cost, by measuring more than one response variable? Or by crossing an additional factor of interest with one of the factors you have already chosen?

Suppose you have now chosen your response(s) and factor(s) of interest. You can think about choosing a design structure in two ways—what you do, and what you get.

a. What you do: the assignment process (What are your choices for units? One size, or two? Should you group them, or regard them as interchangeable?), or

b. What you get: the factor structure. (What are the important nuisance influences? Which should be made into a structural factor for your design? Should the nuisance factor be related to the factor of interest by crossing or nesting?)

Chapter 7 emphasized the first point of view. Table 8.4 shows a summary comparison based on the second.

TABLE 8.4 Using nuisance factors to compare the four basic designs

# structural nuisance factors	Design family	Relation to factors of interest
0	BF	The only nuisance factor corresponds to the units themselves (and thus to residual error). This factor is nested within the factor of interest.
1	CB	The structural nuisance factor (blocks) is crossed with the factor of interest.
1	SP/RM	The structural nuisance factor (blocks) is • nested within the between-blocks factor, and • crossed with the within-blocks factor.
2	LS	Each of the two structural nuisance factors (rows and columns) is crossed with the factor of interest.

Deciding whether to use blocks. Many of the decisions about a design structure come down to questions of blocking. The issues in deciding about blocks depend on how you would get your blocks. Sorting presents one set of tradeoffs. Subdividing/Reusing presents another.

A. Sorting. If your blocks would come from sorting the available units, there are two main considerations:

1. *Will sorting help isolate the effects of interest?* In principle, sorting can only help, but there are two common problems:

   a. Units are very similar already. For example, if you're doing an experiment on potatoes, and you decide to buy them from your local grocery, the potatoes you get will be clones, and so genetically identical. For most experiments, there would be no need to sort the potatoes into blocks. The BF designs work well if your units are relatively uniform. Designs with blocks are generally better if you can sort your units according to a nuisance variable (or variables) that you expect to be closely related to the response. In such cases, blocking better isolates the effects of interest, and leads to smaller residual errors. Sometimes, however, there may be no good way to sort units into groups:

   b. No good measurement to use for sorting. In Kelly's hamster experiment, it would have been nice to measure enzyme concentration before assigning day length, in order to sort hamsters into pairs with nearly equal starting concentrations, then assign one day length to each hamster in a pair. Kelly couldn't do that, however, because she couldn't measure the concentration without sacrificing the hamster. Although she could have sorted the hamsters by weight, there was no reason to expect weight to be related to her response, and so blocking by weight wasn't likely to make the experiment more efficient.

2. *How many units do you have? How many df for estimating chance error size?* If you get blocks by grouping the available units (as opposed to subdividing or reusing, which create more units), and if the number of available units is small, the BF design may be better. The main disadvantage to blocking is that you give up units of information about chance error: degrees of freedom get shifted from residuals to blocks. If df_{Res} is small, the F-test may not be powerful enough to detect treatment differences unless they are quite large. (For this reason, LS designs smaller than 5×5 are not often used.)

B. Sub-dividing/Reusing. If you can reuse or subdivide subjects or material, blocking gives you more units for the same number of subjects or objects. In principle, then, blocking of this sort is a good idea, and a design with blocks is likely to be better than a BF design. In practice, there are two main clusters of possible drawbacks.

1. *Time and effort.* It would take too long to reuse the material.

a. In many agricultural experiments to compare crop varieties or fertilizers or insecticides, it might be theoretically possible to reuse the same plot of farmland, but only if you extend the experiment over more than one growing season.

b. Often in experiments on human or animal subjects it may be possible to have subjects serve as their own control group; doing this offers many advantages, but in some cases it may take too long, or make subjects more likely to drop out of the study.

2. *Linkage and leakage*

a. Sub-dividing: Parts are too closely linked. Often, the nature of your treatments makes it impossible to get units by subdividing. Kelly, for example, couldn't assign a day length to a part of a hamster. You can't assign a teaching method to a part of a person, assign a room temperature to half a rat, or apply fertilizer to half a plant. The effects of a treatment—such as a fertilizer applied to soil—may leak into nearby sub-plots.

b. Reusing: The treatments change the subjects or material. (i) This is true of most medical procedures and teaching methods. For example you couldn't tie off the mammary arteries of the angina patients in Example 4.5 and then go back later and reuse the same patients as controls. (ii) Walking babies: presumably the exercise had an effect that couldn't be removed; certainly there is no way to turn a one-year-old infant who can walk back into a one-week-old infant to reuse as a control. (iii) Comparing teaching methods: if your conditions teach something to your subjects, you can't ordinarily get them to unlearn it completely, despite what some cynical teachers may say.

c. Reusing: The measurement process changes the subjects or material. (i) In many biology experiments, such as Kelly's hamster experiment, or a plant experiment that uses dry weight of the plant as a response, the measurement process kills the animal or plant. (ii) In psychology experiments in which the response is a test score, like the SAT, there is a practice effect: people tend to do better the second time they take the test.

C. **General considerations**. Some issues apply no matter how you would get your blocks.

1. *Interaction*. Don't use a CB design if there's likely to be a block-by-treatment interaction, i.e., if it's reasonable to expect the effect of the treatments to be different for different blocks. Instead, consider a generalized complete block design (Chapter 13), which allows you to estimate block-by-treatment interactions.

2. *Flexibility*. The BF designs are the most flexible; for blocking you need the number of units in a block equal to the number of treatments. Moreover, the BF[1] is particularly flexible: the unbalanced BF [1] is the only unbalanced design that is analyzed in exactly the same way as the balanced design. For LS designs, the requirement that both nuisance factors and the treatment factor all have the same number of levels often makes them impractical. (LS designs larger than 8×8 or 10×10 are rarely used.)

D. **Particular designs**

1. *SP/RM or BF[2]*. The choice of the SP/RM design is often decided by the nature of the conditions and material, e.g., one set of treatments can be applied to parts or time slots, the other can only be applied to entire individuals. However, if you can apply the between-blocks treatment to the smaller units instead of the larger ones, the resulting BF design is better able to detect real differences associated with that factor than the SP/RM.

2. *CB or SP/RM*. If you are considering a CB design, you may be able to get information about an additional factor of interest by applying a second set of treatments to the blocks, or by choosing blocks from two or more populations, i.e., by choosing an SP/RM design instead of the CB.

Choosing a design: Examples

EXAMPLE 8.5 HOSPITAL CARPETS: CHOOSING A DESIGN STRUCTURE

My own experience with hospitals has been that most have tile floors, not carpet, and that most are often noisy. Carpets would soak up some of the sound, but on the other hand might be hard to keep free of germs. If you were a hospital administrator, you might reasonably think that if you installed carpet to keep down the noise, the more restful atmosphere might speed the recovery of some patients. Of course you wouldn't want to do this if it would make the place less sanitary. Suppose you want to compare bacterial levels in hospital rooms with and without carpeting on the floor. How might you design an experiment to do this?

Response: Presumably what matters is not so much the presence of bacteria on the floor or in the carpet, but in the air. One way to measure the presence of airborne bacteria is to pump room air over a growth medium, incubate it, and count the number of colonies of bacteria that you get. The response is the number of colonies per cubic foot of air.

Conditions: Our goal is to compare carpeted and bare floors, under conditions that are in other respects both uniform and representative of what would be typical for a hospital. There is one factor of interest, with two levels.

Material: Suppose that, just as for the study I am basing this on, you have available sixteen rooms in a Montana hospital: #210, 212, 214–217, and 220–229.

Nuisance influences:

1. Exposure: Does the room have windows to the north, south, east, or west? Does it get a lot of sun, or not?
2. Traffic: Is the room at the end of a corridor, or near the nurses' station or an elevator?
3. Patients: You would expect fewer colonies of bacteria on a maternity wing, for example, than in rooms where patients with infectious diseases were staying.

Nuisance factors: All the nuisance influences I listed are observational, built into the rooms, and so they are hard to balance. We might be able to group the rooms into blocks according to how much sun they got, or how much traffic went past them, but it would be difficult to do both at once. Since all the nuisance influences are tied to the rooms, it seems reasonable to take rooms as a nuisance factor.

Relation to factor of interest: There is only one factor of interest, so we have two choices for a design structure. If we make carpeting a between-rooms factor, we get a CR design. Half the rooms, chosen at random, will get carpeting put down; the other half stay bare. The room would be our experimental unit.

On the other hand, if we make carpeting a within-rooms factor, we get a CB design. This means measuring each room twice, once with carpet, and once without. The unit would be a time slot.

Design structure: Which is better, the CR or CB? First, notice that there's more than one way to run the study with rooms as blocks. Plan A: We could measure each room first without carpet, then put down carpet, leave it for a while, and measure a second time. Plan B: We could flip a coin for each room; heads means the room gets carpet first, then the carpet is ripped up and the room is measured a second time with the floor bare; tails means bare first, then carpet.

Plan A confounds the factor of interest with time, but has the possible practical advantage that at the end of the study, all sixteen rooms have carpet on the floor. Plan B gets rid of the confounding, but at a cost: you have to pay to have carpet removed from some of the rooms, before you know whether or not the carpeted rooms are less sanitary. Both CB plans have the disadvantage that they take longer than the CR plan.

If we reject Plan B as too wasteful, and reject Plan A because of the confounding, then we're left with the CR design. That's how the study was actually done. ■

In thinking about studies like the next one, which uses human subjects, there are two somewhat different ways to list your options. The way I have been using emphasizes the factor structure of the design (what you get), but psychologists often use a list that puts more emphasis on what you do with your subjects. According to this second view, you have three choices:

a. *Independent subjects (or between-subjects) design.* Each subject gets assigned to a condition, and contributes one value of the response. This is the same as a BF design.
b. *Matched subjects design.* Each subject gets matched with one or more similar subjects, and groups of matched subjects serve as blocks. In psychology, you might match subjects using some pretest, for example an IQ test, or some test that measures aspects of personality. In biology, you might use littermates as matched subjects, or match them by initial weight. Matching subjects gives a CB design.
c. "*Repeated measures design.*" I've put the phrase in quotation marks because in this context some people use the phrase to mean only that each subject gets measured more than once, and serves as a block of time slots. In this sense "repeated measures" doesn't necessarily refer to what I've called the SP/RM design: there might or might not be a between-subjects factor. If there is no between-subjects factor, the design would have a CB structure, and would be called a **within-subjects** design. If there is a between-subjects factor, the design would have an SP/RM structure, and would be called a **mixed** design.

Notice that (b) and (c) describe two different strategies for creating blocks: by matching subjects and by reusing them. From the point of view of factor structure, (b) and (c) are alike. Both lead to designs with blocks as a nuisance factor, both strategies might be used for a CB design, and both might be used for an SP/RM. From the structural point of view, the difference between a CB and an SP/RM is that the SP/RM has a between-blocks factor, the CB doesn't.

EXAMPLE 8.6 OVERCONFIDENCE IN CASE STUDY JUDGMENTS

"Confirmation bias" is a phrase cognitive psychologists use to refer to a tendency they claim we all fall victim to, a tendency to seek and find evidence that confirms what we think is true, rather than to take a more neutral and objective approach. In part because of this bias, we are often more confident than the facts warrant.

Confirmation bias can be a particular problem for psychotherapists and their patients, because the therapist must continually reevaluate the nature of the patient's problems in order to figure out the best course of therapy, and an overconfident therapist might easily overlook important information. Suppose you want to design a study to measure the extent of this overconfidence in people's judgments based on their information about a clinical case study.

Response: To be overconfident means to be more certain than the evidence justifies. To measure my overconfidence, then, you have to measure both how certain I am, and how justified I am in feeling that certain. Stuart Oskamp, who carried out this study, used a set of 25 multiple choice questions to get his response variables. The questions were all based on a particular published case study, that of Joseph Kidd. Each question, like the sample below, had five possible choices:

20. In conversations with men, Kidd:
    a. Prefers to get them to talk about their work or experiences.
    b. Likes to do most of the talking about subjects with which he is familiar.
    c. Prefers to debate with them about their philosophy or religion.
    d. Likes to brag about his Army days or college exploits.
    e. Confines his discussion mainly to sports, sex, and dirty jokes.

For each question, one of the five choices was actually correct (choice (a), for the sample question) so Oskamp was able to score each answer right or wrong. He asked his subjects not only to answer the questions, but also to rate each of their answers—with a percent—according to how confident they felt. The scale for the percents ranged from 20% to 100%: 20% was the lowest confidence rating, because if you have to guess blindly at an answer, you have a one-in-five chance of guessing right; 100% is the highest rating, for when you're certain.

Oskamp summarized a subject's 25 answers by computing the percent correct, and summarized the confidence rating by computing the average percent confidence. He would call me overconfident, for example, if I had an average confidence of 60%, but I only got 40% of the questions right.

Conditions: Oskamp based his choice of conditions on two ideas related to confirmation bias. One is that once you think you've figured something out, you tend to ignore any new information that ought to make you reconsider. The other idea is that the more information you've seen, the more likely you are to feel confident, regardless of whether the information is relevant or not. Both these ideas suggest that the amount of information available to the subjects ought to be part of what defines the conditions for the experiment. Oskamp's plan was to present the information about Kidd in four stages, and measure the two response variables (percent right, percent confidence) after each stage. Stage 1 told only this much:

> Joseph Kidd (a pseudonym) is a 29-year-old man. He is white, unmarried, and a veteran of World War II. He is a college graduate, and works as a business assistant in a floral decorating studio.

Stage 2 added a page and a half about Kidd's childhood; Stage 3 added information through his college years; Stage 4 covered the years from college to age 29. These four stages serve as levels of the main factor of interest.

Oskamp decided to include a second factor of interest: type of subject, with three levels—clinical psychologist, psychology graduate student, and advanced undergraduate.

Nuisance factors: Subjects.

Relation to factors of interest: The first factor (stages) should be a within-subjects factor, because the hypothesis behind the study predicted that subjects would become more confident as they got more information, and so the experiment ought to compare subjects with themselves as they got more and more data. The second factor, type of subject, can only be between-subjects.

Design structure: With one nuisance factor, one between-blocks factor, and one within-blocks factor, the choice is clear: SP/RM. (Imagine how the data might look in a rectangle, with rows as subjects, in three groups, one for each kind of subject, and with four columns for the four stages. Then visualize the factor diagram.)

The results in Table 8.5, for all subjects combined, speak for themselves:

TABLE 8.5 Confidence grows; accuracy doesn't

	Stage 1	Stage 2	Stage 3	Stage 4
Accuracy (%)	26	23	28	28
Confidence (%)	33	39	46	53

The clinical psychologists tended to be a little less confident, but no more accurate, than the other groups; however, none of the differences between groups of subjects were statistically "significant," i.e. detectable as "real." The main result was the one shown by the table: the subjects did not get any more accurate as they got more and more information, but they did get considerably more confident. ■

The supplement gives examples (S8.3: "Patching up the SAT Study" and S8.4: "Thermal Pollution.") of two designs that I consider workable but less than ideal, and discusses their strengths and weaknesses.

Exercise Set B

1. *Dense tomatoes*. Thirty-six plots of land were used for this study of the effects of planting density on the yield of tomatoes. The goal was to compare the effects of planting density (10, 20, 30 or 40 thousand plants per 10,000 square meters) on three varieties of tomato plants.

   a. What would be a suitable response?

   b. Assume there was no good way to group the plots into blocks.

(i) Tell the units and the conditions, and tell how to assign conditions to units.

(ii) Draw and label a rectangular diagram showing what the data from your design would look like.

(iii) Then summarize the design using the format from Section 1.

2. *Deprived rats*. Suppose you want to study the effects of food deprivation on rats. You choose three levels for your factor of interest: 0 hours of deprivation (control), 24 hours, and 72 hours. After the assigned period of deprivation, you plan to give each rat access to food, and record the amount eaten (in grams) as your response.

a. What design would you use for this study? Tell the units and the conditions, and tell how to assign conditions to units.

b. Draw and label a rectangular diagram showing what the data from your design would look like.

c. Then summarize your design using the format from Section 1.

3–5. *The effect of expectations on biofeedback*. (As you read the following description, think about what sort of experimental designs might be suitable.) Experiments have shown that skin temperature and relaxation are related: the more relaxed you are, the higher your skin temperature. Because of this relationship, doctors sometimes use biofeedback to help tense patients learn to relax. If you were such a patient, the doctor would attach an electronic sensor to the tip of your index finger, and connect the sensor to a machine that measures your skin temperature and plays a musical tone which gets higher when the temperature goes up, lower when the temperature goes down. Many people who practice regularly with such a machine get so they can raise their skin temperature at will.

Suppose you want to design an experiment to study factors which influence how well people learn to raise their skin temperature, and suppose you have twelve human volunteers, people who know nothing about either biofeedback or the scientific literature on control of skin temperature.

It's reasonable to wonder whether patients who believe they can learn to control their temperature will do better than those who believe they probably can't. One way to study the effect of expectation would be to compare subjects who are told that most people can learn to raise their temperature (high expectation) with subjects who are told that hardly anyone is able to do it (low expectation).

It's also reasonable to wonder about the effect of the biofeedback, the changing tone which the machine makes as your temperature changes. Suppose that in addition to expectation, you want to compare three kinds of feedback:

True feedback (TF). The tone goes up when the subject's temperature goes up, and down when the temperature goes down.

False feedback (FF). The machine plays back tones from someone else's biofeedback session. Thus the tone has no relationship to the subject's own temperature.

No feedback (NF). The machine just records the temperature, but doesn't make any sound at all.

Now consider three experimental plans:

3. Completely randomized,
4. Split plot/repeated measures, and
5. Based on Latin squares.

For each plan,

a. Tell how you would use your subjects: tell the units and the conditions, and tell how to assign conditions to units.

b. Draw and label a rectangular diagram showing what the data from your design would look like.

c. Then summarize your design using the format from Section 1.

d. Finally, tell whether you would recommend such an experimental plan: if you would not recommend it, why not? Which of the other plans is better, and why?

6–8. *Exercise, alcohol, and heart rate*. Design and describe an experiment to fit each of the following situations.

a. Tell what you would use as your response and how you would measure it; also

b. Tell who you would use as subjects and how you would get them.

c. Tell the units for your plan, and the name of the design.

d. Draw and label a rectangular format to show what the data from your experiment would look like.

e. Finally, summarize your design using the format from Section 1.

6. Purpose: To compare the effect of exercise on heart rate, for three levels of exercise: sitting, walking, and climbing stairs.

7. Purpose: To see how much athletes and nonathletes differ in how quickly their heart rates return to normal after exercise.

8. Purpose: To see what effect alcohol has on the recovery rate following exercise, and to test the hypothesis that the effect is smaller for athletes than for nonathletes.

9. Pick out a journal in an area that interests you, and page through it until you find an article that looks interesting. Read enough of the abstract to get a sense of what the article is about. Then put the article aside and plan your own experiment on the same topic. Finally, go back to the article and compare your plan with the one the authors used. Here are some journals you might consider:

American Journal of Physiology

Human Factors

Journal of Abnormal Psychology

Journal of Animal Behavior

Journal of Clinical Investigation

Journal of Personality and Social Psychology

---

3. Informal Analysis: Examples

Informal analysis is, like water-witching, a mixture of following ritual and following your nose. At any given point in the analysis, there may be several directions to choose from, and no rules to make the choices for you. Knowing a few standard rituals—the various kinds of plots, for example—may help you see possibilities you might otherwise overlook, but in data analysis, as perhaps in all of life, ritual is no substitute for a good nose. This makes it hard for anyone who tries to write about informal analysis, because, as far as I know, devotees of artificial intelligence have yet to find a good algorithm for following your nose. With a certain diffidence, born of the knowledge that your nose is not my nose, I've decided to show you some of what I've done with two data sets, so you can see not just how I've put some of the graphs to work and what patterns I found interesting, but also how I sometimes wandered a bit before I stumbled onto patterns I might have seen more quickly.

The first of the two examples in this section is fairly straightforward, but the second turned out to be a bit more involved. My source for that example presented it as having a CB design, and I began my analysis without questioning the CB structure, but some of the plots I tried surprised me, and I eventually decided that the most interesting patterns in the data were not the sort that ANOVA would be able to handle. The supplement contains another example of this sort (S8.5: "Feeding Frogs").

Example 8.7 Alpha Waves in Solitary

The data in Table 8.6 come from the study described in Example 8.3. The first ten subjects were kept in solitary confinement, the other ten served as controls. The response is the frequency of alpha waves (measured on days 1, 4, and 7), which psychologists interpret as a measure of arousal.

TABLE 8.6 Frequency of alpha waves

		Days		
Group	Subject	1	4	7
Solitary	1	14	7	6
	2	24	16	14
	3	21	10	7
	4	20	14	13
	5	15	5	2
	6	17	6	3
	7	16	11	9
	8	10	8	5
	9	6	3	0
	10	32	20	19
Control	11	20	20	17
	12	16	15	17
	13	15	17	14
	14	20	17	19
	15	14	16	15
	16	13	13	13
	17	4	6	6
	18	22	22	21
	19	21	21	22
	20	13	12	14

Response = $10 \times (\text{freq} - 9.0)$

I began with a parallel dot graph (Figure 8.3):

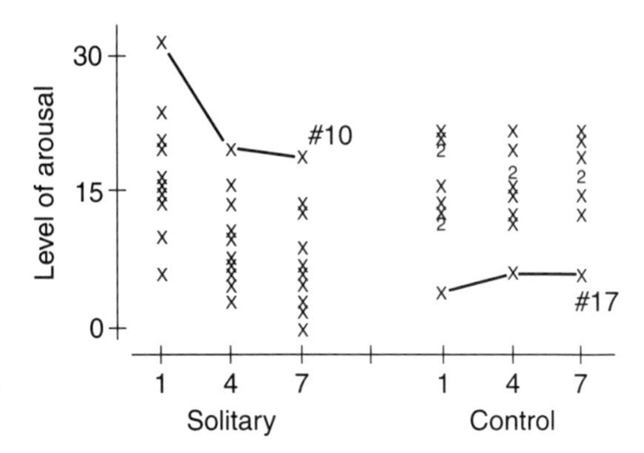

FIGURE 8.3 Parallel dot graph for alpha wave frequencies.

1. *Groups*: The clusters for the prisoners in solitary shift down as you go from day 1 to 4 to 7: the level of arousal, as measured by frequency, drops in response to continued sensory deprivation. For the control group, on the other hand, the three clusters show pretty much the same set of levels for all three days.
2. *Outliers*: Subject #10 is distinctly higher than the other treatment subjects, and Subject #17 is a lot lower than any other of the control subjects.
3. *Spreads*: Roughly equal if you include #17, but much smaller for the control group than for those in solitary if you leave out #17.
4. *Transforming*: not likely to make much difference, because averages and spreads are not related.

The main message from the dot graph is that as the week of the study went by, the two groups showed bigger and bigger differences. To see this pattern more clearly, I computed condition averages and drew an interaction graph (Figure 8.4) using the data in Table 8.7.

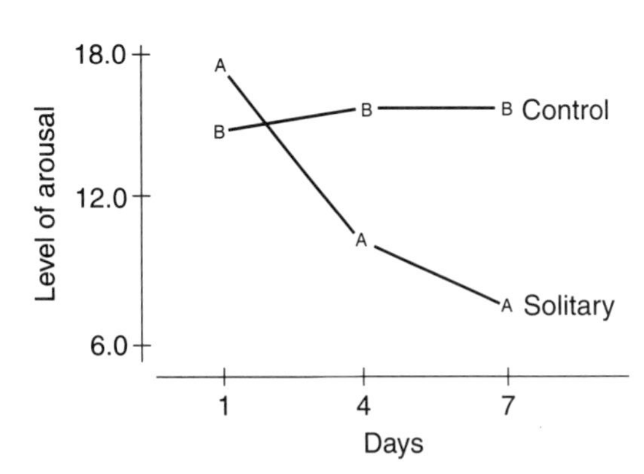

FIGURE 8.4 Condition averages and interaction graph. Alpha frequencies remained stable for the control group, but dropped to about half the starting level for the group in solitary.

TABLE 8.7 Condition averages for the data of Example 8.8

	Days		
	1	4	7
Solitary	17.5	10.0	7.8
Control	14.8	15.9	15.8

The interaction that shows up in the condition averages is the most important message; indeed, the experiment was designed to study that interaction. Nevertheless, there's a bit more to be found in the numbers. For example, Figure 8.5 is a within-blocks scatterplot of levels for Day 7 versus levels for Day 1, using different symbols for the two groups of subjects.

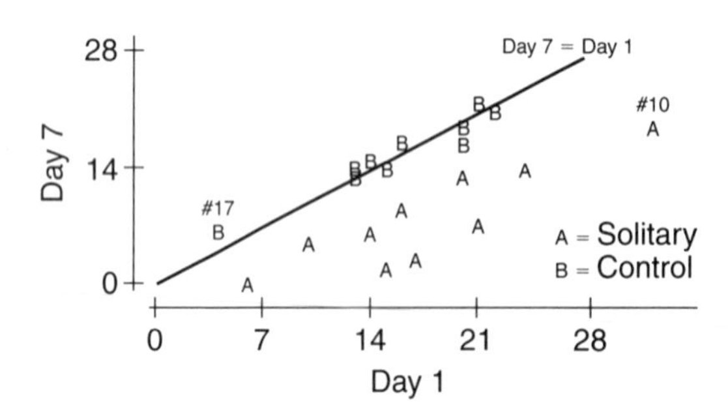

FIGURE 8.5 Scatterplot for the solitary confinement data. The horizontal scale is for arousal level at the beginning of the study, and the vertical scale is for arousal level at the end of the study. The Bs are the control subjects and the As are the subjects in solitary confinement.

Four features of the plot strike me as worth mentioning. First, the As and Bs form separate clusters, with the "control blob" of Bs above the "solitary blob" of As. Each cluster suggests a line whose slope is roughly equal to the slope of the identity line, for which Day 7 = Day 1. Second, the blobs are fairly skinny, indicating a strong relationship between day 1 levels and day 7 levels for the same person. Third, subjects #10 and #17, who stood out from the rest of their groups in the dot graph, also stand out here. Finally, if you leave out subjects 10 and 17, the Bs for the control subjects are quite a bit more tightly bunched than the As for the subjects in solitary: a balloon summary for the Bs would be a lot smaller than one for the As, even though they would have roughly similar shapes.

Apart from the outliers, for which I have no explanation, the other features of the scatterplot make sense if you think about what they mean. It's certainly no surprise to find the strong positive relationships. You'd expect subjects with higher values on one day to have higher values on nearby days, and similarly for subjects with medium or lower values. Nor is it surprising that the blob for the subjects in solitary is more spread out. These subjects were the ones that got something done to them, in a way that the controls did not, and it seems reasonable that the sensory deprivation might have more of an effect on some subjects than on others, and thus would spread out the values for day 7 in the treatment group. I wouldn't have predicted that before I saw the data, so the scatterplot has told me something I didn't already know, but nevertheless, it's no great surprise. Finally, it not only makes sense that the two blobs are separate, but in fact their being separate is related to the main purpose of the study. The Bs for the controls lie near the identity line Day 7 = Day 1, and that's about what you'd expect. The cluster of As lies below the cluster of Bs, and that pattern supports the hypothesis that the study was designed to test, namely, that the day 7 levels of arousal for the subjects in solitary are quite a bit lower than the levels on day 1, before the sensory deprivation began. ■

EXAMPLE 8.8 THE TRANSYLVANIA EFFECT: A CB DESIGN?

The purpose of this observational study, from **Exercise A.16**, was to see whether emergency admissions to a mental health clinic were more frequent when the moon was full than before or after the full moon. The rows (blocks) are for the twelve months from August 1971 to July 1972; the three columns (conditions) are for Before (B), During (D), and After (A) the full moon. The response is admission rate, in patients per day times 10.

The book where I first saw this data set gave only the standard formal analysis for a CB design. That's what I decided to show you first. The data, decomposition, and sums of squares, df, and *F*-tests for months and moon phases are shown in Table 8.8–8.10 and Figure 8.6.

TABLE 8.8 Transylvania data

	B	D	A	Avg
Aug	63	50	58	57
Sep	71	130	93	98
Oct	65	140	80	95
Nov	86	119	77	94
Dec	81	60	111	84
Jan	104	90	130	108
Feb	114	129	135	126
Mar	138	160	131	143
Apr	154	249	158	187
May	157	130	133	140
Jun	117	139	128	128
Jul	158	200	146	168
Avg	109	133	115	119

10 × admissions per day

TABLE 8.9 Decomposition of Transylvania data

	Grand average			Month effects			Moon phases			Residuals		
	119	119	119	−62	−62	−62	−10	14	−4	16	−21	5
	119	119	119	−21	−21	−21	−10	14	−4	−17	18	−1
	119	119	119	−24	−24	−24	−10	14	−4	−20	31	−11
	119	119	119	−25	−25	−25	−10	14	−4	2	11	−13
	119	119	119	−35	−35	−35	−10	14	−4	7	−38	31
	119	119	119	−11	−11	−11	−10	14	−4	6	−32	26
	119	119	119	7	7	7	−10	14	−4	−2	−11	13
	119	119	119	24	24	24	−10	14	−4	5	3	−8
	119	119	119	68	68	68	−10	14	−4	−23	48	−25
	119	119	119	21	21	21	−10	14	−4	27	−24	−3
	119	119	119	9	9	9	−10	14	−4	−1	−3	4
	119	119	119	49	49	49	−10	14	−4	0	18	−18
SS	509,796		+	45,096		+	3,744		+	13,140		
df	1		+	11		+	2		+	22		

TABLE 8.10 ANOVA table

Source	SS	df	MS	*F*-ratio	5% crit. val.
Grand Average	509,796	1	509,796		
Months	45,012	11	4,092	6.85*	2.26
Phases	3,744	2	1,872	3.13	3.44
Residual	13,140	22	597.2		
Total	571,692	36			

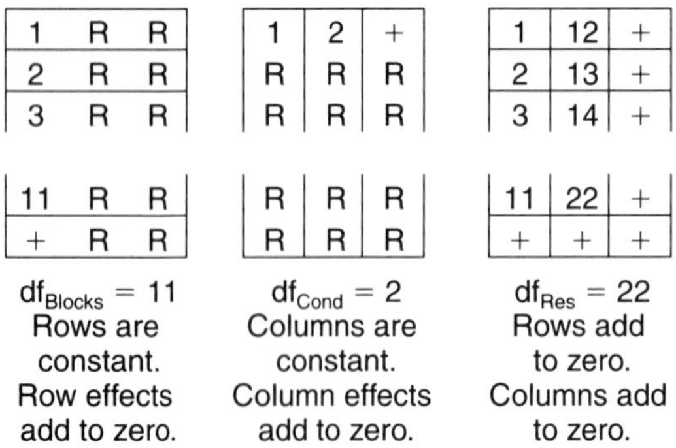

1	R	R
2	R	R
3	R	R
11	R	R
+	R	R

$df_{Blocks} = 11$
Rows are constant.
Row effects add to zero.

1	2	+
R	R	R
R	R	R
R	R	R
R	R	R

$df_{Cond} = 2$
Columns are constant.
Column effects add to zero.

1	12	+
2	13	+
3	14	+
11	22	+
+	+	+

$df_{Res} = 22$
Rows add to zero.
Columns add to zero.

FIGURE 8.6 Counting free numbers.

Discussion: Taking the ANOVA at face value for the moment, we find that the F-ratio for months is significant, but the F-test for phases tells us that phase effects are not big enough to be detected. However, Chapter 11 will use a more focused test to compare rates during the full moon with the average of the rates before and after.

This test *does* find a "real" difference between the admission rate during the full moon and the average of the rates before and after.

How appropriate is this analysis? I'll let a complete answer wait for Chapter 12, on the Fisher assumptions, but for now let's see how much we can learn from an informal analysis.

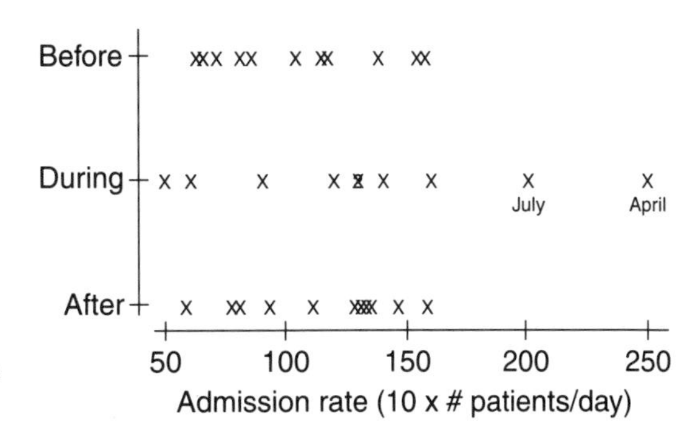

FIGURE 8.7 Parallel dot graph for the Transylvania data.

As Figure 8.7 shows, the averages for Before (109) and After (115) the full moon are nearly equal, and are both quite a bit less than the average for During (133). The same is true of the spreads: SD for Before (36.3) and After (31.0) are distinctly smaller than for During (54.8). Thus large average and large spread go together, small with small.

However, the differences between phases completely disappear if you exclude the two extreme months for During. Without July and April, the average for During drops to 114.7, and the SD drops to 36.2. In order to conclude that there is a "real" effect due to the full moon, we have to assume that the values for July and April should not be excluded. The dot graph itself doesn't give much evidence to support that assumption.

On the other hand, neither the dot graph nor the SDs take into account the effects of the months. For example, the SD for Before gets computed as though the 12 monthly admission rates before the full moon were all made under the same conditions. In fact, as the next plot will show, there is a seasonal pattern to the monthly fluctuations.

Scatterplot of admission rate (y) versus month (x). This data set shows you can't always make a clear distinction between nuisance factors and factors of interest. Here, you can think of months as a nuisance factor, in that what we really want to know about is not the effect of the months, but the effect of the moon phases. The monthly fluctuations are indeed a nuisance, because they make it harder to isolate the effect of the full moon. On the other hand, the levels of a nuisance factor don't usually have the kind of meaningful structure that makes it possible to have systematic relationships. Here, the months have a natural order, so that it makes sense to look for patterns relating the admission rates to the time of year.

The scatterplot in Figure 8.8 shows the average daily rates, by month. For this plot, I've joined the points by lines, to make the time trend easier to see.

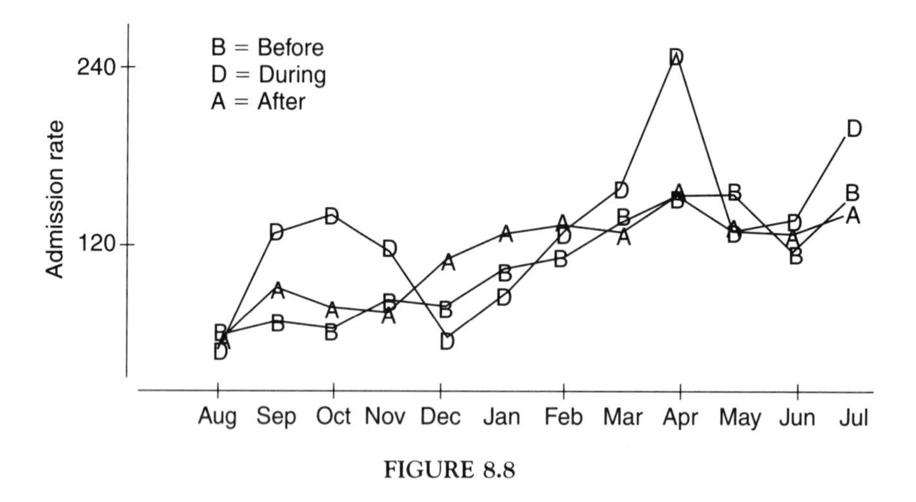

FIGURE 8.8

Admission rates for all three phases are lowest in August (the month when psychiatrists traditionally take long vacations), and rise steadily over the months that follow. The monthly rates for Before and After start close together in August and stay close together throughout the year. The rates for During show a somewhat different pattern, with bigger fluctuations: a sharp rise in September, with high rates continuing in October and November, then a sharp drop in December. From the low point in December, the admission rates climb steeply month after month to a peak in April, before dipping again.

For me, the parallel dot graph doesn't show much evidence for a Transylvania effect, but I find this scatterplot more persuasive. For 7 of the 12 months, the rate for During is higher than the rates for Before and After. (By chance, you'd expect the rate for During to be highest for one third of the months, or 4.) Moreover, for 5 of the months, the rate for During is quite a lot higher than for the other two phases. Finally, the rates for July and April during the full moon, rates which looked suspiciously high to me in the parallel dot graph, don't seem nearly so extreme to me in this scatterplot, which shows them in a different numerical setting. The big increase from March to April looks like part of a sharp upward trend that begins back in December, and the increase from June to July for During roughly parallels similar increases for Before and After.

Scatterplot of rate for During versus average rate for Before and After. I tried one last plot as part of my informal analysis. Since the purpose of the study is to see whether the full moon has an effect, and since the rates for before and after are so close together, I decided to average the before and after rates for each month, then scatterplot the rates for during versus these averages, with the 12 months as points (Figure 8.9).

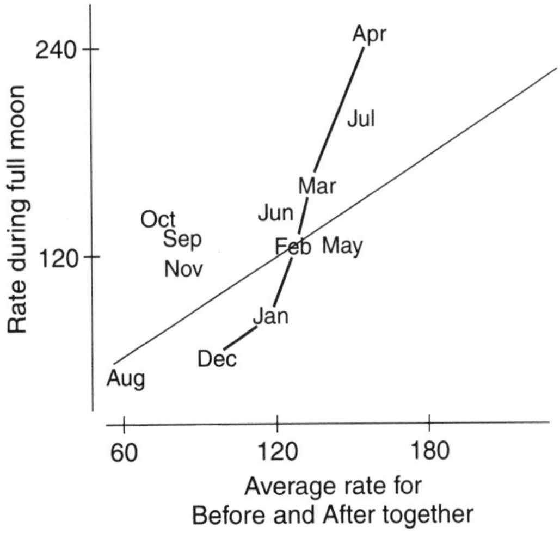

FIGURE 8.9 Admission rate during the full moon vs. average rate for before and after. The 7 months above the diagonal line ($y = x$) have rates that are higher During the full moon than the average rate for Before and After. I drew line segments joining the months December through April to show the time trend.

I see several patterns in this scatterplot. (1) The three fall months (September, October, November) form a tight cluster above the identity line $y = x$, and away from the other months. (2) The point for August is isolated—not close either to the months that come after it or to the months of June and July (of the next year). (3) The points for the other 8 months all fall close to a steeply sloping line. I didn't draw this line, but it would cross the line $y = x$ where the rate is roughly 125 (12.5 patients per day), and would have a slope near 2.5. For months when the admission rate is below 125, the rate for During is lower than the rate for Before and After; but for an increase of 10 in the Before-and-After rate, there is an increase of 25 in the rate for During. (4) If you exclude the two "questionable" months, April and July, then refit the line, your new line would be somewhat less steep than the old one, but not a lot. In other words, in this plot as in the last one, April and July seem part of a pattern that is still present even if you exclude those two months. (5) On the other hand, however, if I cover the points for those two months and look at the other 10 points all at once, ignoring the time sequence, I see a fattish balloon of points that has the line $y = x$ running pretty much through the middle of it.

What do I make of the overall message from the three plots? The data strongly suggest that admission rates are in fact tied to phases of the moon, but the strong suggestion carries two cautions. One: There's no evidence that the full moon *causes* mental illness. It could be, for example, that the moon happened to be full at times that coincided with other events (stressful holidays?) that influenced admission rates. (This would need to be checked.) Two: The pattern relating admission rates and moon phases is not a simple one of the sort you can see clearly just by comparing groups, as in the parallel dot graph or formal ANOVA. The relationship involves the months as well as the moon phases, and would be hard to see without the scatterplots. I'll return to this example in Chapter 12, to show you how it fails to satisfy some of Fisher's assumptions for ANOVA. ■

Exercise Set C

Each of the exercises below asks for an informal analysis of a data set. These exercises are open-ended: there is no one right answer, and no easy way to recognize when you've done enough to quit. To put it more positively, if you think something might be worth a try, then try it. Judge your hunches not by "Is this what they want?" but by "If I do this, will I learn something about the data?"

The first exercise includes several suggestions for things to try. After that, you're on your own, because part of the exercise is to practice deciding what things to try. Exercises 2 and 3 are based on small data sets; Exercises 4 and 5 are based on moderate-sized data sets. Three more exercises of this sort are based on data sets from Section 5, and are given in Set D. The data sets for those exercises have a more complicated structure and offer your imagination more freedom to play.

1. *Hospital Carpets*. Reread Example 8.6; then do an informal analysis of the data given below.

Carpeted floors		Bare floors	
Room #	Colonies/ft^3	Room #	Colonies/ft^3
212	11.8	210	12.1
216	8.2	214	8.3
220	7.1	215	3.8
223	13.0	217	7.2
225	10.8	221	12.0
226	10.1	222	11.2
227	14.6	224	10.1
228	14.0	229	13.7

a. Start with averages and a parallel dot graph; discuss what you find.

b. Then explore the possibility of bias associated with room number: (i) Low versus high. Does one of two treatment groups have more than its share of low numbered rooms? Scatterplot the response (y-axis) against room number (x-axis). Do you see any pattern? (ii) Odd versus even. Does one of the treatment groups have more than its share of even numbered rooms? It could be that even rooms are on one side of the hall, odd on the other side; which side of the hall you are on might determine how much sun you get, and might influence bacteria levels. Is there any evidence for this in the data?

2. *Diabetic dogs: a small data set*. Exercise 7.12 gave data from an experiment to compare two methods to measure the rate of turnover of lactic acid. Do an informal analysis.

3. *Deprived rats*. The experiment described in Exercise B.2 was run as a CB[1] using eight

rats, with each rat providing a block of three times slots as units. Do an informal analysis of the data.

Period of food deprivation	Rat 1	Rat 2	Rat 3	Rat 4	Rat 5	Rat 6	Rat 7	Rat 8
0 hours	3.5	3.7	1.6	2.5	2.8	2	5.9	2.5
24 hours	5.9	8.1	8.1	8.6	8.1	5.9	9.5	7.9
72 hours	13.9	12.6	8.1	6.8	14.3	4.2	14.5	7.9

4. *Pine pruning*. Do an informal analysis of the data given with Exercise A.17.
5. *Emotions and skin potential*. Reread Example 8.3; then do an informal analysis of the data from that experiment, given below.

Subject	Fear	Happiness	Depression	Calmness
1	23.1	22.7	22.5	22.6
2	57.6	53.2	53.7	53.1
3	10.5	9.7	10.8	8.3
4	23.6	19.6	21.1	21.6
5	11.9	13.8	13.7	13.3
6	54.6	47.1	39.2	37.0
7	21.0	13.6	13.7	14.8
8	20.3	23.6	16.3	14.8

4. RECOGNIZING ALTERNATIVES TO ANOVA

ANOVA Is Appropriate If:

1. **You have only one response**
2. **That response is measured on an interval or ratio scale**
3. **Your carrier variables are all categorical**

In this section I will describe and illustrate three kinds of mistakes people sometimes make in choosing an ANOVA model: (a) mistaking multiple measurements for levels of a factor, (b) mistaking the categories of a nominal response for levels of a factor, and (c) mistaking measurements on an interval or ratio scale for levels of a factor. As you read the examples that follow, keep in mind that ANOVA uses averages and deviations to compare two or more groups of units. To help you decide whether a particular ANOVA model is appropriate, think about the averages the model tells you to compute and compare. Are these averages meaningful? Do they give the kinds of comparisons that you want? Is it reasonable to think of the residuals as coming from a chance process?

It is useful to think of statistical methods in families, classified according to the kinds of data for which they are best suited. I've attempted one classification in what follows, to help you see how the various methods are related to each other, and to help you decide when ANOVA or some other family of methods is most suitable.

The classification depends on three pairs of terms: (1) cases and variables, (2) response and carriers, and (3) categorical versus numerical.

1. *Cases and variables*. Crudely, cases are the objects that get "measured," and variables are the "measurements." Consider the solitary confinement data, for example (8.4), and assume for the moment that there is only one response, the alpha wave frequency. Then, although there are only 20 subjects, there are 60 cases that get measured: three different days for each subject. There are 4 variables: the alpha frequency, the environment (solitary or control), the day (1, 4, or 7), and the subject number. Only the first of these, alpha frequency, is what you would ordinarily think of as a measurement (which is why I put quotation marks around the word in my crude summary). The meaning of "variable" as I use it here is somewhat abstract: a variable assigns a value—which might be a number but might just be a label—to each case. The variable "alpha frequency" assigns numbers, but the variable "environment" assigns nonnumerical labels, "solitary" and "control."
2. *Response and carriers*. Crudely, the response variable is the one we want to "explain" in terms of the other, carrier variables, which provide the "explanation." The word "explain" has both a suggestive, everyday meaning, and a specialized statistical meaning. For the hamster data, the response variable is the enzyme concentration, and the other three variables—day length, organ, and hamster—are the carrier variables. We do indeed want to explain, in the everyday sense, why the enzyme levels take on the values that they do, but a statistical analysis can provide only part of an explanation in that sense. The narrow, statistical meaning of "explain" is to find patterns that relate variability in the response to variability in the carriers. For example, an ANOVA for the hamster data splits the total variability, as measured by SS, into pieces associated with various carriers, plus a residual variability. In statistical language, the residual SS is the part not "explained" by the carriers.
3. *Categorical versus numerical.* The basic idea is illustrated by the solitary confinement data: alpha frequency is a numerical variable; environment is a categorical variable. Crudely, a categorical variable assigns categories to cases, a numerical variable assigns numbers. The distinction may seem simple enough, but it can sometimes get tricky, because the distinction depends not just on the variables themselves, but also on how you plan to use them in your analysis. For example, the variable that assigns a subject number to each case does in fact assign numbers, and so might seem to be numerical, but as I mean the words here, the variable is categorical, because we don't plan to use the numbers in our analysis: It makes no difference to our analysis whether we refer to the subjects as #1, #2, #3 or as Rudolph, Vincent, and Horatio. The variable for days is numerical: it assigns the numbers 1, 4, and 7 to cases. However, the variable is nevertheless treated as categorical in our ANOVA

because for that part of our analysis we use of the three days only as ways to group observations. The actual numbers 1, 4, and 7 are irrelevant.

With this vocabulary in mind, we can classify data sets and statistical methods first according to how many response variables there are, and then according to the kind of response(s) and the kinds of carriers. Table 8.11 gives a classification for data sets and methods with just a single response.

TABLE 8.11 A rough classification of univariate methods.

	Response variable is:		
Carriers are:	**Interval/Ratio**	**Ordinal**	**Nominal**
Categorical	ANOVA	Nonparametric ANOVA	Categorical data analysis
Numerical	Regression	Logistic regression and variants	
Some of each	Regression, ANalysis of COVAriance		

You can use this table to identify situations where ANOVA is appropriate: (1) there is only one response, or there is more than one but it makes sense to analyze them one at a time, (2) the response is measured on an interval or ratio scale, and (3) the carriers are all categorical. If any of (1)–(3) are questionable, you should consider alternatives to ANOVA: (1) If you have multiple measurements, then depending on the purpose of your analysis, multivariate methods may be more suitable. (2) If your response is ordinal or nominal, you should consider special methods for counted data. (3) If one or more of your carriers is numerical, you should consider methods described in Chapter 14.

Mistaking Multiple Measurements for Levels of a Factor

Multiple Measurements: One Response Variable or Many?

M1. If you have measurements on different scales, treat each as a different response. (Don't average grams with degrees with milliliters per minute.)

M2. If you have measurements on the same scale, but corresponding to different characteristics of subjects or objects, think about what you want to learn from your analysis:

- **If you want to use all the measurements *jointly* to compare groups of subjects or objects, you need a multivariate model.**
- **If you want to compare the average values for the various characteristics, try ANOVA with the multiple measurements as levels of a factor, but be sure to check the Fisher assumptions.**

EXAMPLE 8.9 WARM RATS: MEASUREMENTS ON DIFFERENT SCALES (M1)

The numbers in Table 8.12 are partial results from a larger student project in a physiology course at Mount Holyoke College, designed to measure the effects of air temperature on the weight, skin temperature, and oxygen consumption of rats. Look over the data set, and try to identify the factors and design structure.

TABLE 8.12 Partial results from a physiology experiment.

		Day 1			Day 3			Day 5		
Air temp[a]	Rat	Wt[b]	Temp[c]	O_2[d]	Wt	Temp	O_2	Wt	Temp	O_2
Warm	1	228	35.4	4.9	238	35.2	4.6	253	35.4	7.1
	2	247	36.2	5.2	254	35.6	5.2	269	36.4	5.6
Cold	3	272	33.2	10.5	275	32.6	9.8	276	33.4	10.9
	5	223	33.4	11.5	228	32.6	11.8	235	32.8	10.7
Control	4	230	34.8	4.6	259	35.4	4.7	269	35.4	5.7
	6	185	35.2	7.5	209	35.2	7.7	221	35.6	7.7

[a]Air temperature in the rat's cage: warm (29°C = 84°F); cold (8°C = 46°F); control (21°C = 70°F).
[b]Weight: Body weight in grams.
[c]Temperature: Skin temperature, in °C. (Core temperature was also measured, but I didn't include it here.)
[d]O_2: Rate of oxygen consumption, in milliliters per minute. ■

I hope you recognize that there are units of two sizes, and that basic design structure is SP/RM. The larger unit is a rat; each rat gets randomly assigned to a temperature. The smaller unit is a time slot for a rat.

What is the structure of the within-rats factor? Each rat was measured nine times, and there are two ways to regard these nine measurements.

a. *Superficially tempting, but wrong:* The nine measurements have a factorial structure that comes from crossing Days (3 levels: 1, 3, 5) with kind of Measurement (3 levels: Weight, Temperature, O_2 consumption).

	Day 1	Day 2	Day 3
Body weight			
Skin temperature			
Oxygen consumption			

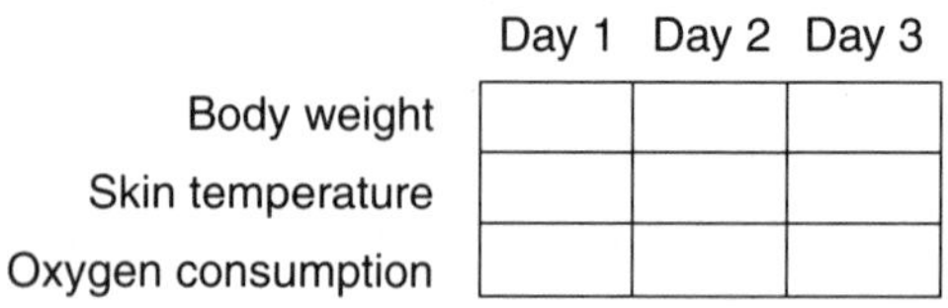

b. *Correct:* there are three different response variables (weight, temperature, O_2 consumption), and only one within-blocks factor (days), with three levels.

One way to spot the flaw in the "tempting but wrong" analysis is to think about using averages to compare groups; in particular, comparing rats using averages for each rat = block = row. These averages are pretty close to meaningless, because they have lumped together weights, temperatures, and oxygen consumption rates, treating them as interchangeable.

There are several reasonable ways to analyze this data set. (1) You could do three separate ANOVAs, one for each of weight, temperature, and oxygen consumption. (2) You could use scatterplots for a graphical analysis of the relationship between, for example, oxygen consumption (y) and temperature (x). (3) You could use a more advanced method, called *analysis of covariance* (Chapter 14), for a formal analysis of the patterns in the scatterplots. (4) You could use an even more advanced method, multivariate ANOVA, to analyze all three responses simultaneously.

Examples like the last one are comparatively easy to recognize, once you've trained yourself to ask whether the various observations are all measured on the same scale, with the same units of measurement. If you see degrees with millimeters, or grams with seconds, presented together as part of the same data set, you'll know you've got more than one response. For such examples, the choice is clear-cut: to think of weights, temperatures, and oxygen consumption as levels of a factor is simply wrong. There are three different response variables.

For a second class of examples, it may not be so clear whether to handle multiple measurements as levels of a factor: this happens if you have measurements on the same scale, but corresponding to different parts or features of your material. For data sets like these, you should pay particular attention to the purpose of your analysis, and to the four Fisher assumptions about chance error (especially the assumption that your chance errors are independent). For factors based on parts or features, no randomization is possible, and your residual errors may not behave the way ANOVA assumes they do. (More in Chapter 12.)

EXAMPLE 8.10 MULTIVARIATE METHODS OR LEVELS OF A FACTOR? (M2)

a. *Kelly's hamsters.* The enzyme in the heart and brain are measured on the same scale, and correspond to features of individual hamsters. If you want to use both concentrations simultaneously to see whether there's a "real" difference between long-day and short-day hamsters, then a multivariate, two-response model is appropriate. If, however, you want to compare heart averages with brain averages for the two groups of hamsters, then it's appropriate to treat the two organs as levels of a factor, provided you're careful to check the assumptions about residual errors.

b. *Sleeping shrews* (*Example 2.4*). Heart rates for the three kinds of sleep were measured on the same scale, and correspond to characteristics of individual shrews. The purpose of the experiment was to compare the three kinds of sleep, one with another. To do this, you have to treat LSW, DSW, and REM as levels of a factor, not as three different response variables.

c. *Murderer's ears* (*Example S7.5*). The purpose of this study was to use two measured characteristics (lengths of the right and left ears) to check for differences among groups of inmates: "ordinary" murderers, "lunatic" criminals, and other criminals. For that purpose you need multivariate methods. (See also Example S5.2: "Fisher's Iris Data.") ■

Mistaking Categories of a Nominal Response for Levels of a Factor

Recall from Chapter 4 that as a rule ANOVA works best for interval and ratio data, and that ANOVA is not appropriate for nominal data. Sometimes it is easy to recognize that your response is nominal (and that rather than try ANOVA, you should choose methods of the sort described in other books, often under the heading "chi-square" or "contingency tables.") Often, however, data sets with a categorical response get summarized in a way that makes the response *appear* to be numerical. Such summaries can mislead you into regarding the categories of the (nominal) response as levels of a factor, and regarding the number of occurrences in each category as a numerical response.

Categories of a Nominal Response Variable

C1. If you can get the data set simply by sorting things into groups, then counting the number in each group, an ANOVA model is not likely to be appropriate.

C2. Don't use an ANOVA model if its factor structure would confound an important interaction with residual error.

C3. Don't use an ANOVA model if any factor averages would be fixed by the design.

Example 8.11 Mothers' Stories

Reread Chapter 4, Review Exercises 7–9, about the use of the TAT test to compare mothers of schizophrenic and normal children. Then look at the way I've arranged the data in Table 8.12, and evaluate the faulty analysis of the design structure that follows the data.

TABLE 8.12 Structure of the data for Example 8.12

		# stories of each type				
Diagnosis of child	Mother	A	B	C	D	IR
Schizophrenic	1	2	2	4	1	1
	2	1	0	2	1	6
	.	.	.	.	.	.
	.	.	.	.	.	.
	20	2	1	2	2	3
Normal	21	8	0	1	0	1
	22	4	0	0	1	5
	.	.	.	.	.	.
	.	.	.	.	.	.
	40	4	0	2	0	4

Faulty analysis: The response is the number of stories told. There are four structural factors:

Factor	*# levels*	*Groups*
1. Diagnosis	2	Rows 1–20, 21–40
2. Mothers	40	Rows
3. Type of story	5	Columns
4. Diag × Type	10	Cells

The nuisance factor (Mothers) is crossed with Type of Story, but nested within Diagnosis. Diagnosis is between-blocks, Type of Story is within-blocks. The two factors of interest are crossed. The design is an SP/RM.

Correct analysis: The numbers in the table are not response values, but counts. They were obtained by classifying stories, and each number tells how many stories there were in a particular group (C1). The true response is nominal, with categories A, B, C, D, and IR.

Notice also that the faulty design structure fixes important factor averages (C3). To assess differences between mothers (= blocks), you would ordinarily compute an average value of the response for each mother, and compare averages. If you compute these averages using the faulty design structure, you find them all equal: the average response for each mother is 2 stories (per category). Similarly, the average response for each level of diagnosis is also 2 stories (per mother per category). The faulty structure makes it impossible to use averages to compare mothers, or to compare levels of diagnosis, the principle factor of interest.

Finally, note that the faulty analysis failed to identify one set of units, which come in two sizes. The larger units are the mothers, and the between-blocks factor is diagnosis; this part of the first analysis is correct. The smaller unit is a time slot for a mother, and the within-blocks treatments assigned to these units are the 10 pictures, which were shown to each mother in a random order.

Here's a summary of the design, following the checklist from Section 1.

A. *Purpose and response:* To compare the kinds of stories (nominal response) told by control mothers and mothers of schizophrenics.

B. *Design structure*

1. Units: There are two sizes of units, mothers and time slots.
2. Grouping:

   a. The mothers are not grouped. (The between-blocks factor is observational, but has the same structural relationship to mothers as for any observational BF design: mothers are sampled from two populations defined by diagnosis. (Structurally, mothers are nested within diagnosis.)

   b. Time slots are grouped into sets of 10 per mother, as in a CB. The 10 pictures are the levels of the within-blocks factor; these get randomly assigned to time slots separately for each mother.

3. Design family: SP/RM.

C. *Treatment structure.* (Table 8.13) There are two crossed factors of interest, the between-blocks factor (diagnosis) and the within-blocks factor (picture).

TABLE 8.13 Summary table of factors: Mothers' stories—an SP/RM[1;1].

Unit	Grouping (Nuisance factors)	Treatment combinations: Main effects	Treatment combinations: 2-Factor interaction
Mothers (20 in all)	(none)	Diagnosis (2)	
Time slots (10 per subj)	Mothers	Pictures (10)	Diagnosis × Pictures

Comment. The best analysis for this data set would use the factor structure I have just listed, and would regard the response as nominal. Nevertheless, you could use ANOVA to analyze certain summaries of the data, thinking of the design as a one-way observational study with diagnosis as the factor of interest, with mothers as the (only) units, and with any of the following as reasonable choices for a numerical response: # type A stories, # type C stories, or (# type A) minus (# type C) stories. ■

EXAMPLE 8.12 SMITH INFIRMARY AND INTRAVENOUS FLUIDS

Decide whether ANOVA is appropriate for the infirmary data (Example S6.1) and the fluids data (Example 5.2). In each case, identify the units.

SOLUTION. a. For the infirmary data, the numbers come from classifying visits into four categories: U (unnecessary), N (requiring a nurse's attention), D (requiring a doctor's attention), and A (requiring admission to the infirmary or local hospital). These four categories appear as columns in the data table presented in Example S6.1, and might be mistaken for levels of a factor. The data table given in Example S6.1 presents the numbers in the same format you would use for a CB design or a two-way factorial with one observation per cell. If you go by the form of the table, without thinking about where the numbers come from, you might make the mistake represented by the following factor diagram.

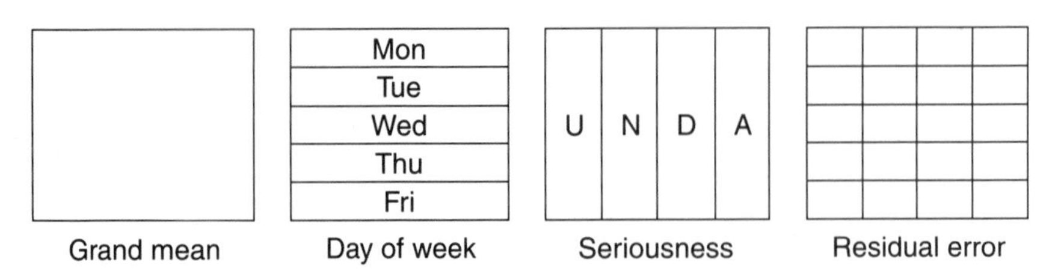

FIGURE 8.10

An analysis of variance based on this factor diagram would be unsound, because the last partition, which I labeled residual error, would also be the partition for the Day x Seriousness interaction (C2). The most interesting feature of these data is the pattern relating seriousness and day of the week, which the factor diagram confounds with residual error. It would be a gross mistake to ignore that pattern by thinking of it as error.

The key here is to recognize that for this data set, the *unit* is a single visit; each unit generates one categorical response value: U, N, D, or A. In my judgment, the best way to analyze these data uses methods for categorical data.

b. For the fluids data, the response is the number of contaminant particles, and the unit is a sample of fluid. As for the infirmary data, the numbers come from counting, but for the fluids data, the things that get counted (individual particles) do not get sorted into groups. (The unit is not a particle.) The response is measured on a ratio scale, and ANOVA is appropriate. ■

Mistaking Numbers on an Interval or Ratio Scale for Levels of a Factor

Because ANOVA uses averages to compare groups of observed values, the analysis treats all the observed values in a group as interchangeable. Moreover, ANOVA treats all the levels of a factor as interchangeable: if the levels of your factor have a meaningful order, or meaningful numerical relationship, ANOVA will ignore such information. Even more important, if the levels of your factor come from multiple measurements on the same subjects or objects, it may be that some of the measurements are better regarded as carrier variables, not as values of the response for different conditions. In such instances, ANOVA may miss the most interesting patterns in your data. It may be more appropriate to use *regression* methods (Chapter 14), which are based on fitting lines or curves to scatterplots.

Regression or ANOVA?

R1. What is the goal of your analysis? If you have multiple measurements on the same subject or object, don't treat them as levels of a factor if some of the measurements should be treated as carrier variables.

R2. If there is a natural way to associate a numerical carrier variable with levels of a factor, or with units that are grouped into levels of a factor, regression analysis may be more informative than ANOVA.

EXAMPLE 8.13 CORN AND PHOSPHORUS

The purpose of this study was to investigate the relationship between the phosphorus content of corn and two sources of phosphorus—from organic compounds and from inorganic compounds—in the soil the corn plants grew in. The data from the study were summarized in a table with 18 rows, one for each of 18 different samples of Iowa soil, by 3 columns, one each for inorganic phosphorus in the soil, organic phosphorus in the soil, and phosphorus in the corn grown from that soil. All phosphorus concentrations were in parts per million.

Faulty analysis: The response is the phosphorus content, each row of the data corresponds to a block (= plot of soil), and the three columns correspond to the three conditions to be compared: organic phosphorus in the soil, inorganic phosphorus in the soil, and phosphorus in the corn. The design is a nonrandomized CB, with multiple measurements as levels of a factor.

An analysis based on this model would tell you whether there were detectable differences between concentrations of phosphorus for three sources. Although such information is not totally worthless, it wouldn't tell you much about how the phosphorus in the corn is related to the two kinds of phosphorus in the soil.

Correct analysis: The phosphorus concentration in the corn is the response, and the two soil concentrations are numerical carrier variables (R1). Regression methods should be used to analyze patterns relating variability in the response to variability in the carriers: To what extent do higher soil concentrations lead to higher concentrations in the corn? ■

EXAMPLE 8.14 COOL MICE AND SOLITARY PRISONERS

a. Mice. Example 7.19 gave pairs of cooling constants for 20 mice. You can think of the data as coming from a CB design, with mice as blocks, and freshly killed versus reheated as the factor of interest. The resulting analysis compares the two conditions by computing an average cooling rate for each condition, then comparing averages. Notice, though, that you could also use the cooling constant for freshly killed to assign a number to each mouse. In other words, there is a meaningful way to associate the levels of the blocks factor with numbers on a line (R2). An analysis of the mouse data using regression turns out to be in some ways more informative than ANOVA. (Example 7.19 showed the scatterplot for such an analysis.)

b. Prisoners. Example 8.3 described an experiment in which 20 prisoners were assigned either to solitary confinement or to a control group. Each prisoner then had the frequency of alpha waves in his brain measured three times, on days 1, 4, and 7. You can analyze the data using the SP/RM design structure given in the example, but notice that you can use the Day 1 scores to associate levels of the subjects factor with numbers on a line (R2). Regression methods might also be useful. In particular, a method called analysis of covariance would let you use each person's starting measurement as a baseline value, and to adjust the measurements for days 4 and 7 before comparing groups. Example 8.8 in Section 3 of this chapter gave a very brief informal beginning. Section 3 of Chapter 14 deals with this method in more detail. ■

EXAMPLE 8.15 TYPE A PERSONALITY AND SELECTIVE ATTENTION

The study described in Exercise 6.B.3 used test scores to sort subjects into two groups, which the investigators called Type A and Type B. The purpose of the study was to compare the behavior of these two groups of people.

Notice that the process of defining the two personality types gives up information: each subject starts with a numerical score, but the design only uses the score to sort people into categories. Those who scored in the top 25% on the test got lumped together in one category (A); those in the bottom 25% went into another category (B). (Those in between weren't used as subjects.) The design throws people who score high in the Type A range together with people who score low in the Type A range, and treats all these people as interchangeable (R2). A regression-based analysis would make it possible to use the actual scores instead of just "high" (Type A) and "low" (Type B). ■

Exercise Set D

1. *The defense of undoing.* The two-way Table 8.14 summarizes the results of the study reported in Example S4.10.

TABLE 8.14 Data for Example S4.10

	Stages			
Sessions	1	2	3	4
1–18	3	1	0	0
19–36	6	5	0	0
37–54	0	2	0	0
55–72	0	4	1	1
73–90	2	0	4	0
91–108	0	2	8	5

a. It would be wrong to analyze the data in the table using ANOVA for either a CB design or a two-way factorial with one observation per cell. Which of the boxed principles from Section 4 apply here?

b. Together Review Exercise 6 of Chapter 4 (portacaval shunt) and Example S6.1 (Smith infirmary data) sketch two alternative approaches that might be appropriate for a data set like this one. List them, and tell whether there is any obvious reason to rule out either of them.

2. *Dandelions and habitats.* Reread Exercises 6.B.5 & 6; then answer the following questions about the data which that exercise summarized in a two-way table. (a) ANOVA is not appropriate. Which of the boxed principles apply to the data table? (b) What is the response? Is it nominal, ordinal, or ratio? (c) What are the units for this study? (d) Imagine that you had a complete list of response values, one value for each unit in the study. Would this list give you more information than you get from the summary table? (e) Is the design structure one-way, or two-way with one observation per cell? Name the factors of interest. (f) Consider the various approaches mentioned in Example S6.1 for the analysis of data sets like the Smith infirmary data. Which one or ones can you rule out as clearly inappropriate for the dandelion data?

3. *More dandelions and habitats.* Reread Exercise 6.B.6, and notice that the numbers in the table are summaries.

a. Suppose you had a complete record of the data used to create the data table in that exercise. (i) What is the unit for the study? (ii) What is the response? (iii) Tell the factor or factors of interest. (iv) What is the design?

b. Now consider the data actually given in the exercise. (i) What model (design) best describes the structure of the data in the table? (ii) How is that model different from the one in part a(iv)? (iii) What would you be able to estimate from the complete data set in (a) that you can't estimate from the data in the table?

4 – 6. For each of the data sets listed below, answer both (a) and (b).

a. Choose one:

i. There is just one response. (Describe it.)

ii. There is clearly more than one response. (List them.)

iii. The number of response variables is ambiguous. (List the variables which correspond ambiguously to the multiple responses and levels of a factor.)

b. Now consider how two kinds of analysis might be useful: (A) ANOVA, for comparing averages for groups of observations that get lumped together and treated as interchangeable; and (R) Regression—methods based on fitting lines or curves to scatterplots. Then tell (i) which groups of observations (which factors) you would compare, if any, using ANOVA; (ii) which pairs of numbers you would scatterplot, if any, for fitting linear or curved relation-ships; and (iii) which approach, ANOVA or regression, seems better suited to analyzing the most important features in the data.

4. *Drunken dogs*. In this experiment, six mongrel dogs were given ethyl alcohol intravenously, one gram of alcohol per kilogram of body weight. The table below shows part of the resulting data. Weight is in kilograms; time is in minutes, after the infusion of alcohol; heart rate (HR) is in beats per minute; mean arterial pressure (AP) is in millimeters of mercury; coronary blood flow (CBF) is in milliliters per minute.

Dog	Weight	Time	HR	AP	CBF
1	18	0	157	111	30
1		15	157	96	42
2	16	0	95	121	31
2		15	101	110	47
3	18	0	144	108	31
3		15	135	92	43
4	20	0	109	128	33
4		15	132	129	56
5	21	0	129	97	35
5		15	131	94	59
6	25	0	154	86	36
6		15	150	84	67

5. *Jobs and hearts*. At the time of this study, jobs at the Hawthorne Works of the Western Electric Company carried a physical demand rating from 1 (least demanding) to 5 (most demanding). The purpose of this study was to test the validity of these ratings by comparing all eighteen of the workers at grade 4 with a random sample of eighteen chosen from the workers at grade 2. (In the table, the headings #1 and #2 refer to Test 1 and Test 2 of the same quantity under the same conditions.)

	Physical Grade 2			
	Caloric expend. (k-cal/min)		Ventilation (liters/min)	
Subj.	#1	#2	#1	#2
1	2.69	2.38	14.23	12.32
2	2.01	2.49	11.07	13.86
3	2.21	2.61	10.54	14.62
4	2.54	2.34	13.56	12.77
5	2.58	2.54	15.50	14.88
6	2.83	2.95	12.38	12.57
7	2.59	2.75	14.69	16.69
8	2.70	2.62	13.09	12.62

	Physical Grade 2			
	Caloric expend. (k-cal/min)		Ventilation (liters/min)	
Subj.	#1	#2	#1	#2
9	2.36	2.51	12.66	13.83
10	2.78	2.88	15.50	14.88
11	2.72	2.25	14.72	12.36
12	2.92	2.87	14.94	15.91
13	2.28	2.37	11.97	12.51
14	2.59	2.51	13.59	12.94
15	2.85	2.73	14.88	13.42
16	3.31	3.18	13.04	12.32
17	3.72	4.36	22.22	23.95
18	2.90	2.90	16.23	17.23

	Physical Grade 4			
	Caloric expend. (k-cal/min)		Ventilation (liters/min)	
Subj.	#1	#2	#1	#2
19	5.59	5.16	25.11	23.59
20	4.42	4.73	19.91	22.53
21	2.99	3.79	13.68	19.22
22	5.63	3.82	30.90	20.22
23	4.89	4.44	22.73	21.49
24	5.32	5.42	21.25	25.49
25	4.74	4.86	24.62	26.53
26	3.26	3.48	18.40	19.67
27	4.68	2.46	24.80	13.23
28	3.42	5.60	18.37	26.60
29	3.39	2.67	16.84	15.19
30	4.05	3.40	17.88	18.70
31	3.33	3.47	15.71	15.96
32	4.18	3.23	19.94	16.74
33	3.22	2.89	14.98	13.29
34	4.62	3.95	21.57	18.56
35	6.43	6.81	27.72	34.60
36	2.73	3.43	16.89	20.05

6. *Pesticides in the Wolf River*. The 60 numbers listed below are concentrations (in nanograms per liter, times 100) of two organic pesticides, aldrin and hexachlorobenzine (HCB), for water samples collected from the Wolf River in Tennessee, downstream from a site used for dumping waste from the manufacture of pesticides. The 60 numbers come in pairs, one

concentration for aldrin and one for HCB, from each of 30 water samples. The samples were all collected within a short interval of time, at a single location, 10 samples from each of three depths.

There were two reasons for measuring at different depths. One: The organic compounds do not have the same density (specific gravity) as river water, and so you could expect different concentrations at different depths. Two: Molecules of the compounds can be expected to cling to sediments, which occur in greater concentrations near the bottom.

Surface:	Aldrin	308	358	381	431	435	440	367	512	517	435
	HCB	374	461	400	467	487	512	452	529	574	548
Middle	Aldrin	517	617	626	426	317	376	476	490	657	517
depth:	HCB	603	655	355	459	377	481	585	574	677	564
Bottom:	Aldrin	481	571	490	535	526	626	376	807	879	730
	HCB	544	688	537	544	503	648	389	585	685	716

Scatterplots and informal analyses

7. *Pesticides in the Wolf River*. Scatterplot aldrin versus HCB for the data of Exercise D.6, separately for each of the three depths. Use your scatterplots to answer the following questions. (a) List any outliers. (b) Compare the variability within each of the three groups: For which depth are the values most spread out? Least spread out? (c) Draw an oval balloon to summarize the plot for each of the three groups, omitting any outliers. Would you describe the relationship between aldrin and HCB concentrations as strong or weak? Positive or negative? Linear or non-linear? (d) The investigators expected to find higher concentrations of the pesticides near the river bottom. Do the data tend to confirm those expectations? (e) Fit a line, by eye, to each scatterplot, omitting outliers. Choose one: (i) The fitted lines are roughly the same: same slopes and intercepts. (ii) The three lines do not all coincide. (Tell which ones differ, and in what way.) (f) Consider the following two approaches to the analysis of these data. (i) ANOVA, to compare groups: Regard the design as an SP/RM with samples as blocks, depths as the between-blocks factor, and pesticide (aldrin versus HCB) as the within-blocks factor. (ii) Regression, to fit lines: Fit lines to the scatterplots for each of the three depths; then compare the fitted lines. Discuss the advantages and disadvantages of each: What kind of information can you get from each that you couldn't get from the other? (g) Average concentrations for the six rows of the data set are given below. Draw an interaction graph and describe the pattern.

	Surface	Middle	Bottom
Aldrin	419	502	602
HCB	480	533	584

(h) Turn back to the data and compare the six rows of numbers. Do average and spread seem related? (If so, a transformation is worth considering.)

8. *Drunken dogs*. Do an informal analysis of the data given in Exercise D.4.

9. *Jobs and hearts*. Do an informal analysis of the data given in Exercise D.5.

SUMMARY

1. *Three preliminary questions to ask when you analyze a design:*

   a. What kind of conditions are being compared? (i) Treatments, which get assigned, (ii) Populations, from which samples are taken, or (iii) Characteristics, which correspond to multiple measurements.

   b. What are the units? (i) Individual subjects or objects, (ii) Groups of subjects of objects, or (iii) Parts or time slots for subjects or objects.

   c. Are there blocks? (i) Groups of subjects or objects (unit = individual), or (ii) Individual subjects or objects (unit = part or time slot).

2. *Deciding whether to use blocks.*

   a. If your blocks would come from **sorting** the available units, there are two main considerations:

   - Will sorting help isolate the effects of interest?
   - Do you have enough units to shift df from residuals to blocks?

   b. If you can **reuse** or **subdivide** subjects or material, blocking gives more units for the same number of subjects or objects. In principle, blocking of this sort is a good idea, but there are two main clusters of possible drawbacks.

   - Time and effort. It would take too long to reuse the material.
   - Linkage and leakage. Parts or time slots are not sufficiently independent.

   c. General considerations.

   - Interaction. Don't use a CB design if there's likely to be a block-by-treatment interaction.
   - Flexibility. BF designs are more flexible than designs with blocks: You can have treatment groups of any size.

3. Three-part checklist for summarizing designs.

   A. Purpose and response.

   B. Design structure.

      1. Units: How many kinds (sizes) of units are there?

      2. Grouping: For each kind of unit, what is the structure that relates treatments to units? (Are the units grouped or arranged?)

      3. What is the design family?

   C. Treatment structure.

   For each kind of unit, what are the treatments or conditions? Is the factor of interest compound? How are the conditions associated with the units: by assigning, selecting, or defining? Was the assigning or selecting randomized?

4. Once you have chosen your response(s) and factors of interest, finding a design structure becomes mainly a matter of identifying nuisance influences and deciding how to deal with them, either by random assignment or by blocking, to make the nuisance influence a structural factor.

   BF: The only nuisance factor corresponds to the units themselves (and thus to chance error). This factor is nested within the factor of interest.

   CB: The structural nuisance factor (blocks) is crossed with the factor of interest.

   SP/RM: The structural nuisance factor (blocks) is nested within the between-blocks factor, and crossed with the within-blocks factor.

   LS: Each of the two structural nuisance factors (rows and columns) is crossed with the factor of interest.

5. A WARNING ABOUT INFORMAL ANALYSIS: You can nearly always find patterns in a bunch of numbers, even if the numbers are purely random. It is dangerous to use a data set to test any hypothesis that grows out of your informal analysis of the same data set. If the hypothesis is important, you should design a new experiment to test it.

6. Three common errors in choosing a model:

M. Mistaking multiple measurements for levels of a factor. (M1) If you have measurements on different scales, treat each as a different response. (Don't average grams with degrees with milliliters per minute.) (M2) If you have measurements on the same scale, but corresponding to different characteristics of subjects or objects, think about what you want to learn from your analysis:

- If you want to use all the measurements jointly to compare groups of subjects or objects, you need a multivariate model.
- If you want to compare the average values for the various characteristics, try ANOVA with the multiple measurements as levels of a factor, but be sure to check the Fisher assumptions.

C. Mistaking categories of a nominal response for levels of a factor. (C1) If you can get the data set simply by sorting things into groups, then counting the number in each group, an ANOVA model is not likely to be appropriate. (C2) Don't use an ANOVA model if its factor structure would confound an important interaction with residual error. (C3) Don't use an ANOVA model if any factor averages would be fixed by the design.

R. Regression or ANOVA? (R1) What is the goal of your analyses? If you have multiple measurements on the same subject or object, don't treat them as levels of a factor if some of the measurements should be treated as carrier variables. (R2) If there is a natural way to associate a numerical carrier variable with levels of a factor, or with units that are grouped into levels of a factor, regression analysis may be more informative than ANOVA.

EXERCISE SET E: REVIEW EXERCISES

1. *Alpha waves and solitary confinement*. A partial ANOVA table for the data of Examples 8.4 and 8.10 appears below. Complete the table, and discuss your results; be sure to keep in mind the informal analysis of Example 8.8.

Source	SS	df	MS	F-ratio	crit.val.
Grand Ave.	11,426.40				
Solitary	248.07				
Subjects	1,610.20				
Days	256.90				
Solitary × Days	260.43				
Residual	867.40				
TOTAL	13,904.00				

2. *Shop stewards*. To what extent does our behavior determine our attitudes? In this study from social psychology, Lieberman compared industrial workers who were promoted either to foreman or to shop steward; he used a questionnaire to measure the percent of issues on which they agreed with the management position. The attitudes were measured at the time of the promotion, and showed no difference between the two groups: both groups averaged about 40% pro-management. A year later, the workers promoted to foreman (a company position)

were up to 70% pro-management; those promoted to shop steward (a union position) were down to 20%. Two years later, the gap was even wider.

Write a summary of the design, following the format from Section 1.

3. *The effect of marijuana smoking on heart rate*. Design an experiment to see whether smoking a marijuana cigarette changes a person's heart rate. (a) Tell what you would use as your response, and how you would choose your subjects. (b) What comparison would you make? Specifically, what would be your control condition? (c) Tell what units to use, and how you would assign conditions to units. (d) Draw and label a rectangular format showing the structure of the data from your experiment.

4. *Pine seeds*. It is a reasonable hypothesis that pine trees living in one environment will be more suited to that environment than will pine trees living in a different environment, and that, moreover, these differences may be passed from one generation to the next. This old study took seeds from four different sources: Louisiana, Texas, Georgia, and Arkansas. Six blocks of land in Louisiana were each divided into four plots, and each plot was planted with seeds from one of the four sources. Fifteen years later, the average height (feet) was computed for the trees in each plot. (a) Summarize the design, following the format of Section 1. (b) Do an informal analysis of the data.

	Block No.					
Seed Source	1	2	3	4	5	6
Louisiana	34.0	29.3	30.6	31.8	34.0	32.7
Texas	27.3	27.6	28.6	29.2	30.2	31.5
Georgia	26.4	25.0	26.6	25.2	27.4	26.2
Arkansas	24.8	24.3	26.0	26.5	25.8	24.2

5. *Extinguished rats*. To study the effect of darkness on the learning behavior of rats, twenty rats were first taught to press a bar in order to get pellets of food. After initial training, once the rat had learned that it got a pellet each time it pressed the bar, all twenty rats were tested. The numbers below, in the Before columns, give the number of bar presses per minute during this first test period.

TABLE 8.15 Bar presses per minute

	Light			Dark	
Rat	Before	After	Rat	Before	After
I	14.87	7.43	XI	15.63	4.80
II	10.10	7.63	XII	4.50	3.43
III	5.80	2.50	XIII	7.43	5.83
IV	12.06	5.00	XIV	17.20	3.87
V	11.03	10.90	XV	10.33	2.20
VI	18.10	12.60	XVI	9.47	3.96
VII	5.73	3.43	XVII	11.50	4.60
VIII	10.63	3.70	XVIII	9.70	2.40
IX	12.93	8.20	XIX	14.96	4.70
X	10.87	5.50	XX	13.96	4.93

After the first test period, all twenty rats were subjected to "extinction of learned behavior": they no longer got rewarded with pellets when they pressed the bar. Ten of the twenty rats had been randomly assigned to the dark condition. They were kept in total darkness during the extinction phase of the experiment. The other ten rats were kept in conditions that were the same except that they had normal lighting during the extinction phase. After extinction, all twenty rats were tested a second time. The numbers in the After columns give the number of bar presses per minute during the second test period.

(a) Summarize the design, following the format of Section 1. (b) Do an informal analysis of the data.

6. *Cool mice #2.* The data set in Table 8.16 is taken from a paper published in the *Canadian Journal of Zoology* more than 25 years ago, before scientists had set up committees to judge whether a proposed experiment was ethical. I find the experiment itself highly objectionable—notice how politely the author's table below tells us that he chilled the mice to the point of death—because I don't think the practical applications of the results are important enough to justify what was done to the mice. Nevertheless, the experiment was done, the data have been published, and it's worth trying to see what the results have to tell us.

TABLE 8.16 Lethal body temperature (°C)

	White mice		Deer mice	
	Colonic	Average	Colonic	Average
	8.2	10.8	10.9	10.5
	12.2	14.4	11.5	12.1
	8.9	10.1	10.0	12.6
	12.0	13.2	11.7	14.7
	14.5	14.1	13.9	11.9
	11.5	14.3	13.5	10.9
	8.0	11.5	0.6	8.7
	5.0	9.3	6.0	7.3
			9.2	9.7
			6.0	7.8
			8.0	9.1
Average	10.0	12.2	9.2	10.5

The purpose of the experiment was to compare two methods for measuring the body temperature of animals. One ("Colonic") is the standard method; the other ("Average") is based on how much the temperature of a water bath increases when the animal is put in the water.

The data set is unbalanced, but in all other respects its structure is one you have seen many times before. (a) Describe the experimental design: Tell the response or responses, the units, factors of interest, and nuisance factors. For each basic factor of interest, tell whether it is experimental or observational. What is the name of the design? (b) Do an informal analysis of the data. (c) The author writes "Results . . . on lethal colonic and average temperatures . . . show a distinct tendency for average temperatures to be higher than colonic temperatures." Evaluate this claim in light of your informal analysis.

Appendix: Supplementary Examples

Comparing and Recognizing Design Structures

Example S8.1a Perch Eggs

For this experiment 120 batches of white perch eggs were randomly assigned to one of 5 salt concentrations and one of 8 temperatures, with 3 batches assigned to each of the 40 possible combinations. The purpose of the experiment was to study the effect of these conditions on hatch rate, and to find the most favorable combination of salinity and temperature. Summarize the design.

A. *Purpose and response(s)*: To compare the effects of temperature and salinity on the hatch rate (response) of perch eggs.

B. *Design structure*

1. Units: Each unit is a batch of eggs. There is only one size.

2. Grouping: The 120 batches are not grouped or arranged in any way before the treatment combinations get randomly assigned. (It is crucial to the design that the treatment combinations are randomly assigned to individual batches. See Example 8.1b below.)

3. Design family: CR.

C. *Treatment structure*. There are 40 different combinations of the two treatments temperature and salinity. The design is a CR[2].

TABLE S8 Summary Table of Factors: Perch eggs—a CR[2].

	Grouping	Treatment combinations	
Unit	**(Nuisance factors)**	**Main Effects**	**2-Factor Interaction**
Batch of eggs (120 in all)	(none)	Temp (8) Salinity (5)	Temp × Salinity

One reasonable rectangular format for the would have 24 rows in 8 sets of 3, one set of rows for each temperature. There would be 5 columns, one for each salt concentration. Thus you would have 40 cells, with 3 observations per cell. A factor diagram would have five boxes, one for each of benchmark, temperature, salinity, interaction, and residual error. ■

EXAMPLE 8.1B PERCH EGGS, VERSION 2

Consider another version of the experiment: Suppose that the batches of eggs were kept in temperature-controlled chambers, 5 batches to a chamber. Then there would be units of two sizes. The chambers (sets of 5 batches) would be the larger units, with temperatures assigned to chambers; the batches would be the smaller units, grouped into blocks of 5, with salt concentrations randomly assigned to batches within each block. The design would then be an SP/RM (Table S9).

TABLE S9 Summary table of factors: Perch eggs—an SP/RM[1;1].

Unit	Nuisance factors)	Main Effects	2-Factor Interaction
Chamber (24 in all)	(none)	Temp (8)	
Batch of eggs (5 per chamber)	Chambers	Salinity (5)	Temp × Salinity

The list has two rows, one for each kind of unit. The first row lists the larger units and the between-blocks factor. The second row lists the smaller units, the within-blocks factor, and its interaction with the between-blocks factor.

You could use the same rectangular format as in Example 8.1a for the data, but whereas before rows of the data table would have no meaning, for this version of the experiment rows correspond to blocks, and are nested within temperatures. ■

EXAMPLE S8.2 CATS AND TETANUS

The purpose of this experiment was to measure the effect of tetanus on the level of the amino acid glycine in the spinal chords of cats. Each of eleven cats was given tetanus in its right side only; the left side served as a control. For each side, the glycine levels were measured separately for the white and gray matter. Summarize the design structure.

A. *Purpose and response(s)*: The purpose is given in the description, and is straightforward. There's a question of whether to think of the data as having one response, glycine concentration, or two: glycine concentration in the white matter, and in the gray matter. Section 5 discusses issues like this one; for now I'll treat the data as though there were only one response, the glycine level.

B. *Design structure*

1. Units: This experiment has units of two different sizes. (The larger unit is the side of a cat, and the smaller unit is a chunk of matter.) Although there are two sizes of units, the design does not belong to the simplest SP/RM family described in Chapter 7, because the larger units get assigned as in a CB.

2. Grouping:

a. The larger unit is a side of a cat; these are the units that get the treatments, tetanus or control. The larger units come neatly grouped in pairs: two sides per cat; a cat is like a block of larger units. Treatments get assigned separately within each cat.

b. The smaller unit is a chunk of matter, gray or white.

3. Design family: A variation on the basic SP/RM. (See Chapter 13.)

C. *Treatment structure*: There are two basic factors of interest. Tetanus versus control gets assigned to the larger units, but note, however, that there is no randomization. Tetanus versus control is completely confounded with right versus left side, although the confounding does not strike me as likely to be a big problem. The second basic factor, color, is observational. The chunks of matter get defined and classified as either white or gray. No randomization is possible. ■

Choosing a Design Structure

EXAMPLE S8.3 PATCHING UP THE SAT STUDY

Examples 4.9 and S7.1 both dealt with how you could evaluate a course that claimed to raise SAT scores, but neither of the designs considered in those examples was a good one. Both the one-way and the complete block plans had problems with confounding. Here's a design which is much better, although not without problems of its own.

Response/conditions/material: In a sense these are all straightforward. The response is combined SAT score, verbal plus quantitative. The conditions to be compared are special course versus no course. The subjects are high school students. The main challenge in planning a good study is in finding a way to isolate the effects of interest.

Nuisance influences: There are two major clusters of nuisance influences, differences among subjects, and the effects of time and practice. These are what undermined the two previous plans.

The one-way design (Example 4.9) simply compared scores for two groups of students, those who took the course, and those who didn't. The selection bias built into this plan confounded the effect of the course with a variety of nuisance influences, such as level of motivation and socio-economic status of the students.

The complete block design (Example S7.1) avoided that kind of confounding by using students as their own controls, comparing SAT scores before and after the course. But for this plan, the confounding was even worse: the effect of the course couldn't be separated from the practice effect.

For both designs the fatal flaw came from making the factor of interest observational rather than experimental. To plan a sound study, you have to find a way to assign conditions using a chance device. Whatever the design, it must contend with the way SAT scores vary from one person to the next. If we regard subjects as a nuisance factor, then we must decide whether to make the factor of interest a between-subjects or within-subjects factor.*Relation to factor of interest*: It would be possible to avoid the confounding using a CR design, which would make the course a between-subjects factor. For example, you might recruit a large number of volunteers who wanted to take the course, and randomly assign half of them to take the course, then take the SATs; the other subjects would take the SATs without the special course. (At this point the experiment would end, but to be fair to the control subjects, you could offer them the special course and a second shot at the SATs, even though you wouldn't need the second set of scores for your study.)

It would be almost as easy to use an SP/RM design, making the course a within-subjects factor. Here, as for the CR, you would take as subjects a set of volunteers who wanted the course, and you would randomly assign half to take the course, half to serve as controls. Both groups would take the SATs to start off. Then the subjects in the treatment group would get the course, then both groups would take the SATs a second time. (Finally, subjects in the control group would be offered the chance to take the course and retake the SATs, although these last scores would not be part of the SP/RM design.)
Design structure: The SP/RM design is much better, because SAT scores vary so much from one person to the next. The CR is less efficient, and would need many more subjects than the SP/RM. (There's still a problem with the units, however: see the discussion following Example S8.4.) ■

Example S8.4 Thermal Pollution

Many people, when they think of pollution, think of toxic chemicals and radioactivity, but environmental biologists know that heat can also be a form of pollution. Living organisms can be sensitive to changes in the temperature of their surroundings; in particular, organisms that live in or near water can be hurt when a new industry that uses water for cooling starts dumping its warm water back into the river or ocean it came from. Designing a study to measure the environmental impact of thermal pollution is not as easy as you might think.

In early 1964 the Hunterston Generating Station, a nuclear power plant located on the north Ayrshire coast of Great Britain, began discharging warm water into the North Sea at a rate of roughly twenty million gallons per hour. Suppose you had known well in advance that the discharge was going to begin, and you wanted to design a study to measure the effects of the thermal pollution on nearby sandy beaches.

Response: There are several choices for response variables. You would certainly want to measure sand temperature; other choices depend on knowing the local biology. Two common inhabitants of that area happen to be *Tellina tenuris* (a mollusk), and *Urothoe brevicornis* (a burrowing sand flea). For the study in question, the investigator chose to measure the density of the mollusk (in thousands per square meter) and the percent gravid (pregnant) among the female sand fleas, although there are a lot of other reasonable choices that could have been used as well.
Conditions: The factor of interest comes from the power plant, before and after.
Nuisance influences: For environmental studies of this sort, two sets of nuisance influences are usually quite marked in their effects: time, and place. The density of a given species, for example, typically shows large fluctuations from one year to the next. This means that you shouldn't rely on before and after measurements at one site only. Suppose, as was in fact the case, that the density of mollusks near the Hunterston plant dropped sharply after the plant began discharging warm water. The drop might be due to the plant, but might just be part of the usual year-to-year variation. One way to deal with this problem is to pick another site or two, not far from the site in question, but far enough away not to be much affected by the plant. For this example, you could use the Millport Marine Station, two miles from Hunterston, as a second site. The idea here is that the two sites are likely to show similar year-to-year variations, except for the effect of the discharge, which would affect Hunterston more than Millport. If the effect of time is the same for both sites (no interaction), you can conclude that the plant has not affected your chosen response variable.

The other nuisance influence to worry about is tied to place: the density of the mollusks will vary from one place to another along the same stretch of beach.
Nuisance factors: There are various strategies for handling the nuisance influences of time and place, but I will focus here on one of them. We can handle the influence of time by making site (Hunterston versus Millport) a factor of interest in the design, crossed with the other factor of interest, before versus after. To handle the local variability, we could sample three places at each of the two sites, making place a nuisance factor.
Design structure: The design I have described is an observational SP/RM, with the six places as whole plots. The two sites are levels of the between-blocks factor, and the two years, before and after, are levels of the within-blocks factor.

The study this example is based on had to rely much more on what data were already available, and the comparisons were not as systematic as would be possible with the design I have suggested here. The investigator concluded, nevertheless, that the thermal pollution had not in fact had an effect that was measurable by his methods. ■

A question about units and confounding. If you think carefully about the treatments and units for the SAT study and the study of thermal pollution, you may recognize the potential weaknesses in both designs.

Critic: For the SAT study, the between-blocks treatment is the special course, which gets taught to an entire class. The unit is a class, not a student. For the pollution study, the between-blocks factor is the power plant, which is present or absent at the what you called a site, Hunterston or Millport. The unit is an entire site, not the smaller unit that you called a place. For each of these designs, you have only two whole plots, one for each level of the between-blocks factor. With only one unit for each condition, you have no way to estimate the size of whole plot error. These designs are unanalyzable.

If you prefer, you could think of the problem in terms of confounding. For your SAT plan, the effects of the course would be confounded with the effects of the teacher and the one class. For your pollution plan, the effects of the power plant would be confounded with the effects of the sites.

Reply: Your point is well taken. I agree that in each case it is possible to improve the design, at least in principle. For the pollution study, it would be better to have several sites, not just one of each kind; for the SAT study, it would be better to have several classes.

But I don't agree that the designs given in the two examples are necessarily worthless. For one thing, although the special course gets taught to the entire class, the students assigned to that group were randomly assigned as individuals. It's not as though I assigned treatments to clusters of people. Also, a lot depends on how much variability there is. I'll describe two opposite extremes for each study.

The SAT study:

Version 1. Suppose there are large class-to-class differences, either differences between groups of students, or differences between teachers. Then I agree—it would be a bad mistake to use a design that ignores these differences, or to assume that the results from one class are typical of the special course in general. You could still analyze the design as an SP/RM, but your conclusion would apply only to the one class you happened to study.

Version 2. Suppose the class-to-class differences aren't much bigger than the student-to-student differences. Then a formal analysis of data from your fancier design with several classes would end up telling you to ignore class-to-class differences, and to use the chance error from student-to-student differences as a yardstick for judging the effect of the course. In other words, you'd end up analyzing the data as though the design were an SP/RM.

The pollution study:

Version 1. If there are large site-to-site differences, then the design I have proposed gives you no way to tell whether differences between Millport and Hunterston are due to the power plant, or are just an instance of site-to-site variability.

Version 2. If the site-to-site differences aren't much bigger than the differences between places within sites, then the design is OK, just as for Version 2 of the SAT study.

Critic: I agree that *if* the true situation is in fact like your Version 2, then the design is acceptable. But your design won't give you the data you need to tell.

Reply: You're right. That's why I agree that designs with more classes or more sites would be better. Unfortunately, though, reality sometimes forces us to substitute assumptions for data. For the SAT study, I know that student-to-student differences are large, and if I can't get data from more than one class, I'm willing to assume that class-to-class differences may not be much bigger. Of course I recognize that the soundness of my conclusions will depend on how well my assumption fits the facts. In particular, if I find a "real" difference between the treatment and control groups, I should realize that the difference might be due to the content of the special course, but might be due more to the teacher than to the course content.

For the pollution study, a lot depends on how far apart "sites" and "places" are. Both terms are a bit arbitrary, in that "places" chosen far enough apart are like "sites." If the places are well spread out, I think the design would be reasonable; but to judge how far apart is far enough, you'd have to know more environmental biology than I do.

Informal Analysis: Examples

EXAMPLE S8.5 FEEDING FROGS

Table S10. shows the data for the experiment described in Example 8.5. The purpose of the experiment was to see whether the amount a frog eats is influenced by having one or more other frogs present.

TABLE S10

	Condition		
Frog	Alone	Paired	Grouped
101	56	32	106
102	50	73	167
103	113	68	157
104	0	36	191
105	55	39	81
106	50	31	98
107	69	64	110
108	0	62	19
109	25	28	41
110	5	12	21
111	12	7	68
112	6	32	140

Total units of food eaten, during observation periods of 200 minutes per condition.

I began with a parallel dot graph, and three condition averages (Figure S7).

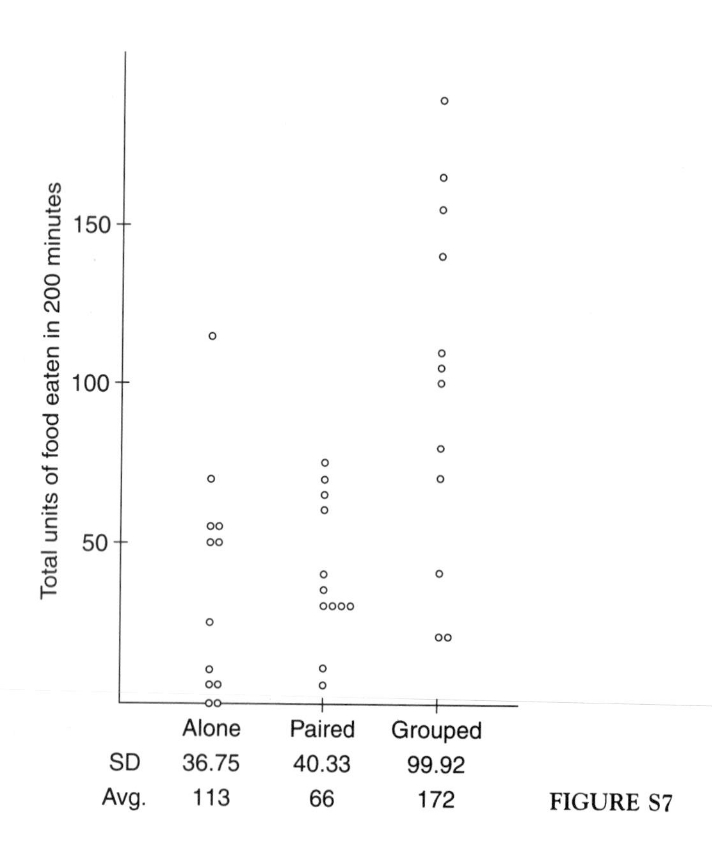

FIGURE S7

1. *Groups*: On average, the frogs ate a lot less when alone (37 units) or paired (40 units) than when grouped with eleven other frogs (100 units). Alone and paired are about the same, although on average the frogs ate about 10% more with another frog present than they ate when alone.
2. *Outliers*: Frog #103 ate a lot more when alone than any of the other frogs did, but other than this, the graph reveals no outliers. If you leave out the 113 units eaten by frog #103 when alone, then the range for alone drops from 113 to 69, and the average drops from 37 to 30. With the 113 omitted, the pattern of the averages is more pronounced: alone—30 units; paired—40 units; grouped—100 units.
3. *Spreads*: There was a lot more variation when the frogs were grouped than in either of the other two conditions. If you leave out #103, the spreads for alone and paired are about the same; with #103 included, alone is a bit more spread out.
4. *Transformation*: The two conditions with low averages also have much smaller spreads, in comparison to grouped. Since the averages and spreads seem related, transforming is probably worth a try. Moreover, Frog #103 is not so far from his peers that I'd want to judge him an exception. Perhaps he was just a bit hun-

grier. Transforming to logs or square roots would bring him more in line with the other eleven.

Next, before trying a transformation, I computed averages by rows, one average per frog. My goal here was to check my impression, just from looking at the table of numbers, that no frog stood out as eating a lot less or a lot more overall. I rounded off these averages to the nearest 10 units of food, and plotted them to see the pattern (Figure S8).

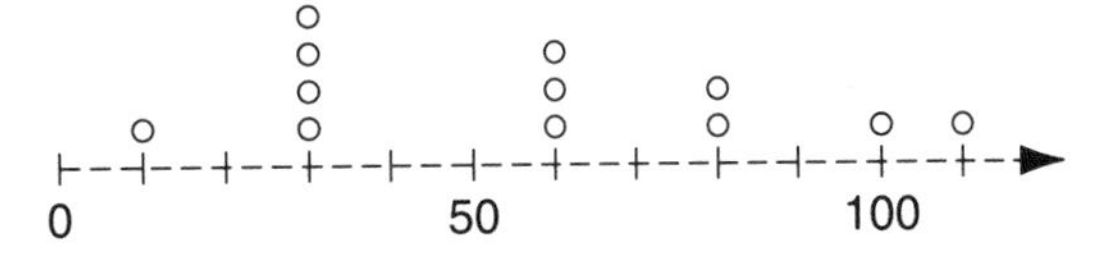

FIGURE S8 Dot graph for the frog (row) averages. There are no odd frogs, if you judge by these averages.

I wondered whether you could use a frog's eating behavior in one condition to predict how much he'd eat in the other two conditions. In other words, I wondered if a scatterplot with frogs as points would show a strong positive relationship between the amounts eaten alone and paired, alone and grouped, or paired and grouped (Figure S9).

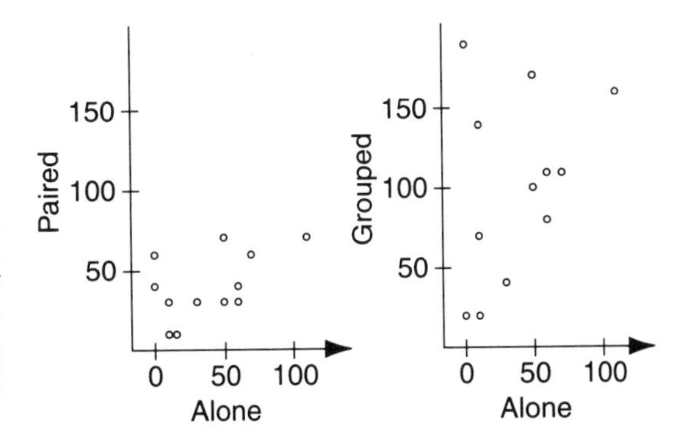

FIGURE S9 Two scatterplots: Paired vs. Alone, and Grouped vs. Alone. Each point corresponds to a frog, and the plots are done in the same way as the within-blocks plots of Chapter 7, Section 5. Both scatterplots show only a mild positive relationship: knowing how much a frog ate while alone doesn't tell you a lot about the other two conditions.

Noisier measurements dominate the average. At this point, I'm going to interrupt my informal analysis long enough to show you another one of the reasons for considering a transformation. As you know, analysis of variance is based on computing averages, and for these data, one meaningful set of averages corresponds to frogs. When you compute such averages, the idea is that each frog's average tells you what is typical for that frog. However, because the numbers for Grouped are so much more spread out than the numbers for Alone or Paired, differences between frog averages will tell you mostly about how the frogs differed while grouped. This is an example of a general fact about averages and related statistical methods, in the same spirit as "the squeaky wheel gets the grease."

The more variable the measurement, the bigger its influence on the average.

To see this in the context of the frog data, compare the two scatterplots in Figure S10.

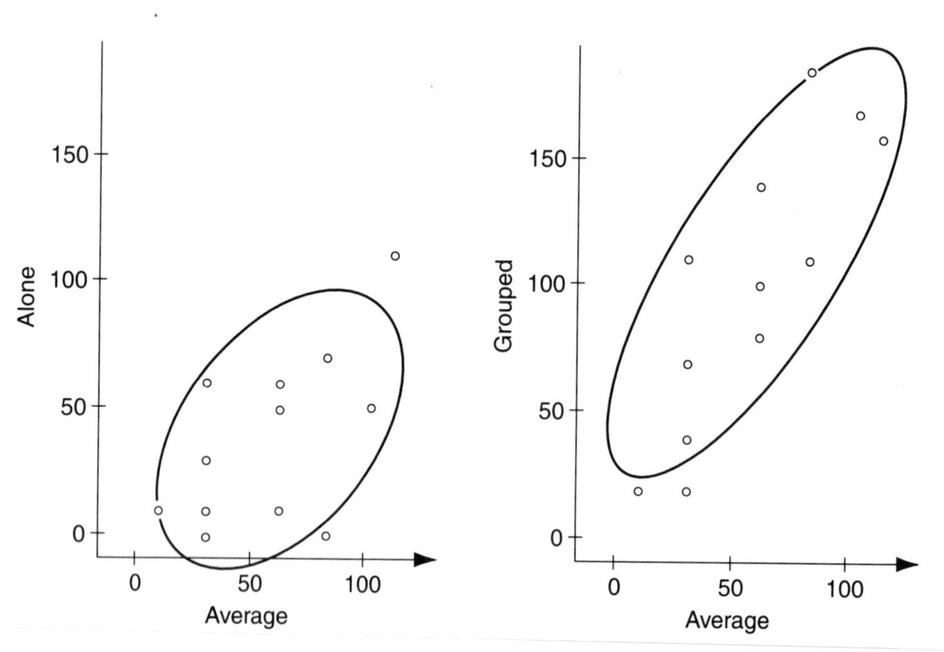

FIGURE S10 Alone versus frog average and Grouped versus frog average. Each point is a frog, just as in the last two scatterplots; for both of these the *x*-value is the average amount eaten by the frog. Notice how the measurement with low variability (Alone) shows a weak positive relationship with the frog average, while the measurement with high variability (Grouped) shows a much stronger positive relationship: the noisier measurement dominates the average. This illustrates another reason for transforming to equalize variability.

For a variety of reasons, then, a transformation seems worth a try, but I wasn't prepared for where it led me. I thought the main message from the data was already clear: on average the frogs ate about 10% more with one other frog present than they ate when alone, and ate almost three times as much when there were eleven other frogs present. I knew transforming wouldn't change that general pattern, and I knew it was more intuitively appetizing to think about how much a frog eats in the original scale than to think about eating roots or logs, so I was half hoping that after trying the transformation, I could forget about it and go back to the original data.

Nevertheless, I decided to try the square root transformation, multiplying by 10 and rounding off to avoid fractions. The dot graph for the transformed data shows the spreads to be more nearly equal this time around, suggesting that ANOVA would be more appropriate for the transformed data than for the original. In the new scale frog #103 no longer looks like such a piggy eater when alone. As expected, the pattern of averages is essentially unchanged (Figure S11).

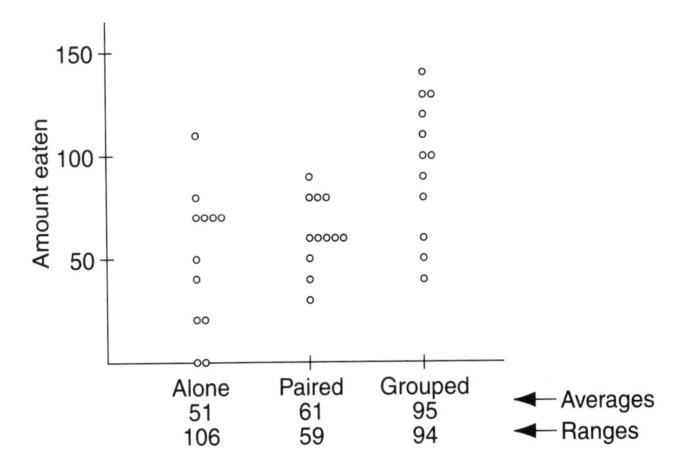

FIGURE S11 Parallel dot graph for the transformed frog data.

I next decided to scatterplot the transformed data, and that's when I got a surprise. In hindsight it's clear that if I'd been more alert, I could have found what I did without doing the transformation first, but as it actually happened, I didn't see it until after I transformed.

Figure S12 shows two scatterplots of Grouped versus Paired. The one on the left uses the original data, the one on the right uses the square roots.

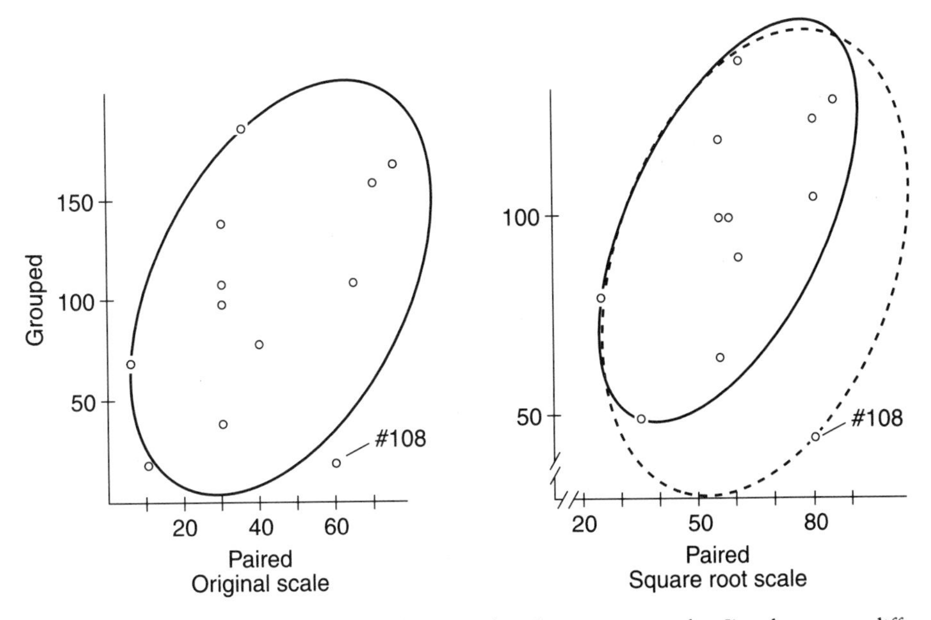

FIGURE S12 Grouped versus Paired in the original and square root scales. I've drawn two different balloon summaries for the right-hand scatterplot, to represent two very different ways to think about the data. The larger of the two balloons pretty much matches what I saw originally in the left-hand scatterplot, and suggests only a weak positive relationship. The smaller balloon not only suggests a stronger positive relationship, it also directs attention to frog #108 as an outlier.

What makes #108 stand out is not so much the *amounts* he ate, but *when* he ate those amounts. He ate no food when alone, but so what: #104 also ate nothing. He ate a lot when paired, but then #102 ate even more. He ate very little when grouped, but so did #110. What sets #108 apart is that he was the *only* frog to eat more when paired than when part of the crowd of twelve (Figure S13).

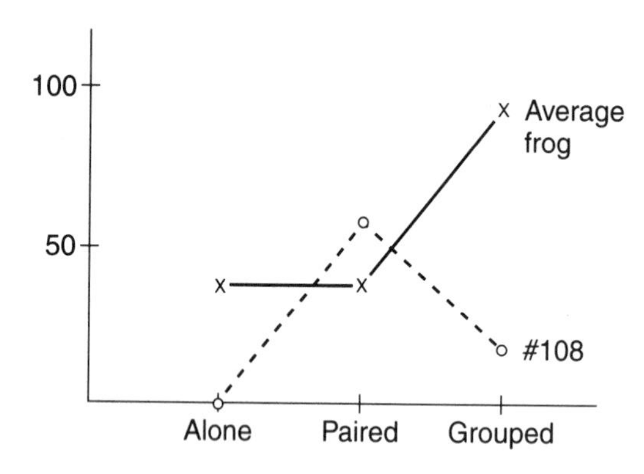

FIGURE S13 Units of food eaten by frog #108 and by the average frog. Frog #108 ate a lot more than average when paired, and a lot less in the other two conditions. All the other frogs ate more when grouped than when paired.

With this pattern now fixed in my in mind, I thought it would be worth going back to look again at each of the other frogs, checking for patterns. What I'm about to show you is the result of my rummaging around in the data, and you should read what follows with an important warning in mind.

WARNING: You can nearly always find patterns in a bunch of numbers, even if the numbers are purely random. It is dangerous to use a data set to test any hypothesis that grows out of your informal analysis of the same data set. If the hypothesis is important, you should design a new experiment to test it.

What I am about to show you is really a bunch of guesses based on patterns that might be nothing more than coincidence. I'm inclined to think the guesses make sense, of course, or I wouldn't waste your time on them, but to test them, *you would have to design a new experiment.*

When I went back to look at the frogs one by one, I ended up thinking of them sorted into five groups (Table S11, Figure S14).

Where does all this leave us? I suggest that it leaves us needing a new experiment. The one we've been looking at is fine for checking out the overall pattern, that frogs tend to eat a lot more when there are eleven other frogs drooling over the same food, but to me, the data suggest there may be more subtle questions of frog sociology involved here. Does how much you eat when paired depend on who you're paired with? Is there a frog hierarchy, with overbearing frogs and deferential frogs? Are the "public gluttons" eating a lot just to assert themselves? Or are they gobbling frantically because they're afraid that if they don't, the other frogs will get it all?

TABLE S11

Group	Frog	Alone	Paired	Grouped
		Condition		
The weirdo (Only eats tête-â-tête)	#108	0	62	19
Picky eaters (Never eat much, but more in a crowd)	#109	25	28	41
	#110	5	12	21
	#111	12	7	68
Moderate eaters (Never picky, but they, too, eat more in a crowd)	#101	56	32	106
	#105	55	39	81
	#106	50	31	98
	#107	69	64	110
Gluttony loves company (Modest in private, but in a crowd, look out)	#102	50	73	167
	#104	0	36	191
	#112	6	32	140
Can't say no (Never turns down a free lunch)	#103	113	68	157

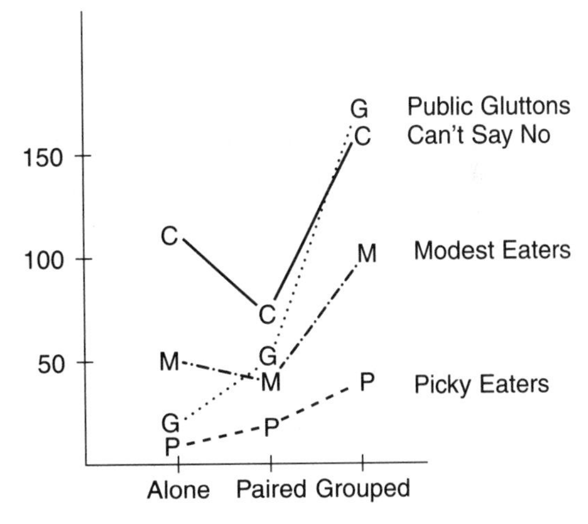

FIGURE S14 Feeding patterns for four of the five groups of frogs. The numbers plotted are the average amounts eaten in each of the three conditions for each of the groups, omitting frog # 108, the outlier, whose pattern was shown in the last graph. All but #108 ate more when grouped than when paired. ■

Or are all these patterns just a numerical shadow, full of random noise, and signifying nothing? It certainly could be that there is simply a lot of variability in the eating behavior of frogs, and the big differences among the five groups are just a result of chance variation. There is no way to tell from this data set whether or not the patterns are meaningful.

CHAPTER 9

EXTENDING THE BASIC DESIGNS BY FACTORIAL CROSSING

Everything in this chapter grows out of a single idea, that you can create new designs from any of the basic four by taking any factor of interest and replacing it with the set of all combinations of levels of two basic factors. If, for example, you do this for the BF[1] design, you get the two-way factorial BF[2] design of Chapter 6. If you then replace one of the factors in the two-way design with combinations of levels of two factors, you get a new design, the three-way factorial BF[3]. Extending the basic designs in this way generates four families of designs, one family for each of the four basic structures.

Designs with three or more factors of interest are quite common both in biology and in psychology. For example, any home gardener can tell you that the health of a plant depends on three essential nutrients, nitrogen, potassium, and phosphorus. To find the best combination of these three nutrients, you might run a three-way factorial experiment, with each nutrient present at each of several concentrations. In a psychology experiment, you might take subjects classified by age and sex (two factors), and measure aspects of their response to several conditions (a third factor).

1. Extending the BF Design: General Principles

This first section does not introduce any new designs; you've already seen several examples of the two-way BF design in Chapter 6. The purpose here is to use a review of that design to illustrate a few general principles that we will apply in the following sections to get new designs.

Extending designs by factorial crossing raises four questions, two about design, and two about analysis.

Design:

1. How do you decide whether to build additional factors into your design?
2. How do you carry out the plan?

Analysis:

3. How do you find the factor structure?
4. How is the informal analysis different from the analysis of the simplest BF, CB, LS, or SP/RM?

Question 1. How Do You Choose the Factors to Cross?

a. Sometimes the purpose of the experiment is to study a particular interaction: the purpose tells the factors.
b. Sometimes you can find additional factors by asking what else might influence your response, or might interact with a factor that is already part of your design.

Example 9.1 Choosing the Factors to Cross

a. *Purpose is to measure interaction.*

i. *Buttercups.* The plant ecologist who planned the experiment described in Example 6.5 began with the hypothesis that buttercups growing in a particular environment (either damp woods, or a sunny field) would adapt to that environment in ways that would make them less successful in the other environment. To test this hypothesis, she had to cross two factors of interest: the original environment, and the new one.

ii. *Classifying habitats* (Exercise 6.B.4). If your goal is to compare a moist habitat with a drier one, and your hypothesis is that seed size matters in the dry habitat (size-beneficial) but not in the moist one (size-neutral), then your experiment has to include both habitat and seed size as factors in the design.

iii. *Kosslyn's imagery experiment* (Example 6.10). Kosslyn designed his experiment to test a particular hypothesis about psychological development: that young children rely much more on images than adults do. This hypothesis leads to a specific prediction, that children answering questions with and without instructions to use a mental image will show a smaller difference in response time than adults will.

To see whether that prediction is correct, you have to have at least two factors of interest: age, and instructions (imagery or not).

b. *Looking for other influences.*

i. *Pigs and vitamins.* The design for Example 6.1 might have started with just one factor of interest (B12) and a goal of measuring the effect of B12 supplements on how fast pigs gain weight. Knowing the biology of a pig's digestion might alert you to the value of including a second factor, antibiotics (present or absent), in your design.

ii. *Kosslyn's imagery experiment.* Remind yourself of the basic comparison in Kosslyn's experiment. He asked his subjects questions like "Does a cat have claws?" and compared how long it took them to answer under two conditions, no imagery instructions (subjects were simply told to answer as fast as they could) and imagery instructions (subjects were told to inspect a mental image). The kind of instruction was one factor of interest, but Kosslyn soon recognized another influence on the response time, and decided to make it another factor of interest. Compare the following two questions:

Does a cat have claws?

Does a cat have a head?

Kosslyn called the first question "High association/low area": we tend to associate claws with cats (high association) and claws are small (low area). The second question is low association/high area: "head" isn't likely to make you think of "cat" the way "claws" might, and a cat's head is comparatively large. Kosslyn thought that questions like the first one are the kind we would more naturally answer just from the words and their meaning, without calling up an image; questions like the second one would not be as easy to answer in the same way. This reasoning led Kosslyn to include the two kinds of questions as a third factor in his design. ■

(For additional examples, see Example S9.1: "Choosing the Factors to Cross.")

Question 2. How Do You Carry Out the Experimental Plan?

a. If all factors are experimental, you use the cells that come from crossing the factors as your treatment combinations, and randomly assign equal numbers of units to each combination.

b. If all factors are observational, you sample and classify your units, sorting them into the cells that come from crossing the factors; then you measure.

c. If you have factors of both kinds, you sample and sort your subjects or material according to the observational factor(s), then randomly assign levels of the experimental factor(s).

Example 9.2 Carrying Out the Plan

a. *Tobacco hornworms* (Exercises 6.A.1 and 6.A.7). Both factors of interest—rearing food (R or C) and test food (R or C)—are experimental. Crossing the two factors gives four cells (RR, RC, CR, CC), so you randomly assign the 24 hornworms to cells, 6 worms to a cell.

b. *Morton's skulls*. The Philadelphia doctor who compared skull sizes (Example 4.4) used a two-way factorial design with sex and ethnic origin as factors defining distinct populations, one for each combination of sex and ethnicity. His samples were nonrandomized convenience samples. He classified his skulls, sorting them into cells, and then measured the volume of each.

c. *Buttercups* (Example 6.5). For this study there was one observational factor (original environment) and one experimental factor. You can think of the experimental plan as like two little CRs in parallel, one for each level of the observational factor: take plants from the woods (level 1) and randomly assign some to the field, some to go back to the woods; then do the same for the plants from the field (level 2). ■

Question 3. How Do You Find the Factor Structure?

Recall from Chapter 6, Section 1, that for a design with compound factors, you can get the factor structure in two steps, which gives you the factor diagram in *reduced* and then *expanded* form. The two steps (and the two forms), correspond to parts B and C of the three-part format for summarizing designs, given in Chapter 8.

Step 1: *Design structure (reduced form)*. Ignore any factorial crossing in the treatment combinations, and analyze the design structure (as in Chapter 8): What are the units, and what is the pattern that relates each set of units to treatment combinations? Use this information to identify the design family: CR, CB, LS, SP/RM (or other). Draw the factor diagram in the usual way.

Step 2: *Treatment structure (expanded form)*. Identify all instances of factorial crossing. Replace any compound factor, everywhere it occurs in your factor list or diagram from Step 1, with three factors: first crossed factor, second crossed factor, and their interaction.

What makes the two-step approach particularly handy is that you can use it for *any* design that has been created by factorial crossing, regardless of whether the basic design is BF, CB, LS, SP/RM, or some other. (For an illustration see Example S9.2: "Factor Structure for the Hornworm Data.")

Question 4. How Is the Informal Analysis Different?

In many ways the analysis is not much different: you can (and should) use any of the graphical methods for simpler data sets, to look for patterns in data from the designs of this chapter. The only new wrinkle is that data sets with three or more factors of interest have a more complicated interaction structure. For example, a plant nutrition experiment with three factors of interest would have four kinds of interactions, three two-factor interactions (nitrogen with potassium, nitrogen with phosphorus, and potassium with phosphorus), and one three-factor interaction. Section 4 illustrates a graphical method for looking at interactions for data sets with more than two factors of interest.

Exercise Set A

Each situation below involves a response and a factor of interest. For each, find a second factor that might be of interest. There is no single right answer, but try to find factors that are likely to be important, given the context of the experiment.

1. *Crying babies*

Purpose: To see whether adults react more strongly to the sound of a baby crying than they do to other sounds.

Response: Change in diameter of the pupil of the eye.

Factor: Four sounds (= levels) superimposed on a tape of classical music: (1) baby crying, (2) church bell ringing, (3) dog barking, and (4) no additional sound (control).

2. *Rating Milgram* (Reread Exercise 5.B.8.)

Response: A rating of how ethical the study is thought to be.

Factor: The three versions of the results.

3. *Dutch elm disease*. In the United Kingdom, Dutch elm disease (a fungus) is transmitted by two species of beetle (*Scolytus destructor* and *S. multistriatus*) which feed between the bark and wood of living elm trees, leaving channels called galleries on the surface of the wood just under the bark. It is easy to tell the species that made a gallery by measuring the diameter: about 2 mm for the larger *S.d.*, about 1 mm for *S.m.*

Purpose: To see whether larger numbers of one species tend to go with large or small numbers of the other.

Response: Mark squares 10 cm × 10 cm on dead elm trees, peel back the bark with a chisel, and count the galleries of each size in the wood underneath.

Nuisance factor: Individual trees.

Factor of interest: Species, *S.d.* or *S.m.*

4. *Snakes and robins*. Robins migrate. They raise their young in northern areas during the spring, but spend winters in southern areas, where snakes are more common.

Purpose: To see whether the presence of a rubber snake model near a robin's nest would prevent the bird from visiting the nest.

Response: Number of occasions when the robin approached the nest, as a percentage of occasions when the robin made an appearance.

Nuisance factor: Nests. (Several different nests were observed.)

Factor of interest: Color of snake model: bright (like a coral snake) or dull brown (like a water snake). (Birds have color vision. Mammals, as a rule, do not.)

5–6. *Before and after structure*. The four conditions of interest for the buttercups example have a "before and after" structure which is quite often used to create factorial crossing. In that experiment, there were two conditions for "before": the original environment was either field or woods. The same two conditions, field or woods, were then used for "after": the new environment was either field or woods. The two examples that follow get their factorial structure in the same way. For each one, (i) list the levels of the compound factor of interest (reduced form), (ii) then list the three factors you get by expanding the compound factor, together with their levels.

5. *Submarine memory*. (Example 7.3.)

6. *Hornworms and cellulose*. (Exercise 6.A.1)

7. *Mental health, memory, and interference*. Reread the description of this experiment (Examples S4.8 and 7.10), and imagine the data arranged in a rectangle with eight columns for conditions, as shown below.

Diagnosis:	**Schizophrenic**		**Schizotypal**		**Borderline**		**Normal**	
Interference:	Yes	No	Yes	No	Yes	No	Yes	No

For the purpose of this exercise, assume that the experimental plan used eight different sets of subjects, one set for each column, and that subjects were not reused: each subject is a unit.

a. Write out the factor diagram in reduced form.

b. Write the factor diagram for the three factors you get by expanding the compound factor of interest.

8. *Kosslyn's imagery experiment*. The table below shows average response times for the six conditions of Kosslyn's experiment (Example 6.10). Draw diagrams for the three factors you get by expanding this compound factor of interest, and compute condition averages for the two basic factors.

	Imagery instructions	**No imagery instruction**
1st graders	2000	1850
4th graders	1650	1400
Adults	1600	800

2. Three or More Crossed Factors of Interest

This section presents two examples, each with three factors of interest. The first example is a three-way CR design, or CR[3]. The second example is an SP/RM with a compound between-blocks factor that comes from crossing two basic factors. (We write SP/RM[2;1] to show that there are two between-blocks factor and one within-blocks factor.) Neither example in this section has a compound within-blocks factor; to find the appropriate factor structure for designs with compound within-blocks factors often requires an extra step, and so I treat such designs together in Section 3.

As you read the examples of this section, focus on what is new—the treatment structure. Once we have identified the design structure (Step 1), our emphasis will be on expanding the compound factors into **main effects** (another name for the basic factors that are crossed), two-factor interactions, three-factor interactions, and so on.

As always, each factor is a partition of the observations into groups whose averages we plan to compare. Instead of showing actual factor diagrams, which tend to be both large and messy, I'll list each factor and describe its groups in words.

EXAMPLE 9.3 COMPETING CRABGRASS, VERSION 1: A 3-WAY CR, OR CR[3]

Competition for scarce resources is an important fact of life and a major focus of research in ecology: animals compete for mates, and all organisms compete for space and food. The way plants compete for nutrients is one aspect of competition that is particularly easy to study systematically with a lab experiment.

The purpose of this experiment was to study the way one species of crabgrass (*Digitaria sanguinalis*) competed with itself and with another species (*D. ischaemum*) for nitrogen (N), phosphorus (P), and potassium (K). (The version of the experiment I give here is simpler than the actual one. I'll give a second version in Example S9.4, and then give the actual experiment in S9.6.)

In this version, bunches of seeds of *D S.* were planted in vermiculite, in 16 Styrofoam cups; after the seeds had sprouted, the plants were thinned to 20 plants per cup. Each of the 16 cups had been randomly assigned to get one of 8 nutrient combinations added to its vermiculite. These combinations are shown as columns in Table 9.1; *n* means normal concentration, — means low concentration. The response is mean dry weight per plant, in milligrams.

TABLE 9.1

Nitrogen	→	*n*	*n*	*n*	*n*	—	—	—	—
Phosphorus	→	*n*	*n*	—	—	*n*	—	—	—
Potassium	→	*n*	—	*n*	—	*n*	—	*n*	—
		96.4	86.1	59.6	68.9	39.7	35.4	36.9	42.2
		73.4	77.8	64.7	47.1	41.4	40.6	45.4	79.0

Step 1: *Design Structure*. There is one kind of unit, a cup. Each cup got randomly assigned to a combination of nutrients. The design is a CR; the compound factor of interest has 8 levels, which correspond to columns in the rectangle.

Step 2: *Treatment Structure*. Expanding the compound factor will give us a total of seven factors (Table 9.2): three main effects (N, P, K), three two-factor interactions (N × P, N × K, P × K), and one three-factor interaction (N × P × K). The number of interactions of each kind is determined by the number of main effects, and I'm sure that with practice you'll soon get so you can use familiar patterns to list these directly. For this first example, however, I'll start with a slower but foolproof way to list them, by applying Step 2 twice.

Step 2 tells us to replace a compound factor with three new ones: first crossed factor, second crossed factor, and their interaction. We can think of getting the compound factor Conditions by crossing the basic factor N with the compound factor

PK, whose levels are the four combinations of concentrations of phosphorus and potassium. Applying Step 2 gives:

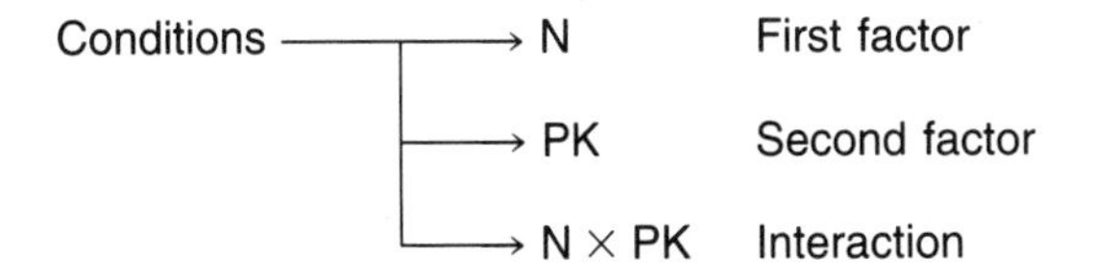

Now we apply Step 2 again, this time expanding the compound factor PK, which comes from crossing P and K. PK appears twice in the factor list above, and Step 2 tells us to expand it everywhere it occurs:

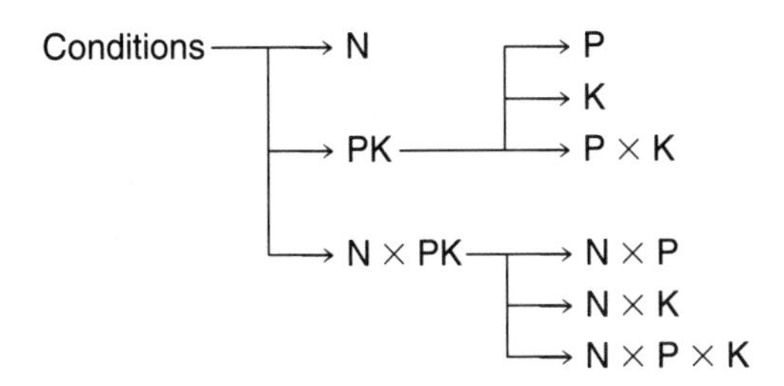

We can summarize the treatment structure by listing factors in columns for main effects, two-factor interactions, and so on:

TABLE 9.2 Summary table of factors.

Unit	Grouping (Nuisance factors)	Treatment combinations		
		Main effects	2-Factor interactions	3-Factor interaction
Cup	(none)	N	N × P	N × P × K
		P	N × K	
		K	P × K	

Rather than draw the factor diagram, which would have seven structural factors, I'll tell the groups in words (Figure 9.1):

Factor	#levels		Groups whose averages are compared
N	2	Columns	1 2 3 4 (n) / 5 6 7 8 (−)
P	2	Columns	1 2 5 6 (n) / 3 4 7 8 (−)
K	2	Columns	1 3 5 7 (n) / 2 4 6 8 (−)
N × P	4	Columns	1 2 (nn)/ 3 4 (n−) / 5 6 (−n)/ 7 8 (− −)
N × K	4	Columns	1 3 (nn)/ 2 4 (n−) / 7 (−n)/ 6 8 (− −)
P × K	4	Columns	1 6 (nn)/ 2 7 (n−) / 3 7 (−n)/ 4 8 (− −)
N × P × K	8	Columns	1 (nnn)/ 2 (nn−) / .../ 7 (− −n)/ 8 (− − −)

FIGURE 9.1 Factor list with partitions: 3-way CR. ■

Example 9.4 Osmoregulation in Worms: A 3-Way SP/RM, or SP/RM[2;1]

Worms that live at the mouth of a river must deal with varying concentrations of salt in their environment as the tides change the mix of salty sea water and fresh river water. Some worms, called osmoregulators, are able to maintain relatively constant concentrations of salt in the body, despite changes in their environment, just as warm-blooded animals can maintain a stable body temperatures. Other worms, the nonregulators, are—like cold-blooded animals—much more at the mercy of environmental changes. The purpose of this experiment was to compare the effects of various mixtures of sea and salt water on the body weight of two species of worms, *Nereis virens* (N) and *Goldfingia gouldii* (G). Eighteen worms of each species were weighed, then randomly assigned in equal numbers to one of three conditions. Six worms of each kind were placed in 100% sea water, another six of each kind were placed in 67% sea water, and the rest were placed in 33% sea water. The worms were weighed after 30, 60, and 90 minutes, then placed in 100% sea water (recovery) and weighed one last time 30 minutes later. The response was body weight, expressed as a percent of initial body weight. Each worm gave four measurements: at 30, 60, 90, and (recovery) 120 minutes after the start of its water bath.

Step 1: *Design Structure*. There are two kinds of units, worms and time slots. The design is an SP/RM, with Species and salt Concentration as between-worms factors, and time as a within-worms factor. (In calling the design an SP/RM, I assume that each worm got its own personal water bath. If, on the other hand, groups of worms were put together in containers of water, with salt concentrations assigned to containers, then a container's worth of worms would be a third kind of unit, and the design structure would be more complicated than an SP/RM.)

Step 2: *Treatment Structure*. There are three basic factors of interest (species, concentration, and time), which together determine the same set of seven factors as in the last pair of examples: three main effects, three 2-factor interactions, and one 3-factor interaction. I want the summary table of factors to reflect the design structure as well as the treatment structure, and so I list the factors in two groups of rows, corresponding to the two kinds of units:

TABLE 9.3 Summary table of factors: SP/RM[2;1].

Unit	Grouping (Nuisance factors)	Treatment combinations		
		Main effects	2-Factor interactions	3-Factor interaction
Worm	(none)	Species Conc.	Species × Conc	
Time slot	Worm	Time	Species × Time Conc × Time	Species × Conc × Time

The data are shown in Table 9.4, together with some group averages. We can use the rectangular layout (Figure 9.2) to describe each of the factors in the summary table in terms of the way it groups observations.

TABLE 9.4

		30 min.	**60 min.**	**90 min.**	**120 min.**
100% sea water					
N	(1)	99.1	99.1	100.46	100.46
	(2)	95.3	90.6	88.6	87
	(3)	100	100	99.6	98.35
	(4)	100.4	100.0	100	101
	(5)	99.02	99.15	98.56	97.15
	(6)	95.4	93.2	93.2	92.5
	Avg.	98.2	97.0	96.7	96.1
G	(7)	100.55	100.00	102.21	99.72
	(8)	96	97	95	90
	(9)	99	88	99.7	91.5
	(10)	108.89	100.8	95.6	104
	(11)	101.15	101.79	101.15	100
	(12)	95.5	92.	92	90
	Avg.	100.2	96.6	97.6	95.9
67% sea water		**30 min.**	**60 min.**	**90 min.**	**120 min.**
N	(13)	114.92	121.63	127.07	115.13
	(14)	103.6	105.6	106.9	98
	(15)	123	125	133	118
	(16)	104.54	110.2	113.6	112
	(17)	106.77	111.76	115.14	111.68
	(18)	116	120	121	112
	Avg.	111.5	115.7	119.5	111.1
G	(19)	113.80	122.17	127.60	118.55
	(20)	102	104	108	101
	(21)	115	117	120	119
	(22)	123.25	134.8	158.1	125
	(23)	113.36	122.14	127.68	121
	(24)	103.01	108	112	105
	Avg.	111.7	118.0	125.6	114.9

TABLE 9.4 (*continued*)

33% sea water		30 min.	60 min.	90 min.	120 min.
N	(25)	123.40	139.50	152.20	142.32
	(26)	117.6	128.3	137.6	124.3
	(27)	134.8	154.7	156.4	149.2
	(28)	108.23	117.6	129.4	118
	(29)	112.04	121.51	128.56	122.3
	(30)	125	112	155	138
	Avg.	120.2	128.9	143.2	132.4
G	(31)	142.76	174.48	188.28	176.90
	(32)	127.3	152	157.7	150
	(33)	149.5	178.6	178.9	177
	(34)	148.27	168.9	175.8	158
	(35)	122.76	129.97	131.89	122.9
	(36)	144	177	191	177
	Avg.	139.1	163.5	170.6	160.3

Factor	#levels	Groups to be compared
Species	2	Rows 1–6, 13–18, 25–30 (N) / 7–12, 19–24, 31–36 (G)
Conc	3	Rows 1–12 (100%) / 13–24 (67%) / 25–36 (33%)
Species × Conc	6	Rows 1–6 / 7–12 / 13–18 / 19–24 / 25–30 / 31–36
Worms	36	Rows 1 / 2 / 3 / ... / 35 / 36
Time	4	Columns 1 / 2 / 3 / 4
Species × Time	8	Rows 1–6, 13–18, 25–50; Column 1 / ... / Rows 7–12, 19–24, 31–36; Column 4
Conc × Time	12	Rows 1–12; Column 1 / ... / Rows 25–36; Column 4
Species × Conc × Time	24	Rows 1–6; Column 1 / ... / Rows 31–36; Column 4

FIGURE 9.2 Factor list with partitions: SP/RM with two between-blocks factors. The factors are grouped as in the summary table, with between-blocks factors first. ■

For additional examples, see S9.2: "Factor Structure of the Hornworm Data," S9.3: "Fear of Rodents, Another CR[3]," and S9.4: "Crabgrass (Version 2), a CR[4]."

Exercise Set B

1. *Memory and mental health.* For the sake of this exercise, assume that the experiment described in Example S4.8 had been run as a CR[3], with conditions determined by crossing the three basic factors below. Write out a summary table of factors.

Diagnosis (observational, 4 levels: schizophrenic, schizotypal, borderline, normal)

Interference (experimental, 2 levels: present, absent)

Length of digit string (experimental, 2 levels: 4 or 5 digits)

2. *Perch eggs*. Consider crossing a third basic factor, turbulence (present or absent), with the 40 treatment combinations for batches of perch eggs described in Example S8.1a. Write out a summary table of factors.

3. *The number of two-factor interactions*. Let A, B, C, ... stand for basic factors in an experiment, and notice that you can use the upper right triangle of a two-way table to get a complete list of two-factor interactions:

2 factors

	A	B
A	-	-
B	-	AB

3 factors

	A	B	C
A	-	AB	AC
B	-	-	BC
C	-	-	-

4 factors

	A	B	C	D
A	-	AB	AC	AD
B	-	-	BC	BD
C	-	-	-	CD

Listing the 2-factor interactions.

a. Construct such a table for an experiment with five basic factors.

b. If an experiment crosses n basic factors, how many 2-factor interactions are there?

c. (If you've studied elementary probability, and know about binomial coefficients.) If you have n basic factors, how many r-factor interactions are there? What is the total number of main effects plus interactions?

4. *Competing crabgrass: A CR[5]*. Suppose you cross a fifth basic factor, Water (normal or low level), with the 32 treatment combinations described in Example S9.4. Give the summary table of factors.

3. Compound Within–Blocks Factors

This section presents two more examples: a two-way complete block design, or CB[2], and a split plot/repeated measures design with one between-blocks factor and two crossed within-blocks factors, or SP/RM[1;2]. For each example, expanding the compound factors of interest will follow the patterns of the last section, and will, I hope, seem straightforward. What makes this section different from the last is that each example has a compound within-blocks factor. There are two ways to analyze such data sets, based on two different factor lists. The first of these is the list you would get by following the examples of Section 2. The second list has all the factors from the first list, plus another set that you get by thinking of the factor for residual error as compound, and expanding it.

For designs with a compound within-blocks factor, there are two alternative factor structures for residual error:

(1) Additive model: Res = Blocks × Within-Blocks Conditions

(2) Nonadditive model: Expand Res by writing the compound factor in terms of basic factors

I'll first show you an example of the two different factor lists, then discuss ways to decide which one to use.

EXAMPLE 9.5 AUTOMATIC PROCESSING: A 2-WAY CB, OR CB[2]

Cognitive psychologists have known for some time that the amount of attention it takes to do something depends on how much practice you've had. Picking a letter like J out of a bunch of numbers is something we can do quickly and easily, but picking J out of a bunch of letters takes longer. Psychologists sometimes classify these two kinds of tasks as *automatic process*, which don't require much attention, and *control process*, which do.

Schneider and Schiffrin reported a series of experiments that demonstrate the difference between the two kinds of tasks. They used a complete block design, with subjects as blocks. Each subject performed a series of tasks, and the tasks themselves had a two-way factorial structure. (The response was the time it took to complete the task.)

Each task had the same form: the subject is told a target letter or number, say J, for example. Then s/he sees twenty cards flashed on a screen, one at a time. If the target letter or number appears on one of the cards, the subject is to say "Yes." If the target doesn't appear on any of the cards, the subject is to say "No."

In all, there were six kinds of tasks, classified according to frame size (1, 2, or 4 symbols on each card) and category condition (same or different). "Same category" meant the symbols on the cards were the same kind as the target, all letters, or all numbers; "different category" meant the target was a letter and the other symbols were numbers, or vice versa.

Frame	Category condition	
size	Same	Different
1		
2		
4		

Factorial structure of the six tasks. Each subject (= block) performed all six tasks. This design has a compound within-blocks factor.

Step 1: *Design Structure.* I've claimed that the design structure for this example is a CB, but my description was vague about the way conditions got assigned to units. It should be clear that each subject is a block, but there are several possible strategies for assigning conditions; only one gives a CB design. We'll assume that each subject's time slots were not grouped ahead of time, and that each slot got randomly assigned to one of the six treatment combinations. This version of the experiment has units of just one size in each block: the design is a (randomized) CB. Figure 9.3 is a standard factor diagram, with Subjects as rows and Conditions as columns.

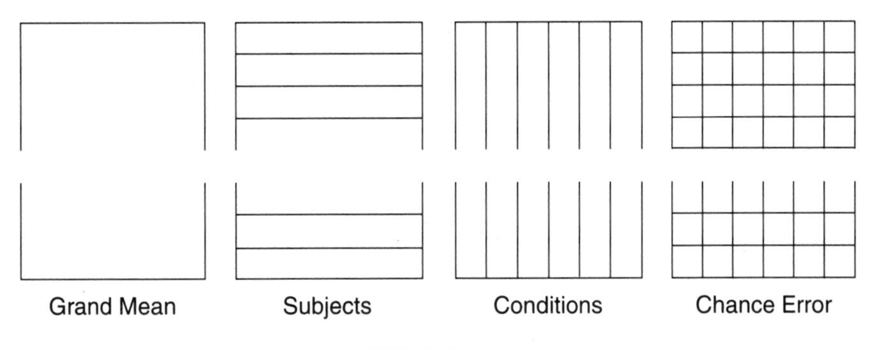

FIGURE 9.3

Step 2: *Treatment structure.* The compound factor, Conditions, comes from crossing two basic factors, Frame Size and Category Condition. If we replace the compound factor with the usual three, we get the following summary table:

TABLE 9.5 Summary table of factors: 2-way CB.

Unit	Grouping (Nuisance factors)	Treatment combinations: Main effects	Treatment combinations: 2-Factor interaction
Time slot	Subjects	Size Categ	Size × Categ

Step 2*: *Expanding the Factor for residual error.* Turn back to the factor diagram from Step 1, and notice that the factor for residual error comes from crossing blocks and conditions. Just as in Step 2, where we replaced the compound factor Conditions with three (Size, Categ, Size × Categ), here we can expand Blocks × Conditions, replacing it with Blocks × Size, Blocks × Categ, and Blocks × Size × Categ. Adding these three new factors at the bottom of the summary table from before gives a factor list which we should consider as an alternative to the first one:

TABLE 9.6 Summary table with chance error expanded.

Unit	Grouping (Nuisance factors)	Treatment combinations: Main effects	2-Factor interaction	3-Factor interaction
Time slot	Subjects	Size	Size × Categ	
		Categ		
		Residual error:	Subj × Size	Subj × Size × Categ
			Subj × Categ	

The bottom group of factors comes from thinking of the factor for residual error as Subj × Cond, and expanding the compound factor.

Two alternative models for residual error structure. When should you use the shorter list, and when should you go on to Step 2*, to expand the factor for error? The two factor lists correspond to two alternative models for the data, called **additive block effects,** and **nonadditive block effects.** Each model describes a way in which the chance variability of a subject's response to the conditions might behave. Each leads to a different set of *F*-tests. The first model assumes that the typical size of the error is the same for all within-blocks factors: there is one typical size (one SD) no matter which set of factor averages you look at. This model leads to *F*-tests that use MS_{Res} in the denominator. The second model assumes that there are, or at least might be, different typical sizes of error depending on the conditions you compare: one typical size when you lump frame sizes together to compare averages for the two category conditions; another size when you lump category conditions together to compare averages for the three frame sizes; and a third typical size when you compare the six condition averages to estimate the interaction effect. According to this model, the right denominator mean square to use in an *F*-test depends on the set of averages you are comparing. To test frame size, you use Subj × Size as the denominator mean square; to test category condition, you use Subj × Categ; to test the Size-by-Category interaction, you use Subj × Size × Categ.

Notice that the assumptions of the second model (nonadditive block effects) include the first model as a special case. If the additive model fits the data, the nonadditive model will automatically fit also. On the other hand, there are many data sets for which the nonadditive model does fit, but the simpler additive model does not. For such data sets, *F*-tests based on the additive model will not be reliable: depending on how badly the model fits, the chances of falsely declaring effects "real" (and in some cases failing to detect "real" effects), will be inflated to an unknown extent.

Since an analysis based on the nonadditive model protects against such inflated error rates, it might seem best to use that model whenever your design has a compound within-blocks factor. However, the extra protection comes at a price. *F*-tests based on the nonadditive model have fewer denominator degrees of freedom, which makes them less powerful: they give you a smaller chance of detecting "real" differences. For this reason you should use the additive model if it fits your data.

How can you tell which model fits better? There are no easy answers, but here are four points to keep in mind:

1. In a CB design, it is not possible to separate block-by-treatment interaction effects from unit-to-unit differences and measurement error. If it is important to estimate block-by-treatment terms, you should use a generalized complete block design, as described in Chapter 13.
2. In general, if you have a compound within-blocks factor, and you expect block-by-treatment interactions to be present, there is no reason to expect the additive model to fit. For example, in Schneider and Schiffrin's experiment (Example 9.5), there is no reason to expect the variation due to Subj × Size and the variation due to Subj × Categ to have the same typical size, as would be required for the additive model. For the data of Example 9.5, I would start my analysis using the nonadditive model, expanding the factor for error.
3. The likely size of block-by-treatment interaction depends on the particular experiment. Within each area of research, scientists build up experience about which model fits better, and there's no general rule that substitutes for knowing how similar experiments have turned out. Nevertheless, it may be useful to ask whether you expect the block-by-treatment interactions to be large enough to matter. For example, compare Schneider and Schiffrin's experiment on automatic processing with a plant nutrition experiment using cups of vermiculite as units and two different starting dates as blocks. In the psychology experiment the blocks are human subjects, and it is not hard to imagine that different people would react differently to different frame sizes, and to different category conditions. By comparison, in the plant experiment, block-by-treatment interaction seems much less likely to be large enough to matter.
4. Sometimes (if you have a large enough number of blocks) you can use your data to test the assumption of the additive model. Section 5 illustrates one such a test. However, if you have only a small number of blocks, the test may be possible in principle, but your data won't supply enough information to make the test worthwhile. Then (as always), you can cross your fingers and substitute an assumption for the information you wish you had, being careful to remind yourself that the more untested assumptions you make, the more tentative your conclusions should be. ■

EXAMPLE 9.6 WORD ASSOCIATIONS AND SCHIZOPHRENIA: AN SP/RM[1;2]

This experiment was designed to test two competing theories about the way schizophrenics think about words. The basic design structure was SP/RM, with 48 subjects as blocks, and psychiatric diagnosis—schizophrenic or normal—as the between-blocks factor. Each subject was measured under four conditions, which correspond to four levels of a compound within-blocks factor. These conditions came from crossing two basic factors, each with two levels.

The experiment was based on word associations: you see a word printed on a 3 × 5 card, and respond with the first word you think of. There were two response variables, called response latency (how long it took to respond), and normative frequency, a measure of how typical the response word is. For example, if you see "boy" and respond "girl," your normative frequency is high, because roughly two out of three people say "girl." If instead, you say "knife" or "finger," as a schizophrenic might, your normative frequency is low, because almost no one responds with those words.

Each value of each response variable was actually an average based on 10 words, and each subject saw the same set of 40 words, 10 for each of the four conditions. Table 9.7 is a two-way table showing the factorial structure of the conditions:

TABLE 9.7

	Kind of words	
Instructions	**Steep slope**	**Flat slope**
Free	10 words	10 words
Idiosyncratic	10 words	10 words

One within-blocks factor was Instructions: for two sets of 10 words, subjects were told to respond with the first word that came to mind (free instructions); for the other two sets of 10 words, subjects were told to respond with a word they were reminded of but thought few people, perhaps no one else, would think of (idiosyncratic instructions). The other within-blocks factor was Slope: "steep slope" refers to words that have one dominant association, words like boy (girl), king (queen), slow (fast); "flat slope" refers to words that do not have a dominant association, words like radish, music, and carry.

Psychologists have known for a long time that in word association tests, schizophrenics often give responses that seem bizarre. (The article in the list of data sources summarizes some of the possible explanations for this fact.) Lisman and Cohen, who designed this experiment, wanted to test the predictions of two competing theories, which they called one-stage and two-stage. Both theories made the same predictions for three of the four conditions: both predict that for flat-slope words, normals and schizophrenics will have equal normative frequencies; both predict that for steep-slope words, with free instructions, normal subjects will score higher than schizophrenics; but for flat-slope words, with idiosyncratic instructions, the one-stage theory predicts that normal subjects will score higher than schizophrenics, whereas the two-stage theory predicts the opposite. I'll discuss the results of the study in Section 4, on interaction graphs for 3-factor experiments.

Step 1: *Design structure.* I've called the design an SP/RM, but be careful. Here, as is often the case when you have compound factors, there are several strategies for assigning conditions, with each strategy corresponding to a different design structure. We'll assume there are four sets of 10 words, 2 steep sets, and 2 flat sets. One set of each kind gets randomly matched with a set of instructions, separately for each subject, and then the subject gets the four instruction/slope combinations in a random order, one set at a time. This strategy gives the SP/RM design, with subjects and time slots as the two kinds of units.

Step 2: *Treatment structure* and Step 2*: *Chance error*.

TABLE 9.8 Summary table of factors: 3-way SP/RM.

Unit	Grouping (Nuisance factors)	Treatment combinations: Main effects	2-Factor interaction	3-Factor interaction
Subject	(none)	Diag		
Time slot	Subj	Instr	Instr × Slope	Diag × Instr × Slope
		Slope	Diag × Slope	
			Diag × Instr	
		Residual error:	Subj × Instr	Subj × Instr × Slope
			Subj × Slope	

■

For additional examples see S9.5: "Submarine memory, a 2-way LS," and S9.6: "Crabgrass, Version 3, a CB[4]."

Exercise Set C

1–2. *Remembering words*. The purpose of this experiment was to compare how easy it is to remember four different kinds of words:

Concrete, frequent: fork, brother, radio, . . .

Concrete, infrequent: blimp, warthog, fedora, . . .

Abstract, frequent: truth, anger, foolishness, . . .

Abstract, infrequent: sloth, vastness, apostasy, . . .

Ten students in a psychology lab served as subjects. For the purpose of this exercise, assume that each subject provided four units (= time slots). During each time slot, the subject heard a list of words of one of the four kinds, and was then tested for recall.

1. Write out a summary table of factors, following the format of Example 9.5. Assume the experiment was done using just one size of unit, and assume the model of nonadditive block effects, that is, expand Error = Blocks × Conditions.

2. Using the data in Table 9.9, compute condition averages for all basic factors of interest, and draw an interaction graph. Is there evidence of interaction between the two basic factors?

TABLE 9.9

		Abstract frequent	Abstract infrequent	Concrete frequent	Concrete infrequent
Subject	I	60	44	60	44
	II	52	32	44	28
	III	20	32	40	24
	IV	36	32	44	36
	V	40	32	52	44
	VI	36	20	28	40
	VII	36	40	32	28
	VIII	44	64	44	64
	IX	36	40	32	56
	X	52	52	64	60
Totals		412	388	440	424

3. *Kosslyn's imagery experiment*. Reread Examples 9.1a(iii) and 9.1b(ii). Suppose Kosslyn had decided to run a small pilot study to see whether there was in fact an interaction between Instructions (imagery or not) and Type of Stimulus (high association/low area or low association/high area). Suppose for his pilot study he decided to use five subjects, all of them volunteers from a psychology class. (For this pilot study, age is not a factor in the design.)

   a. Describe how he could have run the pilot study as a randomized CB, with four conditions. In particular, tell how to assign conditions to subjects.

   b. Draw and label a rectangular format for the data. Then write out the factor diagram in reduced (Step 1) and expanded (Step 2) form.

   c. Write the summary table of factors, assuming the nonadditive model, i.e. expand the compound factor for Error (= Blocks × Conditions).

4. *Submarine memory* (Example S9.5).

   a. Let A–D stand for the four conditions, with the same meaning as before: A = Dry Dry, B = Dry Wet, C = Wet Dry, D = Wet Wet. Suppose the Latin square for subjects V–VIII had been:

		List of words			
		1	2	3	4
Subject	V	A	C	D	B
	VI	C	D	B	A
	VII	D	B	A	C
	VIII	B	A	C	D

Following the format of Example S9.5, draw the factor diagrams for the two basic factors of interest.

   b. Now combine the Latin squares from Example S9.5 and part (a) above, to obtain a design for subjects I–VIII. Sketch the factor diagram, showing factors for benchmark, subjects, word lists, learning environment, recall environment, interaction, and residual error.

c. Make up a situation where you might want to use a Latin square design whose treatments came from crossing two basic factors.

5–8. *Imagery and working memory*. According to the theory of hemispheric specialization, people to use different sides of their brain for visual and verbal tasks. Typically, according to a lot of recent research, we use the right side for visual tasks and the left side for verbal tasks. The purpose of this experiment by R. L. Brooks was to add to such research, by deciding which of two predictions about interaction was more nearly correct. One prediction comes from assuming that our working memory handles verbal and visual tasks separately; the other comes from a theory that working memory must handle both kinds of tasks together. To compare these theories, Brooks devised two kinds of tasks. The verbal task was to scan a sentence like "The pencil is on the desk" and decide whether each word was a noun or not. (The correct response is "No Yes No No No Yes.") The visual task was to scan a block letter like the F shown below, starting at the arrow, and decide whether each corner was an outside corner or not. (The correct response is "Yes Yes Yes Yes No No Yes Yes No Yes.")

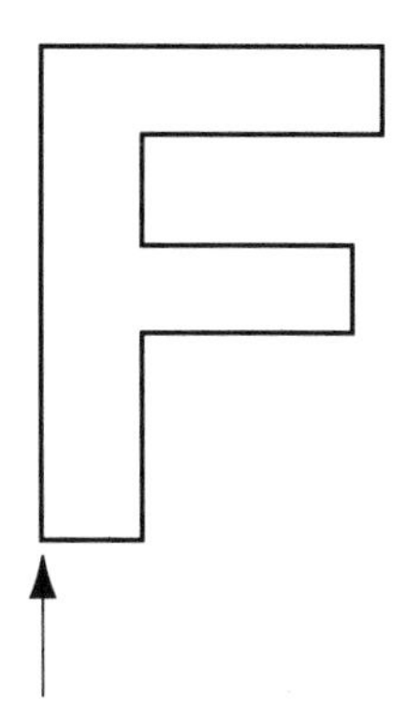

Brooks also devised two ways to report, one verbal and one visual. To report verbally, you would simply say your Yesses and Nos out loud; to report visually, you would point in sequence to a Yes or No on a piece of paper.

If visual and verbal tasks are handled independently, then a visual task with verbal report or a verbal task with visual report would be easier and so would take less time than verbal task/verbal report or visual task/visual report. The theory predicts that when the task and report were of the same kind, they would interfere with each other in memory, and slow the response time.

5. Describe how to run this experiment as a two-way factorial in complete blocks, and write out a summary table of factors. (Expand the factor for error: assume nonadditive block effects.)

6. Suppose you have 20 subjects. Describe how to use a design based on Latin squares to choose the order of the four conditions for each of the 20 subjects, following the example of the submarine memory experiment (Example S9.5).

7. Table 9.10 shows the data from a version of the experiment run by a psychology lab at Mount Holyoke College. Compute condition averages for the basic factors of interest. Then draw an interaction graph and discuss the meaning of the pattern in light of the two predictions. (The response times are in seconds.)

TABLE 9.10

Task:		Visual		Verbal	
Report:		Visual	Verbal	Visual	Verbal
Subject	1	11.60	13.06	16.05	11.51
	2	22.71	6.27	13.16	23.86
	3	20.96	7.77	15.87	9.51
	4	13.96	6.48	12.49	13.20
	5	14.60	6.01	14.69	12.31
	6	10.98	7.60	8.64	12.26
	7	21.08	18.77	17.24	12.68
	8	15.85	10.29	11.69	11.37
	9	15.68	9.18	17.23	18.28
	10	16.10	5.88	8.77	8.33
	11	11.87	6.91	8.44	10.60
	12	17.49	5.66	9.05	8.24
	13	24.40	6.68	18.45	8.53
	14	23.35	11.97	24.38	15.85
	15	11.24	7.50	14.49	10.91
	16	20.24	11.61	12.19	11.13
	17	15.52	10.90	10.50	10.90
	18	13.70	5.74	11.11	9.33
	19	28.15	9.32	13.85	10.01
	20	33.98	12.64	15.48	28.18
	Totals	363.46	180.24	273.78	257.00

8. *Analysis of the data.*

   a. Construct a parallel dot graph. (For the purposes of the graph, you get enough information from the whole-number part of each response value. If you are doing the graph by hand, I suggest you ignore the fractional part, regarding 11.60 as 11, and so on.)

   b. Comment on the way the response times are distributed within each of the four groups.

   c. Each of the four groups contains "straggling" values. Which of the following is more nearly true: (i) There is a small set of atypically slow subjects who are responsible for the extreme values. (ii) Although there are no subjects who (overall) are atypically slow in comparison with the others, there are some (scattered) atypically slow response times that should be regarded as outliers. It would be a good plan to analyze the data twice, both with and without the outliers. (iii) The extreme values don't really seem atypical. The pattern of response times for each condition is bunched at the low end, and spread out at the high end, in a way that suggests the variability is greater for the larger values. Transforming seems like a better strategy than excluding the extreme values.

9. *Dutch elm disease*. Reread Exercise 9.A.3. It is a reasonable hypothesis that the two species of beetles might share elm trees systematically, with one species predominating on the trunk, where the bark is rough, and the other species predominating on the branches, where the bark is smooth. Consider ways to test whether there is an interaction between Species (*S.d.* vs. *S.m.*) and Location (trunk vs. branches).

a. One simple plan would measure two squares from each tree, one from the trunk, one from a branch. Analyze the design structure and treatment structure; give a summary table of factors.

b. A more reasonable design might use more than one square of each kind from each tree. Data from such a design could be arranged in a rectangle like the one in Figure 9.4. The design does not belong to one of the four families, but you can find its summary table of factors by applying the ideas you have used for those other designs. Try it. (Designs like this one are discussed in Chapter 13.)

		Trunk			Branches	
Tree	Square	S.d.	S.m.	Square	S.d.	S.m.
I	1			3		
	2			4		
II	5			7		
	6			8		
III	9			11		
	10			12		

FIGURE 9.4

10. *Kosslyn's imagery experiment*. The diagram in Figure 9.5 shows a rectangular layout for the data from a simplified version of Kosslyn's experiment.

Stimulus type: Instructions:		Hi Assoc./Lo Area		Lo assoc./Hi area	
Age group	Subj	Imagery	No imagery	Imagery	No imagery
1st graders	I				
	II				
	III				
4th graders	IV				
	V				
	VI				
Adults	VII				
	VIII				
	IX				

FIGURE 9.5

a. Describe two strategies for assigning conditions to time slots for subjects. For each, name the resulting design family as CR, CB, LS, SP/RM, or other.

b. Write a summary table of factors for the simpler of the two designs from (a), assuming nonadditive block effects.

c. Draw a partial factor diagram, showing partitions for each of the following factors: Age, Age × Stim Type, Age × Stim × Instr, Blocks × Stim.

11. *Robins and snakes*. The design I'm about to describe is based on an experiment that was part of an honors project at Mount Holyoke. The purpose of this experiment was to determine

which factors influence the behavior of nesting robins in the presence of a snake model. There were several response variables, each concerned with some aspect of the robins' behavior. There were eight conditions of interest, obtained by crossing three basic factors (Table 9.11).

TABLE 9.11 Eight conditions of interest for the robin study.

Condition	Color	Motion	Location
1	Bright	No	Far
2	Bright	No	Near
3	Bright	Yes	Far
4	Bright	Yes	Near
5	Brown	No	Far
6	Brown	No	Near
7	Brown	Yes	Far
8	Brown	Yes	Near

Color: Two snake models were used. The bright one was colored red, yellow, and black like a coral snake. The other was dull brown.
Motion: For half the conditions, the experimenter wiggled the model by pulling on a long thread that was attached to it.
Location: For half the conditions, the model was placed at the base of the tree (far); for the other half the model was 4 feet from the nest (near).

Assume that each of four nests was observed under all eight conditions, and that for each nest, the order of the eight conditions was randomized.

a. Which family does the design belong to?

b. Draw and label a rectangular format for the data.

c. Write a summary table of factors, assuming nonadditive block effects.

12. *IQs and testers' expectations*. (This exercise comes from a final exam given in the Department of Psychology and Social Relations at Harvard several years ago.) "[Imagine that] you have decided to conduct an experiment on the effects of an examiner's expectations on his examinees' IQ tests. You are able to arrange it so that for one of the verbal IQ tests the examiner (E) will expect a very high score while for the other verbal IQ test, E will expect a very low score. Similarly, for one of the reasoning IQ tests E will expect a very high score while for the other reasoning IQ test E will expect a very low score. Four IQ scores are thus to be obtained from each of the 24 subjects. Half of the subjects are male and half are female." Assume, for simplicity, that there is only one (male) examiner. Analyze the design and factor structure of this experiment, and write a summary table of factors.

13. *Crabgrass*. Read, in Example S9.6, my description of a compromise model "in between" the usual models with additive and nonadditive block effects. Use the rectangular format of the data to diagram the partitions for each of Rep × Den, Rep × Nutrients, and Rep × Den × Nutrients.

4. Graphical Methods for 3-Factor Interactions

In this section I will use the experiment of Example 9.6, on schizophrenia and word associations, to illustrate the graphical analysis of interaction for a data set with three crossed factors of interest. Such a data set will have three possible two-factor interactions; for each of these, you should compute condition averages and draw one (or more) interaction graph(s). What you do for the two-factor interactions is the same as for the two-way factorial examples of Chapter 6; the only difference is that you do things three times, once for each pair of main effects.

A three-factor data set also has a three-factor interaction. To study this interaction graphically, you pick one of the three factors, and for each level of that factor, draw an interaction graph using the other two factors to define your conditions. If these graphs all show the same pattern, the data suggest that the two-factor interactions are the same for all levels of the first factor: there is no three-factor interaction. In the example that follows, these graphs do not show the same pattern: there is a three-factor interaction.

EXAMPLE 9.7 SCHIZOPHRENIA AND WORD ASSOCIATIONS

The three factors of interest in Example 9.6 defined a total of $2 \times 2 \times 2 = 8$ conditions. The 8 condition averages are shown in Table 9.12.

TABLE 9.12 Average normative frequency for eight conditions.

Instructions:	Free		Idiosyncratic	
Slope:	Steep	Flat	Steep	Flat
Diagnosis:				
Schizophrenoic	443	86	200	57
Normal	557	100	64	50

Response: The normative frequency measures how typical a subject's response is. Higher scores are for word associations that are common, like responding to "boy" with "girl"; lower scores are for uncommon associations.

Slope: Steep slope words like "boy" have a dominant association; flat slope words like "music" do not.

Instructions: When subjects were in the free condition they were told to say the first word that came to mind; when in the idiosyncratic condition they were told to say a word they thought of but thought other people would be unlikely to think of.

An informal analysis of these condition averages might be based on the factor structure, beginning with the three main effects, and then looking at the three two-factor interactions, and then looking at the three-factor interaction.

A. *Main effects*: diagnosis, instructions, and slope.

1. *Diagnosis*: I computed overall averages for the schizophrenics and for the normal subjects.

Schizophrenic: 196.50
Normal: 192.75

Overall, the two kinds of subjects show roughly the same normative frequency. At first, I found this result a bit surprising, because theory predicts that schizophrenics tend to respond with atypical associations. I expected their average normative frequency to be substantially lower than for the normal controls. However, the interaction patterns that follow suggest an explanation.

2. *Instructions*: I computed averages for all the free responses (columns 1 & 2) and all the idiosyncratic responses (columns 3 & 4):

Free: 296.5 Idiosyncratic: 92.75

The results make sense: the free responses are a lot more typical than the idiosyncratic responses.

3. *Slope*: Here we expect higher average scores for the steep slope words than for the flat slope words, and that's what we get:

Steep: 316.0 Flat: 73.25

B. *Two-factor interactions*. There are three two-factor interactions. For each one, we can find a two-way table of cell averages, and represent those averages with one of the usual interaction graphs.

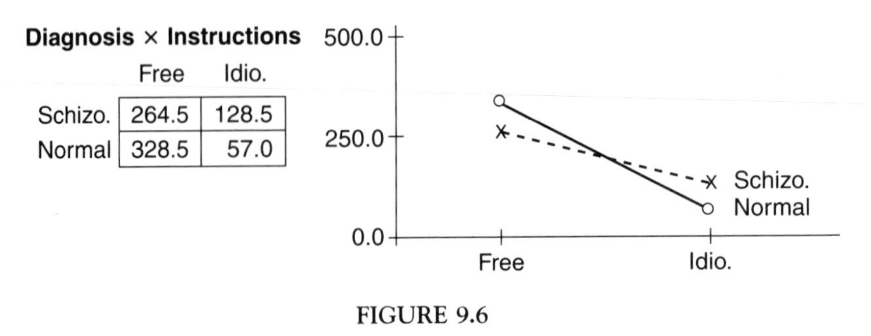

Diagnosis × Instructions

	Free	Idio.
Schizo.	264.5	128.5
Normal	328.5	57.0

FIGURE 9.6

Figure 9.6 shows that the instructions had a bigger effect on the normal subjects than on the schizophrenics. In particular, whereas normal subjects gave more typical associations than the schizophrenics when the instructions were "free," the schizophrenics gave more typical associations when the instructions were "idiosyncratic." As we saw in (A) above, the overall averages for the two groups of subjects were about equal.

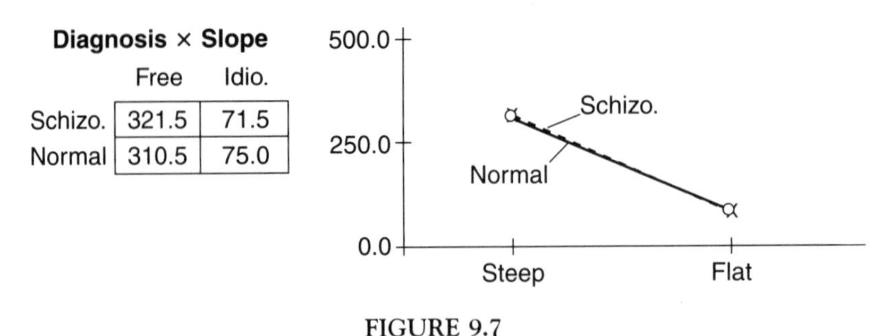

Diagnosis × Slope

	Free	Idio.
Schizo.	321.5	71.5
Normal	310.5	75.0

FIGURE 9.7

Figure 9.7 appears to show no interaction: schizophrenic and normal subjects on average responded in the same way to steep versus flat slope words.

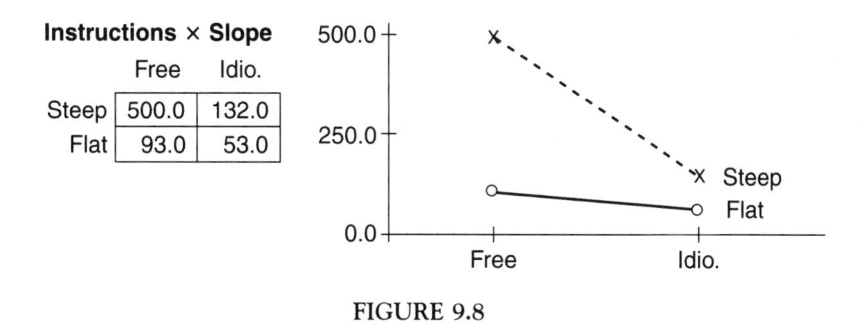

FIGURE 9.8

Figure 9.8 shows a huge interaction effect: instructions made a big difference for steep slope words, and a much smaller difference for flat slope words. Note that this picture shows why the names "steep slope" and "flat slope" make sense for the two kinds of words.

C. *The three-factor interaction.* The last interaction graph showed the interaction of Instructions with Slope, averaged over both kinds of subjects. It is not only reasonable to ask whether this pattern is the same for both kinds of subjects; indeed the main purpose of the study was to test the prediction that schizophrenic and normal subjects would show quite different patterns.

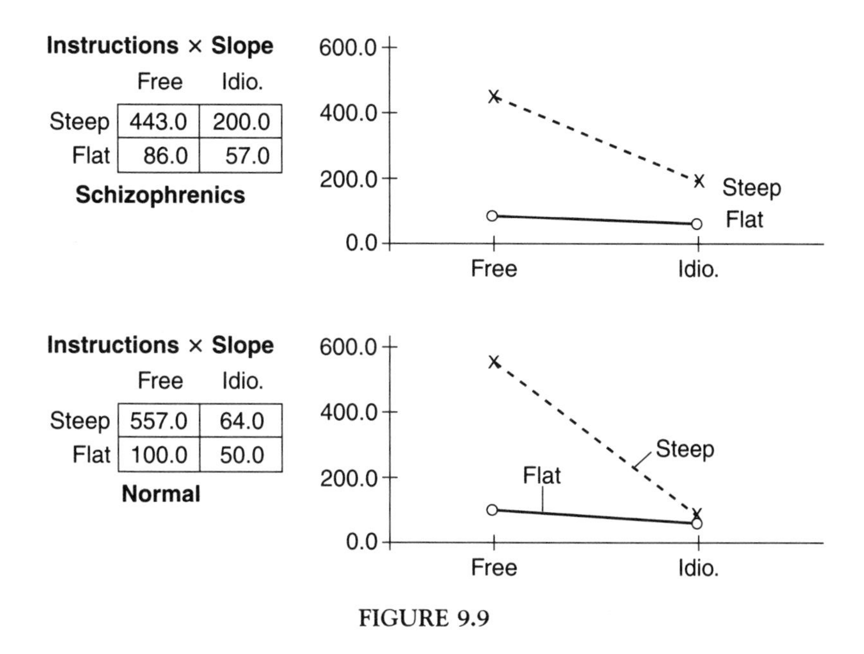

FIGURE 9.9

Although the two graphs in Figure 9.9 show the same general pattern, they are different in one important respect. Both graphs show a small drop for flat slope words, from about 100 (free) to about 50 (idio.), and both show a big drop for steep slope words. However, the effect for the normal controls is much more pronounced: their average drops almost 500 points, from 557 (free) to 64 (idio.), whereas the average for the schizophrenics drops less than 250 points, from 443 (free) to 200 (idio.).

Finally, notice that the data fit the prediction of the two-stage theory: for steep slope words, with idiosyncratic instructions, the schizophrenics score a lot higher on average than the normal controls. ■

We can summarize the main point of this section as follows:

To check for a three-factor interaction, ask whether the interaction pattern for two of the factors is the same for all levels of the third factor.

Exercise Set D

1. *Schizophrenia and word associations*. The experiment of Example 9.6 used two response variables—normative frequency, and response latency, the time it took subjects to respond. Here are the eight condition averages for response latency, which the experimenters transformed by taking reciprocals (1/time) and which I then multiplied by 100 to get rid of decimals. Do an informal analysis, following Example 9.7. Is there evidence of a 3-factor interaction effect?

Instructions:		Free		Idiosyncratic	
Slope:		**Steep**	**Flat**	**Steep**	**Flat**
Diagnosis:	Schizophrenic	78	56	25	25
	Normal	44	28	25	23

Note that a fast response means a small response time, which gives a *larger* value for $100 \times (1/\text{time})$.

2. *Rat phobia*. Here are the eighteen condition averages for the experiment of Example S9.3. Do an informal analysis. Is there evidence of a 3-factor interaction effect?

Expectancy:		Negative			Incorrect Neg.			Free Choice		
Delay:		0	10	40	0	10	40	0	10	40
Fear:	High	9.5	14.7	15.0	11.0	13.0	12.1	11.6	9.9	9.8
	Low	15.0	15.0	15.0	15.0	15.0	15.0	15.0	15.0	15.0

3. *Osmoregulation in worms*. Condition averages for the worm experiment were given along with the data in Example 9.4. Do an informal analysis, and discuss the meaning of any patterns you find. (Be sure to look at the individual response values as well as the averages. I gave you the averages so you wouldn't have to spend time computing them, but there's more to be found in the data set than you can get from the averages alone.) Note that because two

of the basic factors (Time and Conc) are quantitative, your interaction graphs will be more informative if you put levels of one of these factors on the x-axis.

4–10. *Crabgrass*. The data given in Example S9.6 have a very rich structure. A thorough informal analysis would take a lot of time, because there are so many things to look at. Here are several things to try:

4. *Outliers in the grass*. Look over the data given in Example 9.6. Correct the typo, 972.5, to 72.5. Can you identify any other outliers? If so, list them, and tell how you recognized each one as an outlier.

5. *Effects of density*

   a. Compute averages for each of the four basic factors of interest, and also for the nuisance factor Replications.

   b. Using the set of four averages for Density from part (a), graph each average (y) against the density (x), and describe the pattern: Do the plants weigh more when they compete with plants of the same species, or when they compete with plants of the other species?

6. Construct interaction graphs for each of the three nutrient pairs, N × P, N × K, and P × K. Which pair shows the largest interaction effect? The smallest?

7. Repeat #6 another two times, once for Rep I, once for Rep II. Do you find evidence of interaction between Replications and nutrient pairs?

8. Construct interaction graphs for each of N × Den, P × Den, and K × Den. Construct your graphs so that the levels of Den appear on the x-axis, as in 5(b). Write a brief summary of the patterns you find.

9. Construct a set of graphs for showing each of the following three-factor interactions: (a) N × P × K, (b) Den × N × P, (c) Den × N × K, and (d) Den × P × K.

10. Make a list of other things you would try if you had more time.

5. Analysis of Variance

By now I hope you are beginning to develop your skill at decomposing data sets and counting degrees of freedom, so that you can apply these skills to designs you haven't seen before. The same methods you've been using work for more complicated designs, with only one new wrinkle: how you combine mean squares to get F-ratios. (Chapter 13 discusses the logic of choosing denominator mean squares, and gives a single rule that works for all balanced designs. If you go directly to the last part of Section 13.4, on expected mean squares, you can skip this section. In what follows, I'll present a rule that is less general but simpler, as a temporary alternative to learning about expected mean squares.)

F-ratios for Compound Within-Blocks Factors

This section presents two ways to remember how to choose denominator mean squares. The first is based on the general rule you already know: you use MS_{Res} except for testing the between-blocks factor(s) of the SP/RM designs; then you use MS_{Blocks}. The same rule still applies if your design has no blocks, or if the (only) within-blocks factor is basic, or if you have a compound within-blocks factor and assume the additive model. There's a new class of exceptions, however, if you assume the nonadditive model; these occur when you expand the compound factor Blocks × Within to get several components of residual error, and each component has its own mean square. Thus there are several choices for denominator mean squares.

Denominator Mean Squares If You Assume Nonadditive Block Effects

To get the *F*-ratio for testing a within-blocks factor, (or the interaction of a between-blocks factor and within-blocks factor), use for your denominator the MS for the interaction of blocks and the within-blocks factor.

For Example 9.5, Schneider and Schiffrin asked subjects to pick target letters and numbers from arrays of symbols; each subject performed under six conditions defined by crossing Frame Size (1, 2, or 4 symbols) with Category Condition. In Step 2* of Example 9.5, we expanded the factor for error into three parts: Blocks × Frame Size, Blocks × Category Condition, and Blocks × Interaction. The rule says that to compute the *F*-ratio for testing Frame Size, we use as the denominator the mean square for Blocks × Frame Size; to test Category Condition we use the MS for Blocks × Category Condition; to test Interaction we use Blocks × Frame Size × Category Condition.

EXAMPLE 9.8 Remembering Words

Exercise 9.C.1 describes a CB[2] experiment to compare how well we remember four different kinds of words. The four kinds came from crossing two within-subjects factors, abstraction and frequency (each with two levels). Tell how to compute the *F*-ratios for testing abstraction, frequency, and their interaction.

Solution. There is a compound within-blocks factor that comes from crossing two basic factors, Abstraction and Frequency. We expand the factor for residual error and compute separate mean squares for Blocks × Abstraction, Blocks × Frequency, and Blocks × Abstraction × Frequency. A schematic version of the ANOVA table is shown in Table 9.12.

TABLE 9.12

Source	DF	F-Ratio
1. Grand Average	1	
2. Abstraction	1	$MS_{Abstr}/MS_{Bl \times Abstr}$
3. Frequency	1	$MS_{Freq}/MS_{Bl \times Freq}$
4. Abstr × Freq	1	$MS_{Abstr \times Freq}/MS_{Bl \times Abstr \times Freq}$
5. Blocks	9	
6. Blocks × Abstr	9	
7. Blocks × Freq	9	
8. Blocks × Abstr × Freq	9	
Total	40	

■

EXAMPLE 9.9A KOSSLYN'S IMAGERY EXPERIMENT

Refer to the rectangular layout given in Exercise 9.C.10 for a small version of Kosslyn's experiment. Tell how to compute *F*-ratios for testing all basic factors of interest and their interactions.

SOLUTION. This design belongs to the SP/RM family, with one between-blocks factor, Age, and a compound within-blocks factor, which comes from crossing stimulus type (Stim) with instructions (Instr). To get denominators for the various *F*-ratios, we expand the factor for Error = Subj × Within-Blocks into Subj × Stim, Subj × Instr, and Subj × Stim × Instr. For testing the between-blocks factor, we use the MS for subjects in the denominator; for testing Stim and Age × Stim, we use the MS for Subj × Stim; for testing Instr and Age × Instr, we use the MS for Subj × Instr; and for testing Stim × Instr and Age × Stim × Instr, we use the MS for Subj × Stim × Instr (see Table 9.13).

TABLE 9.13

Source	DF	F-Ratio
Grand Average	1	
Age Groups	2	MS_{Age}/MS_{Subj}
Subjects	6	
Stimulus type	1	$MS_{Stim}/MS_{Subj \times Stim}$
Age × Stim	2	$MS_{Age \times Stim}/MS_{Subj \times Stim}$
Subj × Stim	6	
Instructions	1	$MS_{Instr}/MS_{Subj \times Instr}$
Age × Instr	2	$MS_{Age \times Instr}/MS_{Subj \times Instr}$
Subj × Instr	6	
Stim × Instr	1	$MS_{Stim \times Instr}/MS_{Subj \times Stim \times Instr}$
Age × Stim × Instr	2	$MS_{Age \times Stim \times Instr}/MS_{Subj \times Stim \times Instr}$
Subj × Stim × Instr	6	
Total	36	

■

The last two examples treated the nonadditive model as a second exception to the general rule for denominator mean squares. There's another way to think about these examples, however, that combines the two kinds of exceptions in a single rule. (You might want to review Example 7.13, on the product of two factors, before you go on.)

For Designs with Blocks,

$$F = \frac{MS_{Factor}}{MS_{Blocks \times Factor}}.$$

"Factor" stands for any factor of interest or interaction.

Notice that this rule covers the test for the between-blocks factor(s) of the SP/RM designs: the denominator MS is Blocks × Between-Blocks, which is the same as Blocks. (See Example 7.13b.) The rule also works for the CB[1], since for that design the residual factor is the same as Blocks × Treatments.

EXAMPLE 9.9B Kosslyn's Imagery Experiment

Apply the new rule to find denominator mean squares for all factors of interest and their interactions.

Solution. See Table 9.14.

TABLE 9.14

Factor of interest	Factor to use for denominator MS in the *F*-ratio
Age groups	Subj × Age = Age
Stimulus	Subj × Stim
Age × Stim	Subj × Age × Stim = Subj × Stim
Instructions	Subj × Instr
Age × Instr	Subj × Age × Instr = Subj × Instr
Stim × Instr	Subj × Stim × Instr
Age × Stim × Instr	Subj × Age × Stim × Instr = Subj × Stim × Instr

■

If you think of the rule for denominators in two parts ("Res" for BF designs, "Blocks × Factor" for designs with blocks), then this new two-part rule has only one exception—the additive model for designs with compound within-blocks factors. It actually makes more sense to have a rule that makes that one case the exception, because the additive model comes with an extra assumption: that SDs for the various components of residual error are equal. In general, rather than simply assume the SDs are equal, it's better to do some preliminary testing to check the assumption.

Preliminary *F*-tests and Pooling Mean Squares

Note: The next few pages do little more than introduce the basic idea of preliminary tests and pooling. For a more complete treatment of the subject, see Kirk, which also lists several articles that give still more detail.

If your design has a compound within-blocks factor, one strategy for constructing *F*-tests is to assume the additive model, and use the residual mean square for the denominator of your *F*-ratios. A second, more conservative strategy is to assume the model of nonadditive block effects, which identifies and isolates three (or more) different sources of variability, one for each factor you get when you expand Blocks × Within-Blocks = Error. This model leads to *F*-tests that use these components of chance error for denominator mean squares, as illustrated in the previous section. There is also a third, intermediate strategy, which grows out of logic like this: "Instead of assuming we have three or more typical error sizes, why not use the data to do a preliminary test? If the test suggests that the sizes are indeed different, then we'll choose denominator MSs using the nonadditive model and the rules of the previous section. On the other hand, if the test suggests that there's just one size after all, then we can recombine our MSs to get a new, pooled MS to use in the denominator. The resulting *F*-tests will be more powerful, because they will use more information from the data. (The denominator df will be larger.)" The next example illustrates this strategy.

EXAMPLE 9.10 REMEMBERING WORDS: PRELIMINARY TESTS AND POOLING

Example 9.8 gave a table showing how to construct an ANOVA table assuming a non-additive model for the data in Exercise 9.C.1. According to that model, you should split the compound factor for error into three pieces, whose SSs, dfs, and MSs are shown below:

Factor	df	SS	MS
Blocks × Freq	9	1248.0	138.67
Blocks × Abstr	9	385.6	42.84
Blocks × Freq × Abstr	9	518.4	57.60
Blocks × Within	27	2152.0	

There are two preliminary *F*-tests, each with the highest-order interaction MS in the denominator. For these tests, we want to be particularly careful not to end up pooling mean squares whose true error sizes are in fact unequal. To gain extra protection against doing this, we use a 25% critical value, which is quite a bit smaller than the usual 5% value: our preliminary test is more likely to declare differences "real," and correspondingly less likely to miss real differences.

$$MS_{\text{Blocks} \times \text{Freq}}/MS_{\text{Blocks} \times \text{Freq} \times \text{Abstr}} = 138.67/57.60 = 2.41$$
$$MS_{\text{Blocks} \times \text{Abstr}}/MS_{\text{Blocks} \times \text{Freq} \times \text{Abstr}} = 42.84/57.60 = 0.74$$

FIGURE 9.10 Preliminary tests for within-blocks error. For both *F*-ratios, numerator and denominator df are equal to 9. The 25% critical value is 1.59.

a. Blocks × Freq. The *F*-ratio is larger than the critical value. The preliminary test tells us not to pool, but rather to use $MS_{\text{Blocks} \times \text{Freq}}$ for testing the effect of frequency.
b. Blocks × Abstr. The *F*-ratio is less than the critical value: the test tells us to pool variability for Blocks × Abstr and Blocks × Freq × Abstr, to get $MS_{\text{Pooled}} = (385.6 + 518.4) \div (9 + 9) = 50.22$ on 18 df. We then use MS_{Pooled} to test the effects of abstraction and of frequency × abstraction. ■

Exercise Set E

1. *Imagery and working memory*. Exercise 9.C.5 gave data from a CB design with a compound within-blocks factor. Tell which MS to use for testing each of Task, Report, and Task × Report, if you assume nonadditive block effects.

2. *Schizophrenia and word associations*. Refer to Example 9.6. Following the format of Example 9.8, tell how to compute *F*-ratios for testing all basic factors of interest and their interactions. Give the df for each row in the ANOVA table, assuming there were 10 subjects in each of the two diagnostic categories.

SUMMARY

1. You can create new designs from any of the four basic ones by replacing any factor of interest with a compound factor that comes from crossing two (or more) basic factors. If the purpose of your experiment is to study a particular interaction, then the purpose tells which factors to cross. Other times you can find additional factors by asking what else might influence your response or might interact with a factor that is already part of your design.

2. You can find the factor structure systematically in two steps:

First find the design structure (reduced form). Ignore any factorial crossing in the treatment combinations, and analyze the design structure (as in Chapter 8): What are the units, and what is the pattern that relates each set of units to treatment combinations? Use this information to identify the design family: BF, CB, LS, SP/RM (or other). Then find the treatment structure (expanded form): Identify all instances of factorial crossing, and replace each compound factor, everywhere it occurs in your factor list, with three factors: first factor, second basic factor, and their interaction.

3. As an alternative, you can often take a shortcut. List all basic factors and interactions in groups, one group for each kind of experimental unit: First list any nuisance factors for grouping the units, then list main effects, two-factor interactions, three-factor interactions, and so on, for treatment combinations assigned to those units. For within-blocks units (subplots) of an SP/RM design, you should also list interactions of within-blocks factors with between-blocks factors.

4. For designs with compound within-blocks factors, there are two alternative models. The model with additive block effects uses Blocks × Within-Blocks Conditions as the factor for residual error; the model with nonadditive block effects expands that factor into pieces that correspond to the factorial structure of the within-blocks treatment combinations.

5. To check for a three-factor interaction, pick one of the three factors, and for each level of that factor draw a graph showing the interaction pattern for the other two factors, then compare to see whether these interaction graphs show the same pattern.

6. a. Denominator MSs assuming nonadditive block effects: To test a factor of interest or interaction use for your denominator the MS for the product, Blocks × Factor or Blocks × Interaction.

b. Pooling. If a preliminary test using a 25% critical value is not significant, you can get more powerful tests by using $MS_{Pooled} = (MS_1 + MS_2)/(df_1 + df_2)$ in the denominator of *F*-ratios that would otherwise use MS_1 or MS_2.

EXERCISE SET F: REVIEW EXERCISES

1–5. *Speeding rats*. The eight rats who volunteered for the following experiment to study the effect of amphetamines on learned behavior were taught a routine for obtaining milk. After initial training, each rat was measured on 6 different days, each day under a different condition. The 6 conditions, which were assigned in a random order to each rat, were chosen to compare the effects of different doses of two forms of amphetamine: dl-amphetamine and d-amphetamine. Each drug was given (by injection) in three doses: 1 mg per kg of body weight, 2 mg/kg, and 3.2 mg/kg. After the drug had been allowed to take effect (30 minutes), the rats were given a chance to obtain milk. The measured response in the table below is the number of times the rat paused while getting the milk. Table 9.15 gives the data, together with row and column averages and SDs.

TABLE 9.15

		dl-amphetamine			**d-amphetamine**				
Drug dose:		**1**	**2**	**3.2**	**1**	**2**	**3.2**	**Avg.**	**SD**
Rat:	A	7	17	106	24	48	56	43	36.0
	B	3	25	24	34	41	26	25.5	12.8
	C	39	82	68	28	64	22	50.5	24.2
	D	38	5	50	15	98	13	36.5	34.6
	E	1	9	22	1	18	42	15.5	15.6
	F	22	54	21	46	17	50	35	16.7
	G	5	2	1	1	3	0	2	1.8
	H	93	72	92	55	57	63	72	16.9
Avg.		26	33.25	48	25.5	43.25	34	35 = Grand	
SD		31.1	31.6	37.6	19.6	30.7	22.2	average	

1. Describe the experimental design: What are the units, factors of interest, and nuisance factors? Which factor or factors of interest are experimental, which are observational? What is the name of the design? Write a summary table of factors.

2. Draw an interaction graph, and tell in words what the pattern suggests about the effects of amphetamines on learned behavior.

3. Do the data need to be transformed? Give the reason(s) for your answer.

4. Write the complete factor diagram, assuming the nonadditive model for chance error.

5. Give a partial ANOVA table: List each source, give the df, and tell which mean square to use in the denominator of the *F*-test.

6–8. *Puzzled children (continued)*. Exercise 6.C.3 is based on a larger study done at Mount Holyoke College. The goal was to see whether social reinforcement (SR) in the form of a teacher's praise had an effect on how many puzzles a young child would solve in a fixed length of time. There were 40 children in the study, classified by age (3, 4, 5, 6 or 7 years) and sex (M or F) into ten groups of four. Each child performed twice, once under control conditions (no praise; the teacher was "busy" with paperwork), and once with social reinforcement (the teacher praised the child for each puzzle solved). For two of the four children in a group (chosen at random), the control condition came first, then the reinforcement. For the other two children, the order was reversed. The data are given in Table 9.16.

TABLE 9.16 Numbers of puzzles solved by 40 children, with and without reinforcement.

		Age 3		Age 4		Age 5		Age 6		Age 7	
Sex	Order	Con	SR	Con	SR	Con	SR	Con	SR	Con	SR
M	Control	24	26	30	48	18	20	29	57	31	66
	first	19	23	19	26	16	35	41	66	53	90
	SR	48	41	5	26	49	73	28	78	84	51
	first	4	10	21	33	59	57	40	93	61	57
F	Control	20	27	68	73	48	52	58	63	54	69
	first	8	5	32	33	35	70	21	97	66	71
	SR	42	14	5	52	93	73	47	37	103	48
	first	63	52	63	46	64	63	82	74	31	44

6. Describe the experimental design: What are the units, factors of interest, and nuisance factors? Which factor or factors of interest are experimental, which are observational? What is the name of the design? Write a summary table of factors.

7. Give a partial ANOVA table: List each source, give the df, and tell which mean square to use in the denominator of the F-test.

8. Use the table of averages given below as the basis for a graphical exploration of the interaction structure of the data. Then write a summary describing what you consider the most important patterns. (To get the table, I first lumped together the children aged 3 and 4, ignored the children aged 5, and lumped together the children aged 6 and 7; then I computed condition averages and rounded to the nearest whole number.)

	Males				Females			
	Age 3–4		Age 6–7		Age 3–4		Age 6–7	
	Con	SR	Con	SR	Con	SR	Con	SR
Control first	23	31	39	70	32	35	50	75
SR first	20	55	53	70	43	41	66	51

9–11. *Oxygen pressure*. Each of eight patients, while in surgery, had oxygen pressure readings taken in two of their veins, hepatic and portal, under two conditions, control and with the femoral artery clamped. Units of measurement are mm Hg (millimeters of mercury). The data are given in Table 9.17.

TABLE 9.17

Subject	Hepatic vein		Portal vein	
	Control	Clamped	Control	Clamped
1	27	17	72	81
2	33	15	61	46
3	33	22	54	44
4	27	22	36	41
5	36	26	54	66
6	30	23	43	42
7	33	31	33	32
8	25	27	35	26

9. Describe the experimental design, assuming there is only one response, the oxygen pressure: what are the units, factors of interest, and nuisance factors? Which factor or factors of interest are experimental, which are observational? What is the name of the design? Write a summary table of factors.

10. Write the complete factor diagram, assuming the nonadditive model for chance error.

11. Give a partial ANOVA table: List each source, give the df, and tell which mean square to use in the denominator of the *F*-test.

NOTES AND SOURCES

1. Notation for SP/RM designs. The first number tells how many between-blocks (basic) factors there are; the second tells how many within-blocks (basic) factors. Thus the notation SP/RM[2;1] means that the between-blocks treatments have a two-way factorial structure; the within-blocks treatments have a one-way structure.

2. Preliminary F-tests and pooling mean squares. For more information, see Kirk, Roger E. (1982). *Experimental Design*, 2nd ed. Belmont, CA: Brooks/Cole Publishing Company.

Appendix: Supplementary Examples

Extending the BF Design: General Principles

EXAMPLE S9.1 CHOOSING THE FACTORS TO CROSS

a. *Purpose is to measure interaction.*

i. Hornworms (Exercises 8.A.1 and 7). The goal of this experiment was to study the interaction between the food a hornworm was raised on and the test food, as measured by the amount it ate.

ii. Mental health and interference (Example S4.8). This study, too, grew out of a particular hypothesis about interaction: that people with psychiatric problems would be more affected by a distracting voice than normal control subjects would be. To test this hypothesis, the psychiatrist crossed two factors of interest: psychiatric classification, and interference condition.

b. *Looking for other influences.*

i. *Bird calcium.* The experiment described in Exercise 5.B.1 was designed to measure the effect of hormone supplements on the level of calcium in the blood plasma of birds. You can imagine a simple CR plan with one factor of interest: hormone supplement, present or absent. As you consider this plan, if you realize that males and females might respond differently to the hormone, you might decide to cross sex with hormone supplement to get four cells.

ii. *Overconfidence in case study judgments.* Example 8.6 described an experiment by Stuart Oskamp, who gave his subjects four chunks of biographical information about Joseph Kidd, and measured how accurately and how confidently his subjects answered a set of 25 questions as they got more and more information. The main factor of interest was the amount of information the subjects had seen. Oskamp recognized, however, that by using subjects of different kinds—clinical psychologists, graduate students, and undergraduates—he could build a second factor of interest into the design without increasing the total number of subjects. ■

EXAMPLE S9.2 FACTOR STRUCTURE FOR THE HORNWORM DATA (EXERCISES 8.A.1 AND 7)

Step 1: *Design structure*: We ignore the factorial crossing, regard the four conditions as levels of a single compound factor, recognize the design as a CR, and write its factor diagram (Figure S15).

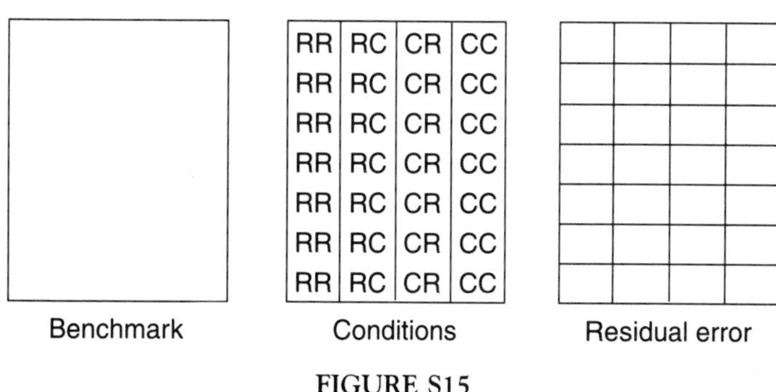

FIGURE S15

Step 2: *Treatment structure*: The compound factor appears only once in the diagram in Step 1. We replace this factor (conditions) with three: first basic factor (rearing food, R or C), second basic factor (test food, R or C), and their interaction (Figure S16).

RR RC CR CC

Reg Cel

Rearing food

RR RC CR CC

Reg	Cel	Reg	Cel
Reg	Cel	Reg	Cel
Reg	Cel	Reg	Cel
Reg	Cel	Reg	Cel
Reg	Cel	Reg	Cel
Reg	Cel	Reg	Cel
Reg	Cel	Reg	Cel

Test food

RR	RC	CR	CC
RR	RC	CR	CC
RR	RC	CR	CC
RR	RC	CR	CC
RR	RC	CR	CC
RR	RC	CR	CC
RR	RC	CR	CC

Interaction

Conditions

FIGURE S16 Expanding the compound factor. The compound factor (conditions) is expanded to give three factors: (1) First basic factor, rearing food, sorts columns (conditions) into groups that correspond to levels of that factor. Columns 1 and 2 go into one group (R); 3 and 4 go in the other (C). (2) Second basic factor, test food, sorts columns a second time: 1 and 3 together (test food = R); 2 and 4 together (test food = C). (3) Interaction. The groups here will always be the same as for the compound factor itself in Step 1. ■

Three or More Crossed Factors of Interest

EXAMPLE S9.3 FEAR OF RODENTS: ANOTHER CR[3]

Would you rather touch a rat, or eat roasted caterpillars? The question may seem bizarre, but it sets the tone for the experiment it came from.

The purpose of this study was to test specific predictions about the effect of mentally rehearsing an unpleasant task. There were 216 subjects, and each was assigned to one of 18 conditions, 12 subjects per condition. The conditions came from crossing three basic factors: Fear, Expectancy, and Delay.

The first factor, Fear, was observational. All 633 female students in the introductory psychology courses at Southern Illinois University took a standard test which was used to measure their fear of white rats. On the basis of the test, 108 "high fear" subjects and 108 "low fear" subjects were chosen. Each of the 216 subjects was eventually given a choice between touching a rat or judging which of two weights was heavier. The choice was recorded, and then the experimenter asked each subject, regardless of her choice, if she would actually be willing to approach a cage at the far side of the room, open the cage, and touch the rat inside. The response variable used a scale from 0 (subject approached no closer than 12 feet) through 13 points (subject approached all the way to the cage), 14 points (subject opened the cage), and 15 points (subject touched the rat).

Each set of 108 subjects, (high fear and low fear) was randomly divided into nine groups of 12, one for each of nine conditions defined by crossing the other two (experimental) factors, Expectancy, and Delay.

Expectancy (3 levels):

1. Negative expectancy (NE): subjects were told that at the end of the experiment, they would be asked to touch a white rat.
2. Incorrect negative expectancy (INE): subjects were shown a plate of three caterpillars, and told that at the end of the experiment, they would be asked to eat the caterpillars. (In fact, the subjects would actually be asked to touch a rat instead, but they didn't find this out until later.)
3. Free choice (FC): subjects were told they'd be given the choice of touching a rat or judging which weight was heavier.

The three levels of Expectancy were chosen in order to compare the behavior of subjects who were told about touching the rat, and who could mentally rehearse, with other subjects who could not rehearse because they were not told the actual task. The third factor, Delay Interval, had 3 levels: 0 minutes, 10 minutes, and 40 minutes, between the initial instructions and the time when the subject was asked to choose a task and then asked if she would touch the rat. The experimenter's hypothesis was that subjects who were told they'd be asked to touch a rat, and who were given time to rehearse mentally, would be more likely to touch the rat, in comparison with other subjects. (Results from this experiment appear in Exercise 9.D.2.)

Step 1: *Design structure*: The unit is a subject, and the design structure is CR, although one of the factors defining the conditions was observational and involved no randomization.

Step 2: *Treatment structure*. The summary table (Table S12) of factors has the same form and pattern as in the crabgrass example (11.3):

TABLE S12 Summary table of factors.

	Grouping	Treatment combinations		
Unit	(Nuisance factors)	Main effects	2-Factor interactions	3-Factor interaction
Subject	(none)	Fear	Fear × Exp	Fear × Exp × Delay
		Exp	Fear × Delay	
		Delay	Exp × Delay	

■

EXAMPLE S9.4 CRABGRASS, VERSION 2: A CR[4]

In this second version of the crabgrass experiment, there were 64 cups, each randomly assigned to get one of 32 treatment combinations. One way to think of these combinations is to take the 8 nutrient combinations from Example 9.3, and cross them with another basic factor, Density, at four levels: 20, 15, 10, and 5. Cups assigned to 20 got 20 plants of *D. sanguinalis*; cups assigned to 15 got 15 plants of *D.s.*, and 5 of the other species, *D. ischaemum*; cups assigned to 10 got 10 of each species; cups assigned to 5 got 5 of *D.s.* and 15 of *D.i.* Although many of the cups contained plants of both species, the response for this data set used only the *D.s.* plants; the response is the average dry weight (per plant) of those plants in the cup.

Step 1: *Design structure*: Treatment combinations get assigned to cups. The cup is the only unit, and the design is CR.

Step 2: *Treatment structure*: You can always fall back on the "brute force" method for expanding compound factors, applying Step 2 over and over, each time splitting off one basic factor, as in Example 9.3 (see Figure S17):

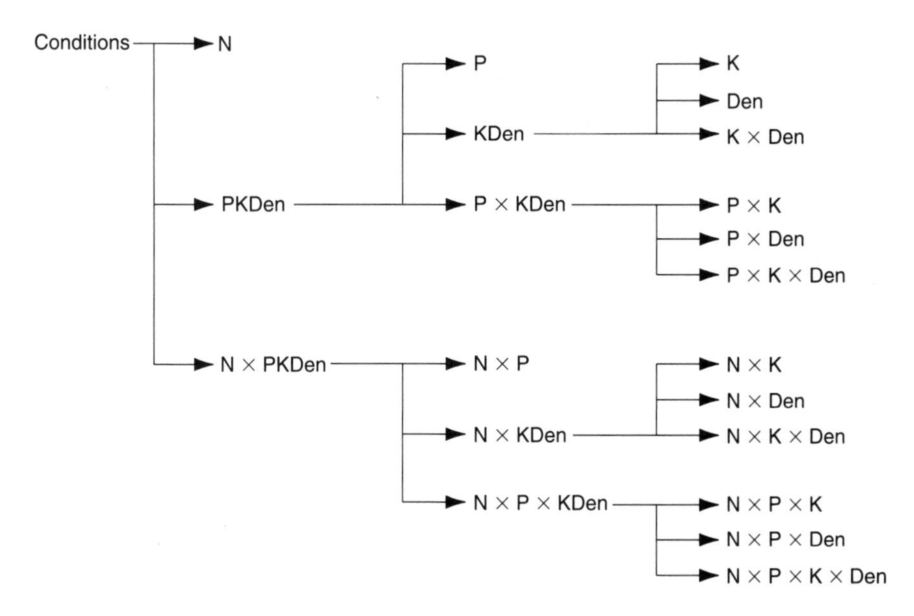

FIGURE S17

We can reorganize the list of factors in a summary table (Table S13):

TABLE S13 Summary table of factors: 4-way CR.

	Grouping	Treatment combinations			
Unit	(Nuisance factors)	Main effects	2-Factor interactions	3-Factor interactions	4-Factor interactions
Cup	(none)	N	N × P	N × P × K	N × P × K × Den
		P	N × K	N × P × Den	
		K	P × K	N × K × Den	
		Den	N × Den	P × K × Den	
			P × Den		
			K × Den		

Every factorial experiment that crosses four basic factors will have four main effects, six 2-factor interactions, four 3-factor interactions, and one 4-factor interaction.

Finding the partitions for these factors would require no additional ideas, just additional patience. ■

Compound Within-Blocks Factors

EXAMPLE S9.5 SUBMARINE MEMORY: A 2-WAY LS, OR LS[2]

The purpose of this experiment, described in Example 6.2, was to study the interaction between learning environment (dry or wet) and recall environment (dry or wet). Crossing these two factors of interest gives four conditions:

Condition	Learn	Recall
A = DD	Dry	Dry
B = DW	Dry	Wet
C = WD	Wet	Dry
D = WW	Wet	Wet

If you had been a subject in this experiment, you would have been asked to learn a list of 36 words, either sitting on the shore (dry), or under twenty feet of water (wet). Then later you would be asked to recall as many of the words as you could, either in the dry or the wet condition. After that, you'd go on to a new list of 36 words, and a new pair of conditions, then to a third list, and finally a fourth, so that in all you would provide a response under each of the four combinations of environments.

The investigators used sixteen subjects, and they had four lists of words; both subjects and lists are nuisance factors. They used Latin square designs to match subjects with lists and conditions. Figure S18 shows a version of a Latin square plan for the first four subjects. Find its factor diagram. (For simplicity, I have presented the design without randomizing. In practice, of course, you should randomize.)

Word List

	1	2	3	4
I	A	B	C	D
II	C	D	A	B
III	D	A	B	C
IV	B	C	D	A

FIGURE S18 Latin square plan for the memory experiment. There was a separate Latin square plan for each set of four subjects, I–IV, V–VIII, IX–XII, XIII–XVI.

SOLUTION:

Step 1: *Design structure*: The unit is a time slot for a subject. The units were grouped by two crossed factors, subject and word list, before conditions were randomly assigned; the design is LS.

Step 2: *Treatment structure*: Expanding the compound factor treatments gives Table S14:

TABLE S14 Summary table of factors: 2-way LS.

Unit	Grouping (Nuisance factors)	Treatment combinations: Main effects	Treatment combinations: 2-Factor Interaction
Time slot	Subjects Lists	Learn Recall	Learn × Recall

Factor diagram. The complete factor diagram will have the two universal factors, the usual two nuisance factors (rows = Subjects, columns = Lists), plus three factors of interest, whose partitions are shown in Figure S19.

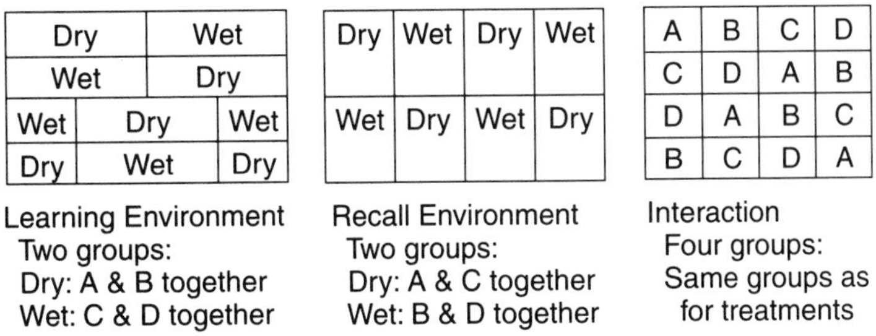

Dry		Wet	
Wet		Dry	
Wet	Dry		Wet
Dry	Wet		Dry

Learning Environment
Two groups:
Dry: A & B together
Wet: C & D together

Dry	Wet	Dry	Wet
Wet	Dry	Wet	Dry

Recall Environment
Two groups:
Dry: A & C together
Wet: B & D together

A	B	C	D
C	D	A	B
D	A	B	C
B	C	D	A

Interaction
Four groups:
Same groups as for treatments

FIGURE S19

Step 2*: Expand the factor for residual error? No: There's not enough information in the data. (However, for the full version of the experiment, with sixteen subjects in four Latin squares, it would be reasonable to expand the factor for error.) ■

EXAMPLE S9.6 CRABGRASS, VERSION 3: A CB[4]

Very often in biology an experiment with factorial structure is run as a series of smaller experiments, called **replications**, with each replication having the structure of a factorial experiment with one observation per cell. The version of the crabgrass experiment that was actually done had two replications, each with 32 cups randomly assigned to the same 32 conditions listed in Version 2. Table S15 shows the data, mean dry weight per plant, in milligrams:

TABLE S15

Nitrogen →		*n*	*n*	*n*	*n*	–	–	–	–	
Phosphorus →		*n*	*n*	–	–	*n*	*n*	–	–	
Potassium →		*n*	–	*n*	–	*n*	–	*n*	–	
Density	20	96.4	86.1	59.6	68.9	39.7	35.4	36.9	42.2	Rep I
	15	95.7	107.3	67.4	70.1	35.0	38.3	43.5	53.7	
	10	117.0	123.4	92.0	72.0	41.8	39.1	47.8	39.8	
	5	224.3	162.7	88.2	54.0	53.3	75.3	47.4	86.6	
	20	73.4	77.8	64.7	47.1	41.4	40.6	45.4	79.0	Rep II
	15	91.0	60.6	972.5	68.6	36.4	50.8	38.4	29.1	
	10	87.4	86.7	66.5	56.0	43.1	53.4	47.3	55.1	
	5	130.0	128.6	82.4	91.0	46.5	50.3	49.5	74.9	

(This data set contains an obvious outlier, with a sad story behind it.)
n: normal level
–: low level

Step 1: *Design structure*: The unit is a cup, but for this version, as opposed to the first two, the cups are grouped into two blocks of 32 cups each before the treatments are assigned. The design is a CB.
Step 2: *Treatment structure*:
Step 2*: *Expanding the factor for residual error*: Although it would be possible to expand the factor for error, there are so few blocks that each of the resulting 15 factors in the expanded form would contain very little information. You could assume the block effects are additive, or, as an alternative, you could use a compromise model, which I prefer. Think of the chance error factor as the result of crossing three factors: Blocks, Density, and Nutrients (a compound factor NPK, with 8 levels). Expand error = Blocks × Conditions to get three factors: Blocks × Density, Blocks × Nutrients, and Blocks × Density × Nutrients. Then plan to use the methods of Chapter 9, Section 5 to test whether the simpler model, with additive block effects, is reasonable. ■

NOTES

Example S9.6. The 972.5 in Rep II at Den = 15 looks like a typographical error, and indeed it is. The graduate student who did the experiment recorded her data in grams rather than milligrams, which made the typo harder to spot. The fifth row of her data looked like this: 0.0910 0.0606 0.9725 0.0686. She fed her data, typo and all, into a computer, and asked for an analysis of variance. She took the results at face value, looking only at a summary of the analysis, without doing any informal analysis. She wrote the sections of her master's thesis dealing with results, conclusions, and discussion based on the ANOVA summaries alone. When the typo was discovered, one week before she was scheduled to defend her thesis, she had to redo her analysis and then spend the next month rewriting about one third of her thesis. (At least her situation could be salvaged. About 35 years ago, NASA lost an 18 million dollar rocket because of a typo: someone left a hyphen out of a computer program. When the computer couldn't find the hyphen, it instructed the rocket to blow itself up.)

CHAPTER 10

DECOMPOSING A DATA SET

OVERVIEW

This chapter presents a new general rule for decomposing a data set. The main reason for learning the new rule is that it shows how the process of decomposition is tied—by way of the assembly line metaphor—to the theoretical model that underlies analysis of variance.

Decomposition as harmonic analysis. According to the ANOVA model, each observed value is a sum of pieces, one piece for each influence on the response—each factor in the design. Our goal is to understand the structure of the data by splitting the numbers apart to get estimates for the various pieces. Back at the beginning of Chapter 2, I described a parallel between decomposition of data and Newton's decomposition of white light into light of various colors. That parallel can in fact be made quite precise. For example, astronomers use a fancier version of Newton's idea to analyze the light from stars. The light is a mixture of several kinds, each with its own wavelength (color), and for a given star, each wavelength is present with its own intensity. The different kinds of light (wavelengths) are like the factors of a design; they correspond to the different workstations of the assembly line. Astronomers measure the intensity at each wavelength; this is like estimating the size of the piece added on at each workstation.

You can think about sound in the same way. When the oboe at a concert plays an A for the orchestra to tune to, the sound is actually a mixture of sounds of various pitches (frequencies). Each of these pitches is present with its own intensity; the pitches are like workstations, and the intensities are like the sizes of the pieces added on. The method called **harmonic analysis** takes a sound wave and decomposes it into a sum of pieces, one piece for each of several wavelengths.

During the last century, mathematicians have found ways to generalize the idea of harmonic analysis so that our current understanding includes the linear decomposition of data sets as one special case. Figure 10.1 shows the linear decomposition of the leafhopper data (Example 5.11) in a graph of the sort that is often used to show the decomposition of a sound wave.

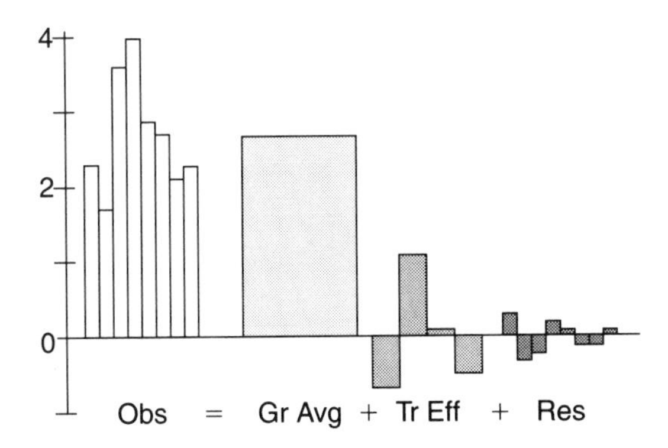

FIGURE 10.1 Linear decomposition as a kind of harmonic analysis. The graph of the leafhopper data (Example 5.11) on the left (Obs), is like a sound wave, which gets analyzed into component waves of different frequencies. The first piece (Gr Avg) is for the factor with the largest groups, and corresponds to the lowest, slowest sound frequency. Next comes the effect of the diets (Tr Eff), for the factor with middle-sized groups. Then comes the set of residuals, for the factor with the smallest groups; its graph bounces around the most, and corresponds to the highest frequency sound.

Disassembly line metaphor. I assume that you have read the sections on decomposition and ANOVA in Chapters 5–7, where I showed you the rule "Estimated effect = Average − Partial fit." Remind yourself how that rule sends you back to the observed values to get estimated effects: each estimated effect answers the question, "How far is it from the average of the observed values to the partial fit?" This approach has the important advantage of directly relating each estimated effect to an average of the observed values.

The new rule, on the other hand, is closer to the idea behind the assembly line metaphor, that each observed value gets "built" one piece at a time, at a sequence of workstations. You can think of the new rule as a disassembly line: Imagine Kelly's entire data set moving on a conveyer belt past a sequence of workstations. At the Benchmark station, you first compute the grand average, then subtract that number from each of the observed values, and send the set of leftovers (Obs − Gr Avg) on down the line. At the next workstation (Day Length), you use the incoming leftovers to compute two averages, one for the long day leftovers, one for short. These averages of leftovers are your estimated effects for day length, and are the same as the numbers you would get using the old rule. Still at the day length station, you subtract the day length effects from the incoming leftovers, to get a new set of outgoing leftovers, which you send on to the next station. At each workstation the process is the same: you sort the incoming leftovers into groups according to some factor, compute an average for each group—these are the estimated effects for the factor—then subtract the averages from the incoming leftovers to get outgoing leftovers for the next step. After you have computed one set of averages for each factor except Error, you'll have a final set of leftovers, the residuals. This decomposition process gives exactly the same results as the decomposition rule you already know.

The next section takes a closer look at how a single step of the new decomposition rule works, and applies that basic step (or SWEEP) to decompose data from BF designs. Then the following section shows a systematic way to combine basic SWEEP steps to decompose data from CB, LS, and SP/RM designs. Because the arithmetic soon gets to be tedious, it makes sense to get a computer to do the grunt work for you, but doing a few by hand is a good way to fix the ideas in your mind.

1. THE BASIC DECOMPOSITION STEP AND THE BF[1] DESIGN

The Basic Decomposition Step [SWEEP]

Here is a summary of the basic decomposition step, in three parts—IN (what you start with), SWEEP (what you do), and OUT (what you get).

The structure of the basic decomposition step:

IN: (1) A table of numbers.
(2) A factor.

SWEEP: (1) Sort the numbers into the groups of the factor.
(2) Compute the average for each group of numbers.
(3) Get a leftover from each number in the table, by subtracting the average for its group.

OUT: (1) A table of averages.
(2) A table of leftovers.

EXAMPLE 10.1A LEAFHOPPER SURVIVAL DATA

Example 5.11 and Exercise 5.C.1 described an experiment to compare the nutritional value of three kinds of sugar (sucrose, glucose, and fructose) for potato leafhoppers. For this CR[1] experiment, eight batches of leafhoppers were randomly divided into four groups of two batches each. One group got a control diet, and each of the other groups got the control diet with one of the three kinds of sugar added. The response was the time in days until half the leafhoppers in a batch had died.

Control	2.3	1.7
Sucrose	3.6	4.0
Glucose	2.9	2.7
Fructose	2.1	2.3

i. Rather than tackle the whole data set at once, let's begin by supposing we had only the control group, with its two values, 2.3 and 1.7. It would be natural to take the average of these two numbers as an estimate of the piece common to both, and use the distance from each number to the average as an estimate of residual error.

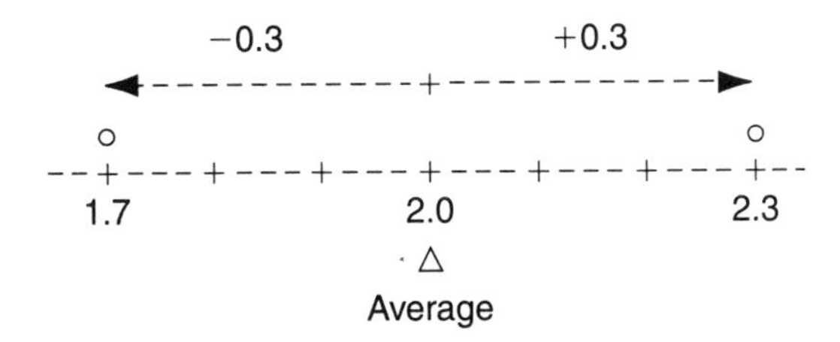

What I have just done gives a particularly simple example of the basic decomposition step.

IN: (1) A table of numbers. (2) A factor.

2.3	1.7

SWEEP: (1) Sort into groups: here there's only one group.

(2) Compute an average for each group:

$$\text{Avg} = (2.3 + 1.7) \div 2 = 2.0$$

(3) Get leftovers by subtracting:

$$\begin{aligned} \text{Obs} - \text{Avg} &= \text{Leftover} \\ 2.3 - 2.0 &= +0.3 \\ 1.7 - 2.0 &= -0.3 \end{aligned}$$

OUT: (1) A table of averages. (2) A table of leftovers.

2.0	2.0

+0.3	−0.3

We can summarize the results of the decomposition step by writing the original table of numbers as a sum of two new tables:

Obs			Ave			Leftover	
2.3	1.7	=	2.0	2.0	+	+0.3	−0.3

The two leftovers are our estimates for the residual errors.

ii. We could carry out the same decomposition for each of the other groups:

	Obs			Avg			Leftover	
Sucrose	3.6	4.0	=	3.8	3.8	+	−0.2	+0.2
Glucose	2.9	2.7	=	2.8	2.8	+	−0.1	−0.1
Fructose	2.1	2.3	=	2.2	2.2	+	−0.1	+0.1

Notice that here, as with any decomposition, (1) the numbers add up according to where they sit in the boxes, and (2) the numbers in each box of leftovers add to zero.

iii. As I'm sure you have guessed, we can put the four little decompositions together, stacked on top of each other, to get a partial decomposition for the whole data set:

	Obs			Avg			Leftover	
Control	2.3	1.7		2.0	2.0		0.3	−0.3
Sucrose	3.6	4.0	=	3.8	3.8	+	−0.2	0.2
Glucose	2.9	2.7		2.8	2.8		0.1	−0.1
Fructose	2.1	2.3		2.2	2.2		−0.1	0.1

Notice that the two properties listed above are still true: the numbers add up according to where they sit in the table, and the leftovers add to zero within each group of the factor.

iv. We could have gotten the same result by applying the basic decomposition step to the whole data set, using the structural factor (Diets) for groups:

IN: (1) A table of numbers. (2) A factor.

Control	2.3	1.7
Sucrose	3.6	4.0
Glucose	2.9	2.7
Fructose	2.1	2.3

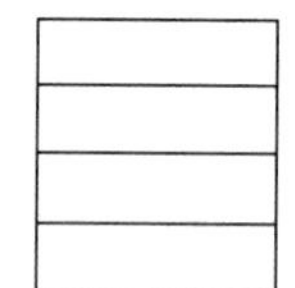

SWEEP: (1) Sort the observations into groups.

(2) Compute the average for each group.

(3) Subtract the averages to get leftovers.

OUT: (1) A table of averages. (2) A table of leftovers.

Control	2.0	2.0
Sucrose	3.8	3.8
Glucose	2.8	2.8
Fructose	2.2	2.2

0.3	−0.3
−0.2	0.2
0.1	−0.1
−0.1	0.1

For this data set, as for any BF, the leftovers from the condition averages are the estimates for the chance errors. ■

Complete Decomposition of a Balanced BF[1] Data Set

The example that follows illustrates two equivalent ways to decompose a BF[1] data set using two decomposition steps. In each case, the complete decomposition will split the observed values into the same three pieces: Grand average + Condition effect + Residual. In each case, also, there will be one decomposition step for each factor except error. For a BF[1] experiment, this means two SWEEP steps, one for the Benchmark, and one for Conditions. The two ways to decompose a BF[1] differ only in the order of the two SWEEPS. One uses conditions first, then benchmark; the other reverses that order. The decomposition you end up with is the same either way, and the same as in Chapter 5.

Each time you do a decomposition step, you have to make two choices: (1) what table of numbers to decompose, and (2) which factor to use. It's a good idea to plan your steps ahead of time, and you can use a flowchart to record your plan, as in the following example.

Example 10.1b Complete Decomposition of the Leafhopper Data, Version 1

We've already done one step of the decomposition, in Example 10.1a.

Step 1: (1) Table to decompose: observed values
(2) Factor: conditions (Diets)

This step broke the observed values into Condition averages + Residuals. For the second step, we'll decompose the box of condition averages, using the benchmark as our factor, to get Grand average + Condition effects.

Step 2: (1) Table to decompose: Condition averages
(2) Factor: benchmark

We can abbreviate our plan for the two steps with a flowchart (Figure 10.2).

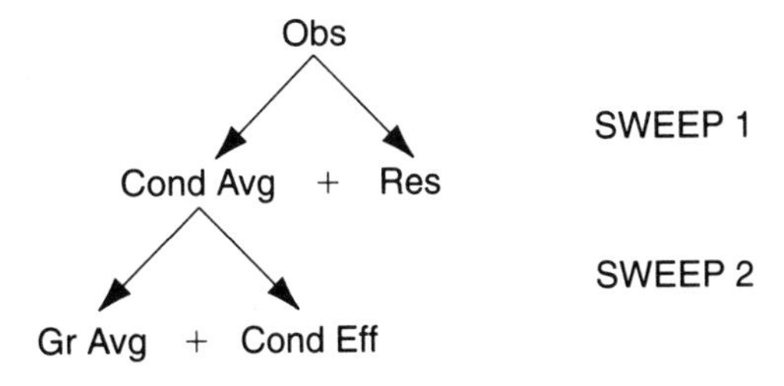

FIGURE 10.2 Flowchart for the two sweeps. The first pair of arrows is for SWEEP 1, and tells us that "Obs" is the table to decompose. The arrows point to what comes out: a set of averages on the left, and the leftovers on the right. (Always put the leftovers on the right.) "Cond Avg" tells us that SWEEP 1 uses Conditions as the factor. The second pair of arrows tells us that SWEEP 2 will decompose "Cond Avg" into "Gr Avg" (so the factor is the benchmark) and "Cond Eff" (as leftovers).

SWEEP 1 corresponds to what we did in Example 10.1a (see Figure 10.3). Before going on, you might want to turn back to part (iii) of that example to see the results of that first decomposition step: Obs = Cond Avg + Res.

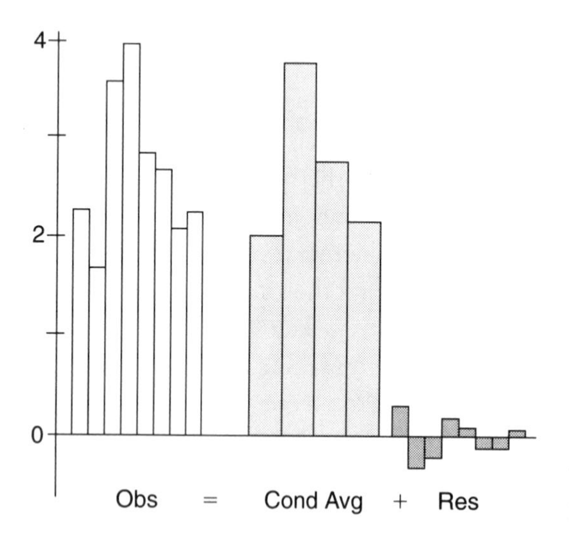

FIGURE 10.3 Observed values decomposed into Condition Averages plus Residuals.

SWEEP 2 first takes the box of condition averages and computes the grand average, using the benchmark as the factor (see Figure 10.4).

$$\text{GrAvg} = \frac{2.0 + 3.8 + 2.8 + 2.2}{4} = \frac{10.8}{4} = 2.7.$$

To complete SWEEP 2, you subtract the grand average from each condition average to get condition effects.

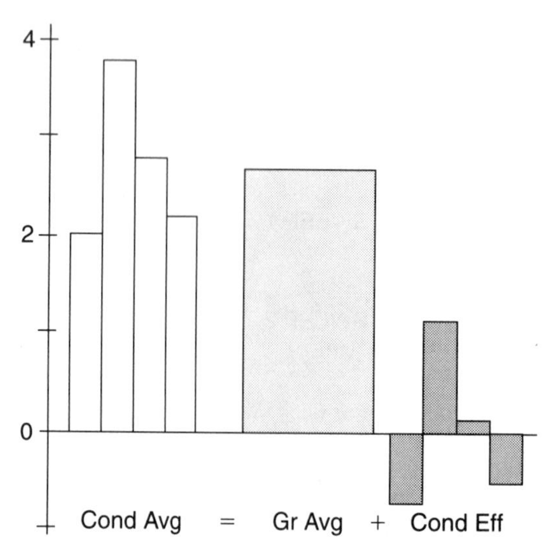

FIGURE 10.4 Condition Averages decomposed into Grand Average plus Condition Effects.

Putting all this in boxes gives the results of SWEEP 2:

	Cond Avg			Gr Avg			Cond Eff	
Control	2.0	2.0		2.7	2.7		−0.7	−0.7
Sucrose	3.8	3.8	=	2.7	2.7	+	1.1	1.1
Glucose	2.8	2.8		2.7	2.7		0.1	0.1
Fructose	2.2	2.2		2.7	2.7		−0.5	−0.5

Notice that the numbers add up according to where they sit in the boxes, and that the condition effects add to zero over the whole box, because we got them by sweeping out the grand average.

Now let's put the two steps together. Start with the 2.3 in the upper left corner of the box of observed values, and follow its progress through the decomposition. SWEEP 1 splits the 2.3 (Obs) into 2.0 (Cond Avg) plus 0.3 (Res) Then SWEEP 2 splits the 2.0 (Cond Avg) into 2.7 (Gr Avg) plus −0.7 (Cond Eff). Together, the two steps have split the 2.3 (Obs) into a sum of three pieces: 2.7 (Gr Avg) plus −0.7 (Cond Eff) plus 0.3 (Res) (see Figure 10.5).

SWEEP 1:	Obs	Cond Avg						Res
	2.3 =	2.0			+			0.3
SWEEP 2:		Cond Avg		Gr Avg		Cond Eff		
		2.0	=	2.7	+	(−0.7)		
Together:	Obs			Gr Avg		Cond Eff		Res
	2.3 =			2.7	+	(−0.7)	+	0.3

FIGURE 10.5 Combining the two decomposition steps.

We can do the same thing for each of the other observations, then put the results in boxes that correspond to the factor diagram. If you do this, you get the same decomposition as in Example 5.12. ■

The decomposition we have just done used factors in the order "conditions first, then benchmark." You can also use the reverse order; each order has its advantage. The approach that uses conditions first makes the arithmetic easier to follow, and is easier to understand at first. However, the order that uses benchmark first (outside factors before inside factors, bigger groups before smaller groups) is easier to generalize to other design structures. With the first order, you have to think about which table to decompose at the next step, averages or leftovers. For a BF this is not hard, but for other designs, it makes life unnecessarily complicated. With the second order, outside factors first, you don't have to think about the choice, because at every step you always decompose the leftovers from the last step. Using this order gives a simple rule that works for all balanced designs.

EXAMPLE 10.1C COMPLETE DECOMPOSITION OF THE LEAFHOPPER DATA, VERSION 2

Just as before, to get a complete decomposition will take two steps, one for each factor except error. This time, we take the benchmark first.

SWEEP 1: Decompose the original data, using the benchmark factor.
This will split the observed values into a grand average and a set of leftovers.

SWEEP 2: Decompose the leftovers from SWEEP 1, using conditions as the factor.
This will split the leftovers into a box of condition effects (the averages) and a box of residuals (the leftovers).

We can summarize this plan with the flowchart in Figure 10.6.

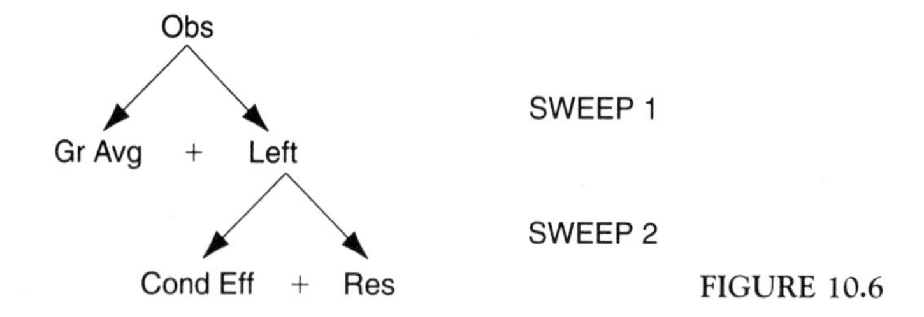

FIGURE 10.6

The first arrows tell us to start with the observed values. The flowchart doesn't say explicitly to use Benchmark as the factor in SWEEP 1, but "Grand Average" refers to the average you get using benchmark as the factor. The second set of arrows says to decompose the leftovers, and the "Cond Eff" on the left reminds us to use Conditions for the factor.

SWEEP 1: Table = Obs, Factor = Benchmark.

	Obs			Gr Avg			Leftover	
Control	2.3	1.7		2.7	2.7		−0.4	−1.0
Sucrose	3.6	4.0	=	2.7	2.7	+	0.9	1.3
Glucose	2.9	2.7		2.7	2.7		0.2	0.0
Fructose	2.1	2.3		2.7	2.7		−0.6	−0.4

To get the grand average, I added up all twelve numbers and divided by 8. The value of the average (2.7) is the same as before. Notice that the leftovers from the grand average don't add to zero across the rows or down the columns, but they do add to zero over the whole box.

SWEEP 2: Table = Leftover, Factor = Conditions.

	Leftover			Cond Eff			Residual	
Control	−0.4	−1.0		−0.7	−0.7		0.3	−0.3
Sucrose	0.9	1.3	=	1.1	1.1	+	−0.2	0.2
Glucose	0.2	0.0		0.1	0.1		0.1	−0.1
Fructose	−0.6	−0.4		−0.5	−0.5		−0.1	0.1

To get the condition effects, I took the averages separately for each row of the box of leftovers from Step 1. For example, the first condition effect, −0.7, equals the average of −0.4 and −1.0. The condition effects are the same as I got doing the steps in the other order. They add to zero because they came from the box of leftovers, which added to zero.

Combining the two steps gives the complete decomposition, which is the same as in Examples 10.1b and 5.12.

This last example (10.1c) illustrates a method that you can use to decompose the data from any balanced design. The next section gives the general rule and shows how to use it for CB, LS, and SP/RM designs. ■

Exercise Set A

1–3. Two small versions of the leafhopper data are given below. First decompose each little BF using the old rule, as in Chapter 5. Then decompose the data a second time, using the new rule and following the flowchart given next to the data set. Check that you get the same decomposition both ways.

1.

Diet 1	1	3
Diet 2	7	11
Diet 3	4	4

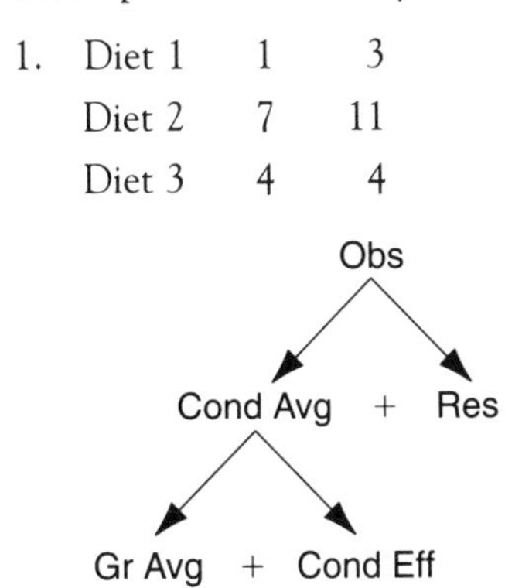

2.

Diet 1	2	4	6
Diet 2	5	1	0

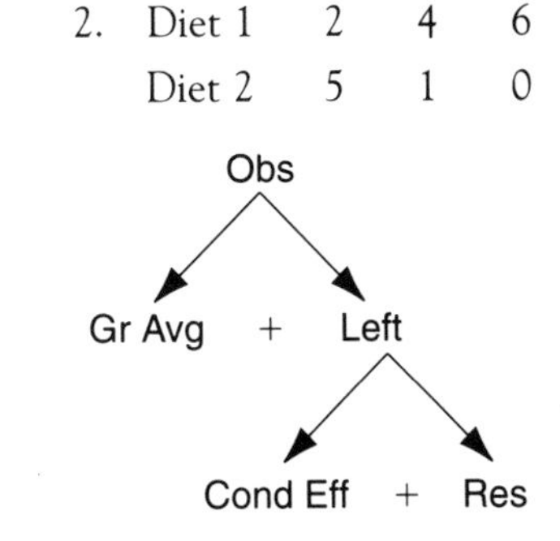

3. Decompose the data set #1 in the other order, that is, according to the flowchart for #2, and check that your final decomposition is the same as the one you got in #1.

4–5. When you decompose a data set, you do one decomposition step for each factor in the factor diagram, except for residuals. Provided your design is balanced, it doesn't matter what order you choose for the steps: you get the same final result no matter what order you use.

Each decomposition step gives you two boxes of numbers, a box of averages, and a box of leftovers. Which box should you decompose at the next step? There are two choices, and what happens if you choose wrong? It turns out there's a foolproof way to tell if you make a wrong choice. That's the point of this exercise.

You've already decomposed the data of Exercise 1 using two different orders. In Exercise 1, you took out condition averages first, then the grand average; in Exercise 3 you reversed the order. Each of Exercises 4 and 5 below shows two flowcharts, one right and one wrong. Do the decomposition steps indicated by the wrong flowchart in each case. Then use your results to tell how you can recognize a wrong choice for which box to decompose. (Use the data from #1.)

4.

5.

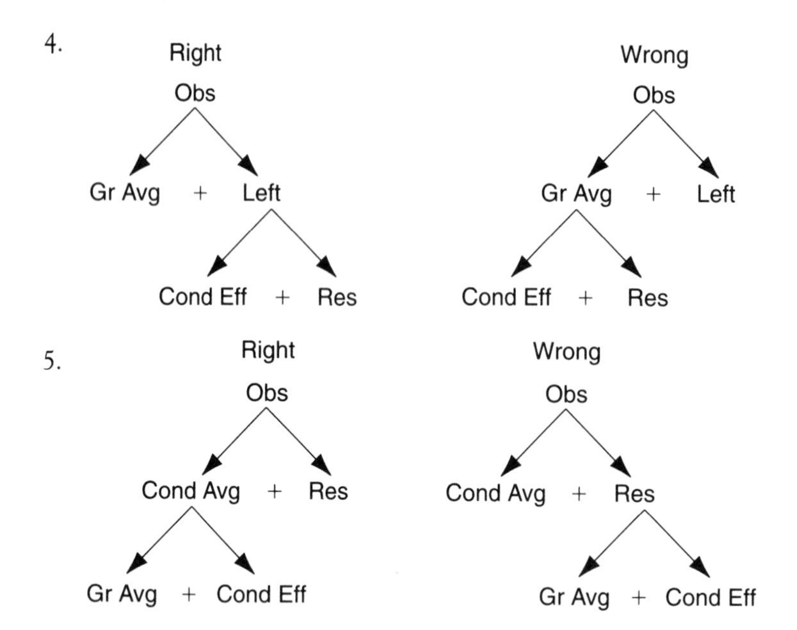

(Don't forget to write out what you discover from doing the arithmetic: How do you recognize a wrong choice?)

6. *The defense of undoing: an unbalanced one-way design*. IN GENERAL, THE FORMAL METHODS OF THIS BOOK (DECOMPOSITION AND ANOVA) DO NOT WORK FOR UNBALANCED DATA SETS, BUT THEY *DO* WORK FOR UNBALANCED BF[1] DESIGNS, such as the walking babies experiment (Example 5.1), the study to rate Milgram's experiment (Exercise 5.B.5) and the study of the defense of undoing (Example S4.10). (If you have a data set that needs only one or two more observations in order to be balanced, you can use the methods of Chapter 12, Section 5 to estimate "replacement" values for these observations, then analyze the data in the usual way. If you have a data set that is not close to being balanced, and not a BF[1], consult a statistician or take a more advanced course: for such data sets, the analysis is not only computationally more involved, but conceptually much more difficult as well.)

Although you can decompose unbalanced BF[1]s using either the method of Chapter 5 or the method of this chapter, many of the shortcuts that work for balanced BF[1]s do not work for unbalanced BF[1]s, as this exercise illustrates. The condition averages for the study of undoing (Example S4.10) are: 1.25 (Sessions 1–18), 1.45 (19–36), 2.00 (37–54), 2.50 (55–72), 2.33 (73–90), and 3.20 (91–108). The grand average equals 2.32. I got this number by adding all the response values and dividing by how many there are. Compute the average of the condition averages, and check that it is not equal to the grand average. Why not?

2. Decomposing Data from Balanced Designs

This section presents the new rule for decomposing balanced data sets and applies it to several designs. There are two main ideas: (1) You do one decomposition step for each factor of the design, except for residual error. (2) For a balanced design, the order of the steps doesn't matter: you end up with the same decomposition no matter what order you choose for using the factors. In practice, though, life is simplest if you "peel the onion": use the factors with bigger groups before the factors with smaller groups.

Using the new rule gives you one more way to count degrees of freedom, and the section ends with a collection of examples of counting this new way.

Decomposition Rule and Examples

Rule for Decomposing a Balanced Data Set

1. **Order your factors from left to right so that no factor is inside a factor to its right.**
2. **Do one decomposition step for each factor except residual error. Start with the box of observed values, and at each later step, decompose the leftovers from the step before.**

To carry out a decomposition step, you have to make two choices: (1) which numbers to decompose, and (2) which factor to use. At any point in the process, you can choose any factor you haven't yet used (except the factor for error). Once you've chosen the factor, there will be one right choice for the box of numbers to decompose. If you did Exercise 10.A.3, you saw there was an easy way to recognize a wrong choice: the decomposition step will give you back the box you started with and a box of zeros. Thus you don't really need a standard order for the factors. You can just play it by ear, using the factors one at a time and guessing which numbers to decompose until you get it right. I mention this in case you decide at some point you want to use some order other than the one(s) in the rule: it's perfectly OK to do that.

The advantage of a standard order is that you never have to guess which numbers to decompose: you always use the leftovers from the step before. To make this work, you have to order your factors according to "big first, then small," just as in peeling layers off an onion.

EXAMPLE 10.2 DECOMPOSING A FAKE BF[1]

Decompose the following fake version of the walking babies data.

Data

Exercise	8	12
Control 1	12	12
Control 2	10	6

Factor structure

SOLUTION. Start by checking the order of the factors: No factor is nested inside a factor to its right in the list.

SWEEP 1: Take out the grand average. Check that the six numbers add up to 60, so their average is 10.

Obs			Gr Avg			Left	
8	12		10	10		−2	2
12	12	=	10	10	+	2	0
10	6		10	10		−3	1

Leftovers add to zero over the whole box

SWEEP 2: Take the condition averages out of the leftovers. It is easy to find the three row averages in your head: 0, 1, −1.

Left			Row Eff			Res	
−2	2		0	0		−2	2
2	0	=	1	1	+	1	−1
−3	1		−1	−1		−2	2

Leftovers add to zero across each row.
Row effects add to zero

Complete decomposition:

Obs			Gr Avg			Row Eff			Res	
8	12		10	10		0	0		−2	2
12	12	=	10	10	+	1	1	+	1	−1
10	6		10	10		−1	−1		−2	2

■

EXAMPLE 10.3 DECOMPOSING A FAKE CB

Think of the numbers below as a fake finger tapping experiment, with only two subjects, and decompose the data.

Data

	Con	Caf	The
I	10	14	12
II	4	10	10

Factor structure

Subjects

Drugs

Residual Error

SOLUTION. The factors are in an acceptable order, with no factor nested inside a factor to its right. There will be three steps to the decomposition: first we'll take out the grand average, then take out row (block) averages, then column (drug) averages. We'll end up with the data split into four pieces, corresponding to the four factors.

SWEEP 1: Take out the grand average. Here, just as in the last example, the six numbers add up to 60, so the average is 10.

Obs				Gr Avg				Left #1		
10	14	12		10	10	10		0	4	2
4	10	10	=	10	10	10	+	−6	0	0

Leftovers add to zero over the whole box.

SWEEP 2: Take the row averages out of the leftovers from SWEEP 1. The first row gives a total of 6, average of 2; the second row gives a total of −6, average of −2. These row effects add to zero over the whole box.

Left #1				Row Eff				Left #2		
0	4	2	=	2	2	2	+	−2	2	0
−6	0	0		−2	−2	−2		−4	2	2

Leftovers add to zero across each row

Notice, before going on to SWEEP 3, that the first two steps have been the same as for the BF: take out the grand average, then take out row averages.

SWEEP 3: Take column averages out of the leftovers from SWEEP 2. The column averages are easy enough to do in your head: −3, 2, and 1. They add to zero.

Left #2				Col Eff				Res		
−2	2	0	=	−3	2	1	+	1	0	−1
−4	2	2		−3	2	1		−1	0	1

Leftovers add to zero across rows and down columns.

Complete decomposition:

Obs				Gr Avg				Row Eff				Col Eff				Res		
10	14	12	=	10	10	10	+	2	2	2	+	−3	2	1	+	1	0	−1
4	10	10		10	10	10		−2	−2	−2		−3	2	1		−1	0	1

■

EXAMPLE 10.4 DECOMPOSING A FAKE SP/RM

Think of the numbers below as a hamster experiment, and decompose the data.

Data

	Hamster	H	B
Long	1	20	10
days	2	20	2
Short	3	10	2
days	4	10	6

Factor structure

Benchmark, Day Length, Hamsters, Organ, Interaction, Residual

SOLUTION. There will be five SWEEPs to the decomposition, to split the data into six pieces, one for each of the factors in the diagram.

SWEEP 1: Take out the grand average. The eight numbers add up to 80, to give a grand average of 10.

Obs			Gr Avg			Left #1	
20	10		10	10		10	0
20	2		10	10		10	−8
10	2	=	10	10	+	0	−8
10	6		10	10		0	−4

Leftovers add to zero over the whole box.

SWEEP 2: Take the day length averages out of the leftovers from SWEEP 1. The top half of the box (long days) adds to 12, for an average of 3; the average for the bottom half works out to −3, making the two day length effects add to zero, as they should.

Obs			Day Eff			Left #2	
10	0		3	3		7	−3
10	−8		3	3		7	−11
0	−8	=	−3	−3	+	3	−5
0	−4		−3	−3		3	−1

Leftovers add to zero over each half,

SWEEP 3: Take the row (hamster) averages out of the leftovers from SWEEP 2. The row totals are 4, −4, −2, and 2; the row averages are 2, −2, −1, and 1. These averages add to zero in pairs: top half (2 − 2 = 0) and bottom (−1 + 1 = 0).

Left #2			Hamsters			Left #3	
7	−3		2	2		5	−5
7	−11		−2	−2		9	−9
3	−5	=	−1	−1	+	4	−4
3	−1		1	1		2	−2

Leftovers add to zero across rows.

SWEEP 4: Take the organ (column) averages out of the leftovers from SWEEP 3. These averages work out to +5 and −5.

Left #3			Organ			Left #4	
5	−5		5	−5		0	0
9	−9		5	−5		4	−4
4	−4	=	5	−5	+	−1	1
2	−2		5	−5		−3	3

Leftovers add to zero across rows,

SWEEP 5: Take the interaction averages out of the leftovers from SWEEP 4.

Left #4			Inter			Res	
0	0		2	−2		−2	2
4	−4	=	2	−2	+	2	−2
−1	1		−2	2		1	−1
−3	3		−2	2		−1	1

Leftovers add to zero across rows, and within cells of Organ crossed with Day Length.

Complete decomposition.

Obs			Gr Avg			Day Eff			Hamsters			Organ			Inter			Res	
20	10		10	10		3	3		2	2		5	−5		2	−2		−2	2
20	2	=	10	10	+	3	3	+	−2	−2	+	5	−5	+	2	−2	+	2	−2
10	2		10	10		−3	−3		−1	−1		5	−5		−2	2		1	−1
10	6		10	10		−3	−3		1	1		5	−5		−2	2		−1	1

Check to see that in fact, the numbers do add up as they should. For example, $20 = 10 + 3 + 2 + 5 + 2 + (-2)$. ■

EXAMPLE 10.5 DECOMPOSING THE FINGER TAPPING DATA (EXAMPLE 7.14)

When I decomposed the finger tapping data back in Chapter 7, I got a grand average of 34, subject effects of −15, 36, −4, −17, and drug effects of −12, 5, 7. The new rule should give these same numbers.

There will be three steps to the decomposition, just as for Example 10.3. The final result will be to split the finger tapping data into four pieces: grand average plus block effect plus condition effect plus residual.

Benchmark Blocks Conditions Residual Error

Factor structure for the finger tapping data.

SWEEP 1: Take out the grand average.

Obs				Gr Avg				Left #1		
11	26	20		34	34	34		−23	−8	−14
56	83	71	=	34	34	34	+	22	49	37
15	34	41		34	34	34		−19	0	7
6	13	32		34	34	34		−28	−21	−3

The grand average is 34, the same as in Example 7.14. Notice that the leftovers for Subject II, the fast one, are all large and positive.

SWEEP 2: Take row averages out of the leftovers from SWEEP 1.

Left #1				Subj				Left #2		
−23	−8	−14		−15	−15	−15		−8	7	1
22	49	37	=	36	36	36	+	−14	13	1
−19	0	7		−4	−4	−4		−15	4	11
−28	−21	−3		−17	−17	−17		−11	−4	15

The box for subjects contains the subject effects, which are the same as the ones I got in Chapter 7. There was one fast subject; the two slowest ones averaged about the same. Check the patterns of adding to zero: The subject effects add to zero over the whole box, because they came from the leftovers from the grand average. The leftovers at this step add to zero across rows, because I got them by taking out row averages.

SWEEP 3: Take column averages out of the leftovers from SWEEP 2.

Left #2				Drugs				Res		
−8	7	1		−12	5	7		4	2	−6
−14	13	1	=	−12	5	7	+	−2	8	−6
−15	4	11		−12	5	7		−3	−1	4
−11	−4	15		−12	5	7		1	−9	8

The box for drugs contains the drug effects, which are the same as in Chapter 7: subjects were on average 12 taps slower than the grand average after the placebo; caffeine and theobromine raised the rate of tapping by roughly equal amounts, 5 and 7. Check that the box of residuals is the same as in Example 7.14, and thus that the complete decomposition is also the same.

The patterns of adding to zero are what they should be: the drug effects add to zero across rows (because I took out row averages in SWEEP 2); the residuals add to zero across rows and down columns. ■

Example 10.6 Decomposing the Milk Yield Data

Look over the numbers below, from Examples 7.6 and 7.15.

	Weeks			
Cow	1–6	7–12	13–18	Diets
I	A 608	B 716	C 845	A = roughage
II	B 885	C 1086	A 711	B = partial grain
III	C 940	A 766	B 832	C = full grain

A complete decomposition will split the numbers into five pieces, one for each factor in the factor diagram.

Benchmark	Rows	Columns	Treatments			Residual Error		
			A	B	C			
			B	C	A			
			C	A	B			

Factor structure for the milk yield experiment.

The decomposition in Example 7.15 gave a grand average of 821, cow effects of −98, 73, 25, column (= time) effects of −10, 35, −25, and diet effects of −126, −10, 136. We should get the same numbers here.

As you read what follows, take a moment to compare the first three steps for this decomposition with the decomposition for the CB in Example 10.5. Notice that these three steps are the same in both cases: first take out the grand average, then row averages, then column averages. The only difference is that for a Latin square, you have a fourth step, taking out treatment averages.

SWEEP 1: Grand average.

Obs				Gr Avg				L1		
608	716	845		821	821	821		−213	−105	24
885	1086	711	=	821	821	821	+	64	265	−110
940	766	832		821	821	821		119	−55	11

SWEEP 2: Rows.

L1				Row Eff				L2		
−213	−105	24		−98	−98	−98		−115	−7	122
64	265	−110	=	73	73	73	+	−9	192	−183
119	−55	11		25	25	25		94	−80	−14

The row effects are the same as before. Cow II produced most; Cow I the least. The row effects add to zero, and the leftovers add to zero across rows.

SWEEP 3: Columns.

L2				Col Eff				L3		
−115	−7	122		−10	35	−25		−105	−42	147
−9	192	−183	=	−10	35	−25	+	1	157	−158
94	−80	−14		−10	35	−25		104	−115	11

The column averages for the box L2 give the column effects. These effects add to zero across rows, because we took out row averages (in SWEEP 2) to get the L2 box; the L3 leftovers add to zero across rows and down columns.

SWEEP 4: Treatments.

L3

A −105	B −42	C 147
B 1	C 157	A −158
C 104	A −115	B 11

=

Tr Eff

A −105	B −42	C 147
B 1	C 157	A −158
C 104	A −115	B 11

+

Res

A 21	B −32	C 11
B 11	C 21	A −32
C −32	A 11	B 21

Check first to see how I got the table of treatment effects: For A, I collected the three A numbers from the L3 box and took their average: $(-105 - 115 - 158) \div 3 = -126$. Similarly for B [$(1 - 42 + 11) \div 3 = -30 \div 3 = -10$] and C. Then I subtracted averages in the usual way to get the residuals. Check that these are the same residuals as we got in Example 7.19. ■

Counting Degrees of Freedom by Subtraction

Once you become comfortable using flowcharts to plan the decomposition of a data set, you can also use them to count degrees of freedom by subtraction. Subtraction by itself will not let you find all the dfs for a decomposition, but it can provide shortcuts, and can also give you a way to check your direct counts.

Each df corresponds to one unit of information about chance error, and a decomposition step neither creates nor destroys that information. Thus:

For every decomposition step, the df for the table of averages plus the df for the table of leftovers equals the df for the table you start with.

EXAMPLE 10.7 COUNTING THE LEAFHOPPER DF BY SUBTRACTION

Figure 10.7 shows a flowchart for decomposing the leafhopper data. (I've written the df in parenthesis under each piece of the decomposition.)

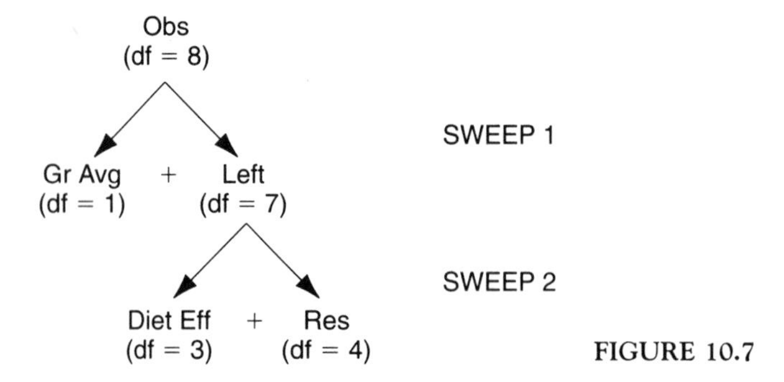

FIGURE 10.7

SWEEP 1: We start with 8 df for the observations, which we split into the grand average plus leftovers. The grand average has 1 df, so the leftovers must have 7.

SWEEP 2: We split the leftovers into diet effects plus residuals. If we count $df_{Diet} = 3$ directly, then we can get $df_{Res} = 4$ by subtraction.

EXAMPLE 10.8A Finger Tapping: A CB Design

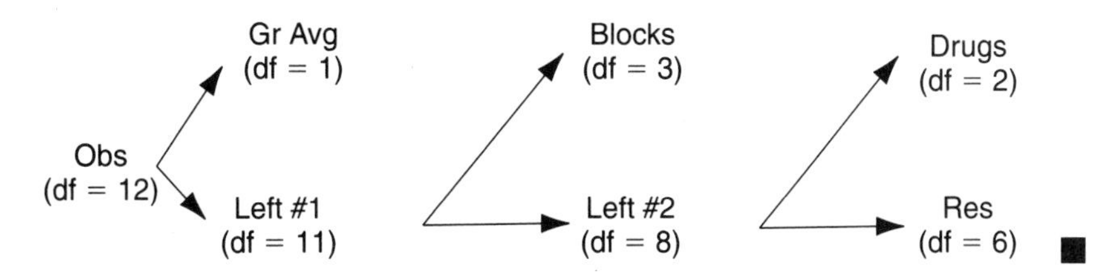

EXAMPLE 10.8B Milk Yields: An LS Design

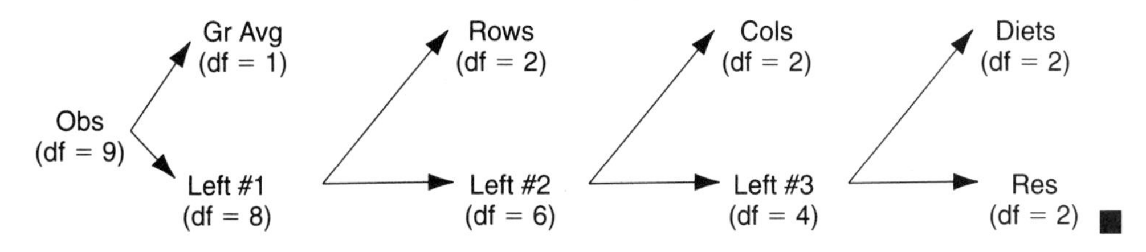

EXAMPLE 10.8C Pigs and Antibiotics: A CR[2] Design

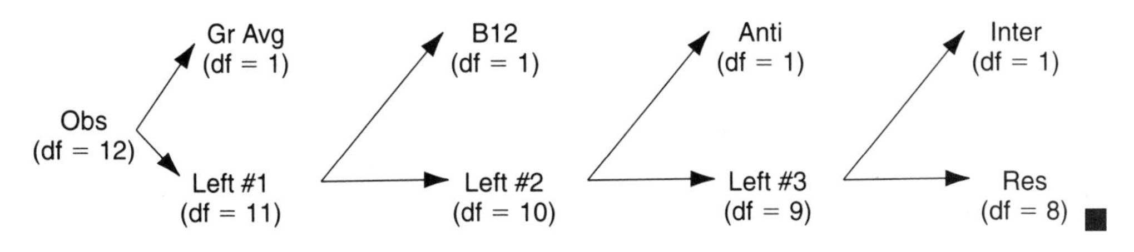

Exercise Set B

1. *Fake data for practice*. You can try the decomposition rule on any of the four versions of the two fake data sets in Exercise 7.D.1–8: #1 & 5 BF[1]; #2 & 6 CB[1]; #3 & 7 BF[2]; and #4 & 8 SP/RM. For each data set, first write out a flowchart showing the steps of your decomposition and the df at each step. Then carry out the decomposition and check that the decomposition you get is the same as you would get using the old rule, as in Chapter 7.

2. Imagine decomposing a CB design. In all you create six boxes: grand average, first leftovers, block effects, second leftovers, treatment effects, and residuals. For each of the following boxes, I give you some of the numbers in the box, together with a flowchart that tells you which of the six possibilities it is. Use this information to fill in the missing numbers; then give the df.

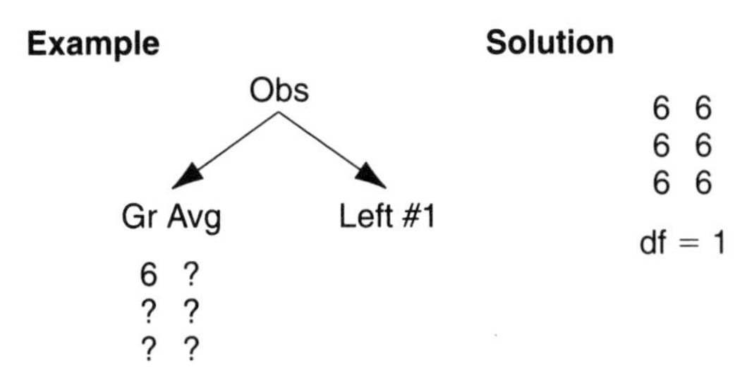

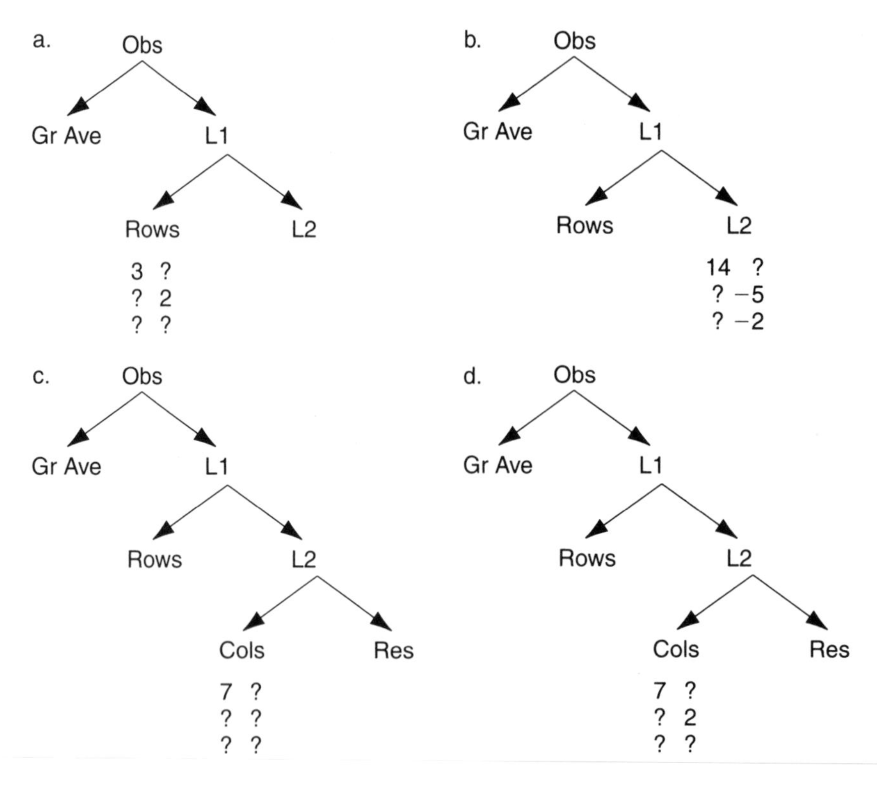

3. *Real data sets for hand computation.* Each of the following data sets has had a few observations changed in the last decimal place to simplify the arithmetic. For each, first write out a flowchart showing the steps of your decomposition and the df at each step. Then carry out the decomposition and check that the decomposition you get is the same as you would get using the old rule, as in Chapter 7.

a. BF[1]: Intravenous fluids (Example 5.2)

b. BF[2]: Puzzled children (Exercise 6.C.3)

c. CB[1]: Beestings (Example S8.3, Exercise 7.D.14)

d. SP/RM: Dopamine and diets (Exercises 7.C.6–9)

4. *Acceptable orders for factors.* Remember that the decomposition rule requires that you order your factors from left to right so that no factor is inside a factor to its right. (a) How many different acceptable orders are there for a BF[1] design? For a CB[1]? (b) For every LS[1] design, there are six possible acceptable orders. List them. (c) For every SP/RM[1;1] design, there are five possible acceptable orders. List them.

Exercise Set C: Data Sets for Decomposition and ANOVA by Computer

1. Cool mice (Example 7.19)
2. Hypnosis and learning (Exercise 7.G.6)

3. Pine pruning (Exercise 8.A.17)
4. Hospital carpets (Exercise 8.C.1)
5. Deprived rats (Exercise 8.C.3)
6. Emotions and skin potential (Exercise 8.C.5)
7. Jobs and hearts (Exercise 8.D.5)
8. Pesticides in the Wolf River (Exercise 8.D.6)
9. Pine seeds (Exercise 8.E.4)
10. Extinguished rats (Exercise 8.E.5)
11. Bloated worms (Example 9.4)
12. Crabgrass (Example S9.6)
13. Remembering words (Exercise 9.C.1)
14. Imagery and working memory (Exercise 9.C.5)
15. Rats and amphetamines (Exercise 9.F.1)
16. Puzzled children (Exercise 9.F.6)
17. Oxygen pressure (Exercise 9.F.9)

SUMMARY

Rule for decomposing a balanced design: First, order your factors from left to right so that no factor is inside a factor to its right.
Then do one decomposition step or "sweep" for each factor except residual error.
Start with the box of observed values, and at each later step, decompose the leftovers from the step before.

The basic decomposition step:

IN: (1) A table of numbers.
(2) A factor.

SWEEP: (1) Sort the numbers into the groups of the factor.
(2) Compute the average for each group of numbers.
(3) Get a leftover from each number in the table by subtracting the average for its group.

OUT: (1) A table of averages.
(2) A table of leftovers.

NOTES AND SOURCES

Arithmetic shortcuts? For many of the examples of decomposition steps, there will be ways to shorten the arithmetic by taking advantage of patterns, but because the point of this chapter is to illustrate a method that works for *all* balanced designs, there is little point in trying to learn shortcuts that only work in particular cases.

CHAPTER 11

COMPARISONS, CONTRASTS, AND CONFIDENCE INTERVALS

OVERVIEW

Comparisons are used to obtain focused information about particular groups of observations.

This chapter introduces two closely related methods, one for testing hypotheses, the other for estimating true values. Whereas the F-tests you have learned so far are designed for general screening—are there differences somewhere among the levels of a factor?—the method of **comparisons** (or **contrasts**) is designed to test hypotheses that are much more focused. For example, consider the leafhopper survival data (Exercise 5.C.1, Example 5.14). The ANOVA F-test tells us only that at least one of the "true" survival times is different from the other three, but doesn't tell us anything more specific about the differences. We can use comparisons to pin down where the statistically detectable differences are. To mention just one possibility, we will use the method of comparisons to test whether there is a "real" effect associated with the number of carbon atoms in the sugars, i.e., a "real" difference between the 12-carbon sugar on one hand and the two 6-carbon sugars on the other. When we do this, we will compute the average for the two dishes of leafhoppers that got the 12-carbon sugar, and subtract the average for the four dishes that got the 6-carbon sugars. If we think of the number we get as an estimate of the corresponding "true" difference, then the method of comparisons tests whether our evidence is strong enough to conclude that the true difference is not equal to zero. As an alternative, we could construct a confidence interval for this true value, an interval of the form "observed difference ± 95% distance," where the 95% distance is chosen to give a 95% chance that the interval will contain the true difference somewhere inside. In a sense, the two methods are equivalent: we conclude that the true difference is different from zero if and only if the confidence interval does not contain zero.

1. Comparisons: Confidence intervals and tests.
2. Adjustments for multiple comparisons.
3, Between-blocks factors and compound within-blocks factors.

1. Comparisons: Confidence Intervals and Tests

This section has five parts. The first two parts give a general overview of the way you use comparisons: one describes and illustrates various strategies for choosing comparisons, and the other illustrates the meaning of the standard error and shows how to use SEs to interpret comparisons. After this two-part overview, the last three parts of the section fill in details: one tells more carefully what it is that a comparison estimates, the next tells how to compute standard errors, and the last tells how to use SEs to construct interval estimates.

Strategies for Choosing Comparisons

A comparison measures the distance between averages for two groups of observed values.

Every comparison arises in the same way: Choose two nonoverlapping groups of observations, compute the average for each group, and subtract one average from the other. The number you get tells how far apart the averages are, so you can think of it as measuring the distance between the two groups.

Example 11.1 Comparison for the Leafhopper Data

Compare the effects of 12-carbon versus 6-carbon sugars for the data of Exercise 5.C.1.

Solution. First identify the groups of observations and find their averages (see Figure 11.1).

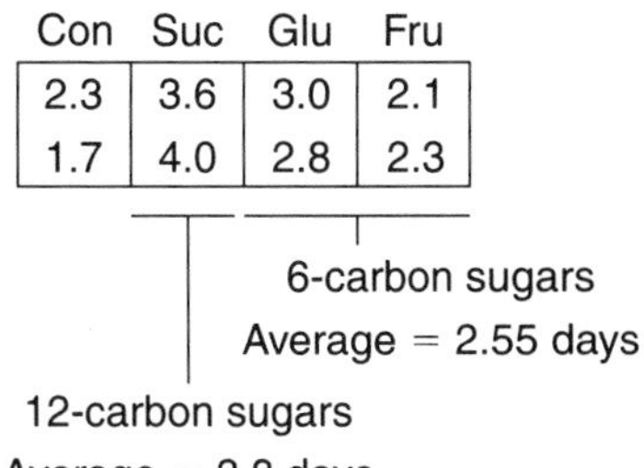

FIGURE 11.1

Next, find the distance between group averages (Figure 11.2):

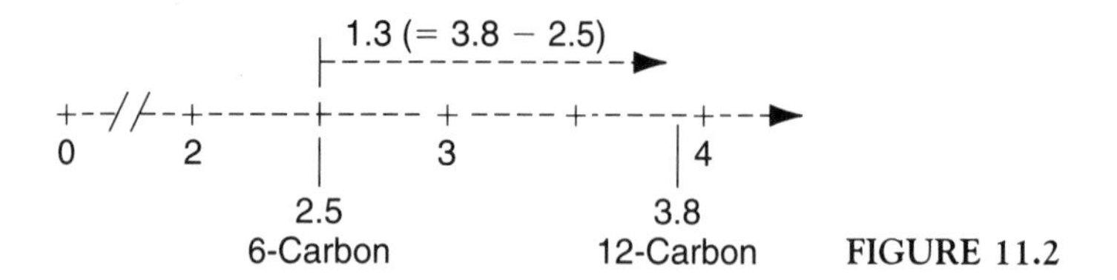

FIGURE 11.2

Compared with the 6-Carbon sugar diets, the 12-carbon sugar diet is worth an extra 1.3 days of life. ■

Statisticians use several strategies for choosing comparisons. What all these strategies have in common is the goal of trying to pin down the nature of observed differences: Which of the various group averages are "really" different from which other group averages?

(a) **All pairwise comparisons.** This is perhaps the simplest and most obvious strategy. For the intravenous fluids data (Example 5.2) for example, there are three drug manufacturers: Abbot, Cutter, and McGaw. It would be natural to compare each drug company with the others, for a total of three comparisons: Abbot versus Cutter, Abbot versus McGaw, and Cutter versus McGaw (Figure 11.3). For the leafhopper data, with four diets, there are six pairwise comparisons; for a data set with five groups, there would be ten pairs.

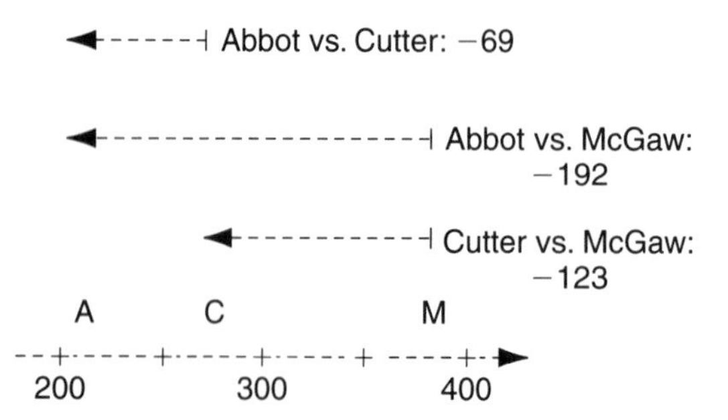

FIGURE 11.3 The three pairwise comparisons for the fluids data. These comparisons show that the average numbers of impurities for Abbot and Cutter are closest together; the averages for Abbot and McGaw are farthest apart.

(b) **Comparisons based on "top down" structure.** Notice that with the leafhopper data set, the groups have a meaningful structure that you don't get with the fluids data. The four diet groups of the leafhopper experiment form natural clusters and subclusters: the three sugar diets form the largest cluster, and the two 6-carbon diets form a subcluster. It makes sense to use this structure based on meaningful clusters to choose the comparisons as in Figure 11.4.

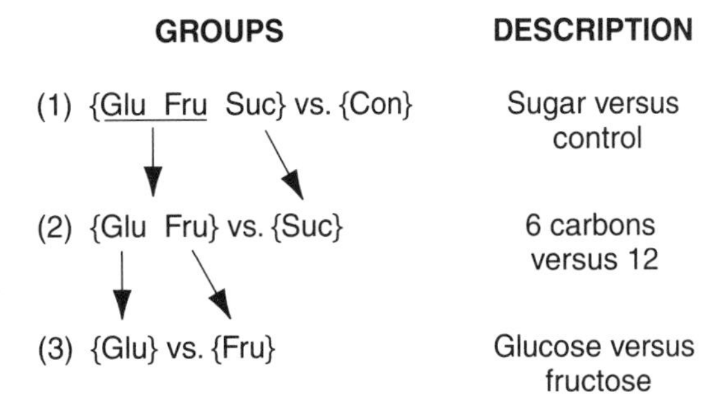

FIGURE 11.4 A set of comparisons based on top-down structure. The leafhopper diets form natural clusters and subclusters in a way that the three drug companies of the fluids study do not. As an alternative to comparing each diet with each other, it makes sense to compare (i) the three sugar diets with the control diet, (ii) the two 6-carbon sugar diets with the 12-carbon sugar diet, and (iii) one 6-carbon sugar with the other.

(c) ***Post hoc* (after the fact) comparisons.** As the name suggests, these are comparisons that you decide to make after looking at your data. For example, remember that for the walking babies data (Example 5.1) there were three control groups: exercise control (C1), no exercise/weekly report (C2), and no exercise/final report only (C3). Before looking at the data, you might expect that among the three control groups, the big difference would be between those who got exercise as part of their placebo treatment (C1) and those who did not (C2, C3). The averages, however, suggest otherwise. The C1 and C2 averages are quite close (11.375 and 11.625 months). Both were about a full month less than the C3 average (12.35 months). These averages suggest that the report method had a bigger effect than the exercise. (Perhaps parents who reported more frequently encouraged their sons to walk sooner.) Based on the data, you might decide to do a formal comparison of the two control groups with weekly reporting (C1, C2) with the other control group.

There's a reason why statisticians have given *post hoc* comparisons a name that sets them apart from comparisons that you plan in advance, before you look at your results: In order to judge whether an observed difference is "real," you must compare the observed difference with a number that tells how big a difference to expect if in fact there are no "true" differences. The expected size of the difference depends on whether you chose your groups ahead of time, or whether you chose them because they turned out to be far apart when you looked at your data. For *post hoc* comparisons, most statisticians say you should not use ordinary confidence intervals or tests. Instead, you should use one of the adjustments discussed in Section 2.

Interpreting the Comparisons Using Standard Errors

To interpret the observed value of a comparison, we rely on our model—the factor structure and Fisher assumptions—as a framework. The observed value is an estimate of an unknown "true" value, true if the model is correct. Assuming that our model gives an accurate description of the process that created our data, we ask "What is the relationship between the observed value of our comparison and the true value it estimates?" The comparison is built by combining individual observations; each observation is—according to the model—a sum, true value plus chance error; so the comparison is also a sum of two parts: a true value, obtained by combining the true values of the observations, plus a chance part, obtained by combining the individual chance errors. The standard error (SE) for the comparison tells the typical size of the chance part, just as the standard deviation (SD) does for a single observations. A later part of this section tells how to compute SEs. Here, I'll illustrate what they tell you.

Think of "doing-your-experiment-and-computing-your-comparison" as a chance process. (More in Chapter 15). Then, roughly, there is a 68% chance that the observed value of your comparison will be within 1 SE of the true value, a 95% chance the comparison will be within 2 SEs, and a near-100% chance for 3 SEs (See Figure 11.5).

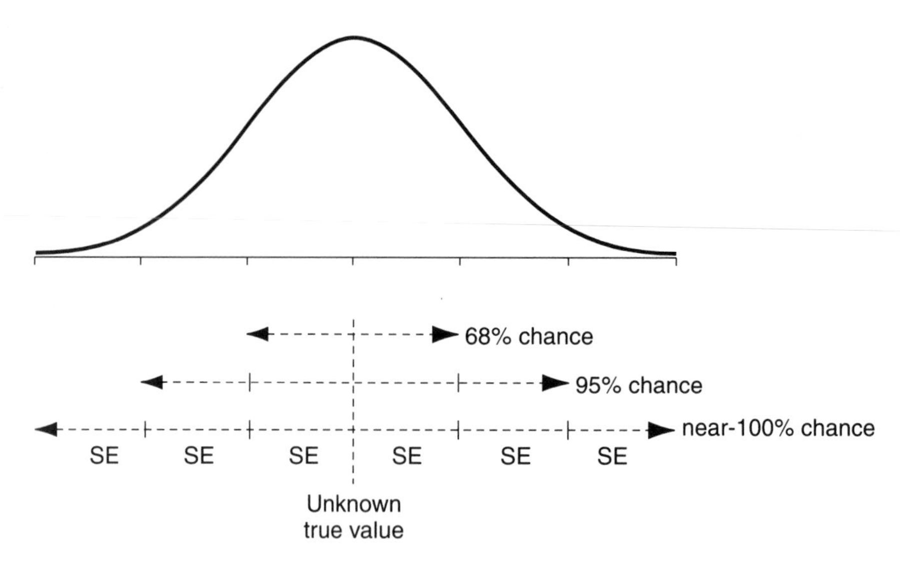

FIGURE 11.5 The SE tells for a comparison what the SD tells for a single measurement. The rough intuitive meaning is that an estimator equals a true value, give or take about one SE. A more detailed version comes from the approximation based on the normal curve. The chances (68%, etc.) are approximate, not exact. The approximations tend to work better when your estimate is based on a large number of observations, and when the individual chance errors follow the normal curve. The approximations don't work as well for small data sets for which the individual chance errors do not fit the normality assumption. Finally, the approximations don't work at all if there is much bias present.

If we focus on 2 SEs and 95%, we can get a preview of the confidence intervals and tests of the next section. (1) There is about a 95% chance that our estimate will fall within 2 SEs of the true value, so there is about a 95% chance that an interval of the form "Estimate ± 2 SEs" will cover the true value: "Est. ± 2 SEs" gives an approximate 95% confidence interval. (2) If an observed difference between group averages is more than 2 SEs from 0, then we regard the difference as "real."

"Quick-and-Dirty" Intervals and Tests

(1) Rough 95% interval: Estimate ± 2 SEs

(2) Approximate test: Differences are "real" if observed difference is more than 2 SEs from 0.

EXAMPLE 11.2 LEAFHOPPERS: COMPARISONS AND SEs

Here are observed values and SEs for the three leafhopper comparisons, first expressed in the original scale (days), and then multiplied by 24 to get the same information in hours:

	Days		Hours	
Comparison	Obs	SE	Obs	SE
Sugar versus Control	0.93	0.22	22.4	5.3
Twelve Carbons versus Six	1.30	0.24	31.2	5.8
Glucose versus Fructose	0.60	0.27	14.4	6.6

Interpret these numbers.

SOLUTION

a. *Sugar versus control.* The observed difference between the two group averages is 22.4 hours, with an SE of 5.3 hours. If we think of the experiment-and-comparison as jointly defining a chance process, then our observed value of 22.4 is like a single measurement with SD = 5.3 and true value unknown, that is, like one draw from a box of 100 numbered tickets, with the average for the box equal to the unknown true value, about 68 tickets within 5.3 of the true value, about 95 tickets within 10.6 of the true value, and no tickets more than 15.9 away from the true value.

Rough confidence interval: If you think of drawing a ticket and constructing the interval "Ticket value ± 10.6," then for 95 of the 100 tickets you might draw, the interval will cover the true value. Our rough interval is 22.4 ± 10.6, or from 11.8 to 33.0 hours.

Approximate test: Suppose (null hypothesis) that the true value is in fact 0, so that 95 of the tickets are within 10.6 = 2SEs of 0. Then it is very unlikely (5% chance) that you'd draw a ticket outside the range −10.6 to 10.6. Therefore, if you *do* get such a ticket—as we did with our 22.4—you know that either (i) the true difference is not 0, or (ii) the true value is 0 and you got a very unlikely result just by chance. Conclusion: The true difference between the control and sugar diets is not 0.

b. *Twelve carbons versus six.* For this comparison, the SE is 5.8, so an approximate box model for the chance process has 100 tickets, 68 of them within 5.8 of the true value = average for the box, 95 within 11.6 of the true value, and none more than 17.4 from the true value. Our observed value of 31.2 hours is like one draw from the box. A rough 95% confidence interval goes from 18.4 (= 31.2 − 11.6) to 42.8: we can be roughly 95% sure that the true advantage of the more complex sugar is worth between 19.6 and 42.8 hours of life to a leafhopper. Since 0 is not in this interval, we declare the difference between 12 and 6 carbons to be "real."

c. *Glucose versus Fructose*. Here the SE is larger, 6.6, and the tickets in the box are more spread out. The rough interval is wider: $14.4 \pm 2(6.6)$, or from 1.2 to 27.6 hours. The interval does not contain 0. Our rough conclusions: Glucose seems to make a "real" difference, but our interval estimate is quite wide, and comes close to 0 at the low end.

We can show the three intervals together on a graph (Figure 11.6):

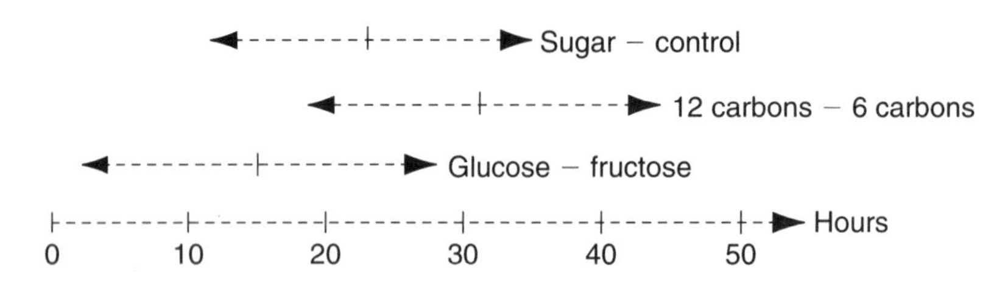

FIGURE 11.6 Quick-and-dirty 95% confidence intervals for the leafhoppers. Each interval appears as a line segment, with the estimate shown in the middle. Together, the intervals neatly summarize the results of the experiment: (a) Measured in hours of extra life, a sugar diet is better than the control by about 22 hours; 12 carbons are better than 6 by 31 hours; and glucose is better than fructose by about 14 hours. (b) All these estimates contain a lot of uncertainty: each of the intervals is about 24 hours wide. (c) All three estimates are about equally imprecise: the three interval widths are about the same. ■

The True Value Estimated by a Comparison

Every statistical model is an imaginary ideal situation, invented in order to describe and understand a more complicated real situation. In what follows, I'll use the leafhopper experiment to illustrate the relationship between our model for the data and certain averages and comparisons.

> **1. The average of a set of observed values equals the average of their true values plus the average of their chance errors.**
>
> **2. The difference of two averages equals the difference of their two true values plus the difference of their two chance parts.**

The observations. We assume that each observed survival time is a sum of two numbers: the unknown "true" survival time, plus a chance error. Think of the (imaginary) set of all possible survival times on the glucose diet as a set of numbered tickets in a box; each observed value corresponds to one draw from the box. The average of all the tickets in the box is the population average, or "true" survival time for the glucose diet. A convenient notation for this theoretical population average is μ_G: the μ is standard notation for an unknown true average, and the G subscript stands for the glucose diet. You can also think of μ_G as the typical survival time for the glucose diet, the number you would observe if conditions were always constant, the leafhoppers were absolutely identical, and there were no chance errors.

Each observed survival time for the glucose diet equals the true μ_G plus an error. For our data set, there were two observations, 2.9 and 2.7 days, each with the same μ_G, but with different error parts. If we use the letter e for error, we could write the two errors as e_{G1} and e_{G2}, and the two observed values as $\mu_G + e_{G1}$, $\mu_G + e_{G2}$.

In the same way, there is a true value for each of the other diets: μ_C for the control, μ_F for the fructose, and μ_S for the sucrose. Using the letter e with subscripts for the errors, we can write the entire data set as in Figure 11.7.

Con	Suc	Glu	Fru
$\mu_c + e_{c1}$	$\mu_s + e_{s1}$	$\mu_G + e_{G1}$	$\mu_F + e_{F1}$
$\mu_c + e_{c2}$	$\mu_s + e_{s2}$	$\mu_G + e_{G2}$	$\mu_F + e_{F2}$

FIGURE 11.7 Notation for the leafhopper data. Each of the eight observations is written as a true value μ plus a chance error e. The chance errors are all different, but for each diet, the two true values are the same. Neither the true values nor the errors can be observed directly.

Averages and comparisons. Any average or comparison will also be a sum of two pieces, a combination of true values plus a combination of errors. To find out how the true value for an average or comparison is related to the true values for the individual observations is just a matter of algebra, although you can often guess the answer.

EXAMPLE 11.3 TRUE VALUES FOR COMPARISONS

Find the true values which are estimated by the following combinations of the observed survival times: (a) the glucose average, (b) the comparison "glucose vs. fructose", (c) the comparison "12 carbons vs. 6", and (d) the comparison "control vs. sugar".

SOLUTION

a. *The glucose average*. This average is just the sum of the two glucose observations, divided by 2:

$$\text{Avg} = \frac{(\mu_G + e_{G1}) + (\mu_G + e_{G2})}{2} = \mu_G + \frac{(e_{G1} + e_{G2})}{2}$$

The true value is μ_G, the same as for a single observation. Thus the average estimates what we want it to, namely, the underlying "true" survival time for leafhoppers on the glucose diet. The chance part is the average of the two chance errors.

b. *Glucose vs. fructose*. This comparison equals (glucose average) − (fructose average). From part (a), the first average is $\mu_G + (e_{G1} + e_{G2})/2$. Following the same steps as in (a), the fructose average works out to $\mu_F + (e_{F1} + e_{F2})/2$.

The true value for the comparison is just the difference of the two true values for the averages: $\mu_G - \mu_F$. This says that the comparison glucose vs. fructose gives an estimate for the true difference in survival times.

The chance part for the comparison is the difference between the two chance parts: $[(e_{G1} + e_{G2})/2 - (e_{F1} + e_{F2})/2]$. (The standard error for this chance part tells the typical size for this particular combination of chance errors.)

c. *12 carbons vs. 6.* This comparison equals (sucrose average) − (average of the four glucose and fructose observations).

First consider the 6-carbon average, which you get by averaging the four observations for the glucose and fructose diets. The true value for the average equals the average of the four true values:

$$\frac{\mu_G + \mu_G + \mu_F + \mu_F}{4} = \frac{\mu_G + \mu_F}{2}$$

The true value for this average is just the average of the true values for the two 6-carbon sugars. Think for a minute about the meaning of this number, $(\mu_G + \mu_F)/2$. We don't assume that μ_G and μ_F are the same, and we can't say that their average is the true survival time for any one particular diet. The number equals the average of two (possibly different) true survival times, which we have lumped together. We lump them together because we think they might be similar, and that their average gives a reasonable one-number summary for 6-carbon sugar diets.

Now consider the comparison 12 carbons vs. 6. The 12-carbon average works out to $\mu_S + (e_{S1} + e_{S2})/2$, and so the true value for the comparison is just the difference of true values for the two averages: $\mu_S - (\mu_G + \mu_F)/2$. You can think of this number as the answer to the question, "The survival time for sucrose is how much bigger than the average of the survival times for glucose and fructose?"

The chance part for the comparison comes from combining chance errors for the observations, and works out to

$$\frac{(e_{S1} + e_{S2})}{2} - \frac{(e_{G1} + e_{G2} + e_{F1} + e_{F2})}{4}$$

The standard error the for comparison tells the typical size for this particular combination of chance errors.

d. *Sugar vs. control.* The true value for the sugar group is the average of the true values for the six dishes that got sugar diets; this average works out to $(\mu_S + \mu_G + \mu_F)/3$. Thus the true value for the comparison is $[(\mu_S + \mu_G + \mu_F)/3] - \mu_C$, the parameter whose numerical value answers the question, "The average of the (true) survival times for the three sugar diets is how much bigger than the survival time for the control diets?" The chance part of the comparison involves all eight chance errors:

$$\frac{(e_{S1} + e_{S2} + e_{G1} + e_{G2} + e_{F1} + e_{F2})}{6} - \frac{(e_{C1} + e_{C2})}{2} \quad \blacksquare$$

The Standard Error for a Comparison

SEs for Averages and Comparison

SE = SD × (leverage factor), where
SD = standard deviation for an observation, and

Leverage factor = $\sqrt{\frac{1}{n}}$ **for an average of n observations**

$\sqrt{\frac{1}{n_1} + \frac{1}{n_2}}$ **for a comparison of averages with n_1 and n_2 observations.**

The standard error for a comparison tells the typical size of its chance part. A large SE means the chance variability is large: the comparison is not very precise. A small SE, on the other hand, means the value you compute for the comparison is likely to be close to the true value you are trying to estimate. Clearly, then, the smaller the SE, the better.

The size of the SE depends on two things: the SD, which tells the typical size of the chance errors for the individual observations, and the numbers of observations for the two averages you compare. Briefly, the smaller the SD, the smaller the SE; the more observations in the averages, the smaller the SE.

Example 11.4 SEs for the Leafhopper Comparisons

For the leafhopper experiment, there were two observations per diet, with SD = 0.274. Find SEs for (a) the control average, (b) glucose vs. fructose, (c) 12 carbons vs. 6, and (d) sugar vs. control.

Solution

a. *Control average*. The average is based on 2 observations.

$$\text{SE} = \text{SD}\sqrt{1/2} = (0.274)(0.707) = 0.194$$

b. *Glucose vs. fructose*. Each average is based on 2 observations.

$$\text{SE} = \text{SD}\sqrt{1/2 + 1/2} = (0.274)(1) = 0.274$$

c. *12 carbons vs. 6*. The first average is based on 2 observations, the second is based on 4.

$$\text{SE} = \text{SD}\sqrt{1/2 + 1/4} = (0.274)(0.866) = 0.237$$

d. *Sugar vs. control*. The first average is based on 6 observations, the second is based on 2.

$$\text{SE} = \text{SD}\sqrt{1/6 + 1/2} = (0.274)(0.816) = 0.224 \ \blacksquare$$

EXAMPLE 11.5 USING THE SE TO CHOOSE A SAMPLE SIZE

"How many measurements do I need?" is an important question that statisticians often get asked. Finding the answer boils down to answering two simpler questions, then doing a little algebra. The two questions: (i) How big is the SD for a single observation? (ii) How small do you want your SE to be?

How many dishes of Leafhoppers?

(i) SD: Suppose you know, either from a pilot experiment or from reading about other people's experiments, that the SD for a leafhopper experiment is likely to be about 0.25 days.

(ii) SE: Suppose you are particularly interested in the 6-carbon comparison, glucose versus fructose, and you would like the SE for this comparison to be fairly small, say 0.10 days. (This would give a confidence interval extending roughly 2 × SE = 0.20 on either side of the estimated difference.)

How many dishes should you use for each diet?

SOLUTION. Let n stand for the number of dishes in each group. Start with the rule for the SE, fill in the numerical values for the SE and the SD, and write the leverage factor in terms of the n you want to find:

$$\mathrm{SE} = \mathrm{SD}\sqrt{\left(\frac{1}{n} + \frac{1}{n}\right)}$$

$$0.10 = 0.25\sqrt{\frac{2}{n}}$$

Now solve for n (Exercise A.13), to get $n = 12.5$. You need 12 or 13 dishes in each group: 12 will give an SE a bit larger than 0.1, and 13 will give an SE a bit smaller than 0.1. ■

Confidence Intervals and Tests for Comparisons

Confidence intervals

Ordinary Confidence Interval = Estimate ± SE × (t-value)

The basic idea for confidence intervals comes from Example 11.2: the observed value of a comparison is like one draw from a box of tickets; the average of the tickets in the box equals the unknown true value, 2/3 of tickets are within 1 SE of the true value, etc. If the box model is right, then an interval of the form "Est ± 2 SEs" has about a 95% chance of containing the true value.

Only one modification is required to convert quick-and-dirty intervals to the more precise ones: Instead of ±(SE × 2) for the interval, you use ±[SE × (t-value from a table)]. Why? The SD you use to compute an SE is itself an estimate of the "true" SD for the chance process. How good the estimate is depends on df_{Res}: the bigger the better. If you have a large number of observations, so that df_{Res} is large, then the SD you estimate from the data will almost surely be close to the true SD for the underlying chance process. In that case, the t-value from the table will be very close to 2. In other words, the "quick-and-dirty" interval you get using ± 2 SEs won't be much different from the one you get using a t-value instead of 2. On the other hand, if you have only a few observed values, so that df_{Res} is small, then the SD you estimate from your data will not be a precise estimate of the true SD for the underlying chance process. Informally, there will be more "slop" in your estimate of the true SD. If your interval is to have a 95% chance of covering the true value, you have to "adjust for the slop," using a wider interval than the one you'd get using ± 2 SEs. The t-value gives the proper adjustment.

Table 11.1 is a typical t-table. (There's also a copy in the back of the book.) The rows of the table correspond to degrees of freedom for residuals, and the columns correspond to chances: 50%, 80%, 90%, 95%, 99%. To get the 95% distance for a confidence interval, you find the row that corresponds to your residual degrees of freedom, and take the number in the 95% column. Unless your data set is tiny, this number will be close to 2.

EXAMPLE 11.6 LEAFHOPPERS

Find 95% confidence intervals for the three comparisons of Example 11.2 and 11.4.
SOLUTION. The standard 95% distance will be the same for all three intervals, because the degrees of freedom for residuals is the same in all three cases, namely 4. From row 4 of the table, we get a standard 95% distance of 2.776. The SD will also be the same in all three cases: 0.274. The only things that change are the observed values for the comparisons and their SEs.

a. *Glucose versus fructose.*

Observed difference $= 0.60$
SE = SD × (lev. fac.) $= (0.274)\sqrt{1} \approx 0.274$
95% distance = SE × t $= (0.274)(2.776)) \approx 0.761$
Confidence interval $= 0.60 \pm 0.761$, or -0.161 to 1.361

b. *Twelve versus six carbons.*

Observed difference $= 1.30$
SE = SD × (lev. fac.) $= (0.274)\sqrt{3/4} \approx 0.237$
95% distance = SE × t $= (0.237)\ (2.776) \approx 0.659$
Confidence interval $= 1.30 \pm 0.659$, or 0.641 to 1.959.

c. *Sugar versus control.*

Observed difference $= 0.93$
SE = SD × (lev. fac.) $= (0.274)\sqrt{2/3} \approx 0.224$
95% distance = SE × t $= (0.224)\ (2.776) \approx 0.621$
Confidence interval $= 0.93 \pm 0.621$, or 0.309 to 1.551. ■

TABLE 11.1

	Table of 2-tailed t-values					
df	50%	80%	90%	95%	98%	99%
1	1.000	3.078	6.314	12.706	31.821	63.656
2	0.816	1.886	2.920	4.303	6.965	9.925
3	0.765	1.638	2.353	3.182	4.541	5.841
4	0.741	1.533	2.132	2.776	3.747	4.604
5	0.727	1.476	2.015	2.571	3.365	4.032
6	0.718	1.440	1.943	2.447	3.143	3.707
7	0.711	1.415	1.895	2.365	2.998	3.499
8	0.706	1.397	1.860	2.306	2.896	3.355
9	0.703	1.383	1.833	2.262	2.821	3.250
10	0.700	1.372	1.812	2.228	2.764	3.169
11	0.697	1.363	1.796	2.201	2.718	3.106
12	0.695	1.356	1.782	2.179	2.681	3.055
13	0.694	1.350	1.771	2.160	2.650	3.012
14	0.692	1.345	1.761	2.145	2.624	2.977
15	0.691	1.341	1.753	2.131	2.602	2.947
16	0.690	1.337	1.746	2.120	2.583	2.921
17	0.689	1.333	1.740	2.110	2.567	2.898
18	0.688	1.330	1.734	2.101	2.552	2.878
19	0.688	1.328	1.729	2.093	2.539	2.861
20	0.687	1.325	1.725	2.086	2.528	2.845
21	0.686	1.323	1.721	2.080	2.518	2.831
22	0.686	1.321	1.717	2.074	2.508	2.819
23	0.685	1.319	1.714	2.069	2.500	2.807
24	0.685	1.318	1.711	2.064	2.492	2.797
25	0.684	1.316	1.708	2.060	2.485	2.787
26	0.684	1.315	1.706	2.056	2.479	2.779
27	0.684	1.314	1.703	2.052	2.473	2.771
28	0.683	1.313	1.701	2.048	2.467	2.763
29	0.683	1.311	1.699	2.045	2.462	2.756
30	0.683	1.310	1.697	2.042	2.457	2.750
40	0.681	1.303	1.684	2.021	2.423	2.704
60	0.679	1.296	1.671	2.000	2.390	2.660
120	0.677	1.289	1.658	1.980	2.358	2.617
	0.674	1.282	1.645	1.960	2.326	2.576

Tests for comparisons

t-Test for a Comparison

Reject the null hypothesis if the observed value for your comparison is more than SE × (t-value) away from 0, that is, if 0 is not in the confidence interval for the comparison.

The basic idea for informal tests based on comparisons appeared in Example 11.2: Think of the observed value for the comparison as a single draw from a box of numbered tickets. The null hypothesis, that there is no "true" difference between the two groups in your comparison, corresponds to a box with tickets whose average is 0. If the null hypothesis is in fact true, then out of 100 tickets, only about 5 are more than 2 SEs from 0. So if the observed value for your comparison is more than 2 SEs from zero, then either (1) the null hypothesis is false, or (ii) you got a very unlikely outcome. Conclusion: The null hypothesis is false.

To convert the informal test to the formal one, you make the same simple modification as you did for confidence intervals: Use a t-value in place of 2. The logic is the same as before. The SE is an estimate of the true SE for the underlying chance process, and using the t-value instead of 2 adjusts for the likely size of the error in estimating the true SE.

What is the null hypothesis being tested? There are various ways to state the hypothesis, some more formal than others. If you think of your comparison as Avg #1 − Avg #2 for two groups of observations, then the null hypothesis is that there is no difference between the "true" averages for the two groups, that is, the observed difference is due just to chance error. More formally, you can define notation for the true values, as in Section 1.3, and write the null hypothesis as an equation involving those true values.

EXAMPLE 11.7 T-TESTS FOR THE LEAFHOPPER COMPARISONS

Using the notation from Example 11.3, write the null hypothesis corresponding to each of the three comparisons in Example 11.6, and test the hypotheses.

SOLUTION

a. *Glucose vs. fructose*. The true value for this comparison (Example 11.3b) is $\mu_G - \mu_F$. The corresponding null hypothesis is that this true difference equals zero:

$$H_o: \mu_G - \mu_F = 0, \text{ or } H_o: \mu_G = \mu_F$$

In words: The true values for glucose and fructose are equal. The 95% confidence interval for $\mu_G - \mu_F$ (Example 11.5c) goes from −0.161 to 1.361. Since this interval includes 0, we can't rule out 0 as a plausible value for $\mu_G - \mu_F$; according to the evidence, μ_G and μ_F are not detectably different.

Notice that the conclusion here is different from the one in Example 11.2c, based on a quick-and-dirty confidence interval for the same comparison. The reasons is that with only four degrees of freedom for error, the SD we compute from the data is not a very precise estimate of the true SD. The approximate interval that uses 2 × SE instead of t × SE is too narrow. The t-interval, which adjusts for having a "squishy" estimate of the SD, uses 2.776 in place of 2, and is more trustworthy. In summary, if df_{Res} is small, the approximate intervals will be *very* approximate: Rely on the t-interval instead.

b. *12 carbons vs. 6*. The unknown true value (Example 11.3c) is $\mu_S - (\mu_G + \mu_F)/2$, and the corresponding null hypothesis is

$$H_o: \mu_S - \frac{(\mu_G + \mu_F)}{2} = 0, \text{ or } H_o: \mu_S = \frac{\mu_G + \mu_F}{2}$$

In words: The true value for the 12-carbon sugar equals the average of the true values for the two 6-carbon sugars. The observed value for the comparison is 1.30 days, which is more than 0.659 ($= t \times SE$) from 0; the 95% t-interval doesn't contain 0. Conclusion: Reject the null hypothesis and declare that μ_S is detectably different from $(\mu_G + \mu_F)/2$.

c. *Sugar vs. control*. Refer to Example 11.3d, and check that the null hypothesis is

$$H_o: \mu_C = \frac{\mu_S + \mu_G + \mu_F}{3}$$

Refer to Example 11.6(c), and check that the 95% confidence interval does not contain 0. Conclusion: Sugar added to the diet makes a detectable difference. ■

Exercise Set A

Quick-and-Dirty Intervals

1. *Walking babies*. Here are (rounded) average times to walk (months) for the four groups of Example 5.1:

Special exercise:	10.1
No exercise, weekly report:	11.6
Exercise control:	11.4
No exercise, final report:	12.4

(a) Compute estimates for the following set of three comparisons based on the top-down approach: (i) Exercise vs. no exercise; (ii) Special exercise vs. exercise control; (ii) Weekly report vs. final report.

(b) The SEs for the comparisons are (i) 0.6, (ii) 0.4, and (iii) 0.9. Construct the corresponding quick-and-dirty intervals.

(c) Show your three intervals on a graph like the one at the end of Example 11.2, and write a summary like the one in the caption for that graph.

(d) Think of the observed value for comparison (i) as like a draw from a box of 100 numbered tickets. Following Example 11.2, describe the tickets in the box. Is the observed difference between the two averages too big to be due just to chance error? Tell, briefly, how you reached your conclusion.

2. *Milk yields*. The average yields, in pounds per week, for the three diets in Example 7.6 are:

Roughage: 695

Partial grain: 811

Full grain: 957.

(a) Use the top-down approach to choose a set of two comparisons; then use the average yields to compute estimates for both.

(b) The SD is 48.78 pounds per week. Use the formula in Section 1.4 to compute leverage factors for your two comparisons; then construct the corresponding quick-and-dirty intervals. (Each of the diet averages was based on 3 observations.

(c) Show your intervals on a graph, with a verbal summary as a caption, as in Example 11.2.

(d) Think of the observed value for the first of your two comparisons as like a draw from a box of 100 numbered tickets. Describe the tickets in the box; then tell whether the observed difference between the two averages is too big to be due just to chance error.

3. *Sleeping shrews*. The average heart rates, in beats per minute, for the three kinds of sleep in Example 7.4 are: 21.0 (light slow wave); 19.1 (deep slow wave); and 19.0 (rapid eye movement). The SD is 1.417 beats per minute.

(a) Use the top-down approach to choose a pair of comparisons, construct the corresponding intervals, and show them on a graph with a verbal summary as a caption. The design was a CB[1] and used 6 shrews.

(b) The data set contains an outlier, the 15.7 beats per minute for Shrew I during REM sleep. With an estimated replacement value of 11.0 in place of the outlier, the REM average drops to 18.22 and the SD drops to 0.938. Construct a new set of intervals for your two comparisons, and show them on a graph with a caption.

(c) Use your results from (a) and (b) to discuss the effect of the outlier.

4. *Beestings*. The average numbers of new stingers for the two conditions in the experiment of Example S7.3 were: 28 (previously Stung) and 16 (Fresh). Each average is based on 9 occasions. The SD is 14.089 stingers. The data set contained an outlier (70 stingers for occasion 9, Stung.) With an estimated replacement of 16 for the outlier, the average for stung drops to 22 stingers, and the SD drops to 6.459. Find quick-and-dirty intervals for "Stung vs. Fresh," first for the original data set, and then with the outlier replaced. Show both intervals on a graph, and write a short caption describing the intervals in words and summarizing the effect of the outlier.

How Sample Size Affects the Size of the SE

5. Suppose you have a CR experiment with n observations in each of three conditions. Match the averages and comparisons (a)–(d) below with the SEs (i)–(iv): (i) SD $\sqrt{1/n}$ (ii) SD $\sqrt{2/n}$ (iii) SD $\sqrt{1/3n}$ (iv) SD $\sqrt{3/2n}$

a. The average of the observations in one group.

b. The grand average of all the observations.

c. The average for one condition minus the average for another.

d. The average for one condition minus the average for the other two.

6. *Extra leafhoppers*. Suppose the leafhopper experiment had been done with eight observations for each condition instead of two. Suppose also that the SD for this larger version of the experiment turned out to be 0.274, just as before. Fill in the rest of following table:

Average or comparison	Two observations SE	Eight observations SE
(a) Control average	$(0.274)\sqrt{1/2} \approx .19$	
(b) Control vs. sugar	$(0.274)\sqrt{2/3} \approx .22$	
(c) 12 vs. 6 carbons	$(0.274)\sqrt{3/4} \approx .24$	

7. *Choosing a sample size.*

   a. Reread Example 11.5. Then solve the equation for n to verify that you need 12 or 13 dishes of leafhoppers.

   b. Suppose you want the SE for glucose versus fructose to be 0.05. How many dishes of leafhoppers will you need for each diet?

   c. Suppose you want the SE for Twelve Carbons versus Six to be 0.1. How many dishes of leafhoppers will you need for each diet?

8–11: *Babies, cows, shrews and bees.*

Adjust each of your sets of intervals in Exercises 1–4, substituting "t for 2," that is, putting a t-value in place of the 2 in 2 SEs. Then show each set of new intervals on a graph. Put a * next to the name of each contrast for which the observed distance is too big to be due just to chance error. Notice that using "t for 2" makes the intervals for the milk yield data a lot wider, but doesn't change the bee intervals very much.

12. *The t-test.* Here is another way to think about tests for contrasts. Since this method gives exactly the same answer as the method in the last section, you might wonder why I bring it up. There are two reasons. One: Many people, and many textbooks, test contrasts this way, so it is worth learning just so you can recognize a familiar thing when you see it. Two: The logic of the test, which I will discuss in Chapter 13, is a little easier to understand if you think of the test in this form.

Alternate Form of the *t*-Test for a Comparison

Reject the null hypothesis if and only if the absolute value of (comparison)/SE is bigger than the *t*-value from the table.

The ratio "(comparison)/SE" is called a ***t*-statistic**, since it is compared with a value from the t-table, or more generally, a **test statistic**, since it is used for testing. (The F-ratios you compute in ANOVA are called ***F*-statistics** and also test statistics.)

Compute the t-statistics for testing each of the comparisons in Example 11.6, and check that your conclusions are the same as the ones you get directly from the confidence intervals.

13. *Carpets and bees: The two-sample problem and the paired-data problem*. Many simple experiments are designed with only two groups to be compared. For example, the hospital carpet experiment (Example 8.7) was designed to compare the bacteria levels for two groups of

rooms, carpeted and bare. The bee sting experiment (Example S7.3), although more complicated (CB rather than CR), was also designed with just one comparison in mind, Stung versus Fresh. I like to regard these single-comparison experiments as special cases of CR and CB designs, and to use F-tests or comparisons to compare the two levels of the one factor of interest. Many scientists, however, would do the test using the version given in Exercise 12.

For the carpet experiment, the average numbers of bacteria colonies per cubic foot of air were 11.2 (Carpeted) and 9.8 (Bare); the SD was almost exactly 3. (Averages and the SD for the bee data are given in #4.) Compute t-statistics for comparing (a) Carpet versus Bare and (b) Stung versus Fresh. Compare these test statistics with the appropriate table values and state your conclusions.

14. *t-test.* Show, by algebra, that the t-test described in Exercise 12 always gives the same result as the test based directly on a confidence interval. That is, show that the absolute value of (estimate ÷ SE) is larger than the table value if and only if the confidence interval for the estimate does not contain zero.

15. *99% confidence intervals; tests at the 1% level.* The choice of 95% confidence is somewhat arbitrary. Scientists sometimes construct 90% intervals or 99% intervals. The 99% intervals are more conservative, in that they are wider than 95% intervals and so have a higher probability of containing the true value. In the same sense, 90% intervals are less conservative. Once you know how to construct 95% intervals, the others are easy: everything is the same except you use the 90% or 99% column of the t-table.

   a. Construct 90% and 99% intervals for the three comparisons in Example 11.6. Then show your 90%, 95%, and 99% intervals on a graph.

   b. Construct 90% and 99% intervals for the comparisons in Exercise 2, and compare your intervals using a graph as in (a).

16. *Finding a sample size.* For the leafhopper experiment, assume SD = 0.25. Suppose you want 95% confidence intervals for the diet averages to be of the form "average ± 0.1." How big should n be?

17. The "root n property." (Fill in the blanks.) Assume SD = 1.

   a. If the number of observations in a group goes from 25 to 100, then the SE for the average of the observations goes from _____ to _____ .

   b. If the number of observations in a group goes from 9 to 25, then the SE for the average of the observations goes from _____ to _____ .

   c. If the number of observations in a group goes from n to $4n$, then the SE for the average of the observations goes from _____ to _____ .

   d. If the number of observations in a group gets multiplied by n, then the SE for the average of the observations gets multiplied by _____ .

2. ADJUSTMENTS FOR MULTIPLE COMPARISONS

When a data set has only one factor of interest, with only two levels, there is only one comparison to make: level one versus level two. For such simple data sets, testing that comparison gives exactly the same result as the ANOVA F-test for the factor of interest. For many data sets, however, including the examples in the last section, there will be two or more comparisons to make. Stat-isticians disagree sharply—on the right way to handle several simultaneous comparisons. In Section 2.1, I'll first sketch two points of view: (1) adjust your comparisons to take into account how any there are, and (2) don't adjust. Then I'll fill in some of the details in the argument. For the most part, I find the argument against adjusting persuasive, but because lots of statisticians don't agree, in Section 2.2 I'll introduce one possible set of strategies for adjusting.

Arguments For and Against Adjusting

I've put the argument below, about whether to adjust, in terms of testing whether the difference between groups means is "real," but a similar argument applies to confidence intervals (since we declare a difference "real" if the corresponding confidence interval doesn't contain zero). Here's an overview of the two points of view:

Adjuster: Each time you do a statistical test to compare two group averages, there's the possibility of a false alarm—declaring a difference "real" when the underlying true values are actually the same. If you have several groups and do several such tests, using the ordinary threshold for declaring a difference real, the chance of a false alarm somewhere among your family of tests can be quite high. In order to keep the overall error rate under control, you should adjust the threshold value for each test to require stronger evidence than you would for just a single test.

Nonadjuster: Adjusting is wrong for several reasons. First, raising the threshold makes it too hard to detect real differences. Granted, you reduce the overall chance of a false alarm, but you pay for that protection by greatly increasing the chance of missing real differences. Second, adjusting encourages the bad idea that you can use the same data set to do two things at once—generate hypotheses and test them at the same time without running a new experiment. And finally, if you adjust your tests, then your decision about whether an observed difference is real will depend in part on information that shouldn't affect your conclusion: Not only will your adjusted threshold depend on the right things—how big your difference is compared to its SE, and the df for error—but also on how many other groups there are in your experiment. Two scientists could reach opposite conclusions based on the same evidence if one studied three groups and the other happened to study five.

The argument in favor of adjusting is based on the concept of **family-wise error rate**. Consider a set of treatments, such as the four leafhopper diets, and a family of comparisons based on the treatment averages. If all the underlying true treatment averages are in fact the same, then any difference declared "real" is actually a false alarm. The **family-wise false alarm rate (or Type I error rate)** is the chance of at least one false alarm somewhere among the set of tests, when all underlying true averages are equal. The reason statisticians worry about the family-wise rate is that it can be much higher than 5% if you are doing several comparisons. As a rough guide, statisticians compute family-wise error rates for sets of comparisons that have the special property of orthogonality, defined in Section 4.2. If you do 2 such tests, and each has a 5% false alarm rate, then the chance of no false alarms is $(0.95)(0.95) = 0.9025$, and the family-wise error rate is 9.75%. If you do 3 such tests, the chance of no false alarms is $(0.95)(0.95)(0.95) = 0.8574$, and the family-wise error rate is 14.26%. With k such tests, the family-wise error rate will be $100\% \times \{1 - (0.95)^k\}$. For $k = 6$ this number is 26.5%, and for $k = 10$ it is 60.1%. Statisticians who adjust their comparisons consider such family-wise error rates un-acceptable. Their adjustments bring the family-wise false alarm rate down to 5%.

The nonadjusters point out that the family-wise false alarm rate only applies when the are no true differences anywhere among your treatment groups—a highly unusual situation. Even then, the expected number of false alarms, if you don't adjust, is just one for every 20 comparisons, which is not really so bad. Moreover, the most common use for large numbers of simultaneous comparisons is to explore data: you don't have a specific set of hypotheses to test, and you're trying to learn which differences are worth attending to. In such situations, the false alarm rate doesn't matter so much, because you're not really drawing conclusions so much as generating hypotheses. You should be more concerned about the chance of a miss than the chance of a false alarm, and adjusting for multiple comparisons will increase the chance of a miss to a level you can't afford.

Many researchers do not find the arguments of the last paragraph persuasive.

A Strategy for Adjusting (If You Must)

Statisticians have developed about a dozen different methods to adjust for multiple comparisons, some methods intended for one kind of situation, some for another. As if that didn't make things complicated enough, even the statisticians who agree that adjusting is a good idea don't always agree on which method to use. I've chosen to present four common methods, together with suggestions about when to use them. For a more comprehensive treatment, you might refer to a more advanced book on design, like Kirk or Milliken and Johnson.

The four methods that follow are the Fisher LSD (for Least Significant Differences), Tukey HSD (for Honest Significant Difference), Scheffé, and Bonferroni. Computationally they are similar in that all four construct intervals of the form "Comparison ± [(SE) × (Adjusted standard distance)]." Both the Comparison and SE are the same as for ordinary intervals. The methods differ in (1) when you use them, and (2) how they adjust the standard distance. In what follows, I'll compare them four different ways: First, in terms of the size of the numerical adjustments; second, based on where the adjustments come from; third, in terms of the situations they are intended to handle; and finally, in terms of the intervals they give for the leafhopper data. Here's how they adjust the distances:

Fisher LSD (Least Significant Difference). No adjustments: Use the same t-value as in Section 1.

Tukey HSD (Honest Significant Difference). Instead of a t-value use $q/\sqrt{2}$, where q comes from a table of values of the Studentized range distribution and depends on the number of groups as well as the error df.

Scheffé. Use the F-table to find the same critical value you use for testing the factor of interest: denominator df (row) equals df for residuals, numerator df (column) equals df for the factor of interest = (# levels − 1). The adjusted distance equals $\sqrt{F \times (\# \text{ levels} - 1)}$.

Bonferroni. Use the t-table, but instead of the 95% column, choose the column for 100% − (5%/#comparisons).

Example 11.8 Comparing the Four Adjusted Distances

Suppose your factor of interest has five levels, and you plan to do tests for all possible pairs of levels. (There are 10 pairs, which you could check by listing them: AB, AC, . . . , DE.) Suppose the residual box of your decomposition has 10 degrees of freedom. Find the four adjusted standard distances.

Fisher LSD. No adjustment. Use the t-value from row 10, column 95%: 2.228.

Tukey HSD. The value of q from row 10 (error df = 10), column 5 (# groups = 5) of the Studentized range distribution is 4.65. The adjusted distance is $4.65/\sqrt{2} = 3.288$.

Scheffé. The F-value in row 10, column 4 is 3.48. The adjusted distance is $\sqrt{(3.48)(4)} = 3.731$.

Bonferroni. 100% − (5%/10) = 99.5%. Use the t-value from row 10, column 99.5%: 3.581.

Notice that in the example, Scheffé gives the biggest distance, Fisher LSD the smallest, and Tukey is in between the two. This illustrates a general property of those three adjustments: Scheffé is most conservative, giving the widest intervals, and requiring the strongest evidence to conclude differences are real; Fisher LSD is at the other extreme, giving the narrowest intervals and requiring the weakest evidence to conclude differences are real; and the Tukey method is in between. The Bonferroni method, which is based on an approximation, doesn't fit neatly into the rankings, except that it never gives narrower intervals than the Fisher LSD. Depending on the number of groups and number of comparisons, Bonferroni will sometimes give wider intervals than Tukey, as in this example, and sometimes narrower. (Bonferroni is narrower than Tukey when the number of comparisons is small enough.) If the number of intervals is large enough, Bonferroni will even be wider than Scheffé, but this is unlikely in practice.

The differences in interval widths reflect differences in where the adjustments come from. Each of the Fisher, Tukey, and Scheffé adjusted distances is computed by a rule that gives a 5% (or 1%) chance that a certain kind of observed difference will be larger than the adjusted distance. (These chances are computed assuming there are no real differences, so that 5% is the chance of a false alarm.) Here are the observed differences used for the three adjustments:

Fisher LSD: (largest group average − smallest group average) for 2 groups.

Tukey HSD: (largest group average − smallest group average) for p groups.

Scheffé: (largest possible weighted average − smallest possible weighted average) from the set of all possible weighted averages of p group averages.

The Bonferroni adjustment is based on a different logic:

Bonferroni. The Bonferroni adjustment is based on an inequality by that name. If each test in a family of k tests has false alarm rate 0.05/k, then the family-wise false alarm rate will be at most 0.05.

The ways the four methods compute their adjustments tells the kind of situations each is intended for. (1) Notice that the Fisher and Tukey adjustments are computed for pairwise comparisons. Fisher sets the threshold for declaring a difference real at the value you would use if you just had two groups. For the Fisher method, the number of groups doesn't affect the adjusted distance. (2) The Tukey distance, on the other hand, is based on the largest among all pairwise comparisons. (3) The Scheffé adjustment considers all possible comparisons, not just those of the form (group average − group average), and the resulting intervals are too wide to use if you have only pairwise comparisons to do. On the other hand, if you are using your data to decide on sets of weighted averages to compare (for example, 6-carbon versus 12-carbon sugar diets), the Scheffé adjustment will keep your false alarm rate at 5%. (4) Finally, the Bonferroni method can in principle be used whenever you have a fixed number of comparisons to make. In particular, it is often used when you have a fairly small number of comparisons that were planned before gathering the data.

1. *Pairwise comparisons: Fisher LSD or Tukey HSD.*

    a. Fisher LSD is designed for situations when (1) the ANOVA F-test for the factor of interest is significant, and (2) you want to test pairwise comparisons of levels, and (3) you are more concerned about missing real differences, less concerned about false alarms.

    b. Tukey HSD is designed for situations where (1) you want to test pairwise comparisons, and (2) you want to keep the family-wise chance of a false alarm at most equal to 5% (or some other small percentage) and you are willing to accept a reduced chance of detecting real differences. (If you are comparing only a few pairs, the Bonferroni adjustment may give narrower intervals than Tukey. If so, use Bonferroni.)

2. *Planned comparisons: Bonferroni*. Use Bonferroni if (1) you have planned a specific set of comparisons before you gather the data, and (2) you want to keep

the family-wise chance of a false alarm at most equal to 5% (or some other small percentage) and you are willing to accept a reduced chance of detecting real differences. (Ordinarily the number of planned comparisons will be fairly small, and often the comparisons will have a special structure like the set of top-down comparisons for the leafhoppers. This property, called *orthogonality*, is defined in Section 4.2.)

3. *Post hoc exploration: Scheffé*. Use Scheffé if (1) you are using the data to construct whatever comparisons look interesting, (2) you are looking not just at pairwise differences, but other comparisons as well, and (3) you want to keep the family-wise chance of a false alarm at most equal to 5% (or some other small percentage) and you are willing to accept a greatly reduced chance of detecting real differences. ■

EXAMPLE 11.9 LEAFHOPPERS

Construct appropriate confidence intervals/tests for the leafhopper data.

SOLUTION. Several approaches are possible here.

1. *Pairwise tests/intervals*.

   a. *Fisher LSD*. The ANOVA F-tests for diets was significant, so it is appropriate to use LSD for all pairwise comparisons, especially if we judge the error of missing a real difference more serious than the error of a false alarm.

   The SD for the data set in Table 11.2 is 0.274, the leverage factor for each pairwise comparison is 1, and the 95% t-value for 4 df is 2.776, so the 95% distance for each pairwise comparison is $(0.274)(1)(2.776) = 0.761$. (Intervals marked * do not contain 0: the difference is statistically significant.)

TABLE 11.2 Fisher LSD intervals

Comparison	Estimate	Interval
Control − Sucrose	−1.8	−2.56 to −1.04*
Control − Glucose	−0.9	−1.66 to −0.14*
Control − Fructose	0.2	−0.56 to 0.96
Sucrose − Glucose	0.9	0.14 to 1.66*
Sucrose − Fructose	1.6	0.84 to 2.36*
Glucose − Fructose	0.7	−0.06 to 1.46

   b. *Tukey HSD*. As an alternative to the Fisher LSD, if we want to control the family-wise chance of a false alarm, and we are willing to take on a higher risk of missing a real difference among the six pairs, we can use the Tukey HSD adjustment. See Table 11.3. The SD and leverage factors are the same as for the LSD, and the 95% q-value from the table of the Studentized range distribution with 4 groups and 4 df is 5.76, which makes $q/\sqrt{2} = 4.073$. This makes the adjusted 95% distance for each com-parison equal to $(0.274)(1)(4.073) = 1.296 \approx 1.30$.

TABLE 11.3 Tukey HSD intervals

Comparison	Estimate	Interval
Control − Sucrose	−1.8	−3.10 to −0.50*
Control − Glucose	−0.9	−2.20 to 0.40
Control − Fructose	0.2	−1.10 to 1.50
Sucrose − Glucose	0.9	−0.40 to 2.20
Sucrose − Fructose	1.6	0.30 to 2.90*
Glucose − Fructose	0.7	−0.60 to 2.00

2. *Planned comparisons: Bonferroni.* Suppose that as part of planning the experiment, we had decided to test three contrasts: control vs. sugar, 12 vs. 6 carbons, and glucose vs. sucrose, just as in Example 11.6. For these tests we could use the Bonferroni adjustment. Since 5%/3 is roughly 2%, we can use as our adjusted standard distance the 98% *t*-value for 4 df, which is 3.747. The leverage factors for the three comparisons are not all the same (Table 11.4).

TABLE 11.4 Bonferroni intervals

Comparison	Estimate	Leverage factor	Distance	Interval
Control vs. Sugar	−0.967	$\sqrt{2/3} = 0.816$	0.838	−1.81 to −0.13*
12 vs. 6 Carbons	1.250	$\sqrt{3/4} = 0.866$	0.889	0.36 to 2.14*
Glucose vs. Fructose	0.700	$\sqrt{1} = 1$	1.027	−0.33 to 1.73

3. *Post hoc comparisons: Scheffé.* Finally, suppose that we had not planned any comparisons in advance, but after running the experiment we noticed that the average for the two 6-carbon sugars was quite different from the average for the 12-carbon sugar, and we decided to test that comparison along with control versus sugar and glucose versus fructose. Although we are only testing three comparisons, we used the data to choose them, and so any comparison at all was a possible candidate for testing. The Scheffé adjustment was designed for just this kind of situation. The F value with 3 (numerator) and 4 (denominator) df is 6.59, which gives a standard distance of $\sqrt{6.59 \times 3} = 4.446$ (see Table 11.5).

TABLE 11.5 Scheffé intervals

Comparison	Interval
Control vs. Sugar	−1.96 to −0.03*
12 vs. 6 Carbons	0.20 to 2.31*
Glucose vs. Fructose	−.52 to 1.92

Table 11.6 compares the various sets of intervals.

TABLE 11.6 Four sets of simultaneous confidence intervals

Comparison	Fisher LSD	Tukey HSD	Bonferroni	Scheffé
Control − sucrose	−2.56 to −1.04*	−3.02 to −0.58*		
Control − glucose	−1.66 to −0.14*	−2.12 to 0.32		
Control − fructose	−0.56 to 0.96	−1.02 to 1.42		
Sucrose − glucose	0.14 to 1.66*	−0.32 to 2.12		
Sucrose − fructose	0.84 to 2.36*	0.38 to 2.82*		
Glucose − fructose	0.06 to 1.46	−0.52 to 1.92		
Control vs. Sugar			−1.81 to −0.13*	−1.96 to −0.03*
12 vs. 6 carbons			0.36 to 2.14*	0.20 to 2.31*
Glucose vs. fructose			−0.33 to 1.73	−0.52 to 1.92

Notice that for the pairwise comparisons, the Tukey tests find only two significant differences among the six possible pairs, compared to four significant differences for Fisher LSD. Similarly, although both Scheffé and Bonferroni find two significant contrasts among the last three, Scheffé comes very close to declaring the first of these not significant.

My interpretation is not that one set of intervals is right and another set wrong, but rather that certain of the differences among group means are not estimated with sufficient precision to support a conclusion that is free from ambiguity. On balance, however, I think that for this data set, the most important summary comes from the F-test for diets (there are detectable differences), together with the Bonferroni adjusted intervals. There is a real difference between control and sugar diets (sugar is better) and between the complex and simple sugars (complex is better); although glucose seems a bit better than fructose, the evidence is not conclusive. ■

Exercise Set B

1. *Walking babies: Bonferroni and Scheffé.*

   a. Use the Bonferroni method to adjust the confidence intervals for your three comparisons in Exercise A.1, and show these new intervals on a graph.

   b. Use the Scheffé method for the same three comparisons, and compare the widths for the two sets of intervals.

   c. Tell when the Bonferroni adjustment is more appropriate than Scheffé.

2. *Sleeping shrews: Bonferroni and Scheffé.* Use the version of the data with the outlier replaced for this exercise. The averages and SD you will need are given in Exercise A.3.

   a. Use the Bonferroni method to adjust the confidence intervals for your two comparisons in Exercise A.3, and show these new intervals on a graph.

b. Use the Scheffé method for the same comparisons, and compare the widths of the two sets of intervals.

3. *Intravenous fluids: Fisher LSD and Tukey HSD*. For the IV fluids data (Example 5.2), the average numbers of contaminant particles were:

Cutter: 274

Abbot: 205

McGaw: 397.

Each average was based on 6 samples of fluid; the SD was 99.165.

a. Use the Fisher LSD method to get a set of confidence intervals for all possible pairwise comparisons, and show these on a graph.

b. Now use the Tukey method for the same set of comparisons, and compare the widths of the intervals.

3. Between–Blocks Factors and Compound Within–Blocks Factors

Many of the designs you have seen have only one kind of error term; these designs include all BFs, almost all LSs, and all CBs for which the additive model applies. For data from these designs, there is only one typical error size (SD), which you estimate as $\sqrt{MS_{Res}}$. Confidence intervals for comparisons based on these designs always have the same form: comparison ± (leverage factor)$\sqrt{MS_{Res}}$ (t-value).

For other designs—all SP/RMs, and all multifactor CBs which use the nonadditive model—there will be two or more kinds of error terms, with a different MS for each kind. Confidence intervals for comparisons have the same form as for the simpler designs, except that the right MS to use for the 95% distance depends on the comparison.

The four parts of this section tells how to choose the mean square to get your 95% distance. The first deals with the SP/RM[1;1] design. For one of the kinds of comparisons based on this design, you have to **pool** (combine) two mean squares to get the one you need, and the second part of this section is about pooling. The third part deals with the CB[2] design. Finally, the last part gives a general rule that applies to all designs from the SP/RM and CB families.

The SP/RM[1;1] Design

All SP/RM[1;1] designs have two error terms: whole plot (between-blocks) and subplot (within-blocks). The right $\sqrt{MS}$ to use as SD depends on your comparison.

Mean Squares for the SP/RM[1;1]

If your comparison is:

within-blocks—use MS_{Res}

between-blocks—use MS_{Blocks}

across-blocks—use MS_{Pooled}

EXAMPLE 11.10 LOSING LEAVES: COMPARISONS FOR THE SP/RM[1;1] DESIGN

Example 7.11 and Exercise 7.D.21 described an experiment designed to compare the effects of four chemical treatments on the number of days before leaf pairs fell off Coleus plants. Sixteen plants served as blocks. The chemical treatments (control, lanolin, auxin at low concentration, auxin at high concentration) were applied to whole plants. Deblading (yes or no) was the within-plant treatment. The condition (cell) averages from the data appear below, along with information about the two error terms. Look at the cell averages in Table 11.7, and think about the kinds of comparisons you might want to make.

TABLE 11.7

	Debladed			Error terms			
	No	Yes	Avg	Source	SS	df	MS
Control	22.0	16.0	19.0	Blocks	483.75	12	40.31
Lanolin	14.0	13.0	13.5	Residual	42.75	12	3.56
Auxin (lo)	24.5	29.25	26.875				
Auxin (hi)	21.0	30.25	25.625				
Avg	20.375	22.125	26.25				

Here are four different kinds of comparisons among levels of a single factor: (a) You might want to compare Deblading vs. No Deblading, averaging over all four chemical treatments; or, because the two factors of interest interact, (b) you might want to compare Deblading vs. No Deblading for just one level of the other factor, for example, looking at the control group only. In the same way, (c) you might want to compare chemical treatments, averaging over both levels of Deblading; or, again because of the interaction, (d) you might compare chemical treatments with the deblading factor "held constant" at a particular level, for example, looking only at the debladed leaves.

As you work through the rest of the example, pay attention to whether and how the differences between plants (blocks) are involved in the comparisons. For the SP/RM[1;1], the right MS to use depends on the role of the block differences.

a. *Debladling vs. no deblading*: Use MS_{Res}.
   Here the comparison is *within-blocks*: each plant (block) gets compared with itself. Differences between plants do not contribute to the comparison, and so the between-blocks errors are not involved. Only the within-blocks errors are part of the estimate: use MS_{Res}.

   Estimate of difference $= 22.125 - 20.375 = 1.75$

   $SD = \sqrt{MS_{Res}} = \sqrt{3.56} = 1.887$

   $SE = SD \times \text{(leverage factor)} = 1.887 \times \sqrt{1/16 + 1/16} = 0.667$

   $95\%\ \text{distance} = SE \times t = 0.667 \times 2.179 = 1.454$

   Confidence interval $= 1.75 \pm 1.45$, or 0.30 to 3.20

b. *Deblading vs. no deblading, control group only*: Use MS_{Res}.
Here again, the comparison is *within-blocks*; the estimate does not include any between-blocks error: use MS_{Res}.

Estimate of difference = 16.0 − 22.0 = −6.0

SE = SD × (leverage factor) = $\sqrt{3.56} \times \sqrt{1/4 + 1/4} = 1.334$

95% distance = SE × t = 1.334 × 2.179 = 2.907

Confidence interval = −6.0 ± 2.9, or −8.9 to −3.1

c. *Auxin vs. control*: Use MS_{Blocks}. This comparison is *between-blocks*: the estimate compares the average for one group of entire blocks with the average for another group of entire blocks. The right SD for a between-blocks comparison is MS_{Blocks}.

Estimate of difference = (26.875 + 25.625)/2 − 19.0 = 7.25

SE = SD × (leverage factor) = $\sqrt{40.31} \times \sqrt{1/16 + 1/8} = 2.749$

95% distance = SE × t = 2.749 × 2.179 = 5.99

Confidence interval = 7.25 ± 5.99, or 1.26 to 13.24

d. *Auxin vs. control, debladed leaves only*: Use MS_{Pooled}. This comparison is neither within-blocks like (i) and (ii), nor between-blocks like (iii). A reasonable name might be **across-blocks**: each average combines observations from parts of a group of blocks. For across-blocks comparisons, the right mean square comes from combining the between-blocks and within-blocks error terms, using a method called pooling, which I'll explain in Example 11.11. It will turn out that for this data set, $MS_{Pooled} = 21.9375$, and $df_{Pooled} = 14$. Take these numbers on faith for the moment, and notice that apart from the pooling, you get your interval in the same way as before:

Estimate of difference = (29.25 +30.25)/2 − 16.0 = 13.75

SE = SD × (leverage factor) = $\sqrt{21.9375} \times \sqrt{1/8 + 1/4} = 2.868$

95% distance = SE × t = 2.868 × 2.145 = 6.152

Confidence interval = 13.75 ± 6.15, or 7.60 to 19.90. ■

Pooling Mean Squares

Computing a pooled mean square is easy: To pool two (or more) mean squares, you simply add the corresponding SSs (sums of squares, not MSs), then divide by the sum of the corresponding dfs. Ordinarily, the df you use to get a t-value from a table is the same as the df in the denominator of your mean square. However, for technical reasons, MS_{Pooled} does not behave like an ordinary mean square, and so for finding a t-value you should use the row in the t-table for the whole number closest to the expression in the box.

$$MS_{Pooled} = \frac{\text{sum of SSs}}{\text{sum of dfs}}$$

$$\text{“df”} = \frac{(\text{sum of SSs})^2}{\text{sum of } (SS^2/df)}$$

EXAMPLE 11.11 LOSING LEAVES: POOLING MEAN SQUARES

For contrast (iv) of Example 11.10, auxin vs. control, for the debladed leaves only, we need to pool MS_{Res} and MS_{Blocks}.

$$MS_{Pooled} = \frac{SS_{Res} + SS_{Blocks}}{df_{Res} + df_{Blocks}} = \frac{42.75 + 483.75}{12 + 12} = 21.9375$$

$$\text{“df”} = \frac{(SS_{Res} + SS_{Blocks})^2}{\frac{(SS_{Res})^2}{df_{Res}} + \frac{(SS_{Blocks})^2}{df_{Blocks}}} = \frac{(42.75 + 483.75)^2}{\frac{42.75^2}{12} + \frac{483.75^2}{12}} = 14.10$$

To get a 95% interval we use $SD = \sqrt{21.9375} = 4.684$, and use df = 14 to get a *t*-value of 2.145. ■

The CB[2] Design

Consider a CB[2] design whose factors are called “A” and “B.” If we assume the additive model, there is only one kind of error term, and we use $SD = \sqrt{MS_{Res}}$ for all comparisons among averages. If, however, we assume the nonadditive model, there are three components of error: Blocks × A, Blocks × B, and Blocks × A × B.

Mean Squares for the CB[2] Design

To compare levels of factor A

by averaging over levels of factor B:	**Use Blocks × A**
by holding a level of B constant:	**Pool Blocks × A and Blocks × A × B**

EXAMPLE 11.12 REMEMBERING WORDS

For the experiment of Exercise 9.C.1, there were four kinds of words to be remembered; the four kinds came from crossing two within-subjects factors, abstraction and frequency. Look over the condition averages (percent remembered) shown below, and think about the kinds of comparisons you might want to make.

	Abstract	Concrete	Avg
Frequent	41.2	44.0	42.6
Infrequent	38.8	42.4	40.6
Avg	40.0	43.2	41.6

Error term	SS	df	MS
Subj × Abstr	385.6	9	42.84
Subj × Freq	1248	9	138.67
Subj × Abstr × Freq	518.4	9	57.6

a. *Comparing levels of one factor, averaging over levels of the other:*

(i) Concrete vs. abstract: Use $MS_{Subj \times Abstr}$.

Estimate of difference $= 43.2 - 40.0 = 3.2$

$SD = \sqrt{MS_{Subj \times Abstr}} = \sqrt{42.84} = 6.546$

$SE = SD \times (\text{leverage factor}) = 6.546 \times \sqrt{1/20 + 1/20} = 2.070$

95% distance $= SE \times t = 2.070 \times 2.262 = 4.682$

Confidence interval $= 3.2 \pm 4.7$, or -1.5 to 7.9

(ii) Frequent vs. infrequent: use $MS_{Subj \times Freq}$.

Estimate of difference $= 42.6 - 40.6 = 2.0$

$SD = \sqrt{MS_{Subj \times Freq)}} = \sqrt{138.67} = 11.776$

$SE = SD \times (\text{leverage factor}) = 11.776 \times \sqrt{1/20 + 1/20} = 3.724$

95% distance $= SE \times t = 3.724 \times 2.262 = 8.423$

Confidence interval $= 2.0 \pm 8.4$, or -6.4 to 10.4

b. *Comparing levels of one factor, holding the other factor constant:*

(iii) Concrete vs. abstract, frequent words only: pool Subj × Abstr and Subj × Abstr × Freq

$$MS_{Pooled} = \frac{\text{sum of SSs}}{\text{sum of dfs}} = \frac{385.6 + 518.4}{9 + 9} = 50.222$$

$$\text{“df”} = \frac{(\text{sum of SSs})^2}{\text{sum of } (SS^2/\text{df})} = \frac{(385.6 + 518.4)^2}{(385.6^2/9) + (518.4^2/9)} = 17.62 \approx 18$$

Estimate of difference $= 44.0 - 41.2 = 3.8$

$SD = \sqrt{MS_{Pooled}} = \sqrt{50.222} = 7.087$

$SE = SD \times (\text{leverage factor}) = 7.087 \times \sqrt{1/10 + 1/10} = 3.169$

95% distance $= SE \times t = 3.169 \times 2.101 = 6.659$

Confidence interval $= 3.8 \pm 6.66$, or -2.86 to 10.46

(iv) Frequent vs. infrequent, concrete words only: pool Subj × Freq and Subj × Abstr × Freq

$$MS_{Pooled} = \frac{\text{sum of SSs}}{\text{sum of dfs}} = \frac{1248 + 518.4}{9 + 9} = 98.133$$

$$\text{“df”} = \frac{(\text{sum of SSs})^2}{\text{sum of } (SS^2/\text{df})} = \frac{(1248 + 518.4)^2}{(1248^2/9) + (518.4^2/9)} = 15.38 \approx 15$$

Estimate of difference $= 44.0 - 42.4 = 1.6$

$SD = \sqrt{MS_{Pooled}} = \sqrt{98.133} = 9.906$

$SE = SD \times (\text{leverage factor}) = 9.906 \times \sqrt{1/10 + 1/10} = 4.430$

95% distance $= SE \times t = 4.430 \times 2.131 = 9.441$

Confidence interval $= 1.6 \pm 9.44$, or -7.84 to 11.04 ■

The CB and SP/RM Families: The "CWIC" Rule for Choosing MSs for Comparisons

Examples 11.10–12 tell how to choose mean squares for two specific kinds of designs, the SP/RM[1;1] and CB[2]. It would be possible to give similar lists of mean squares for a variety of other designs; many books take this approach. However, there is a general rule that works for all designs of the SP/RM and CB families. Once you learn this rule, you won't have to rely on separate lists for the various designs.

To use the rule, you need to practice until you feel at home with two facts about contrasts for the SP/RM and CB families.

(1) There is an error MS for each main effect and interaction. By "error MS" I mean the denominator MS for the *F*-ratio. In Example 11.10, these error MSs are: MS_{Blocks} for the Chemical Treatment, and MS_{Res} for Deblading and Chemical × Deblading. In Example 11.12, the error MSs are: $MS_{Subj \times Abstr}$ for Abstraction, $MS_{Subj \times Freq}$ for Frequency, and $MS_{Subj \times Abstr \times Freq}$ for Abstraction × Frequency.

(2) For each comparison covered by the rule, the factors of interest can appear in any of three ways: comparison factor, absent factor, or constant factor. The comparison factor is the one whose levels you compare. A factor is "absent" if you average over all its levels to get your comparison. A factor is "constant" if you hold its level constant when you compute averages.

Example 11.10 had two factors of interest, Chemical Treatment and Deblading. We looked at four kinds of comparisons. (i) Deblading vs. No deblading. Here the comparison factor is Deblading, and Chemical Treatment is "absent," because the comparison averages over all four treatments. (ii) Deblading vs. No Deblading, control plants only. The comparison factor is Deblading; Chemical Treatment is constant, because the comparison uses observations from one level (Control) only. (iii) Auxin vs. control. Chemical Treatment is the comparison factor; Deblading is absent. (iv) Auxin vs. Control, debladed leaves only. Chemical Treatment is the comparison factor; Deblading is constant.

Example 11.12 also had two factors of interest, abstraction and frequency. Check that for comparisons (i) and (ii), the comparison factors are abstraction (i) and frequency (ii); in each case the other factor was absent. For (iii) and (iv), the comparison factors are again abstraction (iii) and frequency (iv); but this time in each case the other factor is constant.

To get the mean square for a comparison, pool the error MSs for (a) the comparison factor and (b) all interactions of that factor with any factor that is both CONSTANT and WITHIN-BLOCKS.

To remember the rule, just associate it with the acronym "CWIC," for "Constant Within-blocks Interactions with the Comparison factor": these are terms whose error mean squares you pool with the error mean square for the comparison factor. If there are no constant within-blocks factors, the "pooled" MS is just the usual unpooled MS for testing the comparison factor.

Example 11.13 Word Association and Schizophrenia: SP/RM[1;2]

For the experiment described in Example 9.6, there was one between-subjects factor, diagnosis (normal or schizophrenic). There were two crossed within-subjects factors, slope (steep or flat) and instructions (free or idiosyncratic). Use the CWIC rule to choose MSs for the following comparisons: (a) normal vs. schizophrenic, (i) averaging over both within-blocks factors, (ii) averaging over slope, but for free instructions only, (iii) for steep slope words, idiosyncratic instructions; (b) steep vs. flat slope, (i) averaging over both diagnosis and instructions, (ii) averaging over instructions, but for schizophrenics only, (iii) averaging over diagnosis, but for free instructions only, and (iv) for schizophrenics, free instructions only.

Solution. If we assume the nonadditive model, we get four error MSs:

MS_{Subj}	for Diagnosis
$MS_{Subj \times Slope}$	for Slope and Diagnosis × Slope
$MS_{Subj \times Instr}$	for Instructions and Diagnosis ×Instructions
$MS_{Subj \times Slope \times Instr}$	for Slope × Instructions and Diagnosis × Slope × Instructions

a. *Normal vs. schizophrenic*: For these comparisons, the comparison factor is diagnosis, so MS_{Subj} must be included in the MS for the comparison, along with the error MS for the interaction of diagnosis with any constant within-blocks factors.

(i) Averaging over both within-blocks factors. There are no constant within-blocks factors, so you don't need to pool. Use MS_{Subj}.

(ii) Averaging over slope, free instructions only. Slope is absent; instructions is constant. Since Subj × Instr gives the error term for Diagnosis × Instructions, you pool MS_{Subj} and $MS_{Subj \times Instr}$.

(iii) For steep slope words, idiosyncratic instructions. Both within-blocks factors are constant. Pool all four of the error MSs listed above.

b. *Steep vs. flat slope*: For these four comparisons, the comparison factor is slope, so $MS_{Subj \times Slope}$ goes into the MS for the comparison, along with the error MS for the interaction of Slope with any constant within-subjects factor. (Instructions is the only possibility.)

(i) Averaging over diagnosis and instructions. There are no constant within-blocks factors: Use $MS_{Subj \times Slope}$.

(ii) Averaging over instructions, but for schizophrenics only. The one constant factor is not within-blocks. Here, too, there are no constant within-blocks factors: use $MS_{Subj \times Slope}$.

(iii) Averaging over diagnosis, but for free instructions only. Instructions is both constant and within-blocks. The interaction of this factor with the comparison

factor is slope × instructions, so we include the error term for this interaction in the MS for the comparison: Pool $MS_{Subj \times Slope}$ and $MS_{Subj \times Slope \times Instr}$.
(iv) For schizophrenics, free instructions only. Both Diagnosis and instructions are constant, but only instructions is within-blocks. Pool $MS_{Subj \times Slope}$ and $MS_{Subj \times Slope \times Instr}$. ■

Exercise Set C

The SP/RM[1;1] Design

1. *Solitary prisoners: Within-blocks comparisons.* Use the information given below (from Example 8.10) to construct a set of four simultaneous confidence intervals, and show them together on a graph. (Use the Bonferroni method of adjusting the intervals.)

   a. Day 1 vs. days 4 and 7 for the prisoners in solitary.
   b. Day 1 vs. days 4 and 7 for the control group.
   c. Day 4 vs. day 7 for the prisoners in solitary.
   d. Day 4 vs. day 7 for the control group.

	Day 1	Day 4	Day 7		SS	df
Solitary (10 prisoners)	17.5	10.0	7.8	Subjects	1610.2	18
Control (10 prisoners)	14.8	15.9	15.8	Residual	867.4	27

2. *Prisoners: Across-blocks comparisons.*

   a. Compute a pooled MS (pool subjects and residuals) and the approximate df.
   b. Then compute and graph a set of three simultaneous Bonferroni-adjusted intervals for control vs. solitary on day 1, on day 4, and on day 7.

3. *Prisoners: Between-blocks comparison.* What is the between-blocks comparison for this data set? Compute a confidence interval for that comparison. Then tell why that comparison is not particularly meaningful for this study.

4. *Diets and dopamine.* The study described in Exercise S9.C.6–9 compared the effects of two diets (low phenylalanine and normal) on two groups of PKU patients (five with good and five with poor dietary control). The response measured the level of dopamine in the urine. The summary that follows is based on a version of the data set with one outlier replaced.

	Normal	Low		SS	df
Poor control	50	70	Patients	10368	8
Good control	68	160	Residual	8044	7

Four comparisons or pairs of comparisons are listed below. Identify each as within-blocks, between-blocks, or across-blocks. Then find ordinary 95% confidence intervals, and write a short paragraph about the effects of the diets on the two groups of subjects. Should physicians prescribe low phenylalanine diets for PKU patients?

a. Low versus normal phenylalanine diet

b. Good versus poor dietary control

c. Low versus normal for patients with good control

d. Low versus normal for patients with poor control

The CB[2] Design

5. *Left brain, right brain* (Exercise 9.C.5). The numbers below give SSs, dfs, and the average times (in seconds) for twenty subjects to complete four types of tasks.

	Method of reporting	
	Pointing	Saying
Letter task	18.17	13.69
Sentence task	9.01	12.85

	SS	df
Subj × Task	201.91	19
Subj × Rept	248.17	19
Subj × Task × Rept	379.79	19

Find ordinary 95% confidence intervals for

a. The main effects: (i) letter vs. sentence and (ii) pointing vs. saying;

b. Task held constant: (i) letter task: pointing vs. saying and (ii) sentence task: pointing vs. saying;

c. Report held constant: (i) pointing: letter vs. sentence and (ii) saying: letter vs. sentence.

Other Designs: The CWIC Rule

6. *Diets and brains*.

a. For each of your comparisons in #4, identify each of the two factors of interest as either constant, absent, or the comparison factor. Then use the CWIC rule to check that you used the right mean squares for your intervals.

b. Do the same for each of your comparisons in #5.

7. *The SP/RM[1;2]: Word association and schizophrenia (Example 11.13)*. Several comparisons are listed below. For each, identify each of the three factors of interest (diagnosis, instructions, and slope) as constant, absent, or the comparison factor. Then use the CWIC rule to tell which MSs to pool to get the SD for a confidence interval.

a. Schizophrenic vs. normal, for steep slope words (averaging over instructions).

b. Free vs. idiosyncratic instructions

(i) averaging over diagnosis and slope;

(ii) for schizophrenics only, averaging over slope;

(iii) for steep slope words, averaging together both kinds of subjects;

(iv) for schizophrenic subjects only, steep slope words only.

8. *The CB[3]: Automatic processing.* The experiment described in Example 9.10 was in fact a CB[3]: in addition to frame size and category condition, there was a third factor of interest, with two levels: target present or absent. For a CB[3], there are three kinds of comparisons covered by the rule:

   a. Two factors absent;

   b. One factor absent, one constant; and

   c. Two factors constant.

   Give an example of each kind (describe the comparison in words) and use the general rule to tell which MS to use for each.

4. Linear Estimators and Orthogonal Contrasts [Optional]

Linear Estimators

All averages, and all comparisons are **linear estimators**, that is, weighted sums of observed values, of the form "sum of (Weight × Obs)." It can be proved that for linear estimators, the leverage factor for computing the SE equals the square root of the sum of squares of the individual weights.

For a weighted sum of observed values SE = SD × $\sqrt{SS_{Weights}}$.

To use this result, you need to be able to think about comparisons in terms of weighted sums. The next two examples show how to do this and introduce notation for weighted sums.

Example 11.14 Weighted Sums for a Fake Data Set

The main purpose of this example is just to show you the notation and the arithmetic. Think of the data set below as a fake version of the leafhopper data, with three diets instead of four, and with diets as rows. In (a)–(e) below, I'll show you several weighted sums of these observations, using a different set of weights each time. For each set, there will be one weight for each observation.

Control	1	0
Glucose	2	4
Sucrose	3	5

a. *All weights equal 1: The grand total of all the observations.* A particularly simple example is the one with all the weights equal to 1. The weights are shown below, in the box on the left. There is one weight for each of the observations (next box), the weights and observations go together according to where they sit in the boxes. In the middle two boxes, I have multiplied each observed value by its weight. The 15 at the right is the result of adding up each weight·obs.

Since the weights are all 1s, each weight·obs = 1·obs = obs, and the sum of these is the grand total of the observations.

$$\underset{\text{Weight}}{\begin{array}{|cc|}\hline 1 & 1\\ \hline 1 & 1\\ \hline 1 & 1\\ \hline\end{array}} \cdot \underset{\text{Obs}}{\begin{array}{|cc|}\hline 1 & 0\\ \hline 2 & 4\\ \hline 3 & 5\\ \hline\end{array}} = \sum \begin{array}{|cc|}\hline 1\times1 & 1\times0\\ 1\times2 & 1\times4\\ 1\times3 & 1\times5\\ \hline\end{array} = \sum \begin{array}{|cc|}\hline 1 & 0\\ \hline 2 & 4\\ \hline 3 & 5\\ \hline\end{array} = 15$$

Notation for weighted sum (·); Notation for "Compute the sum" (Σ); Multiply the numbers in pairs; Weighted sum = linear estimate (15)

You get $SS_{Weights}$ in the same way you get any sum of squares: square and add. Here, because the weights are 1s, $SS_{Weights}$ = # obs. Indeed, for any sum of observations, the weights will be 1s, and $SS_{Weights}$ will equal the number of observations in the sum.

$$SS_{Weights} = 1^2 + 1^2 + \cdots + 1^2 = 6(1) = 6. \quad SE = SD \times \sqrt{6}.$$

b. *Each weight equals* (1/ # obs): *The grand average.*

$$\underset{\text{Weight}}{\begin{array}{|cc|}\hline 1/6 & 1/6\\ \hline 1/6 & 1/6\\ \hline 1/6 & 1/6\\ \hline\end{array}} \cdot \underset{\text{Obs}}{\begin{array}{|cc|}\hline 1 & 0\\ \hline 2 & 4\\ \hline 3 & 5\\ \hline\end{array}} = \sum \begin{array}{|cc|}\hline 1/6 & 0/6\\ \hline 2/6 & 4/6\\ \hline 3/6 & 5/6\\ \hline\end{array} = \frac{15}{6} = 2.5$$

$$SS_{Weights} = 6(1/6)^2 = 6(1/36) = 1/6. \quad SE = SD \times \sqrt{1/6}.$$

In (a), the weighted sum was a sum, and $SS_{Weights}$ = # obs. Here, the weighted sum is an average, and $SS_{Weights}$ = 1/# obs.

c. *The average for the glucose group.*

$$\underset{\text{Weight}}{\begin{array}{|cc|}\hline 0 & 0\\ \hline 1/2 & 1/2\\ \hline 0 & 0\\ \hline\end{array}} \cdot \underset{\text{Obs}}{\begin{array}{|cc|}\hline 1 & 0\\ \hline 2 & 4\\ \hline 3 & 5\\ \hline\end{array}} = \sum \begin{array}{|cc|}\hline 0 & 0\\ \hline 2/2 & 4/2\\ \hline 0 & 0\\ \hline\end{array} = \frac{(2+4)}{2} = 3$$

$$SS_{Weights} = 2(1/2)^2 = 2(1/4) = 1/2. \quad SE = SD \times \sqrt{1/2}.$$

Here, the average is based on 2 observations, and $SS_{Weights} = 1 / 2$.

d. *The glucose average minus the sucrose average.*

$$\underset{\text{Weight}}{\begin{array}{|cc|}\hline 0 & 0\\ \hline 1/2 & 1/2\\ \hline -1/2 & -1/2\\ \hline\end{array}} \cdot \underset{\text{Obs}}{\begin{array}{|cc|}\hline 1 & 0\\ \hline 2 & 4\\ \hline 3 & 5\\ \hline\end{array}} = \sum \begin{array}{|cc|}\hline 0 & 0\\ \hline 2/2 & 4/2\\ \hline -3/2 & -5/2\\ \hline\end{array} = \frac{(2+4)}{2} - \frac{(3+5)}{2} = -1$$

$$SS_{Weights} = 4(1/2)^2 = 4(1/4) = 1. \quad SE = SD.$$

There are 2 observations in the first average, and 2 in the second. $SS_{Weights} = 1/2 + 1/2$.

e. *The control average minus the sugar average.*

Weight	
1/2	1/2
−1/4	−1/4
−1/4	−1/4

Obs	
1	0
2	4
3	5

$$\text{Weight} = \text{Obs} = \frac{(1+0)}{2} - \frac{(2+4+3+5)}{4} = -3$$

$$SS_{Weights} = 2(1/2)^2 + 4(-1/4)^2 = 3/4 \quad SE = SD \times \sqrt{3/4}.$$

This time there are 2 observations in the first average, and 4 in the second. $SS_{Weights} = 1/2 + 1/4$. ■

EXAMPLE 11.15 LEAFHOPPERS

Here is the real leafhopper data set. The SD is 0.274.

Con	Suc	Glu	Fru
2.3	3.6	3.0	2.1
1.7	4.0	2.8	2.3

(a) Find weights that give the control average, and find the SE.

SOLUTION: The control average equals the average for the first column, (2.3 + 1.7)/2. So the weights for the 2.3 and 1.7 in that column will each be equal to 1/2. None of the other observed values are part of the average, so their weights will all be 0.

Con	Suc	Glu	Fru
1/2	0	0	0
1/2	0	0	0

$$SS_{Weights} = 2(1/2)^2 = 1/2 \quad SE = (.274)\sqrt{1/2} \approx 0.194$$

(Check that the weighted sum does equal 2.0, as it should. Check, too, that there are 2 observations in the average, and $SS_{Weights} = 1/2$.)

(b) Find weights that give control average minus the average of the sugar groups together, and find the SE for this comparison.

SOLUTION. The weights for the control observations will be the same as before. The average of the six sugar observations equals their sum divided by how many there are, which gives weights equal to 1/6. Since we want the weighted sum to subtract the sugar average, we put a −1/6 instead of +1/6.

Con	Suc	Glu	Fru
1/2	−1/6	−1/6	−1/6
1/2	−1/6	−1/6	−1/6

$$SS_{\text{Weights}} = 2(1/2)^2 + 6(1/6)^2 = 2/3. \quad SE = (0.274)\sqrt{2/3} \approx 0.224$$

(Check that the weights work as they should: Compute the two averages in the old way and subtract; then compute the weighted sum and check that it gives the same number. Check that there are 2 observations in the first average, 6 in the second, and $SS_{\text{Weights}} = (1/2 + 1/6.)$

(c) Interpret the weighted sum whose weights are

Con	Suc	Glu	Fru
0	1/2	−1/4	−1/4
0	1/2	−1/4	−1/4

SOLUTION. We can ignore the control group, since its weights are 0s. The two 1/2s go with sucrose, and give us the sucrose average. The four −1/4s go with the glucose and fructose groups, and give us minus the average of these four observations. Since sucrose is the 12-carbon sugar, and the other two are 6-carbon sugars, this weighted sum gives us "12-carbon average minus 6-carbon average". ■

Orthogonal Contrasts (Comparisons)

Definition: **Two linear estimators are** orthogonal **if when you multiply their weights in pairs and add, the sum equals zero.**

Many experiments are designed so that it makes sense to test hypotheses based on several different comparisons, and sometimes you can choose the weighted sums so that testing a set of them is roughly equivalent to the *F*-test for a factor of your design. The purpose of this section is to illustrate a definition that is useful in this context, in order to prepare for exercises here and in later sections. (The definition will not be used except in those exercises.)

Remind yourself of the set of three "top-down" comparisons for the leafhopper experiment: (i) sugar vs. control, (ii) 12 carbons vs. 6, and (iii) glucose vs. fructose. These comparisons are based on meaningful structure, and should seem intuitively appealing. This set of comparisons has two other important properties as well. The two properties can be described formally, but I'll give an informal meaning first: Think of the leafhopper experiment as providing information about the differences among the four diets, and think of each comparison as isolating a piece of that information, three pieces in all. Because of the way I have chosen the comparisons, no two of them have any information in common, and taken together the set of three comparisons uses up all the available information about diet differences. The set of comparisons splits the available information about group differences into independent, nonoverlapping pieces, and no information gets left out. In more formal language, each pair of comparisons is **orthogonal**, and the set of comparisons is **complete**.

The informal meaning, in terms of nonoverlapping pieces of information, becomes easier to see when you look at an example of nonorthogonal comparisons. Consider the three pairwise comparisons for the fluids data: (i) Abbot vs. Cutter, (ii) Abbot vs. McGaw, and (iii) Cutter vs. McGaw. Look back at the graph of these three comparisons (p. 13.2), and check that if you know the observed values for any two of the three comparisons, you can figure out the third. These comparisons are not independent: in fact, any two of the comparisons contain some information in common.

There is a simple way to tell whether two comparisons are orthogonal, that is, have no overlapping information from the observations. Think of the weights for one of your two comparisons as observed values, and use these "observations," together with the weights from the other comparison, to compute a weighted sum. If the sum equals 0, the two comparisons have no information in common, and are orthogonal. The exercises will give you a chance to apply this test.

Exercise Set D

Weighted sums and SEs

1. *Fake data for practice*. Several sets of weights and observed values are shown below. For each pairing of weights and "data," compute the weighted sum. Then compute the sum of squares for each set of weights, and take the square root to get the leverage factor. For four of these sets of weights, one of the shortcuts from Section 1.4 applies. Identify these four, tell which shortcut applies, and check that it gives the same number as the SS. (Example: The weights for a(ii) give the sum of the observations, so $SS_{Weights} = \# \text{ obs} = 3$.)

a.

Obs			Weight (i)	Weight (ii)	Weight (iii)	Weight (iv)
1			1	1	−1	1/3
2	with		2	1	0	1/3
3			3	1	1	1/3

b.

Obs			Weight (i)		Weight (ii)	
1	4		1	0	−1/3	1/3
0	1	with	1	0	−1/3	1/3
2	1		1	0	−1/3	1/3

2. Compute leverage factors for the following sets of weights.

a. The average of two observations:

1/2	1/2

b. The average of three observations:

1/3	1/3	1/3

c. The average of n observations:

$1/n$	$1/n$	$1/n$	$\ldots$	$1/n$

n weights.

3. *Leafhoppers* (Example 11.1).

   a. Write out tables of weights for the average and comparison listed below.

      i. The average of all the 6-carbon observations. (Glucose and fructose)

      ii. Glucose versus fructose.

   b. Which has the smaller SE, the average or the comparison?

4. *Beestings and babies*.

   a. Refer to Exercise A.4. Write out the set of weights for the comparison "stung vs. fresh." Then compute the SS for the weights, and check that it gives the same leverage factor as you get using the shortcut rule of Section 1.

   b. Refer to Exercise A.1. There were 6 babies in each of the first three groups, and 5 in the last group. Write out weights for each of the three comparisons in part (a) of the exercise, and compute the SS for each set of weights. Check that each SS is the same as you would get using the shortcut formulas in Section 1. Finally, take square roots, and check that the SEs I gave in part (b) of A.1 are in fact correct.

5. *Milk yields* (Example 7.9).

   a. Write out weights for the following averages and comparisons for the milk yield data of Exercise A.2:

      i. The roughage average;

      ii. Full grain versus partial grain;

      iii. Roughage versus the average of the two grain diets.

   b. Which of (i)–(iii) has the smallest SE? The largest?

6. *Leverage factor for comparing groups of unequal sizes*.

   a. Suppose you want to compare two group averages, and that each average is based on n observations. Write out weights and find the leverage factor in terms of n.

   b. Suppose the first average is based on n observations, and the second average is based on $2n$ observations. Write out weights, and find the leverage factor in terms of n.

   c. Suppose the first average is based on n observations and the second is based on m observations. Find the leverage factor.

Orthogonal Contrasts

7. Formal definition of a comparison.

Most linear estimators that arise in practice are either averages or else comparisons among averages. Estimates of the second kind have a particular mathematical property, which can be used for a formal definition:

> **Definition: A linear estimator is called a comparison or contrast if the sum of its weights is zero.**

a. Which of the weighted sums (a)–(e) in Example 11.14 are contrasts?

b. Which of the weighted sums (a)–(c) in Example 11.15 are contrasts?

c. According to the formal definition, a weighted sum does not have to be a difference of two averages in order to qualify as a contrast. Give a set of weights for the leafhoppers for a contrast that is not a difference of averages. (This book will only use contrasts that are differences of averages. The more general kinds of contrasts arise in connection with regression methods, which are introduced in Chapter 14.)

8. *Leafhoppers*. Refer back to the set of three comparisons in Example 11.2. (a) Show that every pair of comparisons is orthogonal. (b) Since the number of comparisons equals df_{Diets}, the orthogonality from (a) guarantees that ______________. (Complete the sentence.)

9. *Emotions and skin potential*. Example 8.2 and Exercise 8.C.5 describe an experiment to compare the response of hypnotized subjects who were asked to feel four emotions: fear, happiness, depression, and calmness. One natural comparison is positive emotions (happiness, calmness) versus negative (fear, depression). Find two more comparisons that are orthogonal to each other and at the same time orthogonal to the one I gave.

10. *Rescaling weights*. Consider testing Control versus Sugar for the leafhoppers, using the form of the t-test from Exercises 7 and 8. The point of this exercise is to illustrate a fact about this form of the test: You can multiply all the weights by a constant, and the t-statistic won't change.

a. The weights for Control versus Sugar are

Con	Suc	Glu	Fru
1/2	−1/6	−1/6	−1/6
1/2	−1/6	−1/6	−1/6

If you multiply all the weights by 6, you get rid of the fractions. Do this, then compute the new linear estimate, new leverage factor, new SE, and new t-statistic for Control versus Sugar. Then fill in the blanks below.

b. If you multiply all the weights by the same number c, then the result is to multiply (i) the linear estimate by ________; (ii) the leverage factor by ___________; (iii) the SE by _______; (iv) the t-statistic by ______.

11. *Orthogonal contrasts: Decomposing a sum of squares*. When you test a contrast using a computer program, you typically get several pieces of information:

linear estimate

SE for the estimate

t-statistic = (estimate)/SE

Sum of squares for the contrast = $(\text{estimate})^2/SS_{\text{Weights}}$

F-statistic = $(t\text{-statistic})^2 = SS_{\text{Contrast}}/MS_{\text{Res}}$

You can test the associated null hypothesis using the t-statistic, or you can use the F-statistic, comparing it with the table value with 1 df for the numerator and denominator df the same as for the t-test. The two tests always give the same result. The purpose of this exercise is to illustrate a way to think about the sum of squares for a contrast.

Every linear estimate has df = 1, and has its own sum of squares: you can think of an linear estimate as a specialized way to split off the part of your data that fits the particular pattern represented by the weights. To make this idea precise, you need to know about the method of regression (which I discuss briefly in Chapter 14), so what I say here is necessarily somewhat vague. Nevertheless, you can make a start on the ideas without knowing about regression. The main point is that

You can use a set of orthogonal contrasts to split the sum of squares for a factor of interest into pieces, one piece for each contrast. The number of contrasts will equal the df for the factor.

The leafhopper experiment has one structural factor, diets, with df = 3 and SS_{Diets} = 3.975. In Exercise 8, I asked you to check that the three contrasts (Example 11.2) for the leafhoppers are mutually orthogonal, that is, for every pair of contrasts, the sets of weights are orthogonal.

a. Compute a sum of squares = $(\text{estimate})^2/SS_{Weights}$ for each contrast.

b. Notice that df_{Diets} = 3, and verify that the 3 SSs you computed in (ii) above add up to SS_{Diets}.

12. *Pigs and vitamins: More orthogonal contrasts*. If you regard the pig experiment (Example 6.11) as a CR[1], there is one structural factor, Diets, with df = 3, SS_{Diets} = 4470, so it should be possible to split the SS using three orthogonal contrasts. On the other hand, if you regard the design as a CR[2], there are three structural factors: B12 (df = 1, SS = 2352), Antibiotics (df = 1, SS = 243), and Interaction (df = 1, SS = 1875).

Split SS_{Diets} using the following contrasts:

	Weights			
Contrast	no B12 no Anti	5 mg B12 no Anti	no B12 40 mg Anti	5 mg B12 40 mg Anti
B12 vs. no B12	−	+	−	+
Anti vs. no Anti	−	−	+	+
(Both or neither) vs. (only one)	−	+	+	−

Each group average is based on 6 observations (two sets of 3 observations each), so you could use as weights +1/6 and −1/6. Or you could rescale (see Exercise 10) and use +1, −1.

a. Verify that the contrasts are mutually orthogonal.

b. Verify that their three SSs add to give SS_{Diets}.

c. Verify that the SSs equal the SSs for B12, Antibiotics, and Interaction that you get by regarding the design as a two-way factorial.

SUMMARY

1. *Comparisons and their standard errors.*

   a. A **comparison** or **contrast** is a difference of two group averages. Sometimes the two groups will correspond to levels of a factor, but other comparisons are also useful. The "top down" strategy for choosing a set of comparisons takes advantage of natural clusters among the levels of a factor to split the information about group differences into a complete set of independent pieces. ***Post hoc* comparisons** are those that you choose to do after you have seen your data.

   b. **Standard errors**. Each comparison is an estimate computed from observed values. Just as each observation is part true value, part chance error, in the same way each comparison has a chance part mixed in with the true value. The standard error (SE) for a comparison is an estimate of the typical size of its chance part; the SE has exactly the same meaning for a comparison that the SD has for a single observation. For an average based on n observations, SE = SD $\sqrt{(1/n)}$; for a comparison of two averages based on n_1 and n_2 observations, SE = SD $\sqrt{(1/n_1 + 1/n_2)}$.

2. *Confidence intervals and tests for contrasts.*

   a. Every **confidence interval** has the same form: Estimate ± (95% distance); the 95% distance equals the SE times a standard 95% distance from a t-table. The table of standard distances is constructed so that the confidence interval has a 95% chance of containing the true value estimated by your contrast.

   b. You can use any confidence interval to test a corresponding null hypothesis, that the true value estimated by your comparison equals zero: You reject the hypothesis if zero is not in the interval. In terms of the two groups in your comparison, rejecting the hypothesis means declaring the observed difference between your two group averages to be "real," not just chance.

3. *Adjustments for multiple comparisons.*

   a. Should you adjust? Most experiments lend themselves to several comparisons. Each test has a 5% false alarm rate associated with it, and if you do several tests, the family-wise error rate will be a lot higher than 5% unless you adjust your 95% distances. Some statisticians think the resulting increase in the chance of failing to detect real differences is too high a price to pay, and so they don't adjust. Many other statisticians insist on adjustments.

   b. *Four ways to adjust:*

   Fisher LSD (Least Significant Difference) No adjustments: Use the ordinary 95% t-value.

   Tukey HSD. Instead of a t-value use $q/\sqrt{2}$, where q comes from a table of values of the Studentized range distribution, and depends on the number of groups as well as the error df.

   Scheffé. Use the F-table to find the same critical value you use for testing the factor of interest: denominator df (row) equals df for residuals, numerator df (column) equals df for the factor of interest = (# levels − 1). The adjusted distance equals $\sqrt{F \times (\#\text{levels} - 1)}$.

   Bonferroni. Use the t-table, but instead of the 95% column, choose the column for 100% − (5%/# comparisons).

   c. *Uses of the four methods:*

   Pairwise comparisons. Fisher LSD is designed for situations when the ANOVA F-test for the factor of interest is significant, and you are more concerned about missing real differences, less concerned about false alarms. Tukey HSD is designed for situations where you want to keep the family-wise chance of a false alarm at most equal to 5%, and you are willing to accept a reduced chance of detecting real differences. (If you are comparing only a few pairs, the Bonferroni adjustment may give narrower intervals than Tukey. If so, use Bonferroni.)

Planned comparisons. Use Bonferroni if you have planned a specific set of comparisons before you gather the data, you want to keep the family-wise chance of a false alarm at most equal to 5%, and you are willing to accept a reduced chance of detecting real differences.

***Post hoc* exploration**: Use Scheffé if you are using the data to construct whatever comparisons look interesting, you are looking not just at pairwise differences, but other comparisons as well, you want to keep the family-wise chance of a false alarm at most equal to 5%, and you are willing to accept a greatly reduced chance of detecting real differences.

4. *Between-blocks factors and compound within-blocks factors.*

For comparisons based on any CR design or any CB for which the additive model applies, you use $SD = \sqrt{MS_{Res}}$ to get your 95% distance. However, for SP/RM designs and for multifactor CB designs with the nonadditive model, the right MS to use will depend on the comparison. To get the MS, pool the error MSs for (a) the comparison factor and (b) all interactions of that factor with constant within-blocks factors. MS_{Pooled} = (sum of SSs)/(sum of dfs). For pooled MSs, use "df" = (sum of SSs)^2/[sum of (SS^2/df)] to get your t-value.

EXERCISE SET E: REVIEW EXERCISES

1. *Mothers' stories* (Review Exercise 7 of Chapter 4). For the 20 control mothers, the average (# type A − # type C) stories was 2.15; for the 20 mothers of schizophrenics, the average was 0.00. The SD was 2.588, on 38 df.

   a. Find an ordinary 95% confidence interval for the comparison "Control vs. Schizophrenic."

   b. Think of the observed value for this comparison as like a draw from a box of 100 numbered tickets, and describe the tickets in the box.

2. *Radioactive twins* (Example S7.2). The CB study of the effects of environment on lungs used seven pairs of twins, one rural, one urban in each pair; pair #2 was an outlier. For the complete data set, the average percent radioactivity remaining after an hour was 41.5% for rural and 45.5% for urban; the SD was 7.17%. With pair #2 omitted, the averages were 39.783% (rural) and 47.05% (urban), with SD = 4.108%. Construct and graph two ordinary 95% intervals for urban vs. rural, one for the complete data set, and the other omitting pair #2. Use these to discuss the influence of pair #2 on the results; then state your overall conclusion about the effect of the environment.

3. *Hospital carpets* (Example 8.7 and Exercise 8.C.1). For the carpet experiment, a CR using 16 rooms as units, the average numbers of bacteria colonies per cubic foot of air were 11.2 (carpeted) and 9.8 (bare); the SD was almost exactly 3. However, for one of the 8 bare rooms, the observed value of 3.8 was an outlier. With the outlier replaced (or omitted), the average for bare goes up to 10.657 and the SD drops to 2.503. Construct and graph two ordinary 95% intervals for carpet vs. bare, one for the complete data set, and the other omitting the outlier. Use these to discuss the influence of the outlier on the results; then state your overall conclusion about the effect of the carpeting.

4. *Emotions and skin potential* (Example 8.2 and Exercise 8.C.5). In this CB design, eight hypnotized subjects had their skin conductivity measured after the hypnotist had told them to feel frightened, happy, calm, and depressed (one emotion at a time). The average response values were 27.825 (fear), 25.4125 (happiness), 23.875 (depression), and 23.1875 (calmness); the SD was 3.123.

   a. Use the top-down approach to choose a set of three comparisons.

   b. Use the shortcut rule of Section 1.4 to find the leverage factors for each comparison.

   c. Think of each comparison as a weighted sum of the observations. Tell the weights; then compute the SS for each set of weights and check that you get the same leverage factors as in (b).

   d. Consider a set of intervals for all pairwise comparisons. Compute the interval widths if you use (i) quick-and-dirty intervals; (ii) ordinary 95% intervals; (iii) Fisher LSD intervals; (iv) Bonferroni-adjusted intervals; (v) Scheffé-adjusted intervals. Use these widths to construct five different confidence intervals for fear vs. happiness, and show your intervals on a graph.

   e. Compute and graph simultaneous Bonferroni-adjusted intervals for your set of three comparisons in (a), and write a summary caption as in Example 11.2.

   f. Compute and graph simultaneous Scheffé-adjusted intervals for the four condition averages.

5. *Rats and amphetamines* (Exercise 9.F.1). The averages and SSs below come from a study of the effects of amphetamines on the behavior of rats that had been taught a routine for getting milk. Each of 8 rats was given a total of 6 different injections of amphetamine, one at a time, in random order, on 6 different days. Half an hour after each injection the rat was given a chance to get milk, and the number of pauses was recorded as the response.

	Dose (mg/kg)		
	1	2	3.2
dl-amphetamine	26.00	33.25	48.00
d-amphetamine	25.50	43.25	34.00

	SS	df
Rats × Drugs	2799	7
Rats × Doses	6233	14
Rats × Drugs × Doses	7953	14

Choose a comprehensive set of comparisons, construct and graph the corresponding confidence intervals, and write a short summary of what they tell you about the effect of the three doses of the two drugs.

6. *Darkness and extinction* (Exercise 8.E.5). The averages and SSs below come from a study of the effects of darkness on the extinction of learned behavior in rats. During the conditioning phase ("Before") rats got a food pellet for each bar press; during the extinction phase ("After") bar presses were not rewarded. Half the 20 rats were kept in the dark during the ex-

tinction phase; the other half were kept in control ("light") conditions. I transformed the response to 100 × log(# bar presses per minute).

	Before	After		SS	df
Light	236.3	178.3	Rats	46819.4	18
Dark	237.5	135.9	Residual	14058.2	18

Choose a comprehensive set of comparisons, construct and graph the corresponding confidence intervals, and write a short summary of what they tell you about the effect of darkness on the extinction of learned behavior.

7. *Computer exercises*. Pick one or two of the data sets listed in Exercise Set C of Chapter 10. Choose a set of comparisons you think will best summarize the results of the study, and use a computer to get the corresponding confidence intervals. Show them on a graph with a caption summarizing the intervals in words.

NOTES AND SOURCES

1. **The standard error (SE).** There are two closely related ways to approach the SE, via the data, and via the underlying chance process. If you focus on what you compute from the data, then the SD for an observation and the SE for an average or comparison seem different, because you compute their values in different ways. If, however, you think of every observed value as arising from a chance process, then regardless of whether the observed value is for a single measurement or for an estimate you get by combining measurements, the SD and SE are special cases of the same thing: each provides exactly the same kind of information about its chance process. Many statisticians call them both standard deviations. I'll continue to use SE for estimators and SD for measurements, to remind you that the computing rules are different.

2. **The SE: a simple example.** The following numerical example may help illustrate why averages and contrasts have smaller SEs than single observed values. Imagine computing the average of three measurements: (True + 2), (True + 4), and (True − 3). Adding the three values gives [(3 × True) + 3]; dividing by 3 gives an average of (True + 1). Notice that when you added, some of the chance errors canceled each other, so that the chance part of the average turned out smaller than the individual chance errors.

3. **Additional sources:**

Kirk, Roger E. (1982). *Experimental Design*, 2nd ed., Belmont, CA: Brooks/Cole Publishing Company, Chapter 3.

Milliken, George A. and Dallas E. Johnson (1984). *Analysis of Messy Data*, Volume 1: Designed Experiments. Belmont, CA: Lifetime Learning Publications.

4. **Tukey's method for comparisons among groups of unequal sizes:** see Spjotvoll, E. and M.R. Stoline (1973). "An extension of the T-method of multiple comparisons to include the cases with unequal sample sizes," *Journal of the American Statistical Association*, vol. 68, pp. 975–78.

CHAPTER 12

THE FISHER ASSUMPTIONS AND HOW TO CHECK THEM

OVERVIEW

Note: If you skipped the first three introductory chapters, you will need to read Chapter 2, Section 2, on the Fisher assumptions, before you begin this chapter.

The formal methods of this book (*F*-tests, comparisons, confidence intervals, and *t*-tests) depend on the Fisher assumptions. The focus in this chapter is on ways to check whether particular data sets fit those assumptions, and on some possible remedies for data sets that don't fit. There will be three kinds of examples: data sets that fit the assumptions, data sets that can be made to fit by removing outliers or transforming to a new scale, and data sets that are better analyzed using other models and methods.

Although the other models and methods are for the most part beyond the scope of this book, it is important to keep in mind that good data analysis almost always involves finding an appropriate model, and that this book deals mainly with just one class of models. To help put this one class in perspective, recall from Chapter 8, Section 4 that you can think about choosing a model by first identifying cases and variables, then deciding which of your variables are carriers, which are response variables, and finally, classifying each variable as either categorical or numerical. Analysis of variance is appropriate only for those data sets whose carriers are all categorical and whose single response is numerical. If you have numerical carriers or a categorical response or if you want to analyze two or more response variables at the same time, you should not use ANOVA.

For the most part this chapter will focus on data sets that have categorical carriers and a numerical response. For such data sets, ANOVA will be appropriate provided the Fisher assumptions fit the data.

Recall that you can think of these assumptions in terms of the assembly line metaphor, with random draws from the Berkeley box of numbered tickets providing chance errors at the last workstation. Four of the assumptions refer to the chance errors: they behave as if all the draws come from the same box, whose tickets have *zero average* and the *same SD*, as if each chance error is drawn out *independently* of the others (as if each ticket gets put back and the tickets get mixed before the next draw), and as if the mix of tickets in the box follows the *normal curve*. The other two assumptions refer to the whole assembly line process: the pieces get *added*, and there is one (*constant*) piece for each level of each factor.

The first assumption, that the average of the tickets in the box is zero, cannot be checked directly from the data, because the way we decompose the observations always gives us residuals that add to zero, whether the assumption fits or not. However, if you have planned your study carefully, so that bias is controlled, and if you have chosen the right factor structure, then the assumption will be reasonable.

In the sections that follow, we will consider each of the other five assumptions in turn. It may help you remember them to think of the abbreviations SIN and AC:

S Same SDs (Section 1)

I Independent chance errors (Section 2)

N Normal-shaped set of chance errors (Section 3)

A Adding is the way the effects get combined (Section 4)

C Constant effects for the levels of the factors (Section 4)

For each assumption, I'll say briefly why it is important, then show you one or more things you can do to check whether the assumption is reasonable, then discuss what to do if the assumption doesn't fit. In many cases it will be possible to transform the data to a new scale for which the assumption does fit. In other cases, it may be best to give up on ANOVA in favor of some other method.

The chapter ends with a section on replacement values for outliers.

1. Same SDs

Very briefly, here are two reasons why the assumption is important:

Reasons Why the SDs Should Be Equal

- **When SDs are unequal, the observations with the larger SDs tend to dominate the average (see Example S8.5).**
- **The formal methods (tests and interval estimates) estimate *one* typical size for chance error (or, in the case of some designs, one size for each source of chance error).**

Checking the Assumption

There are two ways you can check to see how well the assumption fits your data. (1) For many data sets, you can divide your data into groups, then compute an SD for each group, and compare SDs. (2) For any data set, you can scatterplot residuals versus fitted values, then look for patterns.

Comparing SDs for several groups

> **Checking the Assumption of Equal SDs**
>
> - **Choose a factor that divides your data into several groups.**
> - **Compute a separate SD for each group.**
> - **Find the ratio of largest to smallest, SD_{Max}/SD_{Min}.**
> - **As a rough rule, if the ratio is bigger than 3, don't assume the underlying true SDs are equal. Try transforming to a new scale.**

EXAMPLE 12.1 WALKING BABIES

The data of Example 5.1 gave the numbers of months it took babies in four groups to learn to walk. Check the assumption of equal SDs.

SOLUTION. The SDs for the four treatment groups are:

Special Exercise:	1.45 months
Control 1 (Exercise):	1.90 months = SD_{Max}
Control 2 (Weekly report):	1.55 months
Control 3 (Single report):	0.96 months = SD_{Min}

The ratio $SD_{Max}/SD_{Min} = 1.90/0.96 = 1.98$, which is less than 3: The assumption is reasonable. ■

In the last example, choosing the groups was easy. With more complicated designs, however, there will be several choices for the groups. Which choices are likely to work better?

> **Choosing Groups for Comparing SDs**
>
> - **Choose a factor for which the response has bigger differences *between* groups, smaller differences *within* groups.**
> - **Try for a medium size number of medium size groups.**

EXAMPLE 12.2 KELLY'S HAMSTERS

The data from the hamster experiment, in mg/ml, were given in Chapter 1, Section 2. First decide which of the structural factors would be best for comparing SDs: Day length, Hamsters, Organ, or Interaction. Then check the assumption.

Solution. Day length is a poor choice: variability within groups is large, variability between groups is small. Hamsters is also a poor choice, for the same reason. Moreover, each group has only two observations, which is too few. Organ is an acceptable choice: variability between groups is large; variability within groups is small. Organ gives only two groups, however, and more would be better. Interaction is the best choice: we get four groups of four observations each, variability between groups is large, variability within groups is small.

Using these groups, we get the SDs in Table 12.1.

TABLE 12.1 Comparing SDs for the hamster data.

Long days	Heart	0.136 mg/ml
	Brain	1.669 mg/ml
Short days	Heart	0.104 mg/ml = SD_{Min}
	Brain	2.985 mg/ml = SD_{Max}

The ratio is huge: $SD_{Max}/SD_{Min} = 2.985/0.104 = 28.7$. This data set has chance errors of two very different sizes. ■

Often you can get around the problem of unequal SDs by reexpressing your data in a new scale. For the hamster data, transforming to log concentrations (then multiplying by 100 to get rid of decimals) works well. In this new scale, using the same four groups as before, $SD_{Max} = 9.06$, $SD_{Min} = 3.16$, and their ratio is 2.86, a tremendous improvement over the 28.7 we got for the original data.

Scatterplots of residuals versus fitted values. A careful analysis of a data set should always include a scatterplot of residuals versus fitted values, to check for outliers and for patterns in the plot. Back in Example 7.21 I showed you such a plot, one that suggests a balloon lying half above the *x*-axis, half below. That pattern is roughly what you should expect if your model and the Fisher assumptions fit your data set.

You can use the scatterplot of Res vs. Fit to check whether your data ought to be transformed. Each fitted value estimates an underlying "true" value; each residual estimates a "true" error. If error size is related to the size of the "true" value, then your scatterplot will show a pattern relating residuals and fitted values. For example, Exercise 7.E.3 gave a scatterplot of Res vs. Fit for the losing leaves data (Exercises 7.D.21–24). This plot suggests a wedge, narrow at the right, wide at the left: the larger residuals go with the smaller fitted values, and vice versa. The leaf data should be transformed.

Here's another example, one that gives a plot with a different pattern:

Example 12.3 Puzzled Children: Scatterplotting Res vs. Fit

Exercise 6.C.3 was based on a study of the effect of a teacher's praise and encouragement on young children. The design for the data set, reproduced below, is an observational BF[2]. Each child in this study was shown how to solve simple puzzles made of blocks, and then told to solve as many as s/he could in a fixed amount of time. The response is the number of puzzles solved.

Look over the data in Table 12.2, and decide whether the sizes of the residuals are related to the sizes of the fitted values. To do this, notice that each fitted value is just the cell average, equal to half the cell total. For each cell, there are two residuals, equal to (±) half the cell difference (larger observation − smaller observation). So to check whether the sizes of the residuals are related to the sizes of the fitted values, look to see if cell differences are related to cell totals (see Figure 12.1).

TABLE 12.2 Numbers of puzzles solved.

	Age									
Sex	3		4		5		6		7	
Male	24	19	30	19	18	16	29	41	31	53
Female	20	8	68	32	48	35	58	21	54	66

Except for the cell in the lower right corner, the data show a pattern: Larger residuals go with large fitted values, small with small. (Larger cell differences occur with larger cell totals. When cell totals are small, cell differences are small also.)

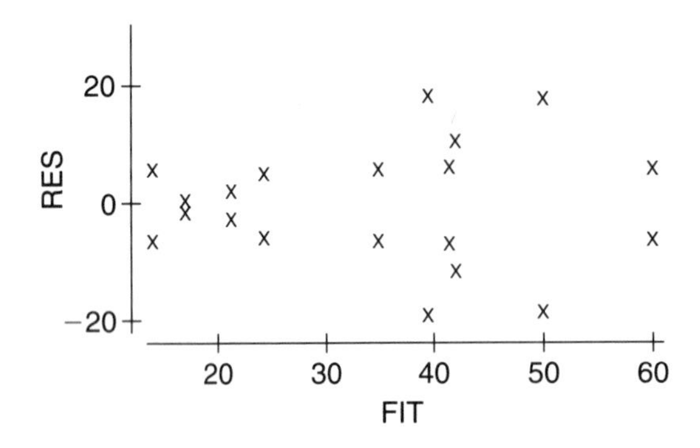

FIGURE 12.1 Scatterplot of Residuals vs. Fitted values for the puzzled kids. The points suggest a wedge, narrow at the left, and wide at the right. The size of the residual (= estimated error) depends on the size of the fit (= estimated true value). These data should be transformed. ■

Finding a Transformation

There are two questions here: What are the choices? How do you choose?

For nonnegative response values—counts and amounts—there is a standard family of transformations, the **power family**, that statisticians usually choose from. Choosing a particular member of that family often involves a mixture of three strategies. (1) Trial and error: It's not unusual to try more than one transformation before you find one that works well. (2) Familiar patterns: Statisticians have learned to recognize kinds of data sets for which particular transformations are likely to work well. (3) Diagnostic plot: There is a specialized scatterplot you can use to help you choose a scale. In what follows, I'll first describe the power family of transformations and give an example that illustrates what they do; then I'll list some of the familiar patterns statisticians rely on; then I'll describe the diagnostic plot and how to use it.

The power family of transformations. Each transformation in this family has the form (obs) $\rightarrow$ (obs)^p: to choose a transformation, you choose a power (p) and raise each observed value to that power. For example, choosing $p = 2$ means you square each observed value to get the transformed data. For most data sets that need transforming, powers between 1 and -1 work best.

Commonly Used Power Transformations

Power	Transformation	
1	Original scale	$x^1 = x$
0.5	Square roots	$x^{0.5} = \sqrt{x}$
0	Logarithms*	$\log x$
-0.5	Reciprocal roots	$x^{-0.5} = 1/\sqrt{x}$
-1	Reciprocals	$x^{-1} = 1/x$

* Even though $x^0 = 1$, not log x, the logarithm transformation behaves in a way that makes it correspond to $p = 0$ in the power family.

Example 12.4 Fake CR for Comparing Transformations

Imagine a CR experiment with four groups and two observations per group. We shall transform using square roots, logs, and reciprocals, and compare the way transforming changes the distances within and between columns.

a. *Raw data: Power* = 1 (Table 12.3).

b. *Square roots: Power* = $\frac{1}{2}$ (Table 12.4).

TABLE 12.3 Raw data.

	A	B	C	D
	1.0	10	100	1000
	1.1	11	110	1100
Average:	1.05	10.5	105	1050
Distance within:	0.1	1.0	10	100
Distance between:	9.45	94.5	945	

Distance within = larger observation in a column − smaller observation

Distance between = right-hand average − left-hand average

For the raw data, each distance is 10 times as big as the one to its left.

TABLE 12.4 Square root transformation.

	A	B	C	D
	1.00	3.16	10.0	31.6
	1.05	3.32	10.5	33.2
Average:	1.025	3.24	10.25	32.4
Distance within:	0.05	0.16	0.5	1.6
Distance between:	2.2	7.0	22	

Each distance is now roughly 3 times as big as the one to its left.

Small distances have gotten a little smaller, big distances have gotten a lot smaller.

c. *Logs: Power* = 0 (Table 12.5).

TABLE 12.5 Log transformation.

	A	B	C	D
	0.00	1.00	2.00	3.00
	0.04	1.04	2.04	3.04
Average:	0.02	1.02	2.02	23.02
Distance within:	0.04	0.04	0.04	0.04
Distance between:	1	1	1	

Each distance is now the same as the one to its left. Large numbers (in the raw scale) have been squeezed together much more than small numbers.

d. *Reciprocals: Power* = -1 (Table 12.6).

TABLE 12.6 Reciprocal transformation.

	A	B	C	D
	1.0	0.10	0.010	0.0010
	0.9	0.09	0.009	0.0009
Average:	0.95	0.095	0.0095	0.00095
Distance within:	0.1	0.01	0.001	0.0001
Distance between:	.855	.0855	.00855	

This transformation reverses directions: the largest value (in the raw scale) is now the smallest, and vice versa. Large numbers (in the raw scale) have been squeezed very close together, much closer than with square roots or even logs.

e. *Comparing the four scales*. Table 12.7 shows the first row of the fake data, in each of the four scales.

TABLE 12.7 Comparing the four scales.

		A	B	C	D
$p = 1$	Raw	1	10	100	1000
$p = .5$	Roots	1	3.16	10.0	31.6
$p = 0$	Logs	0	1.0	2.0	3.0
$p = -1$	Reciprocals	1	0.1	0.01	0.001

Each transformation squeezes the high end of the scale more than the low end. Square roots squeeze less than logs, and logs squeeze less than reciprocals. ■

Common uses of roots, logs, and reciprocals

1. *Square roots: Counted data.* Roots often help equalize SDs when your response counts the number of times something occurs. Examples:
   a. The numbers of wireworms in plots of soil (Example 8.3).
   b. The numbers of particles in intravenous fluids (Example 5.4).

   Note that with both examples, it isn't possible to count nonoccurrences. (For situations where you can count nonoccurrences, a special transformation, not in the power family, is often useful. Exercises A.14–16 illustrate this.)
2. *Logarithms: SDs proportional to averages.* If the size of chance error tends to be proportional to the size of the measurement, taking logs will equalize SDs. Such proportionality is often present for lengths, weights, and concentrations. Example: Enzyme concentrations for the hamster data.
3. *Reciprocals: Waiting times to rates.* Transforming to reciprocals is less common, but is sometimes useful when your response measures how long it takes for something to happen, especially if a few of the times are much longer than the others. Just as the time gives the answer to "How long?," the reciprocal (1/time) gives the answer to "How frequently?" Example: The psychologists who did the word association experiment in Example 9.11 transformed from response time to 1/(response time).

The diagnostic plot for choosing a power. This tool gives you a way to use your data to choose a member of the power family:

The Diagnostic Plot for Choosing a Power

1. **Choose a factor, then compute the average and SD for each group.**
2. **Scatterplot log (SD) = y versus log (average) = x, with one point for each group.**
3. **Fit a line to the scatterplot, by eye; then estimate the slope = (change in y)/(change in x).**
4. **Power = 1 − slope.**

EXAMPLE 12.5 DIAGNOSTIC PLOT FOR THE HAMSTER DATA

Table 12.8 shows the averages, SDs, and their logarithms for Kelly's enzyme concentration data. The diagnostic plot is given in Figure 12.2.

TABLE 12.8

Group		Average	SD	log (avg)	log (SD)
Long days,	heart	1.590	0.136	0.47	−2.00
	brain	8.925	1.669	2.19	0.51
Short days,	heart	1.353	0.302	0.30	−2.26
	brain	13.906	2.985	2.63	1.09

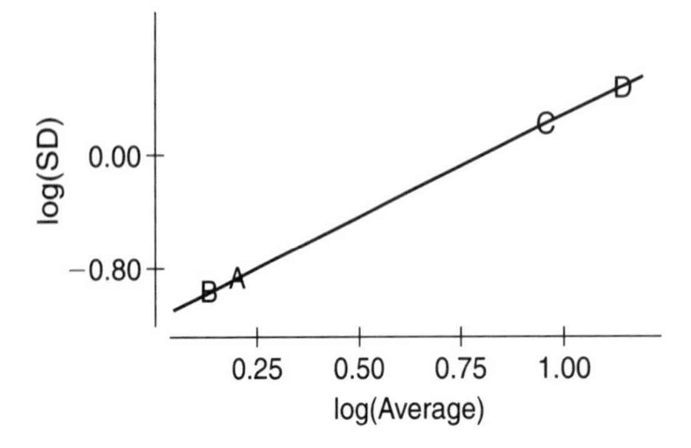

FIGURE 12.2 Diagnostic plot [log(SD) vs. log(avg)] for the hamster data. I drew the line by eye. When your points lie as close to a line as these do, averages and SDs are closely related, and a transformation will be able to bring the ratio SD_{Max}/SD_{Min} close to 1. The slope of the fitted line is very close to 1.5, which suggests transforming to reciprocal square roots.

Table 12.9 shows a comparison of group SDs for the raw concentrations, 100 times the log concentrations, and 100 times the reciprocal square roots of the concentrations:

TABLE 12.9 Three scales for the enzyme concentration data.

Group		Concentration	100 × log (concentration)	$100 \div \sqrt{\text{concentration}}$
Long days,	heart	0.14	3.46	3.22
	brain	1.67	9.02	3.52
Short days,	heart	0.30	3.16	3.27
	brain	2.99	9.06	2.56
SD_{Max}/SD_{Min}		29	2.87	1.35

For the raw concentrations, the group SDs are too unequal for ANOVA. Reciprocal square roots do the best job of equalizing SDs, but the scale is an unfamiliar one. The log concentrations represent a compromise. ■

In practice, choosing a transformation often involves a tradeoff. The power that makes your SDs most nearly equal might be something like $p = 0.22$, and an analysis of data in this scale may seem too unfamiliar to be appealing. Some statisticians would go ahead with $p = 0.22$; others would try more familiar nearby powers, like square roots ($p = 0.5$) and logs ($p = 0$), or possibly cube roots ($p = 0.33$) or fourth roots ($p = 0.25$), in the hope that one of these would work nearly as well. I lean toward using the familiar transformations as long as they work reasonably well; that's why I have chosen to analyze the hamster data using log concentrations instead of reciprocal roots.

For many data sets, you can check the assumption of same SDs by comparing group SDs as in the last example. This check offers the advantage that if your SDs turn out very unequal, you can use the SDs and averages to find a transformation from a diagnostic plot. However, if your design has lots of structure—either many factors, or many levels per factor—there may be no good choice for the groups to compare. For such data sets, indeed for any data set, you can (and should!) scatterplot residuals versus fitted values.

Exercise Set A

1. *Diagnostic plots for simple CR data sets*. Four little data sets are shown in Figure 12.3. Each has a CR design, with three observations per condition. For each one, (a) compute group averages and SDs, and find the ratio SD_{Max}/SD_{Min}; (b) then scatterplot Res vs. Fit; (c) then compute logs of the averages and SDs, plot $y = \log(SD)$ versus $x = \log(avg)$, join your two points with a line, compute its slope, and tell which power ($p \approx 1 -$ slope) you would use for a transformation; (d) finally, use a calculator to carry out the transformation, and replot Res vs. Fit in the new scale. Table 12.10 is a short table of logs to base 10, for finding log(avg) and log(SD):

TABLE 12.10

x	1	2	4	10	20	40	100
$\log(x)$	0.0	0.3	0.6	1.0	1.3	1.6	2.0

a.

A	B
9	90
10	100
11	110

b.

A	B
9	19
10	20
11	21

c.

A	B
8	36
10	40
12	44

d.

A	B
9	16
10	20
11	24

FIGURE 12.3 Data sets.

2. *Diagnostic plots for real data.* Four diagnostic plots are shown in Figures 12.4a–d. For two of them, fit a line by eye, estimate slope = (change in y)/(change in x), and tell which value of p you would use for transforming. The other two plots show points "all over the map," so scattered that no line fits well. Such plots suggest that no transformation from the power family is likely to help much in equalizing SDs. Which two plots are the "hopeless" ones?

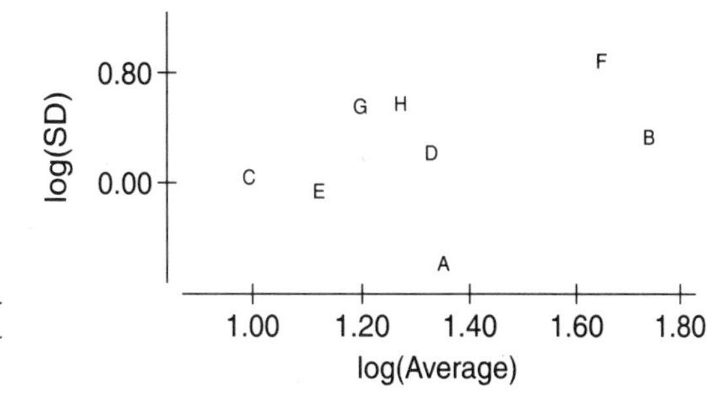

FIGURE 12.4a Emotions and skin potential (Example 8.2). For this plot, I used subjects as groups.

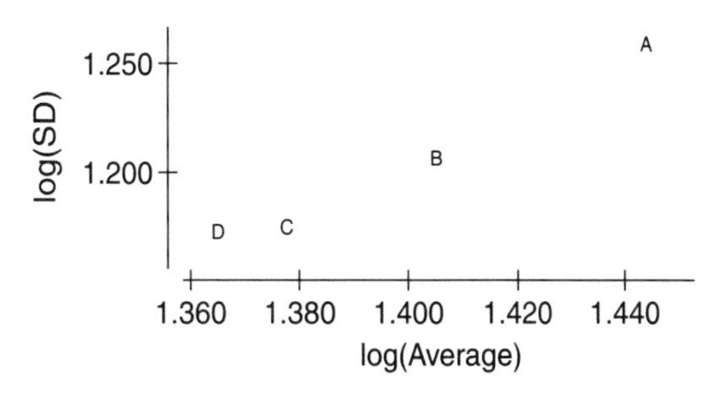

FIGURE 12.4b Emotions and skin potential. For this plot, I used emotions as groups.

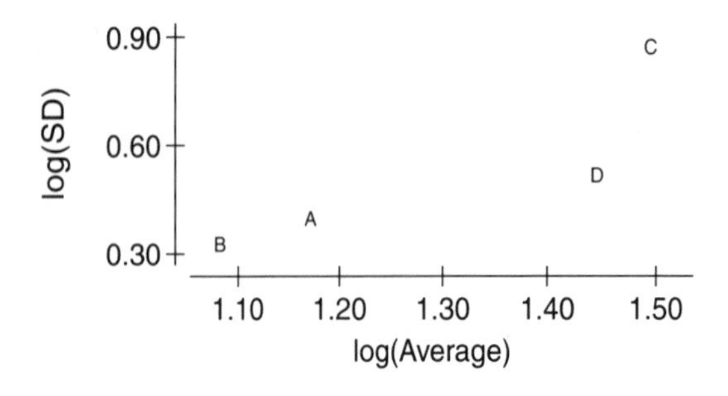

FIGURE 12.4c Bird calcium (Exercise 5.B.1) For this plot, I used the raw data, with cells as groups.

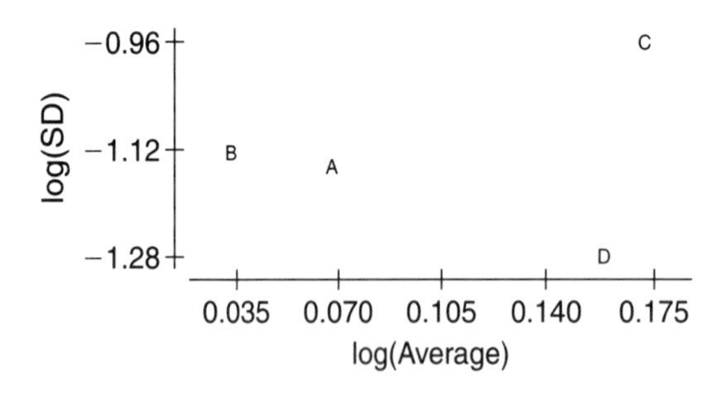

FIGURE 12.4d Bird calcium—log scale. For this plot, the groups are the same as in (c), but the averages and SDs are computed from the log concentrations.

3–5. *Fluids, worms and rats: data sets for practice.* Three data sets are listed below, along with group averages and SDs. For each, decide whether the data should be transformed, and give the reasons for your decision. If you decide a transformation is needed, do a diagnostic plot and use it to decide which member of the power family you'd try first.

3. *Intravenous fluids* (Example 5.2). In the last section, I suggested that because the response variable counts occurrences, the square root transformation might be appropriate. What do the data in Table 12.11 have to say about this suggestion?

TABLE 12.11

Drug Co.	Avg	log(avg)	SD	log(SD)
Cutter	274	2.44	52.1	1.72
Abbot	205	2.31	83.0	1.92
McGaw	397	2.60	141.0	2.15

4. *Tobacco hornworms.* In Exercise 6.C.19 I asked you to estimate a replacement value for the outlier in the RR group of the data from Exercise 6.A.7. The averages and SDs in Table 12.12 are based on the data with the outlier replaced.

TABLE 12.12

Group	Avg	log(avg)	SD	log(SD)
RR	37	1.57	6.63	0.82
RC	107	2.03	10.90	1.04
CR	37	1.57	5.25	0.72
CC	105	2.02	9.63	0.98

5. *Rats and amphetamines*. The data set given in Exercise 9.F.1 has a CB[2] structure, so there are two sets of groups you can use for comparing SDs: Blocks = Rats = Rows, and Conditions = Columns (see Table 12.13).

TABLE 12.13

Blocks as groups					Conditions as groups						
Rat	Avg	log	SD	log	Drug	Dose	Avg	log	SD	log
A	43.0	3.76	36.0	3.58	dl-amph.	1	26.0	3.26	31.1	3.44
B	25.5	3.24	12.8	2.55		2	33.3	3.50	31.6	3.45
C	50.5	3.92	24.2	3.19		3.2	48.0	3.87	37.6	3.63
D	36.5	3.60	34.6	3.54	d-amph.	1	25.5	3.24	19.6	2.97
E	15.5	2.74	15.6	2.75		2	43.3	3.77	30.7	3.42
F	35.0	3.56	16.7	2.82		3.2	34.0	3.53	22.2	3.10
G	2.0	0.69	1.8	0.58						
H	72.0	4.28	16.9	2.83						

6. *Puzzled children (continued)*. The complete version of the experiment on the effects of a teacher's praise was given in Exercise 9.F.6. This data set has so much structure that it would be hard to find suitable groups for comparing SDs. A scatterplot of Res vs. Fit is given in Figure 12.5. Discuss any patterns you find. Does the plot suggest transforming?

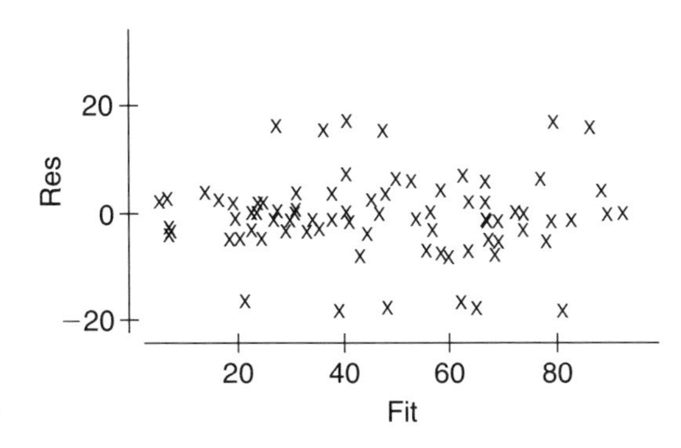

FIGURE 12.5

7. *Sugar metabolism*. Exercise 6.E.5 gave data from a CR[2] experiment designed to study the effects of oxygen concentration on the amount of ethanol produced from two sugars, galactose and glucose, by Streptococcus bacteria. The response was the concentration of ethanol, in micromoles per 0.1 micrograms of sugar. The version of the data given below has some of the observations altered slightly to make the arithmetic easier. Compute cell averages (= fitted values) and residuals. Then scatterplot Res vs. Fit and tell what you conclude from the plot.

	Oxygen concentration (micromoles)			
Sugar	0	46	92	138
Galactose	60* 30	42* 18	21* 23	12 10*
Glucose	26* 4*	14* 2	7 1*	0 2*

8. *A warning about plotting Res vs. Fit.* If df_{Res} is only a small fraction of df_{Total}, you should not rely heavily on patterns in a plot of Res vs. Fit, because the residuals contain too little information about how well your model fits. Professor Francis Anscombe of Yale University suggests a rough rule: df_{Res} should be at least 50% of df_{Total}, and preferably closer to 80%.

a. For the data set in Exercise 7, express df_{Res} as a percent of df_{Total}: does Anscombe's warning apply?

b. For the data set in Exercise 6, think of the design as an SP/RM[1;1] with one (compound) between-blocks factor. (How many levels does this factor have?) Compute df_{Res}, and express it as a percent of df_{Total}. Does Anscombe's warning apply?

9. *Scatterplots of leftovers versus partial fit.* For data sets like the one in Exercise 6, Anscombe suggests the following strategy for obtaining an informative plot. The idea is

"not to estimate small effects in the model, but leave them in the residuals.

Step 1: Perform a standard decomposition of the data and analysis of variance, beginning by taking out the grand average.

Step 2: Repeat the decomposition, beginning by taking out the grand average, but now estimating factor effects (and interactions) in the order of effects with biggest MS first—but don't estimate the interaction of two factors until both main effects have been estimated. At each step calculate the SS for the leftovers divided by the square of the df for leftovers. Continue only as long as this ratio decreases, and also the ratio of df to total number of observations remains fairly high.

Step 3: Finally, plot the leftovers against the corresponding fitted values."

For the data of Exercise 6, Anscombe's strategy leads to fitting the parts of the model listed in Table 12.13:

TABLE 12.14

Source	$SS_{Leftovers}$	$df_{Leftovers}$	$SS/(df)^2$
(Obs)	219,229	80	34.254
Grand Avg	46,155.98	79	7.396
Age	32,550.06	75	5.787
Sex	30,438.55	74	5.558
Cond	28,445.00	73	5.413
Order	27,667.33	72	5.337
Cond × Order	26,564.71	71	5.270
Age × Order	22,756.13	67	5.069
Cond × Sex	22,054.00	66	5.063

Fitting the term with the next largest MS, Cond × Age, causes $SS/(df)^2$ to go up, to 5.150. Fitting the eight factors listed leaves 66 df for leftovers, roughly 80% of the total. A scatterplot of the leftovers versus partial fit is shown in Figure 12.6.

a. Discuss any patterns you find in this plot. Does the plot suggest transforming?

b. Compare this scatterplot with the one in Exercise 6: What similarities and differences do you find?

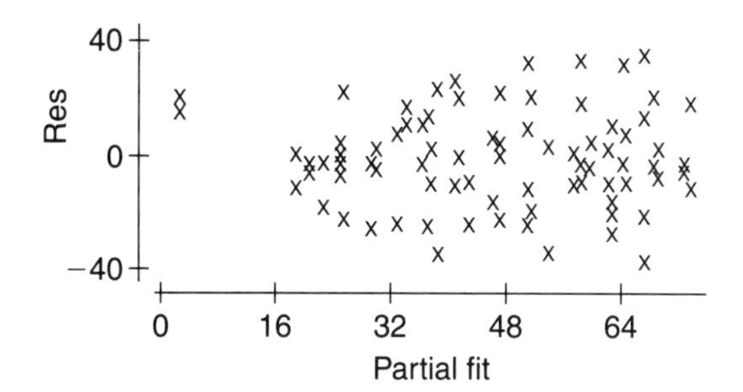

FIGURE 12.6

10–13. *Losing leaves.*

10. Exercise 7.D.23 asked you to find an ANOVA table for the leaf data. Use these numbers to follow Anscombe's strategy (from Exercise 9) to get a set of leftovers and partial fits to scatterplot.

11. Do the plot, and discuss what you find. How does your plot compare with the one I gave in Exercise 7.E.3?

12. Choose a set of groups, compute group SDs, and find SD_{Max}/SD_{Min}. Is your conclusion about transforming the same here as from the scatterplot?

13. Use a diagnostic plot to find a transformation from the power family. Transform and compare group SDs. Keep trying transformations until you either find one you consider satisfactory, or else conclude that you can't find one.

14–16. *Transforming counted data that are bunched at the extremes.*

Question: What do the response variables for the following two experiments have in common? Experiment 1 compares 4 treatments for preventing wheat rust. The response is the number of ears of wheat, out of 20, that show rust. Experiment 2 compares 4 methods of teaching arithmetic. The response is the score on a 20-question multiple choice test.

Answer: For each experiment, you can think of the response as the number of "successes" (no rust, or correct answer) in a fixed number of "trials" (20 ears of wheat, or 20 questions). A response of this sort often shows bunching, with lots of values crowded near 0, or near the maximum, or both. If one of your treatments is highly successful, scores for that treatment will tend to crowd near the maximum. If another treatment is about 50% successful, scores for that treatment will tend to be more spread out, and so the SDs for the two groups will tend to be unequal.

A specialized transformation, the arcsine square root, is designed to spread out the values that bunch near 0 or near the maximum, making the SDs more nearly equal:

Arcsine square root for binomial response:

1. **Compute** $\dfrac{0.5 + (\#\ \text{successes})}{1 + (\#\ \text{trials})}$
2. **Take the square root**
3. **Take the arcsine (inverse sine)**

14. Use a calculator to transform the following fake CR data, with three treatment groups. Each response equals the number of successes in 49 trials. (I chose 49 to make the arithmetic easier.)

Treatment	Scores
Highly successful	49, 48, 48, 47, 47, 46
Moderately successful	26, 22, 25, 25, 28, 29
Unsuccessful	0, 1, 1, 2, 2, 4

15. Construct parallel dot graphs for the raw and transformed data. Compute SD_{Max}/SD_{Min} for both versions of the data.

16–17. *Adjusting the power transformations for tiny response values*. Experience (and theory) shows that when you have tiny response values, a transformation from the power family tends to work a bit better if you make a small modification, such as I did for the wireworms. Here are some common modifications:

Transformation	Modification
Roots ($p = 0.5$)	$\sqrt{1 + \text{Obs}}$ or
	$(\sqrt{\text{Obs}} + \sqrt{1 + \text{Obs}})$
Logs ($p = 0$)	$\log(1 + \text{Obs})$
Reciprocal ($p = -1$)	$1/(1 + \text{Obs})$

16. *Roots*. Transform the following set of numbers twice, once using square roots, and once using the modification: First add 1, then take roots. Plot the raw and transformed data (use three separate axes) and comment on the pattern.

Data: 0 1 2 3 5 8 12

17. *Logs and reciprocals*. If you were to transform the data in (a) using $p = 0$, or $p = -1$, why would you have to use the modified version of the transformation?

18–21. *Building a transformation for the bloated worms*. A check of group SDs for the worm data of Example 9.4 shows that they are not at all the same size. Finding a transformation that equalizes SDs is not easy, however.

18. Use Species × Conc × Time to get groups, and use a computer to get group SDs; compute SD_{Max}/SD_{Min}.

19. Use a computer to do a diagnostic plot. What do you conclude about transforming?

20. a. Notice that group SDs are not so much related to the size of the response values as they are to the size of the distances away from 100%. As a preliminary transformation, subtract 100 from each observation to get percent change.

b. Now notice that because many of your scores are negative, you can't take roots or logs, nor can you do a diagnostic plot, because for many groups, log(avg) doesn't exist. Replace each score with its absolute value. (You may have to do this on the computer in two steps: first square each number, then take square roots.)

c. Use a diagnostic plot to choose a transformation for "absolute percent change" scores from part (b).

d. Finally, replace the minus signs, putting them back in front of those transformed values in (c) that came from negative percent change scores in (a).

21. Using the same groups as in Exercise 18, find group SDs and SD_{Max}/SD_{Min} for the transformed data.

2. Independent Chance Errors (1)

If you draw chance errors one at a time from a box of numbered tickets, replacing each ticket and mixing thoroughly before the next draw, then the chance errors will be independent: knowing the value of any one of them tells nothing about any of the others. The formal methods of ANOVA were designed for data sets whose residual errors are in fact chance-like and independent. The assumption is built into the method for estimating the SD, into the way critical values for test statistics are computed, and into the way standard 95% distances (*t*-values) for confidence intervals are computed.

The assumption also reflects an attitude that is basic in statistical work:

> Any systematic patterns relating observed values to each other should be part of your model.

If you can make such patterns part of the model, and subtract from your observed values the pieces that correspond to those patterns, then your residuals will be free of systematic relationships, and the assumption of independent chance errors is more likely to be appropriate.

The following example illustrates how the failure to recognize systematic patterns can undermine an analysis.

Example 12.6 A Faulty Analysis of the Shrew Data

Example 7.4 gave the heart rate (beats per minute) for six tree shrews, during each of three kinds of sleep. The design was a CB, with shrews as blocks. The factor for blocks in the model reflects our expectation that the three measurements in each block have a piece in common because they come from the same shrew.

Suppose, however, that we ignore the block structure, and think of the design as a BF, as if the 18 heart rates had come from 18 different shrews. According to the BF model, each observed value equals a grand average plus a sleep effect plus a chance error. This incorrect model leaves out the factor for shrews, and as a result, the numbers referred to as "chance errors" will be related by a systematic pattern: they will not be independent.

In this example the faulty model will lead us to grossly over estimate the SD. Any confidence intervals we construct would be much too wide, and the *F*-ratio for testing kinds of sleep would be much too low.

The computations are summarized in Table 12.15. The linear decomposition is the same as the one given in Exercise 7.D.5, except that this time the residual box will equal the sum of the old residual box plus the box of shrew effects.

TABLE 12.15 Failure to fit systematic patterns.

Residuals from					
Correct model			**Faulty model (no blocks)**		
−1.5	−1.5	2.6	−7.0	−7.4	−3.3
1.7	−1.1	−0.6	4.8	2.0	2.5
−0.1	0.7	−0.6	−0.2	0.6	−0.7
−0.4	0.7	−0.3	−2.0	−0.9	−1.9
0.8	−0.1	−0.7	5.0	4.1	3.5
−0.9	1.3	−0.4	−0.6	1.6	−0.1

Correct model	Faulty model (no blocks)
$SS_{Res} = 20.08$	$SS_{Res} = 214.24$
$df_{Res} = 10$	$df_{Res} = 15$
$MS_{Res} = 2.008$	$MS_{Res} = 14.283$
$F = 3.795$	$F = 0.534$

The residuals from the faulty analysis (right) show a pattern: numbers in the same row tend to have the same sign, and to be about the same size. Failure to include that pattern in the model makes MS_{Res} seven times too big. The pattern is not present in the residuals from the correct analysis (left), although the outlier for Shrew I, REM makes the residual in the upper right corner large, and makes the other residuals in row 1 and in column 3 negative. ■

As a general rule, you can check the assumption of independence by checking for patterns that are not already part of your model. The particular tools that are most useful will depend on your data set, but various kinds of scatterplots are nearly always useful. In what follows, I will illustrate two kinds of plots to look for in checking independence: *ad hoc* scatterplots, and within-block scatterplots. For designs with blocks (CB and SP/RM families), you should also consider a third kind of check on the assumption, based on the SDs for within-blocks differences. Even if your residual errors are not independent, ANOVA may still be appropriate provided these SDs are roughly equal.

Ad hoc Scatterplots

I use the phrase "*ad hoc*" to mean that this category is a catch-all. The particular scatterplots to look for depend very much on the data set, but it is always worth thinking about patterns related to place and time.

EXAMPLE 12.7 HOSPITAL CARPETS: RESIDUALS RELATED TO PLACE

For the hospital carpet data of Exercise 8.C.1, we were given the numbers of the 16 rooms used in the experiment. We weren't told where the rooms were located, but it is reasonable to think location would be related to room number. Is there a pattern relating the residuals and room numbers? Figure 12.7 shows a scatterplot.

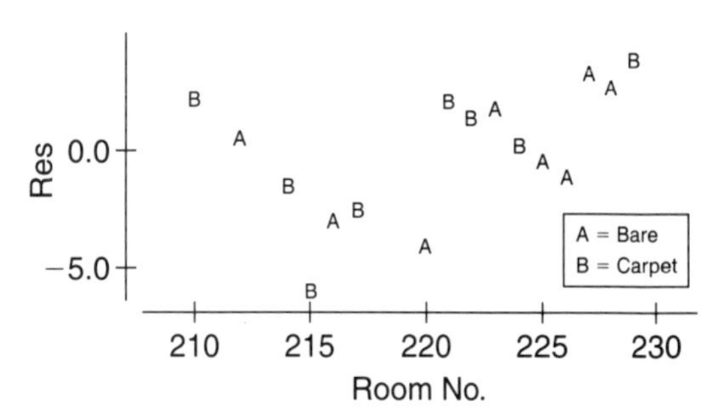

FIGURE 12.7 Residual vs. room number for the carpet data. The scatterplot shows that positive residuals tend to go with higher room numbers, negative residuals with lower room numbers. Depending on how you look, the pattern is either V-shaped or wedge-shaped, with the fat end of the wedge toward the left.

An F-test or t-test fails to find a significant difference between bare and carpeted rooms, but because of the pattern in the scatterplot, I would be reluctant to stop with the test of significance. If the hospital staff thought the observed average difference of 1.4 colonies per cubic foot (between carpeted and bare floors) was not an important difference, then the conclusion seems clear: they can go ahead with plans to carpet the entire hospital. If, on the other hand, a difference of that size is important, it might be worth continuing the data analysis, expanding the model to include room number.

It might also be worth trying to track down the reason for the pattern. The data suggest that higher-numbered rooms tend to be less sterile than lower-numbered rooms. Perhaps a careful study of the rooms used in the experiment could uncover some previously unrecognized influence on bacteria levels. I can at least imagine that such information would turn out to be more valuable than the information the study was planned to give. ■

In the last example we found a relationship between the observations and place. For many data sets, there will be a relationship between the observations and the time they were made. Observations made close together in time may tend to be more nearly equal than observations farther apart in time. A measuring instrument may "drift" during the course of a sequence of measurements. Subjects may get tired, or may get better with practice. Field observations of animals may show time-related effects of temperature, cloud cover, day length, and so on.

If you know the order in which your observations were made, scatterplot observed values or residuals against time to look for possible patterns.

EXAMPLE 12.8 TIME TRENDS IN THE TRANSYLVANIA DATA

The data on admission rates to a mental health clinic (Exercise 8.A.16, Example 8.11) came from a CB design with months as blocks. The scatterplot in Figure 12.8 shows the average daily rate, by month. For this plot, I've joined the points by lines to make the time trend easier to see.

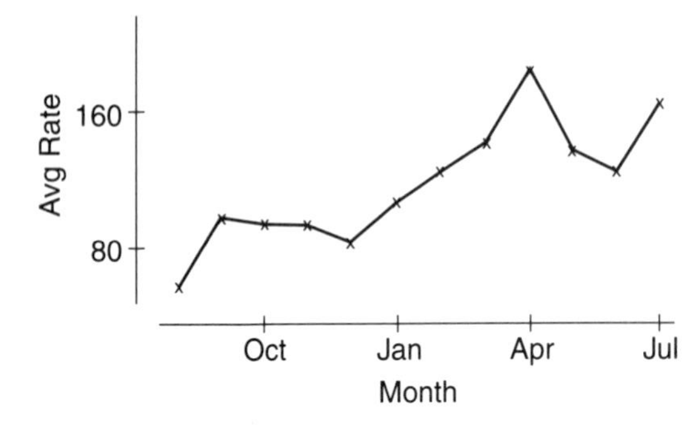

FIGURE 12.8 Average admission rate by month. Rates tend to be more nearly equal for months that are closer together. Admission rate is lowest in August, and increases slowly but steadily to a peak in April.

For the analysis of this data set, including blocks in the model makes it possible to estimate the month effects. One would hope that subtracting estimated month effects would leave the residuals free of time trends. Unfortunately, however, that seems not to be true for this data set. Figure 12.9 shows a scatterplot of residuals versus time.

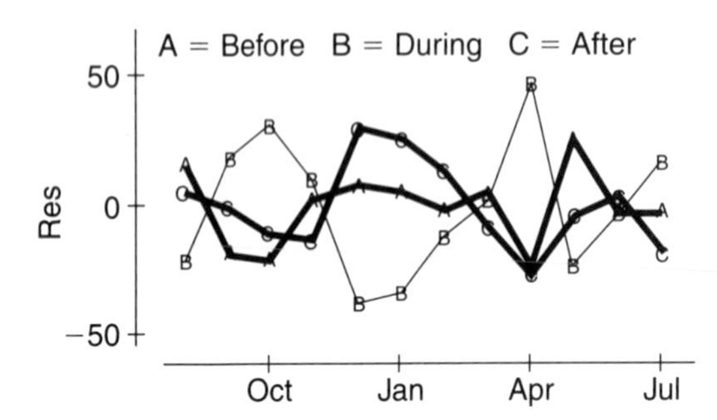

FIGURE 12.9 Residuals versus time. Notice that the plots for Before and After (thicker lines) show the same pattern: up at the same times, and down at the same times. The plot for During (thinner line) shows an "opposite" pattern: down where the other plots go up, and vice versa. If you were to plot month effect versus time, you would find a pattern a lot like the one for During.

The scatterplot makes clear that subtracting month effects has not produced residuals that are free from patterns. The residuals show systematic relationships; the assumption of independence is not appropriate; and MS_{Res} is not a good estimator of the typical size of chance error. ■

Within-Blocks Scatterplots

Within-Blocks Scatterplots

Plot one level of the within-blocks factor versus a second level, with blocks as points. If the model and Fisher assumptions fit, your plot should suggest a line parallel to the identity line $y = x$.

Any pattern that deviates from such a line suggests a systematic relationship that does not correspond to a part of your ANOVA model, and so will not be fitted by that model.

For designs with blocks, there are systematic ways to check the assumption of independence. Some of the more formal methods rely on correlations and SDs, but the methods I'll describe here rely on scatterplots. In Chapter 7, Section 5, I gave an example (7.19) of such a plot: for Hart's study of cooling mice, I scatterplotted the reading for reheated against the reading for freshly killed, with one point for each block (= mouse). If your design has more than two conditions per block, as for the sleeping shrew data (Exercise 8.C.1), you should do scatterplots for all possible pairs of conditions, in that case LSWS vs. DSWS, REM vs. DSWS, and LSWS vs. REM. If the model and Fisher assumptions fit, these plots should suggest lines with slope one.

The lines with slope one represent the part of your data that gets fitted by your model. Everything else becomes part of the residuals. These residuals, because they will not be free of systematic relationships, will not be independent.

I'll start with an example for which the assumption of independence is reasonable, but only after outlying observations have been replaced.

EXAMPLE 12.9 WITHIN-BLOCKS SCATTERPLOTS FOR THE SLEEPING SHREWS

The observed and fitted values for the shrew data are shown in Table 12.16, together with three within-blocks scatterplots (Figure 12.10). The points in each plot correspond to pairs of observed values, one pair per shrew. I've also drawn 45° lines that correspond to the fitted values. To get these lines, I first plotted pairs of fitted values, one

TABLE 12.16

	Observed values			Fitted values		
Shrew	LSWS	DSWS	REMS	LSWS	DSWS	REMS
I	14.0	11.7	15.6	15.1	13.2	13.1
II	25.9	21.7	21.4	24.1	22.2	22.1
III	20.8	19.7	18.3	20.9	19.0	18.9
IV	19.0	18.2	17.0	19.4	17.5	17.4
V	26.1	23.2	22.5	25.2	23.3	23.2
VI	20.5	20.7	18.9	21.3	19.4	19.3

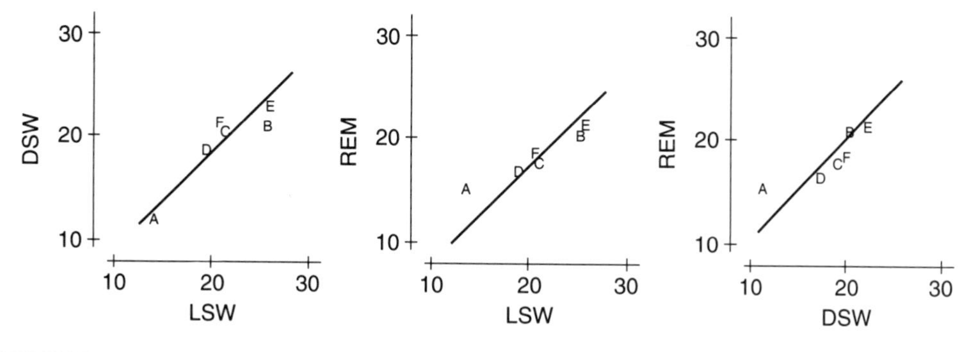

FIGURE 12.10 Within-blocks scatterplots for the sleeping shrews. The letters represent observed values; the lines represent fitted values. Systematic differences between the clusters of points and the lines fitted by the ANOVA model will show up in the residuals. Such differences indicate a failure of the independence assumption.

pair per shrew, then joined the points with a line. (To keep the plots from getting too cluttered, I don't show the individual points for the fitted values; I only show the line.)

I suggest you "read" these plots by looking first at the cluster of points for the observed values, and mentally "fitting" a line, or curve, or balloon—whichever best summarizes the shape. Then compare your fitted shape with the line of fitted values that comes from the ANOVA model. Loosely speaking, the difference between the two fits corresponds to the residuals. Any systematic differences suggest ways the ANOVA model fails as a framework for describing your data. Because the unfitted patterns get included in the residuals, such patterns indicate a failure of the independence assumption.

The first (leftmost) plot shows a good fit: the points lie close to a 45° line, although a curve would give a somewhat better fit. The second and third plots show a poor fit: in each case, if you fit a line by eye, its slope is much less than 45°. However, the y-value of 15.7 for the leftmost point in both plots (for Shrew I, REM) is an outlier, whose estimated replacement value (see Section 5) is 11.0. When I used the replacement value to replot the leftmost points, the revised plots do suggest 45° lines. ■

EXAMPLE 12.10 WITHIN-BLOCKS SCATTERPLOTS FOR THE TRANSYLVANIA DATA

In the last example, once I replaced the outlier, the plots suggested 45° lines. For the Transylvania data, we already know that a CB model fits badly, and we should expect within-blocks scatterplots to show systematic departures from a 45° line (Figure 12.11).

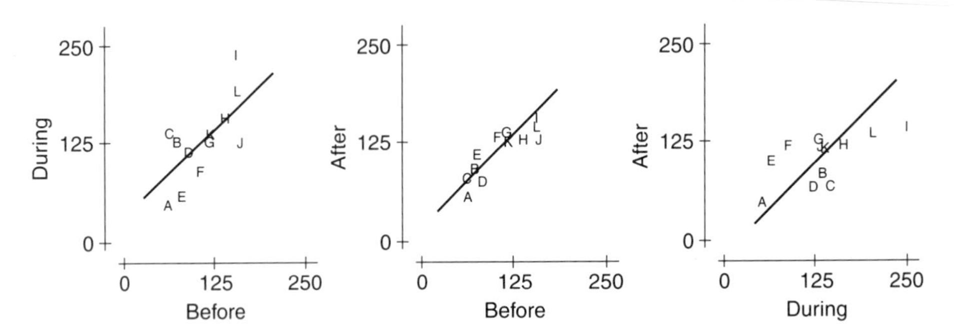

FIGURE 12.11 Within-blocks scatterplots for the Transylvania data. None of the three plots suggests a line with 45° slope. The left plot suggests either a curve or a line much steeper than 45° the middle plot suggests a curve, and the right plot suggests either a curve or a line with slope much less than 45°. A CB model cannot fit these patterns in the data. ■

Plots for the SP/RM designs. You can use within-blocks scatterplots to check the independence assumption for any design with blocks, for the SP/RM as well as for the CB. However, because the SP/RM has a between-blocks factor, you should use different plotting symbols for each level of that factor, and assess the fit separately for each level.

EXAMPLE 12.11 WITHIN-BLOCKS SCATTERPLOTS FOR KELLY'S HAMSTER DATA

I have used the hamster data to illustrate various aspects of the SP/RM model, and I have used the SP/RM model as one framework for describing and analyzing that data set. However, in some ways another model, the bivariate CR[1], gives a better framework. The scatterplot in Figure 12.12 shows what it is about Kelly's data that the SP/RM model cannot handle.

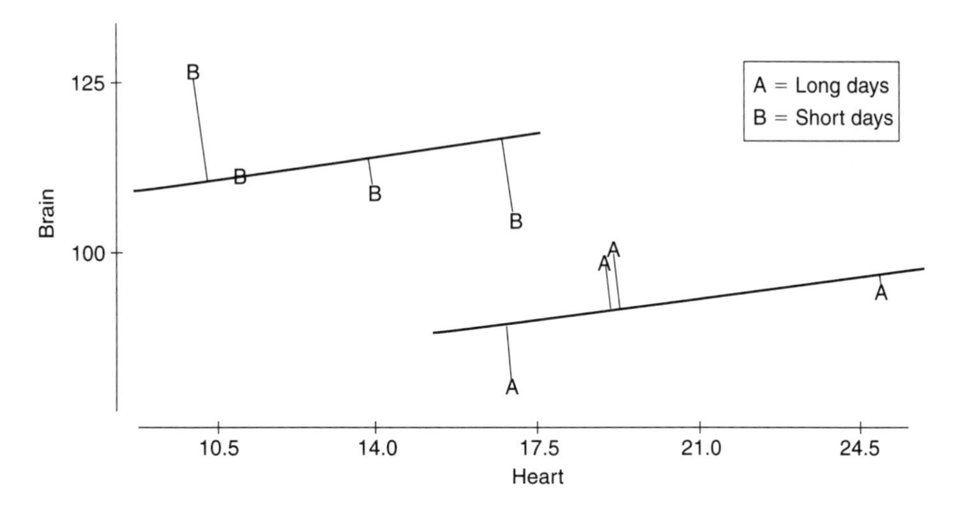

FIGURE 12.12 Within-blocks scatterplot for Kelly's hamster data. There are two sets of points, one for long days (As), one for short (Bs). Each set has its own (thicker) line formed by the fitted values from the SP/RM model. The perpendicular, thinner lines connect points for observed values to points on the thicker line corresponding fitted values. The distances along these lines represent residuals.

If you focus on the observed values (As and Bs), ignoring the rest of the plot, and then mentally exclude the lowest point—an outlier—you can see the pattern which the SP/RM model cannot fit: the seven points suggest a line or pair of lines with negative slope. ■

Judging by the pattern in the plot, the SP/RM has a major flaw as a model for Kelly's data. The strong negative relationship between heart and brain values is one the SP/RM model cannot fit. Because the relationship is present in the data but not present in the fitted values, by default that relationship will be present in the residuals. Thus the pieces of the data estimated by the residuals are not independent. Moreover, the residuals are "too big", in that they contain pieces that ought to be fitted as part of the model. The residual mean square will overestimate the variability due to chance error, in much the same way the faulty analysis of the shrew data in Example 12.6 overestimated chance variability.

Remedies. What should you do if lines y = x do not fit the patterns in your plots? (1) For some data sets, such as the one in Example 12.9, the poor fit may be due to an outlier. If so, you can estimate a replacement value, then reanalyze your data. If you do this, however, you should be careful to identify any conclusions that depend on your decision to replace the outlier. Section 4, on additivity and constant effects, gives examples and illustrates such fitting and subtracting. (2) For some data sets, the patterns depend on your choice of scale. Transforming to a new scale can sometimes "get rid of" the pattern. (More accurately, in the new scale you can fit the pattern and subtract it from the data as part of your linear decomposition.) (3) For many data sets, the patterns in your plots will suggest better models. The Transylvania data can be analyzed using regression-based methods of the sort I sketched in Example 8.11. The hamster data can be analyzed using a bivariate CR[1] model.

The Huynh-Feldt Condition for Designs with Blocks

Checking SDs for Within-Block Differences

For each pair of levels of the within-blocks factor, compute the differences (Level #1 − Level #2), and find the SD for each set of differences. If the SDs are (approximately) equal, the *F*-tests do not require the independence assumption.

Even if your within-blocks scatterplots suggest lines with 45° slopes, the independence assumption can fail in other ways. For example, consider a design with repeated measures, such as the study of prisoners in solitary confinement (Example 8.4). Residual errors for measurements taken close together in time (day 1 and day 4) on the same subject might tend to be related rather than independent. Fortunately, theoretical research has shown that even when the independence assumption fails, the formal *F*-tests may still be valid, provided certain other conditions are met.

EXAMPLE 12.12 CHECKING EQUAL SDs FOR THE TRANSYLVANIA DATA

For the data set in Table 12.17 there are three within-block measurements: Before, During, and After the full moon. Thus there are three sets of differences, and three SDs to check.

TABLE 12.17 Within-blocks differences and SDs.

	D − B	D − A	B − A
	−13	−8	5
	59	37	−22
	75	60	−15
	33	42	9
	−19	−51	−20
	−14	−40	−26
	5	−6	−21
	22	29	7
	95	91	−4
	−27	−3	24
	22	11	−11
	42	54	8
Avg	24.16	18.00	−5.50
SD	38.87	41.90	15.94

For two sets of differences, the SDs are almost equal, but both SDs are much larger than the SD for the third set of differences. The ratio of largest SD to smallest is 2.6, large enough to serve as another indication, along with the within-blocks scatterplots of Example 12.10, that the *F*-test for moon phases should not be taken at face value. ■

EXAMPLE 12.13 CHECKING EQUAL SDs FOR THE HAMSTER DATA

For Kelly's data, regarded as an SP/RM, the within-blocks factor has only two levels, so there is only one kind of difference, Brain − Heart. However, because the SP/RM design has a between-blocks factor, we compute a separate set of differences and SD for each level of that factor (see Table 12.18).

TABLE 12.18

	Differences: Long days	Brain − Heart Short days
	64	96
	82	88
	80	116
	70	100
SD	8.49	11.76

Although the SDs are not equal, they are roughly the same size. The F-tests do not require that the data satisfy the independence assumption. ■

Exercise Set B

1. *Bee stings: Looking for time trends.* Assume that the nine occasions which served as blocks for the experiment of Example S7.3 occurred in the order given, with occasion 1 first, and occasion 9 last. Scatterplot number of stings versus time, and comment on what you find.

2. *Feeding frogs.* A set of within-blocks scatterplots for the frog data (Examples 8.5 and S8.5) is shown in Figure 12.13. The solid lines correspond to fitted values from the CB model. For each plot, fit a line by eye and estimate its slope. For which plots are your lines close to those from the CB model? What do the plots indicate about the suitability of the CB model and the Fisher assumptions?

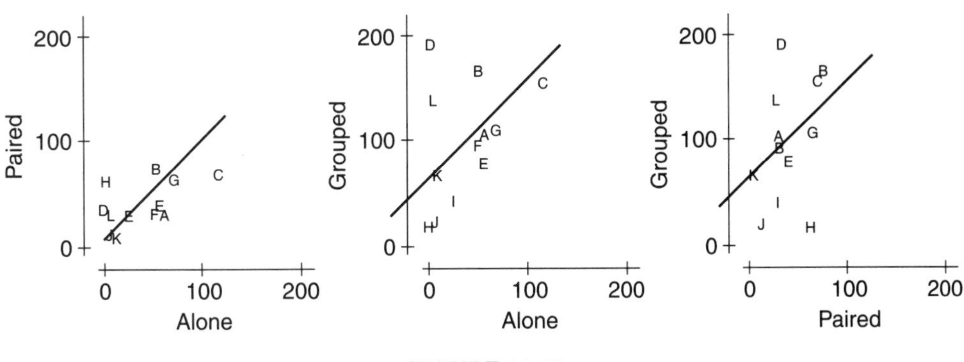

FIGURE 12.13

3. *Losing leaves* (Example 7.11 and Exercise 7.D.21). Describe the patterns in the scatterplot shown in Figure 12.14, and tell what they suggest about the suitability of the SP/RM model and the Fisher assumptions.

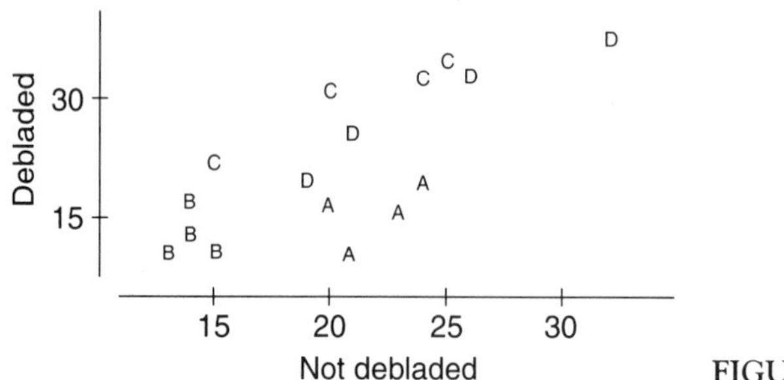

FIGURE 12.14

4–5. *Finger tapping*. The data set from Exercise 7.B.2 and Example 7.14 is a small one, which makes it suitable for practicing the mechanics of within-blocks plots and SDs.

4. Use the data to construct a set of within-blocks scatterplots: For each pair of drugs,

 a. first scatterplot fitted values (use the decomposition from Example 7.14) and join the points with a line; then

 b. scatterplot observed values on the same graph, and compare the pattern of the points with the line of fit.

5. Compute the three sets of differences, and compute an SD for each; then compare SDs. What do you conclude?

6. *Emotions and skin potential* (Exercise 8.C.5). Scatterplot (a) happiness versus fear and (b) calmness versus depression, with the eight subjects as points. (c) Is the assumption of independence appropriate? If not, compute SDs for the two differences corresponding to the two plots: Do the F-tests require independence?

7. *Rats and amphetamines* (Exercise 9.F.1).

 a. Do all three within-blocks plots for the d-amphetamine part of the data set (columns 4–6). What do your plots suggest about the independence assumption?

 b. Compute SDs for the three sets of within-blocks differences that correspond to your three plots, and compare SDs. Are they roughly equal?

 c. The 98 for rat D is an outlier. Use the estimated replacement value of 25 to replot the point for rat D in each of your scatterplots in (a). In what way, if at all, does replacing the outlier change your judgment about the independence assumption?

 d. Without recomputing the SDs for differences, try to judge the effect of replacing the outlier: Will SD_{Max}/SD_{Min} be a lot bigger, about the same, or a lot closer to 1 than before replacing the outlier?

8. *Mothers' stories: The complete data set* (Chapter 4, Review Exercises 7–9). You already know that ANOVA is not appropriate for these data. The purpose of this exercise is to show that within-blocks checks give yet another way to see that the Fisher assumptions don't fit. Two sets of within-blocks differences for the control mothers are shown below. Rather than compute the SD for each set of differences, plot each set on a separate dot graph, use your

graph to estimate each SD, and then estimate—to the nearest whole number – the ratio of the larger to the smaller SD.

(# A – # C): −2 −2 −1 0 0 1 1 2 2 2 2 2 2 3 3 4 5 6 6 7

(# B – # C): −2 −2 −2 −2 −2 −2 −2 −1 −1 −1 −1 −1 −1 −1 −1 0 0 0 0 1

3. THE NORMALITY ASSUMPTION (N)

ANOVA relies heavily on averages and SDs as summaries. These two summaries work very well for data whose residuals follow the normal curve, but often do not work well for non-normal data. (In particular, both the average and the SD are sensitive to the influence of outliers.)

Critical values for F-ratios and standard distances for confidence intervals (t-values) are computed assuming chance errors are normal. For these reasons (and others), you should use plots to check your residuals for normal shape, as part of any ANOVA.

Features of non-normal data. Here are four ways in which residuals may fail to fit the normality assumption:

Four Features to Look for in Residuals

1. **Outliers**
2. **Gaps and lumpiness**
3. **Asymmetry and skewness (lopsidedness)**
4. **Long or short "tails" (sets of extreme values)**

It will usually be easy to spot outliers from a dot graph. As you gain experience, you will find it easier to use a dot graph to check the other features as well. However, these features are somewhat harder to see, and two other kinds of graphs, histograms and normal plots, are often useful. In what follows, I'll introduce these two graphs, and illustrate how to use them to check the four features. Some data sets that violate the normality assumption in a gross way can be transformed to a new scale that makes the assumption more appropriate.

Histograms for Residuals

A histogram represents a list of numbers geometrically in a way that makes it natural to talk about the "shape" of the set of numbers. In particular, we can use histograms to help judge whether a set of residuals has a normal shape.

Histograms for Residuals

1. **Construct a dot graph of the residuals.**
2. **Divide the range into 7 to 15 equal intervals.**
3. **Construct a rectangle above each interval, with height proportional to the number of dots in the interval.**

If your histogram is roughly symmetric and bell-shaped, the normality assumption (N) is reasonable.

The number of intervals to use should depend on the number of residuals. Use fewer intervals if you have only a small number of residuals.

EXAMPLE 12.14 A SET OF RESIDUALS WITH ROUGHLY NORMAL SHAPE

Construct a dot graph and histogram for the following set of residuals, and comment on the shape.

$$0, \pm 1, \pm 2, \pm 4, \pm 6, \pm 8, \pm 11, \pm 14, \pm 18, \pm 23$$

SOLUTION

1. *Dot graph.*

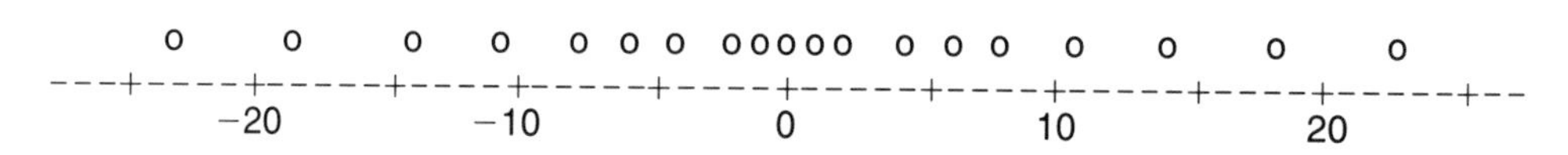

2. *Intervals.* The shape of the histogram depends to some extent on which intervals you choose. Here's one reasonable set of intervals, together with the numbers of residuals in each interval:

```
      1     1     2     3     5     3     2     1     1
--+-----+-----+-----+-----+-----+-----+-----+-----+-----+--
 -27   -21   -15    -9    -3     3     9    15    21    27
```

3. *Histogram.* We construct a rectangle above each interval, with height proportional to the number of residuals in the interval (see Figure 12.15).

   Histograms are good at revealing outliers, gaps and lumpiness, and symmetry.

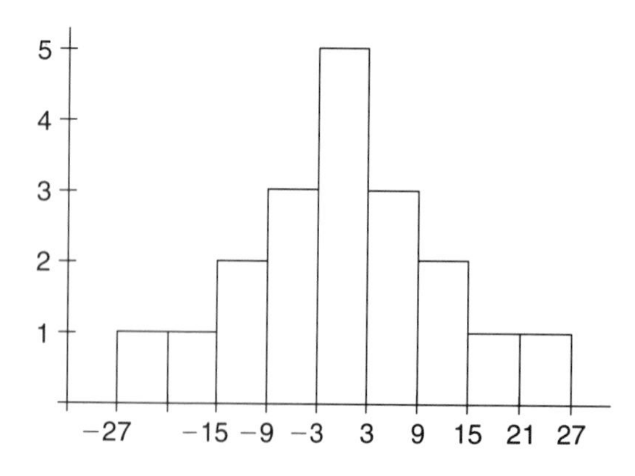

FIGURE 12.15 Histogram for the 19 residuals. The height of each rectangle is proportional to the number of dots in the interval. This histogram is perfectly symmetric, and roughly bell-shaped. The more nearly your histogram has a shape of this sort, the more appropriate the normality assumption is. (You might want to compare it with the normal curve on page 3.7.) ■

They are also useful as a tool for developing your geometric intuition about the shape of a data set. ■

EXAMPLE 12.15 HISTOGRAM FOR THE BIRD CALCIUM DATA

Exercise 5.B.1 gave calcium concentrations (in the plasma) for four groups of birds. A dot graph and histogram of the residuals for this data set are shown in Figure 12.16. Comment on the shape of the histogram.

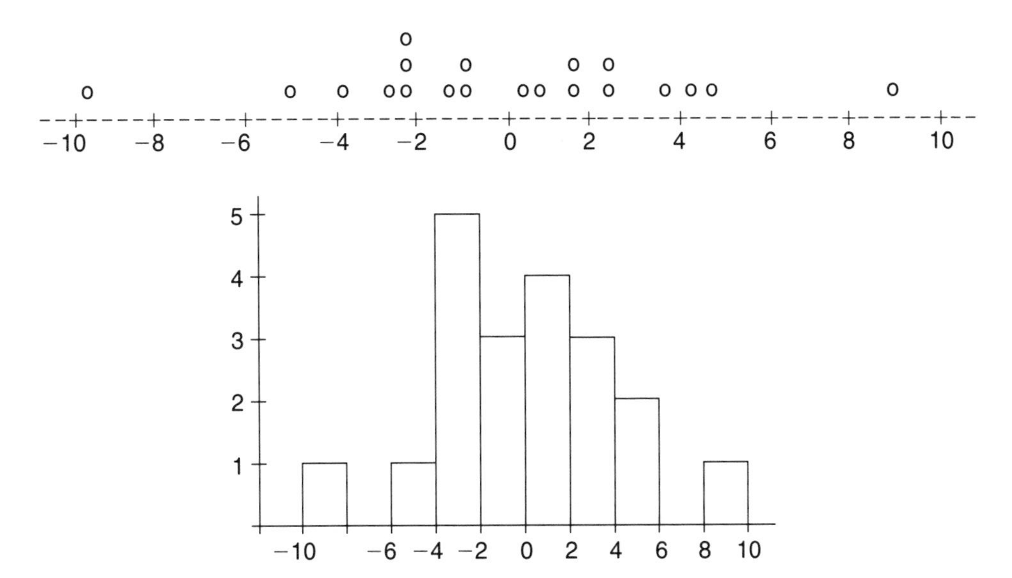

FIGURE 12.16 Histogram for residuals: Bird calcium data. The most striking feature of this histogram is the pair of outliers, one at each extreme, each separated from the rest of the histogram by a gap. The histogram is only somewhat lopsided, and free from big lumps and gaps, except for the gaps created by the outliers. (It is true that both the symmetry and smoothness I claim for this histogram are approximate, but unless you have a lot of residuals, say 40 or more, histograms can only show you coarse features about shape.)

This is a data set whose residuals can be made more nearly normal by transforming to a different scale. In Exercise A.2 I asked you to use a diagnostic plot for these data to find a transformation that would make the SDs for the four groups more nearly equal. Exercises C.10–12 will ask you to check the shape of the residuals for the transformed data.

Normal Plots for Residuals

Although histograms are good at showing coarse features of the residuals like asymmetry or outliers or gross lumpiness, they are not so good at showing other features—most particularly, the relation of the "tails" to the center of a set of residuals. For checking that feature, the normal plot is a better tool.

The **normal plot**, or **rankit plot**, is a specialized scatterplot of observed values or residuals versus certain expected values. The expected values are chosen so that the more nearly the histogram for residuals approximates a normal shape, the closer the points of the scatterplot approximate a line. (These plots are nearly always done by computer.)

EXAMPLE 12.16 NORMAL PLOTS FOR FIVE SETS OF RESIDUALS

a. The roughly normal shaped set of residuals from Example 12.14 (Figure 12.17):

$$0, \pm 1, \pm 2, \pm 4, \pm 6, \pm 8, \pm 11, \pm 14, \pm 18, \pm 23.$$

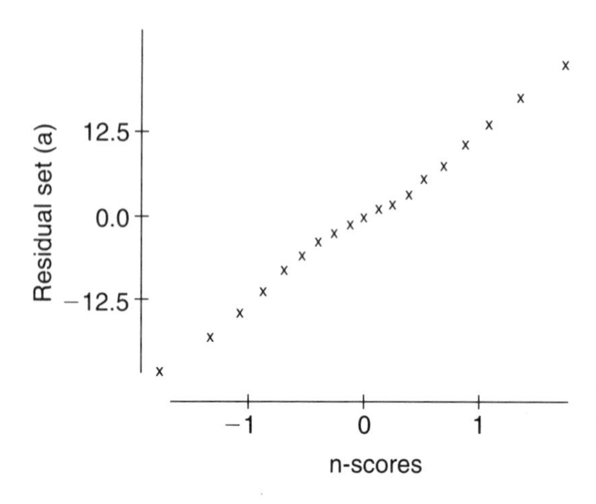

FIGURE 12.17 Normal plot for normal-shaped residuals. The points lie close to a line, indicating normal shape. Each point corresponds to one residual. The y-coordinate is the observed value of the residual; the x-coordinate is the expected value assuming a perfect normal shape (see Note 4 at the end of this chapter for a more detailed description of normal plots).

b. A lumpy but otherwise normal shaped set of residuals (Figure 12.18):

$$-2, -1, -1, -1, -1, 0, 0, 0, 0, 0, 0, 1, 1, 1, 1, 2$$

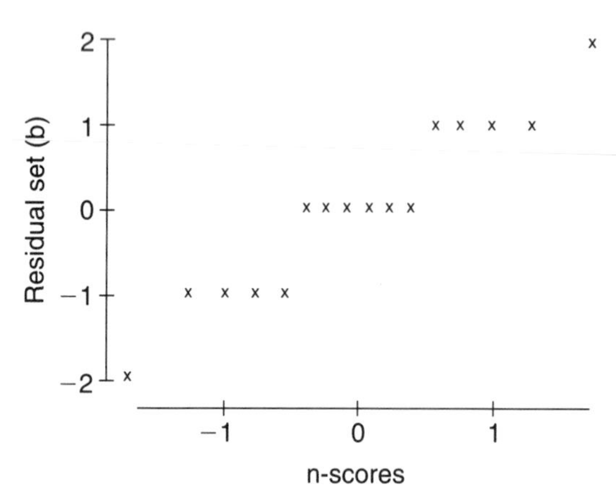

FIGURE 12.18 Normal plot for lumpy but otherwise normal-shaped residuals. Two features stand out that make this plot different from the one in (a). The plot has four big vertical jumps, and three flat stretches.

1. Vertical jumps **in a normal plot correspond to** Gaps **in the set of residuals.**
2. Flat stretches **in a normal plot correspond to** Clusters **of equal or nearly-equal residuals.**

A vertical jump is a jump in y-coordinates, which record observed residuals, so a vertical gap records a gap in residuals. A flat stretch comes from a set of points with equal or nearly-equal y-coordinates, that is, a cluster of nearly-equal observed residuals.

c. A very long-tailed set of residuals (Figure 12.19):

$$0, \pm 1, \pm 2, \pm 4, \pm 6, \pm 8, \pm 11, \pm 14, \pm 30, \pm 50$$

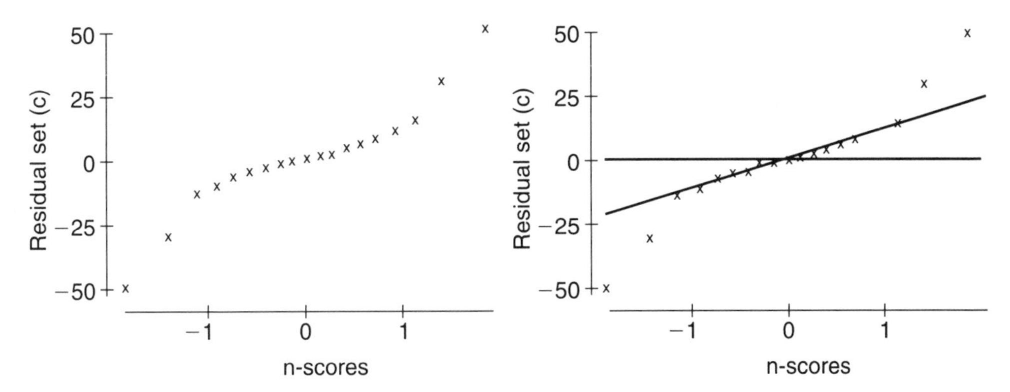

FIGURE 12.19 Normal plot for a long-tailed set of residuals. The normal plot is on the left, and shows two pairs of big vertical jumps, one pair at the left extreme and one pair at the right. The right-hand graph is the same plot, but with two lines added: the x-axis, and a line fitted to the middle set of points. The two pairs of points at the extremes are off the line *away from the x-axis*. This pattern is characteristic of a long-tailed set of residuals.

Long Tails Away, Short Tails Toward.

Fit a line by eye to the middle points of the normal plot, and check where the points at the extreme left and right sit in relation to the line:

***Normal* tail: *near* the line**

***Long* tail: off the line, *away* from the x-axis**

***Short* tail: off the line, *toward* the x-axis**

d. A uniform set of residuals, with short, fat tails (Figure 12.20):

$$0. \pm 1, \pm 2, \pm 3, \ldots, \pm 17.$$

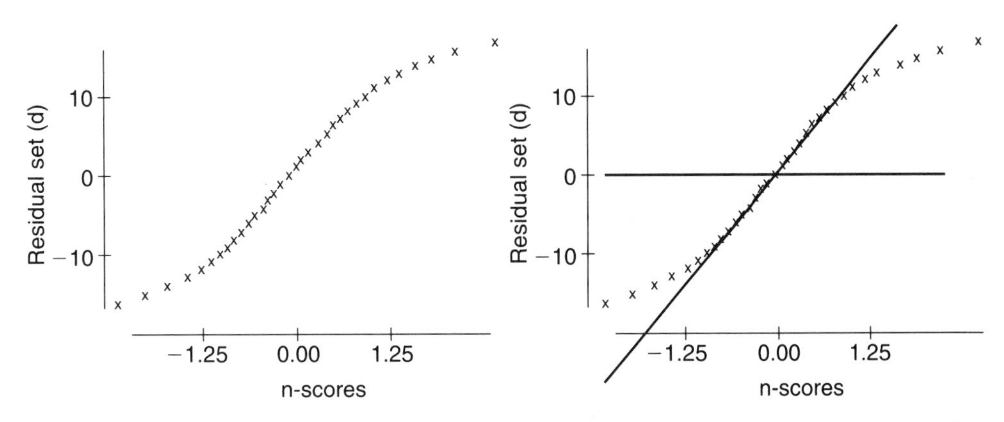

FIGURE 12.20 Normal plot for uniform, short-tailed residuals. The S-shaped plot on the left is characteristic of tails that are shorter than normal. If you fit a line to the middle points, as shown on the right, you find that the points at the extremes are off the line, *toward* the x-axis: short tails.

Note that distances between adjacent residuals are the same. If you order the residuals from smallest to largest, each is 1 bigger than the one before.

e. A highly skewed set of residuals, with a long left tail, and a short, fat right tail (Figure 12.21):

$$-40, -20, -10, -5, 0, 3, 5, 7, 8, 9, 10, 10.5, 11, 11.5$$

long left tail — short, fat right tail

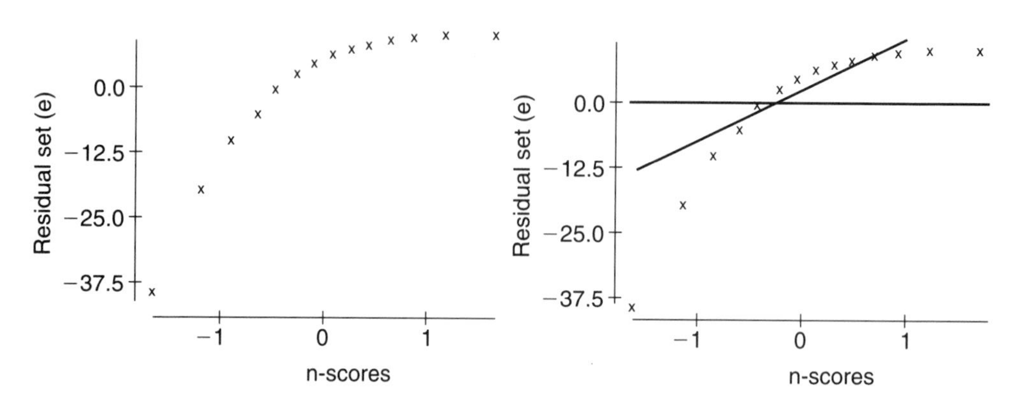

FIGURE 12.21 Normal plot for highly skewed residuals. A normal plot shaped like an arc of an oval (left) is characteristic of skewed residuals. For these data, if you fit a line to the middle points (right), you find the extreme left points off the line *away* from the *x*-axis (long tail), and the extreme right points off the line *toward* the *x*-axis (short tail).

Here, the curve "opens down": the long tail is to the left. If the long tail is to the right, the curve "opens up." ■

Exercise Set C

1–5. *Matching*. Match each of the sets of residuals (1)–(5) with the corresponding verbal description (I)–(V), histogram (A)–(E), and normal plot (a)–(e).

Residuals:

(1) −10, −5, −4, −4, −3, −3, −3, −2, −2, −2, 2, 2, 2, 3, 3, 3, 4, 4, 5, 10

(2) −4, −4, −4, −4, −4, −4, −2, −2, −2, −2, 0, 0, 0, 0, 2, 2, 4, 6, 8, 10

(3) −10, −8, −6, −4, −2, −2, −2, 0, 0, 0, 0, 0, 0, 2, 2, 2, 4, 6, 8, 10

(4) −8, −8, −8, −8, −4, −4, −4, −4, 0, 0, 0, 0, 4, 4, 4, 4, 8, 8, 8, 8

(5) −10, −6, −6, −2, −2, −2, −2, −2, −2, −2, 2, 2, 2, 2, 2, 2, 2, 6, 6, 10

Descriptions:

(I) Long-tailed, symmetric

(II) Symmetric, with gaps and outliers

(III) Lumpy, but otherwise normal-shaped

(IV) Asymmetric, long tail to the right

(V) Uniform (with short, fat tails)

Histograms

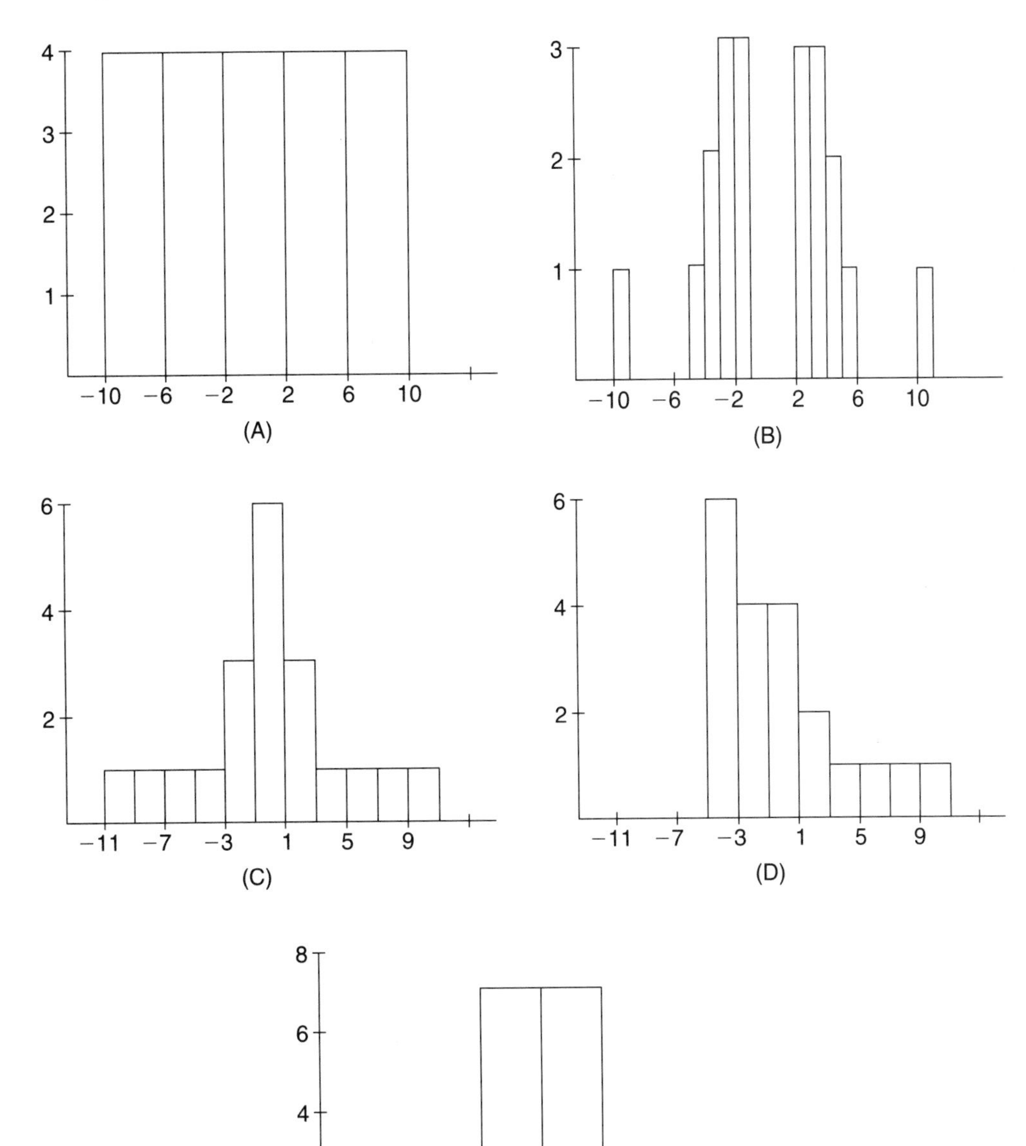

4
3
2
1
−10 −6 −2 2 6 10
(A)
3
2
1
−10 −6 −2 2 6 10
(B)
6
4
2
−11 −7 −3 1 5 9
(C)
6
4
2
−11 −7 −3 1 5 9
(D)
8
6
4
2
−12 −8 −4 0 4 8
(E)

Normal plots:

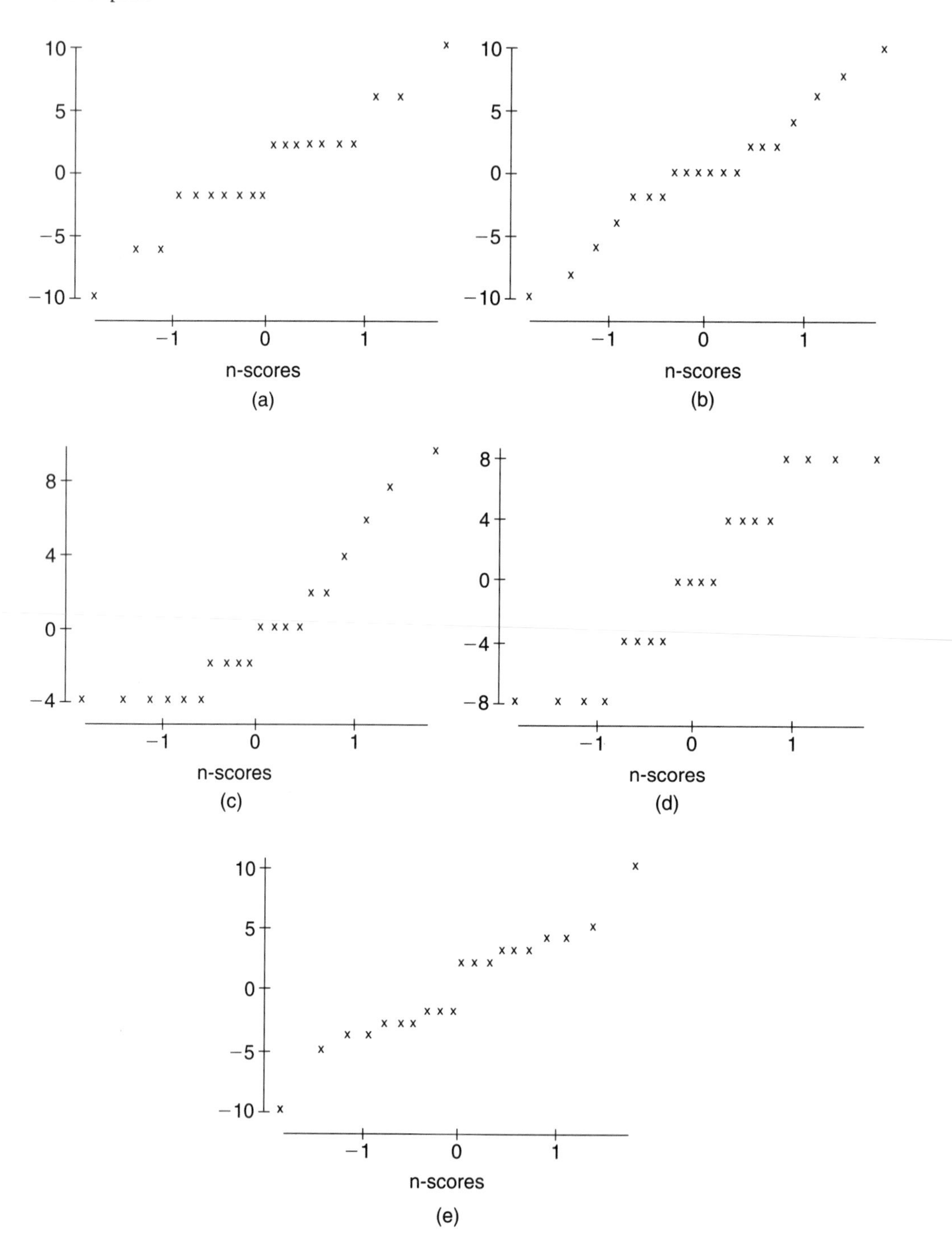

(a) (b) (c) (d) (e)

6–9. *Histograms and normal plots*. For each set of residuals listed, construct a histogram, and then match it with its normal plot.

6. Leafhopper survival data (Example 10.1b): −0.3, −0.2, −0.1, −0.1, 0.1, 0.1, 0.2, 0.3

7. Pigs and vitamins (Example 6.12): −11, −3, −3, −2, −1, 0, 0, 1, 2, 2, 4, 11

8. Finger tapping (Example 10.5): −9, −6, −6, −3, −2, −1, 1, 2, 4, 4, 8, 8

9. Sleeping shrews (Exercise 7.D.17): −1.5, −1.1, −1.1, −0.9, −0.7, −0.6, −0.6, −0.4, −0.4, −0.3, −0.1, −0.1, 0.7, 0.7, 0.8, 1.3, 1.7, 2.6

Normal plots for Exercises 6–9:

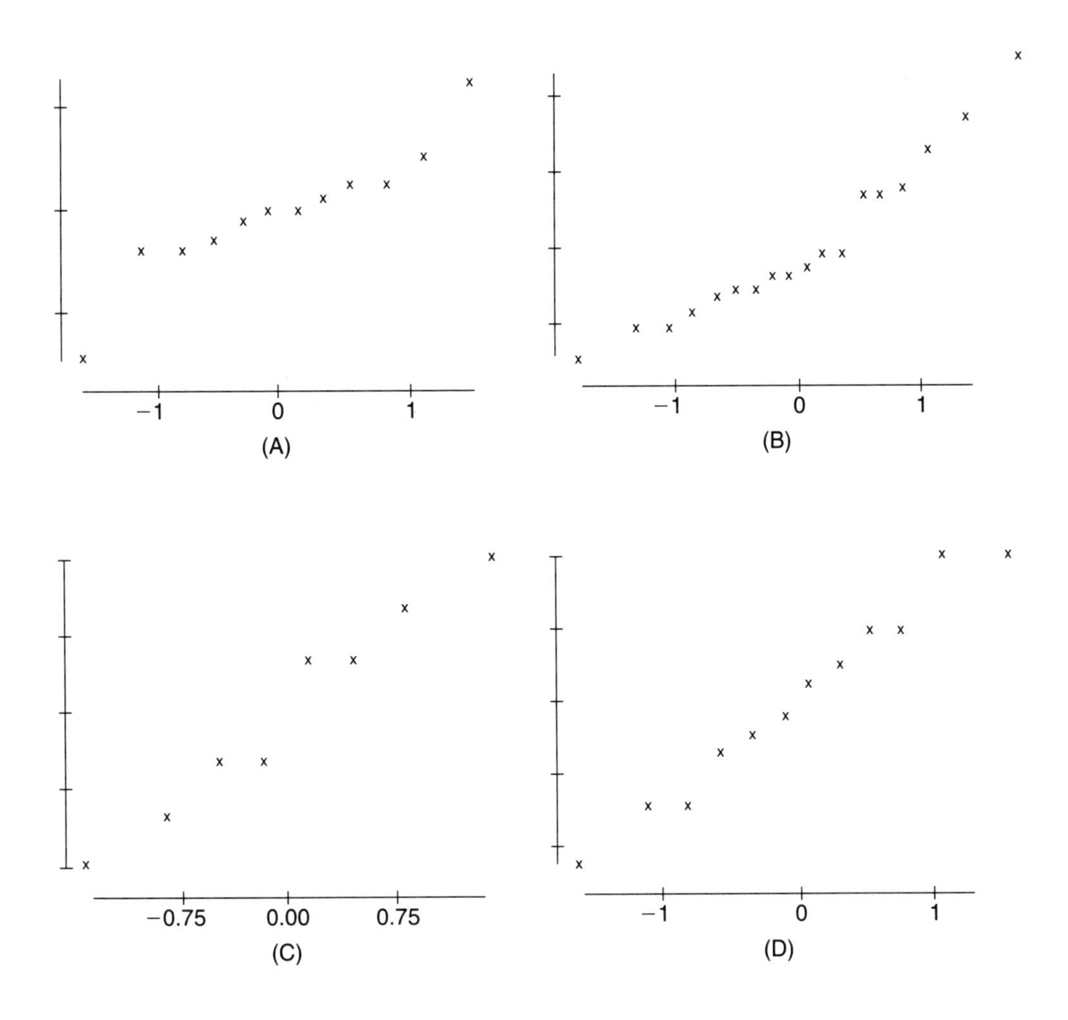

10–12. *Bird calcium* (Example 5.B.1).

10. Use the histogram in Example 12.15 to draw a rough sketch of the normal plot for the residuals from the bird calcium data.

11. Use a computer to get a normal plot.

12. A diagnostic plot for the bird calcium data suggested (Exercise A.2c) transforming to logs. Use a computer to do the transformation. Then decompose the transformed data, and get a histogram and normal plot of the residuals.

---

4. Effects are Additive (A) and Constant (C)

Why the assumptions matter. The model for ANOVA assumes that observed values behave as if they were created by adding numbers associated with the factors of the designs; there is one constant piece for each level of each factor. Both assumptions are needed to justify linear decomposition. Unless we can assume the effects of conditions are additive, it makes no sense to estimate those effects as we do, by adding to get averages and subtracting to get leftovers. Unless we can assume each effect is constant, it makes no sense to use a single number (the average) to estimate "the" effect.

Alternatives to an ANOVA model. This section will concentrate on data sets with categorical carriers and a numerical response, but it is important to keep in mind that what looks like a numerical response may in fact be categorical, and what looks like a categorical carrier may in fact be better regarded as numerical. Sometimes the assumptions fail to fit because ANOVA is simply the wrong approach.

Categorical response. (See Chapter 8, Section 4.) If your response is nominal, a category rather than a number, it makes no sense to assume that the response is built by adding numbers together.

For some studies, even though the data are presented as numbers, the most appropriate analysis is one that regards the response as nominal. An example is the study of mothers' stories: in Example 8.14 I discussed why I thought the best analysis of those data should use the type of story (A, B, C, D, or IR) as the response. If your response is nominal, then it makes no sense to think of your observations (like "A" or "IR") as built by adding numbers together; ANOVA is not appropriate. Instead, the best analysis may be one that seeks to relate the factor structure to probabilities for the various categories for the response.

Numerical carrier. If the groups of a factor come from a test score or measurement, be cautious about assuming constant effects for that factor. Consider an analysis that uses regression-based methods. (See Chapter 15.) For example, Exercise 6.B.3 described a study of personality and selective attention. There were two groups of subjects, called Type A and Type B. (The names refer to personality types, the Type A personality is more hard-driven, competitive, and hostile than Type B.) The response for this study was the time (in seconds) a subject spent looking at cards that s/he thought were descriptions of her/his negative personality traits.

The psychologists who analyzed the data used ANOVA, and so assumed that subjects within each personality group were more or less homogeneous. In particular, ANOVA assumes that the effect of personality type, as measured by the response, is constant within each group (cell). That assumption may in fact be reasonable; the published article doesn't give the information you'd need in order to check. However, the information about how the subjects were classified into groups does suggest that the groups were not uniform.

The subjects in the Type A group were chosen because they had scored in the top 25% on a particular test (The Jenkins Activity Survey); the Type Bs were the ones in the bottom 25%. If the test is valid as a measure of personality, then it seems reasonable to assume that the actual test score provides more information about personality than you get from knowing only "top 25%" or "bottom 25%." An analysis of variance that compares the two groups throws away the extra information by regarding all the people in the top 25% as interchangeable. If there is a relationship between test score and time spent attending to negative information, a scatterplot of those two variables and an analysis based on line fitting would almost surely be more informative than ANOVA.

We can generalize from this example:

When the groups of a factor come from a test score or measurement, be cautious about assuming constant effects for that factor. Consider an analysis that uses regression-based methods (Chapter 14).

Two crossed factors with one observation per cell. For the rest of this section, I'll assume you've already checked the carriers and response, and have decided that an ANOVA model is more reasonable than either of the two classes of alternatives I've just described. If your response is numerical, and your carriers are categorical, then you should be most concerned about additivity and constant effects if your data set has a pair of crossed factors with only one observation per cell, because these are the ones for which the assumptions are most likely to fail. For such data sets, the two assumptions go hand-in-hand: if one fails, the other will also. There are two graphs you can use to check the assumptions: within-blocks plots and scatterplots of residuals versus fitted values. And if the assumptions don't fit, there are two kinds of remedies: often you can transform to a new scale that makes the assumptions reasonable; if not, you can run a new experiment that lets you fit interaction terms.

The following example illustrates all of these ideas.

EXAMPLE 12.17 AREAS OF RECTANGLES: NONADDITIVE DATA

The area of a rectangle equals Base × Height. The data set in Table 12.19 uses Base and Height as two crossed factors, and records the corresponding areas as the response. Although no one would actually plan and carry out this particular study, it has the virtue of being both simple and familiar, and so it makes a good stand-in for messier but more realistic situations where you have to transform to get additivity and constant effects.

TABLE 12.19 Areas of rectangles in sq. in.

	Height		
Base	1"	4"	10"
1"	1	4	10
4"	4	16	40
10"	10	40	100

In this scale, the effects of the two factors, base and height, are multiplicative, not additive.

If there are no measurement errors, as in this example, then you can find the exact area once you know the Base and Height. However, we're thinking of this problem as a stand-in for more realistic situations, where we don't know in advance what the true relationship is. A good analysis ought to allow us to discover a reasonable approximation to the true relationship.

If I didn't know the background, I might begin with a linear decomposition of the areas measured in square inches. What I'd get is shown in Table 12.20:

TABLE 12.20 Linear decomposition of areas (sq. in.).

	Residuals			Row Eff
	16	4	−20	−20
	4	1	−5	−5
	−20	−5	25	25
Col Eff	−20	−5	25	25 = Gr Avg

First notice how the table summarizes the decomposition, by giving the Grand Average (lower right corner), Row Effects (right-hand column, one effect for each row), Column Effects (bottom row, one effect for each column), and Residuals (body of the table).

Then notice how completely misleading the decomposition is, if I take it at face value. The true relationship (area = Base × Height) is impossible to see, and the huge residuals suggest a poor fit of model to data. If we blindly assume that the residuals are estimates of chance errors, and compute *F*-ratios, we find:

$$\frac{MS_{Rows}}{MS_{Res}} = \frac{525}{441} = 1.19 \qquad \frac{MS_{Cols}}{MS_{Res}} = \frac{525}{441} = 1.19$$

If we stop here, and take the *F*-ratios as is, we conclude that length and width have no detectable effect on the areas of rectangles.

Our surface analysis leads to wrong conclusions because the assumptions of additivity and constant effects both fail. *Additivity*: The effects of length and width are multiplicative, not additive. (If you transform to a log scale, however, the effects will be additive.) *Constant effects*: The "effect" of a four-inch length on the area is not constant, but depends on the width (length and width interact, but the model has no interaction terms.) To protect ourselves against reaching such wrong conclusions, we need to check the assumptions. That checking will lead us to a new analysis in the log scale, where the effects are constant and additive. ■

Plots for checking the assumptions

If the Assumptions of Additivity and Constant Effects Don't Fit:

- **A scatterplot of Res vs. Fit will suggest a curve (rather than a horizontal balloon with the x-axis through its middle).**
- **Within-blocks plots will suggest either curves or lines whose slopes are clearly different from the slope of the identity line $y = x$.**

It will often be the case that if your scale makes the assumption of additivity inappropriate, the residuals you get will not fit some of the other key assumptions. You may get residuals of quite different sizes (S won't fit), their distribution may not be normal shaped (N won't fit), or scatterplots of the sort described in Section 2 on the independence assumption will show patterns. For many data sets, if you've chosen a scale that makes these other assumptions reasonable, the assumption of additivity will be reasonable also.

Although you can usually rely on checking SDs, normal shape, and scatterplots in this way, as an indirect check on additivity, there is also a method for checking additivity directly, and for finding a transformation from the power family. As a general rule, if you are using a scale for which effects are not additive, the nonadditivity will create patterns in your residuals. If the levels of your factors have a natural order, as they do for the area example, you may be able to see the pattern in the residual table of your decomposition. For example, look back at the residuals from the first decomposition of the areas—in square inches. We know there are no chance errors for this data set. The residuals are due entirely to nonadditivity, and they show a pattern: large and positive in the northwest and southeast corners, large and negative in the other two corners, and nearer zero elsewhere. (The pattern would be even clearer if I had chosen a data set with four lengths and four widths.)

Regardless of whether the levels of your factors have a natural order, if your scale makes the effects nonadditive, a scatterplot will show a curved pattern relating residuals and fitted values.

EXAMPLE 12.18 SCATTERPLOTTING RESIDUAL VS. FIT FOR THE AREA DATA

Here is the decomposition of the area data into fit (Grand Avg + Row Eff + Col Eff) plus residuals:

$$\begin{bmatrix} 1 & 4 & 10 \\ 4 & 16 & 40 \\ 10 & 40 & 100 \end{bmatrix} = \begin{bmatrix} -15 & 0 & 30 \\ 0 & 15 & 45 \\ 30 & 45 & 75 \end{bmatrix} + \begin{bmatrix} 16 & 4 & -20 \\ 4 & 1 & -5 \\ -20 & -5 & 25 \end{bmatrix}$$

Obs Fit Res

The scatterplot in Figure 12.22 shows residuals on the y-axis and fitted values on the x-axis.

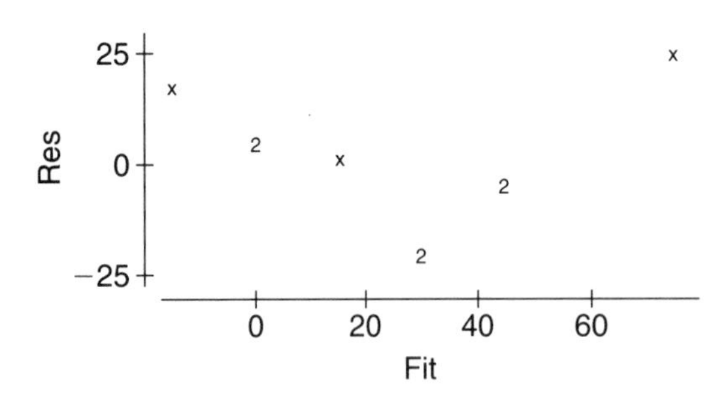

FIGURE 12.22 Residual vs. fit for the area data. Any systematic relationship between residuals and fitted values suggests nonadditivity. Here, the points form a V-shape: residuals are large and positive at the extremes, large and negative in the middle. (A parabola opening upwards would give a reasonable fit to the points.) ■

You can also use within-blocks plots to check the assumptions. If the effects are additive and constant, these scatterplots will suggest lines parallel to the identity line $y = x$. But if the size of the effect varies, the plots will suggest either curves or lines much steeper (or much less steep)

EXAMPLE 12.19 Within-Blocks Plots for the Area Data

Strictly speaking, this data set has no blocks, but we can get a plot that shows the same information as the within-blocks plot by treating each level of the width factor (each row) as a block.

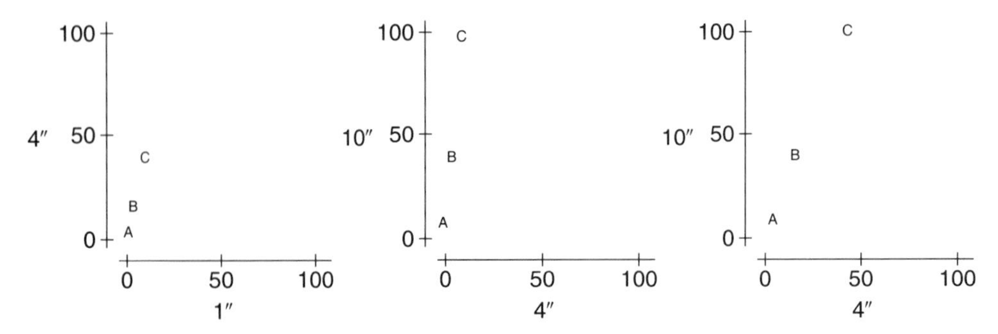

FIGURE 12.23 "Within-blocks" plots for the area data. These plots treat each row (each width) as a block. For the left plot, there is one point for each row (= "block"). The x-coordinate gives the area for a 1" length; the y-coordinate gives the area for a 4" length. The three points fall on a line that is much steeper than the identity line $y = x$, and the other two plots show similar patterns: the two assumptions don't fit. ■

Transforming to make the assumptions fit. Whenever your plot of residual versus fit suggests nonadditivity, it is worth looking for a transformation. The data may be additive, or more nearly so, in a new scale. Depending on the structure of your data set, there are various ways the additivity assumption may fail, and there are various strategies for finding a transformation to a new scale to make the assumption more appropriate. One standard method for finding a transformation is in many ways like the diagnostic plot of Section 1: scatterplot number pairs, fit a line by eye, estimate its slope, and choose a member of the power family with $p \approx (1 - \text{slope})$. The only difference is in the number pairs you plot.

Transforming for Additivity

1. **Compute a comparison value for each residual:**
   Comp = (Row Eff) × (Col Eff) / (Grand Avg)
2. **Scatterplot Res (*y*-axis) vs. Comp (*x*-axis); fit a line by eye; estimate its slope, and take power $p = 1 - \text{slope}$.**

EXAMPLE 12.20 FINDING A TRANSFORMATION FOR THE AREA DATA

Table 12.21 shows the comparison values:

TABLE 12.21 Comparison values for the area data

	Comp = Row × Col/Grand			Row Eff
	16	4	−20	−20
	4	1	−5	−5
	−20	−5	25	20
Col Eff	−20	−5	25	25 = Gr Avg

Each comparison value equals (Row Eff) × (Col Eff)/(Grand Avg)

For this ideal example, each comparison value is exactly equal to the corresponding residual. Because the match is perfect, you don't have to do a scatterplot to fit a line: all the points would lie on the line $y = x$. Because the area example was chosen to be exactly additive in the log scale, the points all lie exactly on a line, its slope is exactly 1, and the power = (1 − slope) is exactly 0, telling us to transform to logs. ■

CB[1] designs. The example I've just given has the structure of a BF[2] with one observation per cell, but you can use the same kind of plot whenever you have two crossed factors. Both the CB[1] and the BF[2] with one observation per cell have the same structure as in the example.

More than one observation per cell. If you have more than one observation per cell, you can always include interaction terms in your model in order to get the additivity assumption to fit. However, it sometimes happens that if you find the right scale, the fitted interaction terms are so small that you don't really need to include them in your model. Instead, you can use the simpler model

$$\text{Obs} = \text{Benchmark} + \text{Row Effect} + \text{Column Effect} + \text{Error}.$$

If there is a transformation that makes this simpler model fit your data, you can use the additivity plot to find it: just apply the method of Example 12.20 to your table of cell averages.

For more complicated data sets, there are also ways to compute comparison values to use for scatterplotting with residuals to find a transformation, but I won't try to explain them here.

EXAMPLE 12.21 THE AREA DATA IN A LOG SCALE

Logarithms convert multiplication to addition. In particular, log(base × height) = log(base) + log(height). Tables 12.22 and 12.23 show the same data, this time expressed as logs to Base 10.

TABLE 12.22 Areas of rectangles, in log (sq. in.)

	Height		
Base	**1"**	**4"**	**10"**
1"	0.0	0.6	1.0
4"	0.6	1.2	1.6
10"	1.0	1.6	2.0

In this new scale, the effects of the two factors are additive. A linear decomposition produces residuals that are all zero.

TABLE 12.23 Linear decomposition of log(area)

	Residuals			Row Eff
	0	0	0	−0.533
	0	0	0	0.067
	0	0	0	0.467
Col Eff	−0.533	0.067	0.467	1.067 = Gr Avg

The effects are additive in a log scale.

■

A new design with interaction terms. If you can't find a scale that makes the assumptions fit, then you'll probably have to rerun your experiment, this time with more than one observation per cell, so that you can include interaction terms in your model. For the kinds of designs we have been considering, with two crossed factors, the assumptions fail because the two factors interact; including interaction terms in the model will make the assumptions appropriate.

If your two crossed factors are both factors of interest, then you can simply repeat your experiment a second time, and analyze the expanded data set as a factorial CB, with each little sub-experiment (or replication) as a block. As an alternative, you could start from scratch, and use a BF design with more than one observation per cell.

If one of your factors corresponds to blocks and you can't find a scale that makes the assumptions fit, then your data set shows a block-by-treatment interaction. For such situations, you should use a **generalized block design**. These designs, which are described in Chapter 14, have the same factor structure as BF designs, and allow you to get separate estimates of interaction terms and residual errors.

Exercise Set D

1–5. *Effects are constant as percents: Finger tapping.* It is not hard to imagine that the effects of a factor might be constant when expressed in percents. Think of the finger tapping experiment (Exercise 7.B.2), and suppose that three subjects had tapping rates of 50, 100, and 150 taps per minute under control conditions (Placebo). Suppose also that the effect of Caffeine is to raise the rate of tapping by 60%. If there were no chance errors, you'd get data like these:

Subject	Placebo	Caffeine	Avg
I	50	80	65
II	100	160	130
III	150	240	195
Avg	100	160	130

1. Within-block plot: Scatterplot the caffeine reading versus placebo, with the three blocks as points. Then fit a line to the points and compute its slope. What is the relationship between the numerical value of the slope and the numerical effect of the caffeine (which raises the rate by 60%)?

2. Decompose the data. The data, in taps/min, are exactly equal to what the model of constant percent effects tells us: chance errors are all zero. Nevertheless, the decomposition gives nonzero residuals.

3. Scatterplot Residual versus Fit. Is there a systematic relationship?

4. Compute a set of comparison values = (Row Eff)×(Col Eff) / (Gr Avg). Then plot Res vs. Comp, fit a line, and use the slope to choose a transformation from the power family.

5. Transform the data and repeat the decomposition.

6. Fill in the blanks: Effects that are constant as percents are not additive in the scale of the raw data, but are ____________. For such data, transforming by taking ____________ makes the effects additive in the new scale.

7–12. *Sugar metabolism.* For this exercise, you can use either the actual version of the sugar metabolism data (Exercise 6.F.5) or the simplified version given in Exercise A.7. The results will be essentially the same either way.

7. For the data expressed as concentrations, the effects of the two basic factors (sugar and oxygen) are not additive. To see this, draw an interaction graph, and notice that the lines are far from parallel. For which sugar does oxygen have the more marked effect?

8. Another way to see the nonadditivity is to think of the interaction effects as residuals you get from the (additive) fit you get by summing the grand average plus the estimated effects for the two basic factors. Decompose the table of cell averages into grand average, sugar effects, oxygen effects, and interaction. Then scatterplot "Res" (= interaction effects) versus "Fit" (= partial fit for interaction). What is it about the shape of the plot that tells you the additive model does not fit the data in the original scale?

9. To see whether transforming might make the additive model appropriate, use the sugar and oxygen effects to compute comparison values for the 8 cells of the data table, and scatterplot "Res" (= interaction effects) versus comparison values. Notice how well a line fits the points of this plot: transforming will make the additive model suitable. Fit a line to the points, estimate its slope, and find power = 1 − slope.

10. Exercise A.7 asked you to plot residuals from the cell averages versus those averages (= fitted values). That plot showed that in the original scale, larger errors go with larger averages, another indication that the data should be transformed. Thus we have two reasons to change our scale: to equalize SDs, and to make the "no interaction" (additive) model fit. Use the information below to do a diagnostic plot for finding a scale to equalize SDs.

Sugar:	Galactose				Glucose			
Oxygen (micromoles):	0	46	92	138	0	46	92	138
log(avg)	1.65	1.49	1.35	1.10	1.15	0.88	0.54	−0.30
log(SD)	1.31	1.26	−0.15	−0.15	1.19	0.89	0.69	−0.15

Notice that your plot shows two outliers. Ignore them, and fit a line to the other 6 points, estimate the slope, and find power = 1 − slope.

11. Compare the transformations suggested by your two plots from (9) and (10). Which member of the power family would you try?

12. Use a computer to transform the data. Then do a complete analysis in the new scale. Include new versions of all the plots you did in the original scale, as well as the formal ANOVA table. What do you conclude about sugar metabolism?

13–16. *Additive or multiplicative?* Several little "data sets" are given below. (Think of each as a 2-way factorial with one observation per cell.) For each, decide whether the effects are exactly additive (A), exactly multiplicative (M), or neither (N).

13.

4	6
6	9

14.

7	9
11	13

15.

9	4
6	6

16.

9	6
6	4

17. *Jobs and hearts*. Exercise 8.D.5 gave data for a study of two groups of workers. Discuss the following plan for a partial analysis of the data: Is the analysis dead wrong? Or is the approach acceptable, though clearly not the best? Or is it basically sound, with no obvious flaws? Give reasons for your answer. If you find flaws in the proposed analysis, what would you suggest as a better approach?

The analysis will focus on factors affecting ventilation rate, using the average of the two tests of oxygen consumption as the response. One factor of interest will be Physical Grade, with two levels, 2 and 4. the other will be Metabolic Rate, with three levels, low, medium, and high. To obtain the levels of this factor, we take the 18 workers in each job grade and sort them into three groups on the basis of their average caloric expenditure: the lowest 6 workers in each grade go into the low groups; the next 6 go into the medium group, and the highest 6 go into the high group. The resulting design is a two-way observational study, with 6 workers in each of 6 cells.

18. *Transylvania effect.* Tell briefly what's wrong with the following statements about the Transylvania data from Example 12.9: "Even if the data did not show the patterns that led us to decide that Assumption I (independence) does not fit the data, it would still be flat wrong to use ANOVA with admission rate as the response, because the data are in fact nominal. Each admission to the clinic gets classified: before, during, or after the full moon. The response is nominal, and so it doesn't make sense to use a model that says response values come from adding numbers together. The additivity assumption (A) doesn't fit. These data should be analyzed using methods for nominal data."

19. *Rogerian therapy and the Q-sort.* Reread Example S4.11 on Carl Rogers' study of psychotherapy. (a) Describe how to use a one-way ANOVA to compare the therapy and control groups. What is the response? (b) Do you consider such an analysis reasonable? (c) Describe an alternative analysis that might be more informative.

5. Estimating Replacement Values for Outliers

As some of the examples in the previous sections illustrate, quite often when one or more of the Fisher assumptions fail, that failure is due to an outlying observation. Statisticians have invented a number of ways to handle outliers, some of them quite elaborate. The approach I'll present here has the advantages of being simple, informal, and close to common sense. Nevertheless, it is only one of the possible approaches.

According to this approach, whenever you identify a deviant value you should do two analyses. First analyze the complete data set, outlier and all. Then do a second analysis: Remove the outlier, regard that observation as missing, estimate a replacement, and analyze the new data set. Comparing your two analyses lets you see the effect of the outlier. You can be confident of any conclusions that are the same for both analyses. For conclusions that are not the same for both analyses, you need to be more cautious. If the outlier is only moderately deviant, the safest conclusion is "not proven." If, however, the residual for an observation is 3 or more SDs away from zero, and the other residuals are normal shaped, then the analysis which replaces the outlier is probably the more trustworthy of the two.

For Designs with One Outlier (or Missing Observation):

1. **Estimate a replacement value, then**
2. **Analyze the data set in the usual way, except**
3. **Reduce df_{Total} and df_{Res} by 1.**

There is a general method for estimating a replacement value for an outlier. I'll first describe the logic of the method, and then describe a shortcut you should use for the actual computations.

The logic. Suppose you have a data set with an outlier. When you decompose your data, estimating all terms in your model and combining them to get fitted values, you'll find that the residual corresponding to the outlier is quite large, because the observed and fitted values are far apart. The fitted value, because it is based on all the observations and not just the one outlier, is a better estimate than the actual observation of what that outlying observation "should have been." So it makes sense to substitute this fitted value in place of the outlier itself, and redo the decomposition. If you do this, you'll get a new fitted value, one whose value depends less than the first one on the outlier, and the corresponding residual will be closer to zero than the first time around. The new fitted value is an improved estimate of what the outlier "should have been." If you do yet another decomposition, this time using the most recent fitted value in place of the outlier, your new fitted value will be an even better estimated replacement for the outlier, and the corresponding residual will be even closer to zero. Each time you repeat the process the residual gets still closer to zero, and the fitted value depends still less on the outlying observation. Eventually you reach a point where additional cycles don't change things. At that point, the residual corresponding to the outlier is zero, and the fitted value is your estimated replacement for the outlier. This fitted value depends solely on the other observations, and is completely independent of the value of the outlier: if you'd started with some other number in place of the outlier, you'd eventually end up with the same fitted value anyway.

The shortcut. Fortunately, if you have a balanced design with only one outlier (or missing value), you don't have to cycle through the decomposition again and again, because there's a shortcut. First, put a zero in place of the bad observation, and estimate fitted values in the usual way. Then multiply the fitted value which corresponds to the zero (the **zero fit**) by (# obs)/(df_{Res}), where df_{Res} is the residual df for the balanced design. It can be shown that the number you get using this shortcut is the same as the one you get by repeated cycles of decomposing and fitting.

$$\textbf{Replacement value} = \textbf{(Zero fit)} \times \left(\frac{\textbf{\# obs}}{\textbf{df}_{\textbf{Res}}}\right)$$

Once you have the replacement value, you can analyze your data set in the usual way, except that you should reduce df_{Res} by one.

Caution. When you estimate a replacement and use an analysis for balanced data, the resulting analysis is approximate, in that for example mean squares that would ordinarily be independent are not, and critical values for F-ratios won't be exactly equal to the values from the F-tables. If you have a lot of data, and estimate only one replacement value, the approximation will generally be good. But if you have a small data set, or several outliers, the approximation may not work so well. In such cases, it would be better to use more advanced methods, based on equation-fitting methods (regression).

My own opinion, which not all statisticians would agree with, is that although regression methods give "exact" answers, they are exact only in an ideal sense, assuming the model is a perfect description of the process that created the data. The degree of approximation that comes from using one replacement value and analyzing your data as though the design were balanced is generally small in relation to the uncertainty introduced by the presence of an outlier. When you keep in mind that you are considering two very different analyses, with and without the outlier, the approximation that comes from pretending your data set is balanced is quite reasonable.

EXAMPLE 12.22 Estimates for Missing Values

In this example we will consider three data sets with outliers. In each case we will regard the outlier as a missing value, and find an estimate to replace it.

a. *BF: Hospital carpets*. The one-way design is a special case, in that the method of decomposition based on averages does not require that the design be balanced. However, decomposition by averages does require balance for factorial designs of the BF family, so it is worth including a BF example.

   Turn back to Exercise 8.C.1, look over the hospital carpet data, and note that the 3.8 colonies per cubic foot for Room #215 is an outlier. Use the rule to estimate a replacement. (Exercise E.1 will ask you to reanalyze the carpet data using the replacement value, and to use your reanalysis to assess the effect of the outlier.)

   Solution. For any BF design, the fitted value is simply the cell average. Putting 0 in place of the 3.8 for Room #215 and computing the average for the rooms with bare floors gives (74.5)/8 = 9.3125. Since # obs = 16 and $df_{Res} = 14$, our replacement value is

$$(\text{Cell Avg})(\#\ \text{obs}/df_{Res}) = (9.3125)(16/14) = 10.64.$$

   Note that the estimate is equal to the condition average of the observations, omitting the one for #215. For a design of the BF family, this always happens: the estimate equals the condition average of the observed values that are present.

b. *CB: Bee stings*. Turn back to the beesting data (Example S7.3), and note that the 70 stings in Row 9, Column 1 (Stung) is an outlier. Estimate a replacement value. (Exercise E.2 will ask you to reanalyze the beesting data using the replacement value, and to use your reanalysis to assess the effect of the outlier.)

SOLUTION. With a 0 in place of the 70 in Row 9, Column 1, we get the following averages and effects:

		Average	Effect
Grand Average:	(326/18)	= 18.11	
Row 9:	(10/2)	= 5.00	−13.11
Column 1:	(182/9)	= 20.22	2.11

The fitted value for Row 9, Column 1 is

$$\text{"Zero fit"} = (\text{Grand average}) + (\text{Row 9 Eff}) + (\text{Column 1 Eff}).$$
$$= \quad 18.11 \quad + \; (-13.11) \; + \quad (2.11) \quad = 7.11.$$

There are 18 observations, and 8 df for residuals, so the replacement value is

$$\text{Replacement value} = (\text{Zero fit})(\#\ \text{obs}/\text{df}_{\text{Res}}) = (7.11)(18/8) = 16.$$

c. *LS: Wireworms.* For the wireworm data of Example 8.3, the 17 wireworms for Row 4, Column 1, Treatment N is an outlier. Estimate a replacement value. (Exercise E.3 will ask you to reanalyze the wireworm data using the replacement value, and to use your reanalysis to assess the effect of the outlier.)

SOLUTION. The number of levels for each of Rows, Columns, and Treatments is 5; $\text{df}_{\text{Res}} = 12$. Putting a 0 in place of the 17 for Row 4, Column 1, Treatment N gives the following averages and effects:

		Average	Effect
Grand Average: (107/25)		= 4.28	
Row 4:	(25/5)	= 5.00	0.72
Column 1:	(17/5)	= 3.4	−0.88
Treatment N:	(18/5)	= 3.6	−0.68

$$\text{"Zero fit"} = \left\{\begin{matrix}\text{Grand}\\\text{Average}\end{matrix}\right\} + \left\{\begin{matrix}\text{Row 9}\\\text{Eff}\end{matrix}\right\} + \left\{\begin{matrix}\text{Column}\\\text{1 Eff}\end{matrix}\right\} + \left\{\begin{matrix}\text{Treatment}\\\text{N Eff}\end{matrix}\right\}$$
$$= \quad 4.28 \quad + \quad (0.72) \quad + \quad (-0.88) \quad + \quad (-0.68) \quad = 3.44.$$

$$\text{Replacement value} = (\text{Zero fit})(\#\ \text{obs}/\text{df}_{\text{Res}}) = (3.44)(25/12) = 7.17. \ ■$$

More than one outlier. If you have more than one outlier or missing value, you can use the same general approach to estimate replacements. However, the two (or more) estimates depend on each other, and this dependence makes it impossible to get the appropriate values by applying the rule just once. Suppose you have two missing values. Start by estimating the first one. Then, using this estimate as though it were an observed value, estimate the second missing value. Now do a second cycle: estimate the first missing value, then the second, each time using the estimate for the other value as though it were an actual observation. Cycle once more, estimating each missing value in turn. Eventually you reach a point where additional cycles barely change your estimates; at this point you stop. Experience shows that usually 3 cycles is enough. When you use this approach to estimate two replacement values, *keep in mind that your analysis, which assumes a balanced data set, is only approximate*. Don't take borderline decisions at face value.

Exercise Set E

1. *Hospital carpets*. In Example 12.22a, we estimated a replacement value of 10.64 for the outlying observation whose actual value was 3.8. Look once again at the data (Exercise 8.C.1) and try to guess how the results of an ANOVA would change if you change the 3.8 to 10.64. Will the average or the residuals change for the carpeted rooms? The bare rooms?

Think about the effect of the change on a confidence interval for the contrast Carpet versus Bare. Will the size of the estimated treatment effect go up or down? What about the sizes of MS_{Res} and the SE?

a. Construct a 95% confidence interval for the contrast, using the original data set as is, including the outlier.

b. Reanalyze the data using the replacement value instead of the 3.8. Give the ANOVA table and a 95% confidence interval for the contrast.

c. Write a short paragraph discussing the effect of the outlier, as shown by the difference between the two confidence intervals.

d. Reread my discussion of the carpet data in Example 12.7. What happens to the scatterplot if you replace the point (#215, 3.8) with (#215, 10.64)? In view of your work in parts (a)–(c) and what you see in the scatterplot, do you think the evidence from the data justifies the effort of designing and carrying out a new experiment? Give your reasons.

2. *Bee stings*. In Example 12.22b, we estimated a replacement value of 16 for the outlying observation for occasion #9, Stung.

a. Look over the original data (Example S7.3) and tell for each of the following quantities whether changing the 70 to 16 will cause the quantity to go up a lot, up a little, down a lot, down a little, or whether the quantity won't change at all: df_{Res}, SS_{Blocks}, SS_{Cond}, average for Stung, average for Fresh, the SD.

b. Residuals from each of two decompositions, one using the original data, the other using 16 in place of the outlier, are listed below. (i) Which set of residuals is for the data set that includes the outlier? (ii) Describe what normal plots for each set of residuals would look like. (iii) Explain why the second set of residuals contains two values very far from zero.

A: 0, 0, 0, 0, ± 2, ± 3, ± 3, ± 4, ± 6, ± 6, ± 6

B: 0, 0, ± 3, ± 3, ± 3, ± 5, ± 6, ± 7, ± 9, ± 24

c. Compute condition averages from the data and use the residuals in part (b) to compute SDs. Fill in the following table:

	Average		
	Stung	Fresh	SD
With the outlier			
Without the outlier			

d. Construct two 95% confidence intervals for the contrast Stung versus Fresh, one with the outlier, one without.

e. This is your last shot at the bee data: Write a short paragraph summarizing your overall conclusions: Are bees more likely to sting targets that have been previously stung?

3. *Wireworms*. The wireworm data set (Exercises 7.B.5–8) has often been used to illustrate transforming to square roots. Consider the following four possibilities for analyzing the data: (A) Do not replace the outlier, do not transform; (B) Do not replace the outlier, but transform to $10\sqrt{1 + \text{Obs}}$; (C) Use the replacement value of 7.17 in place of the outlying observed value of 17 for treatment N in Row 4, Column 1 of the data, but do not transform; (D) Replace the outlier, and transform as in (2).

a. Using treatments as groups, compute two sets of group averages and group SDs, one set using the original data (untransformed), and one with 7.17 in place of 17.

b. Use your results from (a) to compute two values of SD_{Max}/SD_{Min}. Which of the following two approaches do you prefer, and why: (i) Keep the outlier, and analyze the transformed data; (ii) Replace the outlier, and analyze the data in the original scale.

c. If you use the replacement value, then $SS_{Res} = 60.47$; if you keep the outlier then $SS_{Res} = 106.9$. For the data set you get by replacing the outlier, what do you conclude about the effectiveness of the various soil treatments?

d. Tell which, if any, of your conclusions from the data are sensitive to whether or not you replace the outlier.

e. Carry out all four of the analyses suggested at the beginning of the exercise; then compare and discuss what you find,

4. Estimate replacement values for the missing values in the following artificial data sets.

a. Two-way factorial

8	12
12	12
7	10
9	-

b. Complete block

4	5	6
6	-	6
8	7	6

c. Latin square

4A	8B	3C
8B	5C	6A
_C	5A	5B

d. Split plot

6	0
4	2
6	4
2	-

5. *Radioactive twins* (Example S7.2)

a. The 51.8 for Block II, Rural is an outlier. Estimate a replacement value.

b. Construct two 95% confidence intervals for the contrast Rural versus Urban, one using the original data set, the other with the outlier replaced. Use the two intervals to discuss the effect of the outlier.

6. *Pine pruning.*

   a. Draw a parallel dot graph for the data of Exercise 8.A.17, and use the graph to identify the one outlying observation.

   b. Estimate a replacement value.

   c. Scatterplot Treatment Average (y) versus Pruning Height (x), using your estimate in place of the outlier. Sketch a curve that passes through or near your five points. Which pruning height is best for stimulating growth?

   d. Analyze the data with and without the outlier. Does the outlier have much influence on the results?

7. *Walking babies, mothers' stories, hypnosis and pesticides.*

   a. Babies: Exercise 5.F.8 asks you to compare two analyses of the walking babies data, one with outliers, one without. If you have not yet done that exercise, now would be a good time.

   b. Stories: Review Exercise 7, Chapter 4, gives the numbers of Type A minus Type C stories for 20 mothers of schizophrenics and 20 control mothers. The 7 for Subject #6 in the schizophrenic group is an outlier. Estimate a replacement value. Then carry out two ANOVAs, with and without the outlier. Write a short paragraph comparing the two analyses.

   c. Hypnosis and learning: Use within-blocks scatterplots to identify any outliers in the data set given in Exercise 7.G.6. Estimate replacements, and analyze the resulting data set.

   d. Pesticides in the Wolf River: Replace any outliers in the data set of Exercise 8.D.6, and analyze the resulting data.

SUMMARY

The Five Assumptions: Checks and Remedies

S: Same SDs
Check: Compare SDs for levels of a factor; scatterplot Res vs. Fit
Remedy: Transform: power = 1 − slope, for line fitted to a diagnostic scatterplot (log SD vs. log Avg)
Replace outliers

I: Independent chance errors
Check: Scatterplots; SDs for within-blocks differences
Remedy: Replace outliers; transform; change models

N: Residuals are normal-shaped
Check: Histogram and normal plot of residuals
Remedy: Transform; replace outliers

A: Additive effects
Check: (1) Nominal response?
(2) Patterns in residuals? Scatterplot Res vs. Fit
Remedy: (1) Methods for nominal data (Chi-square)
(2) Transform (Scatterplot Residuals vs. Comparison values)

C: Constant effects
Check: (1) Across cells: Patterns in residuals?
(2) Within cells: Do levels of a factor come from a test score or measurement?
Remedy: (1) Transform
(2) Use regression-based methods (Chapter 15)

Tests (and Remedies): A Quick Review

A. *Diagnostic checks to do before you decompose the data*
   1. Parallel dot graph: Check for outliers (replace), unequal spreads and relations between average and spread (transform).
   2. SDs for groups, SD_{Max}/SD_{Min}, plot log SD vs. log Avg (transform).
   3. *Ad hoc* scatterplots, within-blocks plots (replace outliers, transform, new model)

B. *Diagnostic checks to do after decomposing*
   1. Residuals vs. fitted values (replace outliers, transform).
   2. Residuals vs. comparison values (transform).
   3. Normal plot of residuals (remove outliers, transform).

EXERCISE SET F: REVIEW EXERCISES

In an important way, the exercises of this chapter have oversimplified "real life": each set of exercises dealt mainly with a single assumption, and told you which assumption to check. For more realistic practice, you can use any of the 17 data sets listed in Exercise Set C of Chapter 12, but don't try to do too many. Instead, plan to do a thorough job of checking all the assumptions for just one, or two, or at most three. If you find that a data set violates one or more assumptions, try to find a single transformation that brings the data to a scale for which ANOVA would be appropriate. For the data set below, I've made detailed suggestions of things to do. Although these suggestions are based on my own analysis of that particular data set, you might also find them helpful in deciding what to try with the others. (Note, however, that my suggestions below do not include plots of Residuals versus Comparison values—such plots will nevertheless be useful for some of the data sets listed in Exercise Set C, Chapter 12.)

1–11. *Imagery and working memory* (Exercises 9.C.5–8).

1. Start with an interaction graph based on the four condition averages. Does the pattern support the hypothesis of that the two sides of our brains have specialized functions?

2. Do a parallel dot graph to compare the four conditions. Notice how the response times are bunched at the low end, spread out at the high end. Which are the slowest subjects?

3. Draw a dot graph for the subject averages.

4. Compare groups SDs and do diagnostic plots, first using conditions as groups, then using subjects as groups. Do the data suggest transforming?

5. Decompose the data and scatterplot Res vs. Fit.

6. Get a normal plot for Subject × Task × Report, and notice the striking shape: "long tails away."

7. Transform to logs; then redo the interaction graph, the parallel dot graph, the dot graph for subjects, and the diagnostic plots.

8. *Four within-blocks scatterplots.* Plot each of the following: Saying vs. Pointing for the letter task (a) and the sentence task (b); Sentence vs. Letter for Pointing (c) and Saying (d). Are these plots consistent with a model that fits lines parallel to the identity line $y = x$? Notice that one of the plots in particular shows a pronounced wedge shape: even in the log scale, the larger values show more spread than the smaller, suggesting that a transformation with a power less than 0 might have worked better than logs.

9. Decompose the data, and scatterplot Res vs. Fit.

10. Get normal plots for Subject × Condition and Subject × Task × Report. Notice how much straighter these plots are than the one in #6.

11. Get the ANOVA table, and state your conclusions.

12–27. See data sets listed in Exercise Set C of Chapter 12. Follow the instructions given above in the paragraph at the beginning of this exercise set (F).

NOTES AND SOURCES

1. **Normal distribution.** The word "normal" in everyday usage has come to mean "usual," but when statisticians talk of normal shape, they have in mind an older meaning of "normal," closer to "ideal" or "desirable." Don't let the everyday meaning trick you into taking for granted that your residuals have a normal shape.

2. ***The normality assumption.*** Statistical research has found that for critical values, the normality assumption often matters less than the others. Studies of data sets that do not follow the normal curve show that the usual table values often give good approximations to the values computed specifically for those data sets.

3. ***Scale for the hamster data.*** A line fitted to the diagnostic plot had slope very close to 1.5, suggesting a transformation with power equal to −0.5 (reciprocal square roots of the concentrations). In that scale, SD_{Max}/SD_{Min} is only 1.35, less than half as big as in the log scale. I tend to prefer the log scale because reciprocal roots are unfamiliar, whereas logs are quite commonly used for concentrations. However, to be safe, I also did an analysis in the reciprocal root scale. The results were essentially the same as for logs.

4. **Normal plots.** Here's how the expected values (EVs) are chosen for the x-values of a traditional normal plot. Suppose you have 45 residuals, arranged in order from smallest to largest. Imagine drawing 45 times from an ideal chance error box, whose tickets have average equal to zero, SD = 1, and values that follow a normal curve. The smallest of your residuals gets matched with the EV for the smallest of 45 draws from the ideal error box. The next smallest residual gets matched with the EV for the second smallest of 45 draws, and so on up the line. Then each pair (EV, Residual) = (x,y) gets plotted as a point.

Computer programs use a variety of approximations, and they handle ties (two or more observations equal) in different ways. Minitab, for example, gives you equal x-values to go with tied observed values. This means that Minitab's normal plots show repeated observations not as flat stretches in the plot, but as multiple points. The normal plots in this book were generated using DataDesk, which shows repeated values as flat stretches.

The approximations that computers use to get x-values work the same general way: First rank the observed values from smallest to largest. Then convert each rank to a probability P. (Minitab uses the formula $P = (\text{rank} - 3/8)/(\#\ \text{obs} + 1/4)$. DataDesk uses $P = (\text{rank} - 1/3)/(\#\ \text{obs} + 1/3)$. For each value of P, the corresponding x-value (= **normal score**, or **n-score**) is the x-value that divides the standard normal curve so that P is fraction of the area under the curve to the left of the x-value, and $1 - P$ is the fraction of the area to the right.

5. *Test for nonadditivity.* The scatterplot of comparison values versus residuals is closely related to a formal test for nonadditivity, which goes by the name of Tukey's test or Tukey's single degree of freedom for nonadditivity. See, for example, Kirk (1982), pp. 250–3.

Replacements for missing values. For data sets that were designed to be balanced, but end up with one or two missing observations, you can still use the decomposition method (based on averages) for balanced designs to get an approximate ANOVA. Suppose, for example, that you designed your experiment to be balanced, but a lab animal died, or a subject failed to show up, so that your data set has one observation missing. If only you had that one additional number, your design would be balanced. In a case like that, you can estimate a replacement value for the missing number, using the same method as for outliers (Section 5), then analyze the data as though the design were balanced. The same caution from Section 5, about the resulting analysis being approximate, applies.

6. *Same SDs.* Some of the more complicated designs involve more than one source of chance error. For example, the hamster data has two sources, one associated with the hamster differences (between-blocks or whole-plot error), and one associated with the individual measurements. For such designs, the formal methods estimate one typical size for each source.

7. *Exercises A.8–9.* These are based on F. J. Anscombe (1987), "Scatterplots of residuals against fitted values," unpublished notes for Statistics 235a/535a, Yale University.

8. *The Huynh-Feldt condition for designs with blocks.* For more information, see, for example, Section 6.4 in Kirk, Roger E. (1982). *Experimental Design*, 2nd ed., Belmont, CA: Brooks/Cole Publishing Company. In particular, for the condition that standard deviations for within-block differences be equal, see pp. 257–9.

CHAPTER 13

OTHER EXPERIMENTAL DESIGNS AND MODELS

OVERVIEW

Two clusters of extensions. This chapter presents two ways to extend the ideas and methods you have already seen. The two extensions correspond to two parts of an ANOVA model:

The factor structure, which determines the parallel decompositions of the data, df, and SSs.

The assumptions about the underlying process that produced the data, which determine the kinds of conclusions that you can make based on the decomposition.

New factor structures. The first extension deals with **nested factors**. You have already seen nesting, for example, in the SP/RM design: The larger units (blocks) are nested within levels of the between-blocks factor. There are many other kinds of designs with nested factors, but so far we have only considered a few of them, since we have looked only at structures you could get from the four basic designs by factorial crossing. The goal of Section 1 is to enable you to handle all balanced factor structures built using any combination of crossing and nesting.

New assumptions for old factor structures. The second extension, which presents new uses and interpretations for familiar factor structures, deals with the difference between **fixed and random effects**. Although you have in fact already seen both kinds (treatment effects have always been fixed, and block effects have almost always been random), I have put off discussing the difference until now. Section 2 describes the two kinds of effects and why the difference is important. In this section, all the factors in a design will be of the same kind, all fixed or all random. Then, in Section 3, we consider designs with effects of both kinds. For such designs, there can be two kinds of interaction terms, called **restricted** and **unrestricted**, and Sections 3 and 4 are mainly about ANOVA models with these two kinds of interactions.

What else? There are related ANOVA models that are not covered here. Even though this book doesn't present them, you should know that there are other, more sophisticated models and analyses in addition to the ones I present in Section 3. You should also know that choosing the right model when you have both fixed and random effects is an area that statisticians are still actively researching, and that researchers don't always agree about which model is best.

1. New factor structures built by crossing versus nesting
2. New kinds of factors: Fixed versus random effects
3. Models with mixed interaction effects
4. Expected mean squares and F-ratios

1. New Factor Structures Built by Crossing and Nesting

Introduction

As you know, two factors in a design may be related to each other either by crossing or by nesting. In a CB like the finger tapping experiment (Exercise 7.B.2), blocks and treatments are crossed: every combination of subject and drug occurs together in the design. In a CR like the walking babies experiment (Example 5.1), babies are nested within treatments: each baby occurs in the design together with only one of the treatments, and not with any of the others.

Just as you can build new designs from old ones by crossing factors, as in Chapter 9, you can also build new designs by adding nested factors. Your goal should be to learn to work with any balanced factor structure built by crossing and nesting. In what follows, the main focus will be on structural aspects (what you get). We'll start with familiar examples, looking first just at pairs of factors and at ways to tell crossing and nesting apart. With this as background, we'll then look at new structures and more complicated designs.

Crossing Versus Nesting: Familiar Examples, Three Pictures and Four Tests

Familiar examples. So far, we have dealt with only two instances of nested factors: units are nested within treatments in a BF design, and blocks are nested within the between-blocks factor in the SP/RM design.

Example 13.1 Nested Factors

1. *In BF designs, units are nested within conditions.*

   a. Experimental conditions: Walking babies (Example 5.1). Babies are nested within type of exercise: each baby gets only one kind of exercise treatment.

   b. Observational conditions: IV fluids (Example 5.2). Samples (bags) are nested within Manufacturer: each sample was made by one and only one company.

2. *In SP/RM designs, blocks are nested within the between-blocks factor.*

   a. Experimental conditions: Diabetic dogs (Example 7.12). Each dog is a block. The between-blocks factor is Diabetes/Control: half the dogs served as controls, the other half were made diabetic by having their pancreases removed. Dogs are nested within Treatments: each dog belongs to one and only one treatment group. By way of contrast, notice that the within-blocks factor (method of measuring) is crossed with Dogs: each dog gets measured using both methods.

   b. Observational conditions: Diets and dopamine (Exercise 7.C.6). Each subject is a block, and the between-blocks factor corresponds to the two kinds of subjects, those with good dietary control, and those with poor dietary control. Subjects are nested within levels of Dietary Control: each subject is of one and only one kind. In contrast, Subjects are crossed with Diets: each subject gets both a control diet and (at another time) the low phenylalanine diet. ■

Crossing versus nesting: three kinds of pictures. Visually, the difference between crossing and nesting is easiest to see in a picture like the one in Figure 13.1, for the diets and dopamine study:

NESTING	P1	P2	P3	P4	P5	P6	P7	P8	P9	P10
Poor dietary control	□	□	□	□	□					
Good dietary control						□	□	□	□	□

CROSSING	P1	P2	P3	P4	P5	P6	P7	P8	P9	P10
Low phenylalanine diet	□	□	□	□	□	□	□	□	□	□
Control diet	□	□	□	□	□	□	□	□	□	□

FIGURE 13.1 Crossing versus Nesting for the design in Exercise 7.C.6. In the top diagram, the 10 patients (P1–P10) are shown nested within Dietary Control: Each patient is of one kind or the other, but not both. In the bottom diagram, the same 10 patients are shown crossed with Diets: each patient gets both diets, and all possible patient-diet combinations are present in the design.

The format above for nesting is almost never used when showing the actual data from an experiment. Instead, what you see has the same shape as the usual factor diagrams. The diagrams in Figure 13.2 show nesting (IV Fluids, Example 5.2) and crossing (Sleeping Shrews, Example 7.4) using factor diagrams.

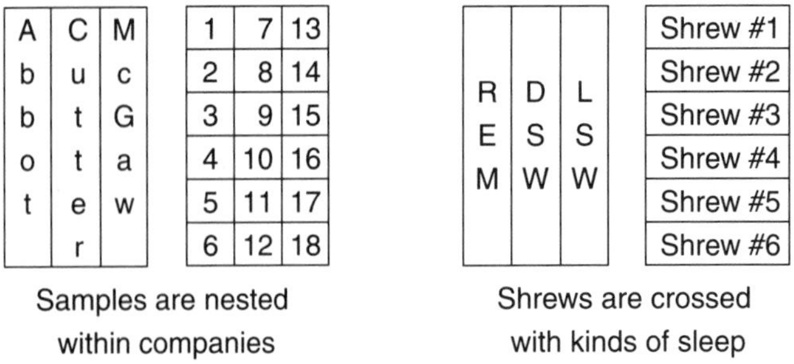

Abbott	Cutter	McGaw
1	7	13
2	8	14
3	9	15
4	10	16
5	11	17
6	12	18

Samples are nested within companies

REM	DSW	LSW
Shrew #1		
Shrew #2		
Shrew #3		
Shrew #4		
Shrew #5		
Shrew #6		

Shrews are crossed with kinds of sleep

FIGURE 13.2

Sometimes the data are arranged so that both factors correspond to sets of rows (or sets of columns) Nesting is easy to show in this format, but crossing is awkward:

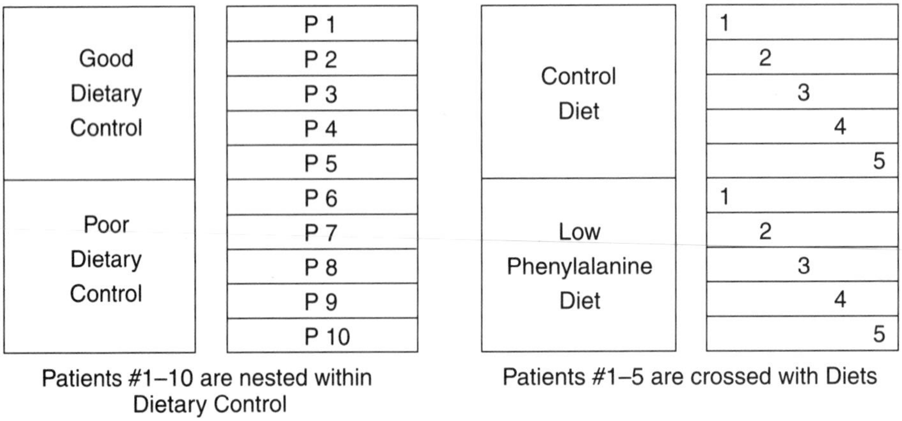

Good Dietary Control	P 1
	P 2
	P 3
	P 4
	P 5
Poor Dietary Control	P 6
	P 7
	P 8
	P 9
	P 10

Patients #1–10 are nested within Dietary Control

Control Diet	1
	2
	3
	4
	5
Low Phenylalanine Diet	1
	2
	3
	4
	5

Patients #1–5 are crossed with Diets

FIGURE 13.3

Crossing versus nesting: four tests. There are four tests you can use to decide whether two factors are related by crossing or nesting. Each test is essentially just a different way of asking the same question: Does each level of the first factor occur with exactly one level of the other factor (nesting), or with all levels of the other (crossing)?

1. Counting test. **How many combinations of levels of A and B appear in the design?**
If A and B are crossed, (# combinations of A and B) = (# levels of A) × (# levels of B)
If B is nested within A, (# combinations of A and B) = (# levels of B)

Example 13.2 Counting Test for Nested Factors

1. *IV fluids* (Example 5.2): There are 15 samples (bags), and 3 manufacturers. There are 15 combinations of Sample and Manufacturer: Samples are nested

within Manufacturers. If the two factors were crossed, there would be 45 combinations.

2. *Sleeping shrews* (Example 5.2): There are 6 shrews, and 3 kinds of sleep. Each shrew was measured during all 3 kinds, so there are 18 combinations of Shrew and Sleep Phase: Shrews are crossed with Phases. If Shrews had been nested within Phases, each shrew would have been measured during just one kind of sleep, with two shrews per kind, and there would have been only 6 combinations of Shrews and Phases.

3. *Diets and dopamine* (Example 5.2): (a) Patients are crossed with Diets: there are 10 patients, 2 diets, and 20 patient-diet combinations. (b) Patients are nested within levels of Dietary Control. There are 10 patients, and 2 levels of control, but only 10 combinations of patient and level of control, one combination per patient. ■

The counting test has a visual version, based on factor diagrams:

> 2. Partition test. **(a) If B is nested within A, then each group in the factor for B fits completely inside a group in the factor for A. (b) If you visualize the factors for A and B on top of each other, what you see is the same as for factor for B, the inside factor, by itself.**

EXAMPLE 13.3 PARTITION TEST FOR NESTED FACTORS

IV Fluids: On the left, (a) each little box for Samples fits inside one of the columns for Manufacturers. (b) If you put the diagram for Manufacturers on top of the diagram for Samples, you get back the diagram for Samples: Samples are nested within Manufacturers.

Shrews: On the right, (a) the rows (for Shrews) do not fit inside the columns (for Kinds), nor do the columns fit inside the rows. Neither factor is nested in the other. (b) If you put the diagram for Kinds on top of the diagram for Shrews, you get a new diagram back, with one cell for each combination of Kinds and Shrews. ■

Quiz yourself: If one factor is nested within another, is it possible to measure the interaction between the two? The answer—No—follows from the partition test, and leads to the interaction test for crossing versus nesting. Recall (from Section 3 of

Chapter 7) that two factors are completely confounded if they correspond to the same partition of the observations. As you just saw above, with the IV fluids example, when B is nested within A, the partition for A × B (which you get superimposing the diagrams for A and B to get all A × B combinations present) is the same as the diagram for B itself. So if B is nested within A, then B and A × B are completely confounded, and it is not possible to measure the A × B interaction. (Caution: Don't confuse this situation with the one where A and B are crossed, but it is not possible to measure interaction because there is only one observation per cell. In that case, the interaction is confounded not with B but with chance error.)

> 3. Interaction test. **If B is nested within A, then the factors for A × B and B are completely confounded. It is not possible to distinguish the AB interaction from the B effects.**

EXAMPLE 13.4 INTERACTION TEST FOR NESTED FACTORS

1. *IV fluids*. If a particular sample was made by Cutter, it cannot also have been made by Abbott. We have no way to measure how that sample might have been different if it had been manufactured by a different company. In fact, it hardly makes sense to ask the question. For this example, as for many examples with nesting, interaction has no meaning.
2. *Sleeping shrews*. Here the two factors are crossed, and the situation is different. Each shrew was measured during all 3 kinds of sleep, and it makes sense to ask, for example, whether the difference between light and deep slow wave sleep is the same from one shrew to the next. Although the data give only one value for each combination of shrew and kind, and so it is not possible to separate the interaction effects from the chance errors, it is possible to get separate estimates for each of Shrews and Kinds of sleep. Interaction is not confounded with a main effect.
3. *Diets and dopamine*. (a) Patient TW has good control, and Patient BR has poor control. We can't get the difference in TW's dopamine readings due to good versus poor control, and so we can't see whether that difference changes from one patient to the next. The Patient-by-Control interactions can't be distinguished from the differences from one patient to the next. Interaction is completely confounded with Patients. (b) On the other hand, each patient gets both diets, and because Patients and Diets are crossed, it is possible to separate the Patient-by-Diet interaction from the main effects for Patients. ■

The interaction test leads to another test, this one based on whether you can interchange data values in pairs of cells without changing the analysis of the data. The cells in the test belong to the partition for A × B, that is, for the set of all combinations of A and B actually present in the design. You get this partition by putting the partition for A on top of the one for B.

4. Interchange test. **Pick two cells of A × B that belong to the same level of B, and consider switching the data values in the first cell with the data values in the second cell. Will the switch affect the analysis of the data? If B is crossed with A, it will; if B is nested in A, it won't.**

EXAMPLE 13.5 Interchange Test for Crossing Versus Nesting

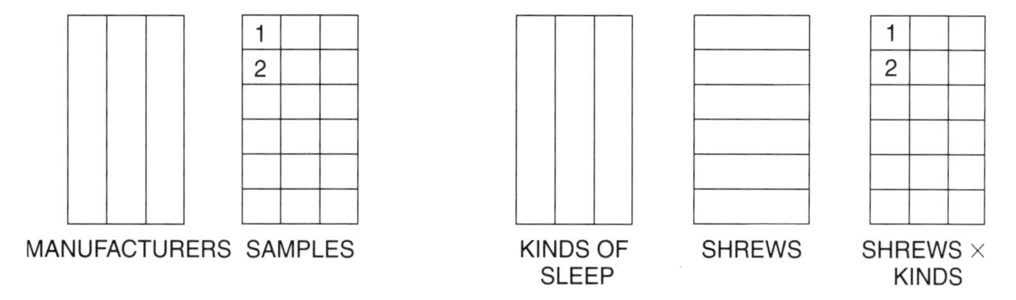

IV fluids: On the left, Samples are nested within Manufacturers. You can interchange the data values in cells 1 and 2, and the switch won't change the meaning of your analysis, because the samples from any one maker are regarded as interchangeable.

Shrews: On the right, Shrews are crossed with Kinds of sleep. If you interchange the data values in cells 1 and 2, you change the analysis. You could switch the entire first row with the entire second row without affecting the analysis, because the analysis regards shrews as interchangeable. However, you can't switch just the two cells in the first column, because then Row 1 would have readings from two different shrews, and so would Row 2. ■

Notation for nesting. It is useful to have a way to name factors that shows the inside/outside relationship for nested pairs. The standard convention is to use parentheses to stand for "nested within." Thus B(A) stands for "B nested within A." In the study of diets and dopamine (7.C.6) we would write Patients(Control) to show that Patients are nested within levels of Dietary Control. (Don't let yourself get confused: It is natural, but exactly backwards, to think that the factor written in parentheses is the inside factor. In fact, the outside factor is *inside* the parentheses. To avoid getting confused, just remember that "(" stands for "within.")

It is also part of the convention that we don't ordinarily show the nesting of the smallest units when they correspond to the factor for chance error.

Examples of Designs with Nested Factors

How does nesting arise in practice? There are essentially two ways, both already familiar to you. (1) Experiments: If you assign one treatment to each unit, then units are nested within treatments. Crossing is a possibility only if you can subdivide or reuse you "units," in which case they are not really units, but blocks of smaller units. You can then cross treatments with *blocks*, but the actual *units* will be nested within treatments. As you can see, the design issues here are the same as in Chapters 7 and 8, on when and whether to use blocks. (2) Observational studies. If you get your data by sampling individuals from different populations, each individual belongs to one and only one population: individuals are nested within populations. In the IV fluids example (5.2), six samples were chosen from each of three populations, Cutter, Abbot, and McGaw. In other studies, instead of sampling from distinct populations, a single sample of individuals is selected, and then sorted into groups afterwards. For example, in a study of rat phobia (Example S9.3), students from a psychology course were chosen as subjects, and then sorted into groups (High Fear and Low Fear) on the basis of a test. Subjects were nested within levels of fear.

Purely hierarchical designs.

The simplest nested designs have no crossed factors, and are called **purely hierarchical**, because *every* pair of factors has an inside/outside relationship.

EXAMPLE 13.6 ELISA TEST FOR HIV

The standard test (called the ELISA test) for whether the AIDS virus is present in a sample of blood uses a spectrophotometer, the same instrument Kelly used in her hamster experiment. Each blood sample is treated with a chemical that reacts with any HIV virus present to reduce the amount of light that can pass through the sample. The test result is positive if the amount of light absorbed by the sample is greater than a certain cut-off value. The proper cut-off value changes, however, from one run to the next, because of the way the spectrophotometer works. Each run must include control samples; the readings for the control are then used to compute the cut-off value for that run. Not only is there variability from one run to another, there is also variability from one control sample to another. Fortunately, it is possible to test several control samples in one run, which makes it possible to measure variability between samples. Finally, there is yet another source of variability. The ingredients for the test will not continue to work properly if they have been kept for too long, and so they are packaged in lots small enough to get used up before they deteriorate. Despite the best efforts of the manufacturer, it is impossible to make the lots perfectly identical, so there is variability between lots.

In all, there are three sources of variability: lots, runs, and samples. Design a purely hierarchical experiment to measure the variability due to each source, and find its factor diagram.

SOLUTION. Each control sample belongs to one and only one run, so Samples will be nested within Runs. Each run uses ingredients from one and only one lot, so Runs will be nested within Lots. In the actual study this example is based on, there were 5 lots (A–E), 5 runs per lot, and 2 samples per run. Figure 13.4 shows a diagram for a smaller version, with only 3 lots, 2 runs per lot, and 2 samples per run.

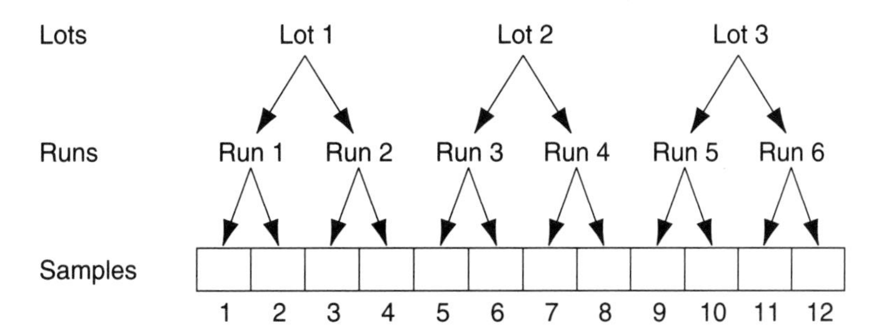

FIGURE 13.4 Hierarchical design: Factors related by nesting. Each sample belongs to only one run; each run uses ingredients from only one lot. No factors are crossed. Each level of the factor Samples occurs with only one level of Runs; each level of Runs occurs with only one level of Lots.

To show the nesting, we can write Runs(Lots) and Samples(Runs(Lots)).

Here's a factor diagram:

Benchmark

Lots

Runs

Samples

What about interactions? Consider Runs × Lots, and remember that to get the partition for the interaction of two factors, we "cross" the partitions for those factors: in terms of the diagram, we superimpose the two boxes for Lots and Runs, putting one on top of the other. If you do that in your mind's eye, you'll see that the box that results gives exactly the same partition as the one for Runs: Runs and Runs × Lots correspond to the same partition of the data, that is, the two factors are completely confounded. ■

EXAMPLE 13.7 ANIMAL BREEDING

Many animal breeding experiments have exactly the same structure as the last example. To make things concrete, imagine that, because the value of a dairy cow depends in part on the butterfat content of her milk, you want to study the variability in the butterfat content due to the genetic background of the cows. A standard design would have three nested factors: Sires, Dams, and Offspring. If each of three bulls were to impregnate each of each cows, on each of two occasions, then the resulting design would be identical to that of the HIV example, with Bulls, Cows(Bulls) and Calves(Cows(Bulls)) instead of Lots, Runs(Lots) and Samples(Runs(Lots)). ■

EXAMPLE 13.8 MULTISTAGE SAMPLING

Are clinical trials that get reported in medical journals based on large enough numbers of patients? One way to study this question would be to take a multistage sample: Start with a list of all medical journals, and choose a random sample of journals. Then, for each journal, list all the issues published in the last ten years, and take a random sample of issues. Finally, for each issue in the sample, list all reports of clinical trials and choose a random sample of reports. The response (one for each report) is the number of patients the report is based on.

The structure of this study is (abstractly) the same as that of the HIV example, with Journals, Issues(Journals), and Reports(Issues(Journals)) as the three nested factors. For a fancier version of the same idea, suppose you wanted to choose a sample of hospital in-patients. It would be too slow and expensive to put together a list of all patients across the country, but multistage sampling would work: First choose a random sample of US states, then cities within states, then hospitals within cities, wards within hospitals, and patients within wards. The same scheme works for sampling elementary school pupils: First choose states, then counties within states, school districts within counties, elementary schools within districts, classes within schools, and finally, pupils within classes. ■

Variations on the SP/RM design. The design I have called the basic SP/RM uses a CR structure to assign one set of treatments to whole plots, and a CB structure to assign a second set of treatments to subplots. For example, the hamster experiment assigns Day Length to Hamsters as in a CR, and associates levels of the observational factor Organ with subplots (parts of hamsters) as in a CB.

In fact, a design can assign treatments to whole plots in any of several ways. The whole plots might sort naturally into groups, for example, so that it would make sense to regard these groups of whole plots as blocks. (The terminology can be confusing if you're not careful. Until now, it was natural to think of "blocks" as another name for whole plots. In what follows, I'll shift to using "blocks" for groups of whole plots.)

EXAMPLE 13.9 CATS AND TETANUS: SP/RM WITH WHOLE PLOTS GROUPED INTO BLOCKS

The purpose of this experiment was to measure the effect of tetanus on the level of the amino acid glycine in the spinal chords of cats. Each of eleven cats was given tetanus in one side only; the other side served as a control. For each side, the glycine levels were measured separately for the white and gray matter. The data are given in Table 13.1. For the purpose of this exercise, assume the assignment of tetanus or control to the two sides was randomized. Analyze the structure of the design.

SOLUTION. There are in fact at least two ways you might choose to analyze this design, but for the purposes of this example, I'll focus on just one (see Table 13.2). The experimental treatment, Tetanus vs. Control, is (randomly) assigned to sides of cats, which serve as whole plots. These whole plots come grouped into blocks (= cats), two sides per cat.

TABLE 13.1

Side:		Control		Tetanus	
Matter:		Gray	White	Gray	White
Cat:	I	5.7	3.0	4.6	3.6
	II	6.1	3.2	5.9	2.7
	III	5.6	2.7	5.3	2.9
	IV	6.1	3.3	5.8	3.4
	V	6.7	2.9	6.6	3.8
	VI	5.4	3.0	5.3	2.8
	VII	5.9	3.3	5.5	3.1
	VIII	5.9	3.9	5.4	3.5
	IX	5.7	2.8	5.2	3.6
	X	4.8	2.6	4.4	2.6
	XI	5.8	2.7	4.9	3.2

TABLE 13.2 Factor list for SP/RM with whole plots grouped.

Unit	Grouping	Treatment combinations	
	(Nuisance factors)	Main effects	2-Factor interaction
Side	Cat	Tetanus/control	
Matter	Side	White/gray	Tetanus × Color

The nesting is Sides(Cats) and Matter(Sides(Cats)).

How should we handle the white and gray matter? (1) If we regard glycine levels for the white and gray matter as two different response values, our model is a bivariate CB[1], with cats as blocks and sides as the only units. For many purposes, this would be the most appropriate model. (2) However, we can also stretch things a bit, just as with the SP/RM model for the hamster data, and regard the white and gray matter in the side of a cat just as we did the hearts and brains of the hamster: the chunks of matter are smaller "units" nested within sides, and White vs. Gray is an observational factor of interest. ■

I hope the last example suggests other variations to you. For example, you could have a design with whole plots arranged in a Latin square, with rows and columns corresponding to two nuisance factors. You could also vary the design by using an LS rather than a CB plan for assigning treatments to subplots. For a crop-yield experiment, for example, you might treat whole plots of ground with various fertilizers, and then assign varieties of wheat to subplots laid out in a Latin square.

Still another way to create variations is to use experimental units of three (or more) different sizes, with different sets of treatments assigned to units of each size.

EXAMPLE 13.10 HAMSTERS: A SPLIT-SPLIT PLOT DESIGN (UNITS OF 3 SIZES)

Before you bid farewell to the hamsters, we can put them to work yet another time. Suppose that Kelly had wanted to compare two different methods for measuring protein concentration, the one she actually used (spectrophotometry) and an alternative. (For example, she might use some chemical to precipitate the protein, and weigh the precipitate.)

The design for her experiment would start out the same as before, but when it came time to measure enzyme concentration, Kelly would need to split each batch of tissue preparation (= subplot) into two halves, then randomly assign one half to each method of measuring. The halves give experimental units of a third size. In the language of design, they would be called sub-subplots, or splits.

Notice that if you ignore the three factors of interest and look at just the three kinds of units, they have a purely hierarchical structure: sub-subplots are nested within subplots, which are in turn nested within whole plots (see Table 13.3).

TABLE 13.3 Factor list for 3-factor split-split plot design

Unit	Grouping	Treatment combinations		
	(Nuisance factors)	Main effects	2-Factor interactions	3-Factor interaction
Hamsters (whole plots)	(none)	Days		
Batches (subplots)	Hamsters	Organ	Days × Organ	
Halves (sub-subplots)	Batches	Method	Days × Method Organ × Method	Days × Organ × Method

■

Finding the List of Factors: A Rule

If your design is a simple one, or is a lot like a design you already know, you may not need a special rule to find the factor structure, but if your design has several factors, with both crossing and nesting, it can help to have a set of steps to follow to find the list of factors. The rule below is based on the same key idea as the interaction test for nesting on page 540: If B is nested within A, then A × B and B are completely confounded.

Finding the List of Factors

Step 1: **Start by pretending all factors are crossed, and write the list of all possible factors and interactions.**

Step 2: **Then "amalgamate" B and AB into B(A):**

a. Cross off every term where B appears without A, and

b. Write B(A) in place of AB in every term containing both A and B.

I'll start with a very simple example, to illustrate why the rule works. (This example will also be important later, if you want to understand how Minitab handles nested factors.)

EXAMPLE 13.11 FACTOR LIST FOR THE IV FLUIDS (EXAMPLE 5.2)

There are two factors: Samples (S) and Manufacturers (M), with Samples nested in Manufacturers.

Step 1: If S and M are crossed, the factor list would have three terms: S, M and SM.

Step 2: We cross off S, and write S(M) in place of SM, to get the list: M and S(M).

Figure 13.5 shows the same two steps using factor diagrams.

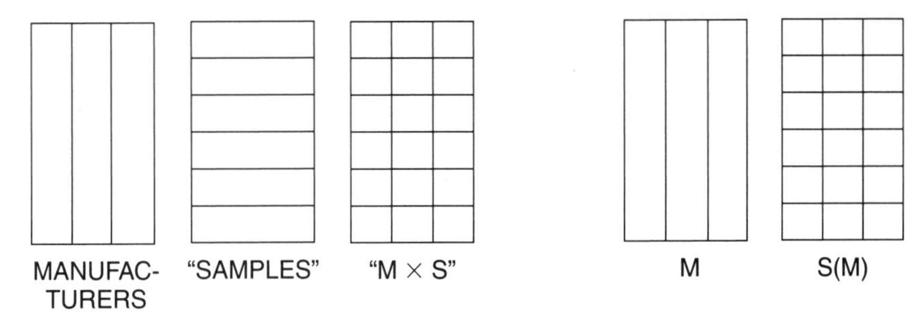

FIGURE 13.5

Step 1: On the left, we pretend that rows correspond to Samples, and that Samples are crossed with Manufacturers, to give three factors in all: M, S, and S × M. *Step 2*: On the right, we cross out the pretend "Samples" factor (rows), and replace "S × M" with S(M). ■

EXAMPLE 13.12 FACTOR LIST FOR DIETS AND DOPAMINE

Use the rule to find the factor list for the study in Exercise 7.C.6, and check that it gives the usual factor list for a basic SP/RM.

SOLUTION. There are three factors: Dietary Control (C), Patients (S), and Diets (D). Patients are nested within Dietary Control.

Step 1: If all factors are crossed, the factor list would have seven terms:

C P D CP CD PD CPD

Step 2: (a) We cross off the inside factor (P) wherever it occurs without the outside factor (C). This eliminates two terms: P and PD.

(b) Wherever C and P appear together, we replace CP with P(C). This gives the final list:

C P(C) D CD P(C)D

For this data set, the interaction P(C)D is assumed to be small enough to ignore, so that we can regard that factor as corresponding to chance error. Thus the four structural factors are the usual four for the SP/RM: Dietary Control = Between, Patients(Control) = Blocks, Diets = Within, and Control × Diets = Interaction. ■

Here's an example complicated enough that you might find the rule useful. Although the structure is in some ways similar to the SP/RM, I think you'll learn more if you try to analyze the design the structure "from scratch" instead of trying to relate it to designs you've already seen.

EXAMPLE 13.13 WATER QUALITY

Suppose that for a study of water quality in the small lakes of northern Maine, you plan to measure the level of toxic heavy metal salts at various places in each lake. One possible design would have four factors: Lakes, Sites, Samples, and Times.

Lakes: Start with a list of all small lakes, say those between 1 and 3 acres in size. Choose 3 at random: these are the 3 levels of Lakes.

Sites: For each lake, use a map of the lake to pick 2 sites at random. In all, you have 6 sites: these are the 6 levels of the factor.

Times: Suppose your research team is large enough to collect water samples from all locations at roughly the same time. You might plan to collect one complete set of samples on each of 3 days, spread over an entire summer: June 15, July 15, and August 15. Time has 3 levels.

Samples: Suppose you plan to take water samples in pairs, two at every site, always at the same fixed depth below the surface. In all you will have $3 \times 2 \times 3 \times 2 = 36$ samples: this factor has 36 levels (Figure 13.6).

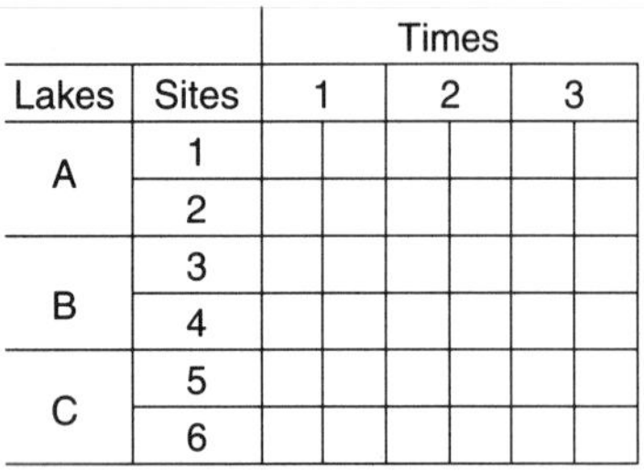

		Times					
Lakes	Sites	1		2		3	
A	1						
	2						
B	3						
	4						
C	5						
	6						

FIGURE 13.6 A rectangular format for the water quality experiment.

Apply the rule to get the factor list for this design.

SOLUTION

Step 1: Start by pretending that the four factors are completely crossed, and make a preliminary list that includes all possible interactions (Table 13.4).

TABLE 13.4 Preliminary list of factor combinations

Main effects	2-Factor interactions	3-Factor interactions	4-Factor interaction
Lakes	Lakes × Sites	Lakes × Sites × Times	Lakes × Sites × Times × Samples
Sites	Lakes × Times	Lakes × Sites × Samples	
Times	Lakes × Samples	Lakes × Times × Samples	
Samples	Sites × Times	Sites × Times × Samples	
	Sites × Samples		
	Times × Samples		

Step 2: Now look for nested pairs, and amalgamate factors.

a. Cross off terms that contain the inside factor but not the outside factor of a nested pair.

- Samples is nested within all other factors. Cross off all terms containing Samples except for the 4-factor interaction. This leaves eight terms:

Lakes Lakes × Sites Lakes × Sites × Times Lakes × Sites × Times × Samples
Sites Lakes × Times
Times
Sites × Times

- Sites is nested within Lakes. Cross off any remaining terms with Sites but not Lakes:

Lakes Lakes × Sites Lakes × Sites × Times Lakes × Sites × Times × Samples
Lakes × Times
Times

b. Replace interactions involving a nested factor:

- Samples is nested within all other factors. Replace the 4-factor interaction with Samples(Lakes × Sites × Times)
- Sites is nested within Lakes: Replace Lakes × Sites with Sites(Lakes), replace Lakes × Sites × Times with Sites(Lakes) × Times, and replace Samples(Lakes × Sites × Times) above with Samples(Sites(Lakes) × Times). This gives the five structural factors listed in Figure 13.6.

Now write out the partitions for the factors that remain on your list. To check that you haven't missed any redundant factors, make sure none of your partitions coincide.

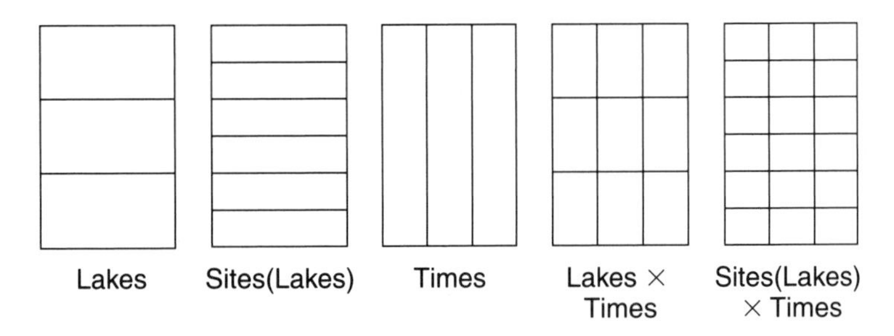

FIGURE 13.6 The five structural factors for the water quality experiment. The complete factor diagram would have two more factors: Benchmark, and Residual Error = Samples(Sites(Lakes) × Times). ■

Decomposing Data Sets with Nested Factors

The decomposition involves no new ideas. Although nesting leads to new factor structures, once you have the factor diagram you can decompose your data, then compute sums of squares, degrees of freedom, and mean squares in the same way as before.

EXAMPLE 13.14 DECOMPOSITION OF THE HIV DATA

Here are actual data for 12 samples in 6 runs using ingredients from 3 lots:

Lot	A				B				C			
Run	1		2		3		4		5		6	
Sample	1	2	3	4	5	6	7	8	9	10	11	12
	1.053	0.977	0.896	1.038	0.996	1.016	1.088	1.280	1.229	1.027	1.118	1.146

a. *Decomposition: Estimated effect equals group average minus sum of estimated effects for all outside factors* (Figure 13.7).

Lot Effect = Lot Average − Grand Average
Run Effect = Run Average − Lot Effect − Grand Average
(= Run Avg − Lot Avg)
Sample Effect = Residual = Observed Value − {Run Effect + Lot Effect + Grand Average} (= Obs − Run Avg)

Grand Avg.	1.072											
Lots	−0.081				0.023				0.058			
Runs	0.024		−0.024		−0.089		0.089		−0.002		0.002	
Samples	0.038	−0.038	−0.071	0.071	−0.010	0.010	−0.096	0.096	0.101	−0.101	−0.014	0.014

FIGURE 13.7

b. *Degrees of freedom: df equals number of levels minus sum of df for all outside factors.*

$df_{Grand} = 1$
$df_{Lots} = \# \text{Lots} - df_{Grand} = 3 - 1 = 2$
$df_{Runs} = \# \text{Runs} - (df_{Grand} + df_{Lots}) = 6 - (1 + 2) = 3$
$df_{Res} = \# \text{Obs} - (df_{Grand} + df_{Lots} + df_{Runs}) = 12 - (1 + 2 + 3) = 6$

As an alternative, you can count df directly, as the number of free numbers (Figure 13.8).

1												Grand Average: df = 1
1				2				+				Lots: df = 2 = # Lots − 1
1		+		2		+		3		+		Runs: df = 3 = # Runs − # Lots
1	+	2	+	3	+	4	+	5	+	6	+	Samples: df = 6 = # Samples − # Runs

FIGURE 13.8 Degrees of freedom for a hierarchical design

As yet a third approach, if you use the alternative decomposition rule of Chapter 10, you can use a flowchart (Figure 13.9).

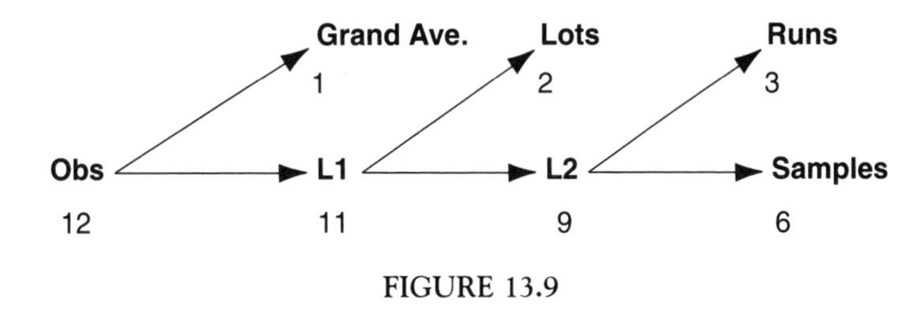

FIGURE 13.9

c. *Sums of squares.* As always for a balanced design, you get the sum of squares for a factor by squaring the numbers in the corresponding table of the decomposition, then adding. This gives:

$SS_{Grand} = 13.79021$

$SS_{Lots} = 0.04182$ $\quad MS_{Lots} = 0.02091$

$SS_{Runs} = 0.03400$ $\quad MS_{Runs} = 0.01133$

$SS_{Res} = 0.05240$ $\quad MS_{Res} = 0.00873$ ■

The interpretation of the mean squares depends on the assumptions we make about how the components in the decomposition behave. That's what Section 2 is about. Before beginning this topic, however, I'll first say something about how to use Minitab to decompose data from designs with nested factors.

Using Minitab. To use the Minitab ANOVA command you need to learn how Minitab handles nested factors for balanced designs. There are two parts to what you need to know: (1) numbering the levels of a nested factor, and (2) showing the model. Both parts are illustrated in Example 9.26, which shows how to use the ANOVA command to analyze the Diabetic Dogs data, from a basic SP/RM design. If you understood that example, then what follows should be familiar. (1) Levels for nested factors: The key is that you have to number the levels of the inside factor as though you had crossing instead of nesting. (2) The model: You need to use parentheses to show nesting when you tell Minitab the model.

EXAMPLE 13.15 USING MINITAB TO DECOMPOSE THE HIV DATA

Table 13.5 shows the columns you would use. Minitab requires you to number the runs as though there were only two, with Runs crossed with Lots. This means that

TABLE 13.5

Row	Response	Lots	Runs	Row	Response	Lots	Runs
1	1.053	1	1	7	1.088	2	2
2	0.977	1	1	8	1.280	2	2
3	0.896	1	2	9	1.229	3	1
4	1.038	1	2	10	1.027	3	1
5	0.996	2	1	11	1.118	3	2
6	1.016	2	1	12	1.146	3	2

Run 3 (Rows 5 and 6) is listed as Lot 2, Run 1, and Run 4 (Rows 7 and 9) is listed as Lot 2, Run 2. Because Samples correspond to the smallest units, Minitab does not require you to number them.

The model is the same as the factor list you get by following the rule on page 546. You need to show nested factors using parentheses to stand for "within," and listing only structural factors. This means you don't include Samples or Samples (Runs(Lots)) in the model, because this factor corresponds to error. There are only two structural factors, Lots and Runs(Lots). To complete the model, you must tell Minitab which factors are fixed, and which are random.

```
MTB > ANOVA > Balanced ANOVA ...
Response:          Response
Model:             Lots Runs(Lots)
Random factors:    Lots Runs(Lots)
```

The difference between fixed and random factors is what Section 2 is about. ■

Exercise Set A

1. *ELISA and HIV: Crossing versus nesting.* Use all four tests to verify that for Example 13.6, Runs are nested within Lots, and Samples are nested within Runs.

2. *ELISA and HIV: Factor lists.* Use the rule of Section 1.3 to find the factor list. (This should be quick and easy. The purpose is to make sure you understand what the rule says before you try to use it for a more complicated example.)

3. *Cats and Tetanus: Crossing versus nesting.* Use all four tests to verify that for Example 13.9, Sides are nested within Cats, and chunks of Matter are nested within Sides, but Color (white/gray) is crossed with Treatment (tetanus/control).

4. *Cats and Tetanus: Crossing versus nesting.* Use the rule of Section 1.3 to verify that the factor list given in Example 13.9 is correct.

5. *Hamsters: Crossing versus nesting.* For each of the following pairs of factors from Example 13.10, use whichever test you like to decide whether the factors are crossed or nested. (Note that you can use the table at the end of Example 13.10 to check your answers.)

   a. Hamsters and Days
   b. Hamsters and Batches
   c. Hamsters and Organ
   d. Halves and Methods
   e. Organ and Method
   f. Halves and Hamsters

6. *Hamsters: Crossing versus nesting.* Use the rule of Section 1.3 to verify that the factor list given in Example 13.10 is correct.

7a. *Dutch elm disease: Crossing versus nesting.* Reread the description of the beetle study in Exercises 9.A.3 and 9.C.9. There are four factors, and so six possible pairs of factors. For each pair, use whatever test(s) you like to decide whether they are related by crossing or nesting.

7b. *Dutch elm disease: Factor list.* Using whatever method you choose, find the factor list for the study of Exercise 9.C.9, and show the factors in a table like the ones used in Examples 13.9 and 13.10.

8. *Acid rain*. Here's a plan to study sources of variability in water pH (response). Choose at random 4 regions in southern Canada; within each region, choose 5 lakes. Within each lake, choose 3 locations, and from each location, collect 2 water samples.

a. Find the factor list, using parentheses to show nesting.

b. Draw and label a rectangular format for the data; then, using that format, write the factor diagram. Analyze the structure as in Exercise 1 above.

9. *Calcium content of turnip leaves*. Notice that the plan of Exercise 8 is easy to transfer to other contexts. Rewrite Exercise 8(a), making the following translations:

Regions → Gardens, Lakes → Turnip plants,

Locations → Leaves, and Samples → Measurements of calcium content.

10. *Gypsy moths*. Using Exercise 8 as a model, describe a plan to study factors that affect the distribution of gypsy moth caterpillars in New England.

11. *Postoperative mortality*. Using Exercise 8 as a model, describe a plan to study factors that affect the distribution of the postoperative mortality rate for gallbladder surgery in the U.S.

12. *An experiment in social psychology*. It's 4 pm on a Friday. You've spent the last 10 minutes in line at a bus station, waiting your turn to buy a ticket home. At last, there are only 3 people ahead of you. Suddenly a person appears out of nowhere and breaks in line, not ahead of you, but right behind you. What do you do?

a. Design an experiment to study factors that affect the behavior of people waiting in line when someone crashes the line. Think about the following questions: Does the kind of line matter? (Is the express line at a supermarket different from the ticket line at a bus station?) Does the person who crashes matter? (An elderly person with a cane versus a scruffy-looking teenager?) Does the position in line matter? What might you use as a response variable or variables? Your design should include at least two nested factors. Make sure your description is precise enough to make the complete factor structure clear.

b. Draw and label a rectangular format for your data. Then draw the factor diagram.

c. Write the degrees of freedom below each box in your diagram.

13. *Tomato leaves*. The data below are real, but in order to make the design structure unambiguous, I've had to guess at a few details of how the experiment was done.

The purpose of the experiment was to study the effect of various factors on the temperature (°F) of the leaves of tomato plants. (The hotter the leaves, the more moisture they lose, which is not good for them.)

The experiment was run in two replications; assume each replication corresponds to a single plant. Two leaves were selected for each plant, one upper leaf, exposed to sun, and one lower, shaded leaf. The temperature of each leaf was measured at 10 a.m., and again at 4 p.m.; at each time the temperature was measured twice, once at the surface of the leaf, and again 1 cm above the leaf:

Distance:	0 cm				1 cm			
Exposure:	Upper		Lower		Upper		Lower	
time:	**10 a.m.**	**4 p.m.**	**10 a.m.**	**4 p.m.**	**10 a.m.**	**4 p.m.**	**10 a.m.**	**4 p.m.**
Replication I	89.9	89.1	81.1	79.3	86.7	85.0	80.3	79.8
II	90.5	94.5	78.9	80.3	89.7	86.2	81.7	82.2

a. List all factors for this design, putting them in a table like the one in Example 14.10 to show units, grouping factors, main effects, and interactions.

b. Draw the factor diagram, and write the df below each factor.

14. *Effective teachers*. Are there personality traits that tend to make a teacher more effective? Here's a design for studying the question: Randomly pick three schools at each of three levels (elementary, intermediate, and high school). At each school, ask the principal to pick the ten most effective and ten least effective teachers. Then ask the principal to fill out special form (the Tuckman Teacher Feedback Form) to rate each teacher on four personality traits: creativity, dynamism, organized demeanor, and warmth and acceptance.

Regard the four ratings as four different responses, and analyze the structure of the design.

2. New Uses for Old Factor Structures: Fixed Versus Random Effects

Overview

Section 2 explains the difference between **fixed and random factors**, gives examples of designs with random factors of interest, and shows how to analyze data from such designs. Fortunately, most of the analysis is the same as for the designs you've already seen. The factor structure involves nothing new, and so finding the decomposition, degrees of freedom, sums of squares and mean squares are all familiar. Only two things are new:

1. How to tell the difference between fixed and random factors, and how this difference affects the meaning of your ANOVA decomposition.
2. How to find the right denominator for an F-test by finding expected mean squares.

Familiar examples. You've actually been working with both fixed and random factors already. In all the examples we have considered so far, the factors of interest have been fixed, and the units have been random; in nearly all the examples of designs with blocks, the blocks have been random. To see the difference, think of a factor as a set of levels, like the four exercise programs in the walking babies experiment (Example 5.1) or the three drug companies in the study of intravenous fluids (Example 5.2). In these examples the levels of the factor are of interest in their own right, and are fixed: we want to know about those four exercise programs or those three manufacturers. The babies and the bags of fluid, on the other hand, were chosen to represent some larger population and are random factors. For designs with blocks, the levels of the blocking factor are also usually chosen to represent some larger population, and so are random. For example, the shrews in the study of heart rate and sleep phases (Examples 8.4 and 8.10) were chosen to represent the population of all tree shrews. This section presents four tests you can use to tell the two kinds of factors apart.

Consequences. Each random factor introduces an extra layer of "things that might have happened but didn't" between the process that generates the data and the numbers you actually observe. This shows up as additional variability in some of the tables of a decomposition, just as the variability due to chance errors is present in all tables of a decomposition. Wherever this extra variability is present in the mean square for a factor, you need stronger evidence in order to conclude that the effects are "real."

For example, consider an F-test for treatments. It turns out (more later) that any random factors *inside* the treatment factor will contribute extra variability to the mean square for treatments. The extra variability means we can't use the residual mean square in an F-test for treatments. (If we did, we wouldn't be able to tell whether a large F-ratio was due to treatments, or to the extra variability from other sources.) We need a denominator that also contains the extra variability, from exactly the same sources as the numerator mean square. This section presents simple rules for finding such denominator mean squares, but the extra variability means that certain effects may be harder to detect.

Fixed and Random Factors

Overview. Fixed and random refer to two alternative assumptions about the levels of a factor. Informally, with random effects we are concerned not only with the levels actually present, but also with levels that might have been chosen but weren't; with fixed effects, we are only concerned with the levels actually present. ("What you see is all there is.")

In what follows, we'll look first at mathematical models for fixed and random effects. In the ideal world of mathematics, the difference is clear and easy to recognize. In practice, however, it may not be so easy to decide which model gives a better fit to reality. To help you decide, I'll give four questions you can ask, and I'll illustrate them with examples. Then I'll give some generalizations about where fixed and random effects typically occur.

Mathematical models for fixed and random effects.

Random = Sample Fixed = Population

A factor is random if its levels are drawn at random from a larger population.
A factor is fixed if its levels correspond to the entire population of interest.

I'll illustrate the difference between the two models using the IV fluids study (Example 5.2). The factor of interest in this study is Manufacturers, with three levels, Cutter, Abbot, and McGaw. In our analysis we treated this factor as fixed: Cutter, Abbot and McGaw were the only manufacturers we cared about, and so they correspond to the entire population of interest.

Model for fixed effects. We are only concerned with the factor levels actually present. We assume there is a fixed underlying "true" value of the response for each level of the fixed factor. For example, with the IV Fluids, we assume that for each of Cutter, Abbot and McGaw there is a fixed "true" reading (equal to the average over all possible samples of fluid) for the concentration of contaminant particles.

$$\mu_C \qquad \mu_A \qquad \mu_M$$

"True" values for the IV fluids

In our analysis, we are interested in this set of three true values, or, equivalently, in the overall mean and the "true" effects:

$$\text{Overall mean: } \mu = \frac{\mu_C + \mu_A + \mu_M}{3}$$

$$\begin{aligned} \text{True effects: } \alpha_C &= \mu_C - \mu \\ \alpha_A &= \mu_A - \mu \\ \alpha_M &= \mu_M - \mu \\ \text{Note that: } \alpha_C + \alpha_A + \alpha_M &= 0 \end{aligned}$$

Fixed effects model for the IV fluids data

With the fixed effects model, we use the F-ratio to test that all the true treatment effects are equal to zero, or, what is the same thing, that all the true means are equal to the overall mean:

$$\begin{aligned} H_0: \ \alpha_C &= \alpha_A = \alpha_M = 0 \\ \mu_C &= \mu_A = \mu_M = \mu \end{aligned}$$

Null hypothesis for the IV fluids data, fixed effects model

Model for random effects. Now consider a different version of the study. Suppose we had said, "I don't particularly care about any one, or two or three manufacturers. Instead, I want to use my data to generalize about the set of all manufacturers of IV fluids. So I'll list them all, and then use a chance device to choose three of them at random." If that's how Cutter, Abbot and McGaw ended up as the three manufacturers in the study, then the factor Manufacturers would be random.

Our model still assumes there is one underlying "true" reading for each manufacturer, but we assume that the set of manufacturers in the study was chosen at random. Abstractly, if we identify each manufacturer with its true value, we assume that the true values were chosen at random as a sample from a larger population of true values (Figure 13.10).

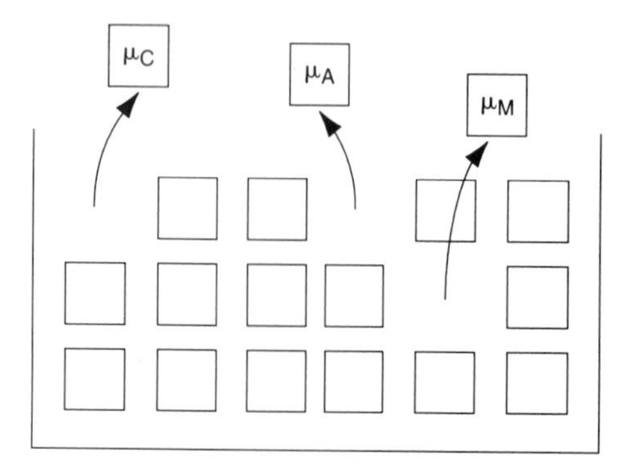

FIGURE 13.10 Random effects model for the IV fluids. The "true" values for Cutter, Abbot and McGaw are a random sample from a population of true values, one for each manufacturer.

We assume that the true values for the levels actually present in the study behave like random draws from a box of numbered tickets. We are interested in these true values mainly for what they can tell us about the set of tickets in the box—the population. We assume that the population of true values follows a normal curve, with mean μ^* and standard deviation σ_M.

Overall mean:	μ^* = average of the tickets in the box
Standard deviation:	σ_M = SD of the tickets in the box
Distribution:	Tickets in the box follow a normal curve
True means:	$\mu_C\ \mu_A\ \mu_M$ are draws from the box
True effects:	$\alpha_C{}^* = \mu_C - \mu^*$
	$\alpha_A{}^* = \mu_A - \mu^*$
	$\alpha_M{}^* = \mu_M - \mu^*$
Note that:	$\alpha_C{}^* + \alpha_A{}^* + \alpha_M{}^*$ does NOT equal zero.

Random effects model for the IV fluids data

With the random effects model, we use the F-ratio to test that all the true treatment effects *for the entire population* are equal to zero, or, what is the same thing, that all the true means in the population are equal to the overall mean. If all the population values are the same, then the population SD is equal to zero, and that is how the null hypothesis is usually stated:

$$H_0: \qquad \sigma_M = 0$$

Null hypothesis for the IV fluids data, random effects model

Two sets of random effects. Notice that the assumptions of the random effects model are almost the same as our standard assumptions for chance errors. In fact, chance errors are themselves random effects. For example, in the IV Fluids study, the chance errors correspond to the individual bags of fluid: chance error = reading for the bag − "true" value for the manufacturer. We assume the six bags for each manufacturer behave like random draws from a population of bags we might have chosen. If we are only interested in Cutter, Abbot and McGaw, then Manufacturers is a fixed factor and Bags(Manufacturers) = Chance Error is a random factor (Figure 13.11).

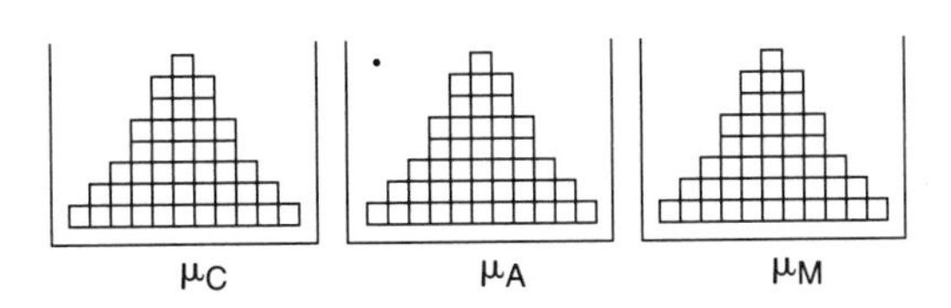

FIGURE 13.11 IV fluids model with Manufacturers fixed and Bags = Error random. The reading for each sample (= bag) behaves like a random draw from a box of numbered tickets. There is one box of tickets for each manufacturer, and the average of that box is the "true" reading for that manufacturer.

For the second version of the study, with manufacturers chosen at random from a list of all manufacturers, both factors are random (Figure 13.12).

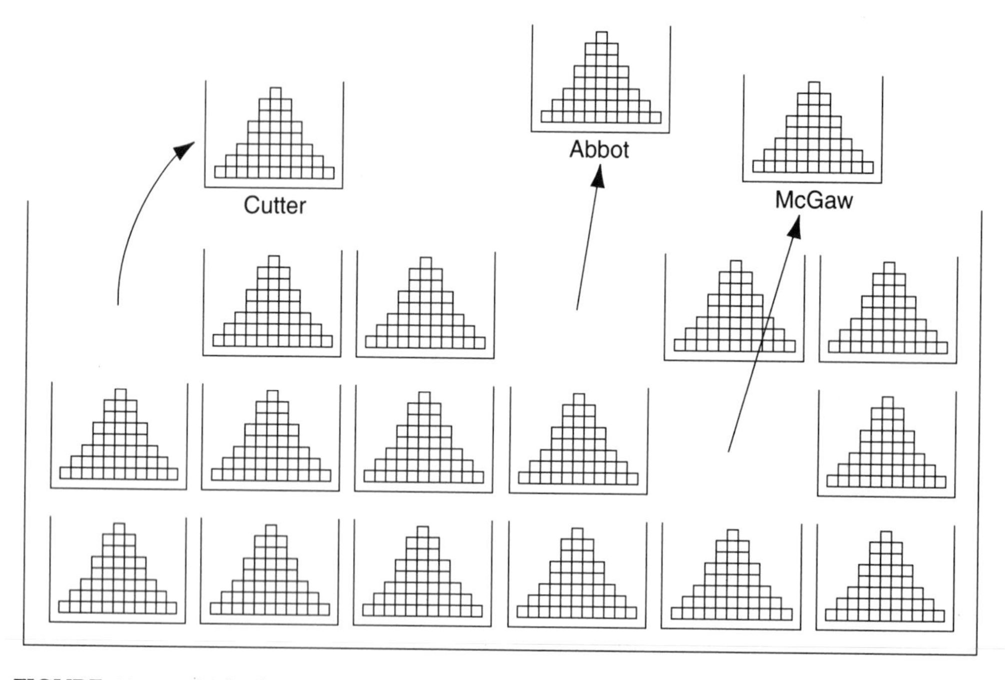

FIGURE 13.12 IV fluids model with both Manufacturers and Bags = Error random. We first choose manufacturers (small boxes) at random from the population of manufacturers (large box), then choose six bags (tickets) at random from the population of bags for that manufacturer (small box).

Deciding which model to use: three tests. Ideally, if a study is well planned, it is easy to decide which effects are fixed and which are random. All you do is ask which is the right assumption about the way the terms in the model behave.

> Assumed behavior: **What do we assume about how the terms in the model behave?**
> **Fixed: Like constants**
> **Random: Like random draws from a box of numbered tickets.**

Sometimes, though, it may not be so clear, possibly because the real situation doesn't quite match either model. In such cases, you can ask yourself a set of three questions: What is the target of your inferences? What fraction of the population of interest is present in your study? If you were to repeat the study, would the levels change or stay the same?

> Target of inference: **What is the target of our inferences?**
> **Fixed: The set of levels actually observed**
> **Random: A much larger set of levels that is represented by the sample of levels actually present.**

When a factor is fixed, your main goal is to estimate the size of the individual effects. For example, in the walking babies experiment we want to know whether the program of special exercises leads the babies to walk earlier than usual, and if so, how much earlier. When a factor is random, your main goal is (ordinarily) to estimate the variability due to that factor. For the study of solitary confinement, it doesn't particularly matter that one individual prisoner's response was above or below average; what matters is how much variability there is among the prisoners as a group.

Sampling fraction: **What fraction of the population of interest have we chosen?**
Fixed: All
Random: Only a tiny fraction

When a factor is fixed, you include all the levels of interest in your study. For example, in the study of sleeping shrews, there were three kinds of sleep (fixed) that were of interest, and all three were used as levels in the actual study. When a factor is random, the levels actually present are only a fraction of the set of levels you are interested in. For example, the goal of the shrew study was to draw conclusions about all tree shrews; the six actually used in the study represent only a tiny fraction of all shrews.

Choice of levels: **How were the levels chosen? If we were to repeat the experiment, would the levels change (random)? Or stay the same as before (fixed)?**

When a factor is fixed, you would use the same levels again if you were to repeat the experiment. For example, consider the study of diets and dopamine (Exercise 7.C.6). If you were to rerun the experiment, you would still use patients of the same two kinds, those with good dietary control and those with poor control. You would also use the same two diets, a control diet and a low phenylalanine diet. Both these factors are fixed. If a factor is random, its levels would be different if you were to repeat the experiment. You wouldn't use the same ten patients in a second run of the dopamine study. Instead, you'd need to get ten more patients, five each with good and poor control. So while Dietary Control is fixed, Patients is random.

EXAMPLE 13.16 FIXED AND RANDOM FACTORS IN THE HAMSTER EXPERIMENT

Classify the factors of the hamster experiment as fixed or random.

SOLUTION

Fixed: Day Length, Organ, Interaction
Random: Hamsters, Chance Error

1. *Target of inferences.* Ask whether the levels of a factor define particular conditions you want to study, like Heart and Brain (fixed), or whether you think of the levels of the factor as a representative set from some larger collection of possibilities, the way you take the eight hamsters (random) to be representative of

the collection of *all* golden hamsters. In other words, ask whether you want your conclusions to apply to the particular levels that appear in your experiment (fixed), or to a larger set of possibilities which (you hope!) your chosen levels represent (random). (Ideally, the levels of a random factor get chosen from the larger set using a chance device to insure that the ones you choose do represent the set; that's why the factor is called random. In practice, however, using the chance device is often not practical, and so you do the best you can to insure that your group of hamsters is representative.)

2. *Sampling fraction*. With the two fixed effects, Day Length and Organ, the levels actually present represent the entire set of levels of interest. With the random effect, Hamsters, however, the 8 hamsters Kelly actually used represent only a tiny fraction of the population of all golden hamsters.
3. *Choice of levels*. Ask what would stay the same and what would change if you or someone else at another lab were to repeat your experiment. The levels of Day Length (fixed) would stay the same, 16 hours and 8 hours. So would the levels of Organ: Heart and Brain. Not so for Hamsters (random), however: you wouldn't use the same hamsters if you reran the experiment. ■

EXAMPLE 13.17 FIXED AND RANDOM FACTORS IN PLANNING EXPERIMENTS

Often, in planning an experiment you have a choice about whether to make a factor fixed or random. Here are examples from ecology, medicine, and psychology.

1. *Ecology*. Whenever you want to measure the abundance of an organism, as in Hurlburt's study (Example 4.11), you should expect the response to vary with time and place. Consider a study with Time and Place as two crossed factors of interest, and Species Density as the response. Each factor can be either fixed or random, depending on how you plan the experiment.

   a. Time: Suppose you decide to take measurements on six different days during the summer months. If you were interested in particular features that changed from one day to the next, such as cloud cover, you might decide to choose (at random) two dates for each of three cloud conditions (completely overcast, partially overcast, full sun). Then Cloud Conditions would be a fixed factor, and Days(Cloud Conditions) would be random. On the other hand, if you thought that a large variety of conditions—not just cloud cover—would affect species density, you might simply use a chance device to choose six dates, and regard them as a representative sample for measuring variability from one day to another. In that case, Days would be a random factor.

   b. Place: Suppose you plan to measure at four different locations. You could choose the locations according to particular characteristics (fixed distances from shore, or fixed depths, for example), or alternatively, you could choose four locations at random.
2. *Medicine*. The HIV testing example (13.6) illustrates the way the results of medical tests can depend on both the sample being tested and the lab doing the test-

ing. Screening for carriers of hepatitis provides another example. A standard test uses the concentration of the enzyme SGPT, which is higher in carriers of hepatitis than in normal controls. Suppose that in order to find out how the levels of SGPT vary from one person to another and from one lab to another, you plan a study with Labs and Patients as two crossed factors of interest.

a. Labs: If you want to compare three particular labs, as you might in order to choose the most reliable, you would take those labs as the three levels of a fixed factor. If, on the other hand, you wanted to estimate the typical size of lab-to-lab variability, you could choose a few labs at random from the population of commercial labs, and regard the labs you chose as levels of a random factor.

b. Samples: You could choose three types of patients (noncarriers, carriers, and those with active hepatitis) and regard the three types as levels of a fixed factor. You could then randomly choose patients of each type, and the nested factor Patients(Type) would be random.

3. *Psychology*. Many psychology experiments rely on human judges to get numerical response values from the behavior of subjects. Consider, for example, the study of tardive dyskinesia (Example S4.5), in which trained raters watched videotapes of subjects and counted the number of facial tics in 5-minutes segments of tape. You could design a study with three judges and four tape segments as crossed factors of interest.

   a. Judges: If you were interested in the three judges as individuals, and wanted to estimate the sizes of the three effects (e.g., is Judge A consistently high in her ratings?), you would analyze the data with judges as a fixed factor. But if you regarded the judges as a representative sample from a population of possible judges and you wanted to estimate the typical size of judge-to-judge variability, you would analyze the data with judges as a random factor. As another example, consider the study of overconfidence in case study judgments (Example 8.6). There the judges were of three types, with different levels of training and experience. Type of judge was a fixed factor, and the nested factor Judges(Types) was random.

   b. Tape Segments: You could choose a sample of segments, and regard the factor as random; alternatively, you could choose tape segments systematically (one with a very high frequency of tics, one moderately high, and one very low) and regard the segments as levels of a fixed factor. ■

Five (imperfect but useful) generalizations. Sometimes, as in the last example, it will be possible for a factor to be either fixed or random, depending on the way you decide to choose the levels. Much of the time, however, the usual choices tend to follow fairly standard patterns, and while there are exceptions, it is nevertheless useful to have the patterns as a way to organize your thinking about examples.

1. *Units should always be random.*

   Example: 5.1 The individual babies in the walking babies experiment

   5.2 The bags of fluid in the IV fluids study

2. *Blocks are usually random*

   Example: 7.4 Shrews, in the study of sleeping shrews, is a random factor

   7.B.2 Subjects, in the finger tapping experiment, is a random factor

   S7.2 Twin Pairs, in the study of tracheobronchial clearance, is a random factor

   S7.3 Occasions, in the beesting study, is a random factor

3. *Nuisance factors are usually random*

   Example: 7.7 Dates, in the rabbit insulin experiment, is a random factor

   7.7 Rabbits, in the same study, is a random factor

   7.6 Cows, in the milk yield experiment, is a random factor

   But: 7.6 Time periods, in the same study, is a fixed factor

   8.9 Months, in the study of the Transylvania effect, is a fixed factor

4. *Nested (inside) factors are usually random*

   Example: 13.6 Runs(Lots), in the HIV study, is a random factor

   13.6 Samples(Runs(Lots)), in the same study, is a random factor

   But: 13.9 Sides(Cats), in the study of cats and tetanus, is a fixed factor

5. *Factors of interest that are experimental are usually fixed*

   Example: 5.1 Exercise, in the walking babies experiment, is a fixed factor

   7.B.2 Drugs, in the finger tapping experiment, is a fixed factor

   7.6 Diets, in the milk yields experiment, is a fixed factor

   7.C.6 Diets, in the study of diets and dopamine, is a fixed factor

Designs with random factors are not as simple as designs with all factors fixed, because not all the levels of the random factors are actually present in the design. In order to generalize beyond the levels actually present, to include all the others that might have been chosen, you need to change your F-tests to take into account the extra variability due to the random factors.

Using Expected Mean Squares to Choose Denominators for F-Ratios

Overview of the logic. If all the structural factors in your model are fixed, then choosing denominators for F-ratios is easy: you always use MS_{Res}. For models with random factors, however, choosing mean squares becomes somewhat more complicated, although there is one general rule that applies to all balanced designs. The rule is based on the following logic:

1. When we decompose a data set, our goal is to get separate estimates for the pieces due to the different factors in the design, but the presence of chance error (and possibly other chance-like components) prevents the decomposition from sorting out those pieces exactly. Each table in the decomposition usually contains

variability from more than one source. As a result, the mean square for a factor will contain variability from exactly those same sources and can be regarded as an estimate for a sum of terms corresponding to those sources.

2. For each factor in our design, we can define a (theoretical) SD that measures the typical size of the true variability in the observations due to that factor. For example, in a CB[1] design each observation contains variability due to treatments, blocks, and error, so we would define three SDs, one for each source of variability. Our goal in an ANOVA is to determine which of these true SDs are detectably different from zero.
3. The SDs are defined and the means square have been constructed so that each mean square estimates a weighted sum of squares of the SDs, one SD for each factor that contributes variability to the mean square. In any given mean square, some $(SD)^2$ terms will be present, others absent.
4. When we construct an F-ratio, we want numerator and denominator MSs to contain contributions from exactly the same set of factors, except that we want the numerator MS to contain one extra term, which corresponds to the factor we want to test. This term will be zero if the effects of the test factor are zero, and positive otherwise. Thus the F-ratio will tend to be near one if effects of the test factor are zero, but greater than one otherwise.

Interpretation of expected value. The rule for choosing F-ratios is usually stated in terms of **expected values of mean squares**, or **EMSs**. You can think of the expected value of a random quantity as the "long-run average" value of that quantity. (The model defines a chance mechanism for creating data sets. For each data set you can compute, say, the mean square for treatments. Now imagine you do this for thousands of data sets, all created according using the same model, and then you take the average of all the mean squares. The resulting number will be very close to the expected value of the mean square, and will differ from it only by a tiny chance amount.) In other words, the expected value of the mean square tells what it is that mean square estimates.

If you write the model for your data set algebraically, as in the appendices of Chapters 5–7, then it is possible to find exact algebraic expressions for all the EMSs, and anyone who wants a thorough understanding of ANOVA should eventually learn to do that, and to understand the probability theory that explains how the process works. Here, however, we will seek a middle ground between such an abstract mathematical approach, and the (defeatist) attitude that you unless you first learn the mathematics, you might as well skip the ideas entirely.

What the EMS tells. Our compromise depends on a fact about ANOVA: *The expected values for all the mean squares for a design are built from the same set of pieces, one piece for each factor*. For a first understanding, it is not essential to find a formula for each piece; what matters is that each of the pieces is a constant times the square of a standard deviation: it measures the true variability in the data due to that factor.

The expected value of a mean square tells which factors in the design contribute to the observed variability in the corresponding table of the decomposition.

The EMS for a factor tells which factors in the design contribute pieces to the mean square for that factor.

Notation: If terms for Error, AB, and A appear in the expected mean square for A, we write

$$\text{EMS(A)} = \text{Error} + \text{AB} + \text{A}$$

EXAMPLE 13.18 INTERPRETING EXPECTED MEAN SQUARES

Consider the dopamine study (7.C.6). It will turn out (Example 13.19) that

a. EMS(Error)= Error
b. EMS(Diets)= Error + Diets
c. EMS(Patients)= Error + Patients
d. EMS(Dietary Control) = Error + Patients + Dietary Control

Interpret these EMSs.

SOLUTION. Start by noticing that the four EMSs are all sums, and that the sums are built from the same four terms: Error, Diets, Patients, and Dietary Control. The term Error appears in all four EMSs, and is the same in all four of its appearances. It measures the variability in the observations due to chance error, and is in fact equal to the square of the true SD for error. Similarly, the term Patients, which appears in the last two EMSs, measures the variability in the observations due to patients, and, apart from a constant, is also a squared SD. The other two terms, Diets and Dietary Control, measure variability in the observations due to each of those two sources.

Now consider the EMSs one at a time.

a. EMS(Error) = Error. This tells us that the only variability in the residual table is due to the errors. None of the other factors in the design contribute variability to the residuals.
b. EMS(Diets) = Error + Diets. The table for Diets in the decomposition has variability from two sources, Diets and Error. Remember that we get this table by computing Diet averages and subtracting the Grand Average, so we want and expect Diet differences (if any) to contribute variability to this table. Because it is impossible to separate the true diet effects from the averages of the chance errors, the table of the decomposition also contains variability due to the chance errors. However, *because of the way we constructed the mean squares, the piece of EMS(Diets) dues to Error is exactly the same as EMS(Error)*. This fact will turn out to be essential to the logic of the F-test.

c. EMS(Patients) = Error + Patients. The table in the decomposition due to Patients has variability from two sources, Error and Patients. Here, too, the piece due to error is exactly the same as EMS(Error), and equal to the piece due to error in EMS(Diets).

d. EMS(Dietary Control) = Error + Patients + Dietary Control. When we decompose the data, we get the table for Dietary Control by subtracting the Grand Average from the two averages for good and poor control. We want and expect this table to contain variability due to differences in dietary control, and we can't avoid variability from two other sources, Error and Patients. Fortunately, however, these two terms in the EMS are exactly the same as the corresponding two terms in EMS(Patients). This fact will allow us to judge the variability due to dietary control by comparing the mean square for Dietary Control with the mean square for Patients. ■

Finding EMSs and choosing denominators for F-ratios. Fortunately, all the EMSs in this section follow the same one general pattern, so it is possible to give a simple rule for finding them.

Random, Inside Rule for EMSs

The EMS for a factor contains a term for the factor itself, plus a term for each RANDOM, INSIDE factor.

Once we've found the expected mean squares for a design, we can use the idea from part (d) of the last example to construct *F*-ratios:

To find the denominator MS for testing a factor, find the MS whose expected value has exactly the same terms as the EMS for the factor you want to test, apart from the test factor itself. Thus
MS_{Den} is the right denominator for testing factor A if EMS(A) = EMS(Den) + A

Here's the logic of the rule: Numerator and denominator MSs will contain contributions from exactly the same set of factors, except that if the effects of the test factor are nonzero, the numerator MS contains an extra contribution due to that factor. Thus the *F*-ratio tends to be near 1 if the null hypothesis is true, and bigger than one otherwise. In symbols, the *F*-ratio estimates

$$\frac{\text{EMS(Den)} + \text{A}}{\text{EMS(Den)}} = 1 + \frac{\text{A}}{\text{EMS(Den)}}.$$

If there is no A effect, the variability due to A will be zero, which makes the term A in the expression above equal to zero, and the *F*-ratio is just an estimate of one. On the other hand, if there is a large A effect, the variability due to A will be large, and the *F*-ratio will tend to be much larger than one.

EXAMPLE 13.19 EMSs AND F-RATIOS FOR THE DOPAMINE DATA

(a) Use the "random, inside" rule to find the EMSs for the study of diets and dopamine (7.C.6). (b) Then use the rule for F-ratios to check that the new method gives the usual denominators for testing the factors of the SP/RM design.

SOLUTION

a. *Expected mean squares.* Start with a factor diagram, and label the random factors:

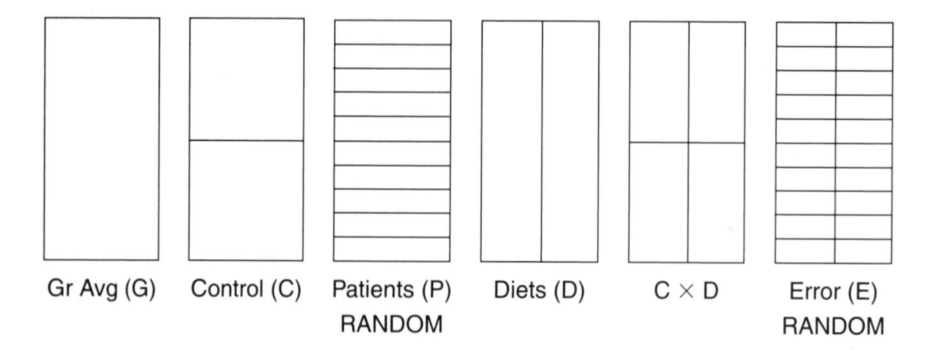

Then go through the factors one at a time, listing the factors that are both inside and random:

Grand average (G): Both Patients (P) and Error (E) are inside and random. EMS(G) = E + P + G

Dietary control (C): Here, too, both Patients and Error are inside and random. EMS(C) = E + P + C

Diets (D): Only Error is inside and random. EMS(D) = E + D

Control × Diet interaction (CD): Only Error is inside and random. EMS(CD) = E + CD

Error: There are no inside, random factors, other than E itself. EMS(E) = E

b. *Denominators for mean squares.* Start with a table of EMSs (Table 13.6).

TABLE 13.6

Source	EMS
Grand Average (G)	E + P + G
Dietary control (C)	E + P + C
Patients (P)	E + P
Diets (D)	E + D
Control × Diets (CD)	E + CD
Error (E)	E

To find the denominator for testing a factor, mentally remove the term for that factor from its EMS, and try to match what's left. For example, to test Dietary

Control (C), remove C from the EMS to get E + P, and look for a factor whose EMS is E + P: Patients gives the right denominator. In the same way, Patients gives the right denominator for the Grand Average. For each of Patients, Diets, and Control × Diets, the EMS contains only two terms, one for the factor in question, and one for Error. The right denominator is Error. ■

EXAMPLE 13.20 FACTORIAL DESIGN WITH ALL FACTORS OF INTEREST FIXED

Use the two rules to find EMSs and choose denominators for a factorial design whose only random factor is Error.

SOLUTION. If the only random factor is Error, and because the Error factor is inside all other factors, the EMS for every main effect or interaction contains just two terms: one for the factor itself, plus a term for Error. Thus the denominator mean square for every *F*-ratio is Error. ■

EXAMPLE 13.21 HIV TESTING

Use the "inside, random" rule to find the EMSs for the HIV test data (13. 6), and find the proper denominators for the *F*-ratios. Summarize the results in an ANOVA table, showing each source, df, EMS and denominator MS.

SOLUTION. As with any purely hierarchical design, the inside/outside relationship is easy, because it puts the factors in order according to group size, from smallest to largest:

$$\text{Samples (S)} \rightarrow \text{Runs (R)} \rightarrow \text{Lots (L)} \rightarrow \text{Grand Average (G)}$$

All three of S, R, and L are random, so the EMS for a factor contains a term for the factor itself, plus terms for all inside factors (Table 13.7).

TABLE 13.7

Source	df	EMS	Denominator MS
Grand Average (G)	1	S + R + L + G	L
Lots (L)	2	S + R + L	R
Runs (R)	3	S + R	S
Samples (S)	6	S	
TOTAL	12		

■

EXAMPLE 13.22 CATS AND TETANUS

Use the "inside, random" rule to find the EMSs for the design in Example 13.9, and find the proper denominators for the *F*-ratios. Summarize the results in an ANOVA table, showing each source, df, EMS and denominator MS.

SOLUTION. (Table 13.8.)

TABLE 13.8 Structure of ANOVA for SP/RM with whole plots grouped

Source	df	EMS	Denominator MS
Grand Ave.	1	Matter + Sides + Cats + Benchmark	Cats
Cats = Blocks	10	Matter + Sides + Cats	Sides
Tetanus/Control	1	Matter + Sides + Tetanus	Sides
Sides (Cats)	10	Matter + Sides	Matter
White/Gray	1	Matter + Color	Matter
Interaction	1	Matter + Inter	Matter
Matter (Cats)	20	Matter	
TOTAL	44		

Notice that apart from the row for Cats, the ANOVA is pretty much the same as for a basic SP/RM with Sides as whole plots. If the 22 sides had not been grouped into cats, there would be no row for Cats, and there would be 20 df for Sides instead of 10 ■.

EXAMPLE 13.23 SPLIT-SPLIT PLOT

Find the EMSs and denominators for the design in Example 13.10, and summarize the results in an ANOVA table.

SOLUTION. (Table 13.9.)

TABLE 13.9 Structure of ANOVA table for a split-split plot design

Source	df	EMS	Denominator MS
Grand Avg	1	Halves + Batches + Hams + Bench	Hamsters
Days	1	Halves + Batches + Hams + Days	Hamsters
Hamsters	6	Halves + Batches + Hams	Batches
Organ	1	Halves + Batches + Org	Batches
Days × Org	1	Halves + Batches + Days × Org	Batches
Batches	6	Halves + Batches	Halves
Methods	1	Halves + Methods	Halves
Days × Methods	1	Halves + Days × Meth	Halves
Organ × Method	1	Halves + Organ × Meth	Halves
Days × Org × Meth	1	Halves + Days × Org × Meth	Halves
Halves = Res	12	Halves	
TOTAL	32		

■

Classifying interactions. Choosing denominator mean squares depends on finding EMSs, and the behavior of EMSs depends in turn on which factors are random. With main effects (basic factors), you can use the four tests from Section 2 to decide, but with interaction effects, there are not just two but three possibilities:

Random = random × random,

Fixed = fixed × fixed,

Mixed = fixed × random.

To handle mixed interaction terms, you need some new ideas and variations on the methods of this section, and so I'll put them off until Section 3. If your design has only fixed or random factors, however, then all the interactions will be of the same type, and you can use the "random, inside" rule to find expected mean squares.

EXAMPLE 13.24 TWO-WAY FACTORIAL DESIGN WITH RANDOM FACTORS
Consider three versions of a factorial design with two crossed factors, A and B:

Version I: A, B both fixed
Version II: A fixed, B random
Version III: A, B both random.

Version II has a mixed interaction, but you can use the "inside, random" rule to analyze the other two. Find all EMSs, and use these to choose denominators for *F*-ratios.
SOLUTION. (Table 13.10.)

TABLE 13.10 EMSs for two versions of a two-way factorial

	Both fixed		Both random	
Source	EMS	Denominator	EMS	Denominator
Gr Avg	E + G	E	E + AB + A + B + G	—
A	E + A	E	E + AB + A	AB
B	E + B	E	E + AB + B	AB
AB	E + AB	E	E + AB	E
Error	E		E	

Both fixed: Tests for A, B, and AB all use MS_{Res}.
Both random: Only the test for AB uses MS_{Res}; tests for A and B both use MS_{AB}. ■

Note that in the last example, when both factors are random, there is no denominator for testing the grand average. Ordinarily, this poses no problem, because we usually know that the overall average will be nonzero, and there will be nothing to learn from testing the hypothesis that it is zero.

As the next example shows, there are other designs involving random factors for which the methods of this section fail to find a suitable denominator.

EXAMPLE 13.25 THREE WAY-FACTORIAL DESIGN WITH ALL FACTORS RANDOM
Consider a factorial design with 3 crossed factors, A, B, and C, all of them random. Find the terms in the EMSs for all factors. Then use the EMSs to choose denominators for the *F*-ratios, and check that there are no denominators suitable for testing any of the main effects.
SOLUTION. (Table 13.11.)

TABLE 13.11 EMSs for a three-way factorial with all factors random

Source	EMS	Denominator
Gr Avg	E + ABC + AB + AC + BC + A + B + C + G	—
A	E + ABC + AB + AC + A	—
B	E + ABC + AB + BC + B	—
C	E + ABC + AC + BC + C	—
AB	E + ABC + AB	ABC
AC	E + ABC + AC	ABC
BC	E + ABC + BC	ABC
ABC	E + ABC	E
Error	E	

There are denominators for testing the two-way interactions (use ABC) and the three-way interaction (use E), but no denominators for testing any of the three main effects. ■

Statisticians have devised a strategy to provide tests when appropriate denominator MSs don't exist. I'll explain how that works in Section 4.

Exercise Set B

1–7. *Fixed versus random.* For each factor listed below, tell whether it is fixed (F) or random (R):

1. Beestings (Example S7.3)
   a. Occasions
   b. Stung versus Fresh
   c. Chance Error
2. Kosslyn's imagery experiment (Example 6.10)
   a. Age
   b. Instructions
   c. Subjects
3. Pigs and antibiotics (Example 6.1)
   a. B12
   b. Antibiotics
   c. Interaction
   d. Chance Error
4. Feeding frogs (Example 8.5)
   a. Frogs
   b. Groupings
5. Emotions and skin potential (Example 8.2)
   a. Subjects
   b. Emotions

6. Dutch elm disease (Exercises 9.A.3 and 9.C.9). All four factors.
7. Effective teachers (Exercise A.14). All factors.

8–10. For each data set listed below, find all EMSs, and tell which denominator MS to use for testing each factor. Assume there are no interactions that involve both fixed and random effects.

8. Emotions and skin potential (Example 8.2).
9. Osmoregulation in worms (Example 9.4)
10. Alpha waves in solitary (Example 8.4)

11–12 For each data set listed below, assume that there are no interactions involving both fixed and random effects.

a. Draw and label a factor diagram.
b. Write the degrees of freedom under each factor.
c. Write F under each fixed factor, and R under each random factor.
d. Construct an "EMS table," using one row for each source (factor), with one column showing EMSs, and another column telling the appropriate denominator to use in the F-test for that factor.

11. Cats and tetanus (Example 13.9)
12. Hamsters (Example 13.10)
13. Hierarchical designs in nursery rhymes.

As I was going to St. Ives,
I met a man with seven wives.
Each wife had seven sacks;
Each sack had seven cats,
Each cat had seven kits.

Analyze the hierarchical structure by listing, in the form of an ANOVA table, Sources, dfs, EMSs, and denominator MSs. Leave out the man: he's not a factor. (Some would say the man should never be a factor.) Assume the effect of Wives is random. (Others would say the effect of wives is always random.) Assume the effects of Sacks, Cats, and Kits are also random.

3. Models with Mixed Interaction Effects

Preview: Restricted and Unrestricted Models for Mixed Interaction Terms

Mixed effects. When a model has both fixed and random main effects, any interaction terms involving factors of both kinds are called **mixed effects.** There are two kinds of mixed effects, called restricted and unrestricted, that correspond to two sets of assumptions about how the mixed interaction terms behave. It will turn out that the unrestricted model is a lot like a random effects model, whereas the restricted model is in some important ways like a fixed effects model. The reason all this matters is that *what we assume about inside factors determines both the meaning and the right denominator for the F-tests for outside factors.*

In what follows I'll first give an example and use it to preview the main differences between the two kinds of models. Next, I'll present the details of the unrestricted model and then the restricted model, before ending with a few more examples.

Example 13.26 Gold Teeth, Version 1: A CB[2] with Mixed Interaction

The goal of this study was to measure the effects of various factors on the hardness of gold fillings. The version I'm about to describe is simplified from a more complicated design.

Fillings need to be hard in order to wear well, and there are various ways to compact a gold filling to change its crystal structure and make it harder. One factor in this study corresponds to three standard ways to do this:

Method 1: Condensing. The dentist uses a special hand tool (a condenser) to pack the gold into the cavity.

Method 2: Hand-malleting. The dentist holds the condenser in place and an assistant taps it with a small hammer.

Method 3: Mechanical malleting is like Method 2, except the hammer is built into the condenser, and the tapping is done by machine.

A second factor corresponds to five dentists chosen from the UCLA School of Dentistry. The two factors were crossed: each dentist used each of the three methods to pack gold into a small cavity drilled in a block of ivory. The hardness was then measured by pushing a pyramid-shaped diamond into the filling and recording the size of the indentation. To keep things simple, assume that Methods and Dentists are the only two factors, and that each dentist filled six cavities, using each method twice. Thus the design is a two-way factorial, with two observations per cell.

Methods is a fixed factor: we are interested in comparing only the three methods actually present in the design. There is no larger population of possible methods we might have chosen. Dentists, on the other hand, can be regarded as a random factor: the five dentists in the study were chosen to represent a much larger population of highly trained dental specialists. The Dentists × Methods interaction effects are mixed, because the interaction involves both fixed and random factors. ■

Restricted versus unrestricted models. There are two commonly used models for mixed effects like the Dentists × Methods interaction. Table 13.12 summarizes the main differences.

TABLE 13.12 Restricted and unrestricted models for mixed effects

	Restricted	Unrestricted
Model		
For each dentist, the interaction terms are	random, but restricted to add to zero across Methods	random and unrestricted
	negatively correlated	uncorrelated
EMSs and denominators		
EMS(Dentist) =	Dentists + Error	Dentists + Inter + Error
Denominator MS for testing Dentist effects is	Error	Interaction
Interpretation		
F-ratio for Dentists tests the null hypothesis that	"true" Dentist averages are equal	for each dentist, the observed response values are uncorrelated

The two models differ in the way the interaction terms are defined. In the restricted model, the terms for each dentist add to zero across methods; in the unrestricted model they do not. The two models lead to different EMSs and *F*-ratios for Dentists, which correspond to different hypotheses being tested.

I'll illustrate the two models using the dental gold example. To keep diagrams and numerical examples simple, imagine a version of the study with just three dentists and two methods. As always, our models assume that each observation is a sum of terms, one for each factor in the design.

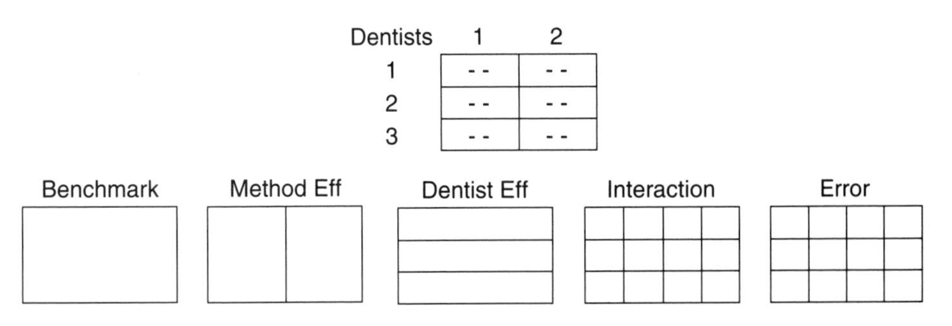

FIGURE 13.13 Factor diagram for the simplest version of the dental gold study. Methods is a fixed factor, Dentists is random, and the D × M interaction is mixed.

The assumptions for Benchmark, Methods, Dentists, and Error are the same for both the restricted and unrestricted models: The Benchmark and the effects for the fixed factor, Methods, are unknown constants, and the Method effects sum to zero over levels of the factor. The chance errors are assumed to behave like random draws from a box whose numbered tickets add to zero and follow a normal curve. Similarly, the random effects for Dentists are assumed to behave like random draws from another box, one whose tickets also add to zero and follow a normal curve, but with a different standard deviation.

The two models differ in what they assume about the mixed Dentist × Method interaction terms.

The Unrestricted Model for Mixed Interaction Terms

Box model for unrestricted interaction effects. Under the unrestricted model, any mixed interaction terms are assumed to behave like any other set of random effects: like random draws from yet another box whose tickets add to zero and follow a normal curve. This box has its own standard deviation, but otherwise is just like the box for the random Dentist effects. Figure 13.14 shows a numerical example. (I've left out the chance errors, and I've made minor adjustments to make the arithmetic cleaner, but apart from that, I got the data using the model I've just described.)

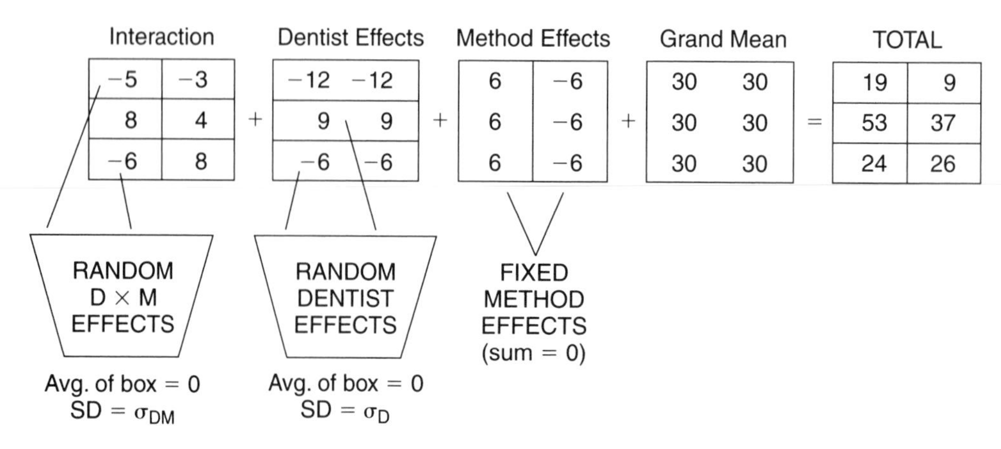

FIGURE 13.14 Numerical example: Unrestricted interaction effects. Note that neither the random Dentist effects nor the unrestricted, mixed Interaction effects add to zero the way the corresponding fixed effects would.

Interpretation of the unrestricted model. We assume the observed hardness of the fillings is a sum of five terms—four "true" values, plus a chance error (not shown). Two of the five terms are constant (the Benchmark = true grand average = 30, and the Method effects of ±6. The chance errors, as always, are assumed to behave like random draws from a box of numbered tickets. For this model, the same is true of both the other two terms:

Dentists: The box of tickets represents the population of dentist effects, one for each of the dentists who might have been chosen. The dentist effects are deviations from the overall average true hardness reading for the whole population of dentists. Some of these effects are negative, for dentists whose fillings give lower hardness than the average for the population; others are positive. Over the whole population, the positive and negative effects add to zero. Choosing three dentists corresponds to choosing three tickets. Once the tickets are chosen, the numbers tell the true dentist effects for those three dentists. In the numerical example, the Dentist effects are -12, 9 and -6. Although the draws come from a box whose average is zero, the draws themselves do not add to zero, and won't in general, except as a fluke.

D × M interaction: Our model includes an overall true effect for Method, and a Dentist effect for each individual dentist, but we also assume that some dentists do better with one method, and other dentists do better with another. To account for these differences, our model includes an interaction term for each combination of dentist and method. In the unrestricted model, we assume these terms, like the random Dentist effects, behave like independent random draws from a box of numbered tickets. In effect, we assume there is a population of possible interaction terms. In the numerical example, the interaction effects for Dentist 1 are -5 and -3. Without interaction, the true hardness readings for Dentist 1 would be $30 + 6 + (-12) = 24$ for Method 1 and $30 + (-6) + (-12) = 12$ for Method 2. Adding in the Interaction effects gives true readings of $24 + (-5) = 19$ for Method 1 and $12 + (-3) = 9$ for Method 2.

Properties of the unrestricted model. The unrestricted model has two properties that are important enough to single out.

1. *Unrestricted random terms are independent.* In the unrestricted model, the random draws from any one box are independent of one another. This means the outcome of one draw has no influence on the outcomes of the other draws, as would be the case if you were always to replace each ticket and mix thoroughly before drawing the next one. In the example, the random effect for Dentist 1 has no influence on the values of the other Dentist effects. Knowing that the Dentist 1 effect equals -12 is of no help in guessing the values of the other effects. (Contrast this with the fixed Method effects: knowing that the first is $+6$ allows you to figure out that the other has to be -6.) Similarly, the unrestricted interaction effects are independent of each other, just as the random effects are.
2. *Unrestricted randomness "leaks out," from inside factors to outside factors.* Effects that are random won't add to zero over the levels actually present. This means, for example, that for Dentist 1, the Method 1 and Method 2 interaction effects (-5 and -3) don't add to zero the way fixed interaction effects do. And that means, in turn, that the true average for Dentist 1 includes a piece equal to the average of the Interaction effects for Dentist 1. Similarly, the true average for each of the other dentists will contain a piece that comes from the average of the interaction effects for that dentist. The randomness of the inside factor (Interaction) has "leaked out" to the true averages for the outside factor (Dentists). In exactly the same way, the randomness of the Dentists "leaks out" to the Grand Mean: The random Dentist effects (-12, 9, -6) don't add to zero over the levels actually present, and the nonzero average of these effects becomes part of the true overall average. (You can see this now in the numerical example, but I'll go over it more carefully in Section 4.)

These two properties of the unrestricted model determine what it is that the *F*-ratio for Dentists is actually testing.

"Dentist effects are zero" means within-dentist correlation is zero. For the fixed effects model, "Dentist effects are zero" means that the true Dentist averages are all equal—any observed differences in the Dentist averages are due to chance error. For the random effects model and the unrestricted mixed effects model, this meaning is no longer correct. "Dentist effects are zero" does *not* mean the there are no differences among the true Dentist averages. Because the random Interaction effects don't add to zero, the true average for Dentist 1 contains a piece due to the average of the Interaction effects, and so can be nonzero even if the Dentist effects are all exactly zero. What, then, does it mean for the effects to be zero?

Figure 13.15 shows a version of the numerical example, with all the true values the same as before, except that the Dentist effects are zero.

Interaction			Dentist Effects			Method Effects			Grand Mean			Total	
−5	−3		0	0		6	−6		30	30		31	21
8	4	+	0	0	+	6	−6	+	30	30	=	44	28
−5	8		0	0		6	−6		30	30		30	32

FIGURE 13.15 Unrestricted model with true Dentist effects zero

Compare the totals for Dentist 1 in this version (31 and 21) with the Dentist 1 totals in the earlier version (19 and 9). Those in the earlier version are lower by 12, because the Dentist effect was −12. The key thing to notice is that when there are Dentist effects present, observations belonging to the same Dentist have a random piece in common—in this case a −12 in common. When the Dentist effects are all zero, observations belonging to the same Dentist have no random pieces in common.

Whenever pairs of observations have a random piece in common, they will be correlated, and one way to see the correlation is by plotting the pairs of points, just as for a within-blocks scatterplot. (Chapter 7, Section 5.) Figure 13.16 shows within-dentist plots for the two versions of the numerical example.

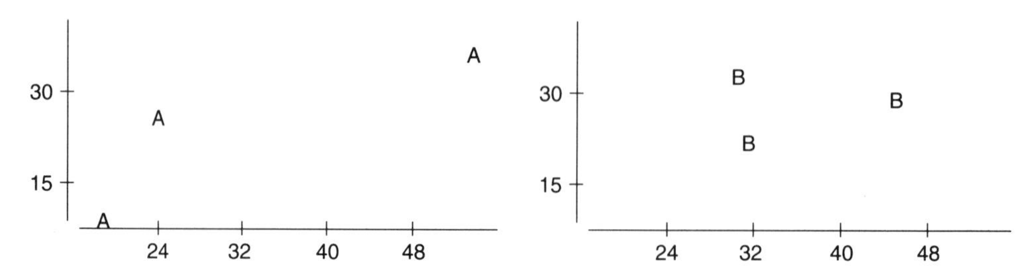

FIGURE 13.16 Within-Dentist plots for data with (left) and without (right) Dentist effects. The points in the left-hand plot, with Dentist effects present, show a positive correlation: they lie near a line that slopes up and to the right. The points in the right-hand plot, with Dentist effects equal to zero, show no correlation: they are scattered about a horizontal line.

The pattern is much clearer with many more points. Figure 13.17 shows plots for 20 data sets of each kind:

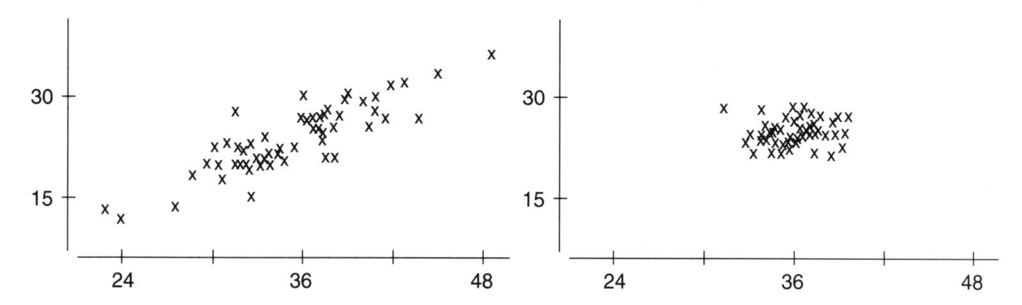

FIGURE 13.17 Plots for 20 data sets with (left) and without (right) Dentist effects. The left-hand plot, which shows a strong positive correlation, is for 20 data sets generated to include random Dentist effects with SD = 5 and random unrestricted interaction effects with SD = 2. The right-hand plot, which shows no correlation, is for 20 data sets generated using the same interaction effects as before, but with SD = 0 for Dentist effects.

There is another useful interpretation for "Dentist effects are zero." (This second interpretation is particularly useful in the context of a class of designs that have the same structure as the simple version of the dental gold study, in which Dentists play the role of blocks. There's more on these designs later in this section.)

"Dentist effects are zero" means Within-dentist variability equals between-dentist variability. If Dentist effects are not zero, then the hardness readings for fillings made by the same dentist will have a random term—the Dentist effect—in common. Readings for the same dentist will tend to be closer together than readings for different dentists. In other words, there is less variability *within* dentists than there is *between* dentists. The extra between-dentist variability is due to the random Dentist effects, and the SD for these effects tells how much extra variability there is. Now suppose the Dentist effects are in fact zero. Then there is *no* extra variability between dentists. This means, for example, that if your goal is to compare methods, it makes no difference whether you use a complete block design with dentists as blocks, or a completely randomized design with dentists as units (as long as you have the same number of fillings).

The Restricted Model for Mixed Interaction Terms

The restrictions. The restricted model for the Dentist × Method interaction terms assumes that for each Dentist, the terms add to zero. More generally, for each level of the random factor, the interaction terms add to zero over the levels of the fixed factor. (If there is more than one factor of each kind in the interaction, then for each combination of levels of the random factors, the interaction terms add to zero over the levels of each fixed factor. As far as the fixed factors are concerned, the mixed terms add to zero in the same ways as in the fixed effects model.)

The "shifted average" version of the restricted model. There are two equivalent box models for restricted interactions. The first uses the same boxes of tickets as the unrestricted model, but shifts and renames the averages of the unrestricted interaction terms to get the restricted terms. To see how this works, look back at the unrestricted interaction terms in the numerical example of Section 3. For the unrestricted model, the terms don't add to zero, but we can split the set of terms for each dentist into an average plus a set of deviations from the average. These deviations, which do add to zero, will be the restricted interaction terms. Figure 13.18 shows the arithmetic for Dentist 1:

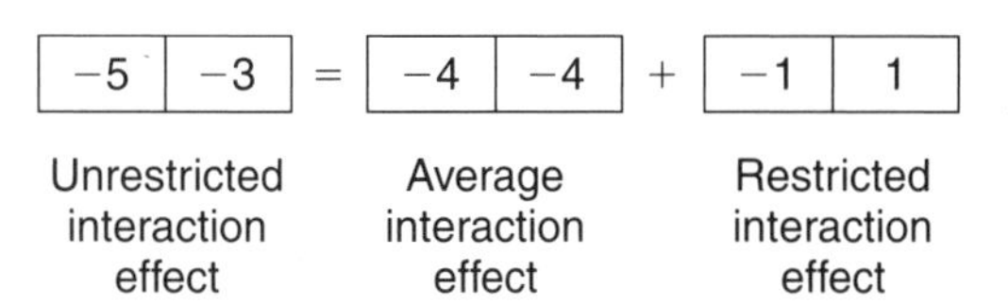

FIGURE 13.18 Unrestricted interactions as average plus deviations. The restricted interaction effects are the deviations of the unrestricted interactions from their average.

To make the models equivalent, we redefine the Dentist 1 effect to include the average unrestricted interaction term (Figure 13.19).

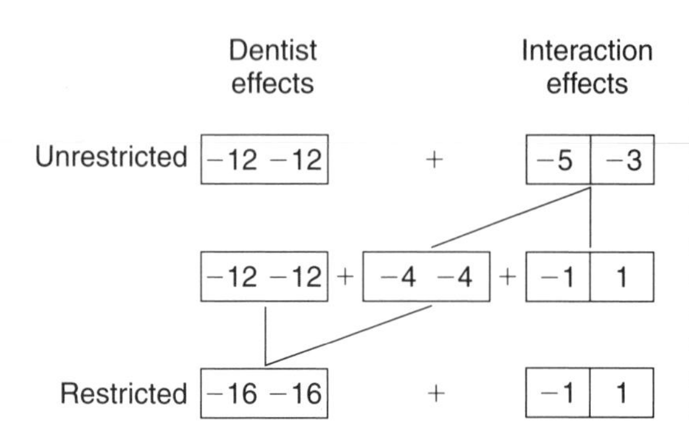

FIGURE 13.19 Shifting the average of the unrestricted interactions. The top row shows the sum of Dentist and Interaction effects for the unrestricted model. In the middle row, the unrestricted interactions have been split into average plus deviations. In the restricted model of the bottom row, the three pieces have been regrouped: the new Dentist effects include the average of the unrestricted Interaction effects, and the new Interaction effects add to zero.

According to this view of the restricted model, all we've done is to rename the pieces, shifting the average of −4 from Interaction effect to Dentist effect (Figure 13.20).

UNRESTRICTED MODEL						RESTRICTED MODEL		
Dentist effect		Interaction effect		SUM		Dentist effect		Interaction effect
−12 −12	+	−5 −3	=	−17 −15	=	−16 −16	+	−1 1
9 9	+	8 4	=	17 13	=	15 15	+	2 −2
−6 −6	+	−6 −8	=	−12 2	=	−5 −5	+	−7 7

FIGURE 13.20 The "Shifted Average" version of the restricted model. In this version of the restricted model, the boxes of tickets are the same as for the unrestricted model, (two left columns), but the averages of the unrestricted interaction effects are "renamed" to be part of the Dentist effects in the restricted model (right two columns).

The "box of deviations" model for restricted interaction effects. There is a second useful way to think about the restricted model. I'll introduce it by way of a comparison with the two models you've already seen. In the numerical example above, the Dentist 1 effect plus the Dentist 1 interactions gave sums of -17 (Method 1) and -15 (Method 2). Here are three different ways to think about getting these totals.

Unrestricted: First, get -12 as a draw from the Dentist box. Then get -5 and -3 as independent draws from the Interaction box, and add each to the -12 to get -17 and -15.

Shifted average (restricted): Start with the same set of draws as for the unrestricted model, but compute the average of the -5 and -3, and add that average of -4 to the -12 to get a Dentists effect of -16. Use the deviations from the average (-1 and 1) as the Interaction effects.

Box of deviations (restricted): First get -16 as a random draw from a box of Dentist effects. Then get the deviations (-1 and 1) as a single draw from a box with an entire "dentist's worth" of deviations written on each ticket (Figure 13.21).

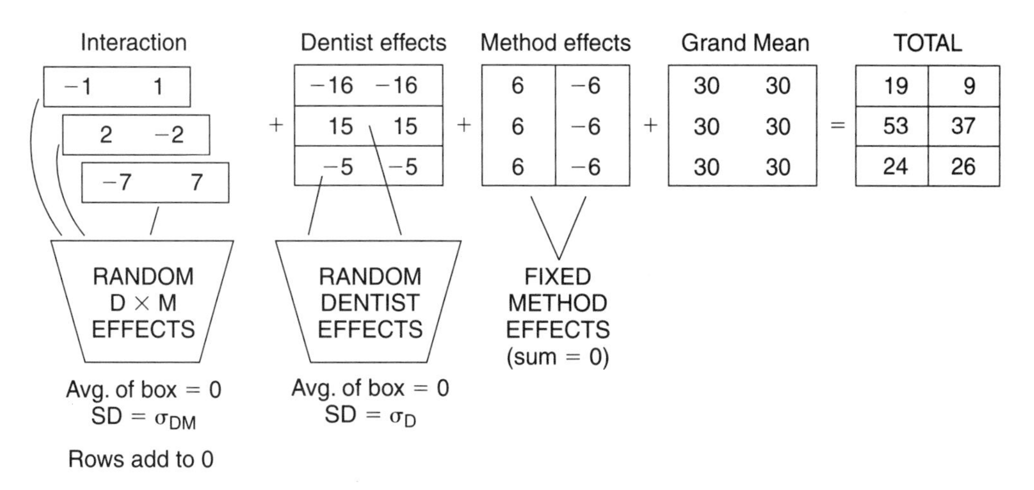

FIGURE 13.21 "Box of Deviations" model for restricted Interaction effects. According to this version of the model, we get all the interaction terms for a dentist with a single draw. Each ticket in this box has one deviation for each level of the fixed factor. The unrestricted, mixed interaction effects add to zero the way the corresponding fixed effects would.

It turns out (you could prove this using probability theory) that for every "shifted average" model, there is an equivalent "box of deviations" model, one whose totals (Dentist effect plus Interaction effects) have exactly the same distribution. You can't always reverse the process though. There are some "box of deviations" models that do not have equivalent "shifted average" versions. For an example, consider a "box of deviations" model with no Dentists effects. For this situation, the within-dentist correlation will be *negative*: hardness readings for fillings made by the same dentists but using different methods will tend to go in opposite directions. A within-dentist scatterplot will show a point cluster that slopes downward.

The "shifted average" model cannot produce negative within-dentist correlations. If the SD for Dentist effects is positive, readings for the same dentist will contain a shared random Dentist effect, and so will have a positive correlation. If the SD for the Dentist effects is zero, then readings for the same dentist will be made up of independent terms, and will have a correlation of zero.

The "limited resources" interpretation of the restricted model. Negative correlations can arise in practice if the response values are linked to some resource whose supply is limited. For a first example, consider Kelly's hamster experiment. Within-blocks scatterplots showed a negative relationship between enzyme concentrations in the heart and brain: the higher the concentration in the brain, the lower the concentration in the heart, and vice versa. This pattern is what you would expect if the hamsters' bodies were shifting either the enzyme itself, or the capacity to produce it, from one organ to another.

For a second example, recall the study of Type A personality and selective attention (Example 8.16). The response was the amount of time a subject spent looking at piles of cards with codes for personality traits written on them. If there were three piles, for positive, negative, and neutral traits, then the more time a subject spent on one pile, the less time there would be for the other two. Within-subject correlations would be negative.

For a third and final example, consider a plant competition study in blocks. Suppose that for each block you put 10 plants, 5 from each of two species, in a styrofoam cup filled with vermiculite and nutrients. Two weeks later, you dry and weigh the plants. Since plant growth depends on the nutrients, which are in limited supply, response values within a cup might well show a negative correlation.

In all these examples, and others like them where negative within-block correlations are a reasonable possibility, the "shifted average" version of the restricted model is not appropriate. You need the "box of deviations" version.

Properties of the restricted model. Both versions of the restricted model have two properties that correspond to those of the unrestricted model:

1. *Restricted random interaction terms are negatively correlated.* Although the draws from the different boxes (Dentists, Interactions, and Errors) are independent, the restricted interaction terms for the same dentist are *not* independent. This is easiest to see if there are only two methods, as in the example above. Then for each dentist the term for one method equals -1 times the term for the other method. Even if there are more than two levels of the fixed factor, the restriction of adding to zero means that if one term is positive, some other term or terms must be negative to compensate.
2. *Restricted randomness only "leaks" into outside factors with the same restrictions as the inside factors.* The restricted interaction terms add to zero within levels of the

random factors, but do not add to zero within levels of the fixed factors. Check this in the numerical example above: Interactions add to zero within rows, i.e., for each dentist, but do not add to zero down columns, i.e., for each method. For Method 1, the interaction terms add to -6, with an average of -2; for Method 2 the sum is 6, and the average is 2. This means that the true average for Method 1 equals the true Method 1 effect (6) plus a -2 from the average of the interaction terms. Check, also, that the random interaction effects only "leak" into the fixed effects, not into the random effects like Dentist, because the interactions add to zero for each dentist.

"Dentist effects are zero" means true Dentist averages are equal. For the restricted model, the interaction effects add to zero for each dentist. This means that the interaction effects do not contribute to the true Dentist averages, the way they do in the unrestricted case. So if the Dentist effects are all equal to zero, the Dentist averages must be equal, just as in the fixed effects model.

Figure 13.22 shows the numerical example, with (true) Dentist effects set to zero. As you can see, the (true) Dentist averages are all equal to the (true) Grand Mean.

Interaction			Dentist effects			Method effects			Grand mean			Total	
−1	1		0	0		6	−6		30	30		35	25
2	−2	+	0	0	+	6	−6	+	30	30	=	38	22
−7	7		0	0		6	−6		30	30		29	31

FIGURE 13.22 Restricted model: Dentist effects zero means Dentist averages equal.

The unrestricted and restricted models serve as two useful ways for thinking about how to analyze a great variety of designs and data sets, as the three following examples illustrate.

Generalized Complete Block Designs and Other Examples

The dental gold study of Example 13.26 illustrates an important design that generalizes the complete block designs of Chapter 7.

Example 13.27 Generalized Complete Block (GCB) Designs

In the CB design, each block provides one and only one unit for each treatment or treatment combination. In those designs, the partitions for interaction and for error are one and the same, which means that it is not possible to separate the effects of interaction and chance error. A natural way around this problem is to use a **generalized complete block design**, or **GCB**. In a GCB, each treatment or treatment combination occurs more than once in each block. In Example 13.26, each dentist is a block, and the methods are the treatments. Each dentist makes two fillings with each method, which makes it possible to get separate estimates for the Dentist × Method interaction effects and chance errors. For another example, consider the finger tapping experiment (7.14), which compares the effects of caffeine, theobromine, and a placebo, using subjects as blocks and time slots as units. For this experiment, each subject provided three units, one per treatment, and the design was a CB. In principle, at least, the experiment could have been run as a GCB instead. If the subjects had been willing to extend their commitment over twice as many days, each one could have provided six time slots, two each for placebo, caffeine, and theobromine. You'd assign treatments to time slots, in random order, separately for each of the four subjects.

The idea is easy to extend to other situations where you might otherwise use a CB design. If you get your blocks by reusing subjects or objects, as in the last example, you can use a GCB design provided you can use each subject twice as many times as for a CB. If you get your blocks by grouping similar subjects or objects, as in a matched pairs design, and if you have enough subjects of each kind to make each group twice as big as for a CB, you can use a GCB instead. If you get your blocks by subdividing, as in the wheat example (9.1), you can use a GCB as long as you can apply your treatments to subdivisions that are only half as big as for a CB.

To analyze the data from a GCB design, you compute exactly the same things as you would for a BF design with blocks as a random factor, although the way you interpret the numbers may be somewhat different for the two designs, since you typically think of blocks as a nuisance factor rather than a factor of interest. The factor or factors for block-by-treatment interaction(s) will be the same as the corresponding interaction factor(s) for the BF design, and the factor for chance error will also be the same for both designs. ■

The meaning of "no block effects" for the GCB design. If Treatments are fixed and Blocks are random, then the Block × Treatment interaction is mixed. Both the unrestricted and restricted models for the GCB design regard each observed value as the sum of five terms, two constants (Grand Average plus Treatment effects) and three random terms (Block effect, Block × Treatment Interaction effect, and Error).

Under the unrestricted model, the three random terms are independent (Property 1). Observations for different treatments in the same block share the random effect for that block, and so tend to be more similar than a pair of observations from two different blocks. In other words, within-blocks observations for different treatments are correlated, which makes comparing treatments within blocks less variable than comparing treatments between blocks. If block effects are zero, then observations for different treatments in the same block have no random terms in common, and so are uncorrelated. The variability for comparing treatments within blocks is the same as the variability for comparing treatments between blocks.

More than two factors. So far, we've relied exclusively on a two-factor example. The same ideas apply to designs with more than two factors, although the details get a bit more complicated, and the number of possible models multiplies rapidly.

EXAMPLE 13.28 GOLD TEETH VARIATIONS

The actual study described in Example 13.26 had a third factor, Type of Gold, crossed with Dentists and Methods: each dentist used each method to make fillings using gold of two types. The first, Gold Foil, consisted of tiny cylinders (500 per ounce) of pure gold. The second type, Goldent, consisted of powdered gold, in a gold foil envelope. The actual data are shown in Table 13.13; each observation is the sum of 10 hardness readings.

TABLE 13.13

		Types		
Dentists	Methods	1	2	
1	1	792	824	
	2	772	772	
	3	782	803	
2	1	803	803	
	2	752	772	
	3	715	707	
3	1	715	724	
	2	792	715	
	3	762	606	
4	1	673	946	
	2	657	743	
	3	690	245	← outlier
5	1	634	715	
	2	649	724	
	3	724	627	

For the purpose of this example, assume that Methods is a fixed factor, but that Dentists and Types may be regarded either as fixed or random, depending on whether you are interested in the levels actually present or in a larger population of levels that might have been chosen. In all, there are fourteen possible models. List them.

SOLUTION. (Table 13.14.)

TABLE 13.14 Each of 2–4 has two mixed interactions, and each mixed interaction can be either unrestricted or restricted, so there are four possible combinations:

a. Both unrestricted
b. First unrestricted, second restricted
c. First restricted, second unrestricted
d. Both restricted

Model	Dentists	Methods	Types	Mixed Interactions
1	Fixed	Fixed	Fixed	(none)
2 a–d	Random	Fixed	Fixed	Dentists × Methods
		Fixed		Dentists × Types
3 a–d	Fixed		Random	Dentists × Types
		Fixed		Methods × Types
4 a–d	Random	Fixed	Random	Dentists × Methods
				Methods × Types

Designs with nested factors. When factors are nested instead of crossed, there will be fewer interaction terms. (Remind yourself that if B is nested within A, then B and B × A are completely confounded, and we regard them as a single factor B(A).) This means that as a rule, designs with nested factors will have fewer mixed interactions and so will present fewer choices between unrestricted and restricted models. In other respects, the use of these models involves nothing new.

EXAMPLE 13.29 WATER QUALITY

Identify fixed, random, and mixed terms in the study of water quality (Example 13.13). For any hypotheses whose meaning depends on the choice between the unrestricted and restricted models, give both meanings.

SOLUTION. Lakes, Sites, and Samples are random factors. Only Times is fixed. There are two sets of mixed interaction effects: Lakes × Times, and Sites × Times.

1. Lakes × Times.

   a. Unrestricted: Under the unrestricted model, the true interaction effects for different times at the same lake are independent, and their (nonzero) average will be part of the true average for that lake. "True Lakes effects are zero" means "within-lakes correlation is zero," or, more precisely, observations taken at the same lake but at different times are uncorrelated.

   b. Restricted: Under the restricted model, the true interaction effects for different times at the same lake add to zero over the three times, and the true average response for a lake does not depend on the Lakes × Times interaction effects. "True Lakes effects are zero" means "true averages for the lakes are equal," that is, any observed differences between lake averages are due to chance variation.

2. Sites(Lakes) × Times. Here, also, there are two models. The two corresponding meanings for "true Sites effects are zero" exactly parallel those for Lakes, with Sites in place of Lakes.

So far, we have looked only at what the possible models are, and what kinds of questions they allow you to answer with hypothesis tests. The next section will show you how to choose F-ratios for carrying out the tests.

Exercise Set C

1. *Gold teeth: A simulation.* The aim of this exercise is to develop your intuitive feel for the difference between the unrestricted and restricted models, by asking you to generate a set of true values using coin tosses instead of random draws from a box. Consider a version of the dental gold study like the one at the beginning of Section 3, with three Dentists (D1, D2, D3) and two Methods (M1, M2). We'll ignore chance errors, and assume that each true value equals the sum of four pieces: Benchmark (fixed) + Method effect (fixed) + Dentist effect (random) + Interaction effect (random). Suppose that the true Benchmark value is 30, and the true Method effects are 6 and −6. To get each random Dentist effect, toss three coins, and compute the effect as 3 × (# heads − # tails).

   a. Unrestricted model: To get each unrestricted Interaction, toss three coins and compute the effect as (# heads − # tails). Write a factor diagram with boxes for each of Benchmark, Method effects, Dentist effects, and unrestricted Interaction, showing the "true" effects for each factor.

   b. Restricted model (shifted average version): Now create the corresponding true values for the restricted model: For each dentist, compute the average of the two unrestricted interaction terms. Add this average to the (unrestricted) Dentist effect to get the (restricted) Dentist effect. Subtract the same average from the two unrestricted interactions for that dentist to get the restricted interactions. Write a new factor diagram showing the true effects under the restricted model.

   c. Comparison: One simple way to compare the two models is to compute a sum of squares for each box of true effects. Compute SSs for the true Dentist and Interaction effects under each model, and write a sentence describing the difference between the two sets of SSs.

2. *Gold teeth variations.* (Refer to Example 13.28.) Consider the three-factor version of the dental gold study, with Methods fixed, Dentists and Types both random. Following Example 13.30 as a guide, identify any hypotheses whose meaning depends on the choice between unrestricted versus restricted, and give both interpretations.

3. *Effective teachers.* Follow the instructions for Exercise 2 above, this time for the study of teachers described in Exercise A.14.

4. *Water quality.* Follow the instructions for Exercise 2 above, this time for the study of water quality described in Example 13.13.

5. *Limited resource models.* Reread the three illustrations of limited resource models on page 580. Invent another example of a situation where it would be reasonable to expect negative within-block correlations.

4. Expected Mean Squares and F-Ratios

Overview

This section comes in five parts. The first presents a rule for finding expected mean squares, and illustrates its use to choose denominators for F-ratios. The second part shows how to use Minitab to analyze balanced designs with any combination of nested and crossed fixed and random factors. The third part gives a way to handle situations like the one in Example 13.24, where there is no mean square with the right expected value to use as the denominator for an F-test. Finally, the last two parts present the logic behind the EMS rule given in the first part.

Once you have learned to find EMSs using the general rule of this section, you can use that rule to choose denominator MSs for all balanced designs. You no longer need to rely on a different set of rules for each design, and you should be able to construct F-ratios and test hypotheses for designs you haven't seen before. In particular, you should be able to find EMSs and F-ratios for all BF designs, regardless of whether the factors of interest are fixed or random, and whether the mixed interactions are restricted or unrestricted.

Finding EMSs and Choosing Denominators

Property 2 for each of the unrestricted and restricted models leads to a simple rule for finding the terms in the EMS for any factor in any balanced design:

The EMS for a factor A contains a term for A itself, plus terms for all random, inside factors whose restrictions, if any, are also restrictions on A.

The last two parts of this section present the logic of this rule. For now, my goal is to show you how to use it.

Constructing EMS diagrams. The rule can be restated as a pair of steps for finding the terms in the EMS for a factor A:

Step 1: Use the factor diagram to list all factors that are both random and inside A.

Step 2: Cross off any factors from the list that are restricted in ways that A itself is not.

There is a useful diagram (the "EMS diagram") for carrying out these steps. The following example shows you how to construct and use one.

EXAMPLE 13.30 GOLD TEETH, BF[2]

Find EMSs and denominators for the following two versions of the dental gold study described in Example 13.26: (a) Dentists random, Methods fixed, Dentists × Methods unrestricted, and (b) Dentists random, Methods fixed, Dentists × Methods restricted. Then (c) tell which F-test differs under the two models.

SOLUTION. Start with a factor diagram (Figure 13.23).

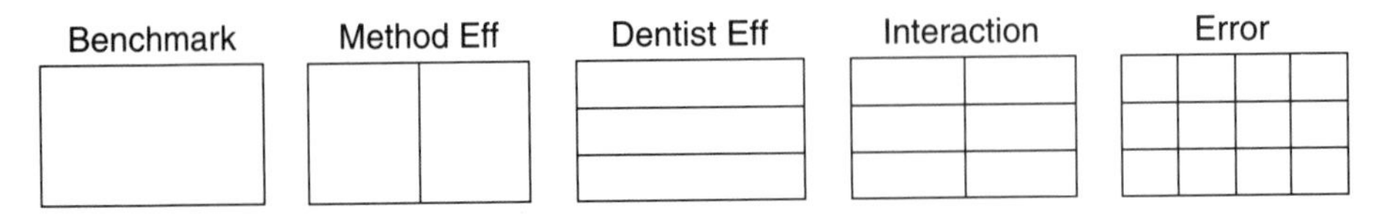

FIGURE 13.23 Factor diagram for the two-way version of the dental gold study

Now replace the factor diagram with an ordered list of (abbreviated) factor names, using arrows to show inside/outside relationships. Circle all random effects (Figure 13.24).

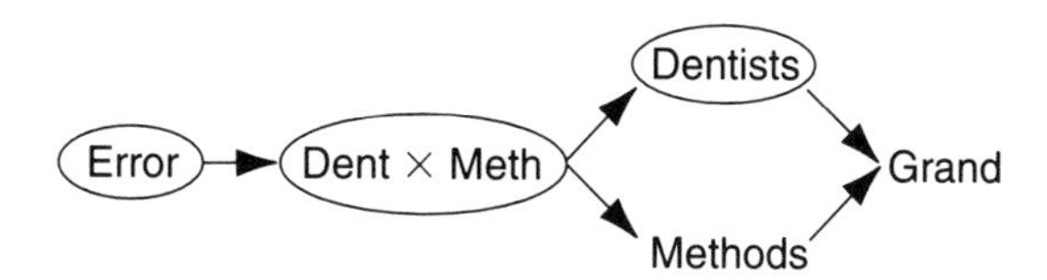

FIGURE 13.24 EMS diagram showing inside/outside relationships and random factors. Random factors are circled. Arrows run from inside factors to outside factors. Randomness (circled factors)"flows" in the direction of the arrows. Thus, reading from left to right, Error contributes to all factors, the Dent × Meth interaction contributes to Dentists, Methods, and the Grand Average, and Dentists contributes to the Grand Average.

a. For the unrestricted model, there are no restrictions, so to find EMSs, you only need Step 1: For each factor A, list all factors that are both random and inside A (Table 13.15).

TABLE 13.15

Source	EMS	Denominator
Grand Average (G)	E + D × M + G	D × M
Dentists (D)	E + D × M + D	D × M
Methods (M)	E + D × M + M	D × M
Interaction (D × M)	E + D × M	E
Error (E)	E	

b. For the restricted model, we can modify the diagram by using rectangles to show restrictions. In this case, any factor that contains Methods is restricted to add to zero over the levels of Methods. So Methods itself is restricted, and the Meth × Dent interaction is restricted with respect to Methods (Figure 13.25).

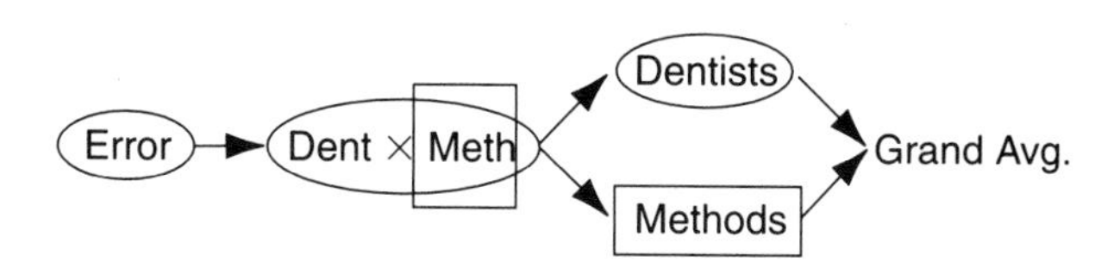

FIGURE 13.25 EMS diagram with boxes to show restrictions. As in the previous EMS diagram for the unrestricted model, randomness (circled factors) "flows" in the direction of the arrows, from inside to outside. With restrictions (boxes) present, restricted randomness can only flow to outside factors that share the restriction(s). Thus Dent × Meth, which is random but restricted with respect to Methods, flows to Methods, which shares the restriction, but not to Dentists or the Grand Average.

Applying Steps 1 and 2 tells us that E is present in all EMSs; D × M is present in itself and in M; D is present in itself and G; M is present only in M itself; and G is present only in itself (Table 13.16).

TABLE 13.16

Source	EMS	Denominator
Grand Average (G)	E + G	E
Dentists (D)	E + D	E
Methods (M)	E + D × M + M	D × M
Interaction (D × M)	E + D × M	E
Error (E)	E	

c. If you compare the two tables of EMSs, you find that only the EMSs and tests for the Grand Average and for Dentists are different under the two models, and ordinarily the test for the Grand Average is of little or no interest. For Dentists, there are two tests, corresponding to the two models:

   Unrestricted: The observed Dentist averages contain a contribution from the averages of the true unrestricted Dentist × Method interaction terms. Thus the mean square for Dentists, and its expected value, contains a contribution due to D × M, and the proper denominator for testing Dentists is the mean square for D × M. The null hypothesis (Dentist effects are zero) is that the Within-dentist correlation is zero, or more precisely, that hardness readings for fillings made by the same dentist but using different methods are uncorrelated.

   Restricted: True interaction effects add to zero within levels of Dentist. The observed Dentist average contains contributions only from the true Dentist effects and the chance errors. The *F*-test for Dentist uses the mean square for Error in the denominator, and tests the null hypothesis that the true Dentist averages are all equal. ■

EXAMPLE 13.31 GOLD TEETH, BF[3]

Find EMSs and denominators for the version of the dental gold study described in Example 13.28, assuming that Methods and Types are fixed and Dentists is random. Do this (a) for the model with both mixed interactions unrestricted and (b) for the model with both mixed interactions restricted. Then (c) compare the two models, as in the example above.

SOLUTION

a. *Unrestricted model*: Here's the EMS diagram. Even though Chance Error and the three-way interaction are completely confounded, I've listed them separately in the diagram (Figure 13.26), in order to see whether any tests cannot use the interaction mean square instead of the error mean square in the denominator.

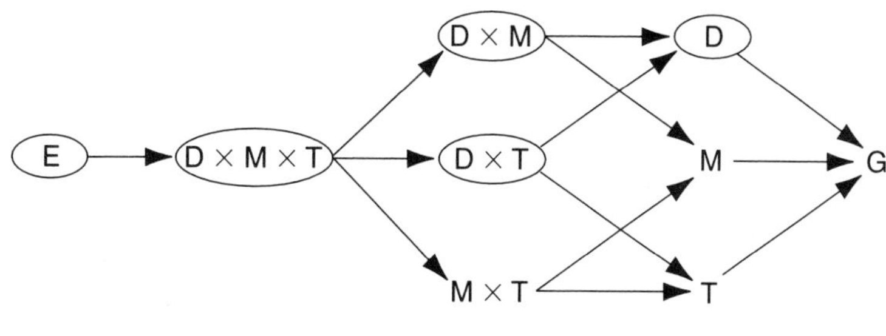

FIGURE 13.26

This diagram leads to the list in Table 13.17.

TABLE 13.17

Source	EMS	Denominator
Grand Average (G)	E + DMT + DM + DT + D + G	D
Dentists (D)	E + DMT + DM + DT + D	—
Methods (M)	E + DMT + DM + M	DM
Types (T)	E + DMT + DT + T	DT
Dent × Meth (DM)	E + DMT + DM	DMT
Dent × Type (DT)	E + DMT + DT	DMT
Meth × Type (MT)	E + DMT + MT	DMT
Dent × Meth × Type (DMT)	E + DMT	E
Error (E)		

Things to notice:

(1) Because DMT and E are completely confounded, there is no mean square for Error, and so no test for DMT. This should come as no surprise. After all, if we can't get separate estimates for DMT and E, we should not expect to be able to use one to test the other.

(2) There is no mean square with the right EMS to serve as a denominator for testing Dentists. This kind of situation requires a special approximate F-test, described later in this section.

(3) For all the other tests, we use the same denominators regardless of whether we assume the three-way interaction DMT is zero, or not. If we don't assume DMT is zero, then the EMSs and denominators are the ones in the table. If we do assume DMT is zero, then DMT drops out of all the EMSs. The EMS for the three-way interaction table is just E, and that mean square gives the right denominator for testing the two-way interactions, even though it is called DMT in the table. For testing M, the right denominator is still DM (since DMT has dropped out of both EMSs). For testing T, the right denominator is DT.

b. *Restricted model*: The EMS diagram is the same as for (a), except that the restrictions are shown as boxes (Figure 13.27).

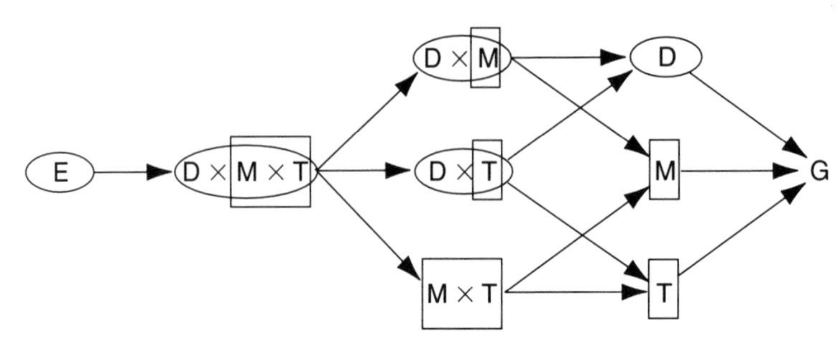

FIGURE 13.27

The restrictions mean that DMT contributes only to itself and MT, because MT is the only outside factor that shares both restrictions. Similarly, DM contributes only to itself and to M; DT contributes only to itself and to T (Table 13.18).

TABLE 13.18

Source	EMS	Denominator
Grand Average (G)	E + D + G	D
Dentists (D)	E + D	E
Methods (M)	E + DM + M	DM
Types (T)	E + DT + T	DT
Dent × Meth (DM)	E + DM	E
Dent × Type (DT)	E + DT	E
Meth × Type (MT)	E + DMT + MT	DMT
Dent × Meth × Type (DMT)	E + DMT	E
Error (E)		

Things to note:

(1) The tests for the fixed effects (Grand Average, Methods, Types, and Methods × Types) use the same denominators as they do under the unrestricted model.

(2) The tests for the random and mixed effects (Dentists, Dentists × Methods, and Dentists × Types) all need MSE for the denominator, but the design does not permit a separate estimate of error variability, and so no tests exist. If, however, we assume the three-way interaction DMT is zero, then the DMT mean square has the proper expected value, and we can use that mean square in place of mean square error in the tests for random and mixed effects.

c. *Comparison*.

(1) If DMT is zero, then the unrestricted and restricted tests are the same for all factors but Dentists. The restricted model uses the DMT mean square as a substitute for the Error mean square, and the F-ratio tests the null hypothesis that the true Dentist averages are all equal. The unrestricted model requires an approximate F-test (see page 596), because no mean square has the right expected value. For this model "no Dentist effects" means that readings for fillings made by the same dentist but using different methods *and* different types of gold will be uncorrelated.

(2) If DMT is not zero, then unrestricted and restricted tests are the same only for the fixed effects. For the random and mixed effects, there are F-tests under the unrestricted model, but no tests (not even approximate ones) under the restricted model. Thus there are no tests that the true averages are equal. Under the unrestricted model, the test for Dentists has the same meaning as above when DMT is assumed to be zero. To find the meaning of the hypothesis tested by the F-ratio for Dentists × Methods under the unrestricted model, note that the expected mean square for that factor is E + DMT + DM. If the null hypothesis is true, and the component due to DM is zero, then the variability in the DM table of the decomposition is due just to Error and the three-way DMT interaction. So observations corresponding to different combinations of DMT will be uncorrelated. The meaning of the hypothesis tested by the F-ratio for Dentists × Types is similar. ■

EXAMPLE 13.32 WHEN CAN YOU TEST FOR BLOCK EFFECTS IN THE COMPLETE BLOCK DESIGN?

The one-way CB design (with just one observation for each block-by-treatment combination) lends itself to ten different sets of assumptions, all for the same factor structure and decomposition. (Blocks can be fixed or random, Treatments can be fixed or random, Interaction can be zero or nonzero, and if it is both nonzero and mixed, it can be either unrestricted or restricted.) In the most common situations, Blocks are random and Treatments are fixed. This still leaves three possible models: Interaction unrestricted, Interaction restricted, and Interaction assumed zero. Find EMSs and denominators for all three models, and compare.

SOLUTION. Start with the EMS diagram for the unrestricted model:

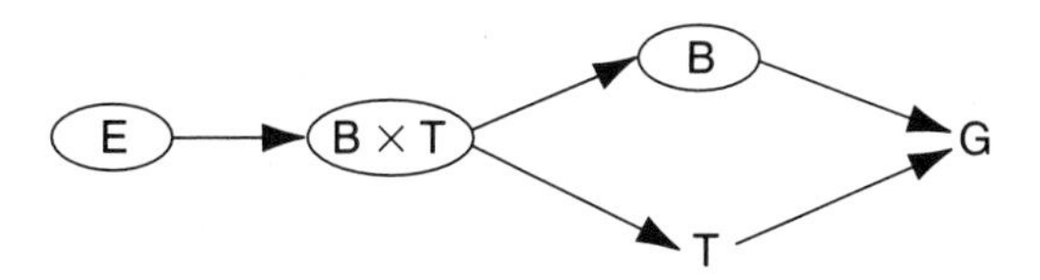

a. *Unrestricted*: Error will be present in all EMSs; Blocks × Treatments in all but the EMS for Error; and Blocks in the EMSs for itself and the Grand Average.

The two fixed factors (Treatments and the Grand Average) will be present only in their own EMSs.

b. *Restricted*: If the Blocks × Treatment interaction is restricted with respect to Treatments, then B × T will be present in the EMS for Treatments, but not in the EMS for Blocks or the Grand Average. So we can get the EMSs for the restricted model from the unrestricted by eliminating BT from the EMSs for Blocks and Grand Average.

c. *Zero*: If the Blocks × Treatment interaction is zero, we can get the EMSs from either model above by eliminating any BT terms from all the EMSs (Table 13.19).

TABLE 13.19

	Unrestricted Interaction		Restricted Interaction		Zero Interaction	
Source	EMS	Denominator	EMS	Denominator	EMS	Denominator
G	E + BT + B + G	B	E + B + G	B	E + B + G	B
B	E + BT + B	BT	E + B	E	E + B	E
T	E + BT + T	BT	E + BT + T	BT	E + T	E
BT	E + BT	E	E + BT	E	E	—
E	E	—	E	—	—	

d. *Comparison*:

(1) There is never a test for B × T interaction. In the first two models there is no estimate of Error, and in the third model, the interaction is assumed to be zero.

(2) All three models use the mean square for B × T as the denominator for testing Treatments, and in all three tests the meaning of the null hypothesis is the same, namely, that the true Treatment averages are all equal.

(3) Tests for Blocks depend on the model. The unrestricted model tests that Within-Block correlations are zero, and uses BT as the denominator mean square. The restricted model cannot test that Block effects are zero (true Block averages are equal) because there is no separate estimate of error variability. For the zero-interaction model, "no Block effects" has both meanings at the same time—true Block averages are equal, and within-block correlation is zero. The test uses the BT mean square as a substitute for mean square error. ■

EXAMPLE 13.33 KELLY'S HAMSTERS: UNRESTRICTED AND RESTRICTED VERSIONS OF THE SP/RM

Our within-blocks scatterplot for Kelly's hamster data showed a negative correlation, which would be consistent with the finite resource version of the restricted model for the Hamster × Organ interaction. Find EMSs and denominators for the hamster experiment assuming this restricted model, and compare the results with those for the unrestricted model and the one that assumes these Interaction effects are zero.

SOLUTION. Here's the EMS diagram for the restricted model; for the unrestricted model, just ignore the boxes. In the diagram, D stands for Day Length, O for Organ, and H(D) for Hamsters within Days.

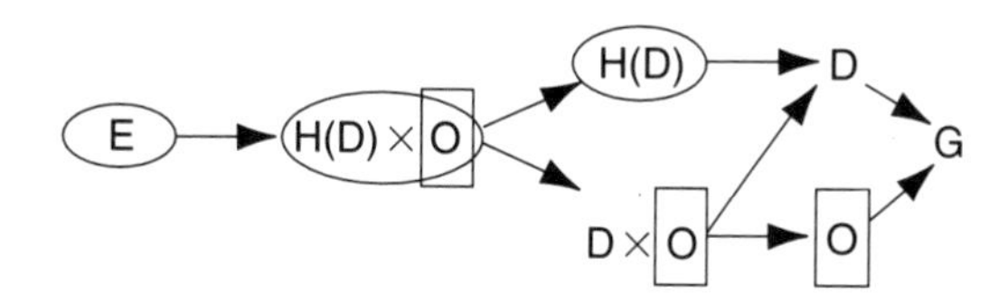

In Table 13.20, I've omitted the parentheses for nesting in order to keep things less cluttered. In this shortened notation, Hamsters within Day Length is just H, and the Hamster × Organ interaction is just HO.

TABLE 13.20

	Unrestricted Interaction		Restricted Interaction		Zero Interaction	
Source	EMS	Denominator	EMS	Denominator	EMS	Denominator
G	E + HO + H + G	H	E + H + G	H	E + H + G	H
D	E + HO + H + O	H	E + H + D	H	E + H + D	H
H	E + HO + H	HO	E + H	E	E + H	E
O	E + HO + O	HO	E + HO + O	HO	E + O	E
DO	E + HO + DO	HO	E + HO + DO	HO	E + DO	E
HO	E + HO	E	E + HO	E	E	—
E	E	—	E	—	E	—

Comparison:

(1) There is never a test for the Hamsters × Organ interaction. In the first two models there is no estimate of Error, and in the third model, the interaction is assumed to be zero.

(2) All three models use the same denominator mean squares for testing Grand Average (Hamsters), Day Length (Hamsters), Organ (Hamsters × Organ) and Day Length × Organ (Hamsters × Organ). Only the denominator for testing the Hamsters effects depends on the model. The first and third models use Hamsters × Organ for the denominator; the second (restricted) model does not permit a test that Hamster effects are zero. Thus it is possible to test that the within-hamster correlation is zero (unrestricted model), but without the additional assumption that the Hamster × Organ interaction is zero, it is not possible to test that the Hamster averages are equal (restricted model). ■

EXAMPLE 13.34 REMEMBERING WORDS: COMPOUND WITHIN-BLOCKS FACTORS
For the experiment in Exercise 9.C.1, assume that Subjects is a random factor, and that Abstraction and Frequency are fixed factors. Find all EMSs, and verify that the unrestricted and restricted models lead to the same tests for all the fixed effects, but different tests for all the random or mixed effects.
SOLUTION. (Figure 13.28, Table 13.21.)

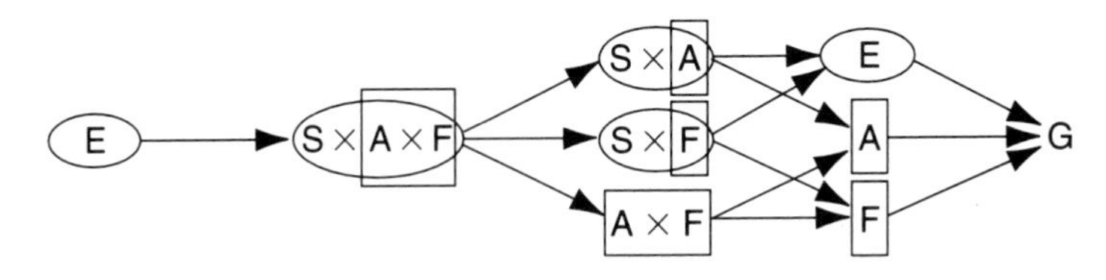

FIGURE 13.27

TABLE 13.21

Source	Unrestricted Interaction EMS	Unrestricted Interaction Denominator	Restricted Interaction EMS	Restricted Interaction Denominator
Gr Avg (G)	E + SAF + SA + SF + S + G	S	E + S + G	S
Subj (S)	E + SAF + SA + SF + S	—	E + S	E
Abstr (A)	E + SAF + SA + A	SA	E + SA + A	SA
Freq (F)	E + SAF + SF + F	SF	E + SF + F	SF
Subj × Abstr (SA)	E + SAF + SA	SAF	E + SA	E
Subj × Freq (SF)	E + SAF + SF	SAF	E + SF	E
Abstr × Freq (AF)	E + SAF + AF	SAF	E + SAF + AF	SAF
Sub × Ab × Fr (SAF)	E + SAF	E	E + SAF	E
Error (E)	E	—	E	—

Comparison:

(1) Tests for fixed effects use the same denominators under both models: S for testing the Grand Average, SA for testing Abstraction, SF for testing Frequency, and SAF for testing Abstraction × Frequency.

(2) Under the unrestricted model, there is no test for Subjects, and tests for the two mixed interactions both use SAF in the denominator. Under the restricted model, tests for Subjects and both mixed interactions require the mean square for error in the denominator, but error is completely confounded with the three-way interaction SAF and cannot be independently estimated. There is no solution to the confounding problem other than a new study using a different design, and so Subjects and the mixed interactions cannot be tested under the restricted model. For testing Subjects under the unrestricted model, however, there is an approximate test, described shortly. ■

Using Minitab

Computer packages. Before using any computer package to analyze data from a design with mixed effects, it is important to find out how to specify whether to regard the mixed effects as restricted or unrestricted, and which choice the package will make for you if you don't specify. You can find a helpful summary comparison of three major packages (BMDP, SAS, and SPSS) in Schwartz (1993).

A reminder. Remember that none of your choices between fixed and random, unrestricted and restricted have any effect at all on the decomposition of your data. The sums of squares, degrees of freedom, and mean squares will be the same regardless of your choices. This means that without learning any new computer commands, you can use whatever method you have already been using to compute almost your entire ANOVA table. The only arithmetic for the table that does depend on your choices is the column of F-ratios, and these are easy to do by calculator once you have the mean squares. On the other hand, it's not hard to get Minitab to compute these for you.

Minitab's ANOVA command. To analyze a balanced design with random effects, choose `ANOVA>Balanced ANOVA` from the `Stat` menu, specify the response and list the structural factors in the model, then list the random terms. To fit the restricted model, choose `Options` and click the box that says `Use restricted form of the mixed model`. If your model contains mixed effects and you don't specify that you want the restricted model, Minitab will assume you want F-tests for the unrestricted model.

EXAMPLE 13.35 A MINITAB ANALYSIS OF THE GOLD TEETH DATA

For the data of Example 13.28, substitute an estimated replacement value of 792 for the outlying value of 245 for Dentist 4, Method 3, Type 2 (but for the purpose of this exercise don't bother adjusting the degrees of freedom). Then use Minitab to carry out the restricted and unrestricted ANOVA with Dentists random, Methods and Types fixed.

SOLUTION. The model contains all combinations of the three factors except the three-way interaction, which is assumed to be zero. Here's how to fit the model with restricted interaction terms:

```
Stat > ANOVA > Balanced ANOVA...
Response: Hardness
Model:    Dentist|Method|Type - Dentist*Method*Type
Random terms:   Dentist
Options: Use restricted form of the mixed model
```

After listing the factors and their levels, Minitab prints the following ANOVA table (Table 13.22).

TABLE 13.22

Source	DF	SS	MS	F	P
Dentist	4	43302	10825	6.19	0.014
Method	2	9193	4597	2.06	0.190
Type	1	4344	4344	0.43	0.548
Dentist*Method	8	17859	2232	1.28	0.369
Dentist*Type	4	40511	10128	5.79	0.017
Method*Type	2	14244	7122	4.07	0.060
Error	8	14000	1750		
Total	29	143453			

To fit the unrestricted model, the choices and commands are the same, except that you don't click the box for the restricted model. For this data set and design, the resulting ANOVA table will also be exactly the same, with one exception: There is no F-test for Dentists under the unrestricted model. There is, however, an approximate F-test, described next. ■

Pseudo F Ratios

The problem: no mean square with the right expected value. In several of the previous examples, there was no mean square to use in the denominator of the F-test for one of the factors in the design, because no mean square had the right EMS. In Example 13.24, the BF[2] with both factors random, there was no denominator for testing the Grand Average. In Example 13.31, the three-way dental gold study with Dentists random, Methods and Types fixed, there was no denominator for testing Dentists under the unrestricted model. In Example 13.34, the CB[2] study of word recall with Subjects (= Blocks) random, Abstraction and Frequency fixed, there was no mean square for testing Subjects under the unrestricted model. In these and similar situations, it is possible to construct approximate tests by combining mean squares.

The solution: Cochran's modification of Satterthwaite's method. I'll use the dental gold example to show you how to construct what is called a "pseudo F-ratio." Here are four EMSs from the unrestricted model for the three-way version of the study in Example 13.31.

Source	EMS
D	E + DMT + DM + DT + D
DM	E + DMT + DM
DT	E + DMT + DT
DMT	E + DMT

For testing Dentists, we need a denominator mean square with EMS = E + DMT + DM + DT (all the terms in EMS(Dentists) except for D itself), but there is no such mean square. Our approximate solution was invented in two stages, the first due to Satterthwaite (1946), and the second to Cochran (1951). Satterthwaite's idea was to combine mean squares, by adding and subtracting, to get a "pseudo mean square" with the right expected value. In our example, $MS_{DM} + MS_{DT} - MS_{DMT}$ has the right expected value:

$$(E + DMT + DM) + (E + DMT + DT) - (E + DMT) = E + DMT + DM + DT.$$

This suggests forming the ratio

$$\text{“}F_S\text{”} = \frac{MS_D}{MS_{DM} + MS_{DT} - MS_{DMT}}.$$

(The quotation marks indicate that the ratio is not really an F-ratio, and the S is for Satterthwaite.) Although Satterthwaite's procedure works reasonably well some of the time, there are also times when it doesn't work very well. For example, if MS_{DMT} is larger than $MS_{DM} + MS_{DT}$, the denominator of the ratio will be negative. Cochran's modified strategy, one that generally works better, is to get rid of any mean squares subtracted in the denominator by adding them to the numerator instead. In our example, since MS_{DMT} appears with a negative sign in the denominator, we add MS_{DMT} to both denominator and numerator to get

$$\text{“}F_C\text{”} = \frac{MS_D + MS_{DMT}}{MS_{DM} + MS_{DT}}$$

To check that Cochran's method gives a test for Dentists, check that the sum of the EMSs for the numerator contains exactly the same terms as the sum of the EMSs for the denominator, except that the numerator also contains a term for Dentists:

$$\begin{aligned} \text{Numerator: } \quad EMS(D) + EMS(DMT) &= (E + DMT + DM + DT + D) + (E + DMT) \\ &= (2 \cdot E + 2 \cdot DMT + DM + DT) + D \\ \text{Denominator: } EMS(DM) + EMS(DT) &= (E + DMT + DM) + (E + DMT + DT) \\ &= 2 \cdot E + 2 \cdot DMT + DM + DT \end{aligned}$$

If the Dentist effects are zero, then numerator and denominator will be estimates of the same quantity, and the ratio will tend to be near one. If Dentist effects are nonzero, the numerator will tend to be larger than the denominator, and the ratio will tend to be larger than one.

Approximate critical values for pseudo F-ratios. Ratios whose numerators and denominators are sums of non-overlapping sets of mean squares, for which the corresponding sums of EMSs are equal except for an extra term in the numerator, are called **pseudo F-ratios**. Pseudo F-ratios do not follow the same distributions as F-ratios do, but Satterthwaite and Cochran found an approximation that usually works well.

Pseudo F-ratio for testing factor A:

$$\text{“}F_C\text{”} = \frac{(MS_1 + MS_2)}{(MS_3 + MS_4)}$$

where $(EMS_1 + EMS_2) = (EMS_3 + EMS_4) + A$

Pseudo degrees of freedom:

$$\text{“}df_{Num}\text{”} = \frac{(MS_1 + MS_2)^2}{(MS_1^2/df_1) + (MS_2^2/df_2)}$$

For “df_{Den}” use subscripts 3 and 4 in place of 1 and 2. If you are using tables instead of a computer, round each “df” down to the nearest whole number less than your computed value.

Approximation: If the pseudo “df” are not too small, then critical values for “F_C” will be close to the critical values for an F-distribution with numerator and denominator df given by the rule above.

EXAMPLE 13.36 PSEUDO F-RATIOS FOR THE GOLD TEETH DATA

After estimating a replacement value for the outlier in the data of Example 13.28, a decomposition gives the following:

Source	df	MS
D	4	10,825
DM	8	2,232
DT	4	10,128
DMT	8	1,750

Construct a pseudo F-ratio for testing the null hypothesis that the Dentist effects under the unrestricted model are zero.

SOLUTION.

$$\text{“}F_C\text{”} = \frac{MS_D + MS_{DMT}}{MS_{DM} + MS_{DT}} = \frac{10825 + 1750}{2232 + 10128} = 1.02$$

$$\text{“}df_{Num}\text{”} = \frac{(MS_D + MS_{DMT})^2}{(MS_D{}^2/df_D) + (MS_{DMT}{}^2/df_{DMT})} = \frac{(10825 + 1750)^2}{(10825^2/4) + (1750^2/8)} = 5.3$$

$$\text{“}df_{Den}\text{”} = \frac{(MS_{DM} + MS_{DT})^2}{(MS_{DM}{}^2/df_{DM}) + (MS_{DT}{}^2/df_{DT})} = \frac{(2232 + 10128)^2}{(2232^2/8) + (10128^2/4)} = 5.8$$

Rounding the “df” down to 5 and 5 gives a 5% critical value of 3.45. Since the computed value of “F_C” is barely larger than 1, we have no evidence to that Dentist effects as defined under the unrestricted model are nonzero. (Look back at Example 13.35, and check that the test for Dentist effects under the restricted model *rejects* the null hypothesis quite decisively. An exercise will ask you to explain how the two different test results for Dentist effects aren't really contradicting each other.) ■

The Logic of the EMS Rule, Part I: Double Decomposition Diagrams

Background: deduction and simulation. Perhaps the biggest challenge facing anyone who tries to learn statistical thinking is this: you never get to see the target of your inferences. Although you do get to see the observed values themselves, the process that generates them remains invisible. Unknown true values are not only unknowable, they are in fact imaginary, in the sense that we can only *assume* they exist.

To understand expected mean squares, you have to confront this challenge. Mean squares themselves are things you compute from observed data, but EMSs are properties of an imaginary chance mechanism that we assume gives a workable model for what actually happens. How, then, can you study an imaginary process? One approach relies on deductive logic: Start with a careful abstract definition of the process, and derive its properties using mathematical theory. A second approach relies on simulation: Create examples of the imaginary process, and study their properties empirically. That's the approach I'll take here, to show you where the EMS rule comes from.

Double decomposition diagrams. To follow the explanation, you'll need to learn to read what I call a double-decomposition diagram. I'll start with a simple example.

Example 13.37 Two Observations with the Same Mean

Consider two observations made with conditions held fixed. Our model and factor diagram has only the two universal factors:

Obs		Gr Avg		Error
	=		+	

Now we'll invent "true" values: Suppose the 'true' Grand Average is 6 and the "true" Chance Errors are 4 and -2. We add these together to get the observed values:

Gr Avg		Error		Obs
6	+	4	=	10
6		−2		4

In practice, of course, we don't get to see the pieces that are added together to give the true values. Instead, we try to estimate the true values by decomposing the observations:

Obs		Gr Avg		Error
10	=	7	+	3
4		7		−3

Our estimates don't quite match the true values: We got 7 instead of 6 for the Grand Average, and we got residuals of 3 and -3 instead of the actual Chance Errors of 4 and -2. The double-decomposition diagram (Figure 13.28) shows how this happens:

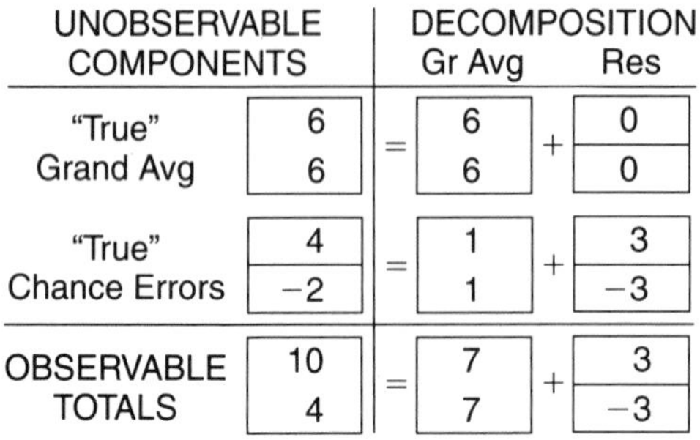

UNOBSERVABLE COMPONENTS			DECOMPOSITION Gr Avg		Res
"True" Grand Avg	6 6	=	6 6	+	0 0
"True" Chance Errors	4 −2	=	1 1	+	3 −3
OBSERVABLE TOTALS	10 4	=	7 7	+	3 −3

FIGURE 13.28 Double-decomposition diagram for a simple data set

The left-hand column shows how the observed values were created, by adding the "true" Grand Average (row 1) and "true" Errors (row 2) to get the observations (bottom left corner). The bottom row, below the line, shows what we actually get to see in a real situation: We start with the Observations (left column), and decompose them into a Grand Average (column 2) plus Residuals (column 3). Thus the left column contains the "true" values, and the bottom row contains the estimates.

The upper right part of the diagram shows how the true values get divided up and added together when we decompose the data. Reading across rows shows how each set of true values gets decomposed:

Row 1: The "true" Grand Average (6) gets decomposed into its overall average (6) plus deviations from that average (0 and 0).

Row 2: The "true" Chance Errors (4 and −2) get decomposed into their overall average (1) plus deviations from that average (3 and −3).

Reading down the columns shows which factors contribute to the tables in the decomposition of the actual observations.

Grand Average (middle column): The observed Grand Average is the sum of two pieces: 6 ("true" Grand Average) plus 1 (average of the "true" Chance Errors).

Residuals (right column): The observed residuals come from adding together 0 (from the "true" Grand Average) plus 3, −3 (from the "true" Chance Errors). Thus, as the diagram shows, the Grand Average does not contribute to the residuals. Only the Chance Errors contribute.

These are precisely the terms that appear in the EMSs:

$$\text{EMS(Gr Avg)} = \text{Gr Avg} + \text{Error}$$
$$\text{EMS(Error)} = \text{Error} \quad \blacksquare$$

EXAMPLE 13.38 A ONE-WAY DESIGN

For a second example, consider a BF[1] design with two treatments and two observations per treatment. We start by inventing "true" values: a Grand Average (10), Treatment Effects (3 and −3) and Chance Errors (−1, 1, −3, and 7). Interpret the double-decomposition diagram that results, and use it to find the EMSs. (See Figure 13.29.)

UNOBSERVABLE COMPONENTS				DECOMPOSITION Gr Avg		Tr Eff		Res
"True" Grand Avg	10 10	=		10 10	+	0 0	+	0 0
	10 10			10 10		0 0		0 0
"True" Treatment Effects	3 −3	=		0 0	+	3 −3	+	0 0
	3 −3			0 0		3 −3		0 0
"True" Chance Errors	−1 −3	=		1 1	+	−1 1	+	−1 −5
	1 7			1 1		−1 1		1 −5
OBSERVABLE TOTALS	12 4	=		11 11	+	2 −2	+	−1 −1
	14 14			11 11		2 −2		1 5

FIGURE 13.29 Double-decomposition diagram for a one-way design

SOLUTION

1. The left-hand column shows how the "true" values are combined to give the observed values (lower left corner).
2. The bottom row shows how the observations get decomposed to give estimates for the true values.
3. The top three rows show how each set of true values would be decomposed. The "true" Grand Average equals itself plus two boxes of zeros, and the "true" Treatment Effects have a zero overall average, and equal themselves plus a box of zeros for the residuals. The Chance Errors, however, which are random, have a nonzero overall average, and also have unequal column averages, so the Chance Errors decompose to give three nonzero tables.
4. The three columns to the right of the dotted line show how the pieces of the true values get recombined to give the estimates. The estimated Grand Average contains contributions from both the "true" Grand Average" and the Chance Errors; the estimated Treatment Effects contain contributions from both the "true" Treatment Effects and the Chance Errors; and the Residuals contain contributions only from the Chance Errors. Thus:

$$\begin{aligned} \text{EMS(Gr Avg)} &= \text{Gr Avg} + \text{Error} \\ \text{EMS(Tr Eff)} &= \text{Tr Eff} + \text{Error} \\ \text{EMS(Error)} &= \text{Error} \end{aligned}$$ ■

With the last two examples as background, we can now use double-decomposition diagrams to compare four models for the simple version of the dental gold study (Example 13.26):

Fixed (Both Dentists and Methods fixed)

Random (Both Dentists and Methods random)

Unrestricted (Dentists random, Methods fixed, Interaction unrestricted)

Restricted (Dentists random, Methods fixed, Interaction restricted)

To make it easier to focus on the differences among the four models, I've omitted the factor for Chance Error, so think of the numbers as showing what would happen if there were no differences between units and no measurement error. In all four examples, I've used a Grand Average of 30. Fixed Method effects are 6 and −6; random effects are 8 and −6. Fixed Dentist effects are −2, 4 and −2; random effects are 7, 4, and −2. Fixed Interaction effects are (by rows) −2 and 2, 3 and −3, −1 and 1; random effects are 7 and −1, 3 and −3, −1 and 1.

EXAMPLE 13.39 FIXED EFFECTS VERSION OF THE STUDY OF DENTAL GOLD

Use the double-decomposition diagram below to find EMSs for the two-way version with Dentists and Methods both fixed. (See Figure 13.30.)

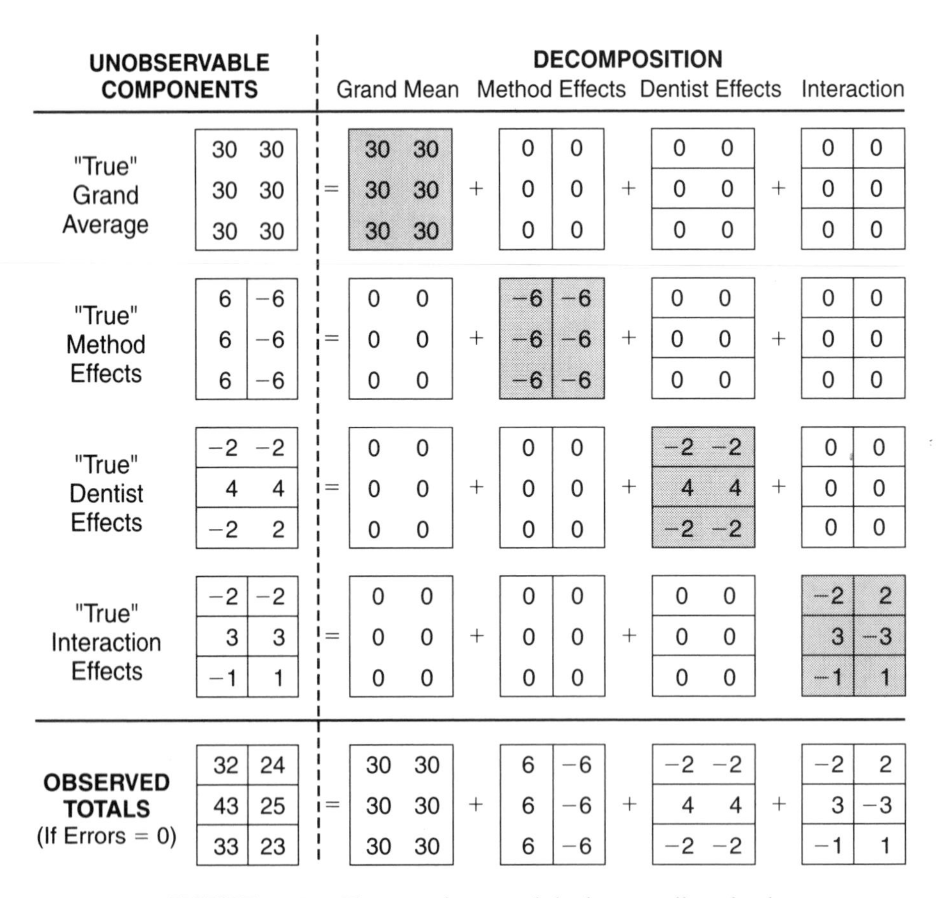

FIGURE 13.30 Two-way design with both main effects fixed

SOLUTION. With all effects fixed, the "true" values add to zero in the same patterns as the estimates do. This means that each set of true values (apart from the Chance

Errors, which are not shown) contributes only to its own counterpart table in the decomposition. In the diagram, nonzero tables in the decompositions of the true values (upper right portion of the diagram) are shaded. To find which sources contribute to a table in the decomposition of the observed values (bottom row), find that table in the decomposition, and "read up" to see which tables above it are shaded. These give the terms in the corresponding EMS:

$$
\begin{aligned}
\text{EMS(Gr Avg)} &= \text{Error} + \text{Gr Avg} \\
\text{EMS(Methods)} &= \text{Error} + \text{Methods} \\
\text{EMS(Dentists)} &= \text{Error} + \text{Dentists} \\
\text{EMS(Interaction)} &= \text{Error} + \text{Interaction} \\
\text{EMS(Error)} &= \text{Error} \quad \blacksquare
\end{aligned}
$$

EXAMPLE 13.40 RANDOM EFFECTS VERSION OF THE STUDY OF DENTAL GOLD

Use the diagram in Figure 13.31 to find EMSs if Dentists and Methods are both random.

UNOBSERVABLE COMPONENTS			DECOMPOSITION: Grand Mean		Method Effects		Dentist Effects		Interaction
"True" Grand Average	30 30		30 30		0 0		0 0		0 0
	30 30	=	30 30	+	0 0	+	0 0	+	0 0
	30 30		30 30		0 0		0 0		0 0
"True" Method Effects	8 −6		1 1		7 −7		0 0		0 0
	8 −6	=	1 1	+	7 −7	+	0 0	+	0 0
	8 −6		1 1		7 −7		0 0		0 0
"True" Dentist Effects	7 7		3 3		0 0		4 4		0 0
	4 4	=	3 3	+	0 0	+	1 1	+	0 0
	−2 2		3 3		0 0		−5 −5		0 0
"True" Interaction Effects	7 −1		1 1		2 −2		2 2		2 −2
	3 −3	=	1 1	+	2 −2	+	−1 −1	+	1 −1
	−1 1		1 1		2 −2		−1 −1		−3 3
OBSERVED TOTALS (If Errors = 0)	52 30		35 35		9 −9		6 6		−2 −2
	45 25	=	35 35	+	9 −9	+	0 0	+	1 −1
	35 23		35 35		9 −9		−6 −6		−3 3

FIGURE 13.31 Two-way design with both main effects random

SOLUTION. Note that the random effects do not add to zero the way the fixed effects do, and that as a result, randomness "leaks out." The "true" values for a random contribute to tables in the decomposition for all outside factors. Here, as in the last example, the nonzero tables are shaded. These give the terms in the EMSs:

$$
\begin{aligned}
\text{EMS(Gr Avg)} &= \text{Error} + \text{Interaction} + \text{Dentists} + \text{Methods} + \text{Gr Avg} \\
\text{EMS(Methods)} &= \text{Error} + \text{Interaction} + \text{Methods} \\
\text{EMS(Dentists)} &= \text{Error} + \text{Interaction} + \text{Dentists} \\
\text{EMS(Interaction)} &= \text{Error} + \text{Interaction} \\
\text{EMS(Error)} &= \text{Error} \ \blacksquare
\end{aligned}
$$

EXAMPLE 13.41 RESTRICTED MIXED VERSION OF THE STUDY OF DENTAL GOLD
Find EMSs if Methods are fixed, Dentists are random, and the mixed Interactions for each Dentist are restricted to add to zero across Methods. (See Figure 13.32.)

	UNOBSERVABLE COMPONENTS			DECOMPOSITION: Grand Mean			Method Effects			Dentist Effects			Interaction	
"True" Grand Average	30	30		30	30		0	0		0	0		0	0
	30	30	=	30	30	+	0	0	+	0	0	+	0	0
	30	30		30	30		0	0		0	0		0	0
"True" Method Effects	6	−6		0	0		6	−6		0	0		0	0
	6	−6	=	0	0	+	6	−6	+	0	0	+	0	0
	6	−6		0	0		6	−6		0	0		0	0
"True" Dentist Effects	7	7		3	3		0	0		4	4		0	0
	4	4	=	3	3	+	0	0	+	1	1	+	0	0
	−2	2		3	3		0	0		−5	−5		0	0
"True" Interaction Effects	7	−7		0	0		3	−3		0	0		4	−4
	3	−3	=	0	0	+	3	−3	+	0	0	+	0	0
	−1	1		0	0		3	−3		0	0		−4	4
OBSERVED TOTALS (If Errors = 0)	50	24		33	33		9	−9		4	4		4	−4
	43	25	=	33	33	+	9	−9	+	1	1	+	0	0
	33	23		33	33		9	−9		−5	−5		−4	4

FIGURE 13.32 Two-way design with Methods fixed, Dentists random, and Interactions restricted

SOLUTION. The first two rows, for Grand Average and Methods, are the same as in the fixed effects version. The third row, for Dentists, is the same as in the random effects version. The fourth row, for Interactions, shows true effects (left column) that add to zero across Methods within each Dentist. These restrictions mean that the Interaction terms do not contribute to the estimated effects for the Grand Average or for Dentists—the two outside factors that do not share the restriction with respect to Methods. The nonzero tables (shaded) give the EMSs:

$$
\begin{aligned}
\text{EMS(Gr Avg)} &= \text{Error} + \text{Dentists} + \text{Gr Avg} \\
\text{EMS(Methods)} &= \text{Error} + \text{Interaction} + \text{Methods} \\
\text{EMS(Dentists)} &= \text{Error} + \text{Dentists} \\
\text{EMS(Interaction)} &= \text{Error} + \text{Interaction} \\
\text{EMS(Error)} &= \text{Error}\ \blacksquare
\end{aligned}
$$

EXAMPLE 13.42 UNRESTRICTED MIXED VERSION OF THE STUDY OF DENTAL GOLD

Find EMSs if Methods are fixed, Dentists are random, and the mixed Interaction terms are unrestricted. (See Figure 13.33.)

UNOBSERVABLE COMPONENTS			DECOMPOSITION: Grand Mean		Method Effects		Dentist Effects		Interaction
"True" Grand Average	30 30 30 30 30 30	=	30 30 30 30 30 30	+	0 0 0 0 0 0	+	0 0 0 0 0 0	+	0 0 0 0 0 0
"True" Method Effects	6 −6 6 −6 6 −6	=	0 0 0 0 0 0	+	6 −6 6 −6 6 −6	+	0 0 0 0 0 0	+	0 0 0 0 0 0
"True" Dentist Effects	7 7 4 4 −2 −2	=	3 3 3 3 3 3	+	0 0 0 0 0 0	+	4 4 1 1 −5 −5	+	0 0 0 0 0 0
"True" Interaction Effects	7 −1 3 −3 −1 1	=	1 1 1 1 1 1	+	2 −2 2 −2 2 −2	+	2 2 −1 −1 −1 −1	+	2 −2 1 −1 −3 3
OBSERVED TOTALS (If Errors = 0)	50 30 43 25 33 23	=	34 34 34 34 34 34		8 −8 8 −8 8 −8	+	6 6 0 0 −6 −6	+	2 −2 1 −1 −3 3

FIGURE 13.33 Two-way design with Methods fixed, Dentists random, and Interactions unrestricted

SOLUTION. The first three rows, for Grand Average, Methods, and Dentists, are the same as in the restricted version. The fourth row, for Interactions, is the same as in the version with both main effects random. The unrestricted random effects "leak out" into all outside factors.

EMS(Gr Avg) = Error + Interaction + Dentists + Gr Avg

EMS(Methods) = Error + Interaction + Methods

EMS(Dentists) = Error + Interaction + Dentists

EMS(Interaction) = Error + Interaction

EMS(Error) = Error ■

The Logic of the EMS Rule, Part II

We can summarize the main idea of the last few examples as follows:

Factor B contributes to the EMS for Factor A if and only if it is possible for changing the "true" B effects to change the estimated A effects.

I won't give a formal proof, but here is an outline of the logic, in two parts. First, suppose there is no way that changing the "true" B effects can change the estimated A effects. Then neither the mean square for A, nor its expected value, can depend on the "true" B effects. On the other hand, suppose that changing the "true" B effects does sometimes change the estimated A effects. Then (for at least some kinds of changes) the mean square for A will be different, and these changes in the mean square will also change its expected value.

Corollary 1: Factor B cannot contribute to EMS(A) if (a) B is outside A or (b) B is crossed with A. Factor B must be inside A.

(a) Illustration: Dentist is outside Dentist × Method, and so Dentist does not contribute to EMS(Dentist × Method). Consider a simple version of the dental gold study, with only two dentists and two methods. Suppose you add 10 to the true effect for Dentist 1. What happens to the estimated interaction effect for Dentist 1, Method 1? All the observations in this cell go up by 10, so the cell average goes up by 10. At the same time, though, all the observations for Dentist 1, Method 2 also go up by the same amount, so the average for Dentist 1 goes up by 10. Thus the increases for the observed average for Dentist 1 and for the observed cell average for Dentist 1, Method 1 are exactly the same, and will cancel when you estimate the interaction effect for that cell. This cancellation will occur no matter what happens to the true effect for the other dentist (Figure 13.34).

	Method 1	Method 2	Avg
Dentist 1	10	10	10
Dentist 2			

FIGURE 13.34 Adding 10 to the true effect for Dentist 1 doesn't change the estimated interaction effect for Dentist × Method, because both cell averages for Dentist 1 go up by the same amount.

(b) Illustration: Dentist is crossed with Method, and so Dentist does not contribute to EMS(Method). Suppose you add 6 to the true effect of Dentist 1 and add four to the true effect for Dentist 2. What happens to the estimated effects for Method? Half the observations for Method 1 go up by 6, and the other half go up by 4. In exactly the same way, half the observations for Method 2 go up by 6, and half go up by 4. Since the observed averages for the two methods go up by exactly the same amount, the distance between the two averages doesn't change, and neither do the estimated effects (Figure 13.35).

	Method 1	Method 2	Avg
Dentist 1	6	6	6
Dentist 2	4	4	4
Avg	5	5	

FIGURE 13.35 Adding 6 to the true effect for Dentist 1 and 4 to the true effect for Dentist 2 doesn't change the estimated effects for Method, because the column averages for the two methods go up by the same amount.

Corollary 2: If B is inside A, and B is restricted with respect to a basic factor that is not part of A, then B cannot contribute to EMS(A).

Illustration: If Dentist × Method is restricted with respect to Method, then Dentist × Method cannot contribute to the expected mean square for Dentist. If Dentist × Method is restricted with respect to Method, then the true interaction effects for Method 1 and Method 2 add to zero for each dentist. Now suppose the true Interaction effects change, and check that the estimated effects for Dentist do not change. For example, suppose that for Dentist 1, the true interactions change by +4 (Dentist 1, Method 1) and −4 (Dentist 1, Method 2), and that for Dentists 2, they change by +6 (Dentist 2, Method 1) and −6 (Dentist 2, Method 2). The averages for the two dentists will be unchanged, and so the estimated effects for Dentists will not change either (Figure 13.36).

	Method 1	Method 2	Avg
Dentist 1	4	−4	0
Dentist 2	6	−6	0

FIGURE 13.36 If true Interaction effects must add to zero across rows, then changing the true Interaction effects cannot change the estimated row effects.

Corollary 3: If B is inside A, and any restrictions on B are also restrictions on A, then B does contribute to EMS(A).

Illustration: If Dentist × Method is restricted with respect to Method, then Dentist × Method does contribute to the expected mean square for Method. Suppose the restrictions and changes are exactly the same as above. What happens to the estimated effects for Method? Half the observations for Method 1 go up by 6, and half go up by 4, so the average for Method 1 goes up by 5. Similarly, half the observations for Method 2 go down by 6, and half down by 4, so the average for Method 2 goes down by 5. The Grand Average doesn't change, and so the estimated effects for Method are changed by exactly the same amount as the column averages: Method 1, up 5; Method 2, down 5. These changes in estimated effects change the mean square for Method, and since the means square for Method depends on the true values for Dentist × Method, the expected values of the mean square does also (Figure 13.37).

	Method 1	Method 2
Dentist 1	4	−4
Dentist 2	6	−6
Avg	5	−5

FIGURE 13.37 If true Interaction effects need not add to zero down columns, then changing the Interaction effects will sometimes change the estimated column effects, and change the MS for Method.

Exercise Set D

1–6. For each data set listed below (a) construct an EMS diagram, (b) construct an EMS table, with one row for each factor in the design, and columns showing for each factor two EMSs and denominator mean squares, one that uses the unrestricted version of all mixed effects and another that uses the restricted version.

1. Water quality (Example 13.13)
2. Dutch elm disease (Exercise 9.C.9)
3. Effective teachers (Exercise A.14)
4. Kosslyn's imagery experiment (Example 9.1b[ii]). Use the full version of Kosslyn's experiment, with two within-subjects factors, Instructions and Stimulus type. Assume nonadditive block effects.
5. Schizophrenia and word association (Example 9.7)
6. Crabgrass, version 3 (Example S9.6)
7. *Dental gold: Explanation for an apparent contradiction*. If we regard Dentists as fixed, Methods and Types random, then under the restricted model (Example 13.35) we reject the hypothesis of no Dentist effects, with p near 0.01, but under the unrestricted model (Example 13.36) a pseudo F-ratio for Dentists is not even close to the 5% critical value. Use the meanings of "Dentist effects are zero" under the two models to explain why the two tests are not contradicting each other. Using both tests together, what do you conclude about the dentists?

8–11. *Pseudo F-ratios*

8. *Two-way factorial with both factors random*. Example 13.24 gives EMSs for a BF[2] with both main effects random, and shows that there is no mean square to use as the denominator of an

F-ratio for testing that the Grand Average is zero. (Although there is usually no reason to test this hypothesis, if your response is the change in some other measurement, such as score after minus score before, then the hypothesis of no overall change corresponds to a Grand Average of zero.) Show that a pseudo F-ratio whose numerator is $MS_{GrAvg} + MS_{AB}$ and whose denominator is $MS_A + MS_B$ is appropriate for testing this hypothesis.

9. *Three-way factorial with all factors random.*

   a. Use the EMSs given in Example 13.25 to decide which mean squares to combine to get a numerator and denominator suitable for testing that the A effects are zero.

   b. Use your answer in (a) to guess which mean squares to use for testing that the B effects are zero, and then check the EMSs to make sure your guess is correct.

10. *Remembering words.* Example 13.34 gives the EMSs for the experiment described in Exercise 9.C.1.

   a. Tell which mean squares to combine to get a pseudo F-ratio for testing Subjects under the unrestricted model.

   b. Example 9.10 gives numerical values of the SSs. Use these to carry out the test in (a).

   c. Now do the F-test for Subjects under the restricted model, and write a sentence summarizing what you conclude from your two tests for Subjects.

11. *Gold teeth variations.* Consider the version of Example 13.28 with Methods fixed, Dentists and Types both random.

   a. Construct an EMS table showing for each factor the EMS and denominator MS assuming the unrestricted model for both mixed effects, and also the EMS and denominator MS assuming the restricted model for both mixed effects. If there is no appropriate denominator mean square, write a *.

   b. For any test in (a) with no suitable mean square, tell how to construct a pseudo F-ratio.

12. *Gold teeth (continuation).* Example 13.35 lists df and MSs for the dental gold study.

   a. Test for Dentists under the restricted and unrestricted models, and tell what you conclude from the results of the two tests taken together.

   b. Test for Methods under both models, and interpret the results.

   c. Test for Types, and interpret.

13–18. *Logic of the rule for EMSs.* The purpose of the next few exercises is to illustrate the ideas you would use to prove that the terms in an EMS correspond to random inside factors with the same restrictions. To make things concrete, we'll work with a BF[2] design with two treatments, A and B, and two observations per treatment:

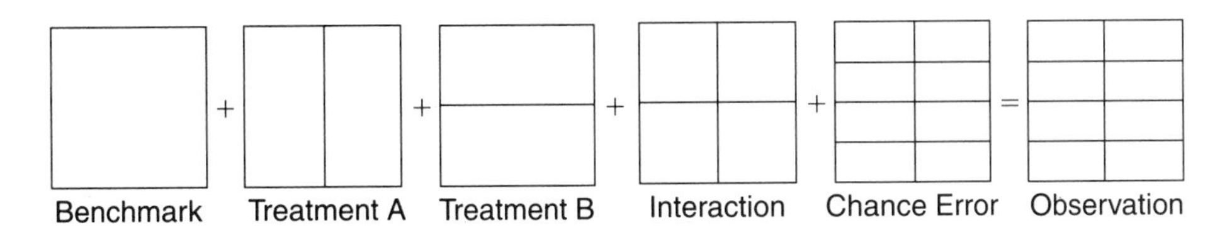

Assume that factor A is fixed, and B is random, and think about the expected mean squares for A and B.

Each of Exercises 13–16 below describes a change in the true values associated with one of the three factors A, B, or A × B. For each, describe how the change affects (a) each of the observed values, (b) each table in the decomposition, and (c) the mean squares for A and B. Example: Add a constant number c to the Benchmark value.

a. Each observed value goes up by c.

b. Gr Avg goes up by c, no other tables of the decomposition change.

c. The MSs for A and B don't change.

13. Fixed treatment effects, which add to zero: For Factor A, add 6 to the first column and −6 to the second column.

14. Random treatment effects, which don't add to zero: For Factor B, add 8 to the first set of rows (top half of the data), and −4 to the second set of rows (bottom half).

15. Restricted mixed interaction effects, which add to zero in the direction of the fixed factor A (across), but not in the direction of the random factor B (down): Add 4 to the upper left cell and −4 to the upper right cell; add −2 to the lower left cell and 2 to the lower right cell.

16. Chance errors (random), which have no patterns of adding to zero: It's enough to think about what happens when you add an error of +8 to the upper left observation, with errors of 0 for all the others. (Of course chance errors don't behave like that, but the key thing here is that every factor is affected.)

17. Use your answers to Exercises 13–16 to complete the following table. For each row, put an X in the columns of the factors for which the MSs change. Then compare the pattern of Xs in each column with the EMS you get using the rule.

True value changed	MSs that change	
	Tr A	Tr B
Benchmark		
13. Tr A (fixed)		
14. Tr B (random)		
15. A × B (mixed)		
16. Error (random)		

18. Unrestricted mixed interaction effects. Repeat Exercises 16 and Exercises 17, this time using unrestricted interactions, which do not add to zero across either rows or columns: Add 4 to the upper left cell and −2 to the upper right cell; add 0 to the lower left cell and 2 to the lower right cell.

19–20. Use Minitab to carry out a careful analysis of each of the data sets listed below. For any mixed effects, fit both the restricted and unrestricted models. Your analysis should end with a summary of your conclusions.

19. Puzzled kids (Exercise 9.F.6)

20. Imagery and working memory (Exercise 9.C.5)

SUMMARY

1. *Four tests for crossing versus nesting:*

   a. **Counting:** How many combinations of levels of A and B appear in the design?

If A and B are crossed, (# combinations of A and B) = (# levels of A) × (# levels of B)

If B is nested within A, (# combinations of A and B) = (# levels of B)

b. **Partition**: (a) If B is nested within A, then each group in the factor for B fits completely inside a group in the factor for A. (b) If you visualize the factors for A and B on top of each other, what you see is the same as for factor B, the inside factor, by itself.

c. **Interaction**: If B is nested within A, then the factors for A × B and B are completely confounded. It is not possible to distinguish the AB interaction from the B effects.

d. **Interchange**: Pick two cells of A × B that belong to the same level of B, and consider switching the data values in the first cell with the data values in the second cell. Will the switch affect the analysis of the data? If B is crossed with A, it will; if B is nested in A, it won't.

2. *Finding the list of factors*. Start by pretending all factors are crossed, and write the list of all possible factors and interactions. Then for each nested pair "amalgamate" B and AB into B(A): cross off every term where B appears without A, and write B(A) in place of AB in every term containing both A and B.

3. *Fixed versus random factors*: A factor is **random** if its levels are drawn at random from a larger population. A factor is **fixed** if its levels correspond to the entire population of interest. Four questions:

a. What do we assume about how the terms in the model behave?
Fixed: Like constants
Random: Like random draws from a box of numbered tickets.

b. What is the target of our inferences?
Fixed: The set of levels actually observed
Random: A much larger set of levels that is represented by the sample of levels actually present.

c. What fraction of the population of interest have we chosen?
Fixed: All
Random: Only a tiny fraction.

d. How were the levels chosen? If we were to repeat the experiment, would the levels change (random)? Or
stay the same as before (fixed)?

4. *Five (imperfect but useful) generalizations*.

a. Units should always be random.

b. Blocks are usually random.

c. Nuisance factors are usually random.

d. Nested (inside) factors are usually random.

e. Factors of interest that are experimental are usually fixed.

5. *Restricted versus unrestricted mixed effects*.

a. An interaction is **mixed** if it involves both fixed and random main effects. Mixed effects are always random, but may be either restricted or unrestricted. **Restricted** effects add to zero over the levels of each fixed factor, and are negatively correlated. **Unrestricted** random terms are independent, and do not add to zero.

b. For a design with a fixed factor ("Treatments") and a random factor ("Blocks"), the meaning of "Block effects are zero" depends on the model.

Restricted: " 'True' block averages are equal," i.e., any observed differences in the block averages are due to chance error. (Under the unrestricted model, each "true" block average will be equal to the (nonzero) average of the unrestricted interaction effects, and so "true" block averages may be different even when block effects are zero.)

Unrestricted: (i) "Within-block correlations are zero," i.e., observations for different treatments within the same block are uncorrelated. (ii) "Within-block variability equals between-block variability," i.e., comparing two treatments in the same block is no more precise than comparing two treatments from different blocks.
Interpretations for other designs are similar.

c. *Two ways to construct the restricted model*:

Shifted average: Under the unrestricted model, the interaction effects behave like independent random draws from a box. For each block, the (nonzero) average of the interaction draws is regarded as interaction, not part of the block effect. Under the restricted model, the average of the unrestricted interaction effects is regarded as part of the block effect.

Deviations: The restricted interactions are defined a block at a time, as a randomly drawn set of deviations. This version corresponds to applications where the response depends on a resource that is in limited supply.

6. *Expected mean squares and denominators for F-tests*.

a. The expected value of a mean square (EMS) tells what the mean square estimates: a sum of squares of SDs, one for each factor that contributes to the variability in the corresponding table of the decomposition.

b. To find the denominator MS for testing a factor, find the MS whose expected value has exactly the same terms as the EMS for the factor you want to test, apart from the test factor itself. Thus MS_{Den} is the right denominator for testing factor A if EMS(A) = EMS(Den) + A.

c. Unrestricted randomness "leaks out" from inside factors to outside factors. Restricted randomness only "leaks" into outside factors with the same restrictions as the inside factors. Thus:

d. The EMS for a factor A contains a term for A itself, plus a term for all random, inside factors whose restriction(s), if any, are also restrictions on A.

NOTES

1. ***Fixed versus random.*** The four tests for fixed versus random skip over some intermediate cases. *Target of inference*: Sometimes the target of our inferences might be the particular set of individuals present in the experiment, but we want to regard these individuals as a sample from a larger population. In such cases the factor in question would be considered random, and the estimates for the individuals of interest would require special methods. *Fraction of population:* Sometimes the fraction is neither one nor tiny, but somewhere in between. The factor is random, just as when the fraction is tiny, but some kinds of estimates require a special correction for a finite population.

2. ***Mixed models.*** Many statisticians refer to models with mixed effects as "mixed models." Unfortunately, this phrase has two quite different meanings. Many psychologists call a design or model "mixed" if it has both within-subject and between-subject factors.

3. ***Other approaches to mixed effects models.*** The treatment of designs with mixed interaction effects in Sections 3 and 4 is somewhat unusual in three ways. First, unlike other books at this level, I present both restricted and unrestricted models; typically an author presents just one kind of model or the other. Second, I don't present the Tukey-Cornfield algorithm for generating expressions for expected mean squares. (See, for example, Ott, 1988, pp. 772–781.) In my opinion, no student learning ANOVA for the first time should spend time on an algorithm as detailed, intricate, and removed from concepts and intuition as this one is, especially these days, when computer packages can do the job. As an alternative, I present a short, simple rule that is both quick to apply and reasonably easy to justify intu-

itively: The EMS for a factor A contains a term for A itself, plus a term for every random, inside factor whose restriction(s), if any, are also restrictions on A. Finally, unlike most authors who treat both restricted and unrestricted models, I do not take the position that you should ordinarily try to decide which model is more nearly correct in a given situation, or even that "restricted versus unrestricted" is a global choice that applies to an entire model. Rather, I agree with what I understand John Tukey to mean when he says, ". . . our focus must be on questions, not models." I see "restricted versus unrestricted" not so much as competing descriptions for a hidden "true" structure, but rather as a useful guide for thinking about ways to answer complementary pairs of questions. Rather than judge one question in the pair "appropriate" and the other "inappropriate," I regard both as potentially worth asking.

Section 3 and 4 present a mix of technique, examples, interpretation and informal exposition of theory. Although some teachers of statistics may disagree with some of the interpretations I present, I hope they will nevertheless find the two sections useful for their exposition of technique and theory, and for the illustrative examples.

REFERENCES

Ott, Lyman (1988). *An Introduction to Statistical Methods*, 3rd. ed. Boston: PWS-Kent Publishing Company.

Schwartz, Carl J. (1993). "The Mixed–Model ANOVA: The Truth, the Computer Packages, The Books. Part I: Balanced Data." *The American Statistician* 47, No. 1, pp. 48–59.

Tukey, John W. (1977). Discussion of J.A. Nelder, "A reformulation of linear models," *J. Royal Statist. Soc.*, Series A 47, p. 72.

CHAPTER 14

CONTINUOUS CARRIERS: A VISUAL APPROACH TO REGRESSION, CORRELATION AND ANALYSIS OF COVARIANCE

INTRODUCTION

Carrier variables, until now, have been categorical. The purpose of this chapter is to give a brief and general overview of the kinds of analyses that are appropriate when your data set has continuous carriers. The methods of this chapter—regression, correlation, and analysis of covariance—are often the subjects of entire courses, and a single chapter can do little more than introduce and illustrate some of the main ideas. All the same, I believe such an introduction is useful, if only as a way to put what you have learned about ANOVA in a larger context. If the only statistical method you know is ANOVA, it is only natural for that to influence the way you think about data. (If all you have is a hammer, everything looks like a nail.) In a sense, this chapter is intended to protect you from developing tunnel vision, and to enable you to think more flexibly about data.

1. REGRESSION

Overview

If both your response and carrier variables are numerical, it is natural to scatterplot response against each carrier. In a sense, this section and the next deal with data sets you can summarize using such plots. However, the two workhorse methods of these sections, regression and correlation, are not suitable for all such data sets: a lot depends on the shape of the plot.

Regression and correlation are suitable for scatterplots whose points lie along a line or form a single oval balloon which lies along a line. Regression fits a line to the points to describe how the response is related to the carrier. Correlation measures the fatness of the balloon to describe how well the fitted line summarizes the relationship.

Introduction to Line-Fitting

Fitting a line to a scatterplot. The scatterplot in Figure 14.1 is for the hamster data, and shows log concentration in the brain versus log concentration in the heart. Apart from the one outlier (omitted from the plot) the points lie near a line. There are several methods for choosing a particular line to summarize the plot: The method of least squares described here has many advantages and is the most common. The simplest method of all is by eye, moving a clear plastic ruler around on the plot until you find a line that you like. For example, for the hamster data a line passing through the points (10,120) and (20,100) comes reasonably close to all the data points (except the outlier), and serves as a workable summary of the relationship between log concentrations in the heart and brain.

EXAMPLE 14.1 HAMSTER DATA: FITTING A LINE BY EYE

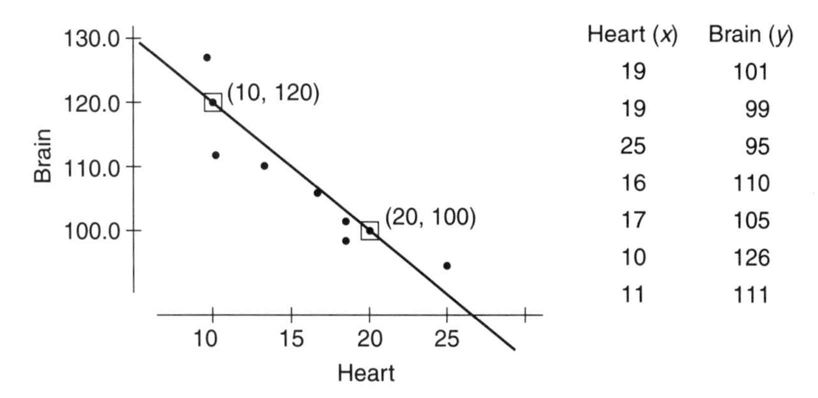

Heart (x)	Brain (y)
19	101
19	99
25	95
16	110
17	105
10	126
11	111

FIGURE 14.1 Scatterplot for the hamster data. The outlier is not shown. A line passing through (10, 120) and (20, 100) gives a workable summary of the observed relationship between brain (y) and heart (x) values. ■

Interpretation: The constant ratio property for lines. If the scatterplot for data pairs (x, y) suggests a line, then the two variables themselves have a particularly simple quantitative relationship.

Constant Ratio Property

For every pair of points on a given line, the ratio "rise over run" or "change in y over change in x" is constant, and equals the slope of the line.

EXAMPLE 14.2 CONSTANT RATIO PROPERTY FOR THE HAMSTER DATA

The fitted line for the hamster data was chosen to pass through the two points (10,120) and (20,100).

	x	y
Point 1	10	120
Point 2	20	100
Change	+10	−20

$$\frac{\text{Change in } y}{\text{Change in } x} = \frac{100 - 200}{20 - 10} = \frac{-20}{10} = -2$$

For any two points on the fitted line, the ratio "change in y over change in x" will equal −2. In other words, the change in brain concentration will be twice as big as the change in heart concentration, but in the opposite direction.

We can use the constant ratio property to find other points on the line, starting with the point (10, 120). For each increase of one unit in heart concentration, there is a decrease of two units in brain concentration. Thus the following points are all on the fitted line: (10, 120), (11, 118), (12, 116), (13, 114), (14, 112) ... (17, 106) ... (19, 102) ... (25, 90). ■

A great many situations, both in science and in everyday life, exhibit the constant ratio property. For example, on any occasion when you buy gas for a car, the price per gallon is a constant ratio: regardless of how many gallons you buy the amount you pay (y) divided by the number of gallons you get (x) is the same. Similarly, the interest paid on a bank account divided by the amount of money in the account is a constant (the interest rate) which doesn't depend on the amount of money in the account. In chemistry, Charles' Law states that if you hold the pressure of a gas fixed, the volume is proportional to the temperature: the ratio of volume to temperature is constant. In all these situations, and many others, a straight line fitted to data pairs gives a good summary.

Regression and ANOVA compared. You can deepen your understanding of how regression works by comparing it with ANOVA. In many ways, regression is like ANOVA with a numerical carrier in place of the factor structure. For ANOVA, the carrier variables define categories; for regression, the carriers are numerical. ANOVA asks "How big a difference in the response is associated with a difference in levels of a factor?" Regression asks, "How big a difference in the response is associated with a difference of any given size in the values of a carrier variable?"

To make the comparison concrete, we'll call on Kelly's hamsters yet again. Think of the enzyme concentrations in the brain as the response, and consider two possible questions: (1) "How big a difference in the brain concentrations is associated with the difference between long and short days?" (2) "How big a difference in brain concentration is associated with a difference of one unit in the heart concentration?" To answer the first question, you would take the categorical variable, day length, as your carrier, and analyze the data using ANOVA. To answer the second question, you would take the numerical variable, heart concentration, as your carrier, and analyze the data using regression. For the first question, you compare group averages by subtracting one from another; for the second you fit a line (or a curve) to a scatterplot, and the steepness of the line tells how the response and carrier values are related.

In many ways, then, regression and ANOVA are quite similar, and are, in fact, two different versions of one general method for analyzing data. (See Exercise A.14.)

EXAMPLE 14.3 HAMSTER DATA: COMPARING ANOVA AND REGRESSION

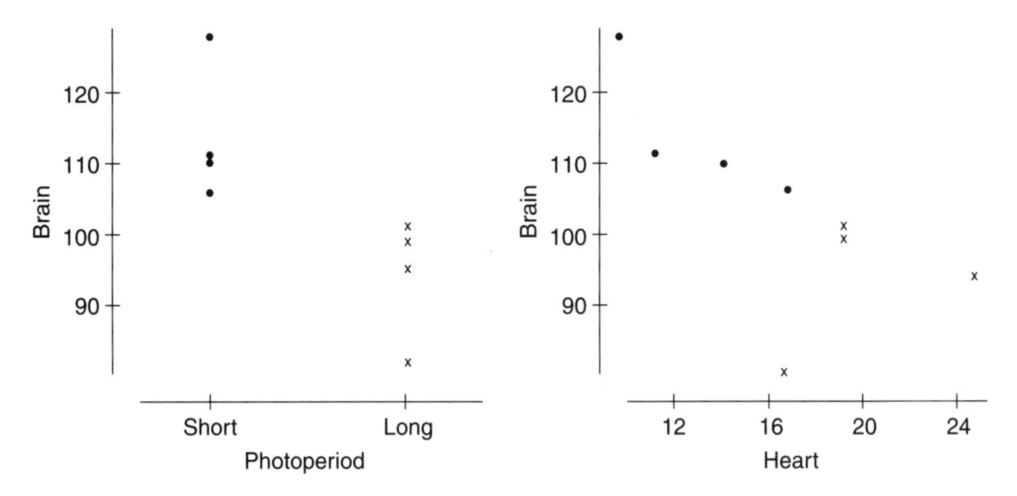

FIGURE 14.2 ANOVA versus Regression for Kelly's hamsters. The parallel dot diagram on the left corresponds to an ANOVA model: the carrier (day length) is categorical, and we compute differences in brain concentration associated with differences in categories. The scatterplot on the right corresponds to a regression model: the carrier (heart concentration) is numerical, and we compute differences in brain concentration associated with differences in heart concentration. ■

Fitted values and residuals: decomposition. Once we have chosen a line to summarize a scatterplot, we can use it to find a fitted value (y) for any x-value (Figure 14.3).

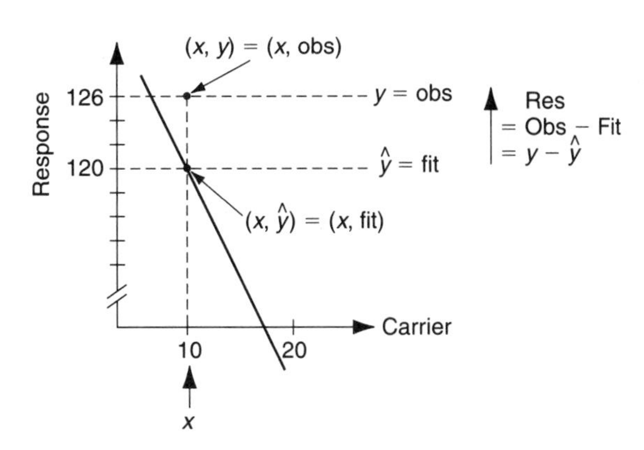

FIGURE 14.3 Using a line to get fitted values and residuals.

Start by locating the x-value on the horizontal axis. Move vertically to the point on the fitted line with that value as its x-coordinate. The y-coordinate of the point is $\hat{y}$, the fitted value. If the y-value corresponds to an observation, the difference Obs − Fit = $y - \hat{y}$ is the residual.

For the hamster data, the observed brain concentration of 126 (= y) corresponds to a heart concentration of 10 (= x). To find the fitted value, locate 10 on the horizontal axis, and move up to the point (10,120) on the line. The fitted value is 120 and the residual is Obs − Fit = 126 − 120 = 6.

We can use a fitted line to find a fitted value ($\hat{y}$) for each observation (y), and then decompose the observations into (Fit + Res), much as with ANOVA (Figure 14.4).

Obs		Fit		Res
101		102		−1
99		102		−3
95		90		5
110	=	112	+	−2
105		106		−1
126		120		6
111		118		−7

FIGURE 14.4 Decomposition of Brain Concentrations into Fit + Res. The fitted values come from the straight line model, as in Example 14.1. As always, Residual = Observed − Fitted. The residual SS gives a measure of how well the line fits.

Uses of the residuals. In this section, I'll describe two important uses of the residuals: for checking the fit, and for inference.

Checking the fit: Just as with ANOVA, you can (and should!) plot Res versus Fit to check for patterns, and you can use SS_{Res} as a measure of how well the straight-line model fits the data. If a straight-line model is appropriate, the plot of Res vs. Fit should show points in a balloon-shaped scatter with the x-axis running through its middle, just as when an ANOVA model fits well. (See Exercise A.4.)

Inference: Under certain conditions that I describe later in this section (if needed assumptions hold, and you get your line by the method of least squares) you can regard the residuals as (estimated) chance errors, and use sums of squares and mean squares much as you would with ANOVA. You can compute SD = $\sqrt{MS_{Res}}$, you can find a confidence interval for the slope of the "true" line relating y to x, and you can compute an F-ratio for testing whether there is an effect due to the carrier variable, i.e., that the slope of the fitted line is nonzero.

Example of a regression analysis. In the following pages, I'll first sketch an example of line fitting, hypothesis testing, and confidence intervals for a real data set, to give you a sense of direction. Then, in the next three parts of this section, I'll back up and go over the same ideas in more detail.

Example 14.4 Mental Activity Level

The data for this example are part of a larger data set obtained to compare the effects of heroin, morphine, and a placebo on the level of mental activity for human subjects.

I'll show you (Table 14.1) only the data for the placebo. There are 24 cases, one for each subject. The response is the score on a 7-item questionnaire, given two hours after a placebo injection; scores range from 0 to 14. There is one carrier variable, the mental activity level just before the injection. (Higher scores mean higher levels of activity.) The carrier values were obtained from the same 7-item questionnaire as the response.

TABLE 14.1 Mental activity data.

Subject	x = Carrier Before	y = Response After	Subject	x = Carrier Before	y = Response After
1	0	7	13	1	0
2	2	1	14	10	11
3	14	10	15	10	10
4	5	10	16	0	0
5	5	6	17	10	8
6	4	1*	18	6	6
7	8	7	19	8	7
8	6	5	20	5	1
9	6	6	21	10	8
10	8	6	22	6	5
11	6	3	23	0	1
12	3	8	24	11	5

Carrier x = activity level before injection.
Response y = level 2 hours after injection.
The number marked * in the data was changed from 2 to 1 to simplify arithmetic.

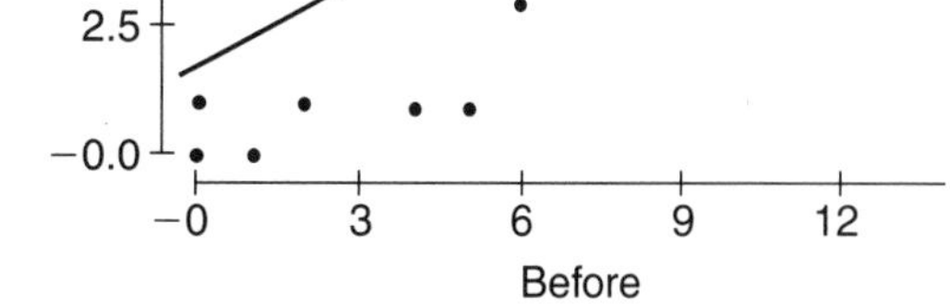

FIGURE 14.5 Scatterplot of mental activity level: After versus Before. For each case (point), the observed response is the y-value of the point, the fitted value is the height of the line directly above or below the data point, and the residual is the vertical distance of the data point above (+) or below (−) the line.

The plot in Figure 14.5 raises several questions:

What is the equation of the fitted line, and what does it tell us about the effect of the placebo on mental activity level?

How well does the model fit?

How appropriate is the straight-line model?
How close do the points come to the fitted line?

How precisely can we estimate the slope of the line?
How precisely can we estimate fitted y-values for various x-values?

The fitted regression line has equation:

$$\hat{y} = \hat{\alpha} + \hat{\beta}x,$$

where $\hat{y}$ = Fitted response = Activity level After
$\hat{\alpha}$ = Intercept = 1.9
$\hat{\beta}$ = Slope = 0.6
x = Carrier = Activity level Before

Thus

Fitted response =	(Intercept)	+	(Slope)(Carrier)
Activity level after =	1.9	+	(0.6) (Activity level before)

The equation tells the estimated effect of the carrier: a difference of 1 in the activity level before the injection corresponds to an average difference of $\hat{\beta} = 0.6$ on the activity level two hours later. (If the placebo had no effect, we would expect "After = Before," which corresponds to intercept 0 and slope 1. The actual slope of 0.6 might seem to suggest a tendency of the placebo to dampen the level of activity but there is another explanation, called the regression effect, described in Section 2.)

The "hats" in the notation ($\hat{y}$, $\hat{\alpha}$, $\hat{\beta}$) indicate that the numbers are estimated from the data. (The values of $\hat{a}$ and $\hat{\beta}$ were computed using a method that in effect considers all possible values and chooses the ones that make SS_{Res} as small as possible. See Exercises A.10 and A.21–23.) For this data set, $SS_{Res} = 147.2$ with $df_{Res} = 22$, so $MS_{Res} = 6.69$. The estimated SD = $\sqrt{MS_{Res}} \approx 2.6$: very roughly, two-thirds of the points fall within ± 2.6 of the line, 95% within ± 5.2 of the line.

The estimates $\hat{\alpha}$ and $\hat{\beta}$ are weighted sums of observed values (Exercise A.19), and the result from Chapter 11 applies: for a weighted sum, SE = $SD\sqrt{SS_{Weights}}$. For the estimated slope, $\sqrt{SS_{Weights}} = 0.055$, and SE = (2.6)(0.055) = 0.142. A 95% confidence interval has the same form as in Chapter 11:

$$\text{Estimate} \pm (\text{SE})(t\text{-value on 22 df})$$
$$0.6 \pm (0.142)(2.074)$$
$$0.6 \pm 0.295$$

The 95% distance is roughly 0.3, and the interval goes from 0.3 to 0.9. The interval does not include zero, so we reject the null hypothesis that the underlying true slope is zero, i.e., that the carrier has no effect on the response.

The regression analysis has decomposed each response value into a sum of three pieces:

$$\text{Obs} = \underbrace{\text{Grand Avg} + \text{Effect of Carrier}}_{\text{Fitted value}} + \text{Residual}$$

Each term in the decomposition has an associated SS and df, which you can compute and interpret the same way you would for ANOVA (Table 14.2).

TABLE 14.2 ANOVA table for mental activity data.

Source	SS	df	MS	F-ratio
Grand Avg	726.0	1		
Carrier	118.8	1	118.8	17.76*
Residual	147.2	22	6.69	
TOTAL	992.0	24		

The *F*-ratio for the effect of the carrier is significant, a result consistent with our confidence interval for the slope. ■

EXAMPLE 14.5 MENTAL ACTIVITY LEVEL: RESIDUALS VERSUS FITTED VALUES
The scatterplot in Figure 14.6 shows Residuals versus Fitted values.

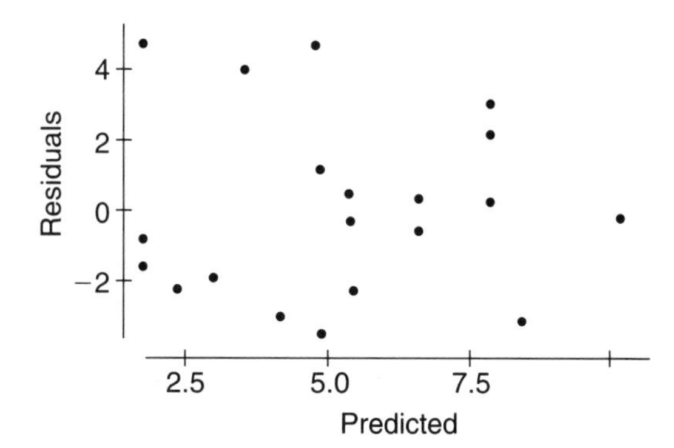

FIGURE 14.6 Residuals versus Fitted values for the mental activity data. Any pattern that departs sharply from an oval balloon lying along the *x*-axis arises doubts about the fitted model and should be investigated. Here, the plot is reasonably balloon-like. ■

Fitting a Regression Line by Least Squares

The least squares principle. For a given data set, each possible fitted line determines a value of SS_{Res}. To get this number, you use the line to get fitted values, split each observation into "fit plus res," square the residuals, and add. Intuitively, it makes sense to want SS_{Res} to be small, since small SS_{Res} means the line comes close to most of the points. The least squares line is the one that makes SS_{Res} as small as possible. (See Exercise A10.)

Least Squares Line

The least squares line for a set of (carrier, response) = (x, y) pairs is the line for which SS_{Res} is as small as possible.

EXAMPLE 14.6 THE LEAST SQUARES LINES MINIMIZES THE SUM OF SQUARED RESIDUALS

Consider fitting a line to the four points $(x, y) = (-2, -2), (-2, -1), (-2, 0)$, and $(6, 3)$. The display in Figure 14.7 shows five possible fitted lines. All five go through the origin, but they have different slopes. Below each fitted line is another version of the same plot, this time showing the residuals from the fitted line. Finally, the bottom panel in each column shows the residual sum of squares for the given line.

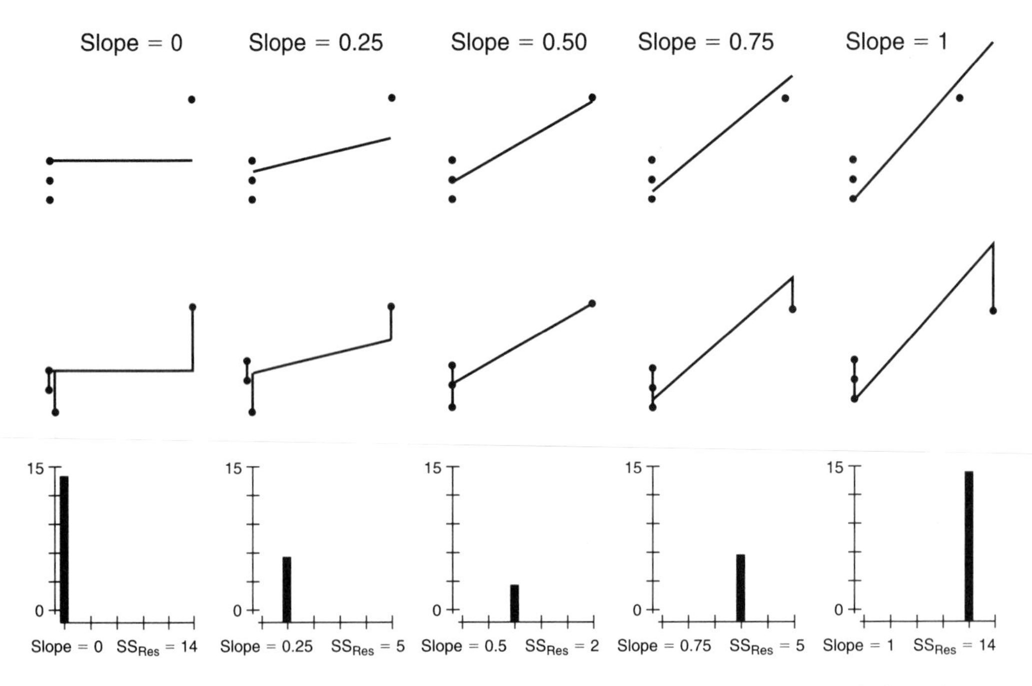

FIGURE 14.7 The least squares line minimizes the sum of squared residuals. The residuals are largest when the slope is either 0 (left column) or 1 (right column). The residuals are quite a lot smaller if the slope is either 0.25 (column to the left of center) or 0.75 (right of center). The line that gives the smallest possible residuals is the one with slope 0.5 (center column). This is the least squares line.

Although there are two disadvantages to using least squares lines, one minor (the calculations are messy if you do them by hand), and one major (the slope of the line can be influenced too much by outliers), the advantages of least squares have made it the standard method for fitting lines, from as far back as the 1700s. There are two main advantages: First, the method is very general, and can be applied to a great variety of situations. (All of ANOVA can be regarded as an application of the method of least squares, for example.) Second, when it is reasonable to assume that your chance errors all have the same SD and follow a normal curve, then the least squares approach brings with it a ready-made theory for hypothesis tests and confidence intervals.

In the next few pages I'll give two descriptions of the method for finding the least squares line. The two descriptions have different goals: The first is to help you understand some of the logic behind the computing rule, and to make the rule seem intuitively reasonable. (If you have good algebra skills, Exercises A.21–23 will show you how to prove that the rule does in fact give the line with the smallest possible SS_{Res}.) The goal of the second description is to give you a simple framework you can use to organize the actual arithmetic for finding the line. We'll start with two facts about lines in general.

Anchor points and slopes. Every line is determined by a point on the line together with the slope of the line. The slope of the line equals "rise over run" or "change in *y* over change in *x*" (Figure 14.8).

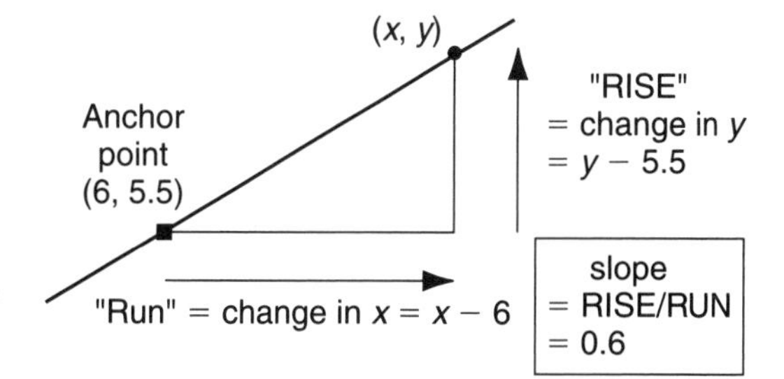

FIGURE 14.8 One point and the slope determine a line. For the mental activity data, the least squares line passes through the point (6, 5.5), and has slope = 0.6.

The first of the two facts gives us a convenient way to think about finding the least squares line: we'll first find an anchor point, then find the slope. The second fact will help us find the slope.

A natural choice for the anchor point is the **point of averages**:

> **The point of averages has *x*-coordinate equal to the average of the observed *x*-values, and *y*-coordinate equal to the average of the observed *y*-values. *The least squares line takes this point as its anchor point.***

There are various reasons to want your fitted line to pass through the point of averages. (1) The point of averages is the center of gravity of the scatterplot, which makes it intuitively appealing as an anchor. (2) Choosing this point as your anchor means that the observed changes in *x* and in *y* are just deviations from averages, and so sum to zero. (3) We want the residuals from the fitted line to add to zero (because we want to use them to estimate chance errors). It is not hard to prove using algebra that the residuals will add to zero if the line passes through the point of averages, but not otherwise. (See Exercise A.21.) Finally, (4) it is also possible to prove that if residuals from a fitted line *don't* add to zero, you can get a new fitted line with a smaller SS_{Res} by moving the old line up or down, keeping its slope the same, until the residuals from the line *do* add to zero. (See Exercise A.22.)

The least squares slope. Of all the lines that pass through the point of averages, we want the one that makes the sum of squared residuals as small as possible.

The Least Squares Slope

$$\hat{\beta} = \frac{\text{weighted average RISE}}{\text{weighted average RUN}}, \text{ where}$$

- **the averages are taken over all the data points**
- **each RISE** $= y - y_{avg}$
- **each RUN** $= x - x_{avg}$
- **each (signed) weight** $=$ **RUN** $= x - x_{avg}$

Exercise 14.23 outlines an algebraic proof that this value for the slope does in fact give the smallest possible SS_{Res}. Even without doing the proof, however, you can convince yourself that the rule is reasonable. Certainly it makes sense that the slope is a form of "rise over run," and computing some sort of average rise and average run over all the data points should seem like a reasonable way to get a fitted slope. The part of the rule that most needs explaining is using the values of $x - x_{avg}$ as signed weights. The explanation is worth reading, because it is linked to an important feature of the least squares line, one that has practical consequences. I'll use an example to illustrate the idea.

Example 14.7 Large $x - x_{AVG}$ Distances Correspond to Influential Points

To see why we use values of "change in x" $= x - x_{avg}$ as weights, consider the effect of changing the slope of a line fitted to the data of Example 14.6. As you can see from the diagram in Figure 14.9, the size of the change in the residual at a point is proportional to the distance along the x-axis from that point to the anchor:

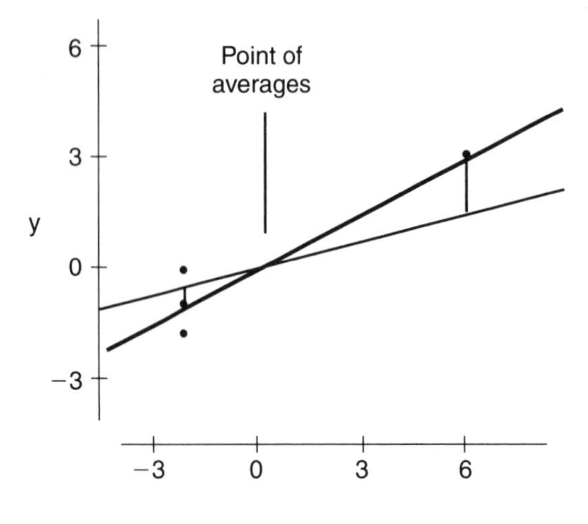

FIGURE 14.9 Change in residual is proportional to distance $x - x_{avg}$ from the anchor. If we change the slope of the line away from the least squares value of 0.5 (heavy line) to a nearby value, say 0.25 (other line), the residual for a point near the anchor (left vertical line) changes only a little. The residual for a point far from the anchor (right vertical line) changes a lot. In fact, by similar triangles, the change must be proportional to the x-distance from the anchor, i.e., $x - x_{avg}$. ■

A change in slope has much more effect on the residual when the point is far from the anchor. If we want to keep SS_{Res} as small as possible, we need to let such points have greater influence on our choice of slope. In other words, we need to give them greater weight when we compute the average values of rise and run. Since the change in residual is proportional to $x - x_{avg}$, we use those values as our weights. There's an immediate consequence:

For the least squares line, points with x-values far from the anchor have large weights, and correspondingly large influence on the value of the fitted slope.

For the mental activity data, there were five subjects whose activity level before the injection was exactly equal to the average of 6. For these subjects, the weight used in the weighted sum is $x - x_{avg} = 0$: the data for these subjects has no effect on the fitted slope. On the other hand, subject 3 had a pre-injection score of 14, for a weight of $14 - 6 = 8$. The fitted slope was more influenced by this subject than by any other. (See Exercise A.4 for more.)

Here's an example where a single influential point has a large effect on the fitted line.

EXAMPLE 14.8 DEATH RATES AND POPULATION DENSITY: AN INFLUENTIAL POINT

Are death rates for the 50 states of the US related to their population densities? If you compare the most densely populated state, New Jersey, with one of the less densely populated states like Iowa, I'm sure you can think of many differences that might be linked to death rates: differences in pollution levels, percent urban population, typical life style, etc. It is reasonable to think density and death rate might be related.

Death rates vary a lot from one state to another. In Alaska, only 400 people out of every 100,000 die each year. In five states, the rate is more than 1,000 per 100,000. Population density varies even more. Alaska is least dense, with an average of only 1 person per square mile, while New Jersey, the densest state, has almost 1000 people per square mile. The graph in Figure 14.10 shows a scatterplot of death rate

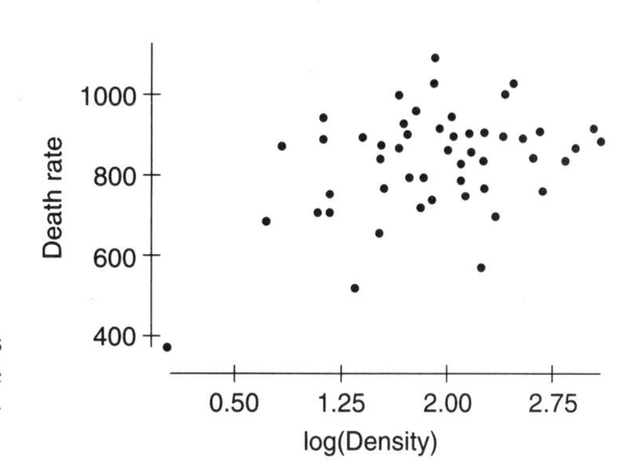

FIGURE 14.10 Death rate (per 100,000) versus the log of population density. The point in the lower left corner is for Alaska, and has a large influence on the least squares line.

against the log of population density for the fifty states. (I used log density because the densities vary over three powers of ten, from 1 to over 1000. The logs go from 0 to just over 3.) Notice in the plot below that one point (Alaska) is far from the others in the x-direction, and so will have a large influence on the least squares line.

The next two graphs (Figure 14.11) compare two least squares lines. The one on the left is fitted to the data for all 50 states. The one on the right is fitted to all but Alaska.

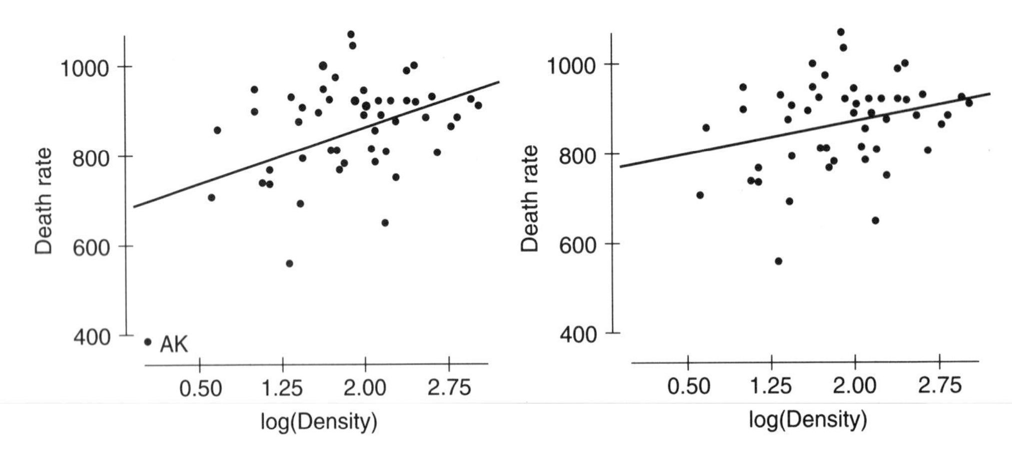

FIGURE 14.11 A point far from the others in the x-direction has large influence. The least squares line on the left is based on all the data, including the influential point, Alaska. The line on the right is based on the same data, except that Alaska is excluded. The change in the line shows the influence of the one point for Alaska. The slope on the left is almost twice as big as the one on the right. ■

Other forms for the equation of the fitted line. The equation of the least squares line can be written in any of three equivalent ways. Each way can be useful, depending on your purpose.

EXAMPLE 14.9 MENTAL ACTIVITY DATA: THREE FORMS OF THE FITTED LINE

For the 24 subjects in Example 14.4, the observed averages are 5.5 (y = After) and 6.0 (x = Before); the fitted slope is 0.6. Here are the three equivalent forms of the equation of the fitted line:

a. **Point-slope:**

$$\frac{y - y_{avg}}{x - x_{avg}} = \text{slope}$$

$$\frac{(y - 5.5)}{(x - 6.0)} = 0.6$$

The point-slope form is useful when you want to emphasize the connections between the equation of a line, the definition of slope as "rise over run," and the rule for computing the fitted slope as "weighted average rise over weighted average run."

b. **Proportional change:**

$$(y - y_{avg}) = (\text{slope})(x - x_{avg})$$
$$(y - 5.5) = (0.6)(x - 6.0)$$

The proportional change form is useful for showing how you compute fitted values for $(y - y_{avg})$, by multiplying slope times $(x - x_{avg})$.

c. **Intercept-slope:**

$$y = [y_{avg} - (\text{slope})(x_{avg})] + (\text{slope})(x)$$
$$y = [5.5 - (0.6)(6.0)] + (0.6)(x)$$
$$y = 1.9 + 0.6x \quad \blacksquare$$

The intercept-slope form is useful if you want to graph the line, or if you want to compare lines with different anchor points, and also because for many people it is the form they learned first in high school.

Computing summary. If you are doing the arithmetic by hand instead of a computer—worth doing only two or three times to learn the method better—it can help to write out your calculations in a standard format.

Step 1: Split the set of ys into average plus deviations: $y = y_{avg} + (y - y_{avg})$; do the same for the xs. The y-deviations are the values of RISE; the x-deviations are the values of RUN.

Step 2: Compute weighted sums, using values of RUN as signed weights:

$$S_{xy} = \text{weighted sum for RISE} \qquad S_{xx} = \text{weighted sum for RUN}$$

Although it is helpful to think in terms of weighted *averages*, the arithmetic is simpler, and leads to the same answer, if you use *sums* instead.

Step 3: Compute the slope $= S_{xy}/S_{xx}$

Step 4: Split the y-deviations (=RISE) into

$$\text{FIT} = (\text{slope})(x\text{-deviation}) = (\text{slope})(\text{RUN})$$
$$\text{plus}$$
$$\text{RES} = \text{RISE} - \text{FIT}$$

EXAMPLE 14.10 REGRESSION MECHANICS FOR A SIMPLE DATA SET

Here, to illustrate the arithmetic, is a simple made-up version of a regression data set.

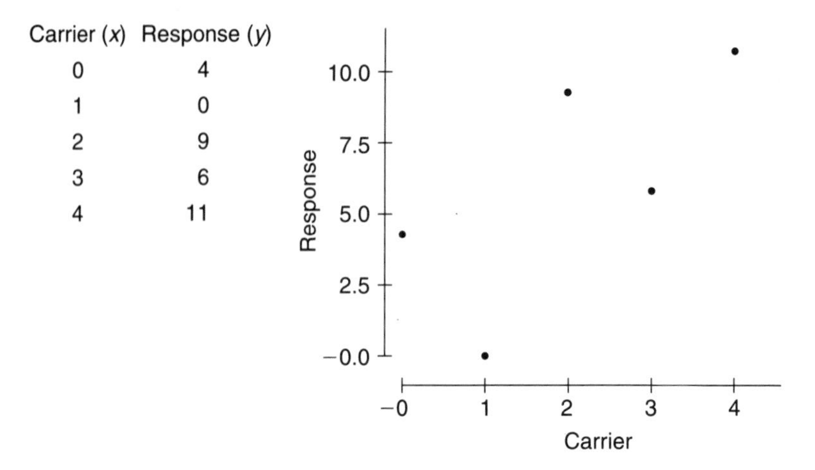

Step 1: Split y into y_{avg} + RISE and x into x_{avg} + RUN.

y		avg(y)		RISE
4		6		−2
0		6		−6
9	=	6	+	3
6		6		0
11		6		5

x		avg(x)		RUN
0		2		−2
1		2		−1
2	=	2	+	0
3		2		1
4		2		2

Step 2: Compute the weighted sums S_{xy} = sum RISE × RUN and S_{xx} = sum RUN × RUN

RISE		RUN		RISE × RUN
−2		−2		4
−6		−1		6
3	×	0	=	0
0		1		0
5		2		10
			Total	20 = S_{xy}

RUN		RUN		RUN × RUN
−2		−2		4
−1		−1		1
0	×	0	=	0
1		1		1
2		2		4
			Total	10 = S_{xx}

Step 3: Compute slope = S_{xy}/S_{xx}

$$\text{Slope} = \hat{\beta} = S_{xy}/S_{xx} = 20/10 = 2$$

Step 4: Compute (slope)(RUN) = FIT and RISE − FIT = RES

(slope)		RUN		FIT
		−2		−4
		−1		−2
[2]	×	0	=	0
		1		2
		2		4

RISE		FIT		RES
−2		−4		2
−6		−2		−4
3	−	0	+	3
0		2		−2
5		4		1

■

Decomposition. The first step (finding the point of averages and the deviations) and the last three steps (finding the slope, fitted values, and residuals) correspond to a two-step decomposition of the observed values (Figure 14.12).

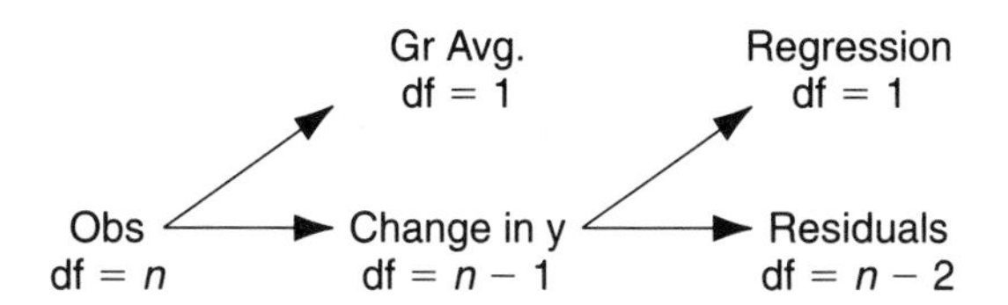

FIGURE 14.12 Flow chart for decomposition. The first step splits the observations (y) into their average ($\bar{y}$) and the leftovers, or change in y ($y - \bar{y}$). The second step splits change in y into a regression fit (= [slope] × [change in x]) plus residuals. To get fitted values, add the average of y ($\bar{y}$) and regression fit ($\hat{\beta}[x - \bar{x}]$). The df are shown in parentheses. If you start with n observations (n df), the first step splits those n into 1 df for the average, and ($n - 1$) for the leftovers. The regression fit = (slope)(change in x) has only 1 df, because once you know the slope, you can figure out all the fitted values. So there are $n - 2$ df for the residuals.

EXAMPLE 14.11 DECOMPOSITION OF A REGRESSION DATA SET

Use Example 14.10 to write out the two-step decomposition that corresponds to the flowchart above. Then summarize the decomposition in an ANOVA table, using the usual format for rows and columns.

SOLUTION. (Figure 14.13, Table 14.3.)

	OBS		Gr Avg		RISE		Gr Avg		Regr		RES
	4		6		−2		6		−4		2
	0		6		−6		6		−2		−4
	9	=	6	+	3	=	6	+	0	+	3
	6		6		0		6		2		−2
	11		6		5		6		4		1
SS	284	=	210	+	74	=	210	+	40	+	34
df	5	=	1	+	4	=	1	+	1	+	3

FIGURE 14.13

TABLE 14.3 ANOVA table for the fake mental activity data

Source	SS	df	MS	F
Grand Avg	180	1		
Regression	40	1	40	3.53
Residual	34	3	11.333	
Total	254	5		

As usual, rows correspond to tables in the decomposition. Each MS = SS/df, and the F-ratio = MS_{Regr}/MS_{Res}. ■

Once you've done the decomposition, computing an F-ratio is quick and easy. However, just as in ANOVA, the F-test is only meaningful and valid under certain assumptions.

Assumptions, Tests, and Confidence Intervals

When do you need assumptions? When you use line fitting to explore your data, or to provide a summary of a pattern, you don't need a model. You do need one, however, if you want to test hypotheses about a fitted line (for example, that the true slope is not zero), or you want to construct confidence intervals. Such tests and intervals are statements about unknown true values, and they make sense only if you have a model that tells how the true values are related to the observed data. For regression, just as for ANOVA, the model describes a mechanism or process that (we assume) created the data. The unknown true values are numerical properties of that process.

Regression model and assumptions. The model and assumptions for regression are almost the same as for ANOVA:

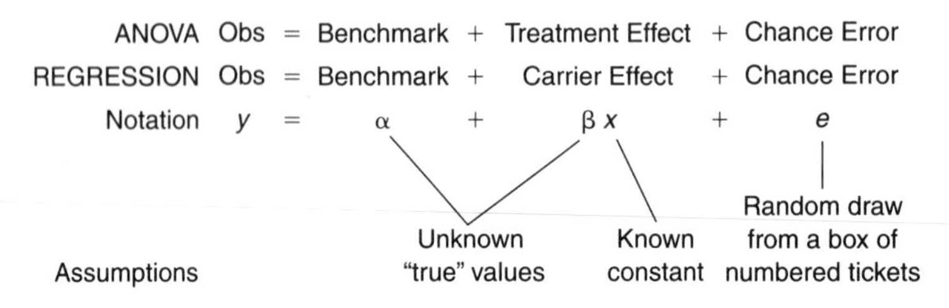

As for ANOVA, we assume:

Additivity: Each observed value is a *sum* of pieces.

Constant effects: The "effect" of a carrier variable x is a constant multiple β of x.

We also assume that each observation equals a "true" part plus a chance error, and that the errors behave like random draws from a box of numbered tickets. The draws need not all comes from the same box, as long as the process has the following properties:

Zero expected value: The average of the tickets in each box is zero.

Same SDs: Each box has the same SD.

Independence: The draws behave either as if they came from different boxes, or, if from the same box, as if you replaced each ticket and mixed thoroughly before the next draw.

Normality: The numbers on the tickets in each box follow a normal curve.

Notice how similar these assumptions are to those for ANOVA: an additivity assumption (A), an assumption of constant "effects" for each carrier value (C), and the same four assumptions about chance errors.

Center, spread, and shape. There's a slightly different way to describe the model that is also useful. For each fixed value of the carrier x, there is a chance process that generates random, independent values of the response y, like random draws from a box of tickets, with a separate box for each value of the carrier x. The assumptions of the model describe the distribution of tickets in each box in terms of the center, spread, relatedness, and shape of that distribution:

Center:	The average of the box for x is $\alpha + \beta x$.
Spread:	The SDs of the boxes are all equal to the same unknown constant σ.
Relatedness:	The draws are independent.
Shape:	The tickets in each box follow a normal curve.

Levels of assumptions. This way of describing the model lets us distinguish four levels of assumptions about the data. Each level is more restrictive than the one before it.

- No assumptions
- Center only: For each x, the values of y behave like draws from a box whose average is $\alpha + \beta x$.
- Center, spread and relatedness: The boxes not only have averages equal to $\alpha + \beta x$, they also have the same SD ($= \sigma$), and the draws are independent.
- Center, spread, relatedness, and shape: The tickets in each box follow a normal curve, with mean $\alpha + \beta x$ and SD $= \sigma$, and the draws are independent.

Levels of inferences. For each set of assumptions, there is a corresponding set of uses of regression. The more you are able to assume about your data, the more you can conclude from your statistical analysis.

- Pure description and exploration; no inferences: If you don't assume anything about the process that generated the data, you have no target for your inferences.
- Unbiased estimation: If, for each x, the true value for y is in fact $\alpha + \beta x$, then the least squares fit gives estimates of α and β that are free from bias.
- "Best" linear unbiased estimation: If, in addition, the observations are independent and all have the same SD, then the least squares estimates will be more precise (have smaller SDs) than any other unbiased estimators that are weighted sums of the observations.
- F-tests and confidence intervals: If, in addition, for each x the distribution of y-values is normal-shaped, then F-tests and confidence intervals will behave as they are supposed to.

In real life, of course, things are not so clear-cut as the list of four levels might suggest. Assumptions are never completely true. At best, they are reasonable approximate descriptions. In practice data sets that fit the assumptions of the third level (center, relatedness, and spread) tend to give a reasonable fit to the normality assumption as well. Thus in practice the important jumps are from the first to the second level (Is there, or is there not, a "true" underlying straight-line relationship?), and from the second to the third (are the SDs equal?).

F-tests for the slope. If the complete set of assumptions is reasonable, then you can use the F-ratio for regression in the usual way, that is, pretty much as in ANOVA (see the end of Example 14.4, and Example 14.9). In ANOVA, the F-ratio for a structural factor (categorical carrier) equals MS_{Factor}/MS_{Error}. In regression, the F-ratio for the numerical carrier equals $MS_{Regression}/MS_{Error}$. In both cases, the F-ratio tends to be large if the effect of the carrier is "real," and tends to be near one if the carrier has no effect. If the computed F is greater than the critical value, you conclude that the carrier effect is detectably different from zero. If not, the verdict is "not proven:" the observed effect could be due just to chance error.

For a straight line model, "no carrier effect" means "slope $\beta = 0$." The F-test is equivalent to constructing a confidence interval for the slope and checking to see whether the interval contains zero. If not, then the hypothesis that $\beta = 0$ is ruled out by the data.

Confidence intervals. Two kinds of intervals are useful: for the fitted slope, and for individual fitted values. Both the slope and the fitted values are weighted sums of observed response values, and so both intervals have the form "weighted sum ± SE · t-value," with SE = SD · (leverage factor), SD = $\sqrt{MS_{Res}}$, and leverage factor = $\sqrt{SS_{Weights}}$. In other words, the intervals are just like those of Chapter 11, except that $SS_{Weights}$ is different. I'll sketch the algebra in the exercises and just give the results here:

Confidence Interval: EST ± (SE)(t-value)

SE = SD$\sqrt{\text{SSWeights}}$, SD = $\sqrt{MS_{Res}}$, and for the fitted slope, $SS_{Weights} = 1/S_{xx}$.

For a fitted response value whose carrier value is x, $SS_{Weights} = \dfrac{1}{n} + \dfrac{(x - x_{avg})^2}{Sxx}$

EXAMPLE 14.12 CONFIDENCE INTERVALS FOR THE MENTAL ACTIVITY DATA

Several summary numbers for the data of Example 14.4 are given below. Use these to find 95% confidence intervals (a) for the slope of the fitted line and (b) for fitted values of "After" corresponding to "Before" values of 0, 3, 6, 9, and 12.

$$df_{Res} = 22 \quad MS_{Res} = 6.69 \quad x_{avg} = 6.0 \quad S_{xx} = 330$$

SOLUTION. All intervals will be of the form EST ± $\sqrt{MS_{Res}}\,\sqrt{SS_{Weights}} \times t$. The 95% t-value on 22 df is $t_{22} = 2.074$, and $\sqrt{MS_{Res}} = \sqrt{6.69} \approx 2.587$ = SD.

a. Slope: 0.6 ± 0.295

$$SS_{\text{Weights}} = \frac{1}{S_{xx}} = \frac{1}{330}, \quad \text{so } \sqrt{SS_{\text{Weights}}} = \sqrt{\frac{1}{330}} = 0.055$$

$$95\% \text{ distance} = SD\sqrt{SS_{\text{Weights}}} \cdot t_{22}$$
$$= (2.587)(0.055)(2.074) = 0.295$$

b. Fitted values:

Each fitted value equals $1.9 + 0.6x$:

x = Before	0	3	6	9	12
$\hat{y}$ = fitted After	1.9	3.7	5.5	7.3	9.1

To get 95% distances, we use the same SD and t as in (a), and use

$$SS_{\text{Weights}} = \frac{1}{\# \text{ obs}} + \frac{(x - \bar{x})^2}{Sxx}$$
$$= \frac{1}{24} + \frac{(x - x_{\text{avg}})^2}{330}$$

(i) Before = 0: Here $x - x_{\text{avg}} = 0 - 6 = -6$, so

$$SS_{\text{Weights}} = \frac{1}{24} + \frac{(-6)^2}{330} = 0.151, \ \sqrt{SS_{\text{Weights}}} = 0.338$$

So $95\% \text{ distance} = SD\sqrt{SS_{\text{Weights}}} \times t_{22}$
$$= (2.587)(0.388)(2.074) \approx 2.1$$

The 95% interval is 1.9 ± 2.1, or -0.2 to 4.0.

(ii) Before = 3: Here $x - x_{\text{avg}} = -3$, so

$$SS_{\text{Weights}} = \frac{1}{24} + \frac{(-3)^2}{330} = 0.069, \ \sqrt{SS_{\text{Weights}}} = 0.263$$

So $95\% \text{ distance} = SD\sqrt{SS_{\text{Weights}}} \times t_{22}$
$$= (2.587)(0.263)(2.074) \approx 2.1$$

The 95% interval is 3.7 ± 1.4, or 2.3 to 5.1.

(iii) Before = 6: Here $x - x_{\text{avg}} = 0$, and $SS_{\text{Weights}} = \frac{1}{\# \text{ obs}} = \frac{1}{24}$, and

$\sqrt{SS_{\text{Weights}}} = 0.204$

The 95% distance is $(2.587)(0.204)(2.074) \approx 1.1$

The 95% interval is 5.5 ± 1.1, or 4.4 to 6.6

The other two cases are like (i) and (ii). We can show the intervals together on a graph (Figure 14.14):

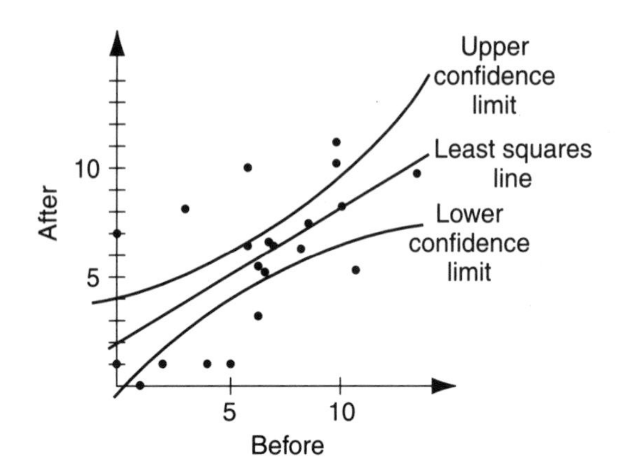

FIGURE 14.14 Upper and lower confidence limits. The graph shows how the width of the confidence interval for a fitted y-value (after placebo injection) depends on how far the corresponding x-value (before injection) is from the observed average of the xs. The narrowest interval is for $x = x_{avg} = 6$. The farther one moves away from the average, the wider the interval. ■

Transformations, Extensions and Limitations

Remember from the introduction that the regression methods of this section (and the correlation methods of the next section) are for scatterplots that suggest a line or an oval balloon. Many data sets yield plots that don't look like lines or balloons. What then?

We'll briefly consider three possibilities: (a) points fall along a curve, (b) points form a wedge, (c) points form more than one cluster. Then we'll end with an example of a data set with more than one numerical carrier.

(a) Points fall along a curve: transforming to get a line. Often, when your scatterplot suggests a curve, changing either the carrier or the response or both to a new scale will make the data set one for which line-fitting is appropriate. Exercise A.15 describes and illustrates a general strategy for choosing a new scale. To keep things brief here, I'll just show you a pair of scatterplots, before and after transforming.

EXAMPLE 14.13 CABBAGE BUTTERFLIES: TRANSFORMING TO GET A LINE

The time it takes a newly hatched butterfly larva to reach pupa stage depends on the surrounding temperature. It seems natural to measure time in days, and to record, for example, that at 20.5°C it took one larva 20 days to reach pupa stage. This scale for time, though natural-seeming, gives a curved plot. If, however, we change from days to 1/days, and record that at 20.5°C, one day is 1/20 of the time it took to reach pupa stage, the data points form a straight line suitable for regression analysis (Figure 14.15).

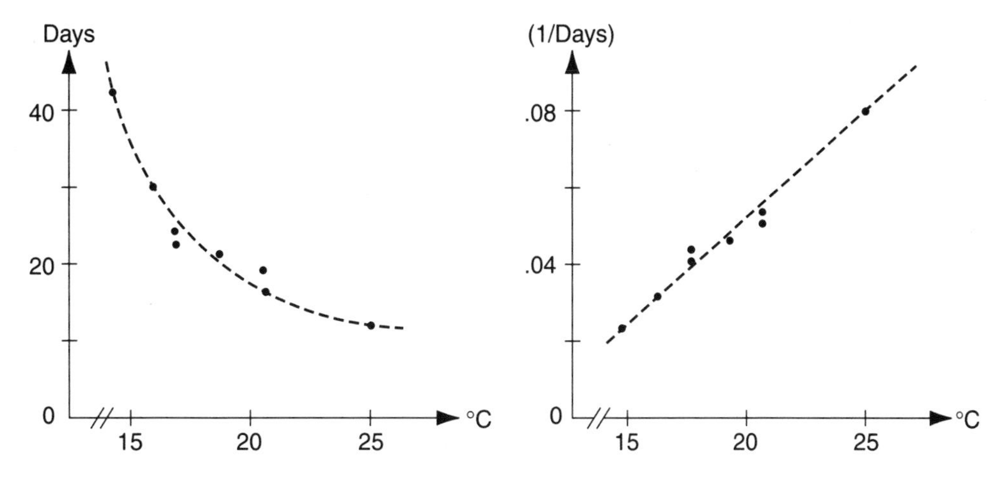

FIGURE 14.15 Sometimes changing scales can straighten a curve. Both plots show the relationship between the time it takes a newly hatched cabbage white butterfly to reach pupa state (y-axis) and the surrounding temperature (x-axis). The graph on the left measures time in days, and shows a curved relationship. Transforming the response to reciprocals (from y = how long, on the left, to y = how fast, on the right) yields a straight-line relationship which can be fit by the regression methods of this section. The straight-line relationship means that, anywhere within the 10° interval of the study, a change of 1° has a constant effect on the speed of development. (See Exercise C.3 for more work with this data set.) ■

(b) Points suggest a wedge: transforming for symmetry. If your plot is asymmetric, with points densely packed in one corner and much more spread out away from that corner, balloon-based methods like regression and correlation will give too much influence to the points farthest from the center of gravity. If the bunching occurs at the low end of the scale, transforming to square roots, logarithms, or reciprocals will usually make the plot more symmetric. (Bunching at the high end of the scale is less usual, except when there is a ceiling on response or carrier values, like scoring 100 on a test. In a case like that, you need special transformations of the sort described in Exercises 12.A.14–16.)

EXAMPLE 14.14 IMAGERY AND WORKING MEMORY: TRANSFORMING FOR SYMMETRY

In the study of imagery and working memory, subjects were timed while they answered a series of yes/no questions. (The questions were either verbal—about the words in a sentence—or visual—about the corners of a block letter. The answers, too, could be either verbal—telling—or visual—pointing.) It is natural to measure time in seconds, and to record, for example, that one subject who took 14 seconds to report by pointing took 10 seconds to report by telling.

A plot of the data in this scale is bunched at the lower left, and more spread out toward the upper right. If we change to 1000/sec, then we record that the subject whose rate when pointing was 1000/14 = 71 reports per thousand seconds had a rate when telling of 1000/10 = 100 reports per thousand seconds. In this new scale, the plot looks much more like a symmetric balloon (Figure 14.16).

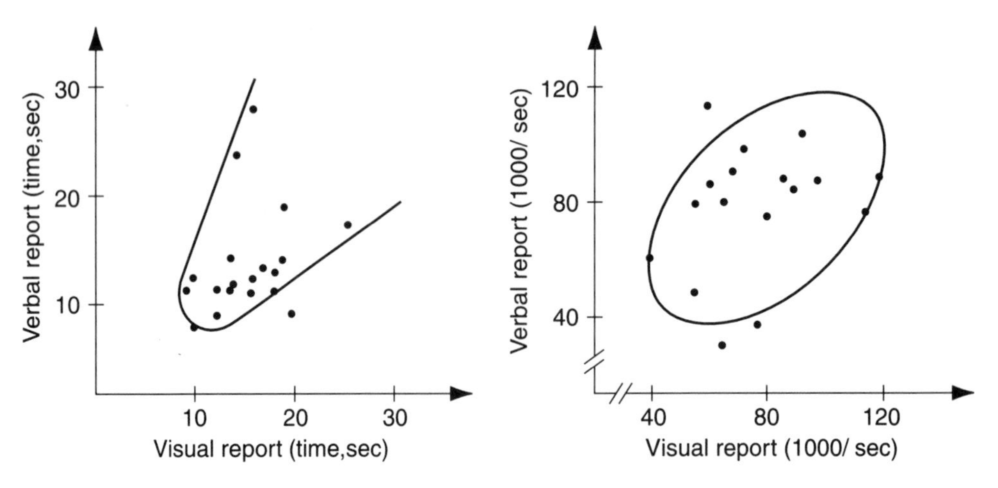

FIGURE 14.16 Transforming a wedge into a balloon. The data for these plots come from Exercises 9.C.5–8, on imagery and working memory. Both plots show versions of the relationship between the time taken for verbal (y) versus visual (x) methods of reporting on a verbal task. The graph on the left shows a wedge, with points (= subjects) densely packed in the narrow, lower-left part, where both x and y values are small, but widely scattered in the upper right, where values are larger. For such a plot, regression and correlation are not suitable. The right-hand plot, however, shows the same relationship expressed in the reciprocal scale, which measures "how often" (reports per second) instead of "how long" (seconds per report). This plot shows the kind of pattern for which regression and correlation are designed: points form a more-or-less balloon-shaped cloud, denser toward the center, and more scattered toward the edges.

In the previous example, the points fell much closer to a fitted line: the developmental behavior of insects is much less variable than the cognitively related behavior of humans. Nevertheless, even here a line gives a reasonable model for the fitable part of the observed relationship. In particular, the 45° line $y = x$ fits about as well as any, suggesting that visual and verbal report rates tend to be equal. Note that this relationship ($y = x$) is much easier to see after transforming. ■

(c) Points form more than one cluster: analysis of covariance. The methods of this section apply to data sets whose points form just one line or cluster. Many data sets show more than one cluster. Nearly always each cluster corresponds to a different set of conditions—different treatments, or different populations. In such cases you need a fancier version of the methods of this section, possibly analysis of covariance (Section 3), or some other regression method that uses both categorical and numerical carrier variables.

EXAMPLE 14.15 PRISONERS IN SOLITARY: MORE THAN ONE CLUSTER OF POINTS

The two clusters in Figure 14.17 show frequency of alpha waves, day 7 versus day 1, for two groups of prisoners: those in solitary confinement (o) and a control group (x). You can use the methods of this section to work with each group separately, for example fitting a line (regression) and measuring how closely the points cluster near it (correlation). But there are also **multiple regression methods** (see (d) below) that could be used first to fit lines simultaneously to both groups, and then to compare the lines as a way of comparing groups.

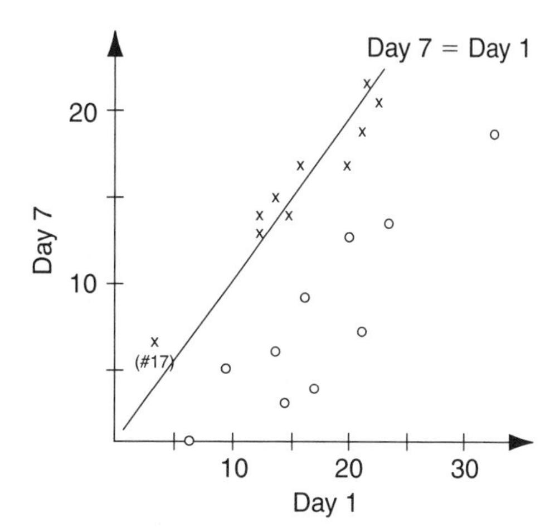

FIGURE 14.17 More general regression methods can handle several clusters.

The pattern here is fit well by two parallel lines with slopes near 1 (45°): any two prisoners in the same treatment group will tend to differ by about the same amount on day 7 as on day 1. The vertical distance between the two lines (roughly 9 units) represents the effect of solitary confinement, which appears to depress the frequency of alpha waves. Notice also that the two point clouds have quite different sizes: prisoners chosen for the treatment group are a more varied lot than the controls. (See Exercises C.13 and C.20 for more work with this data set.) ■

(d) More than one numerical carrier: multiple regression. Just as you can have ANOVA models with more than one factor, you can have regression models with more than one carrier. The models are similar:

Multifactor ANOVA

Observation = Grand Average + Sum of factor effects + Chance Error

Multiple regression

Observation = Grand Average + Sum of carrier effects + Chance Error

In ANOVA, each effect is an unknown (constant) true value. In regression, each effect is an unknown (constant) true value times the carrier value x.

In a regression with two or more carriers, finding and interpreting the least squares fit is simple *only* if the carrier variables have perfect zero correlation with each other. This won't happen unless you can control the values of the carriers, as in a true experiment. Much of the time, you won't have that kind of control, your carriers will be correlated, and the regression model will be hard to interpret.

The next example illustrates a least squares fit to three carriers, and also gives just a glimpse at many of the issues that make interpretation hard.

EXAMPLE 14.16 DEATH RATES BY STATE: A THREE-CARRIER REGRESSION

We know from Example 14.8 that there is a tendency for densely populated states to have somewhat higher death rates than states with fewer people per square mile. Since death rates depend on many factors, it is reasonable to think we could find other variables that we could use, together with population density, to predict a state's death rate. Two relevant variables are the percent of the state's residents with driver's licenses (DLIC%) and the average per capita income, in thousands of dollars (INCOME).

We can use multiple regression to fit an equation of the following form to the data of all fifty states:

$$\text{DEATH} = \alpha + \beta_1 \cdot \text{DLIC\%} + \beta_2 \cdot \text{INCOME} + \beta_3 \cdot \log(\text{DEN})$$

The values of α, β_1, β_2, and β_3 that make the residual sum of squares as small as possibly give the least squares fit:

$$\text{DEATH} = 330 + 10.2 \cdot \text{DLIC\%} - 22 \cdot \text{INCOME} + 132 \cdot \log(\text{DEN})$$

It is nearly always misleading to take an equation like this at face value, for reasons I'll detail shortly. All the same, the literal interpretation is a good place to start, just so long as you don't stop there. If we take the equation at face value, this is what the fitted coefficients tell us:

$10.2 \cdot$ DLIC%	An increase of 1% in the percent of residents with driver's licenses raises the death rate by about 10 per 100,000
$-22 \cdot$ INCOME	Raising the per capita income by $1000 lowers the death rate by 22.
$132 \cdot \log(\text{DEN})$	An increase 1 in log(DEN)—which corresponds to a tenfold increase in the population density—raises the death rate by 132 per 100,000.

If we are careful, the actual message we take from the fitted equation will be somewhat less straightforward, but much less misleading. There are several issues to consider, all of them involving limitations of regression by least squares. ■

Limitations of least squares regression. I've organized my discussion of the limitations of regression under heading corresponding to three main uses of the method: for description, for theory-building, and for prediction.

A. Description: Limitation of least squares lines as summaries

It is always possible, by hand or by computer, to fit a least squares line to a set of number pairs, but *the only way to know whether a line gives a reasonable summary is to look at a scatterplot.*

1. **The relationship may not be linear.**

Example: In the analysis of death rates, it might seem natural to use population density as a carrier instead of log density, but a scatterplot shows that the relationship of death rate to density is not linear. A line would be misleading as a summary.

2. **The fitted line may be too much influenced by extreme cases.**

Example: In Example 14.8, Alaska is an influential observation. The same is true in the multiple regression analysis of death rates. If you leave out that one state, the fitted coefficients change markedly:

All 50 states: DEATH = 330 + 10.2 · DLIC% − 22 · INCOME + 132 · log(DEN)
Without Alaska: DEATH = 360 + 9.6 · DLIC% − 20.5 · INCOME + 122 · log(DEN)

B. **Theory building and interpretation: Association is not causation.**
If a scatterplot shows a strong relationship between a carrier x and a response y, it may be tempting to conclude that a change in the carrier causes a change in the response. *Resist the temptation.* It may be that x causes y, but it might be that y causes x, or that both x and y are caused by a third variable z.

1. **The importance of data production:** The chart in Figure 14.18 lists four kinds of studies: experimental, prospective, retrospective, and cross-sectional. If you want to show causation, an experiment will be most convincing, and a cross-sectional study, least convincing.

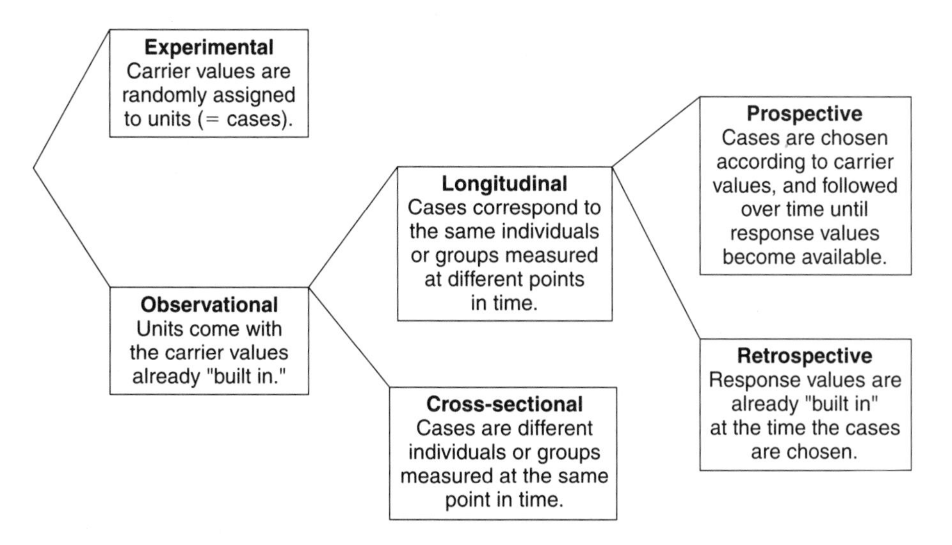

FIGURE 14.18 Different kinds of regression studies. The vertical direction corresponds to how strongly a study can support an inference that changes in the carrier variables cause a change in the response. The experimental study comes out on top: If you randomly assign carrier values to units, then a pattern of association between response and carriers strongly allows you to conclude that changes in the carrier cause changes in the response. Among observational studies, longitudinal is better than cross-sectional for studying causal relationships. Among longitudinal studies, prospective is better than retrospective.

Example: In the analysis of death rates in Example 14.16, I interpreted parts of the fitted equation at face value, saying, "Raising the per capita income by $1000 lowers the death rate by 22." This statement, though handy as a kind of shorthand, is misleading as it stands. The data don't tell us what happens to death rates when incomes change. For that, we'd need to observe actual changes over time: a longitudinal study. Rather, the data we have are cross-sectional. They tell us that states whose average incomes are $1000 apart will have—on average—a difference of 22 in their death rates.

2. The "ecological fallacy:" Patterns that hold for larger groups ordinarily are much less strong (or may not even hold at all) for smaller groups or individuals.

Example: The relationship between incomes and death rates is a pattern relating sets of averages for entire states. If you could look at average death rates and incomes for families instead of states, you'd find much more variability, and much less of a pattern relating the two variables.

3. Lurking variables: A lurking variable is one that has a close relationship to the response but is not one of the carriers. If you fail to recognize such a variable, and so leave it out of your analysis, your fitted equation may be very misleading.

Example: In the study of death rates, Alaska has both a very low death rate and a very low population density. The data almost invite you to speculate: "Perhaps a low population density makes diseases harder to spread, and that's why Alaska has such a low death rate." In fact, deaths from disease are much more often due to cancer, heart disease, and diabetes than to contagious diseases. The actual explanation for Alaska's situation is a lurking variable, age. The percent of people over 65 in Alaska is very low, only 4.2%. (At the other extreme, Florida has a very large percentage, 18.3%, of its population over 65.) It turns out that if you include AGE 65% as a carrier, the entire fitted equation changes drastically. With the information about people over 65 included, the percent of licensed drivers no longer has any value in predicting death rates. Moreover, two other variables that were previously of no value—normal high temperature in January and percent who graduated from high school—now add significantly to the predictive power of the fitted equation. Including the information from the lurking variable has changed the status of three out of the five other carriers I looked at: one no longer belongs in the equation, and two others should be added.

4. Multicollinearity: When carrier variable are correlated with each other, it is misleading to interpret the pieces of a regression equation one at a time.

Example: For the death rate data, a scatterplot of INCOME versus log(DEN) would show a fairly strong relationship. Higher values of one tend to go with higher values of the other. Because of this, the relationship between death rate and the three carriers in the fitted equation of Example 14.16 is far from simple. Here's a table that shows that death rates are highest when both density and income are low, and lowest when density is low and income is high:

Density	Income	
	Low	High
Low	873	823
High	853	837

Average death rates for groups of states

The sense you get from this table, of how death rates are related to incomes and population density, is not at all the same as what you get by interpreting the regression coefficients one at a time.

If the table makes you think of interaction, pat yourself on the back. The reason the regression equation is hard to interpret is the same as the reason why the presence of interaction in a two-way ANOVA makes it hard to say what the effect of either factor is by itself.

C. Fitted values: Limitations of least squares for prediction

If you are using a fitted equation to predict future observed values, issues of interpretation may not matter. (If scores on a placement test give good predictions of how students will do in a course, the fact that the exam makes good predictions is more important than trying to track down the reasons why.) Nevertheless, there are limitations to using least squares for prediction.

1. Extrapolation: It is risky to make predictions for carrier values outside the range covered by the data. Relationships that give good predictions over the range of the data may be quite different for cases whose carrier values fall in a different range.

Example: Washington, DC is often included in almanac listings as though it were a state, and it might seem that we could use our least squares equation to get a good prediction for Washington's death rate. Washington's density, however, is 9883 people per square mile. This is almost ten times as many people per square mile as in the densest actual state, New Jersey. Because of the very high density, DC is in many ways more like a city than a state, and the relationship that gives reasonable predicted death rates for states doesn't work so well for cities. It turns out that a predicted death rate for Washington is way off.

2. Lurking variables: If your response values are affected by a variable that is not part of your fitted equation, predicted values may be unreliable.

Example: Unlike DC, Florida has carrier values that fall in the middle of the range covered by the data. It might seem that we could use the fitted equation to predict death rates for this state, but as we have seen, age is an important lurking variable in the context of the three-carrier fitted equation for death rates. Florida has lots of retirees, with more than 18% of its population over 65, and the fitted value using the equation from Example 14.16 is way too low.

Conclusion. Despite its many limitations, regression is a powerful statistical method. Fitting by least squares is a big subject, in a sense much bigger than ANOVA, since it includes ANOVA as a special case. This section has given only a brief introduction to the main ideas, but some of the exercises give you a chance to go into things more deeply. In particular, there are exercises on the following topics:

The effect of influential points on the slope (A.4)
ANOVA as a special case of regression (A.14)
Transforming (A.15)
More than one carrier variable (A.16–18)

Exercise Set A

1. *Fitting lines by eye: Simple data for quick practice.* Four artificial data sets (a)–(d) are shown below. For each one, (i) plot the pairs (x, y) and fit a line by eye. Then (ii) give the coordinates of two points on the line, (iii) use the coordinates to compute the slope of the line, and write the equation of the line (iv) in the point-slope form, and (v) in the form $y =$ (intercept) + (slope)(x).

   a. (x, y): (0, 1), (0, 3), (4, 1)
   b. (x, y): (0, 2), (4, 2), (4, 4)
   c. (x, y): (0, 2), (0, 4), (3, 0)
   d. (x, y): (0, 3), (0, 4), (2, 0), (3, 1), (3, 2), (4, 2)

2. *Smoking and hearts.* The statistical evidence that smoking increases your risk of heart disease takes various forms. As a general rule the evidence that is easier to get is not as compelling as other evidence that takes a lot more time and effort to collect. This problem is based on evidence of the quick-and-easy kind.

The points in the scatterplot below represent 21 countries. The x-coordinate gives the country's average cigarette consumption per adult per year, and the y-coordinate gives the country's mortality rate from coronary heart disease in deaths per 100,000, for adults age 35–64. The data are from the early 1960s, before cigarette packs and advertisements were required to display a warning from the Surgeon General.

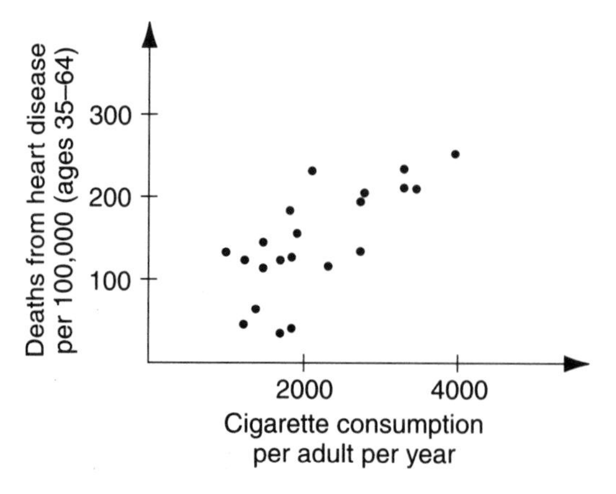

   a. Fit a line by eye to the scatterplot, and give the y-coordinates of the points on your line whose x-coordinates are 0 and 4000.

b. Use the two points from (a) to compute the slope of your line. Be sure to give the units of measurement.

c. According to your fitted line, what would be the death rate for a country in which no one smoked? For a country with average consumption of 2000 cigarettes per person per year? (2000 is about 2 packs per person per week, but the average includes non-smokers as well as smokers.)

d. An increase of 1000 cigarettes per person per year corresponds to an increase of how many deaths per 100,000?

e. People smoke; countries don't. But the points of this plots correspond to countries, not people. In each country, some people smoke a lot, some smoke a little, some not at all. Would you find the data more persuasive if each point corresponded to a group of people who all smoked the number of cigarettes given by the x-coordinate of the point? Why or why not?

3. *Cabbage Butterflies*. Fit a line by eye to the right-hand scatterplot in Example 14.13.

   a. Use your line to find fitted values for $y = (1/\#\text{ days})$ for $x = 10°\text{C}$ and $x = 25°\text{C}$.

   b. Use the two points from part (a) to compute change in y, change in x, and the slope of the line.

   c. Write the equation of the line in point-slope form, with $y = 1/(\#\text{ days})$ and $x = °\text{C}$.

   d. The time it takes, from egg hatch to pupa, is described by an equation of the form

$$(\#\text{ days})(\text{temp}°\text{C} - \text{threshold}) = \text{constant},$$

where "threshold" and "constant" are numbers that depend on the species of butterfly. Use your fitted line to find the values of these two numbers for the cabbage butterfly. (Start with the equation of the line in (c), put (1/# days) in place of y, and then rewrite the equation in a form that puts both °C and # days on the left-hand side.)

4. *Sleeping shrews: The effect of an influential point*. The scatterplot below shows heart rates for REM versus DSW sleep for the shrews of Example 7.4. For Shrew #I, the heart rate for deep slow-wave sleep (the x-value) is far from the average, which means that the point has a lot of influence on the estimated slope.

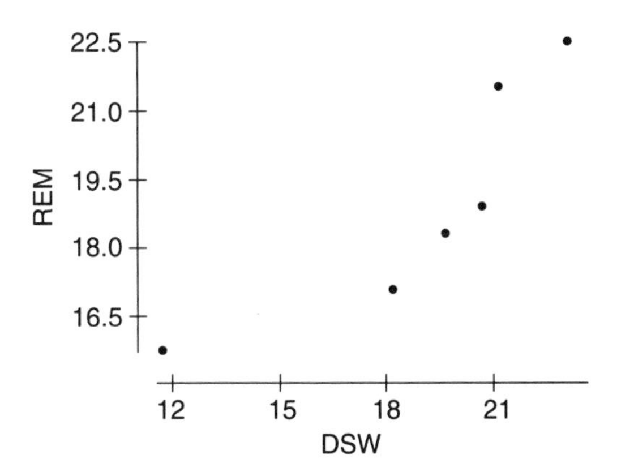

a. Ignore the influential point and fit a line by eye to the other five points. Find the coordinates of two points on your line, and use them to estimate the slope of the line.

b. Now start over again, and this time fit a line by eye to all six points. Find the slope of this new line.

c. Imagine fitting yet another line to the points, this time leaving out a point other than the outlier. Would the slope of this line be much different from the one in (b)?

5. *Interpreting a fitted regression line*. You probably know how much you weigh, but not how much volume you occupy. Weights are easy to measure, but volumes are not; to measure your volume directly, you'd have to measure the amount of water you displace when you submerge yourself in a special tank. Regression methods are often useful for situations like this one, where one of two closely related variables is easy to measure, and so can be used to give fitted values for the other.

In 1933, Edith Boyd published data giving weights (kilograms) and volumes (liters) for 18 children between the ages of 5 and 8. Her data lead to a fitted regression line:

$$(\text{Volume} - 15) = (1.00)(\text{Weight} - 15)$$

a. Write the equation in the form Volume = Intercept + (Slope) (Weight).

b. Find fitted values for the volumes of a child who weighs (i) 10 kg (ii) 15 kg (iii) 20 kg.

c. I'm 50 years old and weight about 65 kg. Find a fitted value for my volume. Explain why you should not trust this number in the same way you would trust your fitted values in part (a).

d. Two of the data points used to fit the line were:

$$(x, y) = (\text{Weight, Volume}) = (15.0, 14.5) \text{ and } (16.0, 15.8).$$

Find the Residual = Obs − Fit for each one.

e. According to the fitted line, two children who differ in weight by 1 kg. will, on average, differ in volume by how many liters? What if they differ in weight by 5 kg? 10 kg?

f. Suppose the data had been expressed in different units of measurement. Find the equation of the regression line if:

	Weights had been measured in	Volumes had been measured in
i.	kilograms	milliliters
ii.	grams	liters
iii.	grams	milliliters

g. Use your results from (f) to fill in the blanks:

i. If you multiply the carrier values by a constant and leave the response values unchanged, the slope of the fitted line gets __________ (multiplied/divided) by that constant.

ii. If you multiply the response values by a constant and leave the carrier values unchanged, the slope of the fitted line gets __________ (multiplied/divided) by that constant.

iii. If you multiply the carrier values by a constant C and multiply the response values by a second constant R, the slope of the fitted line gets multiplied by ________ (C/R, R/C, CR, 1/CR).

6. *Chirping crickets*. Crickets have special organs on their front wings that make a chirping sound when they rub the wings against each other. As a rule, the warmer the air temperature, the faster they rub their wings. The relationship between temperature and chirp rate is well summarized by fitting a regression line, although the particular line depends on the species. Fifteen pairs of measurements for the striped ground cricket (*Nemobius fasciatus fasciatus*) give the following fitted line:

$$(\text{\# chirps per second}) = 7.3 + (0.35)(\text{Temp °C})$$

a. How many chirps per second would you predict at each of the following temperatures: 20°C, 25°C, 30°C?

b. A temperature difference of 1°C corresponds to a difference (on average) of how many chirps per second? What about a difference of 10°C? What about 10°F?

c. Two of the data points were $(x, y) = (22, 16)$ and $(27, 16)$. Find Res = Obs − Fit for each one.

d. Find fitted values for the chirping rates at 0°C and 100°C. What do these numbers mean? What do they tell you about the limitations of line fitting as a method for data analysis?

e. The value of S_{xx} = sum of (change in x)(change in x) for this regression was 238.4.

i. What was the value of sum of (change in x)(change in y)?

ii. $SS_{Res} = 11.95$. Find the SE and 95% confidence interval for the slope.

f. Rewrite the regression equation to correspond to the following units:

	Temperature	**Chirp rate**
i.	°C	Chirps/min
ii.	°K (= °C − 273)	Chirps/sec
iii.	°F	Chirps/sec

7. *Point of Averages = center of gravity*. Use the scatterplots (a)–(d) of Exercise 1 for this exercise. (i) For each one, guess the x and y averages, using the fact that, just as a dot-graph balances at the average, a two-dimensional scatterplot balances at the point of averages, i.e., the coordinates of the center of gravity are (x_{avg}, y_{avg}). (ii) For each graph, write out the x and y coordinates of the points and compute the x and y averages to check your guess in (i).

8. *Fitting lines through the origin: Simple numbers for practice*. Two data sets (a) and (b) are shown below. For each one, the point of averages is the origin (0, 0), so the least squares slope is simply (weighted sum of y)/(weighted sum of x), using the x-values as weights. For each data set, (i) plot the points and use your plot to estimate the slope of the least squares line. (ii) Compute the least squares slope, and compare with your guess. (Some people's guesses tend to overestimate the least squares slope. Did yours?) Notice that (b) is just (a) with x and y switched. Why are the fitted lines not the same?

(a)

x	−2	−1	0	0	1	2
y	−1	−1	−1	1	1	1

(b)

x	−1	−1	−1	1	1	1
y	−2	−1	0	0	1	2

9. *Least squares lines: Simple numbers for practice.* For each of the data sets whose scatterplots are given in Exercise 1 (a)–(d), follow the format of the worksheet in Example 15.14 to compute S_{xx}, S_{xy}, and the least squares slope, the pieces of the decomposition, and their sums of squares.

10. *Estimates minimize SS_{Res}.* The purpose of this exercise is to illustrate an important general fact about the intercept and slope of a least squares line: If you were to use any other line, with a different intercept and/or slope, and compute SS_{Res} for that line, you'd get a bigger number than the SS_{Res} for the least squares line.

Use the data from #1c:

a. Changing the intercept: Keep the slope fixed at the least squares value of −1, and compute residuals and SS_{Res} for lines with slope −1 and intercepts 1, 2, 2.5, 3 (least squares value), 3.5, 4, and 5. Notice that you can save time by finding the residuals geometrically from the scatterplot, as the vertical distances from the data points to the line. For example, the least squares line passes through (0, 3) and (3, 0), so the residuals are +1, −1, and 0.

b. Draw a graph, one point for each of your seven lines in part (a), plotting SS_{Res} on the y-axis versus the intercept on the x-axis. Join the points with a smooth curve, and use the curve to locate the intercept that minimizes SS_{Res}. Is this the same intercept as for the least squares line?

c. Changing the slope: Keep the intercept fixed at 3, and compute residuals and SS_{Res} for lines with intercept 3 and slope equal to −2, −3/2, −1, −1/2, and 0. Here, as in (a), you can save time by finding the residuals geometrically.

d. Graph SS_{Res} versus slope, find the slope that minimizes SS_{Res}, and compare with the slope of the least squares line.

11. *ANOVA table: Simple numerical drill.* For each of the data sets (a)–(d) of Exercise 9, write out the ANOVA table.

12. *Confidence intervals for the slope: Numerical drill.* Use your results from Exercises 9 and 11 to find 95% confidence intervals for the least squares slopes for data sets (a)–(d) of Exercise 1.

13. *Confidence interval for predicted values: Numerical drill.* The width of the confidence interval depends on the x-value for which you want a fitted y-value: the farther x is from the average of the observed xs, the wider the interval will be. Using data set (b) from Exercise 1, find 95% confidence intervals for fitted y-values corresponding to the x-values shown below, and use your intervals to complete the table; then plot "lower vs. x" and "upper vs x" on a set of xy-axes.

			Confidence limits		interval
	x	$x - x_{avg}$	lower	upper	width
0					
1					
2					
3					
4					

14. *A regression approach to two-sample ANOVA: Dummy carriers*. Look back at the hospital carpet data (Exercise 8.D.1), which came from a CR design to compare two groups, corresponding to carpeted and bare floors. Here's an ANOVA table, along with group averages (in # colonies per cubic foot):

Source	df	SS	MS	F	Averages:
Grand Average	1	1764			Carpet: 11.2
Carpet vs. Bare	1	7.84	7.84	0.90	Bare: 9.8
Residual	14	122.58	8.76		
TOTAL	16	1894.42			

The purpose of this exercise is to illustrate an important sense in which regression is a generalization of ANOVA. (In fact, any ANOVA can be done using regression methods.)

Think of the 16 bacteria levels as values of a response variable, and define an "indicator" carrier variable: for each room (case), the carrier equals 1 if the room got carpet, 0 if the room was left bare.

a. Scatterplot response versus carrier, and notice that your scatterplot looks just like a parallel dot graph.

b. Carry out the regression: Fit a least squares line using the methods of Example 14.10 and write the equation of the fitted line. What does the slope tell you about the effect of carpeting?

c. Are the residuals for your regression analysis the same or different from the ANOVA residuals? Are the two SS_{Res}s the same or different?

d. Construct a confidence interval for the slope. This interval is the same as the interval for a particular contrast. Which one?

15. *Regression diagnostics: Clotting blood*. Like ANOVA, regression can be used both informally, to summarize data or look for patterns, and formally, to test hypotheses and construct interval estimates for unknown "true" values. The formal methods depend on assumptions about the underlying chance process that created the data, and there are methods for checking these assumptions, methods a lot like those of Chapter 12. You can find a thorough and readable explanation in a book by Sanford Weisberg, listed in the references. Almost all the methods grow out of a single idea that I hope by now is familiar: Any systematic pattern in your residuals suggests that you need to revise your model and/or transform your data. This exercise gives a brief introduction to transforming.

The time it takes for normal blood to clot depends on the concentration of plasma, as the following data show quite clearly:

y = clotting time (sec)	118	58	42	35	27	25	21	19	18
x = plasma conc (mg/100ml)	5	10	15	20	30	40	60	80	100

a. Scatterplot the data. Why would it be foolish to fit a regression line?

b. Rule of the bulge: Transforming to linearity. If the points of your scatterplot lie near a curve, not a line, you can often get a more-nearly linear relationship by transforming your response or carrier or both, choosing your transformations from the power family (Chapter 12). You can use the "rule of the bulge" to help choose the powers:

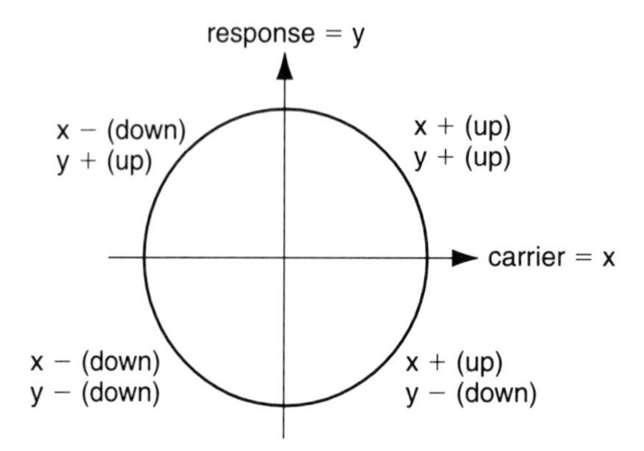

Rule of the bulge for choosing a transformation

Decide which of the four quarter circles best describes your data. Then try transformations in the same "direction" as the bulge in the curve. *Carrier*: The two right-hand curves bulge in the direction of bigger x-values: Try powers of p bigger than 1. The two left-hand curves bulge toward smaller x-values: Try powers less than 1. *Response*: The top two curves bulge in the direction of bigger y-values: Try powers bigger than 1. The bottom two curves bulge down: Try powers less than 1.

Which transformations does the rule suggest for the carrier and response for the clotting data?

c. Here are the data, transformed to log concentrations and clotting rates (= 10/time):

clotting rate	0.085	0.172	0.238	0.286	0.370	0.400	0.476	0.526	0.556
log conc	0.699	1.000	1.176	1.301	1.477	1.602	1.788	0.190	2.000

Which value of p did I use for the carrier? For the response?

d. Scatterplot the transformed data.

e. Fit a regression line, compute SS_{Res}, and find a 95% confidence interval for the slope.

16–18. *Transylvania effect.* An ordinary ANOVA for the Transylvania data (Exercise 8.A.16 and Example 8.11) finds no detectable effect of moon phase on the rate of admission to a mental health clinic. However, in Chapter 12 (Example 12.8) we found such a striking pattern in the residuals that I do not take the results of the ANOVA seriously. The next exercises ask you to carry out a regression-based analysis of the data, an analysis that I find much more sound and informative for these data.

16. The basic idea will be to compare admission rates during the full moon with those not during, that is, before and after. Regard each month as a case, take as your response y the rate during the full moon, and construct a carrier x by averaging the two rates for before and after the full moon.

a. Compute those 12 x-values, one for each month, and use them to fill in the table below.

Month	Aug	Sep	Oct	Nov	Dec	Jan	Feb	Mar	Apr	May	Jun	Jul
y = During												
x = (Before + After)/2												

b. Scatterplot y versus x, using numbers as plotting symbols (8 = Aug, 9 = Sep, . . . , 7 = Jul) so you will be able to see time trends in your plot.

c. I think the patterns in the scatterplot are quite striking. Write a short summary. Be sure to comment on the following:

i. Draw arrows joining the points for the following sequences of months:

Dec → Jan → Feb → Mar → Apr

Jun → Jul

Aug → Sep

Describe the patterns.

ii. If you were to divide the 12 months into two clusters on the basis of what you see in the scatterplots, which points would you put together in each cluster?

iii. Which gives a better summary of the scatterplot, a single line fitted to all 12 points, or separate lines for your two clusters from (ii)? Fit your line or lines by eye, and estimate the intercept(s) and slope(s).

iv. What can you say about the effect of moon phase on admission rate, on the basis of your work in (iii)? (What intercept and slope would you expect for a fitted line if phase had no effect? How do your fitted lines differ from this one? How do you interpret those differences?)

17. *Regression*

a. Compute the slope and intercept for a single regression line fitted to all 12 points; then compute fitted values, residuals, and SS_{Res}.

b. Scatterplot Res versus Fit and comment on the pattern. Remember that any systematic pattern suggests that there is a better model to be found.

c. Compute slopes and intercepts for regression lines fitted separately to each of your two clusters of points from Exercise 16(c). Then compute fitted values, residuals, and SS_{Res} = sum of squares of all 12 residuals.

d. Scatterplot Res versus Fit for each cluster, all 12 points on the same graph, and comment on what you see.

18. *Comparing models*

a. Percent improvement in SS_{Res}: Think of one line (17a) and two lines (17c) as competing models. The first is simpler, the second fits better. Here is one standard way to assess whether the better fit is worth the extra complexity: Start by computing "improvement in fit" = larger SS_{Res} − smaller SS_{Res}; then express this number as a percent (or decimal fraction) of the larger SS_{Res}, to get percent improvement.

b. Two other possible models: So far, you've fit two models: (I) single line, and (II) two separate lines. The first required you to estimate two constants, one intercept and one slope, and would be called a 2-parameter model. The second model is a 4-parameter model: two intercepts, two slopes. Here are two 3-parameter models:

III: Two lines, with different slopes but the same intercept;

IV: Two lines, with same slopes but different intercepts;

Imagine fitting each of these models and computing SS_{Res} for each. (Don't do any of the arithmetic.) Which of models I–IV fits best (smallest SS_{Res})? Next best? Next? Worst? Decide which of II and IV you prefer, and tell why.

c. The general regression significance test for comparing models I and II. Notice that model II (two lines) includes model I as a special case: If the two slopes are equal, and the two intercepts are equal, model II is the same as I. When one regression model includes another as a special case, you can construct an F-ratio to test whether the observed differences between the two models might reasonably be due just to chance error. You compute a ratio of mean squares from the following information:

Model	SS_{Res}	df_{Res}
I	$SS_{(Res\#1)}$	$df_{(Res\#1)}$
II	$SS_{(Res\#2)}$	$df_{(Res\#2)}$

Numerator $SS = SS_{(Res\#1)} - SS_{(Res\#2)}$ = improvement in fit

$df = df_{(Res\#1)} - df_{(Res\#2)}$ = change in df

Denominator $SS = SS_{(Res\#2)}$

$df = df_{(Res\#2)}$

Compute the F-ratio, and test at the 5% level. (Why is $df_{(Res\#2)} = 8$?)

19–23. *Algebraic derivations*. The next five exercises are designed to lead you through the algebra that proves two clusters of results: (1) The line through the point of averages with slope equal to S_{xy}/S_{xx} does in fact minimize the residual sum of squares, and (2) both the fitted slope and the predicted values are weighted sums of the observations.

19. *Regression slope as a weighted sum of the observations*. We got our confidence interval for the slope of the least squares line using a result from Chapter 11: the standard error for a weighted sums of observed values equals $SD\sqrt{SS_{Weights}}$. The goal of this exercise is to show that the regression slope $\hat{\beta} = S_{xy}/S_{xx}$ can be written as a weighted sum of y-values with weights $w = (x - x_{avg})/S_{xx}$. The exercise comes in four parts, starting with a simple concrete example, and building to the general result.

a. Suppose you have only three data pairs (x, y): $(0, y_1)$, $(1, y_2)$, and $(2, y_3)$. Show that the fitted slope S_{xy}/S_{xx} equals $(-1/2)y_1 + (1/2)y_3$.

b. Show that this expression is the same as the sum of (weight)(obs) = $(w)(y)$, where each weight $w = (x - x_{avg})/S_{xx}$.

c. Now suppose there are only two data pairs, but both the x-values and y-values are given as letters (x, y): (x_1, y_1) and (x_2, y_2). Find the regression slope S_{xy}/S_{xx} and show that this is the same as the weighted sum $w_1y_1 + w_2y_2$, with $w = (x - x_{avg})/S_{xx}$.

d. Finally, repeat for the general case of n data points (x, y): (x_1, y_1), (x_2, y_2), . . . , (x_n, y_n).

20. *Predicted values as weighted sums of the observed values*. This exercise is similar to Exercise 19, except this time your goal is to show that any predicted value $\hat{y}$ can be written as a weighted sum of observed values, with weights

$$w_i = \frac{1}{n} + \frac{(x^* - x_{avg})(x_i - x_{avg})}{S_{xx}},$$

where x^* is the carrier value for which we are predicting the response value, and there are n data points (x_i, y_i).

a. To keep things concrete, assume we want to predict y when $x = 3$. The predicted value is

$$\hat{y} = y_{avg} + \hat{\beta}(3 - x_{avg}).$$

As in the last exercises, I'll start with simple special cases, and build to the general result.

a. Suppose the fitted slope is zero. Find the weights.

b. Suppose the average of the y-values is zero. Find the weights.

c. Suppose the average of the y-values is zero, but this time the carrier value for which we want a prediction, instead of being 3, is unspecified and written as x^*.

d. Combine parts (a) and (c) to get the general result.

21. *The least squares line passes through the point of averages.* To show that the least squares line goes through the point of averages takes two steps. The first step is to show that the residuals add to zero if and only if the line goes through the point of averages.

a. Show that the result is true for a simple numerical example with three pairs (x, y): $(-1, 2)$, $(0, 1)$, $(1, -3)$.

b. Now suppose that the x and y values are not specified, but that there are only two pairs: (x_1, y_1) and (x_2, y_2). (i) Assume that when $x = x_{avg}$, the y-value of the fitted line is $y_{avg} + c$. Write the equation of the fitted line in point-slope form, using the point $(x_{avg}, y_{avg} + c)$ and leaving the slope $\hat{\beta}$ unspecified. (ii) First assume that the fitted line goes through the point of averages, which means $c = 0$. Use the equation of the line to find the fitted values that go with x_1 and x_2. Then find the residuals, and show that they sum to zero. (iii) Now go back to the fitted line through the point $(x_{avg}, y_{avg} + c)$. Find fitted values and residuals, set the sum of the residuals equal to zero, and show by solving for c that $c = 0$, i.e., the line goes through the point of averages.

c. Finally, modify your proof in part (b) to fit the general case of n data points (x, y):

$$(x_1, y_1), (x_2, y_2), \ldots, (x_n, y_n).$$

Hints: (1) Residuals will add to zero if and only if the sum of the observed values equals the sum of the fitted values. (2) The line passes through the point of averages if and only if (x_{avg}, y_{avg}) satisfies the equation of the fitted line, i.e., $y_{avg} = \hat{\alpha} + \beta x_{avg}$.

22. (*Continuation* of Exercise 21). The goal of this exercise is to show that if two lines have the same slope, and the first goes through the point of averages but the second doesn't, then the first line has smaller SS_{Res} than the second.

a. Show that $SS_{Res} = n(\text{avg of res})^2 + \text{sum of }(\text{res} - \text{avg})^2 = n(r_{avg})^2 + \text{sum of }(r_i - r_{avg})^2$. Hint: $SS_{Res} = \text{sum of }(r_i)^2$. Put $(r_i - r_{avg}) + (r_{avg})$ in place of (r_i) and use the fact that the sum of $(r_i - r_{avg})$ must be zero.

b. Show that two lines with the same slope may have different values for the average of the residuals, but will have the same values of $(r_i - r_{avg})$.

c. Combine your results from (a), (b), and Exercise 21 to conclude that for any choice of fitted slope, the line through the point of averages has smaller SS_{Res} than any other line with that same slope.

23. $\hat{\beta} = S_{xy}/S_{xx}$ minimizes SS_{Res}.

Exercises 21 and 22 together show that the least squares line must go though the point of averages. This exercise asks you to show that of all lines through that point, the one with slope S_{xy}/S_{xx} leads to the smallest value of SS_{Res}.

a. Show that if a line with slope β passes through the point of averages, then

$SS_{Res} = \text{sum of } [(y - y_{avg}) - \beta(x - x_{avg})]^2$.

b. Show that the sum in (a) can be rewritten as $SS_{Res} = S_{yy} - 2\beta S_{xy} + \beta^2 S_{xx}$, where $S_{yy} = \text{sum of } (y - y_{avg})^2$.

c. Rewrite the expression in (b) as

$$SS_{Res} = \left(S_{yy} - \frac{S_{xy}{}^2}{S_{xx}}\right) + S_{xx}\left(\beta - \frac{S_{xy}}{S_{xx}}\right)^2.$$

Conclude that SS_{Res} takes on its smallest value when $\beta = S_{xy}/S_{xx}$.

2. Balloon Summaries and Correlation

The Correlation Coefficient

Introduction. For scatterplots whose points form an oval balloon, it is possible to summarize the entire plot with numerical answers to just two questions. One: What is the equation of the fitted line that relates y-values to x-values? Two: How fat or skinny is the balloon? That is, how close do the points come to the line? We can answer the first question by giving the slope and intercept of the least squares line. An answer to the second question is given by the **correlation coefficient**. Its value is always between -1 and 1, its sign tells whether the balloon tilts up (+) or down (−), and its absolute value tells the shape of the balloon, with values near 0 for fat, directionless blobs, and values near 1 (+ or −) if all the points lie nearly on a line. I'll give a precise definition of the correlation coefficient later, but first here's a display of several scatterplots to give you a feel for what correlation measures.

EXAMPLE 4.17 Scatterplots and Balloon Summaries for a Range of Correlation Coefficients

The scatterplots in Figure 14.19 have more than 600 points each. Often a data set will have far fewer, and the balloon shape of the plot will be less obvious. Nevertheless, it is reasonable to ask what the plot would look like if you had a lot more points. As you'll see in the next several paragraphs, one useful way to think about the correla-

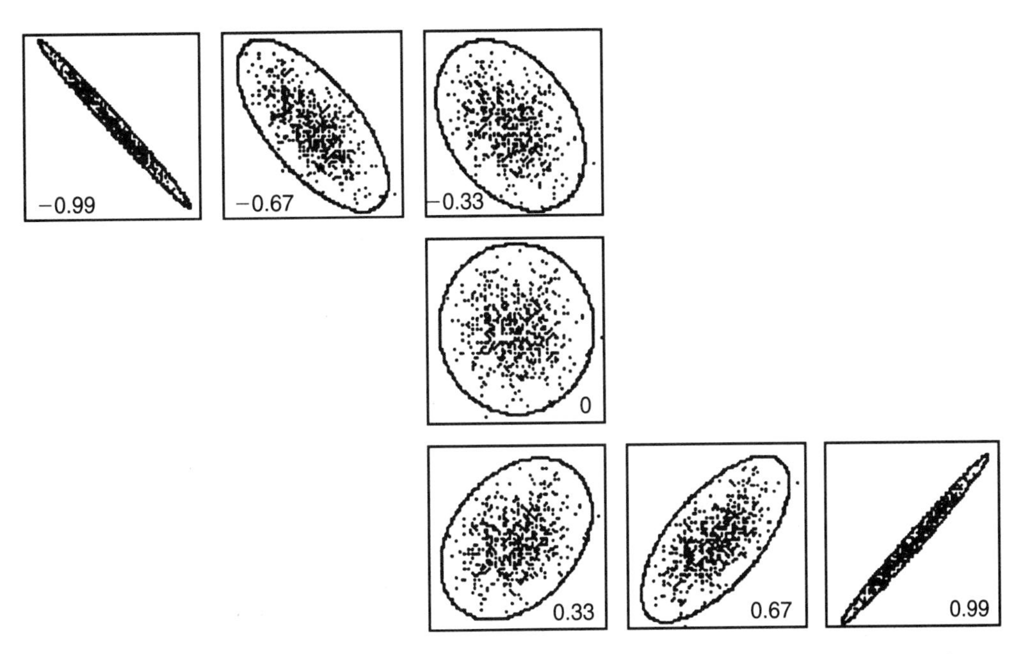

FIGURE 14.19 Correlation tells shape and direction for balloon summaries. For balloon-shaped scatterplots, the correlation tells how well a regression line summarizes the *xy*-relationship. A correlation near 1 or −1 means the "balloon" is practically a line: regression gives very precise predictions (fitted *y*-values) for the response. A correlation near 0 means the balloon is almost round: the direction of the least squares line is not well determined by the data, and the line gives generally poor predictions.

tion is to regard it as a description of the shape of the balloon that summarizes what you think the scatterplot would look like if you had many more points.

Here, for comparison, are four correlations for data sets from this chapter. Notice the range of values, from 0.4 for the fattish balloon of the working memory data up to 0.997 for the almost perfectly linear relationship for the cabbage butterflies.

EXAMPLE 14.18 Correlation Coefficients for Four Data Sets

Example 14.14 Imagery and working memory: correlation = 0.41
x = visual report rate (1000/# sec.)
y = verbal report rate (1000/# sec.)

Example 14.4 Mental activity data: correlation = 0.67
(placebo condition)
x = activity level before
y = activity level after

Example 14.15 Prisoners in solitary: correlation = 0.88
(treatment group only)
x = day 1 frequency of alpha waves
y = day 7 frequency of alpha waves

Example 14.13 Cabbage butterflies: correlation = 0.997
x = temp(°C)
y = rate (1/# days) ■

Like any one-number summary, the correlation has both advantages and disadvantages. A single number takes a lot less space (and attention) than a scatterplot, and so journal articles often report a correlation and don't show the scatterplot. For example, when psychologists report reliabilities, they almost never show the scatterplot.

One-number summaries also make comparison easier. For example, you can use four correlations to summarize the results of Carl Rogers' study of psychotherapy in Example 4.22. There, a higher correlation means that a subject saw her/himself as more like the kind of person s/he wanted to be. The average correlation for subjects who got therapy went from 0.0 before therapy to 0.3 after; the average for subjects in the control group started out at 0.6, and didn't change: before therapy, the way subjects in the treatment group saw themselves was unrelated to the way they wanted to be; after therapy, they saw themselves as somewhat closer to the way they wanted to be, but not as close as subjects in the control group.

Although the correlation is a compact summary that makes comparisons easier, it doesn't tell you about outliers, and what it does tell about the shape of the scatterplot is quite limited: it only tells how close to a line the points fall, and whether the line slopes up or down. If the points lie along a curve, or in two distinct blobs, the correlation coefficient won't tell you that. There's another way to say this:

> Correlations summarize balloons.
>
> **If your plot isn't balloon-shaped, don't use a correlation.**

Balloon-Based Estimates for Scatterplots

For a "balloon-shaped" data set, you can find quick and quite accurate approximations to the least squares line and the correlation coefficient, working directly from a balloon summary. Although the geometric methods are approximate, they offer two advantages that make them worth learning: they are a lot quicker than using a calculator, and they can help you visualize what it is that the messier computing rules refer to. Moreover, in the idealized world of statistical theory, the geometric methods are exactly equivalent to the computing rules: one is as good as the other.

In the rest of this section, I'll use the mental activity data of Example 14.4 to show you three different balloon-based definitions of the correlation coefficient, each with its own interpretation.

Drawing the balloon summary. To get the balloon summary, start with a scatterplot. Then draw a symmetric, oval balloon enclosing most or all of the points (Figure 14.20).

Step 1. Data = scatterplot

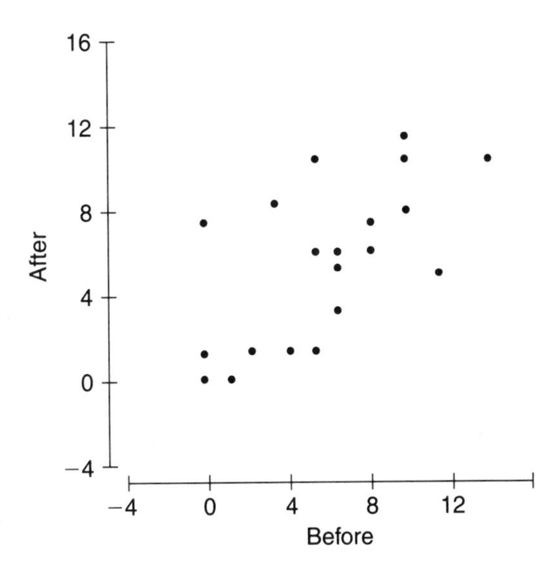

FIGURE 14.20 Scatterplot for the mental activity data of Example 14.4

The scatterplot has far fewer points than the plots in Example 4.17, and the cluster of points shows some irregularities—the plot isn't symmetric, and the density of points doesn't show the same smooth pattern as those plots which are densest in the middle and get gradually less dense as you move away from the center.

What would the plot look like if we had lots more data from the same source—in this case lots more subjects? Here are two possibilities: (1) If the irregularities are due to chance variation, then piling on lots more data would bury the first 24 points in a balloon-shaped cloud of hundreds more. According to this view, the balloon suggested by the 24 points is the message, and the irregularities are just noise. (2) On the other hand, the irregularities might be part of the message: it could be that the "true" relationship between before and after readings under the placebo conditions is not as simple as a balloon summary suggests.

I hope you recognize the two possibilities, and the two ways of looking at the data, as variations on the old theme of "Data = Fit + Residual." A good data analysis should try for a simple fit, but should look at the residuals, too. So keep in mind, as you read the rest of this section, that balloon summaries and correlations correspond to the fit, and don't tell you about the residuals.

Step 2. Balloon = Fit. Draw a symmetric oval balloon, enclosing all or almost all the points (Figure 14.21).

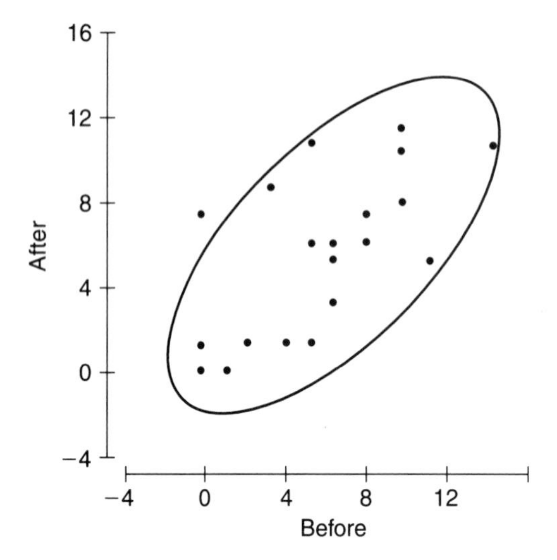

FIGURE 14.21 The balloon should be symmetric and enclose all or almost all the points.

Lines of symmetry. Since our balloon summary assumes that the source of our data tends to produce a symmetric plot, you want your balloon to take advantage of that assumption. If your first try at a balloon isn't symmetric, keep adjusting the shape until you get one that has two lines of symmetry (Figure 14.22).

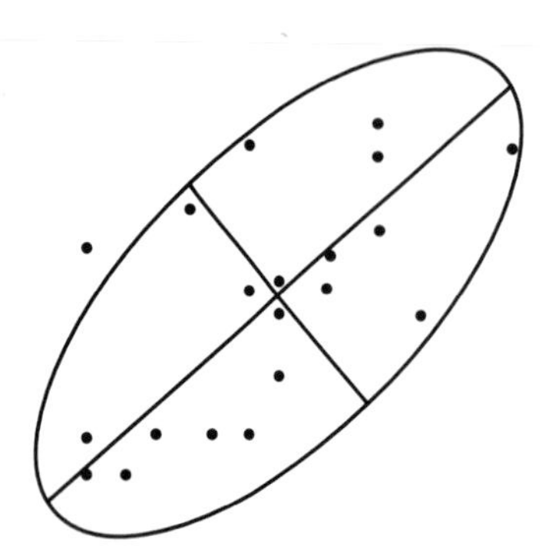

FIGURE 14.22 The balloon should have two lines of symmetry. The two lines of symmetry are the longest possible diameter, called the **major axis**, and the shortest possible diameter, called the **minor axis**. If you were to fold your balloon in half along either line of symmetry, the two halves should match.

It might seem that the major axis ought to be the least squares line, but it isn't. To see why, and to find the least squares line, you need to think about vertical slices.

The Correlation Coefficient and the Regression Effect

Interpretation: vertical slices. Turn back to Example 14.17, and imagine cutting a thin vertical slice through the scatterplot and balloon with correlation 0.67 (Figure 14.23). What does the slice tell us? The slice corresponds to points whose x-values are about the same. The scatter in the vertical direction tells how the y-values are distributed when x is roughly constant. Notice two things about the pattern: the points are densest at the *midpoint* of the slice, and the density falls off in a symmetric way as you move away from the midpoint on either side. A histogram (Chapter 12, Section 3) would look roughly normal: mound-shaped, and symmetric.

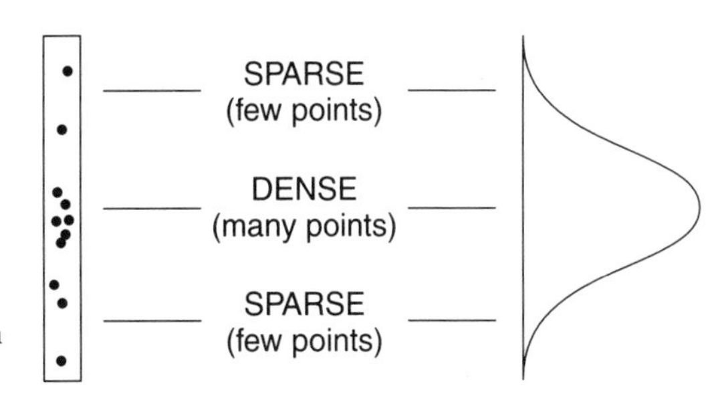

FIGURE 14.23 Distribution of y-values in a slice is roughly normal.

How can we use this information to predict y from x? Suppose you knew that a subject's mental activity level (x) before the placebo injection was 10, and you wanted to predict the level two hours after the injection (y). A reasonable prediction would be the y-value for midpoint of the vertical slice at $x = 10$: If the data behave like the scatterplots in Example 14.17, then the midpoint prediction corresponds to the y-value where the points are densest—the value under the peak of the normal curve.

It can be proved that this is in fact the least squares prediction. Any other choice would give a larger residual sum of squares. In particular, notice that the major axis does not cut the vertical slices at their midpoints (Figure 14.24).

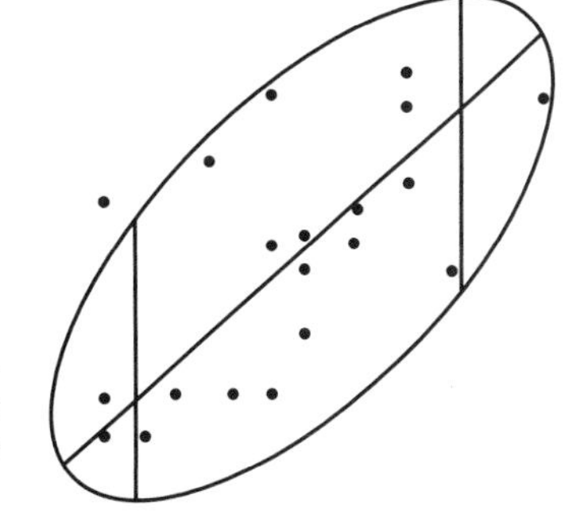

FIGURE 14.24 The major axis misses the midpoints of the vertical slices. Slices to the right of the balloon's center have their midpoints below the line of symmetry. Slices to the left have their midpoints above the line of symmetry.

You can check in Figure 14.25 that the set of midpoints for all possible vertical slices form a line. This line is the least squares line.

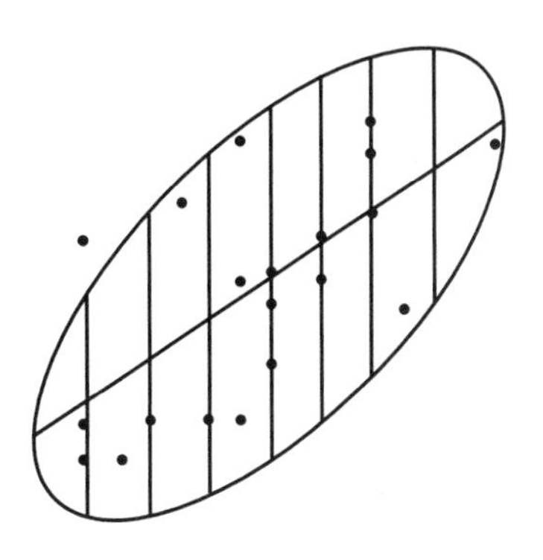

FIGURE 14.25 The line of midpoints is the least squares line.

A shortcut. Notice that in Figure 14.25, the least squares line intersects the balloon at the two points where it is possible to draw vertical tangent lines. To see this, imagine taking vertical slices closer and closer to the right end of the balloon, and notice that the slices get shorter and shorter, shrinking to a single point at the end. The "slice" at that point corresponds to the vertical tangent. To find the line of midpoints, we only need two points on the line, so we can get the least squares line by finding the two points where there are vertical tangents, and joining them with a line.

Step 3. The box of tangents. Draw the two vertical tangent lines, and the two horizontal tangent lines, and connect them to form a box. Join the two points of vertical tangency to get the least squares line (Figure 14.26).

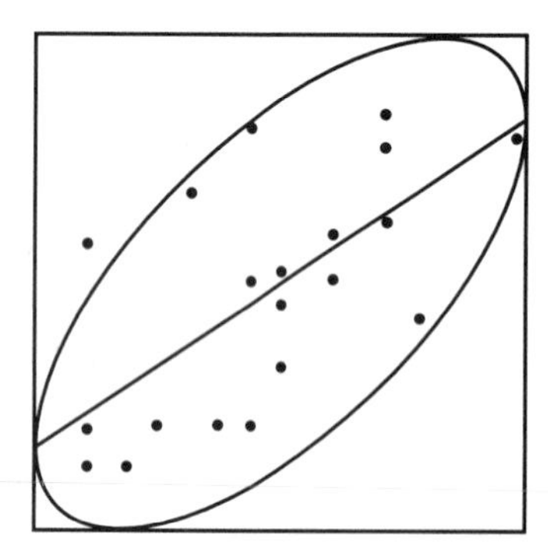

FIGURE 14.26 The box of tangents. The least squares line joins the two points of vertical tangency.

Step 4. Least squares slope = TAN_y/MAX_x. The slope of any line equals "rise over run" or "change in y over change in x," so we can estimate the least squares slope by computing the change in y and the change in x between the two points of vertical tangency (Figure 14.27).

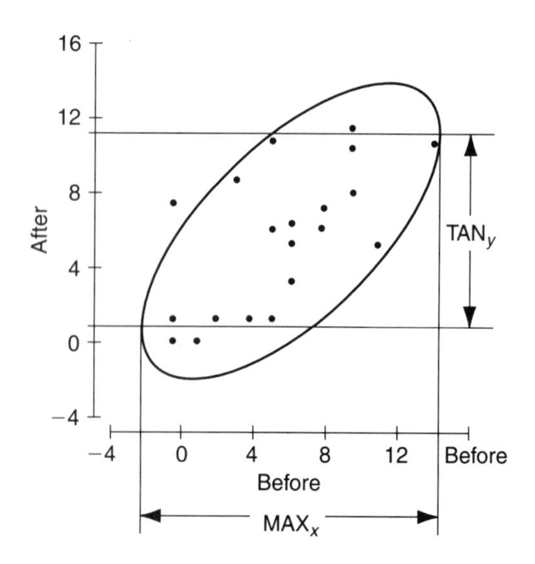

FIGURE 14.27 TAN_y AND MAX_x

Referring to Figure 14.27:

Let TAN_y = rise = vertical distance between the two points of vertical tangency.

Let MAX_x = run = horizontal distance between the same two points = width of the box of tangents.

CAUTION: If your plot uses different scales for y and x, you'll need to measure TAN_y and MAX_x in those different scales. In this example, TAN_y runs from about 0.75 to 10.75 along the y-axis, so $TAN_y \approx 10$. MAX_x runs from about -2.25 up to 14.25, so $MAX_x \approx 16.5$.

Slope = rise over run = TAN_y/MAX_x. In this example, the slope is about $10/16.5 \approx 0.61$. (The actual least squares slope is 0.6.)

Step 5. Correlation = TAN_y/MAX_y. How far will the least squares line be from the diagonal of the box of tangents? Turn back to the plots of Example 14.17, and notice that the answer depends on how fat the balloon is. When the balloon is a perfect circle (correlation = 0), the line joining the points of vertical tangency is horizontal and the two lines are as far apart as possible.

We can use this fact to measure the shape of a balloon summary: We express TAN_y as a fraction of the length of the vertical side of the box. This gives us the correlation coefficient, apart from the sign. If the balloon is a perfect circle, TAN_y will be zero, and so will the correlation. If the "balloon" is actually a line, then TAN_y and MAX_y will be equal, and their ratio will be 1. For the mental activity data, we get a correlation in between (Figure 14.28).

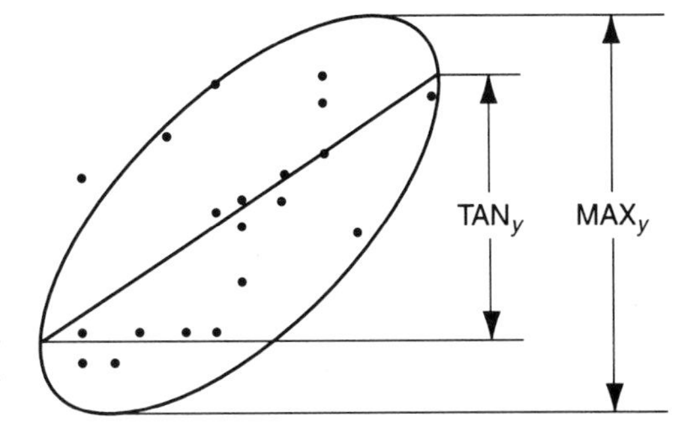

FIGURE 14.28 Correlation = TAN_y/MAX_y. Since TAN_y and MAX_y are both measured vertically, they will always both be on the same scale, and you can measure them with a ruler. In this picture, TAN_y = 27 mm, MAX_y = 40 mm, and the correlation is 27/40 = 0.68.

Interpretation: The regression effect. What sort of predictions would we get if we used the diagonal instead of the least squares line? Since we've defined correlation as a measure of how far apart (or close together) these two lines are, an answer to this question will give you another way to understand correlation.

We'll consider a particularly simple kind of situation—one where the x and y variables have equal averages and equal SDs. Then as you can see in the plots of Example 14.17, the diagonal is the 45° line whose equation is $y = x$. One example of a data set like the one I've just described was studied by the geneticist and statistician Francis Galton. His x-values were the heights of fathers, and his

y-values were the heights of their adult sons. For Galton's data, the diagonal line $y = x$ is the set of points with son's height = father's height. If you know that the father is 6'0", the prediction based on the line of symmetry is that the son will also be 6'0". Galton found that in reality, the average height for sons of six foot fathers was *less* than 6'0". On the other hand, the average height for sons whose fathers were only 5'4" turned out to be *greater* than 5'4". The pattern in Galton's data is similar to what we saw earlier: For vertical slices to the right of the center (fathers taller than the average), the midpoint of the slice (average height of their sons) was below the line of symmetry and here is below the line $y = x$ (less than their fathers' height). For slices to the left of the center (fathers shorter than average), the midpoint of the slice (average height of their sons) was above the line of symmetry and here is above the line $y = x$ (greater than their fathers' height). For Galton's data, the line $y = x$ and the major axis of the balloon are in fact one and the same line. That won't always be the case, but what matters for our purpose is that on both sides of the average, the midpoint of a vertical slice falls in between the diagonal line and the overall mean for the population. Taller fathers tend to have sons who are not quite as tall. Shorter fathers tend to have sons who are not quite as short. Galton said the heights of the sons "regressed toward the mean." (And that's where the name "regression" for line fitting came from.)

Regression to the mean is a very general phenomenon—one that applies whenever a balloon summary is appropriate. It's well know in sports, for example. Teams (or individual players) with extremely good or extremely bad averages early in the season almost always end up drifting back toward the middle of the pack as the season wears on. It's also true of the mental activity data: On average, subjects with very high levels before the placebo injection have somewhat lower levels after; subjects with very low levels before tend to have somewhat higher levels after.

The correlation coefficient measures the size of the regression effect (Figure 14.29). Large regression effects go with correlations near zero. Small regression effects go with correlations near 1 or -1. (Take a minute to verify this for the plots of Example 14.17.)

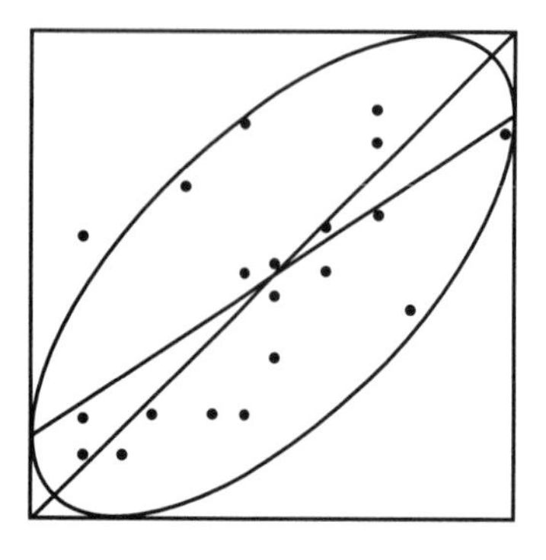

FIGURE 14.29 Correlation measures the size of the regression effect. The least squares line has slope TAN_y/MAX_x and the diagonal of the box of tangents has slope MAX_y/MAX_x. The ratio of the two slopes equals TAN_y/MAX_y, the correlation coefficient.

Correlation and Fraction of Variability "Explained"

Overall and conditional SDs for y. Our interpretation of the correlation coefficient in terms of the regression effect has been based mainly on comparing averages. There's a second interpretation, perhaps even more important than the first, based on comparing variability. This interpretation starts from two ideas you've already seen: (1) A vertical slice through a scatterplot is basically just a dot plot for the y-values when x is fixed at a particular value. (2) The degree of spread in a dot plot is measured by the SD. Putting these two together lets us use the spread of a vertical slice to find the SD.

Extending these ideas will allow us to use the balloon for comparing two different kinds of SDs for the y-values. The first is the overall (or unconditional) SD for y, which I'll write SD_y, the one you get in the usual way, by regarding the whole set of y-values as a single sample. The other SD is the **conditional SD for y given x**, written $SD_{y|x}$, which measures the spread of the y-values in the vertical slice at x. This SD takes the y-values in the slice as the sample, and you would compute it using the deviations from the midpoint of the slice.

Step 6. MID_y tells the conditional $SD_{y|x}$ for y when x is fixed. See Figure 14.30.

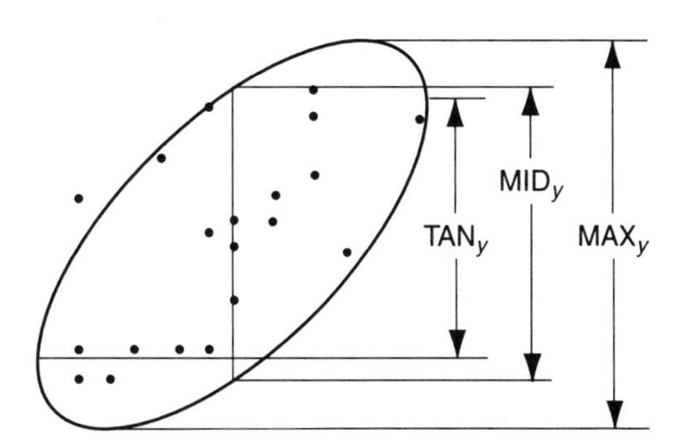

FIGURE 14.30 MID_y tells the conditional $SD_{y|x}$ of y given x.

Let MID_y be the length of the vertical slice at the center of the balloon. MID_y tells the spread of the dot plot of y-values in the vertical slice at the midpoint, and is proportional to $SD_{y|x}$, the SD of the y-values when x is fixed. The more spread out the y-values, the bigger MID_y will be.

A caution. It's an obvious property of the balloon shape that vertical slices are longest at the center of the balloon, and get shorter as the slices get closer to either end of the balloon. Moreover, if you look at slices through one of the plots in Example 14.17, you find that dot plots for the slices extend about as far as the slices themselves. Dot plots for central slices extend farther than dot plots for slices near the ends of the balloon. It would be natural to think this means that the conditional $SD_{y|x}$ for x at the center is larger than $SD_{y|x}$ for x away from the center. Natural, but wrong:

Same SDs

A key assumption for least squares lines and balloon summaries is that $SD_{y|x}$ is the same for all values of x.

On the surface, it looks like the balloon summary and the assumption of same SDs are at odds, but they're not. Here's why it's possible for the assumption to be true and still have a balloon-shaped plot: Even though the dot plot for the central slice of the balloon is more spread out than the plots for slices away from the center, the extra spread results from having *more points* in the center slice. If just 1 of 20 points in a slice is more than 2 SDs from the average of the slice, then a slice with only 10 points might well have all of them within 2 SDs of the center, while a slice with 50 points will probably have 2 or 3 points more than 2 SDs from the average, and so its dot plot will be more spread out even though it has the same SD.

If our assumption holds, and $SD_{y|x}$ is the same for all x, we can use any vertical slice to measure $SD_{y|x}$. What's so special about MID_y? The answer is not obvious, but if you use MID_y to measure the conditional $SD_{y|x}$, then the ratio MID_y/MAX_y has a particularly useful interpretation.

Step 7. $(MID_y/MAX_y)^2$ = fraction of variation "unexplained" by the least squares fit. If MID_y measures the conditional $SD_{y|x}$, what can we use to measure to find the unconditional SD_y? The answer: MAX_y. I won't prove it, but I think you can convince yourself that it makes sense. Turn back yet again to the plots of Example 14.17, turn the page sideways, pick one of the balloons, and imagine that all the points in the scatterplot slide down and pile up along the y-axis to make a dot plot. The spread of this dot plot is what we measure with the unconditional SD_y, and this spread will be proportional to the overall vertical spread of the balloon, that is, to MAX_y.

Using a mathematical derivation based on these same ideas, you can show that the ratio of MID to MAX equals the ratio of the conditional SD to the unconditional SD:

$$\frac{MID_y}{MAX_y} = \frac{SD_{y|x}}{SD_y}.$$

A useful interpretation comes from thinking about predicting y-values. If you don't know x, and you use the average of all the ys as your prediction, then SD_y is the SD for your prediction error, Obs − Pred = $y - y_{avg}$. If you know x, and you use the least squares fit (= midpoint of the vertical slice at x) as your prediction, then $SD_{y|x}$ is the SD for your prediction. If knowing x isn't much help, then $SD_{y|x}$ will be almost as large as SD_y, and the ratio will be near 1. On the other hand, if knowing x allows you to predict y with very little error, $SD_{y|x}$ will be small, and the ratio will be near 0. For technical reasons, it is more convenient to work with the square of the ratio:

$$\left(\frac{MID_y}{MAX_y}\right)^2 = \left(\frac{SD_{y|x}}{SD_y}\right)^2 = \text{fraction of variation not "explained" by the least squares line.}$$

Notice that MID_y can never be larger than MAX_y, so the ratio MID_y/MAX_y will always be between 0 and 1. Values near 1 correspond to fat balloons: almost all the original variation in y remains "unexplained," still in the residuals after fitting the least squares line. Values near 0 correspond to very skinny balloons: only a tiny fraction of the original variation remains in the residuals after fitting the line.

We now have two ways to measure the shape of a balloon: the correlation (TAN_y/MAX_y) and the fraction of unexplained variation $(MID_y/MAX_y)^2$. It is reasonable to expect the two numbers to be related by a formula, since both measure different but related aspects of the shape of an ellipse. In fact, the two are closely related, and the formula relating them will give us the second interpretation of the correlation coefficient.

It can be proved mathematically that for any ellipse, $(TAN_y)^2$ and $(MID_y)^2$ add "like Pythagoras" to give $(MAX_y)^2$:

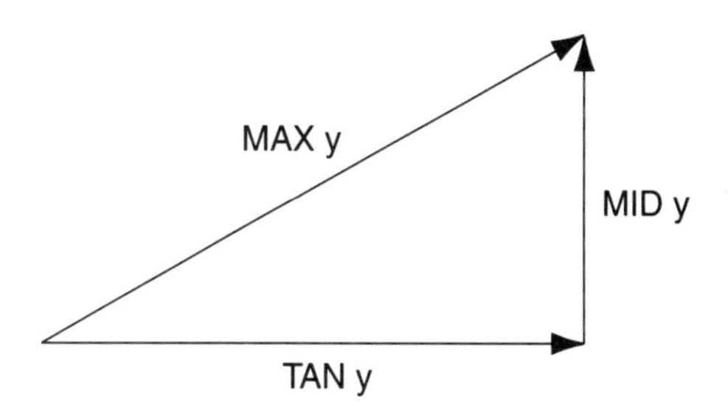

$$(TAN_y)^2 + (MID_y)^2 = (MAX_y)^2.$$

Dividing both sides of the equation by $(MAX_y)^2$ gives

$$\left(\frac{TAN_y}{MAX_y}\right)^2 + \left(\frac{MID_y}{MAX_y}\right)^2 = 1.$$

Since $(MID_y/MAX_y)^2$ is the fraction "*un*explained," then $(TAN_y/MAX_y)^2$ must be (and is) the fraction of the original variation in y "explained" by the regression. The second interpretation of the correlation coefficient, then, is this:

(correlation coefficient)^2 = fraction of variation "explained" by the least squares regression line

The Pythagorean relationship also gives a second way to estimate the correlation coefficient from a balloon summary:

$$\textbf{Correlation} = \pm\sqrt{1 - \left(\frac{MID_y}{MAX_y}\right)^2}$$

(Correlation)^2 = SS_{Regr}/S_{yy}. The balloon-based geometry has an exact parallel in terms of the sums of squares you get from decomposing the data. Remind yourself of the flowchart for the decomposition of response values: first the response y is split into its average plus the "change in y"; then the "change in y" is split into fitted values plus residuals. Just as with ANOVA, the sums of squares adds up (Figure 14.31):

Obs:	"change in y"	=	(slope)(change in x)	+ residual
SS:	S_{yy}	=	SS_{Regr}	+ SS_{Res}

FIGURE 14.31 Decomposition of response and SSs. The "total variation in y," or sum of $(y - y_{avg})^2$, is split into a piece due to regression (SS_{Regr}) and a residual piece (SS_{Res}). The more nearly the points fall along a line, the bigger SS_{Regr} will be, and the smaller SS_{Res} will be.

We can express each of SS_{Regr} and SS_{Res} as a fraction of S_{yy}; the two fractions will then add to 1:

$$\frac{SS_{Regr}}{S_{yy}} + \frac{SS_{Res}}{S_{yy}} = 1 \quad (\text{because } S_{yy} = SS_{Regr} + SS_{Res})$$

The two fractions correspond exactly to $(TAN_y/MAX_y)^2$ and $(MID_y/MAX_y)^2$. Their two sizes tell how well or poorly a line fits the scatterplot (Figure 14.32):

Good fit Points near a line	SS_{Regr} is a large proportion of S_{yy}, i.e. SS_{Regr}/S_{yy} is near 1. SS_{Res} is a small proportion of S_{yy}, i.e. SS_{Res}/S_{yy} is near 0.
Poor fit Points in a fat blob	SS_{Regr} is a small proportion of S_{yy}, i.e. SS_{Regr}/S_{yy} is near 0. SS_{Res} is a large proportion of S_{yy}, i.e. SS_{Res}/S_{yy} is near 1.

FIGURE 14.32 SS_{Regr}/S_{yy} measures how well a line fits the plot.

Since $SS_{Regr}/S_{yy} = (TAN_y/MAX_y)^2$, we can translate the balloon-based rule for computing the correlation coefficient into a rule based on sums of squares:

$$\textbf{Correlation} = \pm\sqrt{SS_{Regr}/S_{yy}} = \pm\sqrt{1 - SS_{Res}/S_{yy}}$$

EXAMPLE 14.19 CORRELATION AS $\sqrt{1 - SS_{Res}/S_{yy}}$.
For the mental activity data, we have

$$S_{yy} = SS_{Regr} + SS_{Res}\ 266 = 118.8 + 147.2$$

Thus the regression accounts for $118.8/266 = 0.45$ of S_{yy}, leaving $147.2/266 = 0.55$ of S_{yy} as residual SS. The correlation is $\sqrt{0.45} \approx 0.67$. ■

Correlation Equals the Standardized Regression Slope

We now, as advertised, have two interpretations for the balloon-based correlation coefficient, one in terms of averages and one in terms of SDs. There is one more important interpretation, in terms of the slope of the least squares line. To set the stage, remember the warning about finding the least squares slope from a balloon summary. If x and y are measured in different scales, you need to use the x scale to measure change in $x = \text{MAX}_x$, and use the y scale to measure change in $y = \text{TAN}_y$. The fitted slope depends on the scales you use. (If you change your x units from feet to inches, all x-values get multiplied by 12, so MAX_x gets multiplied by 12, and slope $= \text{TAN}_y/\text{MAX}_x$ gets divided by 12.)

The correlation coefficient does *not* depend on the scale in this way. If you change your y-units from feet to inches, all of TAN_y, MID_y and MAX_y get multiplied by the same number (12), and the ratios $\text{MID}_y/\text{MAX}_y$ and $\text{TAN}_y/\text{MAX}_y$ don't change.

The final step in setting the stage is to notice that both the least squares slope and the correlation coefficient have the form $\text{TAN}_y/\text{MAX}_{\text{something}}$:

$$\hat{\beta} = \text{least squares slope} = \frac{\text{TAN}_y}{\text{MAX}_x} \qquad \frac{\text{TAN}_y}{\text{MAX}_y} = \text{correlation coefficient} = r$$

Switching from MAX_x in the denominator to MAX_y switches from the least squares slope to the regression coefficient. Algebraically, the switch corresponds to multiplying by the ratio $\text{MAX}_x/\text{MAX}_y$ ($= \text{SD}_x/\text{SD}_y$):

$$\hat{\beta}\left(\frac{\text{SD}_x}{\text{SD}_y}\right) = \left(\frac{\text{TAN}_y}{\text{MAX}_x}\right)\left(\frac{\text{MAX}_x}{\text{MAX}_y}\right) = \left(\frac{\text{TAN}_y}{\text{MAX}_x}\right) = r$$

Note that this gives a way to compute the correlation coefficient numerically. First compute SD_x, SD_y and $\hat{\beta}$; then find r:

Correlation coefficient $r = \hat{\beta}(\text{SD}_x/\text{SD}_y)$.

To get the third interpretation, notice that if $\text{SD}_x = \text{SD}_y$, then their ratio equals 1, and the least squares slope equals the correlation coefficient.

Ordinarily, of course, (x, y) pairs don't come to us with $\text{SD}_x = \text{SD}_y$. But if we compute SD_x and SD_y, then divide each x by SD_x and each y by SD_y, the new, rescaled xs and ys will have $\text{SD}_x = \text{SD}_y = 1$, and so the correlation coefficient r will equal the least squares slope $\hat{\beta}$. Usually, if you're going to divide by the SD, you first subtract the average, then divide by the SD, to get a new **standardized variable,** whose average is 0 and SD is 1. This is the basis of the third interpretation of the correlation coefficient, which also gives a second computing rule:

First standardize x and y: Replace each x-value by $\dfrac{x - x_{\text{avg}}}{\text{SD}_x}$ and each y-value by $\dfrac{y - y_{\text{avg}}}{\text{SD}_y}$.

The least squares slope for the standardized x and y equals the correlation coefficient for x and y.

You can see this relationship between the shape of a balloon and its least squares slope in the plots of Example 14.17. For these plots, although no scales are shown, the x and y values have already been standardized, so that $SD_x = SD_y$, and thus the least squares slope equals the correlation coefficient. Check, for example, that for the first plot in the last row, with correlation of 1/3, $TAN_y = (1/3)MAX_y$, and since $MAX_y = MAX_x$, the slope $TAN_y/MAX_x = 1/3$.

EXAMPLE 14.20 Correlation = Standardized Regression Slope

The standardization is easier to see with artificial data:

						Avg	SS
x	−3	−1	1	1	2	0	2
y	−5	−2	−1	5	3	0	4

For these numbers, $SD_x = \sqrt{16/4} = 2$, $SD_y = \sqrt{64/4} = 4$, $S_{xy} = 27$, and $\hat{\beta} = 27/16 \approx 5/3$. The y-values are twice as spread out as the x-values, which makes the least squares slope twice as big as it would be if the SDs were equal.

To find the correlation, we first divide each x by SD_x (= 2), each y by SD_y (= 4):

						Avg	SS
x	−3/2	−1/2	1/2	1/2	2/2	0	1
y	−5/4	−2/4	−1/4	5/4	3/4	0	1

For these standardized numbers, $SD_x = \sqrt{1} = 1$, $SD_y = \sqrt{1} = 1$, $S_{xy} = 27/32$, and $\hat{\beta} = 27/32 \approx 5/6$. The least squares slope ($27/32 \approx 0.83 \approx 5/6$) is the correlation coefficient (Figure 14.33).

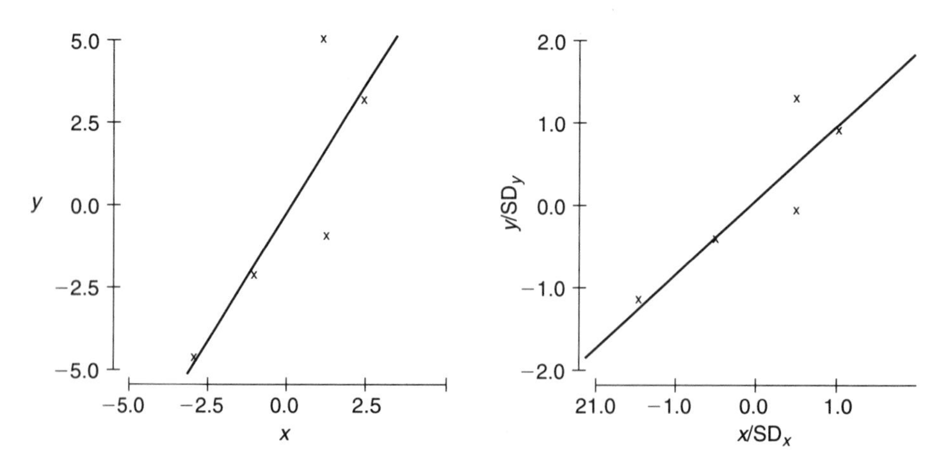

FIGURE 14.33 Correlation coefficient equals standardized regression slope. For the plot on the left, in the original scale, $SD_y = 4$, $SD_x = 2$, and the least squares slope is $5/3 \approx 1.67$. The plot on the right shows y/SD_y versus x/SD_x. For these standardized data, the SDs are equal, and the least squares slope is $5/6 \approx 0.83$, the correlation coefficient. For the left hand plot, the ratio $SD_y/SD_x = 2$, and the slope is twice what it would be if the SDs were equal. The correlation coefficient equals (SD_y/SD_x) times the least squares slope. ■

A final caution: Three traps to avoid. Because the correlation coefficient is so useful as a summary, it can be easy to forget that, like the least squares line, it has limitations. Like regression, the correlation is designed for data pairs whose scatterplot suggests a line or balloon. *The value of a correlation doesn't tell you anything about whether the plot is actually balloon-shaped, and if it isn't, the correlation can be misleading as a summary.* The only insurance is to look at the plot.

Correlation, like regression, helps you study relationships among numerical variables, but *association is not causation.* In particular, a high correlation may be due to a lurking variable. For example, the correlation between the number of deaths in a state and the state's total tax revenue is 0.95. The explanation is not that taxes are life-threatening, but that both variables are roughly proportional to a state's total population. If you divide by population size, to get death *rates* and tax revenue *per capita*, the correlation gets much smaller, and even changes sign, to −0.38.

This example reminds us that correlation isn't causation, and teaches a second lesson as well: *High correlations don't always tell you more than low correlations.* The high positive correlation between deaths and taxes has no value as information because both variables are basically disguised versions of the same one variable—the state's population. The much lower correlation between death rate and per capita tax revenue may or may not be useful, but at least it does provide information that wasn't obvious beforehand.

Exercise Set B

1. Match the following correlations with their scatterplots.

Correlations: (i) 0.56 (ii) 0.03 (iii) −0.90 (iv) 0.34

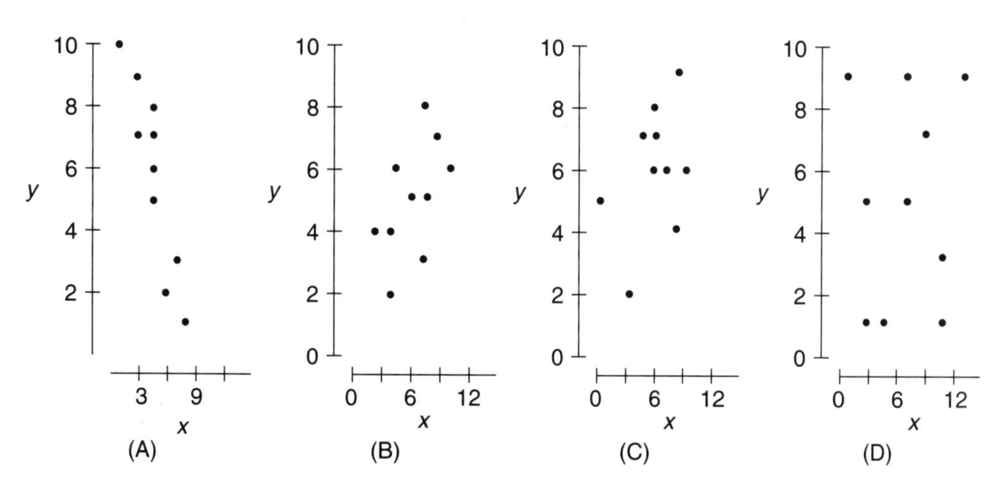

2. *Mother's stories.* The scatterplots in Figure 14.34 are based on the data of Chapter 4, Review Exercise 7. The plot on the left shows the numbers of Type C (overinvolved,

parent-centered) versus Type A (personally involved, child-centered) stories told by mothers of schizophrenic sons. The plot on the right is for mothers in the control group.

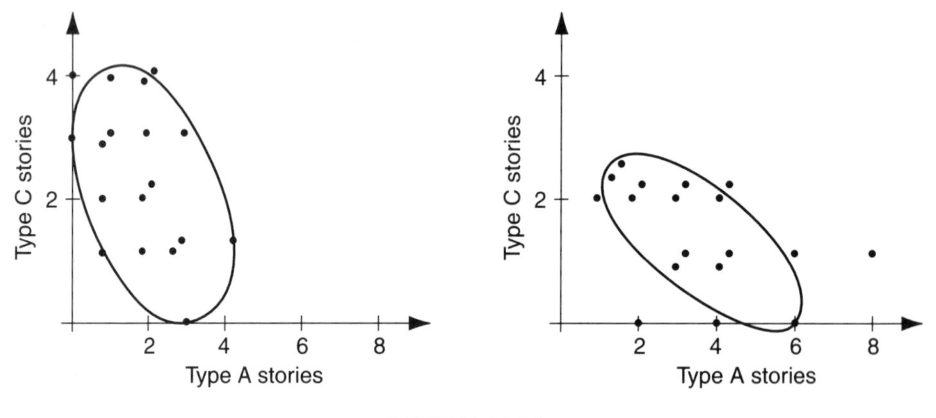

FIGURE 14.34

a. Use the balloon summaries to match each plot with (i) the slope of the least squares line, choosing from −0.4 and −0.6, and (ii) the correlation coefficient, choosing from −0.55 and −0.85.

b. Explain why the fact that each mother told a total of 10 stories all but guarantees the correlations would be negative.

c. Try to find a plausible explanation for why one group of mothers shows a stronger relationship between numbers of stories than the other.

3. Estimate correlation coefficients for each of the balloon diagrams in Figure 14.35. You'll need to measure with a ruler and apply the "balloon rule" from Section 4: don't just guess.

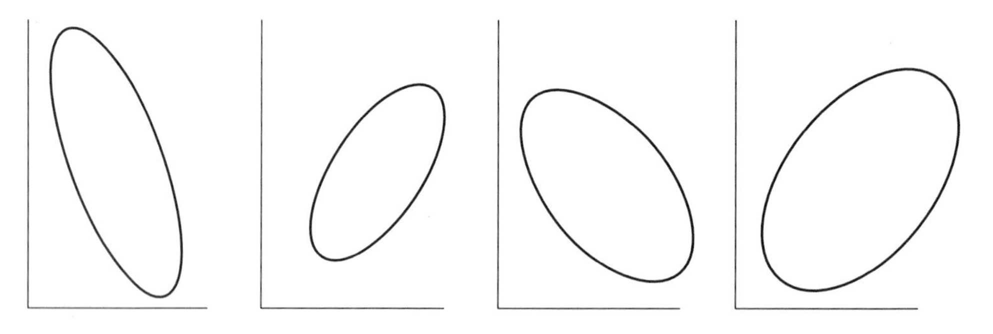

FIGURE 14.35

4. *Shapes of eggs: Drawing a balloon summary.* Would you expect the lengths and widths of a set of eggs to show positive, zero, or negative correlation? The scatterplot in Figure 14.36 shows length versus width for a dozen hens eggs.

a. Draw a balloon summary. You should use a pencil, and you'll probably have to make some changes to your first try in order to get a symmetric balloon. (Here are some things to check: Locate the center of your balloon. The vertical distance up, from center to the edge of the balloon, should be the same as the distance down, from center to edge. Similarly, the horizontal distance right, from center to edge, should equal the distance left.) Once you've got a symmetric balloon, use a ruler to find the following: TAN_y, MAX_y, and MID_y.

b. Estimate the correlation two ways (i) as TAN_y/MAX_y, and (ii) as $\sqrt{1 - RATIO^2}$, where $RATIO = MID_y/MAX_y$.

c. Suppose that instead of a plot for 12 hens eggs, you had a plot for 12 eggs from 12 different birds, ranging in size from tiny to large. What sort of correlation would you expect? (Moral: The value of the correlation can depend a lot on whether your points correspond to units that are similar or different.)

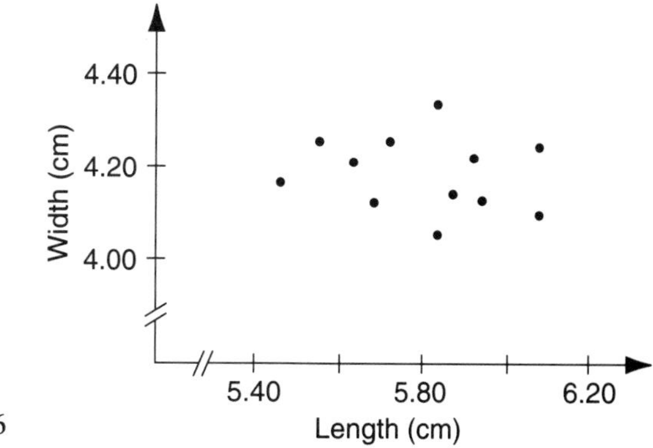

FIGURE 14.36

5. *Smoking and hearts*. Draw a balloon summary for the scatterplot in Exercise A.2. (Use the checks described in Exercise 4(a) to help make sure your balloon is symmetric.)

a. Use a ruler to find TAN_y, MAX_y, MID_y, TAN_x, MAX_x, and MID_x.

b. Estimate the least squares slope as TAN_y/MAX_x, and compare with the slope of the line fit by eye in A.2.

c. Estimate the correlation coefficient two ways, just as in Exercise 4(b).

d. Based on your experience drawing the balloon, which estimate do you have more confidence in? (Which of TAN_y and MID_y do you think is more reliable, given the way you draw balloons?)

6. *Crickets, II*. The data for Exercise A.6 are shown in a scatterplot in Figure 14.37, together with a balloon-summary of the plot. (Note that for this problem, the temperatures are in °F.)

a. Locate the center of the balloon, and use its coordinates to estimate the observed averages for x and y.

b. Use the balloon-method to find the approximate least squares line: Find the points of vertical tangency and use their coordinates to compute the slope of the line which passes through them. Then write the equation of the line in the form $(y - y_{avg}) = (\text{slope})(x - x_{avg})$. (Is your slope roughly the same as the one you got in A.6f(iii)?)

c. Use a ruler to measure TAN_y and MAX_y; then use these to estimate the correlation between chirp rate and temperature.

d. Measure MID_y and use its value, together with the value of MAX_y, to estimate (i) SS_{Res}/S_{yy} (ii) SS_{Reg}/S_{yy}, and (iii) the correlation.

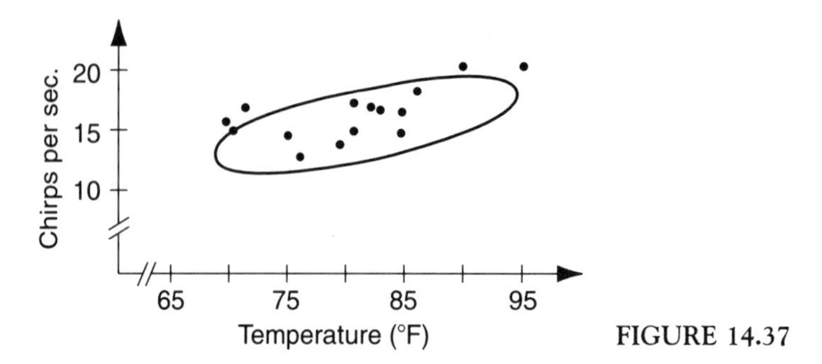

FIGURE 14.37

7. *Crickets, III*. Here are several summary numbers for the data used in the previous problem:

$$S_{xx} = 629.84 \qquad S_{xy} = 133.48 \qquad S_{yy} = 40.56$$

$$x_{avg} = 80.0 \qquad SS_{Gr\ Avg} = 4160.00 \qquad SS_{Res} = 12.28$$

a. Give a complete ANOVA table and test whether there is a "real" effect of temperature on the rate of chirping.

b. Compute a 95% confidence interval for the slope of the least squares line.

c. Compute 95% confidence intervals for the fitted chirping rates at 70°, 75°, 80°, 85°, and 90°. Judging from your intervals, what do you think of people who claim they can tell the temperature from the rate of cricket chirping?

d. Figure 14.38 shows residuals versus fitted values for the cricket data. Comment briefly on whether this plot causes you to reevaluate any of your previous work with this data set, and what you would do next if you were a statistician analyzing the data for a biologist friend.

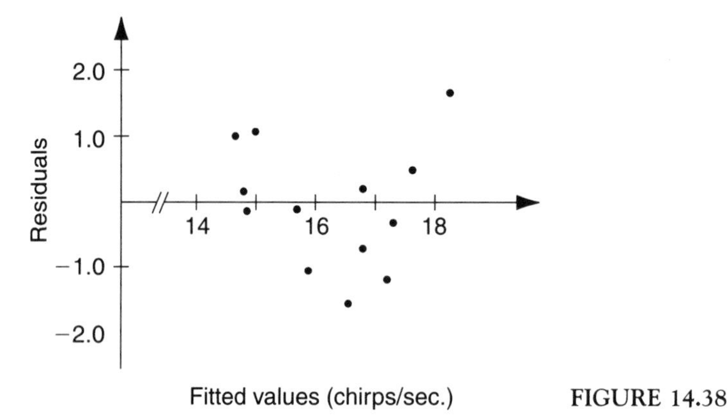

FIGURE 14.38

8. *Jobs and hearts*. It is reasonable to think that how much oxygen you use would be related to how quickly you burn calories, and that the nature of the relationship might depend on the level of physical exertion you are used to. You can use data from Exercise 8.D.5, on workers at the Hawthorne Electric Plant, to explore these relationships.

Three scatterplots are shown below: the first shows ventilation rate, in liters per minute, versus caloric expenditure, in kilocalories per minute, for workers whose jobs were rated at Physical Grade 2. The second plot is similar, but for workers at Grade 4. (Note that the scales for the two plots are not the same.) The third plot (Figure 14.40) shows both sets of workers together on the same graph.

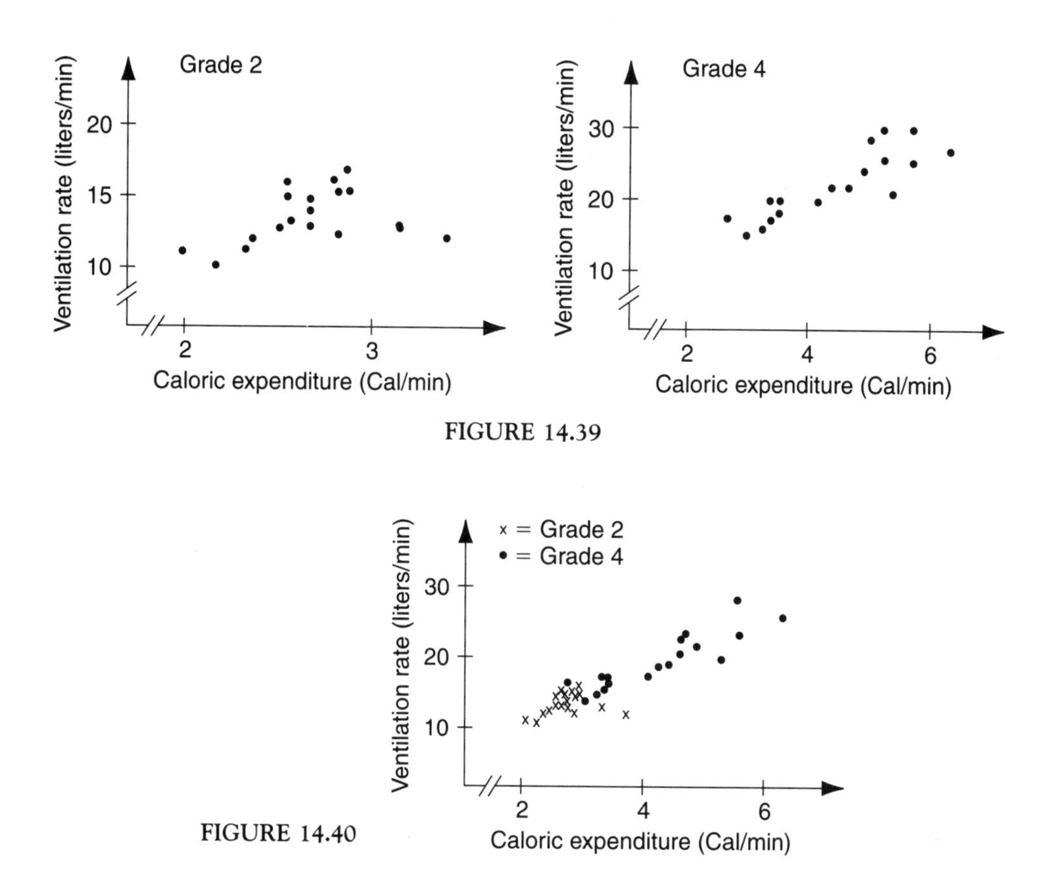

FIGURE 14.39

FIGURE 14.40

a. Outliers: The plot for workers at Grade 2 shows two outliers. True or false: The corresponding two workers seem to be in better physical condition than the other workers at Grade 2. (Omit these outliers in your work on the other parts of this problem.)

b. Slopes: Use balloon summaries for the first two plots to estimate the slopes of the least squares lines for the two groups of workers. (As always, be sure to give the units of measurement for the slopes, along with the numbers.) Can you suggest a plausible explanation for why the slopes are different?

c. Correlations: Use the balloons to estimate correlation coefficients for the two sets of workers. (Use both methods.) Can you suggest a plausible explanation for why one correlation is higher than the other?

d. Scale: Based on the shapes of the plots, do you think a transformation is worth trying? Why or why not?

9. *Correlation: Simple data*. For each of data sets (a)–(d) used in Exercise A.1, first (i) guess the correlation from the scatterplot in A.1, then (ii) compute the correlation as $\sqrt{1 - SS_{Res}/S_{yy}}$, using your work from A.9 and A.11.

10. *Defense of undoing*. In Example S4.10 and Exercise S.A.2, three judges (A, B, and C) each rated 44 instances of "undoing," classifying each instance as stage 1, 2, 3, or 4. Draw balloon diagrams to represent the scatterplots for (a) Judge A versus Judge C: reliability (= correlation) = 0.82; and (b) Judge B versus Judge C: reliability = 0.97.

11. *Rogerian therapy and Q-sort*. In Example 4.22, the therapist Carl Rogers used the Q-sort procedure to compare people's assessments of "who I am" and "who I'd like to be," and used correlations to measure how closely the two assessments agreed. Here is a summary of his results:

	Start	**End**
Therapy group	0.0	0.3
Control group	0.6	0.6

Average correlations

For the therapy group, "Start" and "End" mean "before therapy" and "after therapy."

a. Draw balloon diagrams to represent scatterplots for an average subject in the (i) therapy group, before therapy; (ii) therapy group, after therapy; and (iii) control group.

b. Each balloon you drew in (a) represents a scatterplot with 100 points. Reread Example 4.22, and tell: (i) What does a point correspond to? In other words, what is the pair of numbers that is getting plotted? (ii) What is the range of possible values for numbers plotted on the x-axis? The y-axis?

12–17. What's wrong with this inference? The next several problems each describe evidence—a set of cases and two variables measured on those cases—and a conclusion based on the data. In each problem, the inference is not justified, sometimes because one or more lurking variables affects the correlation. For each problem, tell why the inference is not justified. Identify any lurking variable(s). Then tell what kind of study you would need in order to test whether the conclusion is in fact true.

12. Ham radios:

Cases: Years, from 1920 to 1950

Variables: Number of licensed amateur radio operators registered in Great Britain
Number of officially certified mental defectives registered in Great Britain

Conclusion: The high positive correlation means that amateur radio operators tend to be mentally defective.

13. Money and words:

Cases: The twelve grades of a school district

Variables: Average weekly allowance for all children in the grade
Average vocabulary size for all children in the grade

Conclusion: The high positive correlation means that raising children's allowances leads them to improve their vocabularies.

14. Children and blood pressure:

Cases:	A random sample of mothers, aged 20–45
Variables:	Number of children born to the mother
	Mother's blood pressure
Conclusion:	The positive correlation shows that having children raises your blood pressure.

15. Groceries and cancer:

Cases:	The 100 largest cities in the US
Variables:	Number of grocery stores in the city
	Number of new cases of cancer in a year
Conclusion:	The high positive correlation shows that the food sold at grocery stores tends to cause cancer.

16. Drunk driving:

Cases:	The fifty US states
Variables:	Annual sales ($) of beer, wine and liquor
	Number of arrests in a year for driving under the influence
Conclusion:	Drinking alcohol leads to drunk driving.

17. Videotapes and smoking:

Cases:	The years from 1980 to 1990
Variables:	Percent of US families owning a VCR
	Percent of people over 18 who smoke cigarettes
Conclusion:	The negative correlation shows that as more people were able to entertain themselves watching videotapes, they found it easier to give up smoking.

3. Analysis of Covariance

Analysis of covariance—ANCOVA—is appropriate for designed experiments and some observational studies which include one or more numerical carriers, called **covariates**, along with the usual factor structure. The covariates correspond to nuisance influences that make the experimental or sampling units different from each other, and that therefore make it harder to compare different treatments or populations. In principle there may be any number of covariates, and the factor structure may be any structure for which ANOVA would be appropriate. In this section, however, we will consider only the simplest model, with a single covariate and BF[1] factor structure.

Why and When to Use ANCOVA

The next few examples illustrate what covariance adjustments do, along with some general principles for deciding when to use ANCOVA.

Advantages of ANCOVA

Including covariates in the model:

- **can reduce bias, by adjusting for differences between treatment groups**
- **can reduce residual sum of squares, by fitting and removing systematic variability**

EXAMPLE 14.21 SESAME STREET: A REASONABLE USE OF ANCOVA

(Note: This example is invented, but is based loosely on a real study. See Exercise C3.)

a. *Design*. Suppose you want to design a study to see whether watching Sesame Street for a year will increase four-year-old girls' understanding of number concepts. Since you want to know about *gain* in understanding, it would be natural to give each of your subjects a pre-test before the study begins, and a post-test a year later (Figure 14.41). There are, however, several different ways to use this information.

	Control		"Treatment"	
	PRE	POST	PRE	POST
	4	0	0	2
	8	8	2	12
	10	8	6	4
	10	8	8	14
Avg	8	6	4	8

FIGURE 14.41 Invented data for BF[1] ANCOVA

One possibility would be to use the pre-test scores to define matched pairs, and run the study as a randomized complete block experiment. (Example 14.24 discusses why this often-sensible approach won't work here.) A second approach would be to use the change in score, POST − PRE, as the response. Analysis of covariance is similar in spirit to this second approach, but instead of deciding beforehand how to adjust for the pre-test scores, ANCOVA uses the observed relationship between PRE and POST to choose the adjustment.

The study has a BF[1] structure: subjects were sorted into two groups, those who watched Sesame Street ("treatment") and those who didn't (control). The response is a post-test score. Each subject had been given a pre-test; the pre-test score is the covariate.

Notice two things about the data: (1) The post-test values (response) show a lot of variability, and (2) on average, the two groups have quite different values for the pre-test (covariate). Analysis of covariance can help deal with both problems.

b. *Analysis*

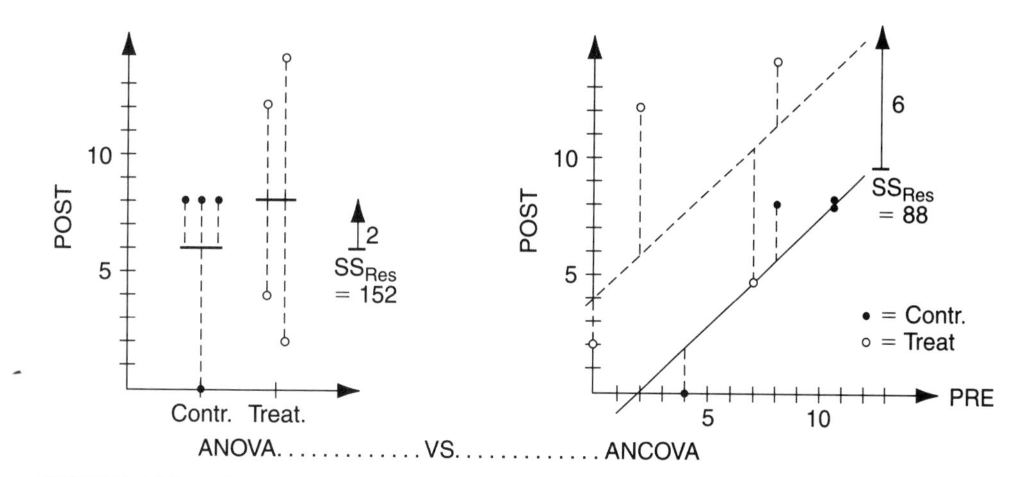

FIGURE 14.42 For ANOVA, on the left in Figure 14.42, the analysis compares group averages for the response values, and uses deviations from these averages as residuals. The difference in averages (treatment minus control) is 8 − 6 = 2, and $SS_{Res} = 152$: ANOVA finds small group differences and large residual variability, i.e., the effect of watching Sesame Street is tiny compared to individual differences. The average difference of 2 is about 40% of the estimated SD = $\sqrt{MS_{(Res\ for\ ANOVA)}}$.

For ANCOVA, on the right, we first scatterplot POST vs. PRE, using different symbols for the two groups of subjects. Then we fit two parallel lines, one for each group, to summarize the relationship between response (POST) and covariate (PRE). For any given value of the covariate, the difference between response values for the two groups is the same, and equals the vertical distance between the two lines. This **adjusted difference** in group averages is 6. The deviations from the fitted values (vertical distances from the lines) are the **adjusted residuals**; $SS_{Res(adj)} = 88$. According to this second analysis, a lot of the variability in post-test scores is associated with the pre-test variability. After adjusting for pre-test differences, the effect of Sesame Street is large compared to individual differences. The adjusted difference of 6 is about 140% of the estimated SD = $\sqrt{MS_{(Res\ for\ ANCOVA)}}$.

In this example, adjusting for the pre-test scores increases the difference in response averages from 2 to 6, and reduces the residual SS from 152 to 88. Here, ANCOVA has two advantages over ordinary ANOVA: it adjusts for the bias from unequal pre-test averages for the two conditions, and it greatly reduces the residual sum of squares. ■

ANCOVA Models May Not Be Suitable

- **if the relationship between response and covariate is not linear**
- **if the relationship is linear, but lines fitted to the groups of points have unequal slopes**
- **if adjusting for group differences violates common sense**

Example 14.22 Mental Activity Level (Continued): A Poor Candidate for ANCOVA.

Sometimes the shapes of scatterplots rule out ANCOVA, even though the design of your study suggests such an analysis might be suitable. The mental activity experiment of Example 14.4 illustrates this. There, 24 subjects provided blocks of time slots in an RCB design to compare the effects of placebo, morphine, and heroin injections on ratings of mental activity. Ratings taken two hours after the injections served as the response, with ratings just before the injection as the covariate. ANCOVA would seem to be the natural method to try.

However, look at the scatterplots of response versus covariate in Figure 14.43:

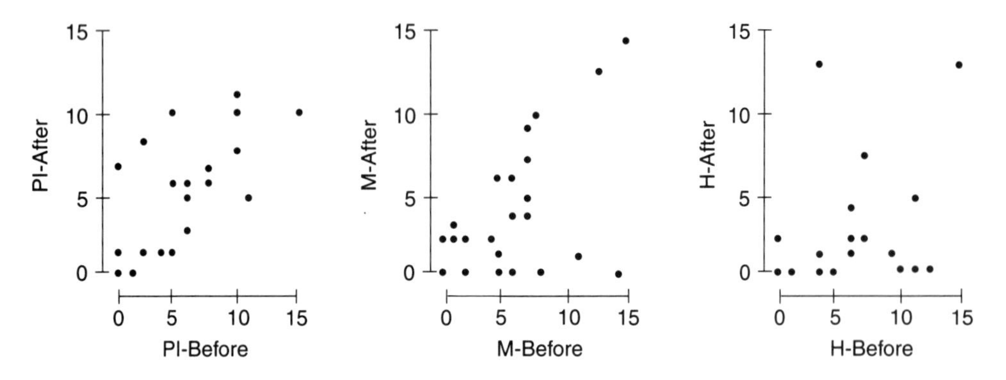

FIGURE 14.43 Scatterplots for the mental activity data

Note that although the plot for the placebo (left) suggests an oval balloon, the plots for morphine (center) and heroin (right) are not balloon-shaped, largely because so many of the points have After = 0. If you ignore the shapes of the plots and fit lines anyway, the lines for morphine and especially heroin are much less steep than the line for placebo. This data set is not a good candidate for covariance analysis. (See Exercise C.6 for more.) ■

In the Sesame Street example, we wanted to compare two groups, but according to the pre-test scores the groups started out unequal: one had an average score that was twice as high as the other. By using analysis of covariance, we could adjust response averages to get the comparison we would have gotten if the two groups had started out with equal pre-test averages.

In the mental activity example, the study was planned to be analyzed using analysis of covariance, but the shapes of the scatterplots told us the ANCOVA model wouldn't fit well, and that we shouldn't use that analysis.

In still other situations, even though the model of parallel lines may give a good fit, a covariance adjustment may violate common sense. Here's an extreme example to illustrate the point.

EXAMPLE 14.23 HEIGHTS AND SHOE SIZES: A SILLY COVARIANCE ADJUSTMENT

Suppose you want to compare the heights of first graders and tenth graders, using shoe size as a covariate. Your two groups start out with very different values of the covariate, and so you use ANCOVA to adjust the differences in average height, to get the difference you would have found if the two groups had started with equal averages for the covariate.

In technical language, this may sound reasonable, but think about it in plain English: The first graders are short and have small feet, while the tenth graders are tall and have big feet. In order to get a more meaningful way to compare heights for the two groups, you "adjust" for shoe size, that is, you find out what the average heights would be if on average the first and tenth graders had shoes of the same sizes. It turns out that your "adjusted first graders" and "adjusted tenth graders" are about the same height!

In principle, you can do this kind of analysis—there's nothing to stop you but common sense—but it makes much more sense to think of the first and tenth graders as two separate populations, and not to use a method that tries to make them equivalent.

MORAL: When the groups you want to compare represent different populations, covariance adjustments may not make sense. ■

ANCOVA or Blocking?

If covariate values are known *before* you assign treatments, using the covariate to define blocks may be better than ANCOVA.

If the conditions you want to compare are experimental, and it is possible to sort your units into blocks with similar values of the covariate in each block, then blocking is ordinarily a better strategy than ANCOVA, because ANCOVA is more restrictive: ANCOVA requires that the relationship between response and covariate be linear, with a single slope for all treatment groups. Blocking works even if the slopes are unequal, or if the relationship is nonlinear. However, blocking may not be an option.

EXAMPLE 14.24 BLOCKING OR ANCOVA?

a. *Sesame Street.* In Example 14.21, it would have been possible to use the covariate to sort the eight children into four blocks of two children each, with pre-test scores paired as follows: 0 & 2, 4 & 6, 8 & 8, 10 & 10. If it were in fact possible to assign the conditions (watch/do not watch the series), then the block design would be better than ANCOVA, because it would not only control for the nuisance influence, but would also insure that we were comparing similar children. The ANCOVA design in the example required us to compare two quite different groups of children.

   However, in reality you could neither force children in the "treatment" group to watch the series, nor prevent those in the control group from watching. In this example, as in many others, the conditions of interest had to be observational, and blocking was not an option. Here, as in most after-the-fact comparisons, ANCOVA was the best available strategy.

   Suppose, just for the sake of example, that the goal had been to see whether taking a vitamin supplement for a year would cause children to gain weight faster. Then the conditions would be experimental, and instead of using ANCOVA with initial weight as covariate, the study could be run as a randomized matched pairs design, with children paired on initial weight.

b. *Mental Activity Level.* In Example 15.11, the treatments were drug injections, and so were experimental. However, the experimental units were time slots, one block's worth per subject, and it was not practical to use the covariate to pre-sort these units into blocks according to the pre-test of mental activity. Here, just as in (a) above, ANCOVA seemed like the best strategy, although the patterns in the data later made such an analysis unworkable. ■

How to Fit the ANCOVA Model: Computing Rules

Fitting an ANCOVA model requires three sets of steps, one set (A) for actually fitting the model, a second set (B) for adjusting the treatment effects, and a third set (C) for testing the hypothesis that treatment effects are zero.

A. Fitting the parallel line model. Your goal is to fit parallel lines, one for each treatment group, to a scatterplot of response versus covariate. Doing this takes two steps, which correspond to (1) finding the point of averages for each group, and (2) finding the common slope.

1. *ANOVA step*: Finding the point of averages. Start by using ANOVA to decompose both the response and then the covariate, using the BF[1] model. Note that this step gives you the treatment averages (= Gr Avg + Tr Eff) for the response and the covariate, for each treatment group. Moreover, this step gives you two sets of residuals (one set for the response, one for the covariate), adjusted for their anchor points.
2. *Regression Step*: Finding the common slope. Because we want the slopes for the two fitted lines to be the same, and the ANOVA step has already adjusted for the anchor points, the slope we want is the same as the one we would get by scatterplotting the two sets of residuals from the ANOVA step and fitting a single line to all the points. In other words, the response residuals serve as "change in y" and the covariate residuals serve as "change in x":

$$\text{slope} = \frac{\underset{\text{of}}{\text{sum}} \begin{pmatrix}\text{response}\\ \text{residual}\end{pmatrix}\begin{pmatrix}\text{covariate}\\ \text{residual}\end{pmatrix}}{\underset{\text{of}}{\text{sum}} \begin{pmatrix}\text{covariate}\\ \text{residual}\end{pmatrix}\begin{pmatrix}\text{covariate}\\ \text{residual}\end{pmatrix}}. = \frac{\Sigma\, y'x'}{\Sigma\, x'x'}, \quad \begin{matrix}\text{where } y' = \text{response residual,}\\ \text{and } x' = \text{covariate residual.}\end{matrix}$$

We now have all we need to draw the parallel lines: the ANOVA step gave us an anchor point for each group, and the REGRESSION step gave us the slope for each line. (Figure 14.44 summarizes the two steps in a flow chart.)

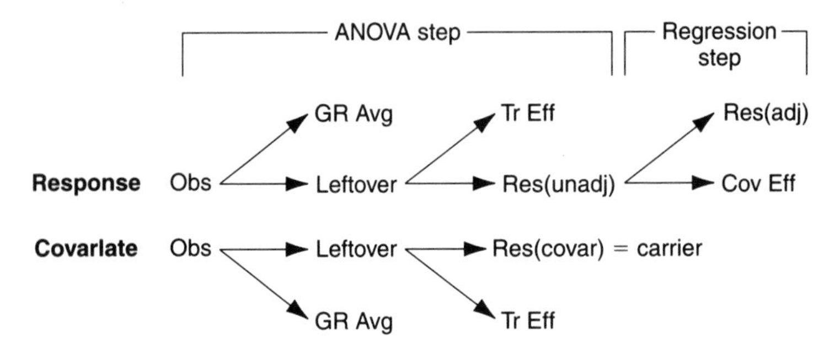

FIGURE 14.44 Fitting the parallel line model. The ANOVA step decomposes both the response (top) and covariate (bottom) in the usual way, with BF[1] factor structure. Then the REGRESSION step splits the response residuals (top) using the covariate residuals (bottom) as the carrier.

EXAMPLE 14.25 SESAME STREET (CONTINUED)

The display in Figure 14.45 shows the steps for fitting a parallel line model to the invented date of Example 14.21, following the flowchart above.

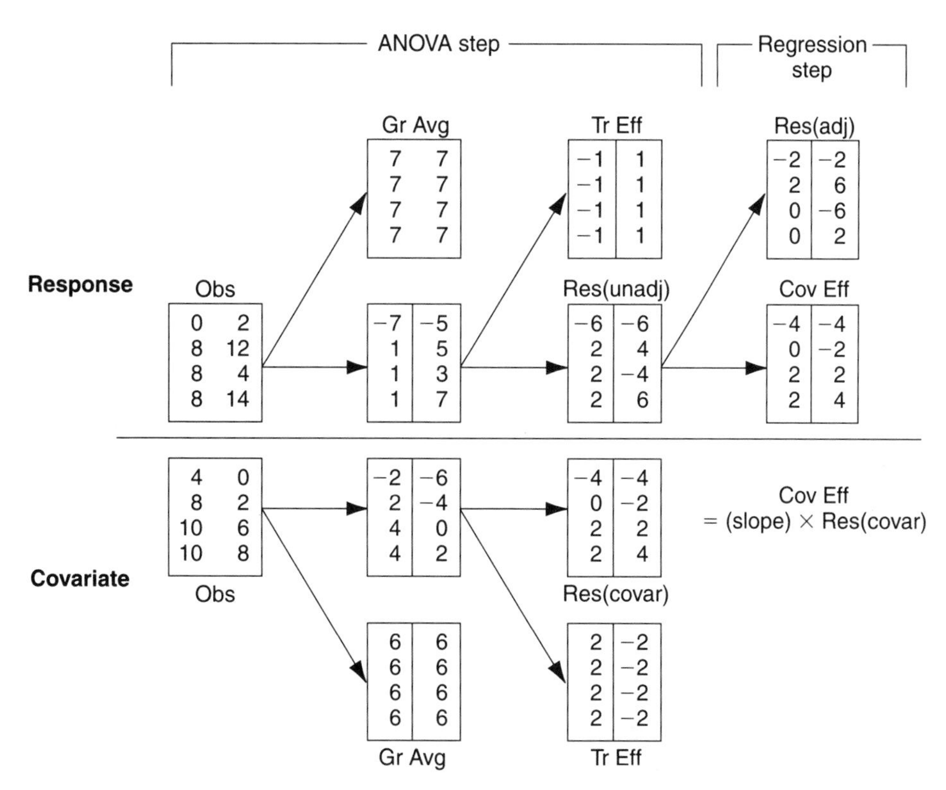

FIGURE 14.45 Computations for fitting the parallel line model. The decompositions here follow the flow charts and description given just before Example 14.25. The response averages (see Example 14.21a) are 8 (treatment) and 6 (control); these give unadjusted treatment effects (top row, third column) of ±1. The covariate averages are 4 and 8, which give covariate treatment effects (bottom row, third column) of −2 and 2. The fitted slope equals 1, so the fitted covariate effects (second row, last column), are equal to 1 times the covariate residuals (third row, third column). ■

B. *Adjusting the treatment effects.* The decompositions of the ANOVA and regression steps tell us all we need to plot the two parallel lines of the fitted ANCOVA model, but numerically, there's one more model-fitting step: adjusting the treatment effects. (If you have only two groups, this next step is equivalent to finding the vertical distance between the two fitted lines.)

The treatment effects we computed in the ANOVA step are based on simple averages, and don't take the covariate into account. (Geometrically, these effects correspond to comparing treatment groups using the y-value from each group's point of averages, ignoring the fact that the points have different x-values.) The adjustment we want corresponds to picking a common x-value for all the groups, locating new anchor points all with this same one x-value, and using the y-values to compare treatment groups. The covariate treatment effects tell how much of a change in x is needed for each group, and multiplying by the common slope gives the corresponding change in y (Figure 14.46).

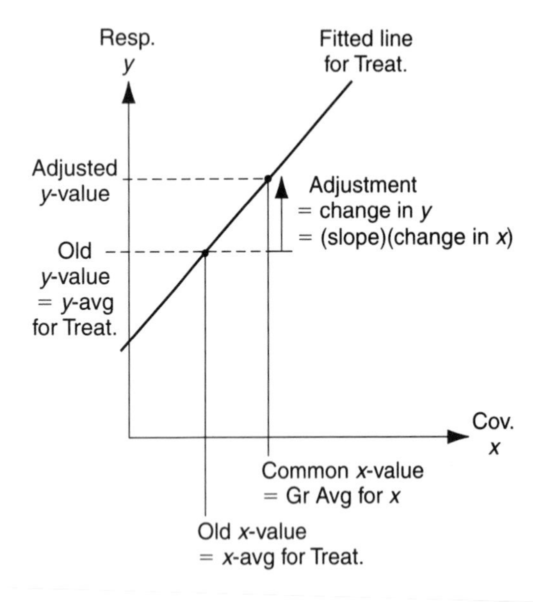

FIGURE 14.46 Adjusting the treatment average. Tr Avg(adj) = Tr Avg(unadj) − (slope)[Tr Eff(covar)]
New y = old y − (slope)[old x − new x]

EXAMPLE 14.26 SESAME STREET (CONTINUED)

Compute adjusted treatment effects for the data of Example 14.21.

SOLUTION: Tr Eff(adj) = Tr Eff(unadj) − (slope)[Tr Eff(covar)]. (See Figure 14.47.) From Example 14.25, the unadjusted treatment effects are 1 and −1, the covariate treatment effects are −2 and 2, and the fitted slope equals 1, so the adjusted treatment effects are ±3: [1, −1] − (1)[−2, 2] = [3, −3].

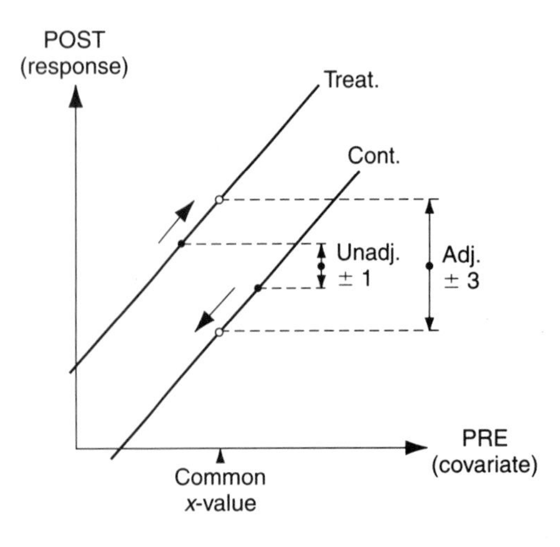

FIGURE 14.47 Adjusting treatment averages. Before adjusting for the covariate, the treatment effects correspond to y-values for the points of averages, shown as solid dots. Adjusting moves these points along their lines until their x-values equal the grand average for the covariate. Comparing y-values for these new points (open circles) gives adjusted treatment effects. ■

C. Testing the hypothesis of zero treatment effects. Unfortunately, the comparatively simple logic for hypothesis testing that works for balanced designs doesn't work for ANCOVA, because the covariate values are not balanced in relation to the rest of the design. This means we have to rely on a less simple but more general logic for hypothesis testing: To test whether the treatment effects are zero, we fit two models, one with treatment effects, one without, and compare the residual sums of squares. The **full model** (with treatment effects) is the parallel line model from before. The **null model** (no treatment effects) corresponds to a single regression line, fitted to all the points of the scatterplot of response versus covariate. (This null model has exactly the same structure as the full model, except that there is no factor for treatments.)

Full model (with): Parallel lines, one line per group
Null model (without): Single line for all groups together

To compare the two models, we compute a residual sum of squares for each one. For the full model, we get $SS_{Res(adj)}$, and use $MS_{Res(adj)}$ for the denominator of an F-ratio. For the null model, the residual sum of squares comes partly from chance error, but—unless the treatment effects are zero—partly from the treatment differences, which were not part of the model. This sum of squares is SS_{T+E}, for "treatment plus error." Figure 14.48 shows a way to visualize the relationships of the various sums of squares:

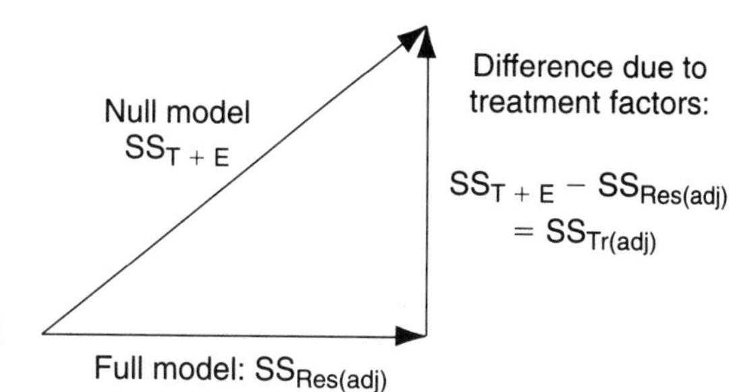

FIGURE 14.48 Comparing the full and null models.

The sums of squares "add like Pythagoras." The null model, which has no treatment factor, has the larger residual SS, and corresponds to the hypotenuse, T + E. The full model, with treatment effects included, has a smaller residual SS, and corresponds to the base, Res(adj). The difference in residual SSs is due to the treatment factor, and corresponds to the remaining side, Tr(adj). The F-ratio for testing treatment effects corresponds to "rise over run": $SS_{Tr(adj)}/SS_{Res(adj)}$. If the treatment effects are large, the null model will not fit nearly as well as the full model, and so will have a much larger residual sum of squares.

Here's how the F-ratio is computed:

Source	df	SS
Treatments(adj)	# tr − 1	$SS_{T+E} - SS_{Res(adj)}$
Residuals(adj)	# obs − # tr − 1	$SS_{Res(adj)}$

$$F = \frac{MS_{Tr(adj)}}{MS_{Res(adj)}}$$

Adjusted F-ratio for ANCOVA

EXAMPLE 14.27 SESAME STREET (CONTINUED)

Test the hypothesis of zero treatment effects.

SOLUTION. Here's an ANOVA table which summarizes the decomposition using the full model:

Source	DF	SS
Gr Avg	1	392
Tr(unadj)	1	8
Covariate	1	64
Res(adj)	5	88
TOTAL	8	552

To fit the null model, we fit a single line to all eight data points. You can check that the point of averages is (7,6), the slope is 0.5, and the decomposition is as follows:

$$\underset{\text{Obs}}{\begin{bmatrix} 0 & 2 \\ 8 & 12 \\ 8 & 4 \\ 8 & 14 \end{bmatrix}} = \underset{\text{Gr Avg}}{\begin{bmatrix} 7 & 7 \\ 7 & 7 \\ 7 & 7 \\ 7 & 7 \end{bmatrix}} + \underset{\text{Cov Eff}}{\begin{bmatrix} -1 & -3 \\ 1 & -2 \\ 2 & 0 \\ 2 & 1 \end{bmatrix}} + \underset{\text{T+E}}{\begin{bmatrix} -6 & -2 \\ 0 & 7 \\ -1 & -3 \\ -1 & 6 \end{bmatrix}}$$

SS	552	=	392	+	24	+	136
df	8	=	1	+	1	+	6

The full model has $SS_{Res(adj)} = 88$ on 5 df, and
the null model has $SS_{T+E} = 136$ on 6 df.

The difference in SS of $136 - 88 = 48$ is the reduction in sum of squares due to treatments ($= SS_{Tr(adj)}$), on $6 - 1 = 5$ df.

Table 14.4 compares the ordinary ANOVA with ANCOVA for this data set:

TABLE 14.4 Adjusting for the covariate

	Unadjusted				Adjusted for covariate			
Source	**df**	**SS**	**MS**	**F**	**df**	**SS**	**MS**	**F**
Grand Average	1	392			1	392		
Treatments	1	8	8.0	0.32	1	48	48.0	2.73
Covariate	—	—	—					
Residual	6	152	25.3	5	88	17.6		
TOTAL	8	552						

The SS for treatments goes from 8 to 48, while the residual SS goes from 152 to 88. The adjusted F-ratio (2.73) is more than eight times as big as the unadjusted F-ratio (0.32). ■

Summary: Fitting the ANCOVA Model

A. Fit a parallel line model

1. ANOVA step: Points of averages for each treatment group. Decompose both the (a) response and (b) covariate using the BF[1] model.

2. REGRESSION step: The common slope

Slope = [sum of (response residual) · (covariate residual)] ÷[sum of (covariate residual) · (covariate residual)]

B. Adjusting the treatment effects

Tr Avg(adj) = Tr Avg(unadj) − (slope)[Tr Eff(covar)]

C. Testing the hypothesis of zero treatment effects

1. Fit a single line (no treatment) model; residual SS = SS_{T+E}

$SS_{Res(adj)}$ = residual SS from A above

2. $SS_{Tr(adj)} = SS_{T+E} - SS_{Res(adj)}$. $F = SS_{Tr(adj)}/SS_{Res(adj)}$

Exercise Set C

1. *Matching scatterplots*. Each scatterplot in Figure 14.49 is for a data set with the same factor structure as in Example 14.21. I've plotted the points in the treatment group as Ts and those in the control group as Cs. Match each scatterplot with one of the following verbal descriptions.

a. The response and covariate show a strong positive relationship, although the slope for the covariance adjustment is not large. The unadjusted SD will be a lot larger than the adjusted SD. Although ordinary ANOVA would find fairly substantial treatment effects, the group differences "go away" if you adjust for the covariate.

b. The response and covariate show almost no linear relationship. There are big treatment differences, whether you adjust for the covariate or not. The SD is pretty much unaffected by the covariance adjustment.

c. Response and covariate show strong linear relationships within groups, and the two groups have very different sets of covariate values. Covariance adjustment will drastically reduce the residual sum of squares and substantially increase the estimated treatment differences.

d. The response and covariate show a strong positive relationship *between* groups, but no linear relationship within groups. Fitted parallel lines would be pretty much horizontal: ANCOVA and ordinary ANOVA would be essentially equivalent.

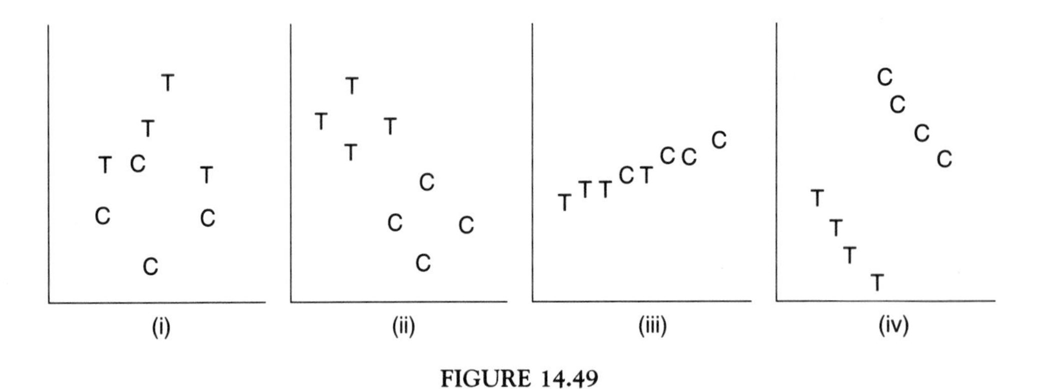

FIGURE 14.49

2. *More fake Sesame Street data, for practicing the computations*. Assume this data set has the same structure as the one in Example 14.21. Analyze the data twice, once using ordinary ANOVA, ignoring the covariate x, and a second time using ANCOVA.

Before you start the arithmetic, look over the numbers and try to judge:

a. If you ignore the covariate, (i) about how big will the (unadjusted) SE be? (ii) Is the effect of A versus B big or small, in comparison to this SD?

b. Scatterplot y versus x, using different symbols for the two groups. Is the relationship between covariate x and response y strong (high correlation) or weak?

c. Is the slope for the covariate adjustment positive or negative? Nearest to 0, ± 1, or ± 4?

d. Guess whether the SD, after adjusting for the covariate, will be closest to (i) the unadjusted SD, (ii) half the unadjusted SD, or (iii) one tenth the unadjusted SD.

Do the computations, and use them to write answers to (a)–(d). (Your regression slopes should be whole numbers.)

A = Treatment		B = Control	
x	y	x	y
6	32	3	34
9	14	6	20
3	44	3	36

3. *Sesame Street—the real story*. Background: When the Sesame Street series was first shown on TV, researchers at Educational Testing Service carried out a very large observational study to evaluate the effect of the series. The data below, which came from that study, are for two groups of four-year-old inner-city girls. The response and covariate are pre- and post-test scores of understanding of number concepts. In between the pre-test and post-test, the girls had the opportunity to watch Sesame Street for a year. Observers scored the frequency of watching, from 1 (= almost never) to 4 (= five times a week). The two groups in Table 14.5 are those

who scored either 2 (less) or 3 (more). Notice that these groups were defined retrospectively, that is, after the data had been gathered.

TABLE 14.5 Pre- and post-test scores for 10 girls

	Watched Sesame Street			
	Less Often		More Often	
	PRE	POST	PRE	POST
	38	42	33	39
	16	22	31	52
	36	42	45	46
	14	28*	23	50*
	21*	36	13	23
Avg	25	34	29	42

(Observations marked * are 1 point higher than actual observations.)

a. The graph in Figure 14.50 shows response (POST) versus covariate (PRE). Use the graphs to answer the following:

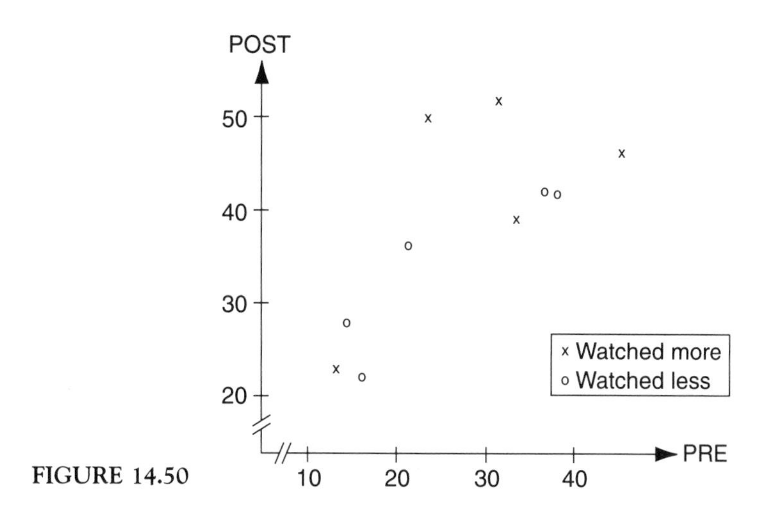

FIGURE 14.50

Covariance adjustment will make the estimated treatment effects (i) ________ (larger/smaller), because the correlation between response and covariate is (ii) ______________ (positive/negative), and the group differences for the response averages go in the (iii) ______________ (same/opposite) direction as for the covariate averages.

b. Covariance adjustment will have a (i) __________ (large/small) effect on the residual sum of squares, because the correlation between response and covariate is (ii) ______________ (near ± 1, close to 0).

c. "Both the pre-test scores and the post-test scores are almost surely correlated with the score used to define the two groups." Either agree or disagree with this statement, give your reasons, and discuss the relationship to conclusions based on ANCOVA.

4–11. *Fat rats.* This experiment was designed to test the effects of a growth hormone extracted from the pituitary gland of cows. Normally, rats stop active growth when they get about 150 days old. Twenty female rats who had stopped growing were injected with the hormone each day for three weeks. Half the rats got daily doses of 0.25 ml, the other half got doses of 0.10 ml. The response y is final weight (in grams); the covariate x is initial weight.

0.25 ml	x	236	228	212	240	220	201	215	207	218	248
	y	255	242	223	254	230	209	223	218	218	255
0.10 ml	x	179	221	218	213	180	231	198	227	228	235
	y	198	245	244	233	203	255	218	250	240	248

4. Scatterplot y versus x using different symbols for the two groups. Describe what you see in the scatterplot, using the descriptions in Exercises 1–3 as a guide.

The next several exercises ask you to carry out several different analyses, in order to compare them.

5. *Ordinary ANOVA.* $SS_{Total} = 1{,}092{,}633$; the two group averages are 232.7 (0.25 ml) and 233.4 (0.10 ml). Use this information to complete an ordinary ANOVA table for the response, and to construct a 95% confidence interval for low dose versus high.

6. *ANOVA for weight gain.* An analysis of difference scores $(y - x)$ corresponds to a covariance adjustment with slope equal to 1. The average initial weights (x) for the two groups are 222.5 (low dose) and 213.0 (high dose). SS_{Total} for the difference is 5632. Use this information to complete an ordinary ANOVA table for the differences, and to construct a 95% confidence interval for Low dose versus High.

7. *ANOVA for weight gain as a percent of initial weight.* Another common seat-of-the-pants adjustment for initial differences is to change the response to After − Before, as a percent of Before. For percent weight gains, $SS_{Total} = 1345.12$, and the group averages are 4.56 (low dose) and 9.70 (high dose). Complete another ANOVA table, and construct another interval.

8. *ANCOVA.* Here are some numbers from an analysis of covariance: $SS_{T+E} = 920.4$, $SS_{Res(adj)} = 429.6$. Use these, in addition to the information in Exercise 5, to complete an analysis of covariance table.

9. The actual slope from the Regression step for the covariance adjustment is almost exactly 1.0. Give the adjusted treatment effects.

10. Suppose that the adjusted treatment effects had been −9.15 (low) and 9.75 (high). What would the slope from the Regression step have been?

11. Write a brief paragraph summarizing what you conclude from comparing Exercises 4–9. Does this data set provide a good advertisement for ANCOVA?

12. *Hospital Carpets.* An ordinary ANOVA for the hospital carpet data (Example 8.7 and Exercise 8.C.1) fails to find a detectable effect of carpeting on the density of bacteria in the air. However, in Example 12.7, a scatterplot of # colonies per cubic foot versus room number revealed a mild positive relationship. Look over that scatterplot once more, and make rough guesses about what an analysis of covariance would show. Then do the analysis, and discuss what you find.

13. *Alpha waves in solitary.* Analyze the data from Example 8.10 as follows. Ignore the Day 4 values. Take the Day 7 values as your response, and Day 1 values as a covariate.

What is the appropriate factor structure? Carry out an analysis of covariance, and discuss the results.

14–19. *More mental activity*. The data for the mental activity experiment (described in Examples 14.4 and 14.22) are given in Table 14.6. In what follows, I'll suggest several things you might do as part of a careful analysis. Here's one question you should think about as you work with the data: A scatterplot for the placebo data (Example 14.4) shows a strong linear relationship between before and after values; yet when you include the data for morphine and heroin (Example 14.22), the slope of the covariance adjustment is essentially 0, as though there were no linear relationship. How can you make sense of the apparent contradiction?

TABLE 14.6 Mental activity data

	Placebo		Morphine		Heroin	
	Before	After	Before	After	Before	After
Subject	x	y	x	y	x	y
1	0	7	7	4	0	2
2	2	1	2	2	4	0
3	14	10	14	14	14	13
4	5	10	14	0	10	0
5	5	6	1	2	4	0
6	4	1*	2	0	5	0
7	8	7	5	6	6	1
8	6	5	6	0	6	2
9	6	6	5	1	4	0
10	8	6	6	6	10	0
11	6	3	7	5	7	2
12	3	8	1	3	4	1
13	1	0	0	0	1	0
14	10	11	8	10	9	1
15	10	10	8	0	4	13
16	0	0	0	0	0	0
17	10	8	11	1	11	0
18	6	6	6	4*	6	4
19	8	7	7	9	0	0
20	5	1	5	0	6	1
21	10	8	4	2	11	5
22	6	5	7	7	7	7
23	0	1	0	2	3*	2*
24	11	5	12	12	12	0

Response y = activity level two hours after injection.
Covariate x = activity level before injection.
The covariate is a numerical carrier. There are also two categorical carriers, which give the experiment the factor structure of a CB design, with Blocks = Rows = Subjects at 24 levels, and Drugs = Columns at 3 levels. (Values marked * were changed slightly to simplify arithmetic.)

14. Construct a parallel dot graph for the response values (the after scores), and write a short summary of what the graph tells you.

15. Use the following numbers to complete an ordinary ANOVA for the response values.

Averages:	Placebo 5.50	Morphine 3.75	Heroin 2.25
Sums of Squares:	Total 2166	Blocks 556.667	

16. Use your parallel dot graph to sketch rough normal plots for each of the three sets of response values. For these data, it's not possible to find a transformation that will straighten the plots much. How can you tell this from the data?

17. Look again at the scatterplot of After (y) versus Before (x) for the morphine data (Example 14.22). Think of summarizing the plot in terms of "balloon plus outliers," that is, a balloon summary for most of the points, plus a list of a few points far from the balloon.

a. Which subjects do you consider outliers? What do they have in common, apart from just being outliers?

b. Fit a line by eye to the other points. Estimate the slope and intercept, and compare these values with the ones for the placebo data Example 14.4).

18. Repeat Exercise 17 using the heroin data.

19. a. Compare the data for the two drugs and placebo by filling in the following table:

	Fitted line		Outliers	
	Intercept	Slope	How many?	What kind?
Placebo				
Morphine				
Heroin				

b. Use your results to argue for or against the following statement:

The data suggest that each subject responds as if s/he has some threshold value for each drug, with different thresholds for different subjects. For some subjects, the drug dose happened to be below the threshold, and had no detectable effect. For other subjects, the dose happened to be above the threshold, and had quite a large effect. For subjects unaffected by the drug, the relationship between Before and After scores is essentially the same as for the placebo condition; but for subjects affected by the drug, the After value shows no relationship to the Before value.

c. If you were to design a new experiment to study the effects of the drugs, how would you change the design from the one that was actually used?

20. *Prisoners in solitary*. Here are three possible analyses of the data shown in Example 14.15:

(i) ANOVA of Day 7 readings, ignoring Day 1 readings

(ii) ANOVA of change in readings, Day 7 − Day 1

(iii) ANCOVA, with Day 7 reading as response, Day 1 as covariate.

a. Treatment effects: Use the scatterplot of Example 14.15 to decide which of the three analyses would find the biggest treatment effects, and which the smallest, or else to explain why all three will be roughly the same.

b. Residual sums of squares: Same question as (a) above, only this time for residual SS.

c. Which analysis do you consider the most suitable, and why?

21. *Prisoners (continued)*. Carry out all three of the analyses (i)–(ii) in Exercise 20, and compare your results with your answers to (a)–(c).

NOTES AND SOURCES

Fraction of variation "explained." The quotation marks are meant to remind you not to assume that technical terms have their everyday meaning. For example, a least squares fit with x = temperature can "explain" a large fraction of the variation in y = frequency of cricket chirps, but that still doesn't explain, in the everyday sense, why faster rates go with higher temperatures.

CHAPTER 15

SAMPLING DISTRIBUTIONS AND THE ROLE OF THE ASSUMPTIONS

Overview

The last several chapters have presented certain formal methods, together with examples of how to use them and interpret the results. This chapter deals with the logic that the methods depend on and why the methods take the forms that they do. I hope you will find that the better you understand the logic of the methods, the easier it will be to see when to use them and when not to, what they can tell you about a data set and what they cannot tell you.

Both confidence intervals and hypothesis tests are methods for **inference**, for using what you can observe directly in your data to draw conclusions about the unseen mechanism that created the data. Both methods are based on **statistics**, numerical summaries you compute from the data: either linear estimators, for confidence intervals, or ratios of mean squares, for hypothesis testing. Until now, you've probably thought of these summaries as fixed numbers. In this chapter, however, you'll need to learn to think of them in a new way, *as rules you can use to define a chance process*.

The logic of inference depends on a careful answer to the question, "Suppose the chance process is defined like so; then what sort of values for the summary statistic would I be likely to get?" Statisticians answer the question by giving the **sampling distribution** of the statistic, which tells all its possible values and how likely they are to occur.

The sampling distribution for a statistic depends on the chance mechanism that created your data. Different mechanisms lead to different sampling distributions, in much the same way that different diseases tend to produce different clusters of symptoms. Just as a doctor tries to use the observed symptoms to figure out the disease responsible for causing them, a statistician tries to use the sampling distribution to figure out the underlying chance mechanism from the observed values of a statistic.

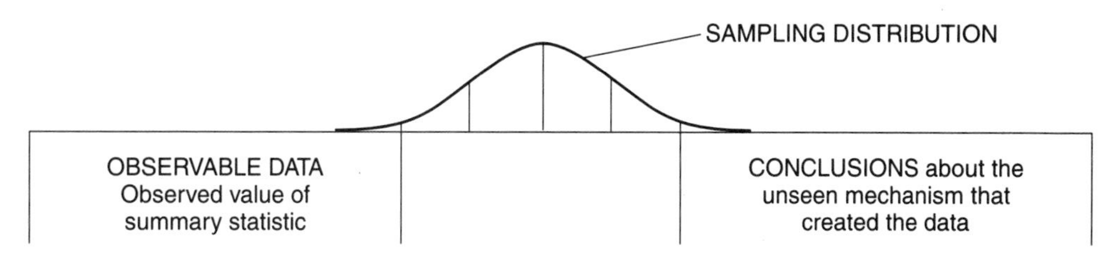

THE SAMPLING DISTRIBUTION SERVES AS A BRIDGE OF LOGIC
It connects the observed data and their underlying chance process.

Somewhat paradoxically, whereas the main ideas about sampling distributions (what one is, and how you use it) are not terribly difficult, the practical details of finding and describing particular sampling distributions can get so messy as to obscure the main ideas. (You may have had this kind of forest-and-trees feeling before, say with linear decomposition: the idea of sorting data into groups and subtracting group averages is simple in principle, but sometimes tricky in practice.) In order to help you focus first on the "big picture" before we go on to the details, I'll start out in Section 1 with an example that involves no numbers, one I took from a detective story.

Then in Section 2 I'll develop several simple numerical examples. For the most part these examples will be too simple to be realistic: their purpose is to illustrate how models and sampling distributions are related, and to introduce various ways to think about the distribution of outcomes from a chance process. With Section 2 as background, Section 3 presents the four most important families of distributions: the normal, chi-square, t and F families. Then the next two sections deal with the sampling distributions of linear estimators (Section 4) and F-ratios (Section 5). Finally, Section 6 discusses some of the reasons why statisticians regard the models we have been using as reasonable.

1. The Logic of Hypothesis Testing

Structure of the Argument

> "Is there any point to which you would wish to draw my attention?"
> "To the curious incident of the dog in the night-time."
> "The dog did nothing in the night-time."
> "That was the curious incident," remarked Sherlock Holmes.
>
> —Sir Arthur Conan Doyle
> *The Memoirs of Sherlock Holmes*

That conversation between Sherlock Holmes and Police Inspector Gregory gives a useful way to think about hypothesis testing in an everyday context where you don't need to compute anything.

In thinking about how hypothesis testing works, I like to break things down into six parts: model, null hypothesis, test statistic, sampling distribution, observed outcome, and conclusion. You've already seen all but one of these. Our standard model for a data set includes the factor structure and Fisher assumptions. The null hypothesis might be that the true effects of the conditions are all equal to zero. Our test statistic would then be an F-ratio of mean squares, and the observed outcome would be the actual F-ratio we compute from our data. If that value is bigger than the critical value, our conclusion would be to reject the null hypothesis.

The only thing in the list of six that I have not discussed in detail is the sampling distribution, the main focus of this chapter. Roughly, the sampling distribution provides a precise description of the way the test statistic behaves. The critical value we use to draw conclusions is computed from the sampling distribution.

In the story Silver Blaze, the detective Sherlock Holmes tries to figure out who has stolen a valuable racehorse (Silver Blaze) just days before the horse was expected to win the Wessex Cup. The crucial clue is that a dog did nothing in the night, and the detective's inference is that because the dog did not bark, he must have known the person who stole the horse. That inference leads Holmes to figure out that Silver Blaze has been stolen by his own trainer. Here's how he might have reasoned.

(1) "I'll start with the assumption that the horse was in fact stolen, by some person, and that the dog was present at the time." (Model)

(2) "Now suppose that the dog did not know the thief (null hypothesis): Are the facts of the case compatible with this working hypothesis?"

(3) "The key piece of information is whether or not the dog barked." (Test statistic)

(4) "I know how dogs typically behave (sampling distribution), and I can use that to deduce what would happen if my hypothesis is true. There are two possible

outcomes, either the dog barks, or he doesn't. If we suppose for the moment that the dog does not know the thief, then we can assign chances to the two outcomes: it is very likely that the dog would bark, and very unlikely that the dog would not bark."

(5) "In fact, the dog did not bark." (Observed value of the test statistic)

(6) "So either my working hypothesis is false, or a very unlikely thing happened. Faced with this choice, I reject my hypothesis, and conclude that the dog must have known the thief." (Conclusion)

Notice how easily the same thinking transfers to analysis of variance, applied, say, to the leafhopper survival data:

(1) "I'll start with the assumption that there is only one structural factor, Diets, and that the Fisher assumptions give a reasonable description of the experiment." (Model)

(2) "Now suppose that the true diet effects are all zero (null hypothesis): Are the results of the experiment compatible with this null hypothesis?"

(3) "The key summary of the data is the *F*-ratio (test statistic), which compares the variability for diets and for residuals."

(4) "I know how the *F*-ratio gets computed, and I can use that to figure out how the *F*-ratio would behave (sampling distribution) if the null hypothesis were true. There are lots of possible values that the *F*-ratio could have, but I can simplify things by using the 5% critical value to define two kinds of outcomes: *F*-ratios that are smaller than the critical value, and those that are bigger. If we suppose for the moment that diet has no effect, then (because of the way the critical value was chosen) it is very likely that the *F*-ratio would be smaller than the critical value (95% chance), and very unlikely that the *F*-ratio would be bigger (5% chance)."

(5) "In fact, the *F*-ratio computed from the data is bigger than the critical value." (Observed value of the test statistic)

(6) "So either the null hypothesis is false, or a very unlikely thing happened. Faced with this choice, I reject the null hypothesis, and conclude that diets have a real effect." (Conclusion)

The six numbered parts in each of the last two paragraphs correspond to the elements of any statistical test:

(1) The **model** provides a general framework of assumptions about the process that gives rise to the data.

(2) The **null hypothesis** gives a more specialized and more tentative assumption that is to be tested within the general framework of the model.

For Sherlock Holmes, the general framework was that Silver Blaze was stolen by a person, and that the dog was present; the null hypothesis to be tested was that the dog did not know the thief. Taken together, the model and null hypothesis describe a mechanism for generating data sets, and you can imagine using that mechanism to create lots of data sets, each one perhaps a little different from the others, but all of them created by the same mechanism.

(3) The **test statistic** is a rule for summarizing any data set generated by the mechanism.

It is important to make a distinction between the *rule* for computing the summary, which you should regard as defining a *process*, and the observed value of the summary that you get from applying the rule to the actual data. The distinction is the same as between the process of finding out whether the dog barked, and the actual outcome (that the dog did not bark). If you regard computing the test statistic as a process, then you're well on your way to understanding (4) its sampling distribution: Parts (1) and (2) specify a mechanism for creating data sets, and (3) specifies a rule for summarizing each data set. Put these together, and you can use them to create a collection of possible outcomes for the test statistic, one for each possible data set. Some outcomes may occur more often than others, so we keep track not just of the possible outcomes, but also how often each one occurs.

(4) The **sampling distribution** of a statistic is the collection of all possible values for the statistic, together with their chances.

We can use the sampling distribution of a test statistic to find the **5% critical value**. In the case of Sherlock Holmes, the process of finding out about barking had two outcomes, with quite different chances: "bark" was likely, "no bark" was unlikely. The sampling distribution tells us what is typical if the null hypothesis is true, and so tells us how to interpret (5) the actual outcome, and (6) what to conclude about the null hypothesis.

(5) The actual **outcome** of a statistic is a single number, computed from the data.

(6) The **conclusion** we draw is based on where the observed outcome falls in relation to the sampling distribution:

Outcome greater than the critical value: Conclude that the null hypothesis is false. (Reject the null hypothesis.)

Outcome less than the critical value: No conclusion. Evidence is not strong enough to rule out the null hypothesis.

Sampling Distribution of a Test Statistic

Now consider a simple ANOVA example, based on the leafhopper survival experiment. My purpose with this example is to show you how you can find the sampling distribution by regarding the model, null hypothesis and rule for the test statistic as defining a chance process, and how you can find the 5% critical value once you have the sampling distribution. Because the details of this example get a bit messy, I'll remind you once more about the purpose of the sampling distribution.

	SHERLOCK HOLMES	ANALYSIS OF VARIANCE
MODEL	The horse was stolen by some person; the dog was present at the time.	Factor structure and the Fisher assumptions
NULL HYPOTHESIS	Dog does not know the thief	True effects are zero
TEST STATISTIC	Whether the dog barked	F-ratio
SAMPLING DISTRIBUTION	If dog doesn't know thief: "Bark" is quite likely. "No bark" is not at all likely	If true effects are zero: "F less than critical value" is quite likely "F greater than critical value" is not
OBSERVED OUTCOME	"No bark"	"F greater than critical value"
CONCLUSION	Reject hypothesis: Dog knew the thief	Reject hypothesis: True effects are not all zero

The structure of hypothesis testing

The sampling distribution of a summary statistic is the collection of possible outcomes together with their chances. The distribution of a test statistic tells how the statistic "behaves" when the null hypothesis is true, and lets us judge whether the observed value of the test statistic is typical or unusual.

EXAMPLE 15.1 FAKE LEAFHOPPERS

Imagine a version of the leafhopper experiment with two diets, control and sugar, and with three observations for each diet. If we know the model and null hypothesis, how can we find the sampling distribution?

(1) The model. Think of the data arranged in two columns, one for each diet, with three observations per column. There are three factors in all: Benchmark, Diets, and Chance Error:

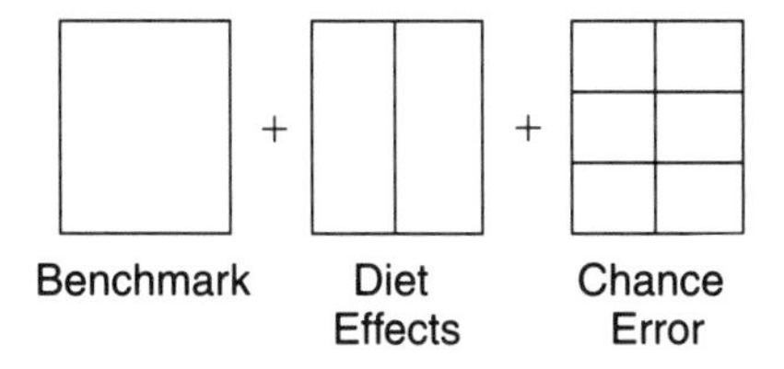

The Fisher assumptions of additivity and constant effects tell us the pieces get added, that there is one benchmark amount for all six observations, and one constant effect for each diet. The assumptions of independence and same SDs tell us that the chance errors behave as if they were drawn one at a time from the same box of tickets, with tickets replaced and mixed between draws. The normality assumption tells us that the tickets follow a normal curve. For my example, in order to keep the arithmetic easy to follow, I'm going to use instead a box of tickets that follows the rule only approximately (Figure 15.1).

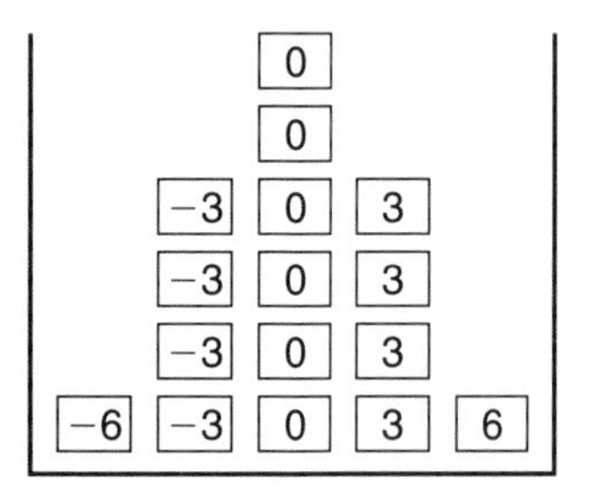

FIGURE 15.1 Chance Error box for the example. There are 16 tickets in all, and their average is zero. We assume that the chance errors behave like draws with replacement from this box.

We will use the model to create data sets: for each observation, we start with the benchmark amount, add on the diet effect, then draw a ticket from the box and add on the number we get as chance error. Although the model tells us how many pieces to add, and how to get chance errors, we still don't know the numerical values of the benchmark amount and two diet effects.

It will turn out that for testing our null hypothesis, we get the same results regardless of the benchmark amount (see Exercise A.6). So we can assume that this number equals zero. Where do we get values for the diet effects?

(2) The null hypothesis: The diet effects are both zero. This is the hypothesis we want to test, within the framework of the model.

(1) & (2): The mechanism for creating data sets. The model and null hypothesis together tell us the mechanism: Start with the benchmark amount (zero), add on the diet effect (also zero), then add on a chance error drawn from the box. For this simple example, if the null hypothesis is true, a data set simply consists of six draws from the error box. I actually drew six errors (−3, 0, 0, −3, 0, −3), and my first data set is shown in Figure 15.2.

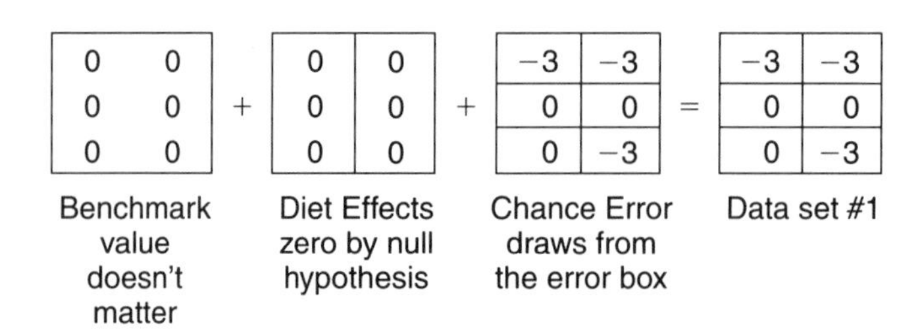

FIGURE 15.2 Data set #1, created using the model

This data set is just one of a huge number of possible data sets I might have gotten. A key idea is that once I've specified the mechanism for creating data sets, I can use that mechanism to create as many as I want. In fact, I will do just that in order to find the sampling distribution of the test statistic.

(3) The test statistic. To test the null hypothesis, we compute an *F*-ratio for Diets: Decompose the data, compute SSs, dfs, MSs, and finally $F = MS_{Diets}/MS_{Res}$ (Figure 15.3).

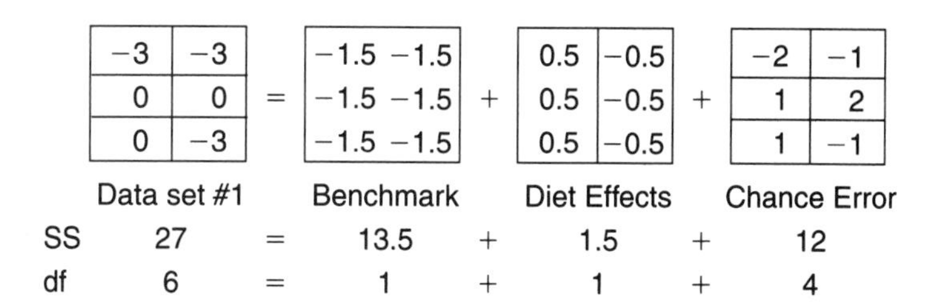

FIGURE 15.3 Computing the *F*-ratio for data set #1.

For this data set, $MS_{Diets} = 1.5/1 = 1.5$, $MS_{Res} = 12/4 = 3$, and $F = MS_{Diets}/MS_{Res} = 1.5/3 = 0.5$. (Note that whereas the true grand average and true diet effects are all zero, the estimates we get by decomposing the data are nonzero, because the process of decomposition "smears" the chance errors over all these factors.)

(4) The sampling distribution. The computations I've just done for this data set could be carried out on any data set created according to the model. The most direct way to find the sampling distribution is to repeat what we've just done, over and over: create a data set, then compute the *F*-ratio. *The sampling distribution is the collection of F-ratios, together with their frequency of occurrence.*

Figure 15.4 shows what I got for data set #2:

Data set #2			Benchmark			Diet Effects			Chance Error	
−3	−6		0	0		−1	1		−2	−7
3	6	=	0	0	+	−1	1	+	4	5
−3	3		0	0		−1	1		−2	2

	Data set #2		Benchmark		Diet Effects		Chance Error
SS	108	=	0	+	6	+	102
df	6	=	1	+	1	+	4

FIGURE 15.4 Data set #2, decomposition and *F*-ratio. I got the data set using the same mechanism as before: six draws with replacement from the error box. Last time the *F*-ratio was 0.5; this time it is $(6/1)/(102/4) = 0.235$.

Here are the *F*-ratios for my first 10 data sets: 0.5, 0.235, 0.182, 0.0, 0.125, 0.571, 1.0, 0.143, 0.2, 4.0

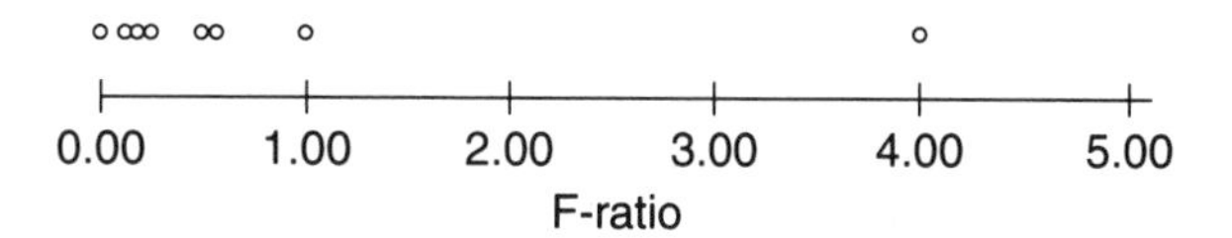

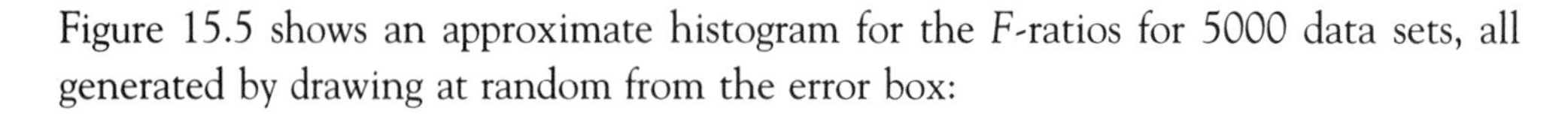

Figure 15.5 shows an approximate histogram for the F-ratios for 5000 data sets, all generated by drawing at random from the error box:

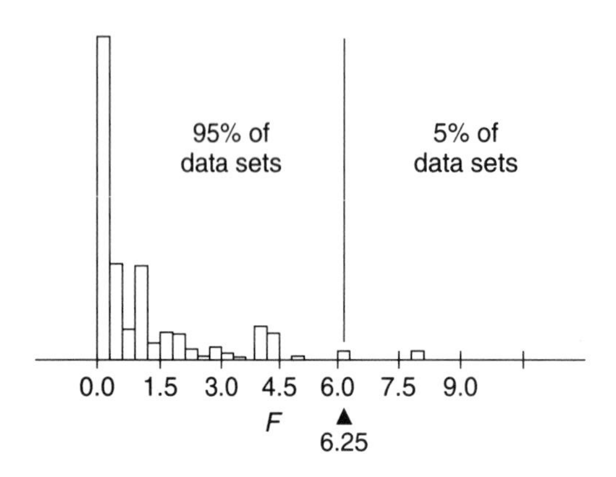

FIGURE 15.5 Histogram for the approximate sampling distribution. The critical value is 6.25: for 5% of the data sets, the F-ratio is greater than or equal to this number.

The histogram summarizes the way the F-ratio behaves if the null hypothesis is true, and so gives us a way to judge whether the F-ratio from an actual data set is typical of what we would expect when there are no real diet effects. In particular, the histogram tells us that 95% of the F-ratios are less than or equal to 6.25, so if the F-ratio we get from actual data is greater than or equal to than 6.25, one of two things must be true: either the null hypothesis is false, or something quite unusual has happened. Faced with this choice, statisticians conclude that the null hypothesis is false. ■

How typical is the example? In some ways the last example is artificial, but in other ways it does illustrate many important ideas about sampling distributions. To help you focus on what is typical, I've tried to think of reasonable questions you might ask about the example:

Question: You used zero for the Benchmark value. In a real situation, you wouldn't know the value, and it almost certainly wouldn't be zero. Doesn't that make your example misleading?

Answer: In fact, you'd get the same values for the MSs and F-ratio no matter what Benchmark you chose. Exercise 6 illustrates this.

Question: Your chance error box had tickets -6, -3, 0, 3, and 6. What if the chance errors had been 10 times as big, say -60, -30, 0, 30, and 60?

Answer: It's an important fact that the bigger the chance errors are, the harder it is for ANOVA to detect "real" effects. However, the sampling distribution we found was based on the working assumption that the true Diet effects were *zero*. It's one of the nice facts about F-ratios that when the null hypothesis is *true*, the sampling distribution does *not* depend on the size of the chance errors. Exercise A.7 illustrates this.

Question: The chance errors in your example didn't follow the rule based on the normal curve. Doesn't that lead to the wrong sampling distribution?

Answer: No: No matter what chance errors you use for your model, you can always find the sampling distribution, at least in principle. However, because the chance errors I used don't follow the normal curve exactly, my sampling distribution is not the same as the one Fisher got assuming the normal curve. The difference shows up in the critical value: my model gives a critical value of 6.25; Fisher's model gives a critical value of 7.71.

Question: Did you really get all your 5000 data sets by drawing tickets from a box? Is that what Fisher did?

Answer: In principle, I *could* have gotten the data sets by drawing tickets from a box, but instead I used a computer to do essentially the same thing. Fisher used mathematical theory. (Using the box model, or using the computer as I did, is called **simulation**.)

Question: Why did you use 5,000 data sets? Why not 10,000? If you'd used 10,000, wouldn't your histogram have been a bit different?

Answer: There's a key distinction involved in these questions, between the true sampling distribution and my estimated version based on a computer simulation. It's essentially the same distinction you've seen before between true values and estimates. My histogram is an approximation to the true one. If I'd used more data sets, I would have gotten a better approximation, but 5000 gave me an approximation I judged to be good enough.

Question: Why do statisticians say to use the F-ratio?

Answer: First of all, it's important to realize that you can find a sampling distribution for any kind of summary based on the data: F-ratio, mean square, linear estimator, or whatever. The summary you choose depends on what you want it to do. A lot of statistical theory deals with how you choose the right summary for the purpose you have in mind. Part of the reason for choosing the F-ratio to test our null hypothesis is contained in my answers to the first two questions: the sampling distribution of the F-ratio doesn't depend on the Benchmark value or on the sizes of the chance errors. Even more important, the F-ratio is used because it tells us about the null hypothesis: when the hypothesis is true, the F-ratio tends to be near 1; when it's false, F tends to be bigger than 1.

Question: Why is the overall model a reasonable one? I don't see much connection between drawing tickets from a box and things like measuring enzyme concentrations or reaction times.

Answer: This is the most important question of all. Sometimes the model is not reasonable, and if that's the case, ANOVA is not reasonable either. There are, however, lots of situations where experience suggests that models of this sort are more or less reasonable. The last section of this chapter discusses some of this in more detail.

Exercise Set A

1. *Testing a coin for bias*. Suppose you want to do a statistical test to decide whether a coin has a 50–50 chance of landing heads. You plan to toss the coin 100 times and count the number of heads. A statistician tells you that if you get fewer than 41 or more than 59 heads, you should conclude that the coin is biased. When you toss the coin, you get 43 heads.

The six parts of the statistical test are listed below in scrambled order. Tell which is which.

a. The actual number of heads you got in 100 tosses: 43.

b. The number of heads in 100 tosses.

c. The coin has a 50-50 chance of landing heads on each toss.

d. The set of chances that 100 tosses of a fair coin would give 0 heads, 1 head, 2 heads, . . . , 100 heads.

e. Do not reject the null hypothesis: the actual results would not be surprising if in fact the coin is fair.

f. The experiment is a sequence of 100 observations. Each observation can be either Heads or Tails. The chance of heads is the same for each observation, and the result of any one outcome does not influence the others.

2. *Testing for ESP*. Identify each of the six parts of a statistical test in this situation: Your friend claims he has ESP, and can guess whether a card is red or black. To test his claim, you shuffle a deck of cards (half red, half black), draw one out at random, and ask him to guess. Then you shuffle, draw, and ask again, repeating this for a total of 20 guesses. You tentatively assume your friend doesn't really have ESP, and is just guessing at random, but if he gets 15 or more right out of 20, you are prepared to conclude that there was more involved than pure guesswork. (Perhaps your friend is a skilled magician.) Just to complete the story, suppose that your friend gets 13 right.

3. *Finding a sampling distribution*. Consider a very simple experiment, one with only one observed value. Assume the model is that the observed value equals true value plus chance error, with chance error determined by tossing a fair coin: Heads $= +1$, Tails $= -1$. The null hypothesis is that the true value equals 10.

a. Suppose your test statistic is (observed value $-$ 10). Find the sampling distribution: Assume the null hypothesis is true, list all possible values for the test statistic, and tell the chances.

b. In (a), what is your conclusion if the actual value of the test statistic is 1? 3?

c. Now suppose your test statistic is (observed value $-$ 10)^2. Find the sampling distribution.

4. *Finding a sampling distribution by simulation*. Consider a somewhat fancier version of the last experiment, this time with two observed values. The model is that each observed value equals (the same) true value plus chance error, and that to get each chance error, you toss three coins together and compute chance error = (# Heads $-$ # Tails). Suppose, as in Exercise 3, your null hypothesis is that the true value equals 10. Take as your test statistic (average of observed values $-10)^2$.

a. Using three fair coins, create 20 data sets, compute the test statistic for each one, and summarize the results. (If you were to do this for 100 data sets instead of just 20, you'd have a decent approximate version of the sampling distribution.)

b. Using your results from (a) to judge, what would you conclude about the null hypothesis if a data set gave a value of the test statistic equal to 0? 16? 4? 9?

5. *Simulating the sampling distribution of an F-ratio*. Actually finding the sampling distribution of an *F*-ratio by experiment is not worth the time and trouble it takes, but doing the first few steps can help you understand the general idea. Consider a very simple CR design, with two conditions, and two observations for each condition. Assume the usual model for a CR experiment, except that each chance error, instead of following the normal curve, is created

by tossing three coins and computing (# Heads − # Tails). Take as your null hypothesis that the true condition effects are zero. You'll need a Benchmark value, and since it won't matter what value you use (see Exercise 6), choose Benchmark = 0, just as I did in the example.

Following the format of Example 15.1, create two data sets, decompose each one, find the SSs, etc., and compute the *F*-ratio. (If you had the patience to repeat this a few hundred more times, you could get a good approximate version of the sampling distribution.) Notice that it is possible to get $MS_{Res} = 0$, which makes it impossible to compute the *F*-ratio. This outcome is not likely, however, and almost never happens with real data.

6. *The Benchmark value doesn't affect the F-ratio.* Go back to the ANOVA example (15.1) of the last section, and using the same two sets of 6 chance errors that I used, create new versions of data sets #1 and #2 using Benchmark = 10 instead Benchmark = 0. Decompose each data set, compute SSs, etc., and compute the *F*-ratio. Then tell which of the following depend on the Benchmark value:

a. The Grand Average
b. SS for the Grand Average
c. df for the Grand Average
d. The estimated condition effects
e. The residuals
f. The MS for conditions
g. The MS for residuals
h. The *F*-ratio

7. Which summaries are influenced by the size of the chance errors?

Go back to the ANOVA example (15.1) once again, and suppose every ticket in the chance error box got its number multiplied by 10, so that an old error of −3 becomes an error of −30, etc. Create the corresponding new versions of my two data sets, decompose them, compute SSs, etc., and *F*-ratios. Then tell which of (a)–(h) in Exercise 6 have changed from the values they had for my original versions of the two data sets.

8. What if the null hypothesis is false?

So far every sampling distribution has been computed assuming the null hypothesis is true. How does the *F*-ratio behave if in fact the true condition effects are not zero? To answer this question, you could in principle find another sampling distribution, using the same methods as before: same model, same test statistic, only a different hypothesis. Go back to the ANOVA example (15.1) one last time, and create versions of the two data sets using −5 and +5 as the true condition effects. Use your results to complete the following table:

	DATA SET #1		DATA SET #2	
	True effects are		True effects are	
	0	5 & −5	0	5 & −5
Estimated Diet Effects				
MS for Diets				
MS for Residuals				
F-ratio				

Now fill in the blanks: If you compare the behavior of the summaries when you change from zero effects to nonzero effects, you find that MS_{Diets} __________, MS_{Res} __________, and the *F*-ratio __________.
(Choose from: goes up, goes down, stays the same.)

2. Ways to Think About Sampling Distributions

Overview

To get a sampling distribution, you need two things: (1) a mechanism for creating data sets (model and null hypothesis); and (2) a rule for computing a summary (test statistic, linear estimator, etc.). Once you've specified these two things, you can find the sampling distribution by simulation, at least in principle: use the mechanism to create lots of data sets, compute the summary for each one, and keep track of the outcomes you get and how often they occur.

Unfortunately, in almost any situation of practical importance, both the mechanism and the rule for the summary will be fairly complicated. This means that while it may be easy to see what you'd need to do to find the sampling distribution by simulation, actually doing the simulation is too messy and too time consuming to be worth it. Fortunately, there are shortcuts that use tools from calculus in place of simulation. Courses in mathematical statistics teach those ways to find sampling distributions. Because my emphasis in this book is on other aspects of statistics, I will be giving you certain key results without showing you the calculus. Because you won't be seeing this key link between our models and the resulting sampling distributions, it will be particularly important to develop other ways of understanding what a sampling distribution is.

This section presents three related ways to think about distributions, as preparation for the key results in Sections 3–5.

One: You can think of any collection of outcomes and chances in terms of drawing just once from the right box model. One way to understand sampling distributions is to practice taking a mechanism and summary, which in general will be complicated, and finding a new box model that gives the same set of outcomes and chances for a single draw.

Two: You can summarize a set of outcomes and chances geometrically, using a dot diagram or a histogram, which present the same information as the single-draw box model. Your main goal in this part of Section 2 should be to develop your intuition so that you can look at a geometric representation and think of the box model for the chance process it represents.

Three: Sometimes you don't need the whole sampling distribution, as long as you know a few key features, such as critical values or 95% distances. It is important to come to associate these key numerical summaries with their underlying chance processes.

Finding an Equivalent Single-Draw Box Model

Example 15.2 Coin Toss

Think about tossing a fair coin twice and counting the number of heads. You can think in terms of a mechanism and summary:

Mechanism: Draw 2 times from

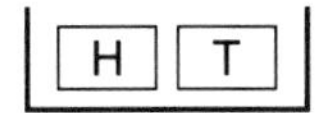

Summary: Count # H

There are four possible "data sets": HH, HT, TH, TT; each has a 25% chance. There are three possible outcomes for the summary: 2, 1, or 0 heads. The outcomes 2 and 0 each have a 25% chance, and the outcome 1 has a 50% chance, because two of the "data sets" have 1 as the summary value.

Here's an equivalent single-draw box model:

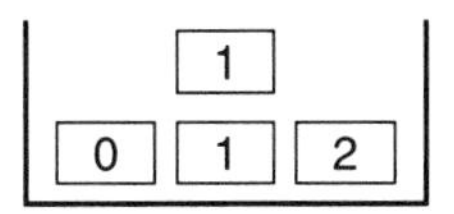

If you draw once from this box, you have a 25% chance of getting 0, a 50% chance of getting 1, and a 25% chance of getting 2: this single-draw model tells you the sampling distribution. ■

EXAMPLE 15.3 SUMMARIES FOR TWO DRAWS

Consider drawing twice (with replacement) from

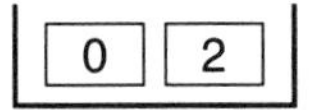

and then computing the following summaries of your data: (a) Average, (b) Sum of squares, (c) SD of the observations. Match each summary with the equivalent single-draw box model.

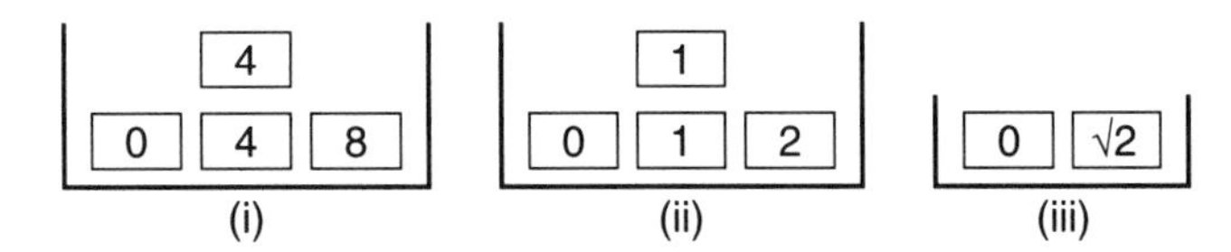

SOLUTION. There are two ways to approach this problem. You can work out the answers systematically, or you can try a shortcut. The systematic approach lists all the possible data sets and computes the summaries:

Data Set	Draw #1	Draw #2	Chance	Average	SS	SD
(1)	2	2	25%	2	8	0
(2)	2	0	25%	1	4	$\sqrt{2}$
(3)	0	2	25%	1	4	$\sqrt{2}$
(4)	0	0	25%	0	0	0

The list tells us that (i) is for SS, (ii) is for the average, and (iii) is for the SD. Notice that even though there are four data sets, a box with only two tickets works for the SD: there are two possible outcomes, each with a 50% chance.

Looking back, it should be easy to see a shortcut here, provided you assume that the boxes I gave show the right chances. The box with 0, 4, and 8 has to be for the SS, because neither of the other summaries could have a value as big as 8. The box with 0, 1, 2 has to be for the average, because the averages of 0s and 2s can't give you 8 or $\sqrt{2}$; and so the last box has to be for the SD. ■

EXAMPLE 15.4 SIMPLE ANOVA

Consider a simple CR experiment, with two observations for each of two conditions:

Observed Data		=	Benchmark		+	Condition Effects		+	Chance Error
			4	4		−1	1		−2 2
			4	4		−1	1		

Part 1. Match the single-draw box models in Figure 15.6 to the summaries: (a) Grand Avg; (b) Cond Avg #2 − Cond Avg #1; (c) SS_{Cond}; (d) MS_{Res}. (The chances given by the box models are correct, but don't try to check this.)

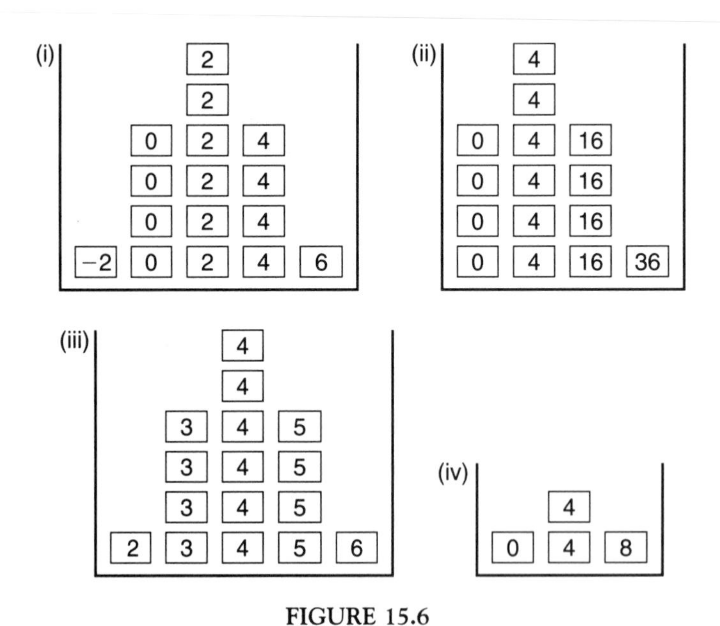

FIGURE 15.6

SOLUTIONS. As a general strategy, we can try to match by looking at typical values, smallest possible values, and largest possible values.

a. Grand Average: Box (iii). The average should tend to be close to 4, the true Benchmark, so (ii), (iii) and (iv) seem like the most reasonable choices. The smallest possible value for Grand Average occurs if all four chance errors equal -2, which gives observed values 1, 1, 3, 3, for an average of 2. (You can check in the same way that the largest possible value for Grand Average is 6.) Box (iii) is for Grand Average.

b. Cond Avg #2 − Cond Avg #1: Box (i). The estimate will tend to be near the true value, which is 2. Box (i) is the only one that makes 2 a likely outcome. Notice also that box (i) has -2 as a possible outcome: you can't have a negative SS or MS, so box (i) can't be for (c) or (d).

c. Sum of squares for conditions: Box (ii). Our choice is between (ii) and (iv). What's the largest possible value for SS_{Cond}? The SS will be largest when the condition averages are as far apart as possible, which happens if the two chance errors for Cond #1 are both -2, and the two errors for Cond #2 are both 2. If you create this data set and compute SS_{Cond}, you get 36, so box (ii) must be the one.

d. MS_{Res}: Box (iv). Consider first the contribution to SS_{Res} you get from looking at just one of the two conditions. If both chance errors are the same, both -2 or both 2, the two observed values will be equal, and the contribution to SS_{Res} will equal 0. If you get one -2 and one 2, the contribution to SS_{Res} will be $(-2)^2 + (2)^2 = 8$. So each half of the data contributes either 0 or 8 to SS_{Res}. If both parts are 0, $SS_{Res} = 0$, and $MS_{Res} = 0$. If both parts are 8, $SS_{Res} = 16$, and $MS_{Res} = 16/2 = 8$. If you get one 0 and one 8, $SS_{Res} = 8$, and $MS_{Res} = 4$. Since the only possible outcomes are 0, 4, and 8, Box (iv) must be for MS_{Res}.

Part 2: Using the sampling distribution. Use the results from (a) to answer the following questions.

a. What is the chance that the Grand Average falls within ±1 of the true Benchmark?

SOLUTION. Box (iii) has 16 tickets, and all but two give values within 1 of the true value. So the chance is $(14/16) \times 100\% = 87.5\%$. ("Grand Average ±1" gives an 87.5% confidence interval for the Benchmark value.)

b. What is the chance that the linear estimate falls within ±1 of the true value (Cond Eff #2 − Cond Eff #1)?

SOLUTION. The true value is 2. Box (i) has 16 tickets, and the only ones that give a value within 1 of the true value are the six 2s, so the chance is $(6/16) \times 100\% = 37.5\%$. (In other words, "estimate ± 1" gives a 37.5% confidence interval.) ■

Representing Distributions Geometrically: Dot Graphs and Histograms

The three families of distributions most often used in analysis of variance are the normal family (for chance errors), the *t* family (for linear estimators), and the *F* family (for ratios of mean squares). All three are often represented by curves called probability histograms, with chances for outcomes given by areas under the curve. Your goal in what follows should be to learn how to use the curves to find chances, critical values, and standard 95% distances.

Why do we need the probability histograms? Why can't we just use box models? A box model gives a useful way to think about a chance process if the number of possible outcomes is small, but not if there are hundreds or thousands of possible outcomes, as there are for most kinds of measurements. Probability histograms use areas to represent collections of tickets, and give a simpler representation than a very long list of tickets.

Probability dot graphs. **Probability dot graphs** are in a sense in between box models and probability histograms, and provide a natural transition.

> **Probability Dot Graph for a Single-Draw Box Model**
> **Plot each ticket as a dot. Think of the dots as representing equally likely outcomes for one draw from the box.**

EXAMPLE 15.5 PROBABILITY DOT GRAPHS FOR EXAMPLE 15.4

Four probability dot graphs are shown in Figure 15.7. Part 1: Match each dot graph with one of the four single-draw box models from Example 15.4. Part 2: Use the dot graphs to find the following chances: (a) the SD is less than or equal to 0; (b) the linear estimator for Condition #1 versus Condition #2 is less than or equal to 0; (c) SS_{Cond} is greater than 4.

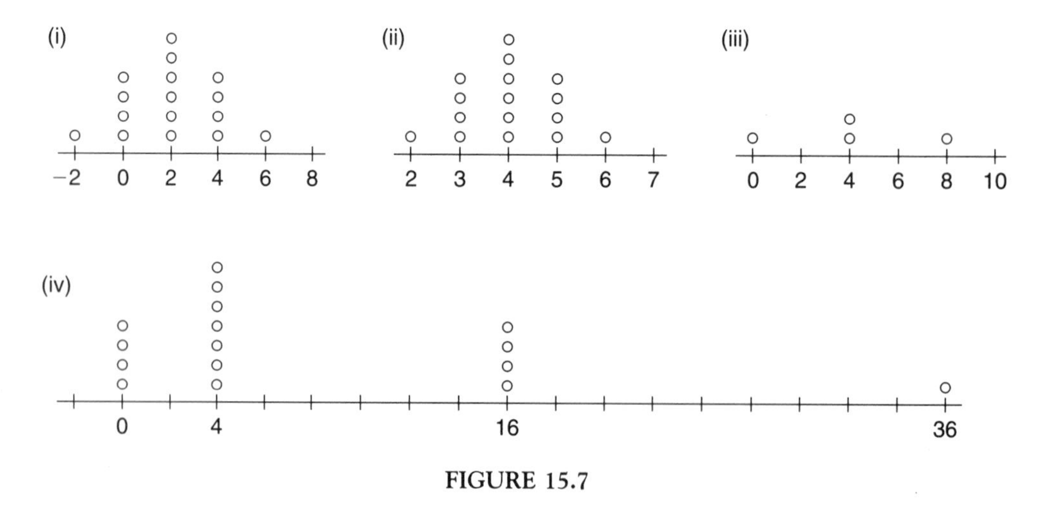

FIGURE 15.7

SOLUTION, PART 1: If you compare the single-draw box models in Example 15.4 with the probability dot graphs above, you should notice how similar they are, and be able to match them easily. Graph (i) is for Condition #1 vs. #2, (ii) is for the Grand Average, (iii) is for the SD, and (iv) is for SS_{Cond}.

PART 2:

a. SD less than or equal to 0. Graph (iii) has one dot (out of 4) at 0, and no dots at values less than 0: chance = 1/4.

b. Estimator for Condition #1 vs. #2 less than or equal to 0. Graph (i) has 5 dots out of 16 at values less than or equal to 0: chance = 5/16.

c. SS_{Cond} greater than 4. Graph (iv) has 4 dots at 16, 1 at 36: chance = 5/16. ■

Probability histograms. Your goal here should be to learn to "read" a histogram, that is, to learn to use it to find chances. The main difference between reading a probability dot graph and reading a histogram is this: Dot graphs represent individual outcomes by dots; to find chances you count dots. Histograms represent intervals of outcomes by areas; to find chances you measure areas.

How to Read a Probability Histogram

To find the chance of an outcome in a particular interval, express the area above the interval and under the curve as a percent of the total area.

EXAMPLE 15.6 A PROBABILITY HISTOGRAM

Use the histogram below to find the chance of an outcome (a) between 2 and 6; (b) less than 2; (c) greater than 2; (d) greater than 8.

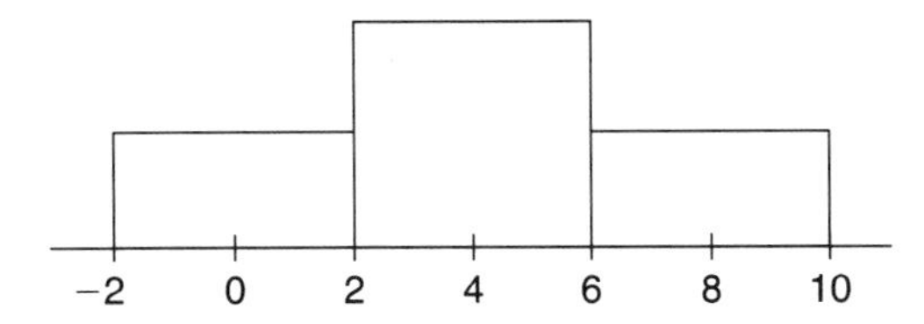

SOLUTION.

a. Between 2 and 6: Half the total area lies between 2 and 6, so the chance is 50%.

b. Less than 2: One-fourth of the total area lies to the left of 2: chance = 25%.

c. Greater than 2: Since one-fourth of the are lies to the left of 2, three-fourths of the area lies to the right of 2: chance = 75%.

d. Greater than 8: The area to the right of 8 equals one-half the block of the histogram between 6 an 8; that block represents one-fourth of the total area, so half of the block has area equal to one-eighth of the total. Chance = 1/8, or 12.5%. ■

EXAMPLE 15.7 PROBABILITY HISTOGRAMS FOR EXAMPLE 15.4

Four probability histograms are shown in Figure 15.8. Match each histogram with one of the four single-draw box models from Example 15.4. (Don't assume there is one histogram for each box model.)

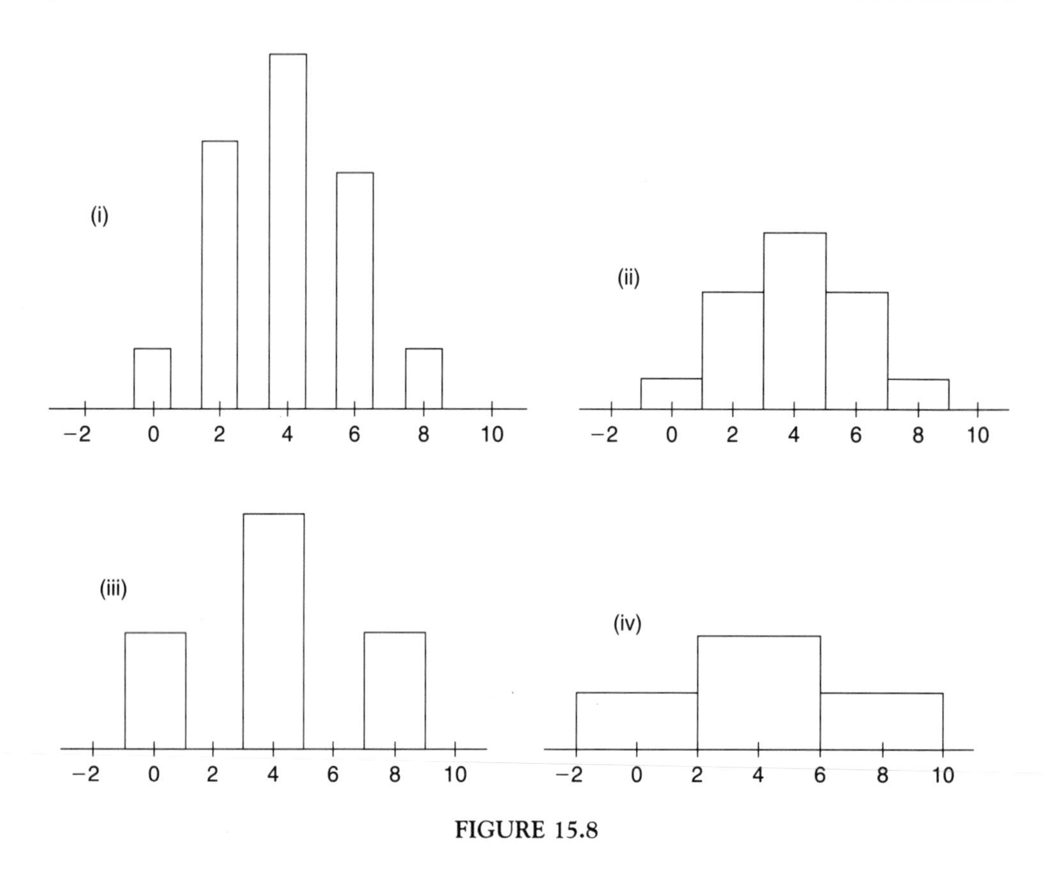

FIGURE 15.8

SOLUTION. Histograms (i) and (ii) both go with the Grand Average; (iii) and (iv) both go with the SD. These pairs illustrate that histograms, unlike probability dot graphs, cannot represent individual outcomes; they can only represent intervals of outcomes. Consider for example, the chance that SD = 4. From the dot graph in Example 15.5, we can easily see that this chance equals 2/4 = 50%. From histogram (iv) above, you can check that half the total area lies above the interval from 2 to 6, and indeed the chance that the SD is between 2 and 6 does equal 50%. Now use histogram (iv) to find the chance of an outcome between 3 and 5. The area above the interval is 1/4 of the total, so the chance is 25%. We know from the box model in Example 15.4 and dot graph in 15.5 that there's a 50% chance that SD = 4, so we seem to have a contradiction. The histogram seems to give the wrong chance.

Actually, the chance is not so much wrong as just a very poor approximation. It's part of the way histograms work that they can only give an approximate way to represent a chance process, and when there are only a few possible outcomes, some of the approximations will be quite crude. However, for a chance process with hundreds or thousands of possible outcomes, a probability histogram will generally give excellent approximations. ■

Smooth probability histograms. The histograms of the last example were built from rectangles, which made finding areas easy. The three families of distributions most often used in analysis of variance are all for idealized chance processes with an infinite number of outcomes, and their histograms are defined by curves. The way you read such histograms is the same as before, chance = area as percent of total area, but the areas are not as easy to find. You can estimate them, as in the example that follows, or if a rough estimate is not good enough, you can use tables that give the areas, as described in Section 3.

EXAMPLE 15.8 PROBABILITY HISTOGRAM FOR A NORMAL CURVE

A normal probability histogram is shown below. Use it to match the following intervals of outcomes with their chances:

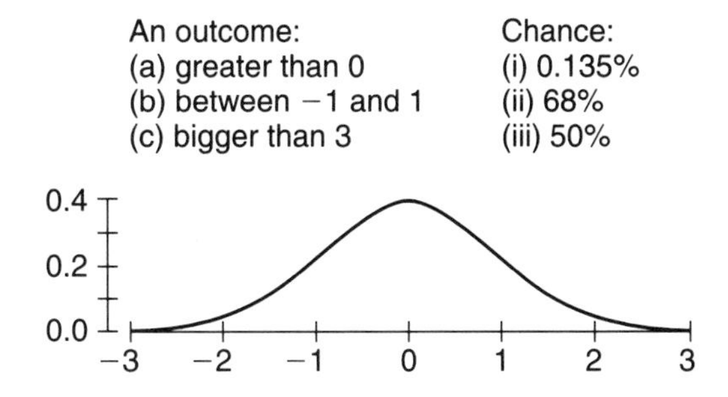

SOLUTION.

a. Greater than 0: The histogram is symmetric with exactly half its area on either side of 0. The chance is 50%.

b. Between −1 and 1: The area above this interval is roughly two-thirds of the total: chance = 68%.

c. There is hardly any area to the right of 3: chance = 0.135%. ■

Key Features: EVs, SEs, and Percentiles

Much of statistics is concerned with finding effective numerical summaries. The correlation coefficient, for example, is a single number that summarizes balloon-shaped scatterplots. The average and SD, taken together, often provide a good summary of a set of observed values. In very much the same way, the expected value (EV) and standard error (SE) often provide a good summary of a chance process, of its single-draw box model, and of its probability histogram. One of your goals here should be to understand how an EV and SE are related to the chance process they summarize.

Although the EV and SE often provide a good summary of the way a chance process behaves most of the time, the two summary numbers don't give much information about extreme and unusual outcomes. To put the same idea geometrically: the EV and SE often provide a good summary of the middle part of a probability histogram, but by themselves don't tell much about the extremes. The information about extreme and unusual outcomes is also important, and statisticians often use percentiles to give that information. (For example, the 95th percentile of a distribution is the number that separates the lowest 95% of the outcomes from the top 5% of them. The 5% critical value for an *F*-ratio is the 95th percentile of the corresponding *F*-distribution.) Table 15.1 summarizes three closely related ways to think about distributions, EVs, and SEs.

TABLE 15.1 Three ways to think about distributions, EVs, and SEs

Distribution	EV	SE
1. As a chance process	Typical value	Typical distance from the outcome to its EV
2. As a single-draw box model	Average of the tickets in the box	SD of the tickets in the box
3. As a probability histogram	Balance point of the histogram	A measure of how spread out the histogram is

For a chance process, the typical outcome equals the EV, give or take one SE or so. You can think of the EV as the typical outcome, in that outcomes (ordinarily) tend to be near the EV. If you represent a chance process as a single-draw box model, then the EV equals the average of the tickets in the box. If you represent the chance process as a probability histogram, then the EV equals the number on the *x*-axis where the histogram balances.

EXAMPLE 15.9 EVs FOR THE BOX MODELS OF EXAMPLE 15.4

For each of the four summary statistics (a)–(d) listed in Example 15.4, first guess the EV of its sampling distribution by looking at the tickets for the single-draw box model; then compute the EV by finding the average of the tickets.

SOLUTION.

a. Grand Average (Box iii): EV = 4, the true Benchmark value. (Notice that the probability dot graph for the Grand Average, (ii) in Example 15.5, balances at 4 = EV.)

b. Cond Avg #2 − Cond Avg #1 (Box i). EV = 2, the difference of the true condition effects. (The dot graph (i) in Example 15.5 balances at 2 = EV.)

c. SS_{Cond} (Box ii): You might be tempted to guess an EV near 4, but notice that the list of tickets "straggles" to the right: most of the tickets are bunched at 0 and 4, but there are a few comparatively large values, four 16s and one 36. The influence of the few large values makes the EV quite a bit larger than 4: EV = 8. (The EV works better as a summary for distributions that are more bell-shaped.) Notice, too, that for this particular chance process, the EV is not one of the possible outcomes. There are no tickets that say 8. Nevertheless, the EV is a "typical" value in that the outcomes tend to cluster on either side of 8. (Dot graph (iv) in 15.5 balances at 8.)

d. MS_{Res} (Box iv): EV = 4. (Dot graph (iii) balances at 4.) ■

EXAMPLE 15.10 EV AS BALANCE POINT

Four probability histograms are shown in Figure 15.9. Match each one with its EV, choosing from 2.5, 4, and 5.

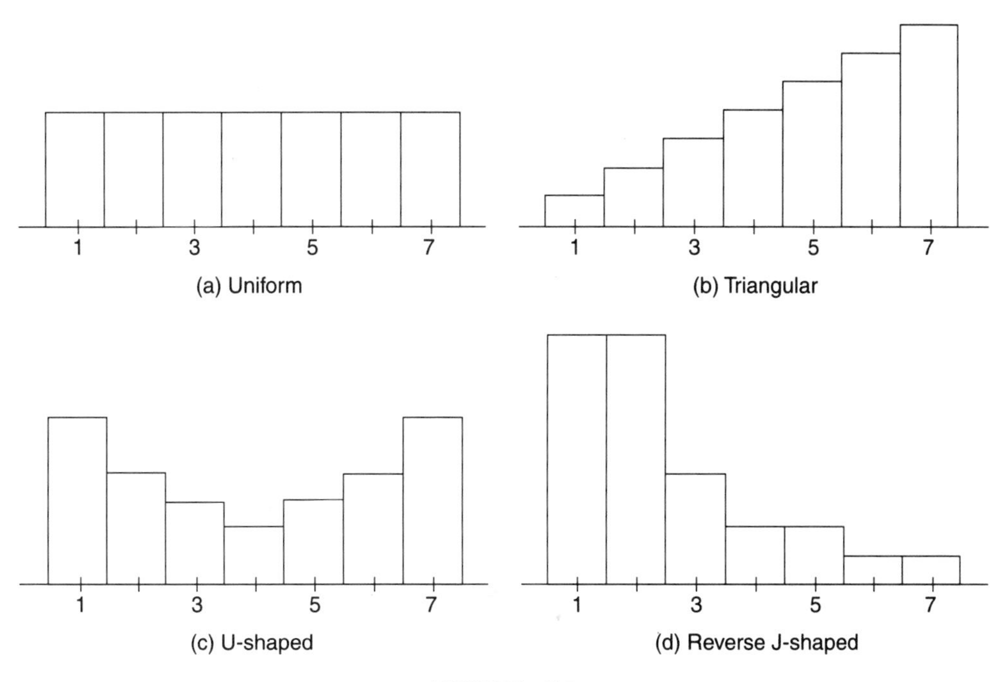

FIGURE 15.9

SOLUTION. The two histograms on the left are perfectly symmetric. For each of these, the EV = balance point is right in the middle, at 4. The triangular histogram (b) puts more weight to the right of 4 than to the left, and balances at 5. The reverse J-shaped histogram (d) put much of its weight at the far left extreme, and balances at 2.5. ■

Standard errors. The SE for a chance process tells the typical distance from an outcome to the EV. If you represent the chance process as a single-draw box model, then you can compute the SE by finding SD* for the numbers on the tickets. I've written SD* instead of SD so the notation can remind you of two ways SD* is different from the SDs you are used to computing. The first difference is conceptual: whereas the SD you compute as $\sqrt{MS_{Res}}$ is an estimate based on observed data, the SD* you compute from the tickets is a parameter, or "true" value, a property of the chance process itself. For most practical situations, you never know SD*; the best you can do is to estimate it from the data using SD. The difference between the SD you compute from the data and the true SD* for a chance process is the same as the difference between an observed average for data and the EV of the underlying chance process. The second difference is computational: whereas to compute an SD you divide a sum of squares by df_{Res}, to compute SD* you divide a sum of squares by the number of squares in your sum, i.e., by the number of tickets in the box.

The Standard Deviation SD* for a Box Model

1. **Decompose each value into EV + Deviation**
2. **$SD^* = \sqrt{MS_{Dev}}$, with $MS_{Dev} = SS_{Dev}/(\#\ tickets)$.**

EXAMPLE 15.11A STANDARD DEVIATIONS FOR EXAMPLE 15.4

Find SD* for the sampling distribution of the first two summary statistics in Example 15.4.

SOLUTION.

a. Grand Average (Box iii).
   1. Deviations: −2, −1, −1, −1, −1, 0, 0, 0, 0, 0, 0, 1, 1, 1, 1, 2
   2. $SS_{Dev} = 16$
   3. $MS_{Dev} = 16/16 = 1$
   4. $SD^* = \sqrt{MS_{Dev}} = 1$

b. Cond Avg #2 − Cond Avg #1 (Box i)
   1. Deviations: −4, −2, −2, −2, −2, 0, 0, 0, 0, 0, 0, 2, 2, 2, 2, 4
   2. $SS_{Dev} = 64$
   3. $MS_{Dev} = 64/16 = 4$
   4. $SD^* = \sqrt{4} = 2$ ■

Just as the SD you compute from observed data estimates SD* for the underlying chance process, the $SE = SD \times \sqrt{SS_{Weights}}$ you compute for a linear estimator estimates the true SE* for its sampling distribution.

For a linear estimator, the true $SE^* = SD^* \times SS_{Weights}$

EXAMPLE 15.11B STANDARD ERRORS FOR EXAMPLE 15.4

Summary statistics (a) and (b) in Example 15.4 are both linear estimators. Compute SD* from the tickets in the chance error box for that example. Then compute $SE^* = SD^* \times \sqrt{SS_{Weights}}$ for the two linear estimators, and check that each SE* is the same as the SD* for the box model in Example 15.11a.

SOLUTION. The chance error box has two tickets, −2 and 2. The EV = 0, $SS_{Dev} = 8$, $MS_{Dev} = 4$, $SD^* = 2$.

For Grand Average,

$$SS_{Weights} = \left(\frac{1}{4}\right)^2 + \left(\frac{1}{4}\right)^2 + \left(\frac{1}{4}\right)^2 + \left(\frac{1}{4}\right)^2 = \frac{1}{4}, \quad \text{so } SE^* = (2)\left(\frac{1}{2}\right) = 1,$$

the same as we got in Example 15.11a.

For Cond Avg #2 − Cond Avg #1,

$$SS_{\text{Weights}} = \left(\frac{1}{2}\right)^2 + \left(\frac{1}{2}\right)^2 + \left(\frac{-1}{2}\right)^2 + \left(\frac{-1}{2}\right)^2 = 1, \quad \text{so SE}^* = (2)(1) = 2,$$

the same as before. ■

If you represent a chance process as a probability histogram, then its SD* or SE* is a measure of how spread out the histogram is. Larger SE*s go with histograms that are flatter, more spread out; smaller SE*s go with histograms that are taller and more bunched together. For histograms that are roughly normal-shaped, about two-thirds of the area is within one SE* of the EV; roughly 95% of the area is within two SE*s.

EXAMPLE 15.12 STANDARD ERRORS AND PROBABILITY HISTOGRAMS

Match each of the probability histograms in Example 15.10 with its approximate SD*, choosing from 1.61, 1.73, 2.00, and 2.28.

SOLUTION. The reverse J-shaped histogram (d) is the most bunched together, and has the smallest SD*, 1.61. The triangular histogram (b) is almost as bunched, and has the second smallest SD*, 1.73. The uniform histogram (a) is quite a bit more spread out than either the J-shaped or triangular histograms, and has SD* = 2. The U-shaped histogram (c) is the most spread out, and has the largest SD*, 2.28. ■

Percentiles. Critical values and standard 95% distances are special cases of percentiles.

Definition

The 95th percentile of a distribution is the value that divides the bottom 95% of the outcomes from the top 5% of the outcomes. Other percentiles are defined in the same way. The 5% critical value for an *F*-ratio equals the 95th percentile of the corresponding *F*-distribution.

EXAMPLE 15.13A CRITICAL VALUES FOR *F*-RATIOS

Three probability histograms for *F*-distributions are shown in Figure 15.10. Match each one with the its 95th percentile = 5% critical value, choosing from 2.53, 4.39, and 9.01.

SOLUTION. At first glance, the three curves look almost identical, but if you look closely at the right tails, you find that the first distribution (denominator df = 2) has a much heavier, longer tail than the others. With only 2 df in the denominator, there's a much greater chance of getting large outcomes, and the 95th percentile is large, 9.01. The distribution with denominator df = 16 has a much lighter tail. Large values are unlikely, and the 95th percentile is comparatively small, 2.53. The distribution with denominator df = 4 is in between.

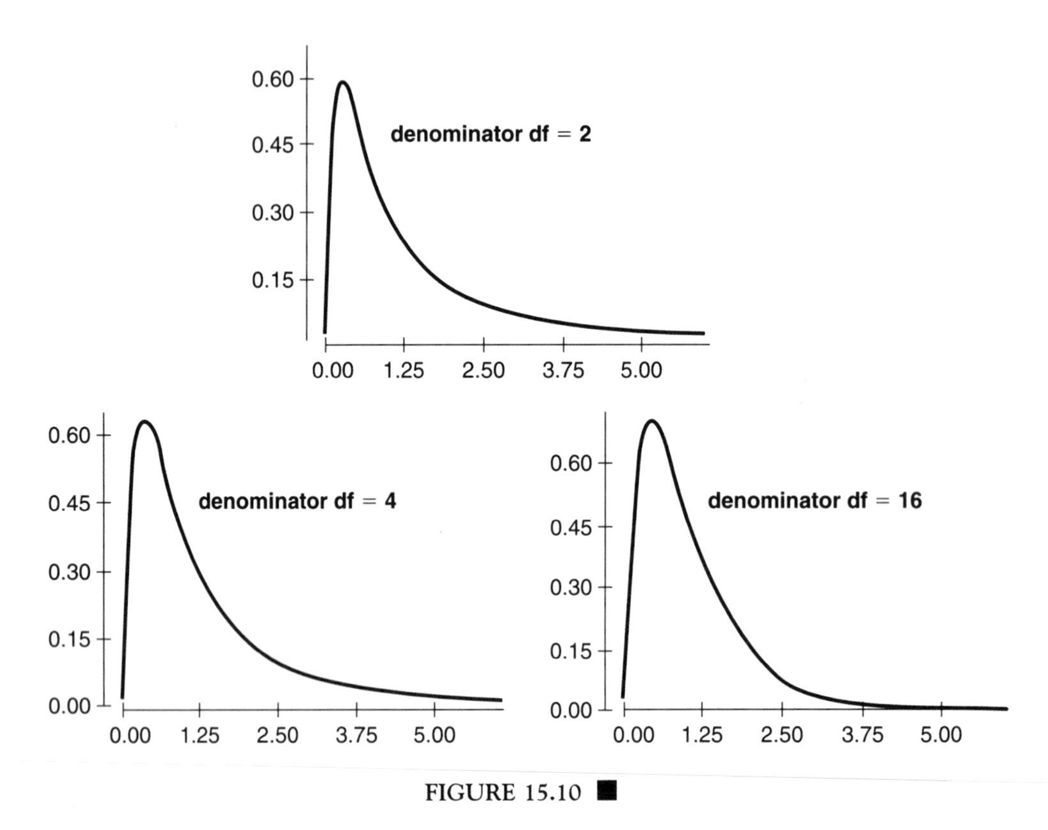

FIGURE 15.10 ■

EXAMPLE 15.13B Critical Values for *F*-Ratios

Look over the table of critical values in the back of the book. Comment on the relation between denominator df and the size of the critical value. (This relationship has an important practical meaning: it tells something about designing a good experiment.)

Comment. If you look at any one row, keeping denominator df fixed, and going from left to right, with the numerator df increasing, you find one of two patterns. For the first two rows (denominator df = 1 or 2) the critical values increase as the numerator df increase. For all the other rows, the pattern is the opposite: critical values decease as the numerator df increase. In other words, if you have at least 3 df for your denominator, then the more groups you compare, the smaller the *F*-ratio you need to reject the null hypothesis of no group differences. In short, more groups make it easier to detect group differences.

If you now pick a column, and run your eyes down the column, keeping the numerator df fixed, you find that the critical value is huge for denominator df = 1, but as the denominator df increases, the critical values decrease, very rapidly at first, and then more slowly. The bigger your denominator df, the more powerful the *F*-test is.

The power of a test is its chance of detecting a true difference.

The power depends on the size of the difference, of course: there's a better chance of detecting differences if they are large. For differences of any fixed size, the power of the F-test will be greater the larger the denominator df. In short: The more observations the better.

The standard 95% distance you get from a table when you construct a confidence interval around a linear estimate is a percentile for a *t*-distribution (Figure 15.11).

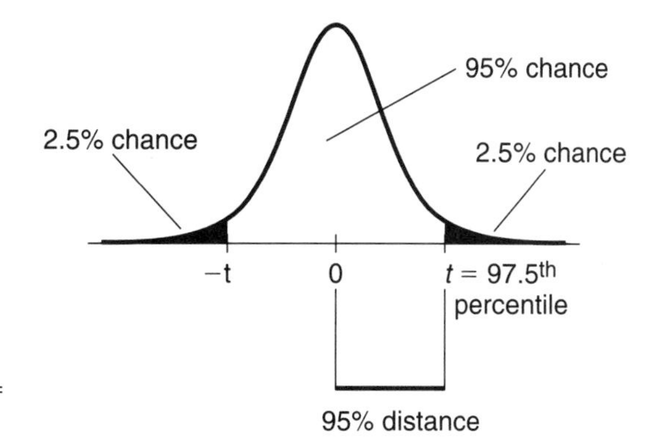

FIGURE 15.11 Standard 95% distance = 97.5th percentile ■

EXAMPLE 15.14 STANDARD 95% DISTANCES AND *t*-DISTRIBUTIONS

The graphs in Figure 15.12 show probability histograms for distributions from the t family, on 1, 2, 10, and 20 df. Match each one with the corresponding 95% distance, choosing from 2.086, 2.228, 4.303 and 12.71. Comment on any patterns.

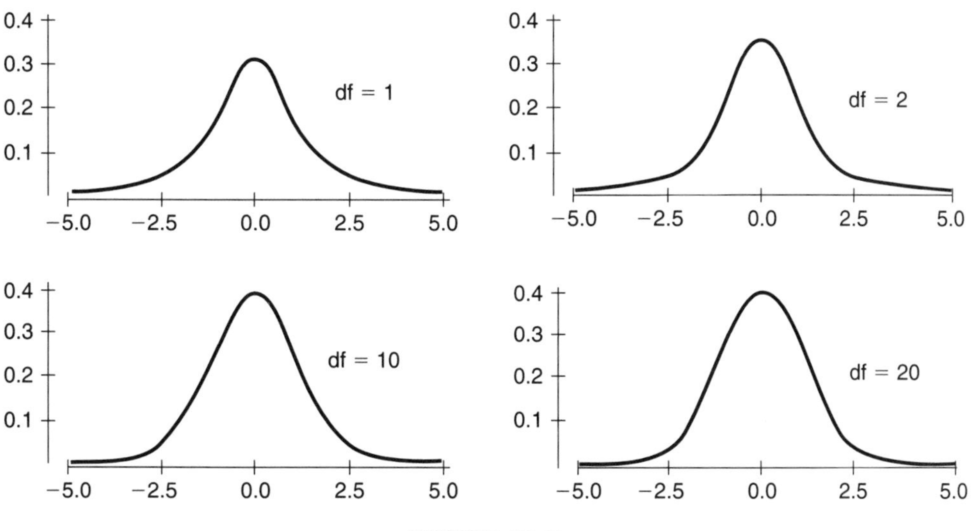

FIGURE 15.12

Comment. Each histogram is symmetric, with center at 0. The larger the df, the smaller the 95% distance. The histogram for df = 1 is the most spread out, with the longest, heaviest tails, and has the largest 97.5th percentile, 12.76. The histogram for df = 2 is much less spread, with lighter tails, and 97.5th percentile at 4.303. For 10 df, the 97.5th percentile is 2.228, and for 20 df, 2.086.

If you turn to the t table in the back of your book, you find the standard 95% distances drop off as you increase the df. The values decrease rapidly at first, then more slowly. The practical message is that the fewer observations you have, the wider your 95% confidence intervals will be. Going from $df_{Res} = 1$ to $df_{Res} = 2$ cuts the width by more than half; going from $df_{Res} = 2$ to $df_{Res} = 10$ cuts the width by roughly half again; but going from $df_{Res} = 10$ all the way to $df_{Res} = 120$ cuts the width by only about 25%. In other words, if your experiment is tiny, adding a few more observations can give you a lot more information, but if your experiment is already moderately large, adding a few more observations under the same conditions probably won't make much difference. ■

Exercise Set B

1. Match each of the box models on the left with the equivalent box model on the right.

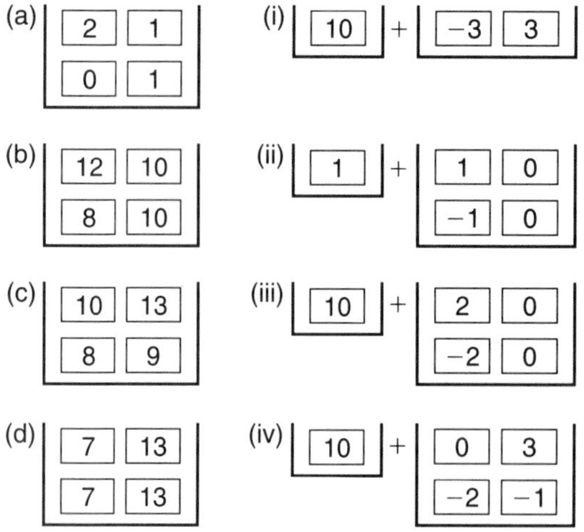

(a) [2 | 1 / 0 | 1] (i) [10] + [−3 | 3]

(b) [12 | 10 / 8 | 10] (ii) [1] + [1 | 0 / −1 | 0]

(c) [10 | 13 / 8 | 9] (iii) [10] + [2 | 0 / −2 | 0]

(d) [7 | 13 / 7 | 13] (iv) [10] + [0 | 3 / −2 | −1]

2. Fill in the blanks to make the box models equivalent.

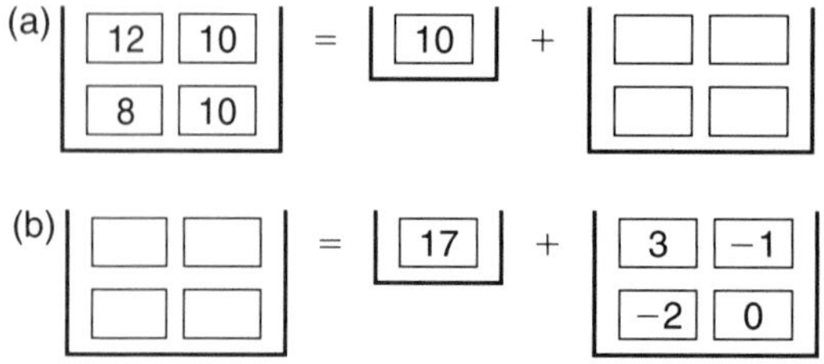

(a) [12 | 10 / 8 | 10] = [10] + [☐ | ☐ / ☐ | ☐]

(b) [☐ | ☐ / ☐ | ☐] = [17] + [3 | −1 / −2 | 0]

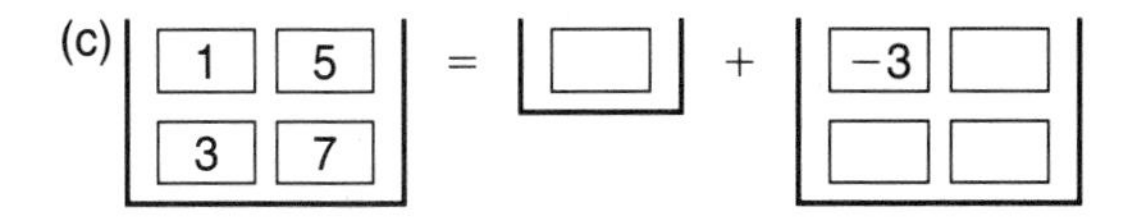

3. *A single-draw box model for # Heads in 3 tosses*. Follow the pattern of Example 15.2 to construct a single-draw box model for the number of heads in 3 tosses of a fair coin:

a. Tell the mechanism and summary.

b. List the eight possible data sets.

c. Give the single-draw box model.

d. What is the chance of getting an even number of heads?

4. *Summaries for three draws*. Consider drawing three times from

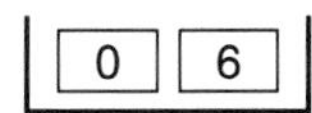

Match each of the following summary statistics with its single-draw box model.

a. Average b. Sum of squares c. SD of the observations

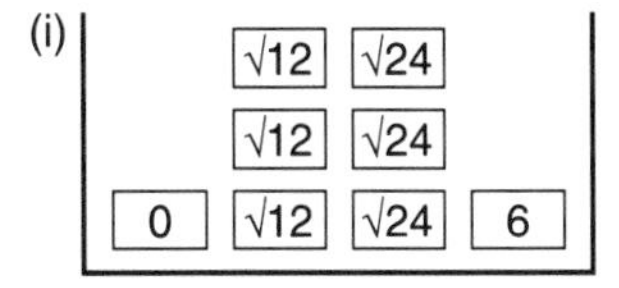

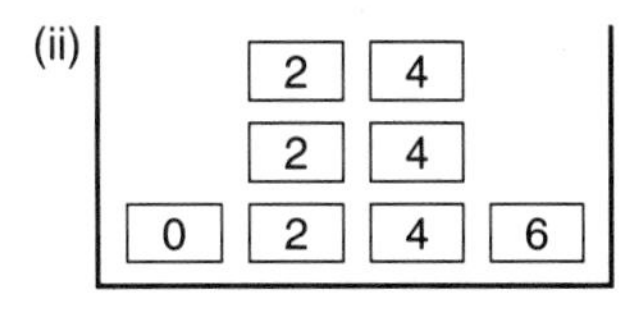

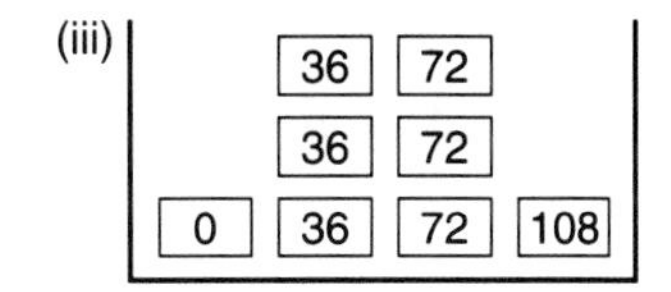

5. Use your solution to Exercise 4 to find the following:

a. The chance that the average for your sample will be within ±1 of the average for the box (= EV).

b. The chance that the sum of squares for the sample will be nonzero.

c. The chance that the SD for the sample will be nonzero.

6. *Simple* ANOVA. Modify the CR experiment in Example 15.4, using 5 as the Benchmark value, ±2 as the Condition Effects, and ±4 as the equally likely Chance Errors. For each of the summaries (a)–(d) in that example, find the appropriate single-draw box model. (For each box the number of tickets of each kind will be the same as for the corresponding box in Example 15.4, but the numbers on at least some of the tickets will be different.)

7. Use your solution to Exercise 6 to find:

a. The chance that the linear estimator for (Cond Eff #2 − Cond Eff #1) will be within

(i) ±1 of the true value

(ii) ±2 of the true value

(iii) ±4 of the true value

b. The chance that the Grand Average you get from your sample will be within

(i) ± 1 of the true value

(ii) ±2 of the true value

(iii) ±4 of the true value

8. Match each mechanism-and-summary (a)–(d) with one of the dot diagrams (i)–(iv).
   a. Mechanism: Toss 2 coins Summary: (# Heads)^2
   b. Mechanism: Toss 3 coins Summary: # Heads − # Tails
   c. Mechanism: Draw twice from Summary: Average of draws

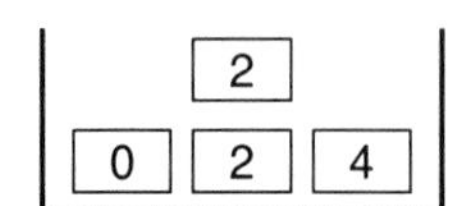

   d. Mechanism: Toss 2 coins Summary: (# Heads − # Tails)^2

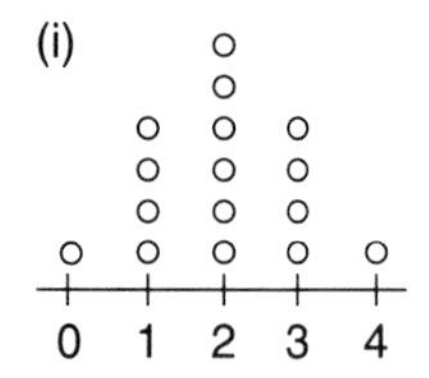

(ii)

0 1 2 3 4

(iii)

0 1 2 3 4

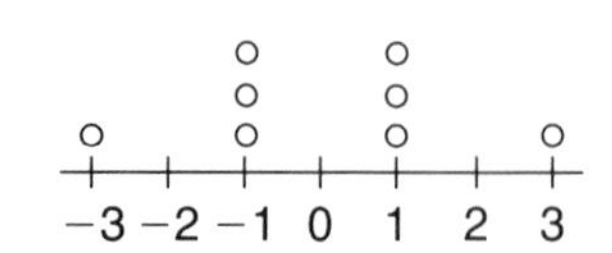

9. Match each mechanism and summary (a)–(d) with its histogram. (i)–(iv)
   a. Mechanism: Toss 2 coins Summary: # Heads
   b. Mechanism: Toss 4 coins Summary: (# Heads)/2
   c. Mechanism: Draw twice from Summary: Average of the draws

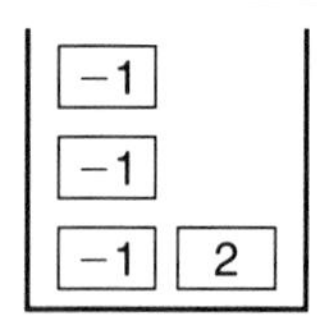

   d. Mechanism: Same as for (c) Summary: SD for the draws

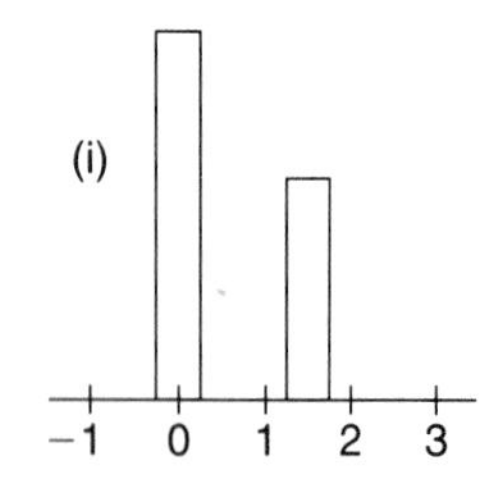

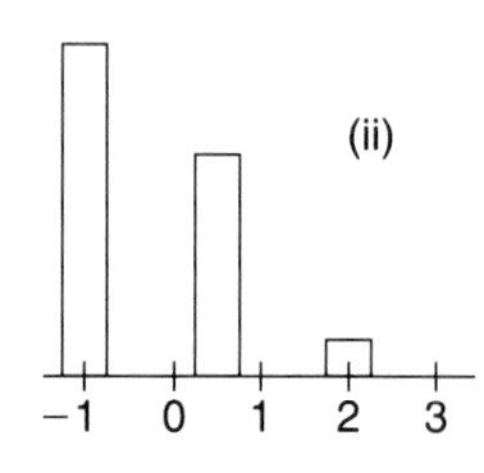

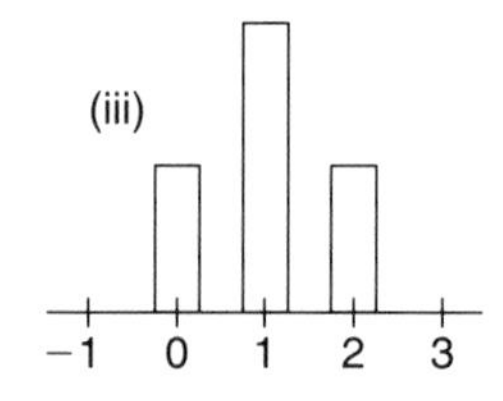

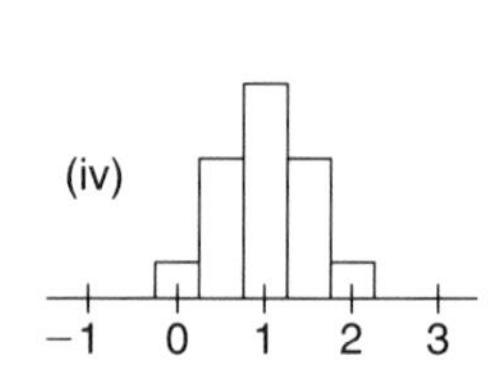

10. Match each of the dot diagrams (i)–(iv) in Exercise 8 with
    a. its EV, chosen from: 0, 1.5, 2, and 2.
    b. its SD*, chosen from: 1, 1.5, $\sqrt{3}$, 2

11. Match each of the histograms (i)–(iv) in Exercise 9 with its EV and SD*, chosen from:
    a. EV = 0.75, SD* = $\sqrt{15/4} \approx 0.97$.
    b. EV = 1.00, SD* = 0.50.
    c. EV = 1.00, SD* = $\sqrt{2/2} \approx 0.71$.
    d. EV = 0.00, SD* $\approx \sqrt{1.5} \approx 1.22$.

12. *Confidence intervals, Part 1*. Four box models for data are shown below on the left. For each, first compute the EV. Then consider the following mechanism-and-summary: Draw once, construct a confidence interval of the form Draw ± 1, and record as your summary either "Hit" if the interval contains the EV, or "Miss" if it doesn't. Match each of the box models (a)–(d) on the left with the appropriate single-draw model (i)–(v) for the Hit/Miss summary on the right.

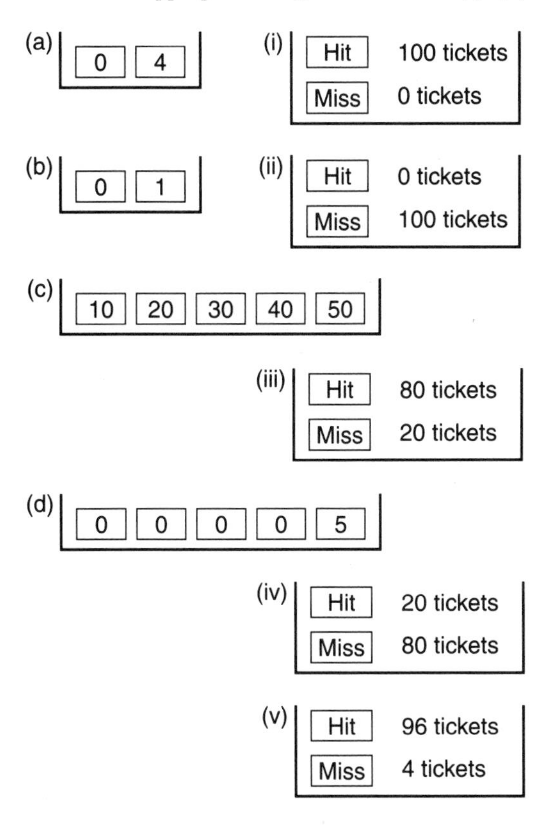

13. *Confidence intervals, Part 2*. Use the same four box models for data as above, but this time use a different mechanism-and-summary: Draw twice, with replacement, compute the average of the two draws, construct the confidence interval as Avg ± 2, and record either "Hit" or "Miss" depending on whether your interval contains the EV for the box. Match each of the box models (a)–(d) on the left with the appropriate single-draw model (i)–(v).

14. *Cool mice: The sign test*. Reread Example 7.19. Suppose you want to test the null hypothesis that for each mouse there is a 50–50 chance that the cooling constant for fresh will be greater than the cooling constant for reheated, just as there is a 50–50 chance that a fair coin lands heads.

a. Think of your null hypothesis as describing a chance mechanism: If the hypothesis is true, then each mouse corresponds to a random draw (with replacement) from a box with two tickets:

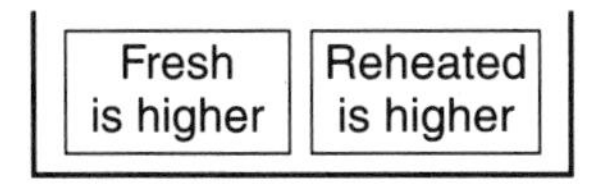

The mouse experiment corresponds to how many draws from the box?

b. Take as your summary statistic the numbers of times Fresh is higher than Reheated.

(i) What is the observed value from the data?

(ii) What is the EV for the summary statistic?

c. Here is a smooth approximate version of the probability histogram for the summary statistic. Guess the SD*.

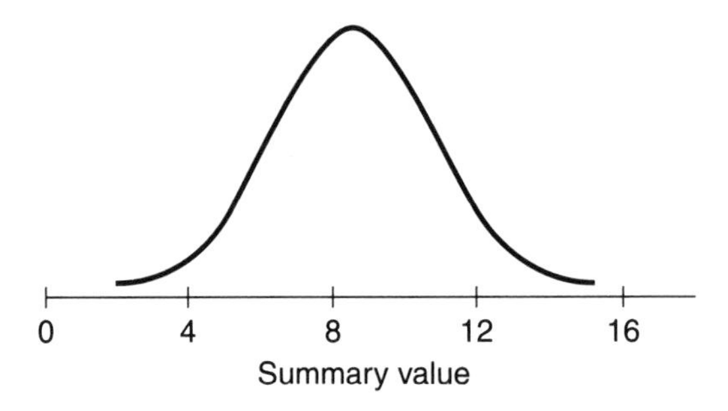

d. Suppose you want to use your summary statistic to test your null hypothesis. You might reason as follows: "If my hypothesis is true, then the histogram gives the sampling distribution of the summary statistic. Outcomes near the center (EV) are likely, and give me no reason to suspect the hypothesis is false. On the other hand, outcomes in the tails, far away from the EV on either side, are unlikely (unless my hypothesis is wrong)."

The 2.5^{th} percentile for the distribution is about 5.2, and the 97.5^{th} percentile is about 13.8. Explain how to use these numbers to complete a test of the null hypothesis. Then do the test and state your conclusion.

e. Notice that you could do an equivalent test by choosing as your test statistic:

Absolute value of (# times Fresh higher − # times Reheated higher).

Explain how you could use simulation to find the 5% critical value.

3. Four Fundamental Families of Distributions

Overview

Most of the statistical methods used by scientists are based on the family of normal distributions. These methods either assume that chance errors themselves have a normal distribution, or else they rely on other assumptions that lead to approximations based on the normal curve. For data sets that satisfy these assumptions, three distribution families derived from the normal are of special importance:

In this section, I will define each of the three families by telling how to construct them from standard normal chance errors. Sections 4 and 5 will then describe how the same families of distributions are related to the data you get from designed experiments.

Three Distribution Families for Normal Data

The chi-square family, for sums of squares;
The *t*-family, for linear estimators; and
The *F*-family, for ratios of mean squares.

The Standard Normal Distribution: A Useful Fiction

Remember that the word "normal" as a name for distributions means "ideal." If you think back to high school geometry, you may remember that what your geometry book called a line was not a thing you could actually draw, but an idealized version that stretched infinitely far in both directions and had no thickness. The standard normal distribution is an idealization in much the same way. No distribution that occurs in nature can be perfectly normal, and it is not possible to create a box of numbered tickets that gives a perfectly normal chance process. Nevertheless, I ask you to pretend that we can construct such a chance error box. Think of it as a useful fiction: For any practical purpose you may have, it is possible to create a box of tickets so close to the ideal normal that the difference won't matter. (See Exercise C.1.)

The standard normal can be by giving the equation of its probability histogram, as in the note at the end of the chapter, or just by the curve itself:

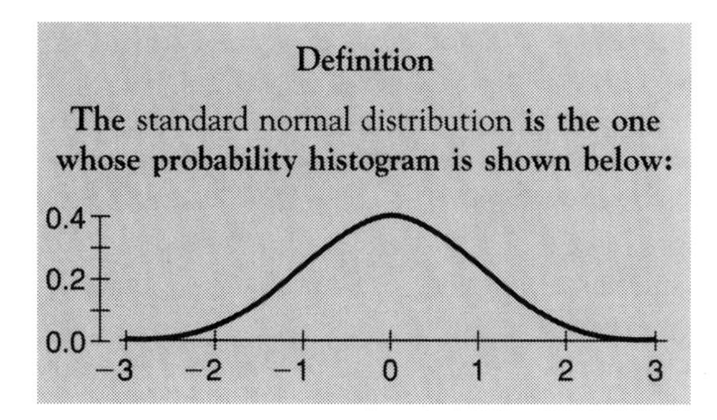

The usual way to find chances associated with the standard normal distribution is to use a computer, or else a table like the one in the back of the book. The next example shows how to use the table.

Figure 15.13 lists pairs of numbers labeled z and $A(z)$. Each z-value refers to a number on the horizontal axis of the standard normal histogram. The corresponding $A(z)$ is the area under the curve between 0 and z, expressed as a percent of the total area.

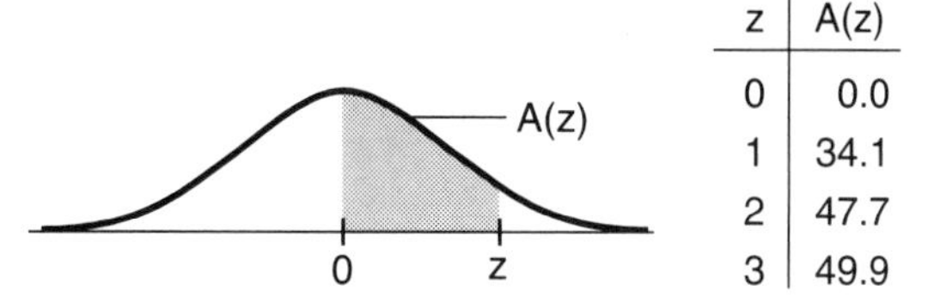

z	A(z)
0	0.0
1	34.1
2	47.7
3	49.9

FIGURE 15.13 A short normal table. Each $A(z)$ tells the area between 0 and z, as a percent of the total area. This number $A(z)$ tells the chance of a value between 0 and z.

Four facts about the standard normal are important enough to memorize:

1. The EV = 0.
2. The SD* = 1.

3. The total area equals 100%.
4. The curve is symmetric: the "right-hand" area between 0 and z equals the "left-hand" area between $-z$ and 0.

EXAMPLE 15.15 Using a Standard Normal Table

Assume your chance errors are N(0,1). Find the chances for each of the outcomes described below. Use the following strategy.

Step 1: First draw a curve. Then shade the area that corresponds to the chance you want to find.

Step 2: Write a "picture equation" that expresses the area in terms of areas listed in the table. Then use the table to find these areas.

a. Outcome: A chance error between -1 and 0.
   SOLUTION.

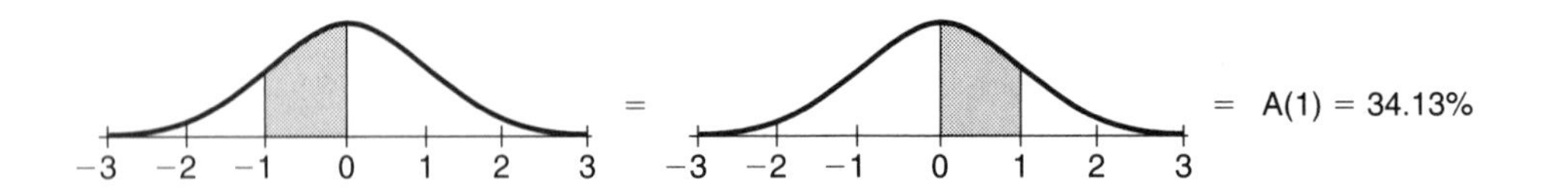

b. Outcome: A chance error between -1 and 1.
   SOLUTION.

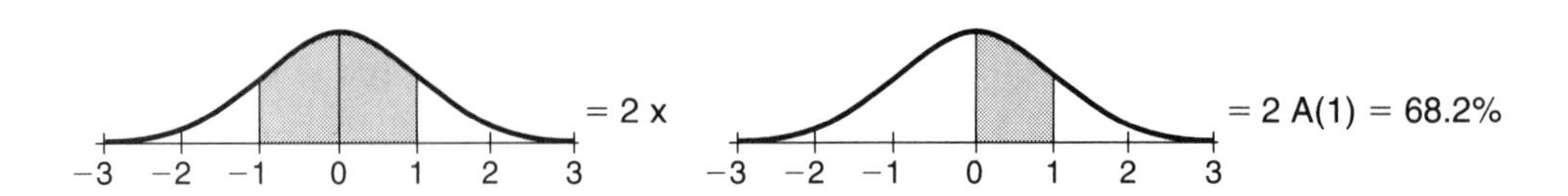

In the same way, you can check that the chance of a value between -2 and 2 is 95.4%; the chance of a value between -3 and 3 is 99.7%.

c. Outcome: A chance error bigger than 0.
   SOLUTION.

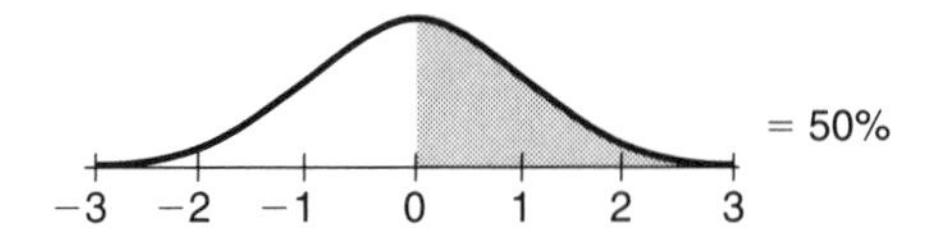

Here I didn't need the table. Because the curve is symmetric, the area to the right of 0 equals the area to the left of 0, and the two areas together add to 100%. So each half equals 50%.

d. Outcome: A chance error bigger than 2.

SOLUTION.

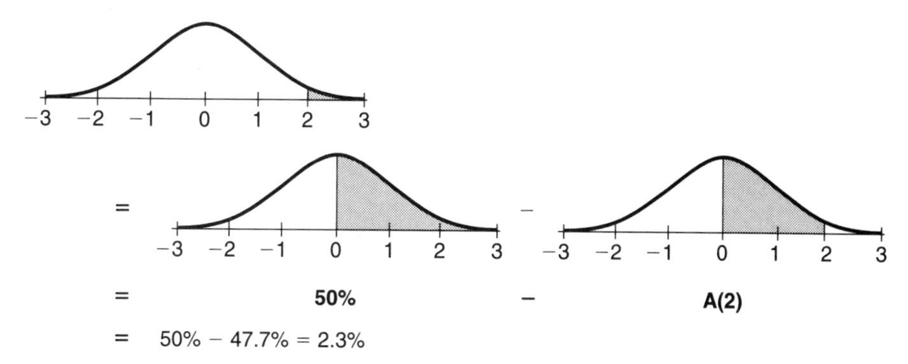

An area like this one is sometimes called a **tail area**. The **"tail" of a distribution** refers to a collection of extreme outcomes, all greater than or equal to some large value, or all less than or equal to some small value.

e. Outcome: A chance error between −2 and 3.
SOLUTION.

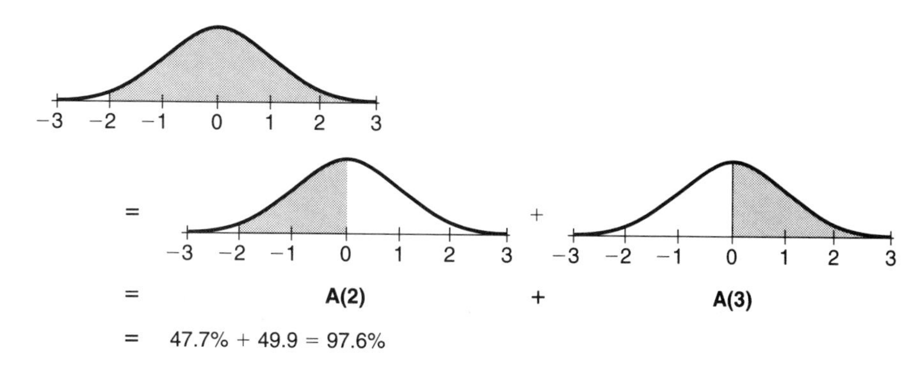

f. Outcome: A chance error bigger than 2 in absolute value.
SOLUTION.

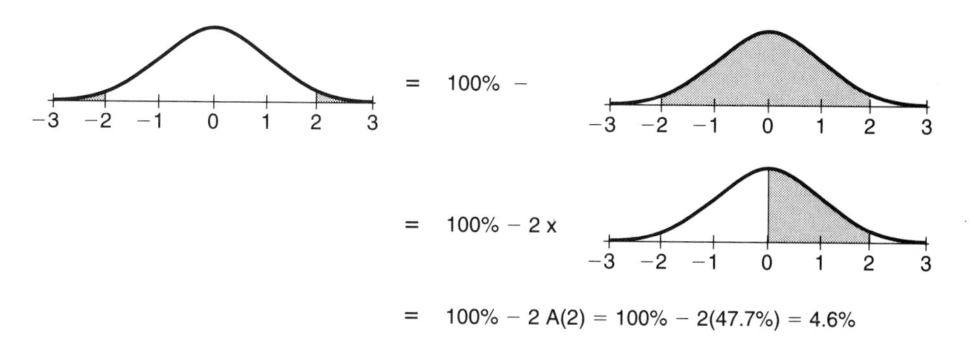

Similarly, the chance or an error bigger than 3 in absolute value equals 0.3%, or 3/1000. ■

For what follows, I urge you to think of the standard normal as a chance process, by imagining that we can draw tickets from an ideal box model, which I'll call the N(0,1) box, or standard normal error box. Statisticians use *z* and Z to refer to the standard normal: *z* to stand for a particular fixed outcome, and Z to stand for a random draw, regarded as a chance process. I'll use the N(0,1) box to define four families of distributions: the normal, chi-square, *t*, and *F* families.

Standard Units and the Family of Normal Distributions

There is one normal curve for each choice of mean = EV and standard deviation = SD*. Together, these curves form the **normal family.** The **standard normal distribution** has EV = 0 and SD* = 1. In practice, it is unusual for an experiment to give measurements with EV ≈ 0, SD* ≈ 1, but the bell shape of the normal distribution is quite common. For example, as a rough approximation, the SAT test was originally constructed to give scores with EV = 500, SD* = 100, and a normal shape.

When statisticians think about histograms, they often focus on three features, as I did in the last paragraph. These features are center, spread, and shape. For the normal distributions, **center** (or **location**) refers to the EV: where is the balance point of the histogram? **Spread** refers to the SD: a large SD for outcomes that are spread out, a small SD for outcomes that are bunched together. **Shape** is harder to summarize, because you can't use numbers the way you can with location and spread; there's no easy way to make sense of "The shape is 3." Instead, statisticians use a process called standardizing to decide whether two distributions have the same shape. For example, a distribution has normal shape if you get the standard normal when you standardize.

Before telling you how to standardize, I'll show you the reverse process, which I'll use to build the family of normal distributions from the standard normal. Suppose you have a standard normal error box, with EV = 0, SD* = 1, and you want to build a model for SAT scores, with EV = 500, SD* = 100. Let Z stand for a random draw from N(0,1). Multiply Z by 100 = SD*, then add 500 = EV, and you've got your model.

If Z is N(0,1), then for any choice of μ = EV and σ = SD*,
Y = $\mu + \sigma$Z is normal, with EV = μ and SD* = σ.

In particular, Y = 500 + 100Z is normal with EV = 500, SD* = 100.

The simplest ANOVA models assume that your observations are built in just this way. There is one unknown but constant SD* for the entire experiment, and there is one unknown but constant EV for each set of conditions or treatment combinations.

[Observed value]	=	[Unknown EV for the particular conditions]	+	[Unknown SD* for the entire experiment]	·	[Random draw from N(0,1) box]
Y	=	μ_{Cond}	+	σ	·	Z

Simple ANOVA model

(Observed value Y) equals (unknown EV σ) plus [(unknown SD* σ) times (N(0,1) draw Z)].

EXAMPLE 15.16 SCREENING BLOOD DONORS FOR HEPATITIS

Although blood transfusions are usually safe, if you get a transfusion you do run a small risk of getting a serious disease from the donated blood. Newspapers and TV have given a lot of attention to the risk of AIDS, partly because the disease is new, and partly because there is no known cure. However, the chance of getting AIDS from a transfusion is tiny in comparison to the chance of getting hepatitis, a disease that is more common, though less often fatal. Fortunately, there is an effective test for screening donated blood, based on the concentration of a particular enzyme, serum glutamic pyruvic transaminase (SGPT). People who are carriers of hepatitis tend to have higher levels of SGPT in their blood than noncarriers.

It turns out that concentrations of SGPT have roughly a normal distribution, provided you measure in log (mg/100 ml). Here's a model

$$\text{Carriers:}\quad \log \text{SGPT} = 1.55 + 0.13Z$$
$$\text{Noncarriers:}\quad \log \text{SGPT} = 1.25 + 0.12Z$$

The log concentrations for carriers are a bit more spread out than they are for noncarriers, but the differences in SDs is pretty small. The difference in EVs is much larger: the expected log concentration for carriers is 0.30 higher than for noncarriers, a difference of more than 2 SDs. In Exercise C.8 I'll ask you to use this model to compare various strategies for screening blood donors. ■

EXAMPLE 15.17 SAT SCORES

Here's a model for one-way ANOVA with verbal SAT score as response and male/female as the factor of interest:

$$\text{Males:}\quad Y = \mu_M + \sigma Z$$
$$\text{Females:}\quad Y = \mu_F + \sigma Z$$

For 1997, σ is about 100, μ_M is about 507, and μ_F is about 503. ■

Suppose you know that the histogram for a large group of SAT scores is normal shaped, with EV = 500, SD* = 1. Roughly what percent of the scores will be above 700? Between 400 and 600? You can answer questions like these by standardizing, or converting to standard units.

Definition

Standard units **tell the distance from an outcome to its EV, measured in SD*s:**

Distance = Outcome − EV   Standard distance = (Outcome − EV)/SD*

EXAMPLE 15.18 SAT Scores and Standard Units

Assume SAT scores have EV = 500, SD* = 100. Convert the following scores to standard units:

a. 500 b. 700 c. 350

Solution.

a. In standard units, EV = 0.

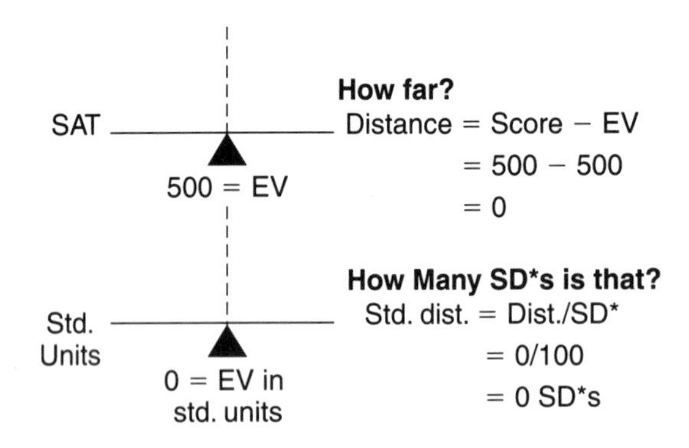

b. Score = 700

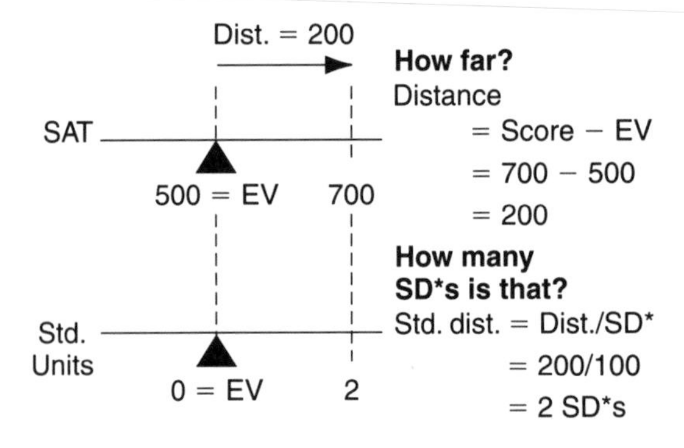

c. Score = 350

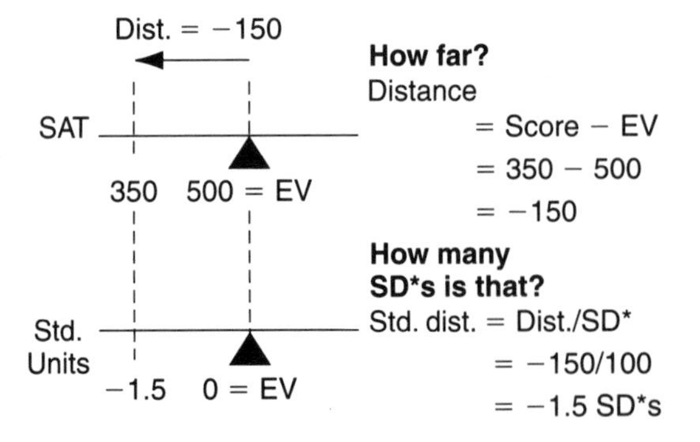

■

A few pages back I said that you could think of standardizing as the opposite of building a distribution from a box model with EV = 0, SD* = 1. Here's the same idea in symbols:

> **$Y = \mu + \sigma Z$ has EV = μ, SD* = σ**
> **if and only if**
> **$Z = (Y - \mu)/\sigma$ has EV = 0, SD* = 1**

To find the chance of an SAT score above 700, first convert to standard units, then use the standard normal table.

EXAMPLE 15.19 CHANCES FOR SAT SCORES

Assume SAT scores are normal, with EV = 500, SD* = 100. Find the chance of a score: (a) above 700; (b) between 400 and 600.

SOLUTION.

a. Above 700

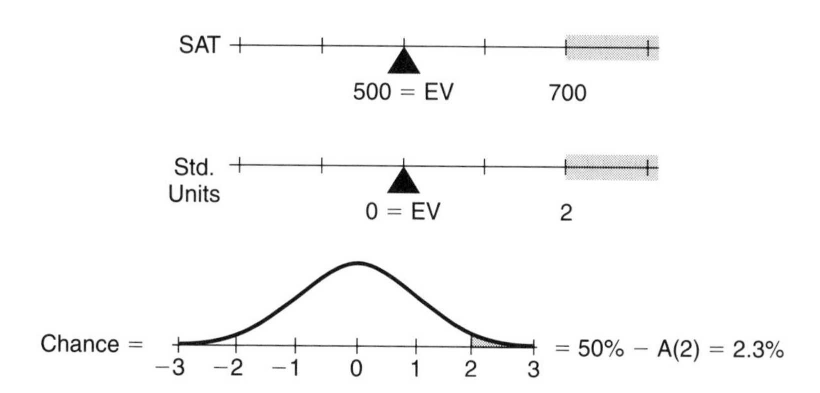

b. Between 400 and 600

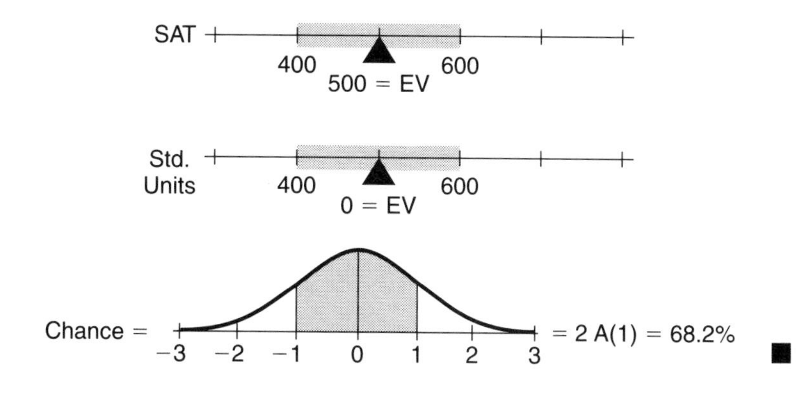

■

A major reason why the normal distributions are so important in statistics is that they often give good approximations to the sampling distributions of linear estimators, even when the estimators are computed from observations whose chance errors are not normal. The more observations you have, the better these approximations tend to be. For the example that follows, the linear estimator is based on just 4 observations, and the chance error distribution is not at all normal: there are only two (equally likely) chance errors: 2 and −2.

EXAMPLE 15.20 Standard Units for a Linear Estimator

Turn back to Example 15.4. The box labeled (i) gives the sampling distribution for the linear estimator "Cond Avg #2 − Cond Avg #1."

a. Express each of the 5 possible outcomes in standard units.

b. Then find the chance of getting a standard distance less than or equal to 0.5, 1.0, 1.5, 2.0, and 2.5.

c. Compare these chances with those for a standard normal distribution.

Solution.

(a) Standard units: First find the EV: The true Condition Effects are −1 (#1) and +1 (#2), so the EV is 1 − (−1) = 2. Next find the SE*: the SD* for the chance error is 2, and $SS_{Weights}$ is $(1/2)^2 + (1/2)^2 + (1/2)^2 + (1/2)^2 = 1$, so the SE* = $SD^*\sqrt{SS_{Weights}} = (2)(1) = 2$. Now convert each possible outcome to standard units:

Outcome	Distance = Outcome − EV	Standard distance = Distance/SE
−2	−4 = −2 − 2	−2 = −4/2
0	−2 = 0 − 2	−1 = −2/2
2	0 = 2 − 2	0 = 0/2
4	2 = 4 − 2	1 = 2/2
6	4 = 6 − 2	2 = 4/2

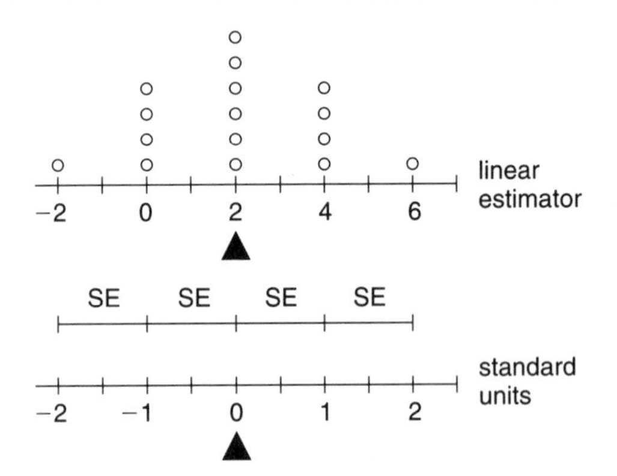

■

A box model for the standard distances will have 16 tickets: one −2, four −1s, six 0s, four +1s, and one +2.

(b) Chances: There are six tickets with distances less than or equal to 0.5 (in absolute value): the chance of getting a standard distance less than or equal to 0.5 is (6/16) × 100% = 37.5%. There are 14 tickets with distances less than or equal to 1.0 (in absolute value): the chance of getting a standard distance less than or equal to 1.0 is (14/16) × 100% = 87.5%. The chance of a standard distance less than or equal to 1.5 is also 14/16, or 87.5%. All the tickets are less than or equal to 2 in absolute value, so the last two chances are both 100%.

(c) Comparison with chances from the standard normal curve:

Standard distance less than or equal to	Actual chance	Normal approximation
0.5	37.5%	35.2%
1.0	87.5%	68.3%
1.5	87.5%	86.6%
2.0	100%	95.4%
2.5	100%	98.8%

Comment. The approximations for 0.5, 1.5, and 2.5 are quite good. The approximation for 1.0 is way off, and the approximation for 2.0 is mediocre at best. Nevertheless, when you consider how lumpy and non-normal the distribution was for the individual chance errors, these results are pretty impressive. (When you use the normal curve to get approximate chances for a distribution like the one in this example, which has only a small number of possible outcomes, you tend to get good approximations for standard distances that fall half-way between outcomes, as 0.5 and 1.5 did here, and poor approximations for standard distances that coincide with outcomes, as 1.0 and 2.0 did.)

In practice, many error distributions are not normal, and if they are not, then the sampling distributions of linear estimators will not be normal either. Moreover, in practice you can't find the true error distribution, because you don't have the option of looking inside a box model to view the tickets. How then can you find the sampling distribution for your linear estimator?

The answer comes from one of the most important results in all of statistical theory, a result called the **Central Limit Theorem**. Roughly, the theorem says that no matter what the true error distribution might be for your observations, if you base your linear estimator on a large enough number of them, then its sampling distribution will be approximately normal. The next example gives an illustration.

EXAMPLE 15.21A SAMPLING DISTRIBUTION OF THE AVERAGE

Twelve histograms are shown in Figure 15.13, arranged in four rows by three columns. Each histogram shows the sampling distribution of an average, in standard units. Each column is for a different distribution for chance error; each row is for a different sample size. Comment on the patterns.

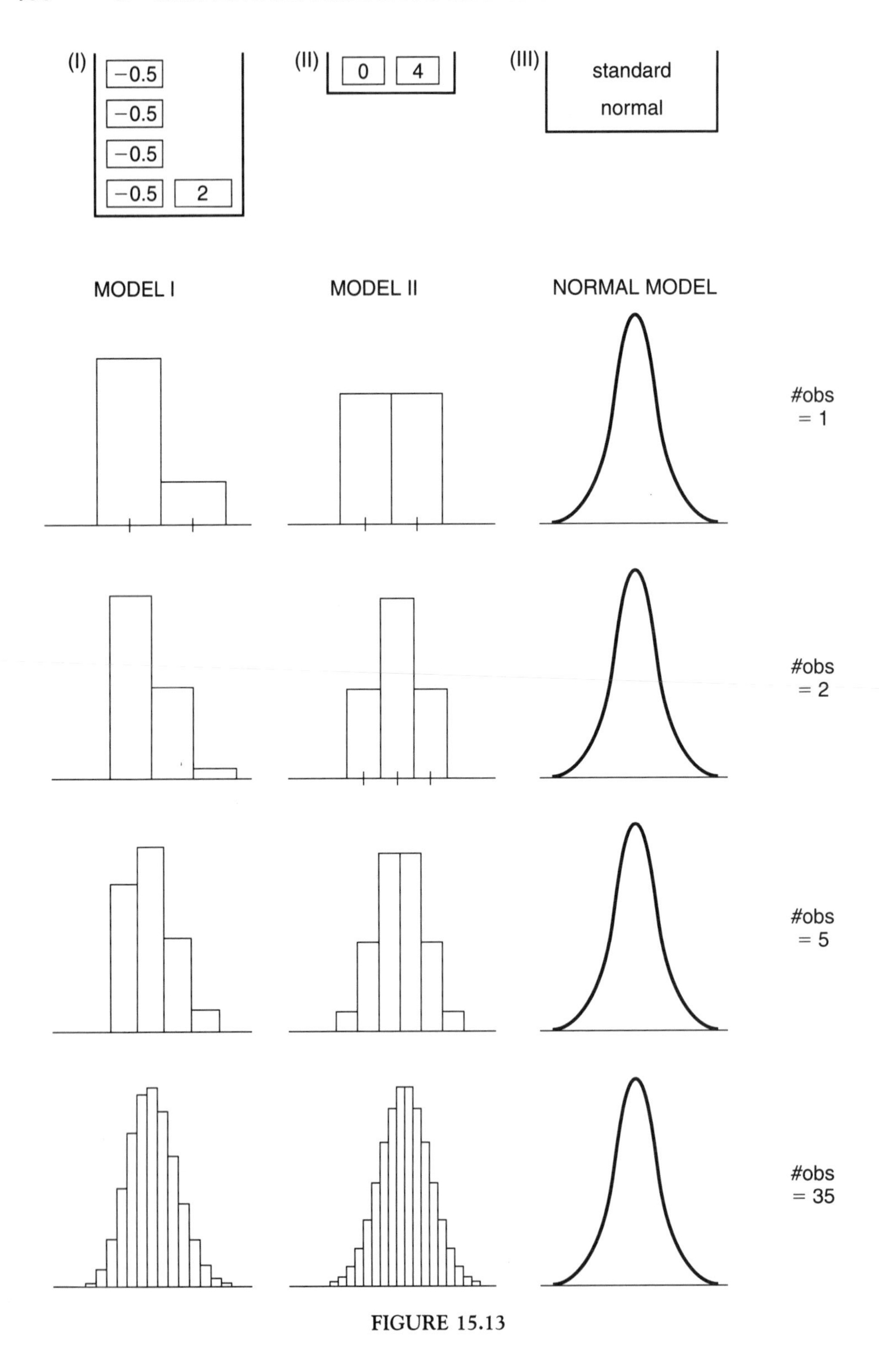
(I)
−0.5
−0.5
−0.5
−0.5
2
(II)
0
4
(III)
standard normal
MODEL I
MODEL II
NORMAL MODEL
#obs = 1
#obs = 2
#obs = 5
#obs = 35

FIGURE 15.13

Comments. The three error distributions have quite different shapes. (I) is skewed (lopsided) and "lumpy" (few outcomes are possible). (II) is symmetric, but also "lumpy." (III) is the ideal situation, normal, and not at all lumpy. Even though both (I) and (II) are far from normal, as you go from 1 to 2 to 5 to 35 observations, the corresponding sampling distributions for the average become more and more like the standard normal. ■

EXAMPLE 15.21B APPROXIMATE AND ACTUAL CHANCES FOR THE AVERAGE

Each of the sampling distributions in Example 15.21(a) was for an average, converted to standard units. If a distribution is perfectly normal, then the chance of getting a standardized average between −1 and 1 is 68.3%. Here is a table showing the actual chances computed from the true sampling distributions.

	Distribution		
Sample size	I	II	III
1	80.0%	100.0%	68.3%
2	64.0%	50.0%	68.3%
5	41.0%	62.5%	68.3%
35	71.1%	68.8%	68.3%

■

EXAMPLE 15.22 CHANCES FOR LINEAR ESTIMATORS

Assume either the ideal model (chance errors have a normal distribution) or that you have a large enough number of observations to rely on the Central Limit Theorem. Find chances for the outcomes described below.

a. Suppose a linear estimator has EV = 8, SE* = 3. Find the chance of an estimate between 5 and 11.

   SOLUTION. First convert the 5 and 11 to standard units:

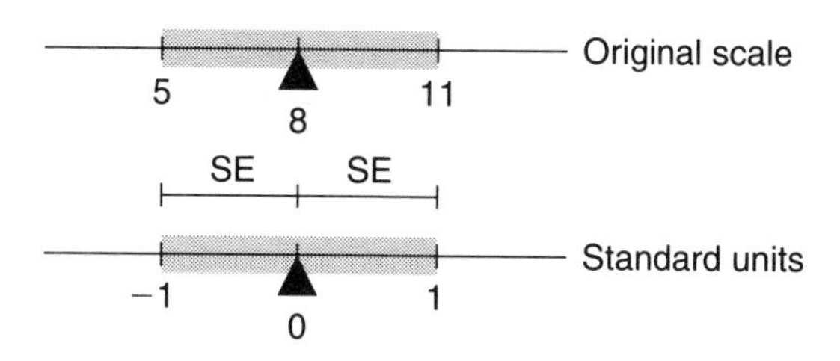

   Then find the chance:

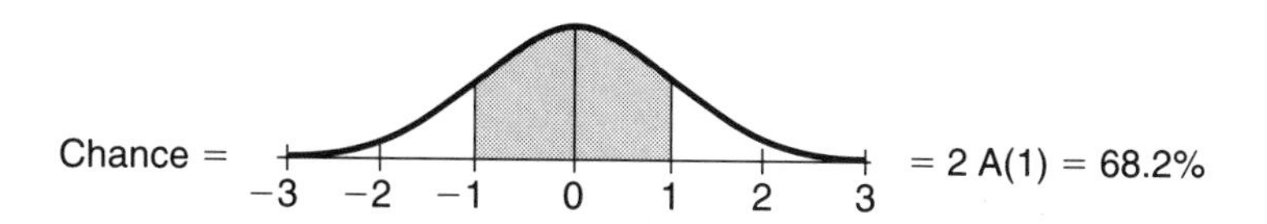

b. Find the chance that an linear estimator is within 2 SE*s of its EV.

SOLUTION. Convert to standard units:

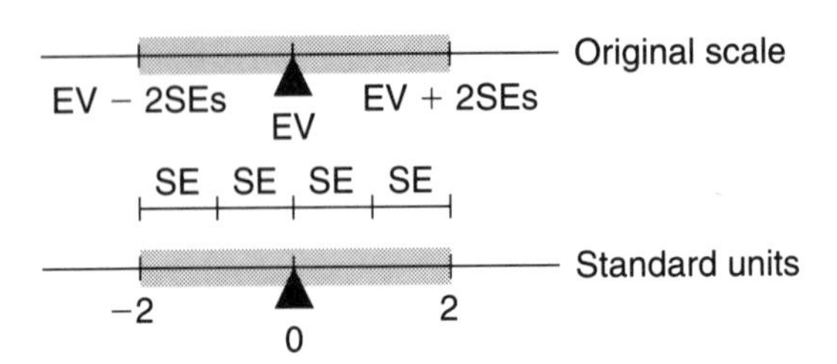

Find the area:

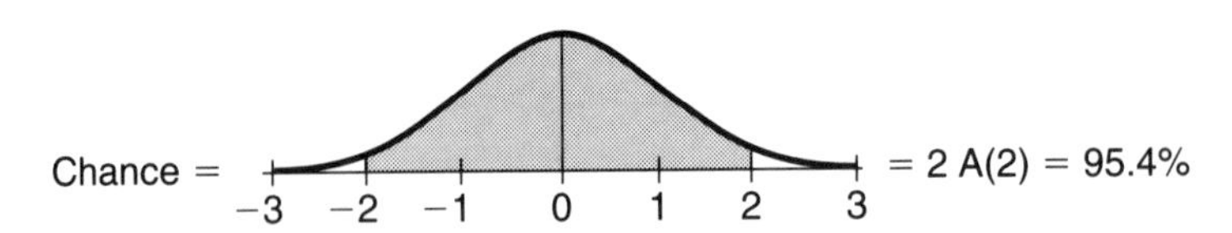

c. Suppose you construct a confidence interval of the form "linear estimator ±(SE*)(2)." Find the chance that your interval will contain the EV.

SOLUTION. "Interval contains the EV" is the same as "estimator is within 2 SE*s of the EV." Part (b) gives the chance: about 95%. ■

The Chi-Square Family: Sums of Squares for N(0,1) Outcomes

The family of normal curves is a two-parameter family: there is one member of the family for each choice of two parameters, μ = EV and σ = SD*. The chi-square family is a one-parameter family, with one distribution for each choice of df.

Definition

The chi-square distribution on n degrees of freedom **is the distribution for the SS of n independent N(0,1) outcomes. In symbols:**

$$\chi_n^2 = Z_1^2 + Z_2^2 + \ldots + Z_n^2.$$

If you had a N(0,1) box, and you wanted to create a single-draw box model for the χ_n^2 distribution, you could create tickets for the chi-square box one at a time by simulation. For each ticket, you'd make n independent draws from the N(0,1) box, compute the sum of squares, and write that number on a ticket for the chi-square box. If you created a few thousand tickets, you'd have a very good approximate single-draw model for $\chi_n{}^2$.

For our purposes, we will use the chi-square family mainly as a stepping stone to the t and F families, so I won't ask you to learn to use chi-square tables. However, before moving on to the next two families, I want to sketch briefly how the chi-square family arises in practical work. The definition is based on the standard normal, and even though real experiments almost never give N(0,1) measurements, the chi-square distributions often provide good approximations for the way ANOVA SSs behave. First suppose that you have independent measurements (Ys) that are normal with known EV = μ and known SD* = σ. If you standardize your observations, you get a set of standard normal outcomes $Z = (Y - \mu)/\sigma$, and the SS for the standardized outcomes will have a chi-square distribution. Now consider the usual situation in ANOVA: you don't know the EV or SD*, and you compute your SSs from the boxes of your linear decomposition. Statistical theory shows that if your model fits, then, for example, the numbers in your residual box have EV = 0, just as for a standard normal. Moreover, even though you don't know the SD*, it turns out not to matter: for *t*-statistics and *F*-ratios, which are built from sums of squares, there will be SD*s in both the numerator and denominator of the ratio you compute, and these will cancel in the same way that (25X)/(25Y) = (X/Y). I'll say more about how this works in the last two parts of this section.

The *t* Family

The family of *t*-distributions is another one-parameter family, with one distribution for each choice of df. A t distribution often gives a good approximate description of the way a ratio of the form (Linear Estimator − EV)/SE behaves.

Definition

The *t*-distribution on *n* degrees of freedom **is the distribution for**

$$t_n = \frac{Z}{\sqrt{\chi_n^2/n}}, \text{ where}$$

1. **Z is N(0,1);**
2. **χ_n^2 is chi-square with df = n;**
3. **Z and χ_n^2 are independent.**

Although it is convenient to define the *t*-distribution by using the chi-square, notice that you could create a single-draw box model for t_n by simulation using just the N(0,1) box. To get each ticket for the t_n box, you'd make $(n + 1)$ independent draws from N(0,1). Compute the SS for the first n draws to get a value for χ_n^2 and use that to compute the value of t_n as (last draw)/$\sqrt{(\text{SS for first } n \text{ draws})/n}$.

The definition of the *t* family uses normal outcomes with SD* = 1. What if you use normal draws from a box with SD* = σ instead of SD* = 1? The distribution will be exactly the same as if you had had SD* = 1. The numerator for each *t*-value will equal σZ. The denominator will also contain a σ: each draw used for your SS will be of the form σZ, each square in the SS will have the form $\sigma^2 Z^2$, and the SS itself will be

$$\sigma^2 Z_1^2 + \sigma^2 Z_2^2 + \ldots + \sigma^2 Z_n^2 = \sigma^2(Z_1^2 + Z_2^2 + \ldots + Z_n^2) = \sigma^2 \chi_n^2.$$

Dividing by n and taking the square root gives you a denominator equal to $\sigma\sqrt{\chi_n^2/n}$, and the σs in the numerator and denominator cancel, leaving you with exactly the same ratio as in the definition.

The behavior of t_n as a chance process depends only on n. Although the probability histograms are not normal shaped, they do look roughly bell shaped, and the larger n is, the more the histogram for t_n is like the N(0,1) histogram. The reason: When n is large, the denominator values for $\sqrt{\chi_n^2/n}$ will be tightly bunched near 1, and for very large n, the ratio t_n = [Z/Denominator] behaves very much as if the denominator were constant and equal to 1.

Section 4 will illustrate how the t family arises in practice, as an approximation to the distribution of (Linear estimator − EV)/SE.

The *F* family

The F family is a two-parameter family, with parameters n = numerator df, d = denominator df. Each F-distribution describes the behavior of a ratio of mean squares.

Definition

The F-distribution with numerator df = n and denominator df = d **is the distribution for**

$$F_{n,d} = \frac{\chi_n^2/n}{\chi_d^2/d},$$ **where** χ_n^2 **and** χ_d^2 **are independent chi-squares.**

Think about how you would create a single-draw box model for $F_{n,d}$ using an N(0,1) box. First make n independent draws and compute Numerator = SS_n/n. Then make d more independent draws for Denominator = SS_d/d. Compute the ratio, Num/Den, and write that number on a ticket for the $F_{n,d}$ box. (Then repeat the whole process a few thousand times!)

I won't go through the argument in detail, but check that if the SD* for your normal draws equals σ instead of 1, both numerator and denominator contain a σ^2, and so the σ^2s cancel when you compute the ratio.

Section 5 will illustrate how the F family applies to ratios of mean squares that you compute from the boxes of a linear decomposition.

Exercise Set C

1. *Box models that are approximately normal.* I've written several lists of numbers below. If you think of each list as the numbers for the tickets of a box model, then each model is closer to the standard normal than the ones that come before it. It would be possible to create new lists of tickets that give models still closer to the normal. Indeed, by using enough tickets,

you could create a box model as close as you like to the standard normal. For each list, do the following things:

i. Notice that the list is symmetric, with positive and negative numbers balancing, so that EV = 0.

ii. Compute SD*.

iii. Notice that the gaps between numbers in the lists get smaller as you go from one list to the next. Each list is "smoother," less lumpy than the ones before it.

iv. Notice that the largest and smallest values get farther from zero as you go from one list to the next.

v. Construct a histogram for each list. Make the height of your rectangle for representing a single value equal to (width of base)/(total # tickets), so that the total area for each histogram will be equal to 1.

a. −1, 1

b. 0 (2x), ± 2 (1x). (The abbreviation means two 0s, and 1 each of +2 and −2: 0, 0, 2, −2.)

c. 0 (6x), ±1 (4x), ±2 (1x): 0, 0, 0, 0, 0, 0, 1, 1, 1, 1, −1, −1, −1, −1, 2, −2

d. ±1/3 (126x), ±3/3 (184x), ±5/3 (36x), ±7/3 (9x), ±9/3 (1x)

2. *Finding chances using the standard normal curve*. Several intervals of outcomes are listed below. By referring to the standard normal curve, match each one with the chance of an outcome of the sort described, choosing from: 1.2%, 13.6%, 45.7%, 86.6%, 95.5%.

a. An outcome greater than 1 but less than 2

b. An outcome between −1.5 and 1.5

c. 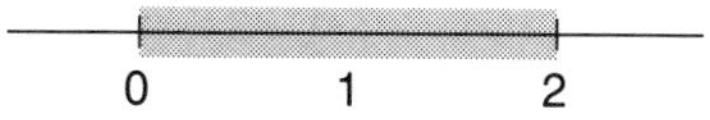

d. An outcome less than 2 in absolute value

e. An outcome greater than 2.5 in absolute value.

3. *Finding chances using standard normal tables*. Follow Example 15.15 to find the following chances for a standard normal outcome Z:

a. Z greater than 1.75

b. Z between −0.5 and 1.6

c. Z between 0.3 and 0.7

d. Z greater than 1.3 in absolute value

4. *Standard units*. Convert each of the following to standard units:

	Obs	**EV**	**SD***
a.	2	1	2
b.	3	−2	2.5
c.	760	500	100

5. Converting back.

Convert each of the following numbers, now expressed in standard units, back to the original scale it came from:

	Std. units	EV	SD*
a.	1.5	0	4
b.	1.5	6	4
c.	1.5	500	100
d.	−1.5	500	100

6. *Matching*. The four numbers 2, 3, 10, and 16, in scrambled order, represent an observed value, EV, SD*, and standardized value.

   a. If 3 is the SD*, identify each of the others.

   b. If 3 is not the SD*, identify each of the four.

7. *SAT scores*. Assume for this problem that SAT scores have SD* = 100. Suppose you have two populations of students, population A, with EV = 550, and population B, with EV = 500. (For 1997, the mean scores nationwide on the quantitative section of the SAT test were 530 for males, 507 for females.)

   a. Social scientists have sometimes argued that a difference in EVs of this size, only half an SD*s worth, matters very little, because for example even though population A has the higher EV, there will be lots of people in population B with scores higher than the EV for population A. What is the chance that

      i. a random person from A scores above 500?

      ii. a random person from B scores above 500?

   b. Other social scientists argue that although a difference of half an SD* doesn't matter much when you compare outcomes near the center of the two distributions, the difference is nevertheless very important when you think about extreme outcomes.

      i. Find the chance that a random person scores above 700 if s/he comes from A, and from B.

      ii. Find chances of a score above 750 for a random person from each population.

      iii. Find the chance of a score below 300 for a random person from each population.

   c. Fill in the following table:

	Chance of score		
Kind of score	**Population A**	**Population B**	**Chance for A/Chance for B**
Above 550			
Above 700			
Above 750			
Below 300			

   d. Suppose the populations are the same size. Fill in the blanks:

      i. For every B score above 700, there are ________ A scores above 700.

      ii. For every B score above 750, there are ________ A scores above 750.

      iii. For every A score below 300, there are ________ B scores below 300.

8. *Screening blood donors for hepatitis*. (Refer to Example 15.16.) Suppose you run a blood bank, and you want to use blood concentrations of SGPT to screen out hepatitis carriers. Your idea is to choose a cut-off value: anyone with log SGPT above that value you'll call a carrier, and not accept their blood; anyone with log SGPT below that level, you'll call a noncarrier, and allow them to donate.

   a. Two types of errors. For each true condition, carrier and noncarrier, the test based on SGPT can either be right or wrong. Fill in the blanks to describe the two kinds of errors (choose from carrier, noncarrier, greater than and less than):

   A Miss occurs if a __________ has log SGPT __________ than the cut-off, and gets called a __________;

   A False Alarm occurs if a __________ has a log SGPT __________ than the cut-off, and gets called a __________.

   b. For each of the possible cut-off values listed below, find the chance that (i) a carrier, and (ii) a noncarrier will be correctly classified. Cut-off values: 1.34, 1.40, 1.45.

   c. In your role as blood-banker, tell which kind of error from part (a) is more serious. Which of the three cut-off values in part (b) would you use if you ran the blood bank?

9–12. *Simulating distributions from the four families*. You can get a very rough approximation to a standard normal chance process by tossing four coins and computing (# Heads − 2). The purpose of these problems is to develop your intuition about the way distributions from the normal, chi-square, *t*, and *F* families are related to the standard normal.

9. *Not quite normal: mound-shaped distributions with* EV = 0, SD* = 1.

   a. There are 2 × 2 × 2 × 2 = 16 equally likely outcomes for four tosses of a fair coin. (HTHH and HHTH count as different outcomes.) List the 16 outcomes, and use your list to check that the distribution for (# Heads − 2) is the same as the one in Exercise 1(c), which has EV = 0, SD* = 1.

   b. Get four coins, and toss them 24 times to create 24 values for (# Heads − 2). List them in the order you get them; then summarize your results in a histogram.

   c. SAT scores. Use your 24 values from part (b) to get 24 normal outcomes with EV = 500, SD = 100. List the outcomes; then summarize your results in a histogram.

10. *Not quite chi-square: Sums of squares of not-quite normals*.

   a. Chi-square, df = 1: Square each of the numbers in Exercise 9(b) to get values for an approximation to the χ_1^2 distribution. Draw a histogram.

   b. Chi-square, df = 3: Use your numbers from Exercise 9(b) in sets of 3, computing a SS for each set, to get 8 values for an approximation to the χ_3^2 distribution. Draw a histogram, and write a sentence comparing it with your histogram in (a). (Since you have only 8 outcomes, your histogram may not be very typical, but I thought you'd rather not generate as many as 50 outcomes. Of course, if I'm wrong, be my guest . . .)

11. *Not quite a t-distribution*.

t, df = 2: Use your numbers from Exercise 9(b) in sets of 3 to generate 8 values from an approximate t_2 distribution: for each set of 3, compute

$$t_2 = \frac{\text{first outcome}}{\sqrt{(\text{SS for 2}^{\text{nd}} \text{ and 3}^{\text{rd}} \text{ outcomes})/d}}$$

Are these values more or less spread out than those in (b)?

12. *Not quite F: Ratios of not-quite chi-squares.*

a. "$F_{2,2}$." Use your numbers from Exercise 9(b) in sets of 4: for each set, compute

$$\text{“}F_{2,2}\text{”} = \frac{\text{MS for 1}^{\text{st}}\text{ and 2}^{\text{nd}}\text{ outcomes}}{\text{MS for 3}^{\text{rd}}\text{ and 4}^{\text{th}}\text{ outcomes}}$$

b. "$F_{1,2}$."

i. Tell how to use the numbers from Exercise 9(b), in sets of 3, to get values from an approximate $F_{1,2}$ distribution: follow the form of (a) to write an equation $F = \ldots$ that tells how to get the values.

ii. If you were actually to follow your instructions in (i), how would your 8 values be related to the 8 values in Exercise 11?

13–16. *Using Minitab to simulate the distributions based on coin tosses.*

13. *Not-quite normal.*

a. To generate 100 random numbers that behave like the random numbers in Exercise 9(b), follow these two steps:

1. Enter the following sixteen numbers in `c1` of a new worksheet:

−2, −1, −1, −1, −1, 0, 0, 0, 0, 0, 0, 1, 1, 1, 1, 2

2. Choose: `Calc > Random Data > Sample from Columns`

Fill in the dialog box as follows:

`Sample [100] rows from column(s): [c1]`

`Put samples in: [c2]` `[X] Sample with replacement`

b. Now choose: `Graph > Histogram` and fill in the dialog box to get a histogram of your values in (a). In what way does your distribution look like a standard normal? In what way does it look different?

c. To get simulated SAT scores like those in Exercise 9(c), choose:

`Calc > Calculator`

and fill in the dialog box with:

`Store result in variable: [c3]`

`Expression: [500+100*c2]`

Now follow the steps in (b) to get a histogram for these scores.

14. *Not quite chi-square.* Get histograms for the following two variables:

a. "Chi-square," df = 1: Follow the same steps as in Exercise 13(c), except when you fill in the dialog box, tell Minitab to make `c4` your new variable, and use the expression `c2**2`.

b. "Chi-square," df = 3: First create three new sets of not-quite normals, in columns `c5`, `c6`, and `c7`, by repeating step 2 from Exercise 13(a) three times. Then create a new variable `c8` as in (a) above, this time using as your expression `c5**2 + c6**2 + c7**2`.

15. *Not quite t.* Get a histogram for the following variable:

Put the variable in `c9`, and use the expression `c2/(c5**2 + c6**2)`.

16. *Not quite F.* Get histograms for:

a. "$F_{2,2}$" = `c10` = `(c2**2 + c5**2)/(c6**2 + c7**2)`

b. "$F_{1,2}$" = `c11` = `c2**2/(c6**2 + c7**2)`

17–20: The following four problems have exactly the same structure as the previous four, except this time, instead of using simulated coin tosses, you use Minitab's simulated random normal numbers.

17. Simulated normal distribution.

Repeat problem 13, except that this time, get your observations for `c2` by choosing:

`Calc > Random Data > Normal` and filling in the dialog box to get 100 rows of data, stored in `c2`, with mean of 0.0 and standard deviation of 1.0.

18–20. Repeat Exercises 14–16, this time using the normal data from Exercise 17. When you repeat Exercise 14(b), create three new columns of normal data, by following Exercise 17.

4. Sampling Distributions for Linear Estimators

This section deals with the relationship between the *t*-distributions and 95% confidence intervals based on linear estimators. On the surface, it is quite a remarkable claim that for so many different ANOVA models, and so many different linear estimators, the form of a confidence interval is always the same: Estimate ± SE · *t*-value. The claim rests on an equally remarkable fact, that for any of those models and estimators, if the Fisher assumptions fit exactly, then the sampling distribution of (Est − True)/SE depends only on the df used to compute the SE, and belongs to the *t* family. In the first part of what follows, I'll ask you to take that fact on faith, and I'll show you how it leads to confidence intervals of the form you've been using. Then in the second part, I'll focus on the fact itself. The *t*-distribution was defined (Section 3) not in terms of an ANOVA model, but in terms of independent draws from a standard normal box model. I'll first sketch the logic that leads from the ideal ANOVA model to the *t*-distribution. Then, because the ideal model is at best a good approximation to a real situation, I'll discuss the conditions that must be satisfied if the approximation is to be a good one.

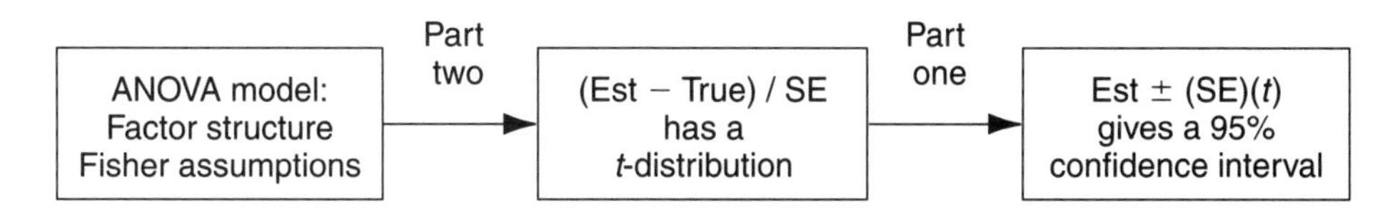

The Meaning of 95% Confidence, and Its Relation to the *t* Family

To understand where 95% confidence intervals come from, I think you first have to have a very clear sense of what 95% confidence means. Every confidence interval you construct either scores a "Hit" (the interval contains its target true value = EV) or a "Miss." If you think of an ideal ANOVA model as a chance mechanism for creating data sets, and think of recording Hit or Miss as your summary outcome for each data set, then for a properly constructed interval, 95% refers to the sampling distribution of your summary: 95% is the chance of a Hit. A single-draw box model for the sampling distribution of your Hit/Miss summary would have 100 tickets, 95 that say Hit, and 5 that say Miss. Although with a real data set, you never know whether your interval is a Hit or a Miss, nevertheless—provided, as always, that the assumptions fit—you can be certain that if you used the same model and assumptions to create lots of data sets and intervals, over the long run 95% of your intervals would score a Hit.

The Meaning of 95% Confidence

Mechanism: **ANOVA model**
Summary: **Hit or Miss, for 95% confidence interval**
Distribution: **Chance of Hit = 95%, Chance of Miss = 5%**

At this point, my last paragraph is only a claim, one that remains to be justified by logic. In what follows, I'll sketch part of that logic, by showing how the claim is related to another, more basic claim, that I'll also state in terms of a chance mechanism, summary outcome, and its sampling distribution:

Mechanism: **ANOVA model**
Summary: **(Linear estimator − True value)/SE**
Distribution: ***t* family.**

Notice how much the summary outcome looks like the standardized (Outcome − EV)/SD* from Section 3. The linear estimator is the outcome, the "true" value is the EV, and the SE is very much like the SD*. The only difference is that SD* is a true value from the model, whereas the SE is an estimate: $SE = SD\sqrt{SS_{Weights}}$, and $SD = \sqrt{MS_{Res}}$ gets computed from your data. I'll call (Est − True)/SE an **estimated standard distance,** and say that the linear estimator has been converted to **estimated standard units**.

Estimated Standard Units Tell the Distance to the EV, Measured in SEs.

Distance	**= Value − EV**
Standard distance	**= Distance/True SE***
Estimated standard distance	**= Distance/Estimated SE**

We can use this language to put the claim about the *t*-distribution more compactly:

For any linear estimator from an ideal ANOVA model, the estimated standard distance has a *t*-distribution.

The logic that links the t-distribution to 95% confidence comes from translating what it means to score a Hit:

> **Two Equivalent Ways to Score a Hit**
>
> **For any positive number t:**
> **The interval Est ± SE · t covers the true value**
> **if and only if**
> **(Est − True)/SE is less than or equal to t in absolute value.**

According to this translation, the chance of a Hit equals the chance that (Est − True)/SE is less than or equal to t in absolute value. If (Est − True)/SE has a t-distribution, then we can use that distribution to choose the number t to make the chance of a Hit equal to 95%.

Although it is not hard to show by algebra that the translation is valid (see Exercise C.8), it may help your intuition to see the same kind of translation in a more familiar context.

EXAMPLE 15.23 Two Ways to Score a Hit

Imagine that you live on an interstate highway at milepost 100, that you drive at 50 miles an hour, and that you're now at milepost 175. Are you within 2 hours of home? Here are two equivalent ways to find the answer:

(1) Interval covers the target:

"I'm now at mile 175 (observed value). In 2 hours I can drive 50 × 2 = 100 miles in either direction (SE · t = 95% distance). So every milepost in the interval 175 ± 100, or from 75 to 275, is within 2 hours' drive. In particular, mile 100 (my target) is in the interval: Yes, a Hit." (See Figure 15.14.)

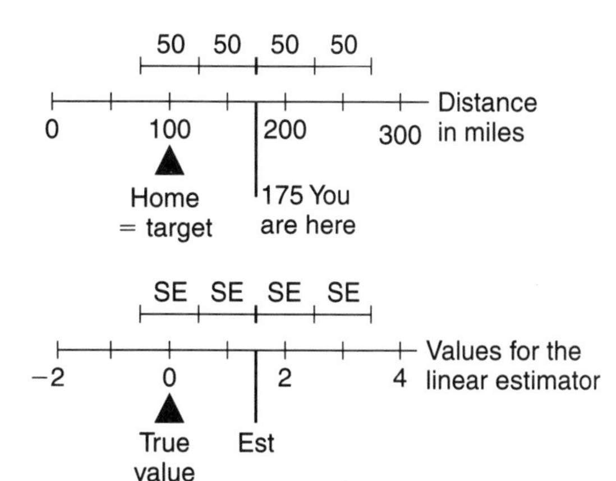

FIGURE 15.14 Hit, version 1: Interval contains the true value. If you're at mile 175, and you drive 50 mph, then in 2 hours you can reach any point in the interval 175 ± 50 · 2. In particular, the interval contains your target value, 100.

(2) (Est − True)/SE less than *t*-value:

"I'm at 175, the target is 100, so my distance from the target is 75 miles (Distance = Est − True). At 50 mph, 75 miles takes 75/50 = 1.5 hours (Distance/SE). In this new scale, the distance is less than 2 hours (*t*-value): Yes, a Hit." (See Figure 15.15.)

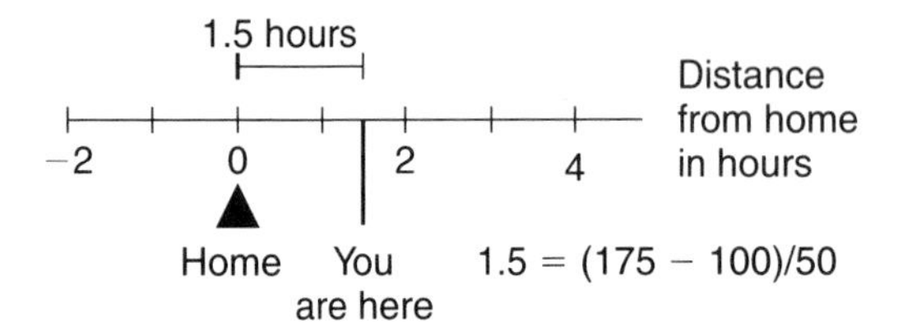

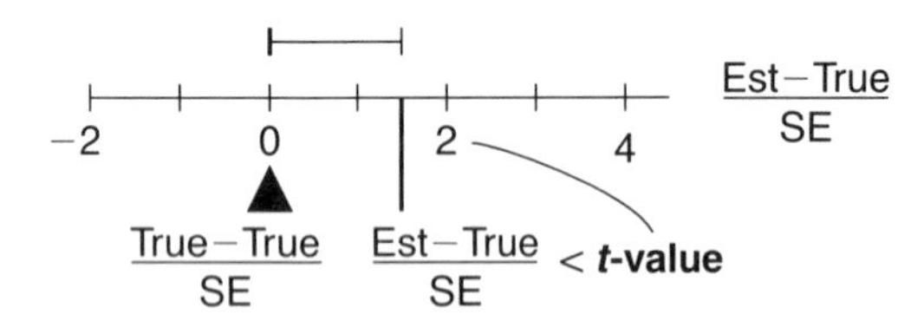

FIGURE 15.15 Hit, version 2: (Est − True)/SE less than *t*-value. If you're at mile 175, your distance from home is 75 miles. At 50 mph, that distance takes 1.5 hours, which is less than 2 hours. ■

If we know "Hit" means "(Est − True)/SE is less than or equal to *t* in absolute value", and we know that (Est − True)/SE has a *t*-distribution, how can we choose *t* (= 95% distance) to make the chance of a Hit equal 95%? The answer: Take *t* equal to the 97.5^{th} percentile of the *t*-distribution (Figure 15.16).

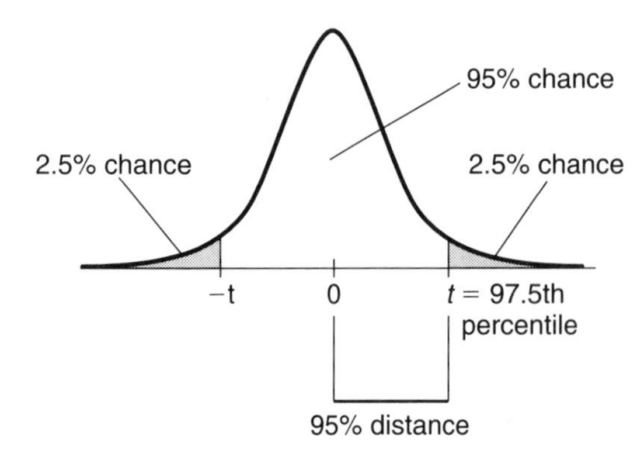

FIGURE 15.16 95% distance = 97.5^{th} percentile. A box model for the *t*-distribution of (Est − True)/SE will have 2.5% of the tickets greater than $t = 97.5^{\text{th}}$ percentile, 2.5% less than $-t$, and 95% in between, that is, less than or equal to *t* in absolute value.

We can summarize the last few pages in an "If . . . then . . ." form:

If (Est − True)/SE has a *t*-distribution, then there is a 95% chance that an interval of the form Est ± SE · (97.5^{th} percentile of the *t*-distribution) will contain the true value.

The next few pages deal with the "if" part: When will (Est − True)/SE have a *t*-distribution, at least approximately?

The Sampling Distribution of (Est − True)/SE

Our goal now is to see how the Fisher assumptions lead to the conclusion that (Est − True)/SE has a *t*-distribution. In a book on mathematical statistics, the connection would be stated in the form of a theorem, and proved rigorously. What follows is essentially an informal overview of such a proof.

Start by reminding yourself that we defined the *t*-distributions in terms of independent draws from a standard normal distribution, with χ^2_n as the sum of squares for *n* draws:

$$t_n = \frac{Z}{\sqrt{\dfrac{\chi^2_n}{n}}}$$

To prove that some other chance outcome has a *t*-distribution, we must first write that outcome as a ratio, Num/Den, and then prove three things:

i. Num has a N(0,1) distribution.

ii. $n(\text{Den})^2$ has a χ^2_n distribution.

iii. Num and Den are independent.

I won't actually prove any of the three for (Est − True)/SE. Instead, I'll simply point out the connections that link them to the Fisher assumptions.

You might think that because (Est − True)/SE is already in the form of a ratio, we should use (Est − True) as our numerator, and SE as our denominator, but those choices won't work. We want our numerator to have SE* = 1, and our linear estimator has $\text{SE}^* = \text{SD}^*\sqrt{\text{SS}_{\text{Weights}}}$. To get around this problem we take

$$\text{Num} = (\text{Est} - \text{True})/\text{SE}^*$$
$$\text{Den} = \text{SE}/\text{SE}^*.$$

The SE*s cancel, so that Num/Den equals (Est − True)/SE.

i. Num = (Est − True)/SE* is N(0,1).

Because the numerator is expressed in standard units, EV = 0 and SD* = 1, as required. Whether the shape of the distribution is normal depends directly on the normality assumption for the chance errors: If the chance errors for the observed values have a normal distribution, so will any linear estimator.

ii. $n(\text{Den})^2$ is χ^2_n.

Remember that *n* tells the degrees of freedom for estimating SD*, so if we use $\text{SD} = \sqrt{\text{MS}_{\text{Res}}}$, then $n = \text{df}_{\text{Res}}$. You can use a bit of algebra (Exercise D.6) to prove that $n(\text{Den})^2$ is just a disguised form of $\text{SS}_{\text{Res}}/(\text{SD}^*)^2$.

We would next need to prove that $\text{SS}_{\text{Res}}/(\text{SD}^*)^2$ has a χ^2 distribution with $\text{df} = \text{df}_{\text{Res}}$, that is, that $\text{SS}_{\text{Res}}/(\text{SD}^*)^2$ behaves like the sum of squares of $n = \text{df}_{\text{Res}}$ independent standard normal draws. The proof is complicated, because SS_{Res} comes from squaring and adding the residuals. The number of residuals is bigger than df_{Res} (there is one for each observed value), the residuals are not independent, and although each (Res/SD*) has EV = 0, the true standard deviation for (Res/SD*) is not 1. Nevertheless, the result can be proved, and the proof uses all five of the Fisher assumptions.

iii. Num and Den are independent.

If the individual chance errors are independent, and the factor structure is correct, then it can be proved that SS_{Res} will be independent of any condition average or comparison of averages, so that Num and Den will be independent.

With this very informal outline of a proof as background, let's review the Fisher assumptions one at a time, so I can point out where they are needed.

[Z]: If your chance errors do not have EV = 0, then your estimate will be biased, and what you think is the true value will not be the actual true value. This means the numerator in (i) above will not have EV = 0.

[S]: If your chance errors do not have the same SD*, then two links in the proof cannot be made. You can't rewrite (Est − True)/SE as Num/Den in the way I did, because the canceling that we relied on won't be valid if there's more than one SD*. Moreover, $SS_{Res}/(SD^*)^2$ won't have a χ^2 distribution.

[I]: If your chance errors are not independent, $SS_{Res}/(SD^*)^2$ may not have a χ^2 distribution, and Num and Den may not be independent.

[N]: If chance errors do not have a normal distribution, then neither will your linear estimator, and $SS_{Res}/(SD^*)^2$ will not be χ^2 .

[A & C]: If these assumptions don't fit, then the averages you compute in order to decompose your data will not estimate what they should, and neither will any summaries you compute from your decomposition.

Although it is true that you need all six assumptions to guarantee that (Est − True)/SE has a *t*-distribution, and also true that, in practice, at least some of the assumptions will not fit your data exactly, nevertheless, the *t*-distribution can provide a workable approximation much of the time.

I think it helps to make a distinction between gross and minor violations of the assumptions. By gross violations I mean those you can detect using the methods of Chapter 12: systematic patterns that tell you to transform to a new scale, or add new terms to your model, or give up on ANOVA in favor of correlation-based methods or methods for counted data. If you have chosen a scale and a model that make your data free of such gross violations, then you can be reasonably confident that the *t*-distributions will give workable approximations.

In particular, you needn't worry if your chance errors depart somewhat from a normal curve, unless your experiment is tiny. One of the most important results in statistical theory is a theorem that says, in effect, that the more observations you have, the more nearly the sampling distributions for averages and other linear estimators will have normal shapes. (Remember Example 15.21.)

Exercise Set D

1. Give the sampling distribution for a properly constructed 90% confidence interval by describing a box model with tickets of two kinds: What are the two labels? How many of each kind are there?

2. Take an ideal ANOVA model as your chance mechanism, and imagine computing a linear estimator. Several summaries based on the estimator are listed below. For each one, tell whether its sampling distribution depends on (i) the EV (= true value) for the estimator, (ii) the SE* for the estimator, and (iii) the df_{Res} used for estimating the SE.

   a. The Hit/Miss summary for a 95% confidence interval

   b. The linear estimator itself

   c. The distance = Est − True

   d. The estimated standard distance = (Est − True)/SE

3. *The key translation: Two ways to score a hit.* For each of the following situations, (i) find the interval of mileposts within 2 hours drive, and tell if you score a Hit; then (ii) find your distance from home, measured in hours, and tell if you score a Hit.

	Home	You are here	Driving speed
a.	150	75	50
b.	100	170	40
c.	200	60	75

4. Show by algebra that the interval Est ± SE · t contains the true value if and only if the estimated standard distance is less than t in absolute value. Draw a picture to illustrate your proof.

5. *Estimated standard units for real data.*

   a. For the leafhopper survival experiment, the SD is about 0.25. The average survival times in days, each based on 2 observations, are: Control: 2.0 Sucrose: 3.8 Glucose: 2.9 Fructose: 2.2

If you take as a tentative hypothesis that all the corresponding EVs are zero, express each average in estimated standard units.

   b. Assume the EVs for Glucose and Sucrose are unknown but equal. Express the comparison Glucose vs. Sucrose in estimated standard units.

6. *Proof that* $n(Den)^2 = SS_{Res}/(SD^*)^2$.

Rewrite $n(\text{Den})^2$ using the following facts:

$$n = df_{Res}, \quad \text{Den} = \text{SE}/\text{SE}^*, \quad \text{SE} = \sqrt{\text{SDSS}_{Weights}},$$
$$\text{SD} = \sqrt{\text{MS}_{Res}}, \quad \text{SE}^* = \text{SD}^*\sqrt{\text{SS}_{Weights}}, \quad \text{MS}_{Res} = \text{SS}_{Res}/df_{Res}\,.$$

Then simplify your expression—you should get a lot of terms canceling—to get $SS_{Res}/(SD^*)^2$.

7–13. *Class simulation experiments.*

The purpose of the next few exercises is to give you a chance to study the behavior of confidence intervals when your data sets are created by a chance mechanism that does not fit some of the Fisher assumptions. Each exercise begins by describing a chance mechanism, then asks you to create several data sets by simulation, to construct a confidence interval for each, and to check whether your interval contains the true value. It would be quite tedious for one person working alone to create enough data to show meaningful patterns, so I hope you will be able to combine your results with those of several other people.

7. *Groups have unequal SD*s*. Consider a basic CR model: Obs = Benchmark + Condition Effect + Chance Error. To keep arithmetic as simple as possible, assume the true Benchmark = 0, and that the true Condition Effects are also 0. The Benchmark value could be any number, and your results wouldn't be affected, so we might as well use 0. The same is true of the Condition Effects: the chance that a confidence interval contains its EV doesn't depend on the value of the EV, but only on the way the chance errors behave.

You'll create chance errors for your data sets by tossing coins in sets of four, and computing Chance Error = (# Heads − 2). These chance errors will be independent, with EV = 0, SD* = 1, and a probability histogram that is lumpy but roughly normal-shaped.

a. The basic experiment:

For this experiment, you have 2 groups (conditions), 3 observations per group, and SD* = 1 for both groups; there are 6 observations in all, each equal to (# Heads − 2) for a set of four tosses.

i. Create a data set, and construct an ordinary 95% confidence interval for the contrast Condition 1 versus Condition 2. The EV for the contrast is 0, so your interval scores a Hit if it contains 0.

ii. Repeat part (i) 4 more times, for a total of 5, and summarize your results: How many Hits out of 5? (When you combine your results with those of others, you should find the combined percentage of Hits near 95%.)

b. Unequal SD*s:

This time use SD* = 1 for the three Condition 1 values: Obs = (# Heads − 2); but use SD* = 10 for the Condition 2 values: Obs = 10 × (# Heads −2). Create 5 data sets, and record the number of Hits.

c. Variations:

The effect of unequal SD*s has been shown to depend on several things, among them the ratio of the group SD*s, the number of observations per condition, and the number of conditions for each of the two averages in your comparison. Here are some variations on the basic experiment. (Unless you are able to combine your results with a lot of others, I don't suggest doing them "by hand." However, if you know a little bit of computer programming, you could easily get 1000 repetitions for each variation.)

i. Same as (b), but with 10 observations per condition instead of 3.

ii. Same as (b), but with four conditions instead of two, and using Condition 1 vs. other three conditions as your comparison. Use SD* = 1 for Condition 1, and SD* = 10 for the others.

iii. Same as (ii) above, but use SD* = 10 for Condition 1, SD* = 1 for the others.

d. A factorial experiment:

Design a multifactor CR experiment for studying the effect of unequal SD*s on the chance of a Hit. Use the ideas from (c) above to help you choose your factors. (If you can program a computer, you might want to consider carrying out your experiment.)

8. *Observations are not independent.* Start with the same CR model as in Exercise 7: 3 values of Obs = (# Heads − 2) for each of 2 conditions.

a. Observations positively correlated:

To introduce positive correlations, use coin tosses in overlapping sets of 4:

```
Condition 1     Condition 2
H T T H T T     T H H T T H      ← 12 tosses in all
\_____/         \_____/
# H = 2         # H = 2
  \_____/         \_____/
  # H = 1         # H = 2
    \_____/         \_____/
    # H = 1         # H = 2
```

The observed values from this set of 12 tosses would be

Cond 1: 0, −1, −1   Cond 2: 0, 0, 0

Create 5 data sets, and record the number of Hits.

b. Weak positive correlation:

In (a), there was a lot of overlap, and the positive correlation was strong. For weaker correlation, use less overlap:

```
H T H T T T H T      Chance error:
\_____/
# H = 2                   0
    \_____/
    # H = 1              −1
        \_____/
        # H = 1          −1
```

Create 5 data sets, and record the number of Hits.

c. Negative correlation:

To get a negative correlation between adjacent observations, use the same method as in (a), then change the sign of every other observation. In part (a), the first 3 values were 0, −1, −1. To get negative correlations, switch the sign of the second (fourth, sixth, etc.) value: 0, +1, −1.

Create 5 data sets, and record the number of Hits.

d. Variations:

The effect of the correlation depends on the number of observations per condition and on the number of conditions in each average for your contrast. Design a factorial CR experiment to study these effects.

9. *Chance errors not normal shaped.* Start with the same model as in Exercise 7: 3 values of Obs = Chance Error for each of 2 conditions. This time, use a different rule for finding your chance errors:

Chance Error = (# tosses until first Head − 2).

These chance errors will have EV = 0, SD* = $1/\sqrt{2} \approx 0.7$, and a very non-normal shape. A rough sketch of the probability histogram looks like Figure 15.17.

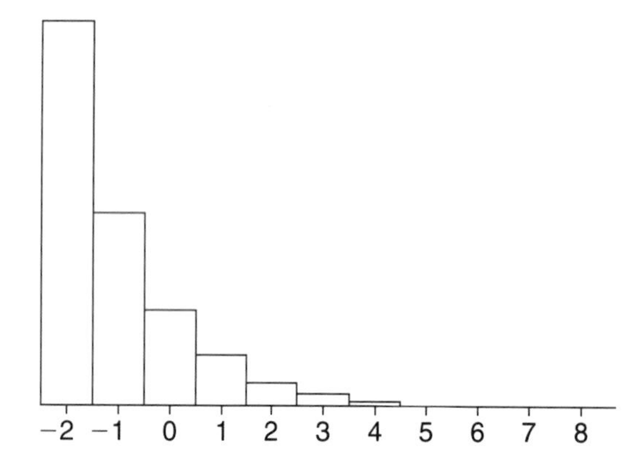

FIGURE 15.17 Histogram for non-normal chance errors. The distribution is highly skewed, with a very long right tail. (Although the histogram doesn't show it, there is positive probability for all integer values to the right of 4.)

a. Create 5 data sets, and record the number of Hits.

b. Variations:

i. Create 5 data sets with 10 observations per condition instead of 3, and record the number of Hits.

ii. Create 5 data sets with 3 observations for each of 4 conditions and use as your contrast Condition 1 versus the other 3.

c. Yes/No data:

A non-normal histogram can have any of a variety of shapes. The one for parts (a) and (b) was long-tailed and highly skewed. Many experiments have a response with only two possible values: Yes or No, Died or Survived, Right or Wrong.

For this experiment, use the same model as in Exercise 7, but get each of your chance errors from a single toss:

$$\text{Chance Error} = 1 \text{ if Heads}, -1 \text{ if Tails}.$$

These chance errors have EV = 0, SD* = $\sqrt{2} \approx 1.4$, and a symmetric histogram:

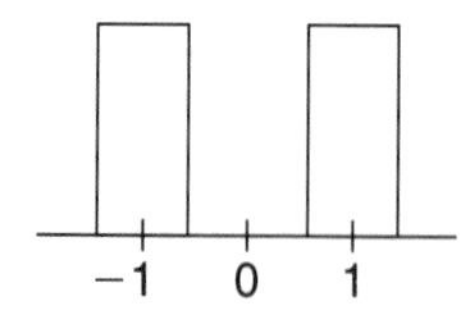

i. Create 5 data sets and record the number of Hits.

ii. Repeat (i), but with 10 observations per condition.

iii. Repeat (ii), but with 4 conditions, and Condition 1 versus other 3 as your contrast.

10. *The basic experiment.* In this exercise you'll use Minitab to simulate the experiment described in Exercise 7(a). For each of 1000 repetitions, you'll generate six observations, three for each of two conditions. Next, you'll compute the two condition averages, and their difference, which is the estimated value for the contrast. Then you'll compute the residual sum of squares, the SD, the SE, and the 95% distance. Finally, you'll create a summary variable, with value 1 if the interval misses the true value of zero, −1 if the interval scores a hit.

Step 1. Put the sixteen equally likely outcomes in c1. Start with a new worksheet, and enter the data for `c1` just as in Exercise C.13: −2 (1x), −1 (4x), 0 (6x), 1 (4x), 2 (1x).

Step 2. Generate observed data in `c21`–`c26`. Each row will correspond to one repetition of the experiment. Columns `c21`, `c22`, and `c23` will hold the three observations for Condition 1; `c23`–`c26` will hold those for Condition 2. To create `c21`, follow Exercise C.13a(2): Choose `Calc > Random Data > Sample` from `Columns` and fill in the dialog box to sample 1000 rows from `c1`, putting the result in `c21`, and sampling *with* replacement. Create `c22`–`c26` in the same way.

Step 3. Condition averages: Label `c31` as `'Ave1'` and `c32` as `'Ave2'`. Then choose `Calc > Calculator`. Use the dialog box to define `Ave1` as `(c21 + c22 + c23)/3` and `Ave2` as `(c24 + c25 + c26)/3`.

Step 4. Estimated contrast. Label `c33` as `'Est'` and define it as `'Ave1' - 'Ave2'`.

Step 5. Residual sum of squares: Label `c34` as `'SS(Res)'` and define it as `(c21 - 'Ave1')**2 + (c22 - 'Ave1')**2 + (c23 - 'Ave1')**2 + (c24 - 'Ave2')**2 + (c25 - 'Ave2')**2 + (c26 - 'Ave2')**2`

Step 6. Standard deviation: Label `c35` as `'SD'` and use `Calc > Calculator` to define it as `Sqrt('SS(Res)'/4)`

Step 7. Standard error: For this contrast, there are six weights, each equal to $\pm(1/3)$, so $SS_{Weights} = 6(1/3)^2 = 2/3$. Label `c36` as `'SE'` and define it as `'SD' * Sqrt(2/3)`.

Step 8. 95% distance: The 95% *t*-value on 4 df is 2.776. Label `c37` as `'Dist'` and define it as `'SE' * 2.776`.

Step 9. Hit or Miss: The interval scores a Hit if and only if it contains the true value, which for this experiment is 0. The interval will contain zero if and only if `'Est' - 'Dist'` is negative and `'Est' + 'Dist'` is positive, which will happen if and only if the product of the upper and lower values is negative. So to find out whether the interval misses, we can look at the sign of the product. To do this, label `c38` as `'Miss'` and use the calculator to define it as `Signs(('Est' - 'Dist')*('Est' + 'Dist'))`. For each of your hundred data sets, this variable will equal 1 if the interval misses zero, and −1 if it scores a Hit.

Thus the average of this column lets you find the fraction of misses: Fraction = (1 − Avg)/2.

11. *Unequal SDs*. To repeat the simulation with SD* = 1 for the three observations of Condition 1 and SD* = 10 for the other three observations, define `c2` as `10*c1`. Create `c21`–`c23` by sampling from `c1`, and `c24`–`c26` by sampling from `c2`. All the other steps are the same. Carry out the simulation and compare your results with those when the SD*s were equal.

12. *Dependent observations*. To create data sets with correlated observations, as in Exercise 8, follow the same steps as in Exercise 10, except for the first two:

Step 1. To simulate individual coin tosses, change `c1` to contain just two values, 0 and 1. Then put the set of twelve tosses, as described in Exercise 8, into `c2`–`c13`: for each of these twelve columns, sample 1000 rows with replacement from `c1`.

Step 2. Generate the observed values, for `c21`–`c26`, as overlapping sums of the simulated coin tosses:

```
c21:  (c2 + c3 + c4 + c5 - 2)        c24:  (c8 + c9 + c10 + c11 - 2)
c22:  (c3 + c4 + c5 + c6 - 2)        c25:  (c9 + c10 + c11 + c12 - 2)
c23:  (c4 + c5 + c6 + c7 - 2)        c26:  (c10 + c11 + c12 + c13 - 2)
```

What effect does the correlation appear to have on the Hit rate for the "95%" intervals?

13. *Non-normal shape*.

    a. Skewed: To create data sets with a highly skewed error distribution much like the one in Exercise 9, repeat the steps of Exercise 10, with just one change:

    Step 1: For `c1` values, use: −1 (16x), 0 (8x), 1 (4x), 2 (2x), 3(1x), 5 (1x).

    This distribution has EV = 0, and SD* = $\sqrt{31/16} \approx 1.4$. It is not at all symmetric. How does this appear to affect the Hit rate for the confidence intervals?

    b. Long-tailed: To create data sets with a much higher percentage of outliers than the normal, repeat the steps of Exercise 10, except for the first one:

    Step 1. For `c1` values, use: −6 (1x), −1 (4x), 0 (6x), 1 (4x), 6(1x).

    This distribution has EV = 0, and SD* = $\sqrt{5} \approx 2.2$. The two extreme values of ±6 are almost 2.7 SD*s from the average, and together they account for 12.5% of the outcomes. For a normal distribution, less than 1% of the outcomes are that far from the mean. How does this appear to affect the Hit rate for the confidence intervals?

5. Approximate Sampling Distributions for F−Ratios

Whenever you compare an observed *F*-ratio of mean squares with a table value in order to test a null hypothesis, in effect you are assuming that your ratio of mean squares has a sampling distribution that belongs to the *F* family. This section deals with one fundamental question: Under what conditions is that assumption reasonable? The mean squares used to define the *F* family in Section 3 were computed using independent draws from a standard normal box model. The mean squares you use to get your *F*-ratio, on the other hand, come from the boxes of a linear decomposition. Under what conditions will a ratio of mean squares computed from a decomposition have the same sampling distribution as a ratio computed from standard normal draws?

A short answer is this: If your model is exactly right (your factor structure and the Fisher assumptions describe your experiment exactly), and your null hypothesis is true, then your *F*-ratio has an *F*-distribution. I'll give a more detailed answer in two parts, dealing with two sets of requirements.

Expected mean squares. One requirement is that the two mean squares in your ratio must have the same expected values. The expected value of any mean square can be written as a sum of pieces; each piece is a number that measures the variability associated with some factor in your design. If the ratio of two mean squares is to have an *F*-distribution, it is essential that the two mean squares have exactly the same pieces in their expected values. For example, in Chapter 3, Section 4, the *F*-ratio for testing the effect of Organ was MS_{Organ}/MS_{Res}. The denominator MS_{Res} has only one piece, a measure of the variability due to chance error. The numerator MS_{Organ} is a sum of two pieces, one that measures the variability due to Organ plus a second that measures the variability due to chance error. If the true Organ effects are zero (null hypothesis true), then the true variability due to Organ is zero, and the expected value for MS_{Organ} has only the one term that measures variability due to chance error. Thus MS_{Organ} and MS_{Res} have the same EVs, which must be true if MS_{Organ}/MS_{Res} is to have an F-distribution. (If the null hypothesis is false, then the EVs will not be equal, and the ratio of mean squares will not follow an F-distribution. The F-test boils down to judging whether an observed ratio might have come from an F-distribution: if we judge not, we reject the null hypothesis.)

The requirement of equal EVs is a major reason you can't just take any old pair of mean squares and get a ratio that has an F-distribution. You have to choose the denominators for your F-tests so that they have the right expected values. As long as you choose your two mean squares according to the principles and rules of Chapter 13, the requirement of equal EMSs will be satisfied. There is a second set of requirements, however, that you may have less control over. This set has to do with the model, most particularly the Fisher assumptions.

The role of the model. In Section 3, I defined the F-distributions in terms of independent draws from a N(0,1) error box. The F-ratios you compute for an ANOVA appear to be quite different, since they come from your linear decomposition, not from independent standard normal outcomes. Nevertheless, it can be proved as a theorem that if your model is exactly right (both the factor structure and Fisher assumptions fit your data exactly), and if you have the right denominator MS, and if your null hypothesis is true, then your F-ratio has an F-distribution. Of course, no model is ever exactly right. Thus to get from a real-life experiment to an F-distribution, there are two gaps that must be bridged, first from your experiment to the ideal ANOVA model, and then from that model to the F-distribution as defined in Section 3. The theorem I mentioned serves to bridge only the second gap. Bridging the first was the topic of Chapter 12.

To prove the theorem, you would have to show how the Fisher assumptions insure that an ANOVA F-ratio has the same sampling distribution as a ratio of mean squares computed from independent standard normals. I won't present a proof, but I think it is worthwhile to look more closely at what a proof would involve, as a way to focus on why the usual ANOVA assumptions are important, and on how they are related to the definition of the F-distributions. You can also use various properties from the definition to help judge when and whether the ideal situation leads to a reasonable approximation.

In the definition, the ideal mechanism creates observed values as independent draws from a standard normal error box. Thus there are four requirements, corresponding to center (EV = 0), spread (SD* = 1), shape (normal), and relationship (independence). A proof must show that the mean squares you compute in an ANOVA behave as if they were built from terms with these four properties. We can consider the four one at a time.

Center: EV = 0. In a balanced design, the estimated effects for every factor other than the grand mean will always add to zero. Thus in practice, the way the linear decomposition is done makes the numbers we square behave as if EV = 0.

Spread: SD = 1.* In practice, the SD* for a data set is unknown, and almost surely not equal to 1; but provided the chance errors behave as if the SD* is *constant* (which you can check using the methods of Chapter 12, Section 1), then it isn't necessary that the constant be equal to 1. The reason is that we compute a *ratio* of mean squares to get the F-statistic. Whatever the actual SD* may be, it appears in both the numerator and denominator, and gets canceled out in exactly the way that canceling 25s makes $(25 \times MS_1)/(25 \times MS_2) = MS_1/MS_2$.

Shape: Normal. In practice shapes are never exactly normal, but are often close enough. You can check by looking at residuals, using the methods of Chapter 12, Section 3. Even if the individual residuals do not approximate a normal pattern very well, the more observations you have, the less you need to worry about the chance errors being normal. (Look again at example 15.22.)

Relationship: Independence. In the definition, the draws were independent. In practice, the numbers from a decomposition that you square and add to get a sum of squares are not independent, but are correlated. Informally, the correlation reflects a kind of overlap in the information contained in the numbers you square and add. Dividing your sum of squares by the df, instead of by how many numbers get squared, makes just the right adjustment for the overlap.

In the definition, because the draws are independent, the two mean squares are also independent. In practice, if your design is balanced, and if your data set does not show the kinds of correlation patterns discussed in the Chapter 12, Section 2, then the mean squares you compute will be independent of each other, as required.

Relation to Fisher Assumptions. We can use the four properties to summarize the relationship between the Fisher assumptions and the form of the F-test.

[Z]: If your chance errors do not have EV = 0, then some of the estimated effects you get when you decompose your data will be biased. This bias will make the corresponding mean square larger than it would be otherwise, and you risk concluding than an effect is "real" when in reality what you've detected is part of the bias.

[S] If your chance errors do not have the same SD*, then the canceling of SD*s that is needed when you compute the F-ratio will not occur, and the ratio will not behave as if SD* = 1, as in the definition.

[I]: If you chance errors are not independent, then your mean squares will not have the right expected values. Moreover, the numerator and denominator may not be independent, as required by the definition. For both these reasons, the critical values from an F table will not be the right ones for your data set.

[N]: If your residuals aren't roughly normal-shaped, then the numbers you square to get SSs will not be normal-shaped in the sense required by the definition. In particular, if your percentages of large residuals are higher than those given by the rule based in the normal curve, the squares of these large residuals will have too big an effect on the behavior of the mean squares, and the distribution of your ratio will not be close enough to an actual F-distribution to make the critical values trustworthy.

[A & C]: If the true effects are not constant, or are not additive, the decomposition you use to compute your mean squares will not split the observations into the right pieces. Tables in your decomposition will contain contributions from factors that aren't supposed to contribute, and so the expected mean squares will be off. In particular, the residuals may be "contaminated" by these nonconstant or nonadditive factors, and so will give poor estimates for the chance errors.

Exercise Set E

Class simulation experiments. The next few exercises are very similar, both in spirit and in structure, to the simulation exercises at the end of the Set D. I assume you have at least read the description of those exercises.

Each exercise will describe a CR model. Use each model to create 5 data sets. For each data set, compute the F-ratio for testing the null hypothesis that the true condition effects are all 0. Record the number of times you reject the hypothesis. For a model that satisfies all 5 assumptions, if the null hypothesis is true, the chance of rejecting is 5%. (Because these models are all chosen so that the null hypothesis is in fact true, rejecting it corresponds to a "False Alarm.")

1. *Unequal SD*s.*
   a. Use the CR model from Exercise D.7 part (a). For this model, SD*s are equal.
   b. Use the model from Exercise D.7 part (b).
   c. Variations: Use each of the three models from Exercise D.7 part (c).

2. *Observations not independent.* Create 5 data sets for each of the models in Exercise D.8, parts (a)–(c), and for each, record the number of False Alarms.

3. *Chance errors not normal shaped.* Create 5 data sets for each of the models described in Exercises D.9, and record the number of False Alarms for each model.

4. *Power of the F-test.* All the models in Exercises 1–3 had true condition effects equal to 0; rejecting the null hypothesis in such cases is an error. The **power** of a statistical test is the chance of rejecting the null hypothesis, computed for data sets that have nonzero true effects. The power depends on the size of these effects: the bigger they are, in comparison to the SD*, the more likely you are to reject the null hypothesis. The purpose of this exercise is to explore the relationship between the size of the effects, the number of observations, and the chance of rejecting the null hypothesis.

   Use the CR model from Exercise D.7(a) as a starting model, and create variations according to the following SP/RM factorial design:

	True condition effects			
	0	1 & −1	3 & −3	5 & −5
3 observations per group				
10 observations per group				

Create one set of chance errors for each row of the table; use each set of chance errors to create four data sets, one for each column. These four data sets will differ only in the condition effects; the chance errors will be the same for all four. Record for each data set whether or not you reject the null hypothesis. Then combine your results with those from the rest of the class, in order to see the patterns.

Minitab: Any of the simulations in Exercises 1–4 can be carried out in Minitab, using the instructions in the exercises at the end of Sets C and D as a guide.

6. WHY (AND WHEN) ARE THE MODELS REASONABLE?

Overview

In this section I give examples of three justifications for box models and the Fisher assumptions: (a) sampling from a population, (b) measurement error, and (c) randomization. Each justification is to some extent idealized, in that the situations that arise in practice rarely fit exactly to the conditions of any one justification. Here, as elsewhere, statistics uses idealized models as approximations to a more complicated reality.

In what follows, I'll illustrate each of the three justifications for the case of a one-way design. The justifications for other designs are similar, although the technical details are more complicated.

Sampling from a Population

The first justification is built from the ideas of population and simple random sample.

Reminder: The population **is the entire group of individuals you want to know about. The** sample **is the group of individuals you actually observe. For a** simple random sample**, each individual in the population has the same chance of being in the sample. All possible samples are equally likely.**

The ideal example of simple random sampling is the box model itself. The population is a box of numbered tickets. The sample is the set of draws. If you draw at random, putting each ticket back and mixing before the next draw, you get a simple random sample.

When a design calls for you to select individuals, the more you can make the process of selection behave like drawing tickets from a box, the better the standard model and assumptions will fit your data.

As an example, suppose you wanted to choose 100 students from your college to take part in an experiment. You might find it easiest to track down 100 people you know from your classes, or run an ad in the school paper and take the first 100 volunteers, or try to get 100 volunteers from a single large lecture class. However, none of these three selection methods is much like drawing tickets from a box. To get a simple random sample, you'd need to get a complete list of all students enrolled in your college, number the students, and then use a random number table to choose 100 of them for your study.

This last little example involves only one population. For a one-way design, each level of your factor of interest corresponds to a different population, and you get your observations by taking simple random samples separately from each population. For example, suppose you wanted to compare freshmen, sophomores, juniors, and seniors, regarding each class as one level of your factor of interest. You'd need separate lists for each of the four classes; using a random number table, you might choose 25 students from each class.

How does simple random sampling lead to the standard model and assumptions? I've created an artificial example to show you. Assume there are just two populations, with just four individuals in each one. Assume, too, that there's no randomness or ambiguity in the measurement process: for the idealized sampling model, all the randomness comes from the process of choosing individuals from the populations.

EXAMPLE 15.24 A SIMPLE ONE-WAY SAMPLING SCHEME

The factor of interest has two levels, A and B, corresponding to two populations. Population A has four individuals with response values 3, 4, 4, and 5. Population B also has four individuals, with values 5, 6, 6, 7. The sampling plan calls for choosing two individuals at random from each (see Figure 15.18).

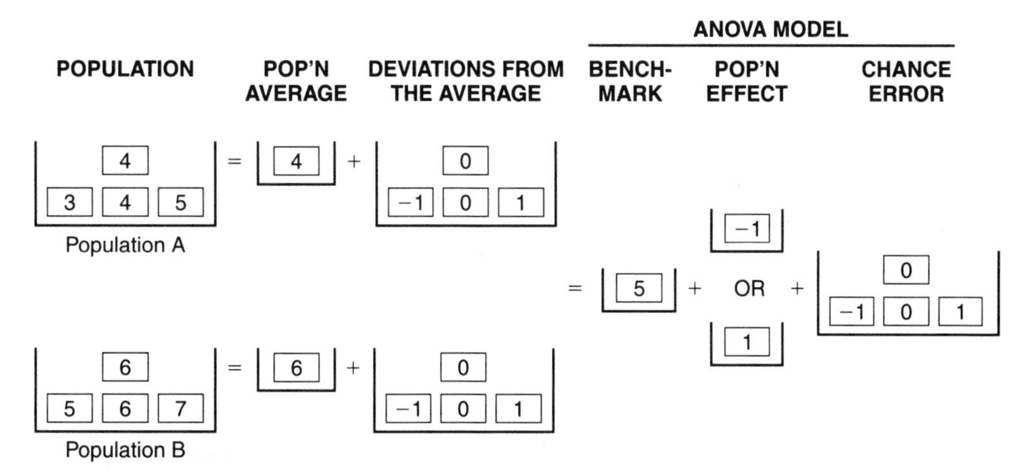

FIGURE 15.18 One-way ANOVA: Simple random sampling from two populations. If the observations are chosen at random, with replacement, from the two left-hand populations A and B, then the data fit the standard ANOVA model on the right, which satisfies the assumptions of zero EVs, same SD*s, independence, additivity, and constant effects. If the distribution of response values in each population had been normal, then all six assumptions would be satisfied. ■

There are several questions you might want answered before you'd be willing to regard an example like this one as a justification for the standard model. I've tried to cover the most important questions and objections:

Objection: Your two populations are exactly the same size. Surely you don't expect me to believe that happens in practice.
Answer: No, I don't. Most populations of interest are quite large, and if they are, the actual size of the population makes no practical difference as far as the standard model is concerned.

Question: Why do you insist on sampling with replacement? Isn't it silly to allow the possibility of choosing the same individual more than once?
Answer: If your populations are large, as they almost always will be, it makes almost no difference whether you sample with or without replacement. For sampling without replacing, the chance of getting the same individual more than once is zero, of course. If you sample with replacement, and your population is large, the chance of a repetition will be so close to zero that the difference won't matter. (See Exercise F.5.) The main reason for assuming that you sample with replacement is that it makes the theory a lot simpler. For large populations, that simpler theory gives a good approximation even if you actually sample without replacement.

Objection: In practice, you can't measure everyone in the whole population, so you don't know how the population values are distributed. Yet one of the Fisher assumptions is that chance errors follow a normal distribution.
Answer: There are two answers. First, experience has shown that for many kinds of situations, if you choose the right scale for your response, the distribution in the population will be at least roughly normal. Second, even if the distribution is not normal, the standard assumptions nevertheless very often lead to reasonable approximations for critical values, etc., for most data sets that are not tiny.

Objection: You set up your two populations to have exactly the same set of values for the deviations from the average. You don't expect me to believe that happens in real life, do you?
Answer: No. What's important is that the SDs be roughly equal. You can check that, as you know, and try transforming if the SDs are not equal.

Although the questions and objections I've just discussed are reasonable things to think about, there are other issues of much greater practical importance. Here's a more realistic example we can use to illustrate them.

EXAMPLE 15.25 Hepatitis and Blood Transfusions

In Example 15.16 and Exercise 15.C.8 I described a test that can be used to screen blood donors for hepatitis, by measuring the concentration of an enzyme, serum glutamic pyruvic transaminase (SGPT). People who are carriers of hepatitis tend to have higher levels of SGPT in their blood than noncarriers.

Imagine how researchers might have developed this test for hepatitis. Suppose you're a doctor, and your experience with hepatitis patients has led you to notice that often their levels of SGPT seem a bit higher than normal. You'd like to design a careful study to determine what range of SGPT levels should be considered normal, and whether carriers of hepatitis do in fact tend to have higher levels. There are two groups you want to compare: hepatitis carriers and noncarriers. A reasonable plan might be to choose 50 individuals of each kind, then measure their SGPT concentrations. Just how reasonable the plan would be depends very much on how you choose the individuals. To see how important the method of choosing is, it helps to look ahead to the way you might analyze your data.

Your plan for the analysis might include a one-way ANOVA, with Normal versus Carrier as the factor of interest, and log SGPT concentration as the response (For SGPT, you could easily discover from the data that log concentrations have a normal distribution; the untransformed concentrations do not.) The standard model for one-way ANOVA assumes that each value of the response equals a Benchmark value, plus an effect due to the factor of interest, plus a chance error. The chance errors are assumed to be independent and normally distributed, with the same SD. How might you justify these assumptions for the case of the SGPT study? How does the way you choose your 100 individuals relate to the standard box model?

For this study there are two populations, which I'll call the Normals and the Carriers. These are the people you want your results to tell you about. If we assume that you want your results to apply to all U.S. residents over 18, then the population of Normals is the set of all U.S. residents over 18 who are not carriers of hepatitis; the carriers over 18 are your second population.

Now assume that you have available to you a complete list of all the individuals in each of your two populations. You could use a table of random numbers to choose 50 people from each population, then measure their log SGPT, and record your data. If you chose your samples in that way, the standard ANOVA model would describe your data pretty well. The distribution of log SGPT concentrations would be very close to normal for each of your populations. You chose your individuals by simple random sampling, so each of your samples is like a set of random draws from a box of tickets. It happens to be true that the SD for the Carrier population is a bit larger than for the Normals, but not enough to make much difference in an ANOVA. Overall, the standard model is quite reasonable, provided you chose simple random samples.

There's the rub. Who has a complete list of all the hepatitis carriers over 18 in the U.S.? (If you could get such a list, you wouldn't have to measure SGPT to screen blood donors. You could just use the list.) Even if you had the list, and were willing to go to the trouble of using random numbers to choose people at random, could you afford to track them all down? ■

These questions illustrate why it is rare for investigators to use simple random samples from the actual populations they want to study. It's easy to see similar problems in other contexts: Who has a complete list of all the silver maples in the U.S., all the king snakes in Georgia, all the heroin addicts in Chicago, or all the psychotics in the State Department? Only Caesar Augustus or the Bureau of the Census can insist that all individuals of interest go forth to be enrolled.

Practical difficulties of this sort have led statisticians to emphasize a crucial distinction, between a *population* and a *frame*.

Reminder: The population **is the entire group of individuals you want to know about. The** sampling frame **is the list of individuals you actually choose from.**

If you use a frame to choose a random sample, then formal statistical methods give you a logical basis for generalizing from your sample to the *frame*. There are no *formal* methods to carry you the extra mile from the frame to the population. For that, you have to use your judgment about how well the frame represents the population.

Let's return to the hepatitis example. A practical plan might be to sample from the patients at one large metropolitan hospital over the course of five years. Because hepatitis is, fortunately, not common, you might decide to use *all* patients with a diagnosis of hepatitis as your sample from the carrier population. For this part of your sample, there would be no randomness—in the statistical sense—involved in the selection process. You'd need to worry about whether patients diagnosed with hepatitis really represented the much larger group of carriers, many of whom escape diagnosis. For your normal sample, you might choose individuals at random from patients who met three conditions: (i) no diagnosis or history of hepatitis, (ii) a blood donation within the last five years, (iii) no suspicion that the person who received the donated blood had developed transfusion hepatitis. As you can see, the practical considerations can easily overwhelm the statistical goal of choosing a simple random sample from the population of interest.

To summarize, here are four ways in which the ideal model of simple random sampling from a large population is likely to be beyond the reach of what is feasible in practice.

Frame not equal to population. The main danger here is bias. The group of individuals that is easiest to list and select from may for that very reason be atypical of the group you really want to know about.

Samples are not random. Studies in psychology often depend on volunteers. Lab studies use animals purchased from (and chosen by) a supply house, not chosen at random. Field studies in biology are based on who is hungry enough, or careless enough, to walk into your trap; or on who happens to fly or swim into your net.

Problems that are more technical: SDs not equal, distributions not normal. These you can check using your data. Often transforming to a new scale provides a reasonable solution.

Populations are small, not large. This is actually more of an advantage than a problem. If the actual populations of interest are small, then sampling without replacing is more efficient than sampling with replacing, and you can get by with smaller samples. To take full advantage, you would need to make appropriate adjustments to the methods described in this book. Any text on sampling tells what adjustments to make. For most studies in biology and psychology, however, the populations of interest will be large.

Measurement Error

The justification of assumptions that comes purely from random sampling puts all the randomness in the process of *choosing* the individuals to study; there is no randomness in the process of *measuring*. In practice, few measurements are completely free of uncertainty. For many kinds of measurements, the process of measuring behaves a lot like sampling from a population.

For the sampling model, the objects of interest were collections of numbers—the populations—and the process of obtaining data involved choosing at random from these collections. For the pure version of the model based on measurement error, we regard the individuals as already chosen, and in that sense fixed. The randomness comes from the measurement process.

Viewed abstractly, the sampling and measurement models are the same. For each, the objects of interest are fixed collections of numbers, and the process of getting observations corresponds to choosing at random from the collections. For the model based on measurement error, the fixed collection for each individual is the (hypothetical) set of all possible values for the measurement process.

There are two extreme versions of this model, one in which the set of values comes from unpredictable changes in the individual, and one in which the values come from unpredictable changes in the measurement process itself. Most situations for which the model of measurement error is appropriate fall somewhere between the extremes.

Consider first unpredictable changes in the individual. Suppose I wanted to test the claim that scientists have made recently, that having my pet dog nearby lowers my blood pressure. My blood pressure, like everyone else's, is far from constant. I've had it measured lots of times, and it's never exactly the same. You could imagine designing an experiment to compare my blood pressure under two conditions, sitting in a doctor's office with my dog beside me, and sitting in the same chair, with no dog. For each of the two conditions, there is a set of possible blood pressure readings, and experience has shown that it is reasonable to think of each set of blood pressures as a (hypothetical) population, and to think of taking a particular reading as like sampling at random from a population. No one knows all the details of why the readings behave like a random sample, although the basic idea is generally agreed on. There are so many influences on blood pressure, and their combined effect is so complicated, that even when you try to keep conditions fixed, there are unpredictable fluctuations from one reading to the next. In principle, it might be possible to learn more and more about the what makes one reading higher than another, just as in principle, you might be able to learn what it is about the way you flip a coin that makes it land heads instead of tails. In practice, though, it's not necessary. A sequence of coin tosses behaves like a set of random draws from a box with two tickets labeled Heads and Tails. Viewed collectively, a set of blood pressure readings taken under controlled conditions very often behaves like a random sample from a population.

Now consider the opposite extreme. For this version of the model of measurement error, the number to be measured is completely fixed, but the measuring instrument and process introduces unpredictable fluctuations. One rough but intuitive example comes from the Olympics. In diving, or skating, or gymnastics, each performance gets rated several times. One fixed performance generates several ratings, one by each of several "instruments," in this case human judges. In a case like this, there may be some systematic variability, but after making appropriate allowances for such patterns, the residual differences in ratings of the same performance behave much like a random sample from a population.

In the Olympic example, differences in the measurements were easy to associate with concrete and visible differences in the instrument: each rating came from a different judge. Often, however, there is nothing so concrete to associate with the observed differences in the response. A lovely example is described in the book *Statistics*, by Freedman, Pisani and Purvis. In this example, the objects that get measured are weights used by the National Bureau of Standards (now the National Institute of Standards and Technology) to calibrate all official scales in the U.S. Each weight gets weighed about once a week. It's the same weight each time, the same people doing the weighing, and the same instrument. Nevertheless, the results of the weighings are different from week to week, and they behave very much like a random sample from some (hypothetical) population of possible outcomes for the measurement process.

Randomization

For the justification based purely on sampling, we regard the populations as the fixed objects of study and the selection of individuals as a chance process. For the justification based purely on measurement error, we take the individuals as given, along with their factor levels; the chance process is in the measuring. For the third and last justification, based purely on randomization, we take not only the individuals but also the measurement errors as given. The chance part of the model comes from using a chance device to assign individuals to levels of the factor of interest. The assumptions that support this third justification are quite different from the usual Fisher assumptions. Nevertheless, there is considerable evidence that the critical values from an F table often provide good approximations to the critical values based purely on randomization. Before giving a brief summary of this evidence, I'll first show you a small, simple, artificial example that illustrates the basic ideas of the randomization test.

EXAMPLE 15.26 Randomization Test for Simple Data

Remind yourself of the experiment with hospital carpets (Example 8.5 and Exercise 8.C.1). Then imagine a very simple version of the experiment, using just four rooms. We use a chance device to pick two rooms that get carpet put down; the other two floors are left bare. Suppose we measure bacteria levels and get the following results:

	Carpet		Bare	
Bacteria level:	10	8	0	2
Room #:	201	204	202	203

If we assume the usual factor structure, decompose the data, and compute an F-ratio, we get the results in Figure 15.19.

$$\begin{array}{l} \text{Carpet} \\ \text{Bare} \end{array} \begin{array}{|cc|} \hline 10 & 8 \\ \hline 0 & 2 \\ \hline \end{array} = \begin{array}{|cc|} \hline 5 & 5 \\ \hline 5 & 5 \\ \hline \end{array} + \begin{array}{|cc|} \hline 4 & 4 \\ \hline -4 & -4 \\ \hline \end{array} + \begin{array}{|cc|} \hline 1 & -1 \\ \hline -1 & 1 \\ \hline \end{array} \qquad \begin{array}{l} MS_{Cond} = 64 \\ MS_{Res} = 2 \end{array}$$

$$\begin{array}{lccccccccl} \text{SS} & 168 & = & 100 & + & 64 & + & 4 & \qquad & F = 32 \\ \text{df} & 4 & = & 1 & + & 1 & + & 2 & & \end{array}$$

FIGURE 15.19

Our null hypothesis is that the true effects for Carpet versus Bare are both zero. We need to find the sampling distribution of the F-ratio, starting from the assumption that the null hypothesis is true.

If the treatment effects are zero, then each observed value equals Benchmark + Error. The floor covering has no effect on the observed values. Suppose we ignore the process that created the four errors, and assume that because carpeting has no effect, Room 201 was going to give a response of 10, regardless of what we did to the floor. Similarly, Room #202 would give a 0, #203 a 2, and #204 an 8, regardless of treatment. Then, still assuming the null hypothesis true, and taking these four values as given, the value we get for the F-ratio will be completely determined by the chance process we use to assign the four rooms to the two groups. According to this view, the chance is not in the errors themselves, but rather in the process of random assignment.

There are six different ways to sort the four rooms into a carpet pair and a bare pair. If we use a chance device to do the sorting, these six ways are equally likely (Figure 15.20).

	#1	#2	#3	#4	#5	#6
Carpet	10 8	10 2	10 0	8 2	8 0	2 0
Bare	0 2	0 8	2 8	0 10	2 10	8 10
SS_{Cond}	64	4	0	0	4	64
SS_{Res}	4	64	68	68	64	4
F-ratio	32	0.125	0	0	0.125	32

FIGURE 15.20 Six equally likely data sets and F-ratios. Each of the six ways to sort the 10, 8, 0, and 2 has the same 1-in-6 chance, because the sorting was done randomly. There are three possible values for the F-ratio: 0, 0.125, and 32. Each has chance $2/6 = 1/3$. This sampling distribution for the F-ratio comes from the random assignment of observed values to groups. ■

This simple example illustrates two general properties of sampling distributions based purely on randomization:

(1) $SS_{Cond} + SS_{Res}$ is constant. (For the example, the two sums of squares always add to 68.) To see why this always happens for the CR design, check that because the set of observed values is the same no matter how you randomize, both SS_{Total} and $SS_{Gr\ Avg}$. are constant, and so their difference, $SS_{Total} - SS_{Gr\ Avg}$, is also constant. But $SS_{Total} - SS_{Gr\ Avg} = SS_{Cond} + SS_{Res}$, so $SS_{Cond} + SS_{Res}$ is constant, too. For the CB and LS designs, there are similar results.

(2) $EMS_{Cond} = EMS_{Res}$. For the example, there are six equally likely ways to sort the data, so the expected value for each mean square equals the ordinary average of the six values for that mean square; here, both EMSs equal 22.67. Theoretical statisticians have proved that the same result, $EMS_{Cond} = EMS_{Res}$, is true in general for properly randomized CR, CB, and LS designs for which the null hypothesis is true.

Notice that we got a sampling distribution for the F-ratio without assuming anything about how the errors got created. We only used the null hypothesis and the chance process for assigning rooms. It's quite true that the sampling distribution we got is not at all like the F-distribution derived from the Fisher assumptions, but if we were to use the actual carpet data, with 16 observations instead of only 4, the 5% critical value would be quite close to the usual one from the F table.

In general, the justification based on randomization requires only that you have assigned conditions at random. If that assumption fits, then it is straightforward, though tedious, to find the exact sampling distribution of the F-ratio by first listing all possible ways to sort the observations into groups and then computing a value of the F-ratio for each. As a "quick-and-dirty" alternative, you can assume that the usual critical value from an F table will give an adequate approximation to the exact value based on randomization; for most data sets except very small ones. There is considerable evidence that such an assumption is often reasonable.

For one thing, statisticians have proved that the expected mean squares behave as they should: When the null hypothesis is true, EMS_{Cond} and EMS_{Res} will be equal, just as for sampling distributions derived from the Fisher assumptions. Second, statisticians have also proved that under certain specified conditions, the two sampling distributions will themselves be very nearly the same. Finally, in addition to the theoretical evidence, there is also numerical evidence from simulated experiments.

Summary. In all, this section has given three justifications for the standard box models. In the first, the randomness is assumed to come from the process of choosing the individuals to measure. In the second, the randomness is assumed to come from the process of measuring the individuals, regardless of how they are chosen. In the third, the randomness is assumed to come from the process of assigning individuals to conditions, regardless of how the individuals are chosen or how the measurements are made.

In practice, most situations don't quite fit any one of the three idealized justifications. Nevertheless, for many if not most well-designed studies, some mixture of the three justifications is likely to be appropriate. Unless the methods of Chapter 12 give you reason to question the assumptions, you are probably safe in regarding them as reasonable approximations.

Exercise Set F

1. List three possible sources of chance variation that may be present in an experiment, and for each, tell the name of the corresponding ideal model.

2. *Sampling from a population*. Suppose you want to do a one-way ANOVA to compare three populations of five individuals each: A [3,4,5,6,7], B [8,9,10,11,12], and C [4,5,6,7,8].

   a. Draw a diagram like the one in Example 15.24 to illustrate how drawing tickets at random from the three populations leads to an ANOVA model which satisfies five of the Fisher assumptions.

   b. Which assumption is not satisfied by the model?

3. Which of the following statements about the ideal sampling model are true?

   a. The ideal model assumes the SDs for the populations you sample from are equal.

   b. The pure sampling model ignores any randomness that comes from the measurement process.

   c. If you take simple random samples, then normality assumption (N) is automatically satisfied, no matter what the populations are.

4. *Departures from the ideal sampling model*. On page 758, I listed four ways in which a sampling scheme often falls short of ideal. For each of the situations described below, tell which one(s) of these four ways—call them (i)–(iv)—would be major concerns.

   a. Crabgrass (Exercise 4.A.5)

   b. Mental rotation (Exercise 4.E.2)

   c. Phytoplankton (Example 4.11)

   d. Memory and meaning (Example 4.7)

e. The defense of undoing (Example 6.21)

f. Mothers of schizophrenics (Exercise 4.E.7)

g. Bird calcium (Exercise 5.B.1)

h. Pigs and antibiotics (Example 6.1)

5. *Sampling with and without replacement*. The purpose of this exercise is to illustrate two facts: that if your population is small, there is a big difference between sampling with and without replacement, but that if your population is large, the two sampling methods behave pretty much alike.

a. Suppose you draw two tickets at random from a box with just two tickets, labeled 1 and 2. As you think about the list of all possible samples, consider samples like (First draw = 1, Second draw = 2) and (First draw = 2, Second draw = 1) to be different.

i. How many different samples (of size 2) are possible if you sample without replacing tickets? List the samples.

ii. If you sample with replacement? List the samples.

iii. If all possible samples in (ii) are equally likely, what is the chance that your sample contains the same individual twice?

b. Suppose your population has 3 different tickets, (1, 2, 3) and you draw a random sample of size 2. Answer (ii)–(iii) above.

c. Suppose your population has 4 different tickets, and you draw a sample of size 2. Answer (iii) above.

d. Suppose your population has N tickets. Answer (iii) above. Use your answer to find the chance of getting the same individual twice for populations of size 5, 10, 50, 100, 1,000, and 1,000,000.

6. *Measurement error*. Which of the following statements are true of the pure measurement model?

a. The pure model ignores any chance variation that comes from the sampling process of choosing subjects or material.

b. The "populations" for the measurement model are hypothetical constructions.

c. If you plan to appeal to the measurement model to justify the Fisher assumptions, then you don't need to use a chance device to assign conditions to material.

d. The pure measurement model applies only to genuine experiments, not to observational studies.

e. Experience has shown that for most measurements, the Fisher assumptions are likely to be true, regardless of the scale you use for your measurements.

7. For the pure randomization model which of the following statements are true?

a. None of the chance in the model (as opposed to the actual experiment) comes from the way individuals are sampled.

b. You can't apply the randomization model unless your chance errors follow the normal curve, at least approximately.

c. The null hypothesis that there are no real differences among levels of the factor of interest is equivalent to the assumption that all ways of sorting the observed values into groups are equally likely.

d. The exact sampling distribution (for the F-ratio) that you get from the model is the same as the one derived from the Fisher assumptions.

e. Because the randomization model ignores the sampling process that relates the individuals in a study to the populations they came from, an F-ratio that is significant according to the randomization model tells you nothing about the sampling process.

f. The randomization model is equally appropriate for genuine experiments and for observational studies.

8. *Randomization*. Suppose you randomly sort four individuals into a treatment and control group, and obtain the following results:

Treatment:	14	6
Control:	4	8

Follow the pattern of Example 15.26 to find the sampling distribution of the F-ratio that results from the assumption that the six possible ways to sort 14, 6, 4, and 8 into two groups are equally likely.

9. *Randomization model for an RCB design*. Review the finger tapping experiment (Exercise 7.B.2 and Example 7.14), and think about how you would construct a randomization model. Take the three response values from each subject as given, and assume that all the chance comes from the random assignment of those values to Placebo, Caffeine, and Theobromine.

a. The three values for Subject I were 11, 26, and 20. List and count all possible ways to assign these values to the three conditions.

b. Each subjects' values could be assigned in the same number of ways as Subject I's. Taking all four subjects together, how many ways are there, in all, to assign values to conditions?

c. The largest possible F-ratio occurs when one condition is assigned the largest value for each subject, and a second condition is assigned the smallest value for each subject. Here is one such assignment:

	Placebo	Caffeine	Theobromine	
I	11	26	20	
II	56	83	71	F-ratio = 8.46
III	15	41	34	
IV	6	32	13	

(i) How many other assignments give the same F-ratio?

(ii) Use your answers to (i) and (b) to compute the chance of an F-ratio this big.

(iii) Is this F-ratio significant, as judged by the exact sampling distribution? Is it significant, as judged by the usual critical value from an F table?

10. *Randomized beestings*. For the beesting data (Example S7.3),

a. How many different ways are there to assign observed values to conditions for (i) Occasion I? (ii) Occasion II? (iii) Each one of the seven other occasions?

b. How many different assignments are possible in all, considering all 9 occasions together?

c. The F-ratio for Fresh vs. Stung equals 3.11. It turns out that 20 of the possible assignments give an F-ratio this larger or larger.

(i) If you assume all possible assignments are equally likely, what is the chance of getting an F-ratio 3.11 or larger?

(ii) Is the observed F significant, if judged by the exact sampling distribution?

(iii) Is it significant, as judged by the usual critical value?

11. For each of the situations listed below, tell which of the three ideal models seems least appropriate.

a. The defense of undoing (Example S4.10)

b. Tracheobronchial twins (Example S7.2)

c. Pigs and antibiotics (Example 6.1)

d. Walking babies (Example 5.1)

e. Intravenous fluids (Example 5.2)

f. Leafhopper survival times (Exercise 5.C.1 and Example 5.11)

g. Submarine memory (Example 6.2)

h. Kosslyn's imagery experiment (Example 6.10)

NOTES AND SOURCES

1. ***Percentiles*** can be used to describe not only the extremes of a distribution, but also its center: the *median* is the 50th percentile, for example. For our purposes, however, we will use percentiles for the extremes, and summarize the center part using the EV and SE.

2. ***Denominator for SD****. One way to think about the reason for dividing by the number of tickets is this: The deviations for an SD* are independent, and so each provides a unit of information about chance error. The number of units of information equals the number of tickets, and so you divide SS_{Dev} by # tickets. The residuals you square to get an SD are not independent, but are correlated. Any two residuals contain some overlapping information about chance error, and dividing by df_{Res} instead of # residuals makes the right adjustment for the overlap.

3. ***The equation of the standard normal curve $f(z)$:***

$$f(z) = \frac{1}{\sqrt{2\pi}} \exp(-z^2/2)$$

4. ***The pure randomization model.*** From Hoaglin, Mosteller, and Tukey (1991), *Fundamentals of Exploratory Analysis of Variance*, New York: John Wiley & Sons, pp. 260–1: "**Randomization theory** (see Chapter 7) shows that the distribution of the ratio of mean squares will resemble the tabulated F-distribution quite closely, even when the distribution of the data is quite far from Gaussian." Also see Scheffe, Henry (1959). *Analysis of Variance*, New York: John Wiley & Sons, p. 324.

TABLES

Table 1. Random numbers

	1	2	3	4	5	6	7	8	9	10	11	12	13	14	15	16	17	18	19	20
1	37	64	41	84	55	51	41	40	04	91	85	24	49	21	04	03	95	29	20	70
2	68	88	02	58	36	58	89	46	24	05	14	89	93	88	93	25	69	69	15	05
3	37	56	97	70	02	80	75	28	61	60	77	43	92	47	04	68	09	72	44	65
4	88	66	90	30	30	53	90	41	22	01	97	86	57	05	02	53	69	53	30	44
5	59	44	28	28	29	46	48	61	28	98	08	62	02	82	11	01	04	65	78	48
6	63	63	12	88	82	21	83	44	00	53	54	81	59	24	19	21	96	44	80	55
7	95	50	14	78	75	18	21	94	50	13	65	18	91	18	30	48	62	39	70	95
8	84	84	59	25	83	76	84	62	67	50	16	19	88	18	96	52	62	07	71	61
9	65	90	44	85	00	23	79	63	74	38	82	07	88	51	12	06	89	33	42	64
10	68	19	13	68	90	96	94	04	29	57	44	61	96	03	36	37	13	55	08	74
11	65	75	92	25	81	95	65	56	66	94	93	50	86	35	23	79	48	92	55	58
12	24	87	60	25	71	63	67	80	92	16	73	66	63	12	55	00	20	28	71	79
13	83	50	13	42	18	40	50	91	00	39	74	36	15	73	46	70	47	31	72	08
14	61	04	35	98	89	43	98	79	42	46	36	30	57	09	52	52	04	41	43	97
15	25	58	09	06	95	66	70	58	67	39	89	27	99	32	22	52	63	88	39	78
16	80	83	18	96	85	49	52	18	07	97	04	89	35	60	50	89	25	80	10	92
17	50	55	34	23	73	90	53	43	30	89	44	27	72	25	01	54	98	94	46	96
18	14	52	40	71	97	48	43	81	55	51	55	08	09	92	33	75	05	72	91	43
19	83	76	09	56	24	11	31	24	97	14	45	36	92	87	09	29	79	15	77	95
20	04	44	73	80	35	56	54	06	22	47	94	75	39	55	72	99	54	53	55	65
21	10	82	64	87	10	99	12	25	44	24	16	17	25	47	38	32	67	68	74	53
22	11	03	09	15	43	95	55	37	74	04	39	47	53	10	17	21	25	99	14	64
23	78	06	57	52	83	26	79	00	82	34	19	85	47	04	77	25	53	17	72	41
24	76	77	45	41	07	56	75	43	89	38	53	81	31	63	72	05	91	95	01	16
25	35	00	13	53	35	86	56	09	01	47	76	97	78	37	24	88	36	52	29	20
26	42	73	82	94	87	13	55	83	15	55	23	36	64	53	25	99	20	99	11	90
27	80	12	59	49	01	03	53	34	58	09	29	55	24	62	48	75	84	45	45	53
28	23	87	61	61	26	63	14	04	64	43	41	53	02	88	34	70	08	61	13	36
29	29	62	89	67	52	29	18	35	36	33	45	40	43	02	09	82	35	09	89	05
30	79	59	60	04	73	58	33	67	67	71	69	50	46	58	63	74	38	64	24	97
31	49	37	99	48	98	98	33	40	22	23	05	02	25	57	47	58	90	73	93	04
32	81	58	67	94	40	05	42	08	13	80	67	45	91	82	25	45	98	39	56	68
33	25	05	62	29	62	60	31	79	98	99	91	76	90	72	59	05	71	39	49	18
34	86	02	93	89	65	35	87	42	15	12	41	84	14	52	57	46	03	02	79	47
35	70	68	41	40	38	50	08	66	26	91	42	59	69	67	13	34	51	25	48	52
36	75	79	12	61	06	19	25	30	55	41	02	07	45	29	45	04	45	81	40	12

Table 1. Random numbers (*continued*)

	1	**2**	**3**	**4**	**5**	**6**	**7**	**8**	**9**	**10**	**11**	**12**	**13**	**14**	**15**	**16**	**17**	**18**	**19**	**20**
37	09	66	07	24	18	09	28	45	05	33	88	12	02	48	17	98	70	09	43	20
38	72	69	93	00	48	75	67	27	08	31	55	18	84	87	20	14	09	79	23	31
39	70	54	75	11	69	33	96	84	51	07	36	81	70	55	95	13	66	36	19	34
40	15	44	22	40	98	40	93	86	74	55	54	78	21	77	17	06	85	55	45	19
41	45	56	99	21	50	25	94	31	12	43	81	32	84	59	83	75	45	40	80	51
42	44	33	53	63	20	03	01	67	27	62	52	39	48	23	78	25	54	55	74	13
43	50	67	56	86	67	92	69	72	08	65	49	13	76	88	93	30	62	19	00	32
44	85	95	73	83	26	29	97	21	54	90	00	10	14	52	20	07	21	08	05	26
45	77	24	19	50	53	43	35	76	37	54	93	03	84	20	77	82	78	74	07	22
46	28	20	35	09	03	71	56	41	13	37	80	36	70	10	06	94	32	08	62	33
47	76	16	69	93	45	52	25	99	10	94	30	57	42	20	89	68	87	67	04	62
48	66	11	98	74	30	14	45	11	82	65	64	71	24	21	19	59	82	82	00	79
49	04	35	85	55	55	96	92	39	11	32	35	70	69	11	43	51	04	13	53	76
50	86	66	32	66	07	84	98	97	52	32	30	80	87	43	03	45	99	97	91	92

Table 2. Areas A(z) between 0 and Z under the standard normal curve

z	0.00	0.01	0.02	0.03	0.04	0.05	0.06	0.07	0.08	0.09
0.00	0.0000	0.0040	0.0080	0.0120	0.0160	0.0199	0.0239	0.0279	0.0319	0.0359
0.10	0.0398	0.0438	0.0478	0.0517	0.0557	0.0596	0.0636	0.0675	0.0714	0.0753
0.20	0.0793	0.0832	0.0871	0.0910	0.0948	0.0987	0.1026	0.1064	0.1103	0.1141
0.30	0.1179	0.1217	0.1255	0.1293	0.1331	0.1368	0.1406	0.1443	0.1480	0.1517
0.40	0.1554	0.1591	0.1628	0.1664	0.1700	0.1736	0.1772	0.1808	0.1844	0.1879
0.50	0.1915	0.1950	0.1985	0.2019	0.2054	0.2088	0.2123	0.2157	0.2190	0.2224
0.60	0.2257	0.2291	0.2324	0.2357	0.2389	0.2422	0.2454	0.2486	0.2517	0.2549
0.70	0.2580	0.2611	0.2642	0.2673	0.2704	0.2734	0.2764	0.2794	0.2823	0.2852
0.80	0.2881	0.2910	0.2939	0.2967	0.2995	0.3023	0.3051	0.3078	0.3106	0.3133
0.90	0.3159	0.3186	0.3212	0.3238	0.3264	0.3289	0.3315	0.3340	0.3365	0.3389
1.00	0.3413	0.3438	0.3461	0.3485	0.3508	0.3531	0.3554	0.3577	0.3599	0.3621
1.10	0.3643	0.3665	0.3686	0.3708	0.3729	0.3749	0.3770	0.3790	0.3810	0.3830
1.20	0.3849	0.3869	0.3888	0.3907	0.3925	0.3944	0.3962	0.3980	0.3997	0.4015
1.30	0.4032	0.4049	0.4066	0.4082	0.4099	0.4115	0.4131	0.4147	0.4162	0.4177
1.40	0.4192	0.4207	0.4222	0.4236	0.4251	0.4265	0.4279	0.4292	0.4306	0.4319
1.50	0.4332	0.4345	0.4357	0.4370	0.4382	0.4394	0.4406	0.4418	0.4429	0.4441
1.60	0.4452	0.4463	0.4474	0.4484	0.4495	0.4505	0.4515	0.4525	0.4535	0.4545
1.70	0.4554	0.4564	0.4573	0.4582	0.4591	0.4599	0.4608	0.4616	0.4625	0.4633
1.80	0.4641	0.4649	0.4656	0.4664	0.4671	0.4678	0.4686	0.4693	0.4699	0.4706
1.90	0.4713	0.4719	0.4726	0.4732	0.4738	0.4744	0.4750	0.4756	0.4761	0.4767
2.00	0.4772	0.4778	0.4783	0.4788	0.4793	0.4798	0.4803	0.4808	0.4812	0.4817
2.10	0.4821	0.4826	0.4830	0.4834	0.4838	0.4842	0.4846	0.4850	0.4854	0.4857
2.20	0.4861	0.4864	0.4868	0.4871	0.4875	0.4878	0.4881	0.4884	0.4887	0.4890
2.30	0.4893	0.4896	0.4898	0.4901	0.4904	0.4906	0.4909	0.4911	0.4913	0.4916
2.40	0.4918	0.4920	0.4922	0.4925	0.4927	0.4929	0.4931	0.4932	0.4934	0.4936
2.50	0.4938	0.4940	0.4941	0.4943	0.4945	0.4946	0.4948	0.4949	0.4951	0.4952
2.60	0.4953	0.4955	0.4956	0.4957	0.4959	0.4960	0.4961	0.4962	0.4963	0.4964
2.70	0.4965	0.4966	0.4967	0.4968	0.4969	0.4970	0.4971	0.4972	0.4973	0.4974
2.80	0.4974	0.4975	0.4976	0.4977	0.4977	0.4978	0.4979	0.4979	0.4980	0.4981
2.90	0.4981	0.4982	0.4982	0.4983	0.4984	0.4984	0.4985	0.4985	0.4986	0.4986
3.00	0.4987	0.4987	0.4987	0.4988	0.4988	0.4989	0.4989	0.4989	0.4990	0.4990
3.10	0.4990	0.4991	0.4991	0.4991	0.4992	0.4992	0.4992	0.4992	0.4993	0.4993
3.20	0.4993	0.4993	0.4994	0.4994	0.4994	0.4994	0.4994	0.4995	0.4995	0.4995
3.30	0.4995	0.4995	0.4995	0.4996	0.4996	0.4996	0.4996	0.4996	0.4996	0.4997
3.40	0.4997	0.4997	0.4997	0.4997	0.4997	0.4997	0.4997	0.4997	0.4997	0.4998
3.50	0.4998	0.4998	0.4998	0.4998	0.4998	0.4998	0.4998	0.4998	0.4998	0.4998
3.60	0.4998	0.4998	0.4999	0.4999	0.4999	0.4999	0.4999	0.4999	0.4999	0.4999
3.70	0.4999	0.4999	0.4999	0.4999	0.4999	0.4999	0.4999	0.4999	0.4999	0.4999
3.80	0.4999	0.4999	0.4999	0.4999	0.4999	0.4999	0.4999	0.4999	0.4999	0.4999

Table 3. Critical values of the F distribution

denominator df	p-value	numerator df														
		1	2	3	4	5	6	7	8	9	10	15	20	40	80	120
1	0.100	39.86	49.50	53.59	55.83	57.24	58.20	58.91	59.44	59.86	60.19	61.22	61.74	62.53	62.93	63.06
	0.050	161.4	199.5	215.7	224.6	230.2	234.0	236.8	238.9	240.5	241.9	245.9	248.0	251.1	252.7	253.3
	0.010	4052	4999	5404	5624	5764	5859	5928	5981	6022	6056	6157	6209	6286	6326	6340
2	0.100	8.53	9.00	9.16	9.24	9.29	9.33	9.35	9.37	9.38	9.39	9.42	9.44	9.47	9.48	9.48
	0.050	18.51	19.00	19.16	19.25	19.30	19.33	19.35	19.37	19.38	19.40	19.43	19.45	19.47	19.48	19.49
	0.010	98.50	99.00	99.16	99.25	99.30	99.33	99.36	99.38	99.39	99.40	99.43	99.45	99.48	99.48	99.49
	0.001	998.4	998.8	999.3	999.3	999.3	999.3	999.3	999.3	999.3	999.3	999.3	999.3	999.3	999.3	999.3
3	0.100	5.54	5.46	5.39	5.34	5.31	5.28	5.27	5.25	5.24	5.23	5.20	5.18	5.16	5.15	5.14
	0.050	10.13	9.55	9.28	9.12	9.01	8.94	8.89	8.85	8.81	8.79	8.70	8.66	8.59	8.56	8.55
	0.010	34.12	30.82	29.46	28.71	28.24	27.91	27.67	27.49	27.34	27.23	26.87	26.69	26.41	26.27	26.22
	0.001	167.06	148.49	141.10	137.08	134.58	132.83	131.61	130.62	129.86	129.22	127.36	126.43	124.97	124.22	123.98
4	0.100	4.54	4.32	4.19	4.11	4.05	4.01	3.98	3.95	3.94	3.92	3.87	3.84	3.80	3.78	3.78
	0.050	7.71	6.94	6.59	6.39	6.26	6.16	6.09	6.04	6.00	5.96	5.86	5.80	5.72	5.67	5.66
	0.010	21.20	18.00	16.69	15.98	15.52	15.21	14.98	14.80	14.66	14.55	14.20	14.02	13.75	13.61	13.56
	0.001	74.13	61.25	56.17	53.43	51.72	50.52	49.65	49.00	48.47	48.05	46.76	46.10	45.08	44.57	44.40
5	0.100	4.06	3.78	3.62	3.52	3.45	3.40	3.37	3.34	3.32	3.30	3.24	3.21	3.16	3.13	3.12
	0.050	6.61	5.79	5.41	5.19	5.05	4.95	4.88	4.82	4.77	4.74	4.62	4.56	4.46	4.41	4.40
	0.010	16.26	13.27	12.06	11.39	10.97	10.67	10.46	10.29	10.16	10.05	9.72	9.55	9.29	9.16	9.11
	0.001	47.18	37.12	33.20	31.08	29.75	28.83	28.17	27.65	27.24	26.91	25.91	25.39	24.60	24.20	24.06
6	0.100	3.78	3.46	3.29	3.18	3.11	3.05	3.01	2.98	2.96	2.94	2.87	2.84	2.78	2.75	2.74
	0.050	5.99	5.14	4.76	4.53	4.39	4.28	4.21	4.15	4.10	4.06	3.94	3.87	3.77	3.72	3.70
	0.010	13.75	10.92	9.78	9.15	8.75	8.47	8.26	8.10	7.98	7.87	7.56	7.40	7.14	7.01	6.97
	0.001	35.51	27.00	23.71	21.92	20.80	20.03	19.46	19.03	18.69	18.41	17.56	17.12	16.44	16.10	15.98
7	0.100	3.59	3.26	3.07	2.96	2.88	2.83	2.78	2.75	2.72	2.70	2.63	2.59	2.54	2.50	2.49
	0.050	5.59	4.74	4.35	4.12	3.97	3.87	3.79	3.73	3.68	3.64	3.51	3.44	3.34	3.29	3.27
	0.010	12.25	9.55	8.45	7.85	7.46	7.19	6.99	6.84	6.72	6.62	6.31	6.16	5.91	5.78	5.74
	0.001	29.25	21.69	18.77	17.20	16.21	15.52	15.02	14.63	14.33	14.08	13.32	12.93	12.33	12.01	11.91
8	0.100	3.46	3.11	2.92	2.81	2.73	2.67	2.62	2.59	2.56	2.54	2.46	2.42	2.36	2.33	2.32
	0.050	5.32	4.46	4.07	3.84	3.69	3.58	3.50	3.44	3.39	3.35	3.22	3.15	3.04	2.99	2.97
	0.010	11.26	8.65	7.59	7.01	6.63	6.37	6.18	6.03	5.91	5.81	5.52	5.36	5.12	4.99	4.95
	0.001	25.41	18.49	15.83	14.39	13.48	12.86	12.40	12.05	11.77	11.54	10.84	10.48	9.92	9.63	9.53
9	0.100	3.36	3.01	2.81	2.69	2.61	2.55	2.51	2.47	2.44	2.42	2.34	2.30	2.23	2.20	2.18
	0.050	5.12	4.26	3.86	3.63	3.48	3.37	3.29	3.23	3.18	3.14	3.01	2.94	2.83	2.77	2.75
	0.010	10.56	8.02	6.99	6.42	6.06	5.80	5.61	5.47	5.35	5.26	4.96	4.81	4.57	4.44	4.40
	0.001	22.86	16.39	13.90	12.56	11.71	11.13	10.70	10.37	10.11	9.89	9.24	8.90	8.37	8.09	8.00
10	0.100	3.29	2.92	2.73	2.61	2.52	2.46	2.41	2.38	2.35	2.32	2.24	2.20	2.13	2.09	2.08
	0.050	4.96	4.10	3.71	3.48	3.33	3.22	3.14	3.07	3.02	2.98	2.85	2.77	2.66	2.60	2.58
	0.010	10.04	7.56	6.55	5.99	5.64	5.39	5.20	5.06	4.94	4.85	4.56	4.41	4.17	4.04	4.00
	0.001	21.04	14.90	12.55	11.28	10.48	9.93	9.52	9.20	8.96	8.75	8.13	7.80	7.30	7.03	6.94
11	0.100	3.23	2.86	2.66	2.54	2.45	2.39	2.34	2.30	2.27	2.25	2.17	2.12	2.05	2.01	2.00
	0.050	4.84	3.98	3.59	3.36	3.20	3.09	3.01	2.95	2.90	2.85	2.72	2.65	2.53	2.47	2.45
	0.010	9.65	7.21	6.22	5.67	5.32	5.07	4.89	4.74	4.63	4.54	4.25	4.10	3.86	3.73	3.69
	0.001	19.69	13.81	11.56	10.35	9.58	9.05	8.65	8.35	8.12	7.92	7.32	7.01	6.52	6.26	6.18
12	0.100	3.18	2.81	2.61	2.48	2.39	2.33	2.28	2.24	2.21	2.19	2.10	2.06	1.99	1.95	1.93
	0.050	4.75	3.89	3.49	3.26	3.11	3.00	2.91	2.85	2.80	2.75	2.62	2.54	2.43	2.36	2.34
	0.010	9.33	6.93	5.95	5.41	5.06	4.82	4.64	4.50	4.39	4.30	4.01	3.86	3.62	3.49	3.45
	0.001	18.64	12.97	10.80	9.63	8.89	8.38	8.00	7.71	7.48	7.29	6.71	6.40	5.93	5.68	5.59

Table 3. Critical values of the F distribution (*continued*)

denominator df	p-value	numerator df 1	2	3	4	5	6	7	8	9	10	15	20	40	80	120
13	0.100	3.14	2.76	2.56	2.43	2.35	2.28	2.23	2.20	2.16	2.14	2.05	2.01	1.93	1.89	1.88
	0.050	4.67	3.81	3.41	3.18	3.03	2.92	2.83	2.77	2.71	2.67	2.53	2.46	2.34	2.27	2.25
	0.010	9.07	6.70	5.74	5.21	4.86	4.62	4.44	4.30	4.19	4.10	3.82	3.66	3.43	3.30	3.25
	0.001	17.82	12.31	10.21	9.07	8.35	7.86	7.49	7.21	6.98	6.80	6.23	5.93	5.47	5.22	5.14
14	0.100	3.10	2.73	2.52	2.39	2.31	2.24	2.19	2.15	2.12	2.10	2.01	1.96	1.89	1.84	1.83
	0.050	4.60	3.74	3.34	3.11	2.96	2.85	2.76	2.70	2.65	2.60	2.46	2.39	2.27	2.20	2.18
	0.010	8.86	6.51	5.56	5.04	4.69	4.46	4.28	4.14	4.03	3.94	3.66	3.51	3.27	3.14	3.09
	0.001	17.14	11.78	9.73	8.62	7.92	7.44	7.08	6.80	6.58	6.40	5.85	5.56	5.10	4.86	4.77
15	0.100	3.07	2.70	2.49	2.36	2.27	2.21	2.16	2.12	2.09	2.06	1.97	1.92	1.85	1.80	1.79
	0.050	4.54	3.68	3.29	3.06	2.90	2.79	2.71	2.64	2.59	2.54	2.40	2.33	2.20	2.14	2.11
	0.010	8.68	6.36	5.42	4.89	4.56	4.32	4.14	4.00	3.89	3.80	3.52	3.37	3.13	3.00	2.96
	0.001	16.59	11.34	9.34	8.25	7.57	7.09	6.74	6.47	6.26	6.08	5.54	5.25	4.80	4.56	4.48
16	0.100	3.05	2.67	2.46	2.33	2.24	2.18	2.13	2.09	2.06	2.03	1.94	1.89	1.81	1.77	1.75
	0.050	4.49	3.63	3.24	3.01	2.85	2.74	2.66	2.59	2.54	2.49	2.35	2.28	2.15	2.08	2.06
	0.010	8.53	6.23	5.29	4.77	4.44	4.20	4.03	3.89	3.78	3.69	3.41	3.26	3.02	2.89	2.84
	0.001	16.12	10.97	9.01	7.94	7.27	6.80	6.46	6.20	5.98	5.81	5.27	4.99	4.54	4.31	4.23
17	0.100	3.03	2.64	2.44	2.31	2.22	2.15	2.10	2.06	2.03	2.00	1.91	1.86	1.78	1.74	1.72
	0.050	4.45	3.59	3.20	2.96	2.81	2.70	2.61	2.55	2.49	2.45	2.31	2.23	2.10	2.03	2.01
	0.010	8.40	6.11	5.19	4.67	4.34	4.10	3.93	3.79	3.68	3.59	3.31	3.16	2.92	2.79	2.75
	0.001	15.72	10.66	8.73	7.68	7.02	6.56	6.22	5.96	5.75	5.58	5.05	4.78	4.33	4.10	4.02
18	0.100	3.01	2.62	2.42	2.29	2.20	2.13	2.08	2.04	2.00	1.98	1.89	1.84	1.75	1.71	1.69
	0.050	4.41	3.55	3.16	2.93	2.77	2.66	2.58	2.51	2.46	2.41	2.27	2.19	2.06	1.99	1.97
	0.010	8.29	6.01	5.09	4.58	4.25	4.01	3.84	3.71	3.60	3.51	3.23	3.08	2.84	2.70	2.66
	0.001	15.38	10.39	8.49	7.46	6.81	6.35	6.02	5.76	5.56	5.39	4.87	4.59	4.15	3.92	3.84
19	0.100	2.99	2.61	2.40	2.27	2.18	2.11	2.06	2.02	1.98	1.96	1.86	1.81	1.73	1.68	1.67
	0.050	4.38	3.52	3.13	2.90	2.74	2.63	2.54	2.48	2.42	2.38	2.23	2.16	2.03	1.96	1.93
	0.010	8.18	5.93	5.01	4.50	4.17	3.94	3.77	3.63	3.52	3.43	3.15	3.00	2.76	2.63	2.58
	0.001	15.08	10.16	8.28	7.27	6.62	6.18	5.85	5.59	5.39	5.22	4.70	4.43	3.99	3.76	3.68
20	0.100	2.97	2.59	2.38	2.25	2.16	2.09	2.04	2.00	1.96	1.94	1.84	1.79	1.71	1.66	1.64
	0.050	4.35	3.49	3.10	2.87	2.71	2.60	2.51	2.45	2.39	2.35	2.20	2.12	1.99	1.92	1.90
	0.010	8.10	5.85	4.94	4.43	4.10	3.87	3.70	3.56	3.46	3.37	3.09	2.94	2.69	2.56	2.52
	0.001	14.82	9.95	8.10	7.10	6.46	6.02	5.69	5.44	5.24	5.08	4.56	4.29	3.86	3.62	3.54
21	0.100	2.96	2.57	2.36	2.23	2.14	2.08	2.02	1.98	1.95	1.92	1.83	1.78	1.69	1.64	1.62
	0.050	4.32	3.47	3.07	2.84	2.68	2.57	2.49	2.42	2.37	2.32	2.18	2.10	1.96	1.89	1.87
	0.010	8.02	5.78	4.87	4.37	4.04	3.81	3.64	3.51	3.40	3.31	3.03	2.88	2.64	2.50	2.46
	0.001	14.59	9.77	7.94	6.95	6.32	5.88	5.56	5.31	5.11	4.95	4.44	4.17	3.74	3.50	3.42
22	0.100	2.95	2.56	2.35	2.22	2.13	2.06	2.01	1.97	1.93	1.90	1.81	1.76	1.67	1.62	1.60
	0.050	4.30	3.44	3.05	2.82	2.66	2.55	2.46	2.40	2.34	2.30	2.15	2.07	1.94	1.86	1.84
	0.010	7.95	5.72	4.82	4.31	3.99	3.76	3.59	3.45	3.35	3.26	2.98	2.83	2.58	2.45	2.40
	0.001	14.38	9.61	7.80	6.81	6.19	5.76	5.44	5.19	4.99	4.83	4.33	4.06	3.63	3.40	3.32
23	0.100	2.94	2.55	2.34	2.21	2.11	2.05	1.99	1.95	1.92	1.89	1.80	1.74	1.66	1.61	1.59
	0.050	4.28	3.42	3.03	2.80	2.64	2.53	2.44	2.37	2.32	2.27	2.13	2.05	1.91	1.84	1.81
	0.010	7.88	5.66	4.76	4.26	3.94	3.71	3.54	3.41	3.30	3.21	2.93	2.78	2.54	2.40	2.35
	0.001	14.20	9.47	7.67	6.70	6.08	5.65	5.33	5.09	4.89	4.73	4.23	3.96	3.53	3.30	3.22
24	0.100	2.93	2.54	2.33	2.19	2.10	2.04	1.98	1.94	1.91	1.88	1.78	1.73	1.64	1.59	1.57
	0.050	4.26	3.40	3.01	2.78	2.62	2.51	2.42	2.36	2.30	2.25	2.11	2.03	1.89	1.82	1.79
	0.010	7.82	5.61	4.72	4.22	3.90	3.67	3.50	3.36	3.26	3.17	2.89	2.74	2.49	2.36	2.31
	0.001	14.03	9.34	7.55	6.59	5.98	5.55	5.24	4.99	4.80	4.64	4.14	3.87	3.45	3.22	3.14

Table 3. Critical values of the F distribution (*continued*)

denominator df	p-value	numerator df 1	2	3	4	5	6	7	8	9	10	15	20	40	80	120
25	0.100	2.92	2.53	2.32	2.18	2.09	2.02	1.97	1.93	1.89	1.87	1.77	1.72	1.63	1.58	1.56
	0.050	4.24	3.39	2.99	2.76	2.60	2.49	2.40	2.34	2.28	2.24	2.09	2.01	1.87	1.80	1.77
	0.010	7.77	5.57	4.68	4.18	3.85	3.63	3.46	3.32	3.22	3.13	2.85	2.70	2.45	2.32	2.27
	0.001	13.88	9.22	7.45	6.49	5.89	5.46	5.15	4.91	4.71	4.56	4.06	3.79	3.37	3.14	3.06
30	0.100	2.88	2.49	2.28	2.14	2.05	1.98	1.93	1.88	1.85	1.82	1.72	1.67	1.57	1.52	1.50
	0.050	4.17	3.32	2.92	2.69	2.53	2.42	2.33	2.27	2.21	2.16	2.01	1.93	1.79	1.71	1.68
	0.010	7.56	5.39	4.51	4.02	3.70	3.47	3.30	3.17	3.07	2.98	2.70	2.55	2.30	2.16	2.11
	0.001	13.29	8.77	7.05	6.12	5.53	5.12	4.82	4.58	4.39	4.24	3.75	3.49	3.07	2.84	2.76
40	0.100	2.84	2.44	2.23	2.09	2.00	1.93	1.87	1.83	1.79	1.76	1.66	1.61	1.51	1.45	1.42
	0.050	4.08	3.23	2.84	2.61	2.45	2.34	2.25	2.18	2.12	2.08	1.92	1.84	1.69	1.61	1.58
	0.010	7.31	5.18	4.31	3.83	3.51	3.29	3.12	2.99	2.89	2.80	2.52	2.37	2.11	1.97	1.92
	0.001	12.61	8.25	6.59	5.70	5.13	4.73	4.44	4.21	4.02	3.87	3.40	3.15	2.73	2.49	2.41
60	0.100	2.79	2.39	2.18	2.04	1.95	1.87	1.82	1.77	1.74	1.71	1.60	1.54	1.44	1.37	1.35
	0.050	4.00	3.15	2.76	2.53	2.37	2.25	2.17	2.10	2.04	1.99	1.84	1.75	1.59	1.50	1.47
	0.010	7.08	4.98	4.13	3.65	3.34	3.12	2.95	2.82	2.72	2.63	2.35	2.20	1.94	1.78	1.73
	0.001	11.97	7.77	6.17	5.31	4.76	4.37	4.09	3.86	3.69	3.54	3.08	2.83	2.41	2.17	2.08
120	0.100	2.75	2.35	2.13	1.99	1.90	1.82	1.77	1.72	1.68	1.65	1.55	1.48	1.37	1.29	1.26
	0.050	3.92	3.07	2.68	2.45	2.29	2.18	2.09	2.02	1.96	1.91	1.75	1.66	1.50	1.39	1.35
	0.010	6.85	4.79	3.95	3.48	3.17	2.96	2.79	2.66	2.56	2.47	2.19	2.03	1.76	1.60	1.53
	0.001	11.38	7.32	5.78	4.95	4.42	4.04	3.77	3.55	3.38	3.24	2.78	2.53	2.11	1.86	1.77
240	0.100	2.73	2.32	2.11	1.97	1.87	1.80	1.74	1.70	1.66	1.63	1.52	1.45	1.33	1.25	1.22
	0.050	3.88	3.03	2.64	2.41	2.25	2.14	2.05	1.98	1.92	1.87	1.71	1.61	1.44	1.33	1.29
	0.010	6.74	4.69	3.86	3.40	3.09	2.88	2.71	2.59	2.48	2.40	2.11	1.96	1.68	1.50	1.43
	0.001	11.10	7.11	5.60	4.78	4.26	3.89	3.62	3.41	3.24	3.09	2.65	2.40	1.97	1.71	1.61

Table 4. 2-tailed t-values

df	Probability between -t and t					
	50%	80%	90%	95%	98%	99%
1	1.000	3.078	6.314	12.706	31.821	63.656
2	0.816	1.886	2.920	4.303	6.965	9.925
3	0.765	1.638	2.353	3.182	4.541	5.841
4	0.741	1.533	2.132	2.776	3.747	4.604
5	0.727	1.476	2.015	2.571	3.365	4.032
6	0.718	1.440	1.943	2.447	3.143	3.707
7	0.711	1.415	1.895	2.365	2.998	3.499
8	0.706	1.397	1.860	2.306	2.896	3.355
9	0.703	1.383	1.833	2.262	2.821	3.250
10	0.700	1.372	1.812	2.228	2.764	3.169
11	0.697	1.363	1.796	2.201	2.718	3.106
12	0.695	1.356	1.782	2.179	2.681	3.055
13	0.694	1.350	1.771	2.160	2.650	3.012
14	0.692	1.345	1.761	2.145	2.624	2.977
15	0.691	1.341	1.753	2.131	2.602	2.947
16	0.690	1.337	1.746	2.120	2.583	2.921
17	0.689	1.333	1.740	2.110	2.567	2.898
18	0.688	1.330	1.734	2.101	2.552	2.878
19	0.688	1.328	1.729	2.093	2.539	2.861
20	0.687	1.325	1.725	2.086	2.528	2.845
21	0.686	1.323	1.721	2.080	2.518	2.831
22	0.686	1.321	1.717	2.074	2.508	2.819
23	0.685	1.319	1.714	2.069	2.500	2.807
24	0.685	1.318	1.711	2.064	2.492	2.797
25	0.684	1.316	1.708	2.060	2.485	2.787
26	0.684	1.315	1.706	2.056	2.479	2.779
27	0.684	1.314	1.703	2.052	2.473	2.771
28	0.683	1.313	1.701	2.048	2.467	2.763
29	0.683	1.311	1.699	2.045	2.462	2.756
30	0.683	1.310	1.697	2.042	2.457	2.750
40	0.681	1.303	1.684	2.021	2.423	2.704
60	0.679	1.296	1.671	2.000	2.390	2.660
120	0.677	1.289	1.658	1.980	2.358	2.617
normal	0.674	1.282	1.645	1.960	2.326	2.576

DATA SOURCES

- *Abundance of phytoplankton:* Hurlbert, S. J. (1983). "The unpredictability of the marine phytoplankton," *Ecology*, v. 64, pp. 1157–1170.
- *Agressive mice:* Leshner, Alan I. and John A. and Moyer (1975). "Androgens and agonistic behavior in mice: relevance to aggression and irrlevance to avoidance of attack," *Physiol. and Behavior*, v. 15, no. 6, pp. 695–699.
- *Attribution in troubled marriages:* Fincham, Frank D. (1985). "Attribution preocesses in distressed and non-distressed couples: 2. Responsibility for marital problems." *J. Abnormal Psych.*, v. 94, no. 2, p. 183.
- *Automatic processing:* Schneider W. and R. M. Shiffren (1977). "Controlled and automatic human information processing: I. Detection, search, and attention," *Psychological Review*, vol. 84, pp. 1–66 (via Anderson).
- *Automatic processing:* Shiffren, R. M and W. Schneider (1977). "Controlled and automatic human information processing: II. Perceptual learning, automatic attending, and a general theory," *Psychological Review*, vol. 84, pp. 127–190 (via Anderson).
- *Bee stings:* Free, J. B. (1961). "The stinging response of honeybees," *Animal Behavior*, v. 9, pp. 193–196.
- *Behavior therapy for alcoholics:* Okulitch, Peter V. and G. Alan Marlatt (1972). "Effects of varied extinction conditions with alcoholics and social drinkers," *J. Abnormal Psych.*, v. 72, no.2, pp. 205–211.
- *Body weight, body volume:* Boyd, Edith (1933). "The specific gravity of the human body," *Human Biology*, v. 5, p. 62 (via Larsen and Marx).
- *Buttercups:* Lovett-Doust, Lesley (1980). Personal communication based on research conducted in the Department of Biological Sciences, Mount Holyoke College.
- *Cabbage butterflies:* Gilbert, N. (1984). "Control of fecundity in Pieris rapae II. Differential effects of temperature," *J. Animal Ecol.*, v. 53, pp. 589–597 (via Begon, Harper and Townsend).
- *Cats and tetanus:* Sema, T. and N. Kano (1969). "Glycine in the spinal chord of cats with local tetanus rigidity," *Science*, v. 174, pp. 571–572 (via Srivastatva and Carter).
- *Chick thyroids:* Based on a student laboratory project in endocrinology, Department of Biological Sciences, Mount Holyoke College.

- *Chirping crickets:* Pierce, George W. (1949). *The Songs of Insects*. Cambridge, MA: Harvard University Press, pp. 12–21 (via Larsen and Marx).
- *Circadian rhythms:* Boon, D. A., R. E. Keenan, and W. R. Slaunwhite, Jr. (1972). "Plasma testosterone in men: variation but not circadian rhythm," *Steroids*, v. 20, no. 3, pp. 269–278.
- *Classifying habitats:* Southwood, T. R. E. (1977). "Habitat: the templet for ecological strategies?" *J. Animal Ecol.*, v. 46, pp. 337–365 (via Begon, Harper and Townsend).
- *Clotting blood:* Hurn, M. W., N. W. Barker, and T. D. Magath (1945). "The determination of prothrombin time following the administration of dicumarol with special reference to thromboplastin," *J. Lab. Clin, Med.*, v. 30, pp. 432–447 (via Bliss, 1970).
- *Competing crabgrass:* Maruk, Katherine Ann (1975). "The effects of nutrient levels on the competitive interaction between two species of *Digitaria*." Unpublished master's thesis, Department of Biological Sciences, Mount Holyoke College.
- *Cool mice:* Hart, J. S. (1951). "Calorimetric determination of average body temperature of small mammals and its variation with environmental conditions," *Canadian J. Zool.*, v. 29, pp. 224–233.
- *Dandelions:* Solbrig, O. T. and B. B. Simpson (1974). "Components of regulation of a population of dandelions in Michigan," *J. Ecol.*, v. 62, pp. 473–486 (via Begon, Harper and Townsend).
- *Dandelions:* Solbrig, O. T. and B. B. Simpson (1977). "A garden experiment on competition between biotypes of the common dandelion (*Taraxacum officinale*)," *J. Ecol.*, v. 65, pp. 427–430 (via Begon, Harper and Townsend).
- *Death rates and density: The World Almanac and Book of Facts* (1989). Equivalent current data can be found in the 1997 edition, pp. 211 (licensed drivers), 383 (density by state), 662–87 (income), and 963 (death rate).
- *Deaths and climate: The World Almanac and Book of Facts* (1997), p. 963.
- *Defense of undoing:* Sampson, Harold, Joseph Weiss, L. Mlodnosky, and Edward Hause (1972). "Defense analysis and the emergence of warded off mental contents," *Arch. Gen. Psychiat.*, v. 26, 524–532.
- *Dense tomatoes:* (1976). "Effects of plant density on tomato yields in western Nigeria," *Exper. Ag.*, pp. 43–47 (via Devore and Peck).
- *Deprived rats:* (1972). "The relation between differences in level of food deprivation and dominanace in food getting in the rat," *Psych. Sci.*, pp. 297–298 (via Devore and Peck).
- *Diabetic dogs:* Forbath, N., A. B. Kenshole, and G. Hetenyi, Jr. (1967). "Turnover of lactic acid in normal and diabetic dogs calculated by two tracer methods," *Am. J. Physiol.*, v. 212, pp. 1179–1183.
- *Diets and dopamine:* Krause, Wilma, Margaret Halminski, Linda McDonald, Philip Dembure, David Friedes, and Louis Elsas (1985). "Biochemical and neurophysio-

logical effects of elevated plasma phenylalanine in patients with treated phenylketonurea," *J. Clin. Invest.*, v. 75, pp. 40–48.

- *Dutch elm disease:* Lewis, Trevor and L. R. Taylor (1967). *Introduction to Experimental Ecology*. London: Academic Press, pp. 164–166.
- *Effective teachers:* Tuckman, B., J. Steber, and R. Hyman (1979). "Judging the effectiveness of teaching styles: The perceptions of principals," *Educational Administration Quarterly*, via James P. Stevens (1990). *Intermediate Statistics: A Modern Approach*, Hillsdale, NJ: Lawrence Erlbaum Associates.
- *Elaboration and memory:* Hyde, T. S. and J. J. Jenkins (1973). "Recall for words as a function of semantic, graphic, and syntactic orienting tasks," *J. Verbal Learning and Verbal Behavior*, v. 12, pp. 471–480 (via Anderson).
- *Emotions and skin potential:* (1963). "Physiological effects during hypnotically requested emotions," *Psychosomatic Medicine*, pp. 334–43 (via Devore and Peck).
- *Estradiol and uterine weight:* Based on a student laboratory project in endocrinology, Department of Biological Sciences, Mount Holyoke College.
- *Federalist papers:* Mosteller, Frederick and David L. Wallace (1984). *Applied Bayesian and Classical Inference: The Case of the Federalist Papers*. NY: Springer-Verlag/Wallace.
- *Feeding frogs:* Haubrich, Robert (1961). "Hierarchical behavior in the South African clawed frog Xenopus laevis Daudin," *Animal Behavior*, v. 9, pp. 71–76.
- *Finger tapping*: Scott, C. C. and K. K. Chen (1944). "Comparison of the action of 1–ethyl theobromine and caffeine in animals and man," *J. Pharmacol. Exptl. Therapy*, v. 82, pp. 89–97 (via Bliss, 1967).
- *Fisher's iris:* Fisher, R. A. (1936). "The use of multiple measurements in taxonomic problems," *Ann. Eugenics*, v. 7, Pt. II, pp. 179–188.
- *Freud's stages:* Kline, Paul (1981). *Fact and Fantasy in Freudian Theory*, 2nd ed. London: Methuen.
- *Gold teeth:* Brown, Morton B. (1975). "Exploring interaction effects in ANOVA," *Applied Statistics, vol. 24*, pp. 288–298 (via Katherine Taylor Halvorsen, "Value Splitting involving more factors," in Hoaglin, Mosteller and Tukey).
- *Hens' eggs:* Dempster, A. P. (1969). *Elements of Continuous Multivariate Analysis*. Reading, MA: Addison-Wesley, p. 151.
- *Hepatitis screening:* Prince, A. M. and Gershon, R. K. (1965). The use of serum enzyme determinations to detect anicteric hepatitis. *Transfusion*, v. 5, p. 120 (via Colton).
- *Hibernating hamsters: specific activity:* Acampora, Kelly Ann (1978). "The photoperiodic effects of Na+, K+-adenosinetriphosphatase activity in the golden hamster," undergraduate honeors thesis, Department of Biological Sciences, Mount Holyoke College.
- *HIV testing:* Schmid, Christopher H. (1991). "Value spliting: taking the data apart," in Hoaglin, Mosteller and Tukey.

- *Hornworms and cellulose:* Dorfman, Katherine (1980). Personal communication based on research conducted in the Department of Biological Sciences, Mount Holyoke College.
- *Hospital carpets:* Walter, W. G. and A. Stober (1968). "Microbial air sampling in a carpeted hospital," *J. Environ. Health*, v. 30, p. 405 (via Larsen and Marx).
- *Hypnosis and learning:* Liebert, Robert M. Norma Rubin, and Ernest R. Hilgard (1965). "The effects of suggestions of alertness in hypnosis on paired-associate learning", *J. Personality*, v. 33, pp. 605–612.
- *Imagery and working memory:* Brooks, L. R. (1968). "Spatial and verbal components of the act of recall," *Canadian J. Psych.*, v. 22, pp. 349–368.
- *Intravenous fluids:* Turco, Slavatore, and Neil Davis (1973). "Particulate matter in intravenous infusion fluids—Phase 3," *Am. J. Hospital Pharm.*, v. 30, p. 612 (via Larsen and Marx).
- *IQ and expectations:* Example is based on Rosenthal, Robert (1973). Unpublished final examination, Department of Psychology and Social Relations, Harvard University.
- *Jung's word association test:* Jung, Carl (1918). *Studies in Word Association.* Trans. by M. D. Eder. London: W. Heinemann, Ltd.
- *Kosslyn's imagery experiment:* Kosslyn, S. M. (1980). *Image and Mind.* Cambridge, MA: Harvard University Press.
- *Leafhopper survival:* Dahlman, Douglas (1963). "Survival and behavioral responses of the potato leafhopper, *Empoasca Fabae* (Harris), on synthetic media," MS thesis, Iowa State University (via David M. Allen and Foster B. Cady, *Analyzing Experimental Data by Regression*, Belmont, CA: Lifetime Learning (Wadsworth).
- *Losing leaves:* Jennifer Early (1986). "Studies in abscission: the inhibiting effect of auxin," unpublished paper based on data provided by Karen Davidson (1983), Department of Biological Sciences, Mount Holyoke College.
- *Lung cancr:* Moses, Lincoln E. (1986). *Think and Explain with Statistics.* Reading, MA: Addison-Wesley, p. 159 (no primary source given).
- *Mammary ligation:* Dimond, E. Grey, C. Frederick Kittle, and James E. Crocket (1960). "Comparison of internal mammary artery ligation and sham operation for angina pectoris," *Am. J. Cardiol.*, v. 5, pp. 483–486.
- *Memory and meaning:* Wanner, H. E. (1968). "On remembering, forgetting, and understanding sentences. A study of the deep structure hypothesis." Unpublished doctoral dissertation, Harvard University.
- *"Memory, mental health, and interference":* Based on research by William Edell, Department of Psychology, University of Massachusetts, 1982.
- *Mental activity:* Smith, G. M. and H. T. Beecher (1962). *J. Pharm. and Exper. Therapy*, v. 136, p. 47 (via Snedecor and Cochran).
- *Mental rotation:* Cooper, L. A. and R. N. Shepard (1973). "Chronomeric studies

of the rotation of mental images," in Chase, W. G. (Ed.) *Visual Information Processing*, New York: Academic Press (via Anderson).

- *Milgram compliance:* Milgram, Stanley (1975). *Obedience to Authority: An Experimental View*, New York: Harper and Row, p. 35.
- *Milk yields:* Cochran, W. G., K. M. Autrey, and C. Y. Cannon (1941). *J. Dairy Sci.*, v. 24, p. 931.
- *MMPI:* Graham, John R. (1987). *The MMPI: A Practical Guide*, 2nd ed., New York: Oxford University Press.
- *Morton's skulls:* Gould, S. J. (1981). *The Mismeasure of Man*. New York: W. W. Norton & Company, p. 54.
- *Mothers' stories:* Werner, M., J. B. Stabenau, and W. Pollin (1970). "Thematic Apperception Test method for the differentiation of the families of schizophrenics, delinquents, and normals," *J. Abnormal Psych.*, v. 75, pp. 139–145.
- *Muderers' ears:* Griffiths, G. B. (1904). "Measurements of one hundred and thirty criminals," *Biometrika*, v. 3, pp. 60–63 (via Srivastava and Carter).
- *Osmoregulation in worms:* Mitchell, Katherine (1985). Unpublished student project, Department of Biological Sciences, Mount Holyoke College.
- *Overconfidence in case study judgments:* Oskamp, Stuart (1982). "Overconfidence in case study judgments," in Daniel Kahneman, Paul Slovic, and Amos Tversky (eds), *Judgment Under Uncertainly: Heuristics and Biases*. New York: Cambridge University Press.
- *Oxygen pressure:* Tygstrup, Niels, Kjeld Winkler, Kresten Mellengaard, and Mogens Andeassen (1962). "Determination of the hepatic arterial blood flow and oxygen supply in man by clamping the hepatic artery during surgery," *J. Clin. Invest.*, v. 41, no. 3, pp. 447–454
- *Parsnip webworms:* Hendrix, S. D. (1979). "Compensatory reproduction in a biennial herb following insect defloration," *Oecologia*, v. 42, pp. 107–118 (via Begon, Harper and Townsend).
- *Personality and selective attention:* Lifshitz Cooney, Judith and Amos Zeichner (1985). "Selective attention to negative feedback in Type A and Type B individuals," *J. Abnormal Psych.*, v. 94, no. 1, p. 110.
- *Pesticides in the Wolf River:* Jaffe, P. R., F. L. Parker, and D. J. Wilson (1982). "Distribution of toxic substances in rivers," *J. Envir. Eng. Division*, v. 108, pp. 639–649 (via Rovert V. Hogg and Johannes Ledolter (1987). *Engineering Statistics*, New York: Macmillan.
- *Pigs and vitamins:* Iowa Agricultural Experiment Station (1952). Animal Husbandry Swine Nutrition Experiment No. 577 (via Snedecor and Cochran, p. 345).
- *Pine pruning:* Stoate, T. N. and C. E. Lane-Poole (1938), "Application of statistical methods to some Australian forest problems," *Commonwealth Forestry Bureau Australia Bull. No. 21* (via Bliss, 1967).

- *Pine seeds:* Wakely, P. C. (1944). "Geographic sources of loblolly pine seed," *J. Forestry*, v. 42, pp. 23–32 (via Bliss, 1967).
- *Portacaval shunt:* Grace, N. D., H. Meunch, and T. C. Chalmers (1966). "The present status of shunts for portal hypertension in cirrhosis," *J. Gastroenterol.*, v. 50, pp. 646–691.
- *Premature infants:* Moses, Lincoln E. (1986). *Think and Explain with Statistics.* Reading, MA: Addison-Wesley, p. 193–194 (no primary source given).
- *Premenstrual syndrome:* Based on research by Kim Kendall, Department of Psychology, University of Massachusetts, 1982.
- *Projective tests:* Jensen, A. R. (1965). "Review of the Rorschach," in Buros, O. K. (Ed.) *The Sixth Mental Measurements Yearbook*, Highland Park, NJ: Gryphon Press.
- *Projective tests:* Machover, K. (1949). *Personality Projection in the Drawing of the Human Figure: A Method of Personality Investigation*, Springfield, IL: Charles C. Thomas, Publisher (via Stanley and Hopkins).
- *Projective tests:* Murray, H. A., et al. (1938). *Explorations in Personality*, New York: Oxford University Press (via Stanley and Hopkins).
- *Puzzled children:* Haegel, Kathleen M. (1969). Undergraduate honors thesis, Department of Psychology and Education, Mount Holyoke College.
- *Pygmalion effect:* Rosenthal, Robert and Lenore Jacobson (1968). *Pygmalion in the Classroom: Teacher Expectation and Pupils' Intellectual Development*. New York: Holt, Rinehart, and Winston.
- *Rabbit insulin:* Young, D. M. and R. G. Romans (1948). "Assay of insulin with one blood sample per rabbit per day," *Biometrics*, v. 4, pp. 12–131 (via Bliss, 1967).
- *Radioactive twins:* Camner, Per and Klas Phillipson (1973). "Urban factor and tracheobronchial clearance," v. 27, p. 82 (via Larsen and Marx).
- *Rating Milgram:* DiMatteo, Mary Ann (1972). "An experimental study of attitudes toward deception," Unpublished manuscript, Department of Psychology and Social Relations, Harvard University.
- *Remembering words:* Data from student laboratory project, Department of Psychology and Education, Mount Holyoke College.
- *Rogerian therapy and Q-sort:* Stephenson, W. (1973). *The Study of Behavior: Q-technique and its Methodology*, Chicago: University of Chicago Press (via Stanley and Hopkins).
- *Salk polio vaccine trials:* Meier, Paul (1989). "The Biggest Public Health Experiment Ever: The 1954 Field Trial of the Salk Poliomyelitis Vaccine," In Tanur, Judith M., et al., *Statistics: A Guide to the Unknown*, 3rd. ed., Pacific Grove, CA: Wadsworth & Brooks/Cole.
- *SAT scores: The World Almanac and Book of Facts* (1997), p. 221–222.
- *Semantic differential:* Osgood, C. E., C. J. Suci, and P. H. Tannenbaum (1957). *The Measurement of Meaning*, Urbana, IL: University of Illinois (via Stanley and Hopkins).

- *Sesame Street:* Stevens, James (1992). *Applied Multivariate Statistics for the Social Sciences*, 2nd ed. Hillsdale, NJ: Lawrence Erlbaum Associates, pp. 578–585. (No primary source given.)
- *Shop stewards:* Lieberman, S. (1956). "The effects of changes in roles on the attitudes of role occupants," *Human Relations*, v. 9, pp. 385–402 (via Myers).
- *Sleeping shrews:* Berger, R. J. and J. M. Walker (1972). "A polygraphic study of sleep in the tree shrew," *Brain, Behavior, and Evolution*, v. 5, pp. 62 (via Larsen and Marx).
- *Smith College infirmary:* Hodge, Mary Beth (1977). "Statistical Analysis of Smith College Infirmary Data," undergraduate honors thesis, Department of Mathematics, Smith College.
- *Snakes and robins:* Mostrom, Alison Mary (1982). "The response of six local avian species to a snake model." Unpublished undergraduate honors thesis, Department of Biological Sciences, Mount Holyoke College.
- *Solitary prisoners:* Gendreau, Paul, et al. (1957). "Changes in EEG alpha frequency and evoked response latency during solitary confinement," *J. Abnormal Psych.*, v. 79, pp. 54–59.
- *Sperling's partial report procedure:* Sperling, G. A. (1960). "The information available in brief visual presentation" *Psychological Mongraphs*, v. 74, Whole No. 498 (via Anderson).
- *Sponge cells:* Williamson, Craig (1977). "Algal symbiosis in the freshwater sponge *Spongilla lacustris*." Unpublished master's thesis, Department of Biological Sciences, Mount Holyoke College, and personal communication.
- *Submarine memory:* Godden, D. R. and A. D. Baddeley (1975). "Context-dependent memory in two natural environments: On land and under water," *British J. Psych.*, v. 66, pp. 325–331 (via Anderson).
- *Sugar metabolism:* (1985). "Effects of oxygen concentration on pyruvate formatelyase in situ and sugar metabolism of *Streptocucoccus mutans* and *Streptococcus samguis*," *Infection and Immunity*, pp. 129–134 (via Devore and Peck).
- *Tardive dyskinesia:* Based on research by Nancy Keuthen and Patricia Wisocki, Department of Psychology, University of Massachusetts, 1982.
- *Thymus surgery:* Barnes, Benjamin A. (1977). "Discarded operations: surgical innovation by trial and error," in John P. Bunker, Benjamin A. Barnes, and Frederick Mosteller, *Costs, Risks, and Benefits of Surgery*. New York: Oxford University Press, pp. 109–123.
- *Tomato leaves:* Waggoner, P. E. and R. H. Shaw (1952). "Temperature of potato and tomato leaves," *Plant Physiol.*, v. 27, pp. 710–7124 (via Bliss, 1970).
- *Transylvania effect:* Olvin, J. F. (1943). "Moonlight and nervous disorders," *Am. J. Psychiatry*, v. 99, pp. 578–584 (via Larsen and Marx).
- *Varieties of wheat:* Fisher, R. A. (1935). *The Design of Experiments*, Edinburgh: Olvier and Boyd, p. 56.

- *Walking babies:* Zelazo, Phillip R., Nancy Ann Zelazo, and Sarah Kolb (1972). "Walking in the Newborn," *Science*, v. 176, pp. 314–315 (via Larson and Marx).
- *Warm rats:* Clapp, Karen (1980). "The acute effects of an ambient temperature change on the body temperature and oxygen consumption of rats," unpublished student project, Mount Holyoke College.
- *Wireworms:* Cochran, W. G. (1938). *Emp. J. Exp. Agric.* v. 6, p. 157 (via Snedecor and Cochran).

SECONDARY SOURCES

- Anderson, John R. (1985). *Cognitive Psychology and Its Implications.* New York: W. H. Freeman.
- Begon, Michael, John L. Harper, and Colin R. Townsend (1986). *Ecology: Individuals, Populations and Communities*, Sunderland, MA: Sinauer Associates.
- Bliss, C. I. (1967). *Statistics in Biology*, Vol. One, New York: McGraw Hill.
- Bliss, C. I. (1970). *Statistics in Biology*, Vol. Two, New York: McGraw Hill.
- Colton, Theodore (1974). *Statistics in Medicine.* Boston: Little, Brown and Company.
- Devore, Jay and Roxy Peck (1986). *Statistics: The Exploration and Analysis of Data.* St. Paul, MN: West.
- Hoaglin, David C., Frederick Mosteller and John W. Tukey, eds. (1991). *Fundamentals of Exploratory Analysis of Variance*, New York, John Wiley & Sons.
- Myers, David G. (1983). *Social Psychology*, New York: McGraw-Hill.
- Snedecor, George W. and William G. Cochran (1967). *Statistical Methods*, Ames, IA: The Iowa State University Press.
- Srivastava, M. S. and E. M. Carter (1983). *An Introduction to Applied Multivariate Statistics*, New York: North Holland.
- Stanley, Julian C. and Kenneth D. Hopkins (1972). *Educational and Psychological Measurement and Evaluation*, Englewood Cliffs, NJ: Prentice-Hall.

ADDITIONAL REFERENCES

- Cochran, William G., and Gertrude M. Cox (1957). *Experimental Designs*, 2nd ed., New York: John Wiley & Sons.
- Cox, D. R. (1958). *The Planning of Experiments.* New York: John Wiley & Sons.
- Kirk, Roger E. (1982). *Experimental Design: Procedures for the Social Sciences.* Belmont, CA: Brooks/Cole.
- Milliken, George A. and Dallas E. Johnson (1984). *Analysis of Messy Data, vol. 1.* New York: Van Nostrand Reinhold.
- Ott, Lyman (1988). *An Introduction to Statistical Methods and Data Analysis.* Boston: PWS Kent.

DATA INDEX

SUBJECT INDEX

R

S

T